Studies in Computational Intelligence

Volume 1056

The series "Studies in Computational Intelligence" (SCI) publishes new developments and advances in the various areas of computational intelligence—quickly and with a high quality. The intent is to cover the theory, applications, and design methods of computational intelligence, as embedded in the fields of engineering, computer science, physics and life sciences, as well as the methodologies behind them. The series contains monographs, lecture notes and edited volumes in computational intelligence spanning the areas of neural networks, connectionist systems, genetic algorithms, evolutionary computation, artificial intelligence, cellular automata, self-organizing systems, soft computing, fuzzy systems, and hybrid intelligent systems. Of particular value to both the contributors and the readership are the short publication timeframe and the world-wide distribution, which enable both wide and rapid dissemination of research output.

Indexed by SCOPUS, DBLP, WTI Frankfurt eG, zbMATH, SCImago.

All books published in the series are submitted for consideration in Web of Science.

Muhammad Alshurideh · Barween Hikmat Al Kurdi · Ra'ed Masa'deh · Haitham M. Alzoubi · Said Salloum
Editors

The Effect of Information Technology on Business and Marketing Intelligence Systems

Volume 1

Springer

Editors
Muhammad Alshurideh
Department of Management, College of Business Administration
University of Sharjah
Sharjah, United Arab Emirates

Department of Marketing, School of Business
The University of Jordan
Amman, Jordan

Barween Hikmat Al Kurdi
Department of Marketing, Faculty of Economics and Administrative Sciences
The Hashemite University
Zarqa, Jordan

Haitham M. Alzoubi
Skyline University College
Sharjah, United Arab Emirates

Ra'ed Masa'deh
Management Information Systems Department, School of Business
University of Jordan
Aqaba, Jordan

Said Salloum
School of Comupting, Science and Engineering
University of Salford
Salford, England

ISSN 1860-949X ISSN 1860-9503 (electronic)
Studies in Computational Intelligence
ISBN 978-3-031-12381-8 ISBN 978-3-031-12382-5 (eBook)
https://doi.org/10.1007/978-3-031-12382-5

This Springer imprint is published by the registered company Springer Nature Switzerland AG
The registered company address is: Gewerbestrasse 11, 6330 Cham, Switzerland

Contents

Business and Data Analytics

Innovation, Entrepreneurship and Leadership

Knowledge Management

Social Marketing and Social Media Applications

Assessment of the Perception of Usage of Facebook as a Business Tool in SMEs Through the Technological Acceptance Model (TAM) and Structural Equation Modeling

Mohammed T. Nuseir, Ahmad I. Aljumah, Ghaleb A. El Refae, Muhammad Alshurideh, Sarah Urabi, and Barween Al Kurdi

Abstract The implementation of novel technologies such as social networks is altering the landscape of contemporary Small and Medium Enterprises (SMEs) including the ones located in UAE. The purpose of this research is to define the extent of technological acceptance of Facebook in business activities through the Technology Acceptance Model (TAM). To achieve this, the structural equation modeling

M. T. Nuseir (✉)
Department of Business Administration, College of Business, Al Ain University, Abu Dhabi Campus, Abu Dhabi, UAE
e-mail: mohammed.nuseir@aau.ac.ae

A. I. Aljumah
College of Communication and Media, Al Ain University, Abu Dhabi Campus, P.O. Box 112612, Abu Dhabi, United Arab Emirates
e-mail: Ahmad.aljumah@aau.ac.ae

G. A. El Refae
Department of Business Administration, College of Business, Al Ain University, Al Ain Campus, Al Ain, UAE
e-mail: ghalebelrefae@aau.ac.ae

M. Alshurideh
Department of Marketing, School of Business, The University of Jordan, Amman, Jordan
e-mail: m.alshurideh@ju.edu.jo; malshurideh@sharjah.ac.ae

Department of Management, College of Business Administration, University of Sharjah, Sharjah, United Arab Emirates

S. Urabi
Abu Dhabi Vocational Education and Training Institute (ADVETI), Abu Dhabi, UAE
e-mail: sarah.urabi@adveti.ac.ae

B. Al Kurdi
Department of Marketing, Faculty of Economics and Administrative Sciences, The Hashemite University, Zarqa, Jordan
e-mail: barween@hu.edu.jo

M. Alshurideh et al. (eds.), *The Effect of Information Technology on Business and Marketing Intelligence Systems*, Studies in Computational Intelligence 1056,
https://doi.org/10.1007/978-3-031-12382-5_1

is employed with the application of 130 surveys. Empirical evidence displays positive significance in most of the constructs such as Perceived Usefulness, Easy to Use, Intended Use, Attitude of Use and Current Use and also suggests the lack of a culture regarding the use, handling, and acceptance of Facebook as a marketing tool in most businesses consulted.

Keywords Facebook · Technology acceptance model · SME · SEM

1 Introduction

The existence of millions of users on social media sites is characteristically attractive for companies because it opens up novel means to interact with customers and other important stakeholders, like employees and suppliers (Aljumah et al., 2021a; Al-Maroof et al., 2021a; Nuseir et al., 2021b; El Refae et al., 2021). For this reason, different types of organizations around the sphere are aiming to assimilate social media in their numerous facets of business operations and processes (Al Khasawneh et al., 2021a, b; Al Kurdi & Alshurideh, 2021; Bhanot, 2012). Culnan et al. (2010) stated that social networks are changing the ways traditional communication of human beings, who have advanced with the usage of novel channels to becoming a tool which is founded on the creation, generalized trust and shared knowledge.

In the view of Ahmad et al. (2019), the main activities of the Internet in UAE are the use of social networks, which occupy the first place, while statistical data company Global Media Insight shows that UAE has 9 million users on Facebook, which represents an area of opportunity for companies because of the large number of potential customers (GMI, 2018; Nuseir et al., 2021a; Radcliffe et al., 2019).

However, despite the benefits and uses that social network can provide to the organizations (Alhamad et al., 2021; Al-Hamad et al., 2021; Alshurideh et al., 2016, 2019c), there are many managers who have not decided to venture into this type of technology (Ahmad et al., 2019; Al Kurdi et al., 2021a; Alshurideh et al., 2021a, b; Culnan et al., 2010). Therefore, the purpose of this investigation is to examine the degree of technological acceptance of Facebook in SMEs (trading and services firms), located in Dubai International Financial Centre in Dubai and Khalifa Industrial Zone in Abu Dhabi, UAE, through the Technology Acceptance Model (TAM) proposed by Davis (1989), Akour et al. (2021), and Almazrouei et al. (2020).

To achieve this goal, a literature review was conducted and the application of a survey of 130 SMEs was performed. Afterward, the data obtained were analyzed by the technique of Partial Least Squares (PLS). Finally, this paper is structured as follows: the first section reviews the literature, and the next one provides the description of the method and the proposed model; followed by analysis of the results, and finally in the last section the main contributions to the knowledge obtained are shown.

2 State of the Art

2.1 Social Networks

Social networks are altering the participation and commitment of users; consequently, businesses have concentrated on their benefits, since the extensive usage of the Internet and its tools has meant the advent of a communication channel that permits to collaborate with a lot of people on a daily basis (Al-Dmour et al., 2021a; Al-Khayyal et al., 2021; Guesalaga, 2016). Increasingly, more and more individuals are using online format to interrelate and share experiences (Al Kurdi et al., 2021b; Alshurideh et al., 2012b; Hamadneh & Al Kurdi, 2021; Joghee et al., 2021; Radcliffe et al., 2019). Some studies contemplate the usage of social networks as one of the fundamentals of change in the manner SMEs compete in the global world, and that its examination can be an excellent tool for gaining pertinent information (Aburayya et al., 2020a, b, c, d; Al Suwaidi et al., 2021; Al-Khayyal et al., 2020; Baró & Costa-Sánchez, 2017; Radcliffe et al., 2019).

Moreover, social networks are component of the digital marketing as they propose exchange of ideas amongst participants and allow the advertisement of products, services or brand openly to existing and potential customers (Abu Zayyad et al., 2021; Alaali et al., 2021; Alshurideh et al., 2015b; Elsaadani, 2020; Sweiss et al., 2021). As per Baró and Costa-Sánchez (2017), there are different social networking platforms such as blogs, online communities like Twitter, Facebook, Wikis, forums, etc. which can be employed by SMEs to increase awareness and share content or information to surge the visibility, prestige and popularity of a brand or company. These platforms have the benefit of permitting direct collaboration with consumers without the requirement for a considerable investment; so, assist in decreasing in marketing costs (Agha et al., 2021; Al Kurdi et al., 2020a; Al-Dmour et al., 2021b; Kurdi et al., 2020a; Madi Odeh et al., 2021; Prantl & Mičík, 2019); lower the costs of technical services (Aljumah et al., 2021b, c; Felix et al., 2017) or even promote the formation of emotional ties and reinforce the levels of loyalty to the services, products or business (Zhang et al., 2019). As a result, creating a marketing campaign effectively on social networks saves on advertising, as well as generates a kind of advertisement, which is better received by customers, since it is likely to create a closer relationship with them, eventually creating brand loyalty (Al-Dmour et al., 2014; Alshurideh et al., 2012a, 2019b; Alshurideh, 2016a, b, c, 2017, 2019; Obeidat et al., 2019). To attain good results, it is significant that SMEs interrelate day-to-day in social networks and display an interest in taking part in these communication platforms, together with nurturing consumer relations (Al Kurdi et al., 2017; Alshurideh et al., 2014, 2020b; Alyammahi et al., 2021; Alzoubi et al., 2020; Guesalaga, 2016; Hayajneh et al., 2021; Zhang et al., 2019).

According to a study by Lin and Kim (2016), Facebook is the social network with the largest brands which reflects that 59% of those who follow a brand in social networking channels have been affected in their procuring choices. Moreover, Bogea and Brito (2018) state that social networking platforms enhance business prospects, yield and stay in the market as well as bringing it closer to consumers and potential customers, networking with companies and people, counting its clients in developing their position and business itself as a standard in its sector.

2.2 Technology Acceptance Model

At present, the optimum usage of IT within SMEs is a requirement, keeping in view its effect on the generation of services and goods. This, together with social networks, is turning out to be more affordable in the market; nevertheless, proper management is more important than the investment (Alshurideh et al., 2019a, d; Lin & Kim, 2016; Nuseir et al., 2021c). Consequently, in contemporary years, various studies have concentrated on recognizing the factors which affect acceptance behaviors in the case of technology by the end-users (Alshurideh et al., 2017; Nikou & Economides, 2017; Scherer et al., 2019).

This resulted in the existence of different models and theories to measure the acceptance of technology; among which the Technology Acceptance Model introduced by Davis (1989) is most significant. It is an extremely effective model in forecasting the usage of IT (Kurdi et al., 2020b; Lin & Kim, 2016). The model has its origins in the Theory of Reasoned Action (TRA) but replaces several measures of attitude of the TRA model to predict behavior, acceptance and intended use of the technologies by individuals (Abuhashesh et al., 2021; Al-Maroof et al., 2021b; Lai, 2017).

As per Al Kurdi et al. (2020b), Alshurideh et al. (2020a), Amarneh et al. (2021), and Lai (2017), the model attempts to predict the technology acceptance based on two main variables: the *perceived usefulness*—the extent to which an individual believes that employing a specific system enhances performance and the *perceived ease of use*—the extent to which an individual believes that employing a specific system, one can free oneself from the effort it involves performing the work.

Moreover, in the view of Al-Qaysi et al. (2020), the model proposes that when customers are presented with an innovative technology, several factors affect their decision on how and when to employ. According to Davis (1989), the primary aim of the TAM is to elucidate the reasons of approval of technologies by users. Therefore, the model suggests that views of an individual in perceived usefulness and perceived ease of use of technology are categorical to define their intent to employ it (Al Kurdi et al., 2020b; Alkitbi et al., 2021; Al-Qaysi et al., 2020; Bettayeb et al., 2020; Lin & Kim, 2016).

Moreover, the literature review allowed us to detect the lack of studies related to social and TAM networks in UAE; however, it was found that in other dimensions exist, but in a limited amount. One of these studies examined the adoption and use of social networking by users including model aspects such as trust and perceived risk to them, the results of this study support a relationship and positive influence among the variables (Biswas, 2016). Similarly, another study was carried out to recognize the variables that affect the inclination to use the social network for their purchasing decisions. The outcomes of this study specify that the anticipated usage of Facebook in purchasing decisions is affected by the attitude, social influence, and perceived usefulness. In addition, this study recognizes that perceived enjoyment could have a vital part, even more than the perceived usefulness in defining an individual's attitude to Facebook as a tool for discovering info about the good to purchase (Aissani & Awad ElRefae, 2021; Alshurideh et al., 2021c; Singh & Srivastava, 2019).

3 Method

This study presents a conceptual model in which it is proposed that the attitude of use is affected by the ease of use and perceived usefulness, the latter also inferred by the ease of use. In the same way, both perceived usefulness and attitude of use are direct antecedents of intended use, which influences the perception of the current use of certain technology. Figure 1 shows the model and the causal relationships among the variables.

Moreover, the assumptions of the testing model for the proposed research are summarized in Table 1.

Regarding the process of creation of the instrument, it was based on the literature review, which allowed us to detect different models of operationalization of the variables related to the subject, which have been used successfully in different studies. Therefore, the first draft of the instrument was sent for review to a group of researchers and practitioners in the area. Each expert was received their opinion on the consistency, relevance and clarity of the items-factors proposed. Afterward, a pilot study was conducted and the questionnaire consisting of 29 items, of which 8 were multiple and 21 were Likert scale responses of 5 points (5. Strongly agree and 1. Strongly disagree) and were sent to different SMEs through e-mail. The responses were received back within the period of 30 days.

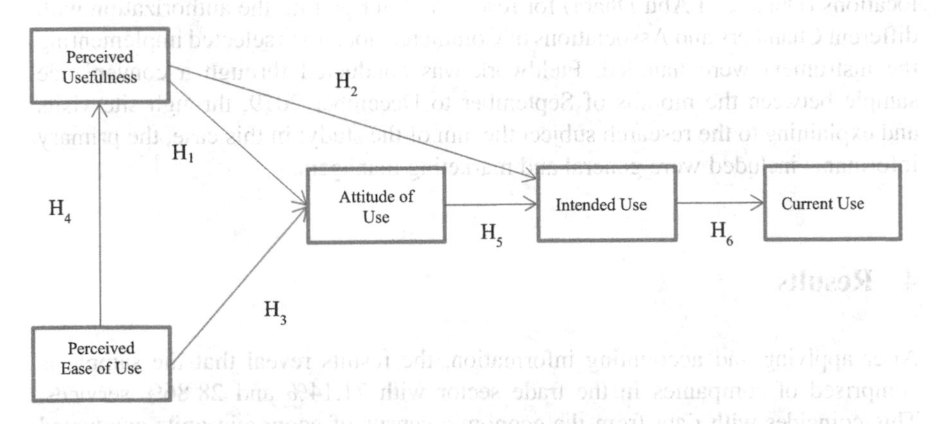

Fig. 1 Theoretical model to test

Table 1 Summary of hypotheses

Variables	Hypotheses	References
Perceived usefulness	H1. Perceived usefulness of the SMEs regarding social network Facebook has a positive influence on the attitude of use of social networking as a marketing tool	Davis (1989)
	H2. Perceived usefulness of the SMEs regarding social network Facebook positively influences the intention of use of social networks as a tool for marketing	Davis (1989)
Ease of use	H3. Perceived ease of use Facebook as a marketing tool has a positive effect on the attitude of use in SMEs	Davis (1989)
	H4. The effect of the ease of use Facebook perceived by SMEs is positive in the perception of usefulness	Lane and Coleman (2012), George et al. (2014)
Attitude of use	H5. Affects the attitude of entrepreneurs positively regarding intended use of Facebook as a marketing tool	Biswas (2016)
Intended use	H6. The intended usage of Facebook as a promotion tool by SMEs positively influences their current use of this social network	Sell et al. (2012), Lin and Kim (2016)

The sample was obtained by the UAE Chamber of Commerce and Industry for the year 2019 which registered 2,800 SMEs located within the established geographic locations (Dubai and Abu Dhabi) for research. After getting the authorization with different Chambers and Associations of Commerce, locations selected implementing the instrument were handled. Fieldwork was conducted through a convenience sample between the months of September to December 2019, through site visits and explaining to the research subject the aim of the study; in this case, the primary informants included were general and marketing managers.

4 Results

After applying and accounting information, the results reveal that the sample is comprised of companies in the trade sector with 71.14% and 28.86% services. This coincides with data from the economic census of economic units conducted by GMI (2018), which states that in Dubai and Abu Dhabi, commercial enterprises represent the most economic units. Also 49% of SMEs have 0–10 employees, 30% of SMEs have 11–30 employees and 21% of SMEs have 31–100 employees. The research participants were also asked to indicate the positions they play within the organization, obtaining the highest percentage of respondents were owners (44%).

Moreover, to check the model, structural equation modeling (SEM) was applied using the statistical technique called PLS through MS Excel 2010. This requires technical evaluation of the quality of the model before obtaining structural validation (Queiroz & Fosso Wamba, 2019). Therefore, the researcher conducted tests of the main criteria of quality, beginning with the analysis of individual reliability of the item, which states that to accept an indicator as part of a reflective construct, this must have a factor loading (λ) or 0.707 or above simple correlations (Al-Maroof et al., 2021a). Taking the above acceptance criteria ($\lambda \geq 0.707$), the FU3 and FU4 indicators were eliminated. Table 2 shows the results.

Continued with the quality tests of the model, the next phase is to determine the reliability of the construct, which is evaluated by Cronbach's alpha (α) and the coefficient of composite reliability (ρc), in both cases, their performance is similar. Therefore, guidance provided by Achjari (2004) who suggests 0.7 as a reference point is used. Table 3 shows the results and as noted, all constructs have a satisfactory internal consistency and are reliable.

Moreover, the researcher also calculated the Average Variance Extracted (AVE). This ratio indicates the amount of variance that a reflective construct gets from its indicator's comparative to the amount of variance because of the measurement error and its value must be greater than 0.5 (Manerikar & Manerikar, 2015) and as shown in Table 4 all AVE measures are valid.

Table 2 Reliability of single items

Attitude of use (AU)				Intended use (IU)				Current use (CU)			
Item	λ		T-statistic	Item	λ		T-statistic	Item	λ		T-statistic
AC1	0.81	***	17.50	INT1	0.84	***	23.38	UA1	0.93	***	53.07
AC2	0.91	***	43.85	INT2	0.92	***	68.89	UA2	0.95	***	110.03
AC3	0.91	***	38.95	INT3	0.88	***	28.64	UA3	0.75	***	9.09
AC4	0.91	***	49.93	INT4	0.92	***	69.52				

Ease of use (EU)				Perceived usefulness (PU)			
Item	Λ		T-statistic	Item	Λ		T-statistic
FU1	0.74	***	9.94	UP1	0.81	***	18.90
FU2	0.78	***	14.90	UP2	0.90	***	46.14
FU3	**0.59**	***	7.54	UP3	0.87	***	27.80
FU4	**0.61**	***	6.17	UP4	0.90	***	35.70
FU5	0.82	***	16.90	UP5	0.88	***	36,15
FU6	0.77	***	12.75	UP6	0.88	***	38.40

***Value t > 3.31 (p < 0.001), **p value > 2.58 (p < 0.01), *p value > 1.96 (p < 0.05)

Table 3 Reliability of the construct

Constructs	Cronbach's alpha (α)	Composite reliability (ρc)
Perceived usefulness	0.942	0.949
Easy to use	0.807	0.865
Intended use	0.919	0.943
Attitude of use	0.911	0.938
Current use	0.874	0.916

Table 4 Validity convergent

Constructs	Factor
Perceived usefulness	0.775
Easy to use	0.565
Attitude of use	0.792
Intended use	0.805
Current usage	0.787

Finally, the values were analyzed, and the correlation matrix constructs made by the square root of the coefficient of each AVE obtained, indicating that these values should be higher than the rest of the same column (Manerikar & Manerikar, 2015). As shown in Table 5, constructs meet this criterion.

Once it has been verified that the constructs are valid and reliable, the researcher proceeded to calculate the magnitude and weight of relations between different variables, using the above two indices proposed by Radcliffe et al. (2019): (i) The Explained Variance (R^2)—which determines the predictive power of the model; therefore, their values must be equal to or greater than 0.1 since little information; and (ii) the standardized coefficients path (β), which show the strength of relations between independent and dependent variables, so their values need to reach the least significant to 0.2 and as seen in Table 6 the values obtained for R^2 are within suitable ranges.

Table 5 Discriminant validity

Constructs	Attitude of use	Ease of use	Intended use	Current use	Perceived usefulness
Attitude of use	**0.890**				
Easy to use	0.438	**0.751**			
Intended use	0.848	0.396	**0.897**		
Current use	0.363	0.350	0.423	**0.887**	
Perceived usefulness	0.801	0.481	0.768	0.444	**0.880**

Table 6 Explained variance (R^2)

Constructs	Explained variance
Perceived usefulness	0.232
Attitude of use	0.645
Intended use	0.741
Current use	0.179

Table 7 Results of the structural model

	Hypothesis		Coefficient (β)	t values	p values
H1	Perceived usefulness -> Attitude of use	+	0.768	13.618	0.000
H2	Perceived usefulness -> Intended use	+	0.247	2.941	0.003
H3	Easy to use -> Attitude of use	+	0.069	1.160	0.246
H4	Easy to use -> Perceived usefulness	+	0.481	6.223	0.000
H5	Attitude of use -> Intended use	+	0.650	8.037	0.000
H6	Intended use -> Current use	+	0.423	7.272	0.000

***Value t > 3.310 (p < 0.001), **p value > 2.586 (p < 0.01), *p value > 1.965 (p < 0.05)

Note that resorted to non-parametric technique Bootstrap, a resampling process with replacement, considering 130 cases with 5000 samples, which is suggested for concluding findings (Simar & Wilson, 2000), the above values were obtained using Student t-test and significance (p). For a t distribution, two-tailed with n degrees of freedom, where n is the number of samples to be considered in the Bootstrap technique, values that define the statistical significance are: t (95%) = 1.96* T (99%) = 2.58** t (99.9%) = 3.31***. As seen in Table 7, only the hypothesis H3, not significant.

Finally, the value of the Residual Standardized Root Mean Square was calculated (SRMR), which is interpreted as the average difference between the covariance and observed based on the standard error of the residual; consequently, it can be deliberated as a measure of goodness of fit for PLS-SEM models (Hair et al., 2014). Values must be between 0.0 (perfect fit) and less than 0.08 to be considered as valid (Wong, 2013), in this sense, the value obtained for SRMR research model proposed is 0.069, indicating an adequate level of fit. After obtaining the results of the PLS modeling, proceed to accept the coefficient path and by extension confirm the assumptions made that are significant.

To this end, Fig. 2 shows the results obtained.

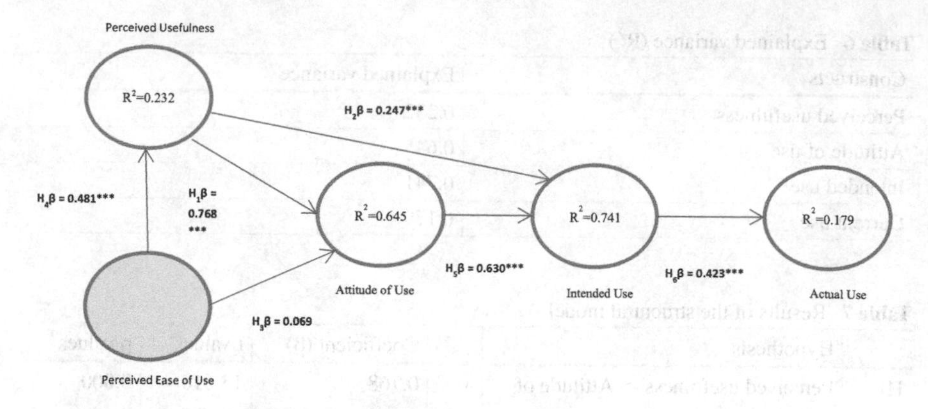

Fig. 2 Coefficients path and statistical significance

4.1 Results Analysis

From Fig. 2, it can be inferred that: the perceived utility of the SMEs regarding the social network Facebook has a positive and significant influence (H_1: $\beta = 0.768$***) in the attitude of use of networks social as a marketing tool having a direct causal relationship, so the H1 hypothesis is accepted, this is comparable to the investigation of Ahmad et al. (2019) in which their results indicate that the *Perceived Usefulness* of users influence their attitude of using e-learning, and also this statement is checked in TRA theory, which states that attitudes toward behavior are influenced by relevant beliefs.

While the relationship between *Perceived Usefulness* and *Intended Use* (Hypothesis H2) is accepted, because the *Perceived Usefulness* of SMEs regarding Facebook has a positive impact on the intended use of social networks as marketing tool ($\beta =$ 0.247**), which proves that the easier it is to interact with technology, the greater the satisfaction of the user and therefore shows a greater intention to use (Guesalaga, 2016), also they are related to investigations of Nikou and Economides (2017), Prantl and Mičík (2019) since *Perceived Usefulness* significantly influences the *Intended Use* for the system.

Now, regarding the usability of Facebook and inference in the construct *Attitude of Use*, as shown there is no significant relationship in attitude towards their use in the company (H3: $\beta = 0.069$); therefore, the hypothesis is rejected. The result is very similar to that obtained in the investigation of Lin and Kim (2016) where he also investigated this relationship in the field of social networks, and it was also rejected. This may be due to the fact that most respondents perceived easy handling and personal use; however, at workplace, most activities are directly reflected in their attitude to the use and management of this network.

With regard to the issues raised in the H4, there was a significant effect on the *Ease of Use* on *Perceived Usefulness*; therefore, this hypothesis is accepted demonstrating that the effect of the ease of use of Facebook perceived by SMEs is observed it is positive in the *Perceived Usefulness* (H4: β = 0.481). In this way, *Ease of Use* contributes to improving the performance of the task, which saves effort, so more and better results with the same effort can be attained (Lin & Kim, 2016).

Moreover, the relationship between *Attitude of Use* and *Intended Use* was accepted (hypothesis H5), as it shows that this construct has an essential influence on the intended use of this social network (β = 0.650***). This demonstrates that the intention is determined by the attitude of the employer using Facebook. This result is consistent with Biswas (2016) who mentions that the attitude towards social networking virtual positively and significantly influences the intention to use them.

Regarding the H6, a causal relationship between the intended use and the current use of Facebook was determined, obtaining a positive and significant effect on end-use level (β = 0.423***); therefore, it shows that the intended use acts as an intermediary between the varying effect exerted by perceptions (usability and perceived utility) and the end-use of the individual, which is similar to that obtained in the investigation of Lane and Coleman (2012) as the results obtained by the authors show a significant positive correlation between aspects of *Intended Use* and *Current Use*, and similarly are related with Guesalaga (2016) because the results of this study support a relationship and positive influence among the variables used as intended and current use.

5 Conclusions and Limitations

Regarding the use of Facebook in business trade and service sector related SMEs in UAE, it has been found that 56% have incorporated this tool to their activities while 44% not used or hardly used, representing nearly half of the sample did not know the use and therefore the benefits of this media. Also, through data processing it has been identified that only 3% of companies carry out strategic planning to carry out marketing activities, representing an alarming fact (Alshurideh et al., 2015a; Alshurideh, 2022; ELSamen & Alshurideh, 2012; Robertson & Kee, 2017).

Another result obtained in relation to the current use of Facebook in SMEs is the time that they invest to make their publications advertising, promotions, sales, and general marketing activities specifically in this social network, denoting that most companies invest little time.

This is also an area of opportunity to incorporate into higher education programs for future business leaders possess this knowledge. On the other hand, regarding the results of utility employers perceive this social network, most are agreed that this tool is useful for your organization and for those companies that have not joined Facebook as its perceived usefulness is high.

As a general conclusion, research and statistical indicators show that the dissemination strategy in social networking is an exceptional tool for the market positioning of SMEs as social networks have generated a significant effect on customers. Especially businesses look to be recognized by them and to maintain sound corporate image of the firm in the eyes of customers. It can be assumed that by employing set of tactical actions like planning in managing social networking platforms, there can be a positive impact on the positioning of a company in a market.

Furthermore, this paper has some limitations that should be taken in account before simplifying the findings, the first relates to the validity of a model, which cannot be recognized on the base of a single research, since the data signify a snapshot in the time. Second is associated to the geographic area (Dubai and Abu Dhabi); consequently, future studies ought to employ a greater sample and assume a longitudinal tactic and can also comprise and explore the relationships between individual or organizational variables and their impact on the usage of Facebook. Consequently, criticism of the cause-effect perceived must be conducted with cautiousness.

References

Abu Zayyad, H. M., Obeidat, Z. M., Alshurideh, M. T., Abuhashesh, M., Maqableh, M., & Masa'deh, R. (2021). Corporate social responsibility and patronage intentions: The mediating effect of brand credibility. *Journal of Marketing Communications, 27*(5). https://doi.org/10.1080/13527266.2020.1728565.

Abuhashesh, M. Y., Alshurideh, M. T., & Sumadi, M. (2021). The effect of culture on customers' attitudes toward Facebook advertising: The moderating role of gender. *Review of International Business and Strategy*.

Aburayya, A., et al. (2020a). Critical success factors affecting the implementation of TQM in public hospitals: A case study in UAE hospitals. *Systematic Reviews in Pharmacy, 11*(10). https://doi.org/10.31838/srp.2020.10.39.

Aburayya, A., et al. (2020b). An empirical examination of the effect of TQM practices on hospital service quality: An assessment study in UAE hospitals. *Systematic Reviews in Pharmacy, 11*(9). https://doi.org/10.31838/srp.2020.9.51.

Aburayya, A., Alshurideh, M., Alawadhi, D., Alfarsi, A., Taryam, M., & Mubarak, S. (2020c). An investigation of the effect of lean six sigma practices on healthcare service quality and patient satisfaction: Testing the mediating role of service quality in Dubai primary healthcare sector. *Journal of Advanced Research in Dynamical and Control Systems, 12*(8), 56–72.

Aburayya, A., Alshurideh, M., Albqaeen, A., Alawadhi, D., & Al A'yadeh, I. (2020d). An investigation of factors affecting patients waiting time in primary health care centers: An assessment study in Dubai. *Management Science Letters, 10*(6). https://doi.org/10.5267/j.msl.2019.11.031.

Achjari, D. (2004). Partial least squares: Another method of structural equation modeling analysis. *Jurnal Ekonomi dan Bisnis Indonesia, 19*(3), 238–248.

Agha, K., Alzoubi, H. M., & Alshurideh, M. T. (2021). Measuring reliability and validity instruments of technologically driven cognitive intrusion towards work-life balance. In *The International Conference on Artificial Intelligence and Computer Vision* (pp. 601–614).

Ahmad, S. Z., Bakar, A. R. A., & Ahmad, N. (2019). Social media adoption and its impact on firm performance: The case of the UAE. *International Journal of Entrepreneurial Behavior & Research, 25*(1), 84–111.

Aissani, R., & Awad ElRefae, G. (2021). Trends of university students towards the use of YouTube and the benefits achieved—Field study on a sample of Al Ain University students. *AAU Journal of Business and Law* مجلة جامعة العين للأعمال والقانون, *5*(1), 1.

Akour, I., Alshurideh, M., Al Kurdi, B., Al Ali, A., & Salloum, S. (2021). Using machine learning algorithms to predict people's intention to use mobile learning platforms during the COVID-19 pandemic: Machine learning approach. *JMIR Medical Education, 7*(1), 1–17.

Al Khasawneh, M., Abuhashesh, M., Ahmad, A., Alshurideh, M. T., & Masa'deh, R. (2021a). Determinants of e-word of mouth on social media during COVID-19 outbreaks: An empirical study. In *The effect of coronavirus disease (COVID-19) on business intelligence* (Vol. 334).

Al Khasawneh, M., Abuhashesh, M., Ahmad, A., Masa'deh, R., & Alshurideh, M. T. (2021b). Customers online engagement with social media influencers' content related to COVID 19. In *The effect of coronavirus disease (COVID-19) on business intelligence* (Vol. 334).

Al Kurdi, B., et al. (2017). Investigating the impact of communication satisfaction on organizational commitment: A practical approach to increase employees' loyalty. *International Journal of Marketing Studies, 9*(2). https://doi.org/10.5539/ijms.v9n2p113.

Al Kurdi, B., Alshurideh, M., & Al Afaishata, T. (2020a). Employee retention and organizational performance: Evidence from banking industry. *Management Science Letters, 10*(16), 3981–3990.

Al Kurdi, B., Alshurideh, M., & Salloum, S. (2020b). Investigating a theoretical framework for e-learning technology acceptance. *International Journal of Electrical and Computer Engineering, 10*(6), 6484–6496.

Al Kurdi, B. H., & Alshurideh, M. T. (2021). Facebook advertising as a marketing tool: Examining the influence on female cosmetic purchasing behavior. *International Journal Online Marketing, 11*(2), 52–74.

Al Kurdi, B., Alshurideh, M., Nuseir, M., Aburayya, A., & Salloum, S. A. (2021a). The effects of subjective norm on the intention to use social media networks: An exploratory study using PLS-SEM and machine learning approach. In *Advanced Machine Learning Technologies and Applications: Proceedings of AMLTA 2021* (pp. 581–592).

Al Kurdi, B., Elrehail, H., Alzoubi, H. M., Alshurideh, M., & Al-adaileh, R. (2021b). The interplay among HRM practices, job satisfaction and intention to leave: An empirical investigation. *24*(1), 1–14.

Al Suwaidi, F., Alshurideh, M., Al Kurdi, B., & Salloum, S. A. (2021). *The impact of innovation management in SMEs performance: A systematic review* (Vol. 1261). AISC.

Alaali, N., et al. (2021). The impact of adopting corporate governance strategic performance in the tourism sector: A case study in the Kingdom of Bahrain. *Journal of Legal, Ethical and Regulatory Issues, 24*(Special Issue 1).

Al-Dmour, H., Alshuraideh, M., & Salehih, S. (2014). A study of Jordanians' television viewers habits. *Life Science Journal, 11*(6), 161–171.

Al-Dmour, A., Al-Dmour, H., Al-Barghuthi, R., Al-Dmour, R., & Alshurideh, M. T. (2021a). Factors influencing the adoption of e-payment during pandemic outbreak (COVID-19): Empirical evidence. In *The effect of coronavirus disease (COVID-19) on business intelligence* (Vol. 334).

Al-Dmour, R., AlShaar, F., Al-Dmour, H., Masa'deh, R., & Alshurideh, M. T. (2021b). The effect of service recovery justices strategies on online customer engagement via the role of "customer satisfaction" during the Covid-19 pandemic: An empirical study. In *The effect of coronavirus disease (COVID-19) on business intelligence* (Vol. 334).

Alhamad, A. Q. M., Akour, I., Alshurideh, M., Al-Hamad, A. Q., Kurdi, B. A., & Alzoubi, H. (2021). Predicting the intention to use google glass: A comparative approach using machine learning models and PLS-SEM. *International Journal of Data and Network Science, 5*(3). https://doi.org/10.5267/j.ijdns.2021.6.002.

Al-Hamad, M. Q., Mbaidin, H. O., Alhamad, A. Q. M., Alshurideh, M. T., Kurdi, B. H. A., & Al-Hamad, N. Q. (2021). Investigating students' behavioral intention to use mobile learning in higher education in UAE during Coronavirus-19 pandemic. *International Journal of Data and Network Science, 5*(3). https://doi.org/10.5267/j.ijdns.2021.6.001.

Aljumah, A., Nuseir, M. T., & Alshurideh, M. T. (2021a). The impact of social media marketing communications on consumer response during the COVID-19: Does the brand equity of a university matter. In *The effect of coronavirus disease (COVID-19) on business intelligence* (Vol. 334, pp. 384–367).

Aljumah, A. I., Nuseir, M. T., & Alam, M. M. (2021b). Traditional marketing analytics, big data analytics and big data system quality and the success of new product development. *Business Process Management Journal, 27*(4), 1108–1125.

Aljumah, A. I., Nuseir, M. T., & Alam, M. M. (2021c). Organizational performance and capabilities to analyze big data: Do the ambidexterity and business value of big data analytics matter? *Business Process Management Journal*.

Al-Khayyal, A., Alshurideh, M., Al Kurdi, B., & Aburayya, A. (2020). The impact of electronic service quality dimensions on customers' e-shopping and e-loyalty via the impact of e-satisfaction and e-trust: A qualitative approach. *International Journal of Innovation, Creativity and Change, 14*(9), 257–281.

Al-Khayyal, A., Alshurideh, M., Al Kurdi, B., & Salloum, S. A. (2021). Factors influencing electronic service quality on electronic loyalty in online shopping context: Data analysis approach. In *Enabling AI applications in data science* (pp. 367–378). Springer.

Alkitbi, S. S., Alshurideh, M., Al Kurdi, B., & Salloum, S. A. (2021). *Factors affect customer retention: A systematic review* (Vol. 1261). AISC.

Al-Maroof, R., et al. (2021a). The acceptance of social media video for knowledge acquisition, sharing and application: A comparative study among YouTube users and TikTok users' for medical purposes. *International Journal of Data and Network Science, 5*(3). https://doi.org/10.5267/j.ijdns.2021.6.013.

Al-Maroof, R. S., Alshurideh, M. T., Salloum, S. A., AlHamad, A. Q. M., & Gaber, T. (2021b). Acceptance of Google Meet during the spread of Coronavirus by Arab university students. *Informatics, 8*(2), 24.

Almazrouei, F. A., Alshurideh, M., Al Kurdi, B., & Salloum, S. A. (2020). Social media impact on business: A systematic review. In *International Conference on Advanced Intelligent Systems and Informatics* (pp. 697–707).

Al-Qaysi, N., Mohamad-Nordin, N., & Al-Emran, M. (2020). A systematic review of social media acceptance from the perspective of educational and information systems theories and models. *Journal of Educational Computing Research, 57*(8). https://doi.org/10.1177/0735633118817879.

Alshurideh, M., Masa'deh, R., & Al Kurdi, B. (2012a). The effect of customer satisfaction upon customer retention in the Jordanian mobile market: An empirical investigation. *European Journal of Economics, Finance and Administrative Sciences, 47*(12), 69–78.

Alshurideh, M., Nicholson, M., & Xiao, S. (2012b). The effect of previous experience on mobile subscribers' repeat purchase behavior. *European Journal of Social Sciences, 30*(3), 366–376.

Alshurideh, M., Shaltoni, A., & Hijawi, D. (2014). Marketing communications role in shaping consumer awareness of cause-related marketing campaigns. *International Journal of Marketing Studies, 6*(2), 163.

Alshurideh, M., Alhadid, A., & Al kurdi, B. (2015a). The effect of internal marketing on organizational citizenship behavior an applicable study on the University of Jordan employees. *International Journal of Marketing Studies, 7*(1), 138.

Alshurideh, M., Bataineh, A., Al Kurdi, B., & Alasmr, N. (2015b). Factors affect mobile phone brand choices—Studying the case of Jordan universities students. *International Business Research, 8*(3), 141–155.

Alshurideh, M. T. (2016a). Exploring the main factors affecting consumer choice of mobile phone service provider contracts. *International Journal of Communications, Network and System Sciences, 9*(12), 563–581.

Alshurideh, M. (2016b). Is customer retention beneficial for customers: A conceptual background. *Journal of Research in Marketing, 5*(3), 382–389.

Alshurideh, M. (2016c). Scope of customer retention problem in the mobile phone sector: A theoretical perspective. *Journal of Marketing and Consumer Research, 20*, 64–69.

Alshurideh, M., Al Kurdi, B., Vij, A., Obiedat, Z., & Naser, A. (2016). Marketing ethics and relationship marketing—An empirical study that measure the effect of ethics practices application on maintaining relationships with customers. *International Business Research, 9*(9), 78–90.

Alshurideh, M. (2017). A theoretical perspective of contract and contractual customer-supplier relationship in the mobile phone service sector. *International Journal of Business and Management, 12*(7), 201–210.

Alshurideh, M., Al Kurdi, B., Abu Hussien, A., & Alshaar, H. (2017). Determining the main factors affecting consumers' acceptance of ethical advertising: A review of the Jordanian market. *Journal of Marketing Communications, 23*(5). https://doi.org/10.1080/13527266.2017.1322126.
Alshurideh, M. (2019). Do electronic loyalty programs still drive customer choice and repeat purchase behaviour? *International Journal of Electronic Customer Relationship Management, 12*(1), 40–57.
Alshurideh, M., Al Kurdi, B., & Salloum, S. A. (2019a). Examining the main mobile learning system drivers' effects: A mix empirical examination of both the expectation-confirmation model (ECM) and the technology acceptance model (TAM). In *International Conference on Advanced Intelligent Systems and Informatics* (pp. 406–417).
Alshurideh, M., Alsharari, N. M., & Al Kurdi, B. (2019b). Supply chain integration and customer relationship management in the airline logistics. *Theoretical Economics Letters, 9*(02), 392–414.
Alshurideh, M., Salloum, S. A., Al Kurdi, B., & Al-Emran, M. (2019c). Factors affecting the social networks acceptance: An empirical study using PLS-SEM approach. In *PervasiveHealth: Pervasive computing technologies for healthcare* (Vol. Part F1479). https://doi.org/10.1145/3316615.3316720.
Alshurideh, M., Salloum, S. A., Al Kurdi, B., Monem, A. A., & Shaalan, K. (2019d). Understanding the quality determinants that influence the intention to use the mobile learning platforms: A practical study. *International Journal of Interactive Mobile Technologies, 13*(11). https://doi.org/10.3991/ijim.v13i11.10300.
Alshurideh, M., Al Kurdi, B., Salloum, S. A., Arpaci, I., & Al-Emran, M. (2020a). Predicting the actual use of m-learning systems: A comparative approach using PLS-SEM and machine learning algorithms. *Interactive Learning Environments*, 1–15.
Alshurideh, M., Gasaymeh, A., Ahmed, G., Alzoubi, H., & Al Kurd, B. (2020b). Loyalty program effectiveness: Theoretical reviews and practical proofs. *Uncertain Supply Chain Management, 8*(3), 599–612. https://doi.org/10.5267/j.uscm.2020.2.003.
Alshurideh, M. T., et al. (2021a). Factors affecting the use of smart mobile examination platforms by universities' postgraduate students during the COVID 19 pandemic: An empirical study. *Informatics, 8*(2), 32.
Alshurideh, M. T., Al Kurdi, B., & Salloum, S. A. (2021b). The moderation effect of gender on accepting electronic payment technology: A study on United Arab Emirates consumers. *Review of International Business and Strategy.*
Alshurideh, M. T., Hassanien, A. E., & Masa'deh, R. (2021c). *The effect of coronavirus disease (COVID-19) on business intelligence*. Springer.
Alshurideh, M. (2022). Does electronic customer relationship management (E-CRM) affect service quality at private hospitals in Jordan? *Uncertain Supply Chain Management, 10*(2), 1–8.
Alyammahi, A., Alshurideh, M., Kurdi, B. A., & Salloum, S. A. (2021). *The impacts of communication ethics on workplace decision making and productivity* (Vol. 1261). AISC.
Alzoubi, H., Alshurideh, M., Kurdi, B. A., & Inairat, M. (2020). Do perceived service value, quality, price fairness and service recovery shape customer satisfaction and delight? A practical study in the service telecommunication context. *Uncertain Supply Chain Management, 8*(3). https://doi.org/10.5267/j.uscm.2020.2.005.
Amarneh, B. M., Alshurideh, M. T., Al Kurdi, B. H., & Obeidat, Z. (2021). The impact of COVID-19 on e-learning: Advantages and challenges. In *The International Conference on Artificial Intelligence and Computer Vision* (pp. 75–89).
Baró, B. F., & Costa-Sánchez, C. (2017). Corporate communication and social media. Spanish companies' communicative activity index on the audiovisual social networks. In *Media and metamedia management* (pp. 189–194). Springer.
Bettayeb, H., Alshurideh, M. T., & Al Kurdi, B. (2020). The effectiveness of mobile learning in UAE universities: A systematic review of motivation, self-efficacy, usability and usefulness. *International Journal of Control and Automation, 13*(2), 1558–1579.
Bhanot, S. (2012). Use of social media by companies to reach their customers. *SIES Journal of Management, 8*(1).

Biswas, A. (2016). Impact of social media usage factors on green consumption behavior based on technology acceptance model. *Journal of Advanced Management Science, 4*(2).
Bogea, F., & Brito, E. P. Z. (2018). Determinants of social media adoption by large companies. *Journal of Technology Management & Innovation, 13*(1), 11–18.
Culnan, M. J., McHugh, P. J., & Zubillaga, J. I. (2010). How large US companies can use Twitter and other social media to gain business value. *MIS Quarterly Executive, 9*(4).
Davis, F. D. (1989). Perceived usefulness, perceived ease of use, and user acceptance of information technology. *MIS Quarterly*, 319–340.
El Refae, G. A., Nuseir, M., & Alshurideh, M. (2021). The influence of social media regulations boundary on marketing and commerce of industries in UAE. *Journal of Legal, Ethical and Regulatory Issues, 24*(Special Issue 6), 1–14.
Elsaadani, M. A. (2020). Investigating the effect of e-management on customer service performance. *AAU Journal of Business and Law مجلة جامعة العين للأعمال والقانون*, *3*(2), 1.
ELSamen, A., & Alshurideh, M. (2012). The impact of internal marketing on internal service quality: A case study in a Jordanian pharmaceutical company. *International Journal of Business and Management, 7*(19), 84–95.
Felix, R., Rauschnabel, P. A., & Hinsch, C. (2017). Elements of strategic social media marketing: A holistic framework. *Journal of Business Research, 70*, 118–126.
George, J., Dietzsch, N., Bier, M., Zirpel, H., Perl, A., & Robra-Bissantz, S. (2014). Testing the perceived ease of use in social media. In *International Conference on Computers for Handicapped Persons* (pp. 169–176).
GMI. (2018). *UAE Facebook usage statistics: 2018 (infographics)*.
Guesalaga, R. (2016). The use of social media in sales: Individual and organizational antecedents, and the role of customer engagement in social media. *Industrial Marketing Management, 54*, 71–79.
Hair, J. F. J., Sarstedt, M., Hopkins, L., & Kuppelwieser, V. G. (2014). Partial least squares structural equation modeling (PLS-SEM): An emerging tool in business research. *European Business Review, 26*(2), 106–121. https://doi.org/10.1108/EBR-10-2013-0128.
Hamadneh, S., & Al Kurdi, B. (2021). The effect of brand personality on consumer self-identity: The moderation effect of cultural orientations among British and Chinese consumers. *Journal of Legal, Ethical and Regulatory Issues, 24*(Special Issue 1), 1–14.
Hayajneh, N., Suifan, T., Obeidat, B., Abuhashesh, M., Alshurideh, M., & Masa'deh, R. (2021). The relationship between organizational changes and job satisfaction through the mediating role of job stress in the Jordanian telecommunication sector. *Management Science Letters, 11*(1), 315–326.
Joghee, S., et al. (2021). Expats impulse buying behaviour in UAE: A customer perspective. *Journal of Management Information and Decision Sciences, 24*(1), 1–24.
Kurdi, B. A., Alshurideh, M., & Alnaser, A. (2020a). The impact of employee satisfaction on customer satisfaction: Theoretical and empirical underpinning. *Management Science Letters, 10*(15). https://doi.org/10.5267/j.msl.2020.6.038.
Kurdi, B. A., Alshurideh, M., Salloum, S. A., Obeidat, Z. M., & Al-dweeri, R. M. (2020b). An empirical investigation into examination of factors influencing university students' behavior towards elearning acceptance using SEM approach. *International Journal of Interactive Mobile Technologies, 14*(2). https://doi.org/10.3991/ijim.v14i02.11115.
Lai, P. C. (2017). The literature review of technology adoption models and theories for the novelty technology. *JISTEM-Journal of Information Systems and Technology Management, 14*, 21–38.
Lane, M., & Coleman, P. (2012). Technology ease of use through social networking media. *Journal of Technology Research, 3*, 1.
Lin, C. A., & Kim, T. (2016). Predicting user response to sponsored advertising on social media via the technology acceptance model. *Computers in Human Behavior, 64*, 710–718.
Madi Odeh, R. B. S., Obeidat, B. Y., Jaradat, M. O., Masa'deh, R., & Alshurideh, M. T. (2021). The transformational leadership role in achieving organizational resilience through adaptive

cultures: The case of Dubai service sector. *International Journal of Productivity and Performance Management*. https://doi.org/10.1108/IJPPM-02-2021-0093.

Manerikar, V., & Manerikar, S. (2015). Cronbach's alpha. *Aweshkar Research Journal, 19*(1), 117–119.

Nikou, S. A., & Economides, A. A. (2017). Mobile-based assessment: Integrating acceptance and motivational factors into a combined model of self-determination theory and technology acceptance. *Computers in Human Behavior, 68*, 83–95.

Nuseir, M. T., Aljumah, A., & Alshurideh, M. T. (2021a). How the business intelligence in the new startup performance in UAE during COVID-19: The mediating role of innovativeness. In *The effect of coronavirus disease (COVID-19) on business intelligence* (Vol. 334).

Nuseir, M., El Refae, G. A., & Alshurideh, M. (2021b). The impact of social media power on the social commerce intentions: Double mediating role of economic and social satisfaction. *Journal of Legal, Ethical and Regulatory Issues, 24*(Special Issue 6), 1–15.

Nuseir, M. T., Ghaleb, A., & Aljumah, A. (2021c). The e-learning of students and university's brand image (Post COVID-19): How successfully Al-Ain University have embraced the paradigm shift in digital learning. In *The effect of coronavirus disease (COVID-19) on business intelligence* (Vol. 334, p. 171).

Obeidat, Z. M., Alshurideh, M. T., Al Dweeri, R., & Masa'deh, R. (2019). The influence of online revenge acts on consumers psychological and emotional states: Does revenge taste sweet? In *Proceedings of the 33rd International Business Information Management Association Conference, IBIMA 2019: Education Excellence and Innovation Management through Vision 2020* (pp. 4797–4815).

Prantl, D., & Mičík, M. (2019). Analysis of the significance of eWOM on social media for companies.

Queiroz, M. M., & Fosso Wamba, S. (2019). Blockchain adoption challenges in supply chain: An empirical investigation of the main drivers in India and the USA. *International Journal of Information Management, 46*, 70–82.

Radcliffe, D., & Abuhmaid, H. (2020). *Social media in the Middle East: 2019 in review* (Available SSRN 3517916)

Robertson, B. W., & Kee, K. F. (2017). Social media at work: The roles of job satisfaction, employment status, and Facebook use with co-workers. *Computers in Human Behavior, 70*, 191–196.

Scherer, R., Siddiq, F., & Tondeur, J. (2019). The technology acceptance model (TAM): A meta-analytic structural equation modeling approach to explaining teachers' adoption of digital technology in education. *Computers & Education, 128*, 13–35.

Sell, A., de Reuver, M., Walden, P., & Carlsson, C. (2012). Context, gender and intended use of mobile messaging, entertainment and social media services. *International Journal of Systems and Service-Oriented Engineering, 3*(1), 1–15.

Simar, L., & Wilson, P. W. (2000). A general methodology for bootstrapping in non-parametric frontier models. *Journal of Applied Statistics, 27*(6), 779–802.

Singh, S., & Srivastava, P. (2019). Social media for outbound leisure travel: A framework based on technology acceptance model (TAM). *Journal of Tourism Futures*.

Sweiss, N., Obeidat, Z. M., Al-Dweeri, R. M., Mohammad Khalaf Ahmad, A., Obeidat, A. M., & Alshurideh, M. (2021). The moderating role of perceived company effort in mitigating customer misconduct within online brand communities (OBC). *Journal of Marketing Communications*, 1–24. https://doi.org/10.1080/13527266.2021.1931942.

Wong, K.K.-K. (2013). Partial least squares structural equation modeling (PLS-SEM) techniques using SmartPLS. *Marketing Bulletin, 24*(1), 1–32.

Zhang, H., Zhang, Y., Ryzhkova, A., Tan, C. D., & Li, F. (2019). Social media marketing activities and customers' purchase intention: The mediating effect of brand image. In *2019 IEEE International Conference on Industrial Engineering and Engineering Management (IEEM)* (pp. 369–373).

Digital Marketing Strategies and the Impact on Customer Experience: A Systematic Review

Mohammed T. Nuseir, Ghaleb A. El Refae, Ahmad Aljumah, Muhammad Alshurideh, Sarah Urabi, and Barween Al Kurdi

Abstract The aim of this study is to explore the contemporary digital marketing strategies and tools and the role played by these in various marketing activities or areas. The study also explores the market segmentation in the digital era to improve customer experience tools. In addition, the study determines which strategy has been suggested to be most optimized for enhancing Customer Experience. The research method comprises a Systematic Literature Review (SLR), which included choosing the key publications, data extraction and synthesis, quality assessment for the chosen

M. T. Nuseir (✉)
Department of Business Administration, College of Business, Al Ain University, Abu Dhabi Campus, Abu Dhabi, UAE
e-mail: mohammed.nuseir@aau.ac.ae

G. A. El Refae
Department of Business Administration, College of Business, Al Ain University, Abu Dhabi Campus, Abu Dhabi, UAE
e-mail: ghalebelrefae@aau.ac.ae

A. Aljumah
College of Communication and Media, Al Ain University, Abu Dhabi Campus, P.O. Box 112612, Abu Dhabi, United Arab Emirates
e-mail: Ahmad.aljumah@aau.ac.ae

M. Alshurideh
Department of Management, College of Business Administration, University of Sharjah, Sharjah, United Arab Emirates
e-mail: m.alshurideh@ju.edu.jo; malshurideh@sharjah.ac.ae

Department of Marketing, School of Business, The University of Jordan, Amman, Jordan

S. Urabi
Abu Dhabi Vocational Education & Training Institute (ADVETI), Abu Dhabi, UAE
e-mail: sarah.urabi@adveti.ac.ae

B. A. Kurdi
Department of Marketing, Faculty of Economics and Administrative Sciences, The Hashemite University, Zarqa, Jordan
e-mail: barween@hu.edu.jo

M. Alshurideh et al. (eds.), *The Effect of Information Technology on Business and Marketing Intelligence Systems*, Studies in Computational Intelligence 1056,
https://doi.org/10.1007/978-3-031-12382-5_2

publications, and assessing and presenting the results. This systematic review is important for marketing professionals as it emphasizes the significance of selecting appropriate digital marketing strategies as per marketing activities to maximize customer experience. The results of SLR identify different contemporary marketing strategies such as eWOM, emailing, affiliate marketing, search engine optimization, social media marketing, and corporate blogging. Among these strategies, Social Media Marketing is found to be most effective for the brands in their endeavor to maximize customer experience. It is because of the personalization, customization as well as an interaction which this digital strategy offers to the customers.

Keywords Digital marketing · Strategies · Tools · Customer experience

1 Introduction

The development of digital technology has allowed the emergence of a new environment of social interaction that facilitates and demands, at the same time, a profound transformation of the marketing strategies (Alshurideh et al., 2021; Alzoubi et al., 2021). Digital marketing is currently one of such marketing strategies in which many companies are dedicating greater investment (Lee et al., 2022; Tariq et al., 2022). This new marketing paradigm has been focused, specifically, on the careful management of the relationship between the customers and the company to have information about the customers, their characteristics, needs, and preferences (Edelman, 2010; Alkitbi et al., Alkitbi et al., 2020; Alsharari & Alshurideh, 2020).

Digital marketing is a fundamentally set of approaches, strategies, and tools to promote services and products on online platforms: emails, blogs, social networks, websites, mobile, SEO, etc. (Almaazmi et al., 2020; Nuseir et al., 2021). The achievement of digital marketing initiates with the on-going process of converting leads into loyal customers with a positive customer experience (Cook, 2014). It offers a set of strategies, tools, techniques, and operations coordinated through the Internet to increase the sales of a product or service (Aljumah et al., 2021; Sweiss et al., 2021). As per Parise et al. (2016), digital marketing varies from the conventional marketing manner through the methods and channels. It can be asserted that digital marketing helps to monitor things like conversions, what content works and what does not; how many individuals visit the business web page, collaborate with its social networks, search it in the enormous world of web, etc., and in short can offer the measurement of customer experience in real-time (Al Kurdi & Alshurideh, 2021; Alshurideh et al., 2016).

In the current era, novel technologies and particularly the potential of the Network, have been consolidated as an ideal complement for the establishment of digital marketing activities that include notices on Websites, e-mailing or mass mailings, search marketing, the usage of social networking platforms, and blog marketing among others (Alshurideh et al., 2019; Kurdi et al., 2021). A service or product with an appropriate digital marketing strategy or tool can generate a positive customer experience (Alshurideh, 2022; Edelman, 2010). Therefore, companies need to establish an adequate digital marketing strategy to generate traffic on the Web, capture potential customers and speed up effective communication with them by providing them with answers or solutions to their needs. In addition, companies need to recognize different types of customers as an essential part of generating a successful customer experience program having strategies that bring about a personalized experience for each segment.

2 Research Objectives

This research aims to explore the contemporary digital marketing strategies and tools and the role played by these in various marketing activities or areas. The study also explores the market segmentation in the digital era to improve customer experience tools. Lastly, the study determines which strategy has been suggested to be most optimized for enhancing Customer Experience.

3 Literature Review

3.1 Customer Experience

The customer experience is defined as perceptions of consumers or users, conscious and subconscious of their relationship with the brand as a result of all interactions during the life cycle of this.

As per Meyer and Schwager (2007), the customer experience is now more than ever relevant for three main reasons. First, the customer experience is playing a critical role in the ability of companies to differentiate themselves from their competitors. Second, the expectations of the customers are changing, many times because of the appearance of new businesses (many of them with a relevant digital component) or new companies in the market that have managed to break "established barriers" (such as WhatsApp, breaking the concept of traditional communication between people). Last, customers no longer expect an incredible customer experience only from large multinationals; they expect it from any company, including small and medium-sized enterprises, since it is not a matter of scale, but to develop the qualities and capabilities required for it (Meyer & Schwager, 2007).

Literature in customer experience management has identified three key levers to offer a superior customer experience: employees, organization, and "detail management" (Verhoef et al., 2009; Alshurideh et al., 2012; Verma et al., 2012; Rawson et al., 2013; Alzoubi et al., 2022). Personal employees are one of the main points to change the customer experience; they are the first line of contact with consumers and should be skilled and exceed the needs of customers (2020b; Kurdi et al., 2020a; Verhoef et al., 2009). Another aspect is the organization, which means that the customers at the center of all decisions, creating a responsibility at the management team level and making the customer experience a very powerful function within the business (Verma et al., 2012). Last but not least, leaders in customer experience management invest more time in understanding in greater depth the needs of the different segments of clients, or even of individual clients, to be able to offer them more personalized services (Alameeri et al., 2020; Rawson et al., 2013). However, how can a company go from offering a good customer experience to an outstanding one? From the literature point of view, everything is related to offering a personalized experience with a clear brand image and end-to-end vision at any point of contact. For many companies, this can be achieved by implementing digital marketing strategies. In the view of Cook (2014), the ultimate goal of exceeding the client's expectations via digital marketing would attract the client and turn him into a loyal fan of the brand and the product.

3.2 Digital Marketing

Digital marketing uses mobile devices, social media, the internet, search engines, and other mediums to get to customers. Specific marketing experts like Ryan (2016) and Chaffey and Ellis-Chadwick (2019) regard digital marketing as a novel endeavor that necessitates a new method of approaching consumers and comprehending how they behave in contrast to traditional marketing. The implementation of digital tools, together with traditional communication between customers and the company to achieve marketing objectives, is named digital marketing (Chaffey & Ellis-Chadwick, 2019, p. 20). It is an advanced way of advertisement to present the customers with the information materials they require through diverse digital tools. Eventually, traditional marketing and digital marketing have no huge difference in the industry. Nevertheless, the manner of making contacts and delivering info to the customers are more innovative in the latter (Tiago & Veríssimo, 2014).

4 Research Method

The research was carried out in line with the Hair et al. (2008) "*Essentials of marketing research,*" which included tasks such as developing the procedure or review, recognizing and choosing the main publications, data extraction and synthesis, quality assessment for the chosen publications, and assessing and presenting the results.

The researcher used NVivo software to achieve the systematic mapping of the carefully chosen publications. One main advantage of NVivo is that it can distinguish the same papers spontaneously, lessening the assessors' effort during the selection stage.

4.1 Research Question

RQ1: What are the various marketing activities/areas, and how digital marketing strategies and tools play a role in these activities/areas?
RQ2: What are the contemporary strategies and tools in digital marketing?
RQ3: How Market Segmentation can be achieved in the Digital era to improve Customer Experience?
RQ4: Which digital marketing strategies are most effective in enhancing Customer Experience?

4.2 Search Process

The search process for this SLR began with a systematic search, together with the assessment of references for a number of studies. The digital databases used for the SLR were: ASCE Library, Google Scholar, ScienceDirect, and ASCE Library.

Systematic exploration was carried out on publications' full text and restricted to "Digital Marketing" and "Customer Experience" via the search engine filter. For every database, different keywords and search strings were used for the research questions, and the findings were combined (see Table 1).

4.3 Selection Process

The selection procedure included five stages. The first stage was carrying out a string search on databases and attaining the findings. The second stage was eliminating dual publications via NVivo. The third was investigating the headings and abstracts of studies from each digital database in comparison to the exclusion and inclusion benchmarks and eliminating the irrelevant studies. The fourth stage was revising the whole text of all the chosen studies that remained after the second stage again on the base of inclusion and exclusion benchmarks. The fifth stage was once more looking for the repetition of studies by examining articles having the same authors and including related research areas. In this condition, the up-to-date publications were selected.

The researcher inspected the citations of all main publications remained after the fifth stage to distinguish any other publication not by now included in the designated main studies. After that, these studies had to go through the same selection stage. Figure 1 shows the selection process for the studies and the number of selected

Table 1 Keywords and search strings for searching the studies

Search Strings	Google scholar	ASCE library	ASCE library	Science direct	Total
Digital marketing tools	4	5	1	2	12
Social media marketing + Customer Experience	3	3	2	4	12
Targeting customers in digital era	4	7	4	9	24
Digital customer experience	6	9	6	5	26
Contemporary digital marketing strategies	7	7	2	4	20
Omni-channel marketing strategies + Customer experience	5	3	4	2	14
	29	34	19	26	**108**

studies at each stage. The first stage shows the total publications gotten from each digital database, which was 108 altogether. The circles on the left-side include the number of remaining studies after each stage for the SLR. The circles on the right-side contain the systematic of outstanding studies after each selection stage for the citation search, which was carried out via 10 papers gotten through systematic search. The last sample included 39 studies.

4.4 *Exclusion and Inclusion Benchmarks*

For the studies to be on a sample, it should be in English and handle digital marketing and customer experience. The researcher intended to include studies that explore contemporary digital marketing approaches to enhance the customer experience even if the authors do not assume clearly that digital marketing is the ultimate way for a positive customer experience. On the other hand, several types of papers and articles were eliminated: papers that do not take into account digital marketing and rather focus on traditional marketing for customer experience; publications that summarize results from prevalent studies like roadmaps, surveys, and reviews; repeated papers and finally, articles that cannot be downloaded from digital databases.

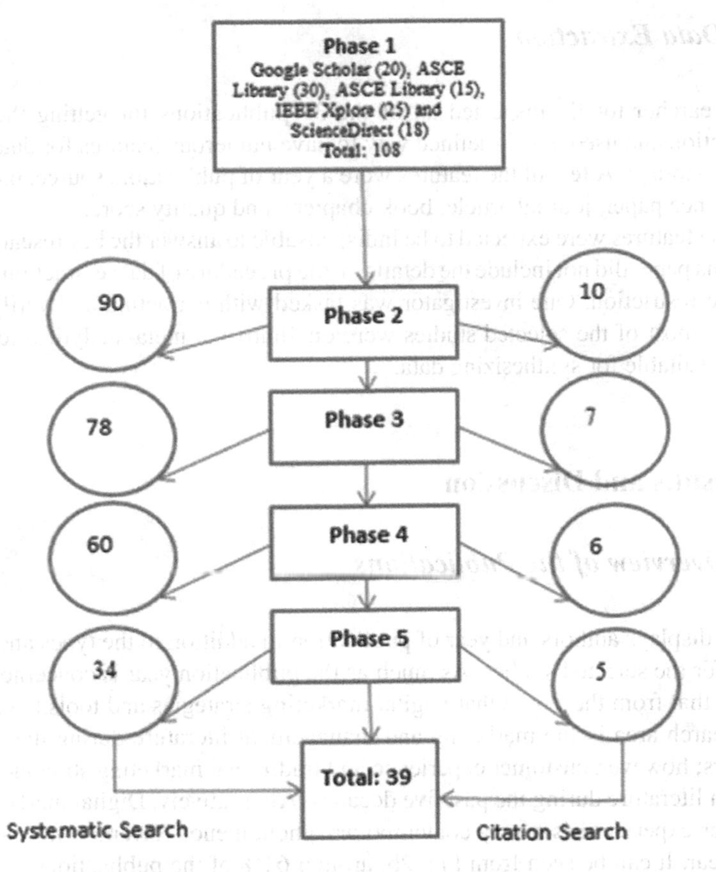

Fig. 1 The selection procedure for the studies and the number of selected studies at each stage

4.5 Quality Evaluation

The quality evaluation is a subjective process as different researchers value different aspects of the research. In this systematic review, the researchers chose the inclusion and exclusion benchmarks for the evaluation of quality. The quality evaluation is based founded on the specified by Hair et al. (2008). Some of the key points were: The aims of the publication are defined in a clear manner (Yes = 1 or No = 0); The publication efficiently describes the matrices (Yes = 1 or No = 0); The data gathering methods are defined well (Yes = 1 or No = 0); The findings are presented in a clear and an unambiguous way (Yes = 1 or No = 0); The findings are supported by empirical evaluation (Yes = 1 or No = 0); The threats to the paper validity are reflected (Yes = 1 or No = 0).

4.6 Data Extraction

The researcher totally inspected all of the 39 publications for getting the crucial information and used a well-defined way to have numerous features for data extraction and storage. A few of the features were a year of publication, source, paper type (conference paper, journal article, book chapter), and quality score.

These features were expected to be indispensable to answer the key research question. This paper did not include the details on the procedure of data extraction because of space restriction. One investigator was tasked with extracting and verifying the data. As most of the selected studies were qualitative, a meta-analytical technique was not suitable for synthesizing data.

5 Results and Discussion

5.1 Overview of the Publications

Table 2 displays authors and year of publication in addition to the types and quality scores for the selected studies. As much as the publication year is concerned, it can be seen that from the Fig. 2 that Digital marketing strategies and tools have been a hot research area in the marketing and management literature during the previous ten years; however, customer experience and traditional marketing strategies can be found in literature during the past five decades. Alternatively, Digital marketing and customer experience is a very contemporary phenomenon. Averagely, 3 published every year. It can be seen from Fig. 2b, around 61% of the publications were journaled articles, while 31% were reported. On the other hand, 11 publications got the maximum score which meant that they defined their objectives, metrics, and data collection methods effectively. When classifying the publications in Table 2, the researcher coded journal articles as 'J,' reports as 'R,' and conference papers as 'C.'

It can also be seen that the researcher restricted the date of publication in the search stage to nine years, i.e., 2010 to 2019. The reason for choosing 2010 was that as this was the year when the most relevant study was published. This shows that there is much room for growth within this area of marketing literature.

For the quality score, the researcher used the aforementioned 6-point criteria (0–1 score) to evaluate the sample. The higher the score, the detailed and clearer the study is to the present SLR. The lowest score acquired by the study comprised in this SLR was 1. Only 2 studies got this score. These publications failed to include any empirical methods efficiently. Most of the publications achieved either a 4 or 5 score, which means that the general sample was strong in nature as the publications backed empirical evidence. This rationalizes the effort to carry out the SLR in this emerging area of marketing.

Table 2 Overview of the publications as per year, quality score, and type

Sr. number	Authors and year	Quality score	Publication type
1	Edelman (2010)	5	J
2	Chaffey (2010)	4	J
3	Peterson et al. (2010)	2	J
4	Munro and Richards (2011)	4.5	J
5	Heller and Parasnis (2011)	6	C
6	Hollebeek (2011)	3	J
7	Gopalani and Shick (2011)	4	C
8	Smith (2012)	6	C
9	Verma et al., (2012)	1	J
10	Hudson et al. (2012)	5	R
11	Mitic and Kapoulas (2012)	5	J
12	Dumitrescu et al.(2012)	4	J
13	Rawson et al. (2013)	5	J
14	Nash et al. (2013)	6	J
15	Tiago et al. (2014)	6	J
16	Cook (2014)	2	J
17	Solis et al., (2014)	5	R
18	Ryan (2014)	5	R
19	Schneider (2014)	6	R
20	Dennis et al. (2014)	3	J
21	Dörner and Edelman (2015)	4	R
22	Edelman and Heller (2015)	6	R
23	Bennett and El Azhari (2015)	6	R
24	Kane et al. (2015)		J
25	Lemon and Verhoef (2016)	6	J
26	Parise et al. (2016)	4	J
27	Ryan (2016)	4	R
28	Jackson and Ahuja (2016)	6	J
29	Bilgihan (2016)	6	R
30	Gill and VanBoskirk (2016)	4	J
31	Payne et al. (2017)	4	J
32	Homburg et al. (2017)	5	J
33	Bolton et al. (2018)	5	J
34	Bughin et al. (2018)	3	R
35	Iankova et al.(2018)	4	R

(continued)

Table 2 (continued)

Sr. number	Authors and year	Quality score	Publication type
36	Petit et al. (2019)	1	J
37	Goodman (2019)	2	R
38	Chou (2019)	4	J
39	Flavián et al. (2019)	3	J

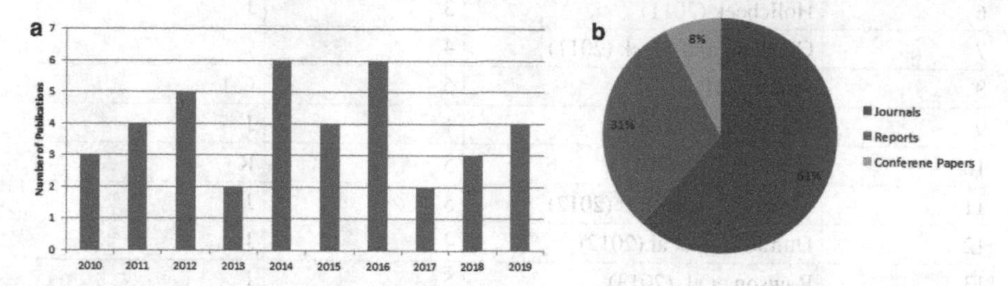

Fig. 2 **a** Number of publications per year; **b** Types of publications

5.2 *Marketing Activities and Related Digital Strategies*

The systematic review of the sample showed that digital marketing strategies vary as per marketing activities. The researchers identified different marketing activities and related digital marketing strategies presented below.

Market research: In this area, reference is made to all the tools designed to understand better the market in which it competes and the targets in the ones to focus on to obtain leads that can easily become customers. An example of a digital marketing tool in this area is electronic surveys (Edelman, 2010) and online focus groups (Smith, 2012).

Brand: It refers to all actions aimed at improving product or service awareness (brand recognition) among the public of interest. Some of the best electronic tools to improve the top of mind (positioning) and the share of voice (SOV or voice participation) are, for example, corporate blogs and direct search engine marketing (Cook, 2014).

Product: Actions oriented to a product or a line of them. Some of the proposed tools in the publications are digital product verifiers or online product testing (Peterson et al., 2010; Al-Dmour et al., 2021; Al-Khayyal et al., 2021; Khasawneh et al., 2021a).

Price: This area includes the actions related to the price variable for improving the marketing mix. The digital world offers important contributions in this area, such as digital price timing or electronic auctions (e-auctions) (Gill and VanBoskirk, 2016).

Communication: This area and those of promotion and advertising are integrated into the model of the 4P's; however, in practice, it is appropriate to treat the actions separately of communication in which the company does not pay for appearing in a third-party media (advertising) or encourages the purchase of a direct form (promotion). As stated by Homburg et al. (2017), virtual communities and RSS marketing (Content Syndication Systems) are an explanatory example from the electronic point of view.

Promotion: The point of sale is one of the marketing scenarios in which the digital revolution has been noticed more because it is detached in a way totally from the Internet like the other forms of electronic promotion such as digital coupons or re-merchandising (Al-Dmour et al., 2014; Dennis et al., 2014; Ryan, 2016).

Advertising: Online advertising is commercial and informative information expressed with creativity about a product or service on the network (Alshurideh et al., 2017). As an example, Homburg et al. (2017) cited buzz marketing or contextual advertising.

Distribution: Like, the price and the product, in this area, the correspondence between theory and practice is clear and direct. Examples of tools are infomediaries (Mitic & Kapoulas, 2012) or affiliate marketing (Bolton et al., 2018).

Marketing: Without a doubt, the ultimate goal of marketing is selling, so that in practice, there must be an area that groups together all the tools to make it more profitable and productive, which from the electronic channel they get without going away the marketplaces and the e-commerce portals (Alzoubi et al., 2022; Kane et al., 2015).

Control: It is necessary to measure the effectiveness of digital marketing tools to improve both the selection and how to use it (strategy). The Customer Relationship Management (CRM) which is based on knowing the customer as much as possible and identify to produce what customers want to buy, and the Gross Rating Point (GRP), which corresponds to the metrics par excellence used to measure the impact of spots determining how much of the target the company has reached. The CRM (Heller and Parasnis, 2011) and the electronic GRP (Bolton et al., 2018) constitute excellent audit tools for e-marketing.

The following table summarises the findings for the various marketing activities and related digital marketing strategies and tools.

5.3 Digital Marketing Strategies

5.3.1 Viral Marketing or Buzz Marketing

Viral marketing or buzz marketing is one of the most examined digital strategies in the chosen publications. Also called the electronic word of mouth (eWOM), this strategy encourages individuals to quickly convey a commercial message to others seeking to create exponential growth in the exposure of said message (Edelman, 2010). As per Chaffey (2010), it deals with commercial communication that propagates by itself. It is a tool of communication on the Internet that allows the diffusion of a message, starting from a small emitter core that is multiplied by the collaboration of the receivers in their transmission and diffusion, generating a pyramidal effect with geometric growth (Nash et al., 2013). With the viral marketing advertising campaigns spread like a virus through the network, and the users are the ones who share and transmit the advertising message, the costs associated with this strategy are considerably low or zero.

Nash et al. (2013) maintained that any eWOM campaign on the network should be feasible to be shared or transmitted from one user to another. The ease that users have to pass a message to others will directly influence the scope and spread of the campaign. So, if a firm publishes a very interesting article, it should allow with just a press of a key to be forwarded to a friend; or if a video is published, it must be ensured that it can be download in a format compatible with most users and that additionally can be easily forwarded to other users; if an image is loaded; that it can be downloaded and forwarded easily; if an application is created; that it can be installed in any operating system or if it is a document then it can be added to the social networks (Edelman, 2010).

5.3.2 Email Marketing

This digital marketing strategy consists in the use of electronic mail as a loyalty strategy; that is, it proposes to attract new customers and retain customers already acquired through the sending of messages to email accounts to maintain a continuous dialogue with the client throughout the entire commercial relationship and finally feed a certain base of data (Peterson et al., 2010; Smith, 2012). The foundation of this strategy is focused on the fact that more than one trillion emails are sent every day. This behavior makes e-mail marketing one of the most effective, fastest tools and to carry out advertising campaigns directly through electronic mail system because it is an ideal means of carrying out any type of one to one marketing action (Smith, 2012).

In the view of Parise et al. (2016), e-mail marketing is a good complement to offline marketing. Some of the benefits highlighted in the studies are interactivity, customization, low intrusiveness, savings, reduction of times and distances; it also avoids inconveniences related to legal and brand risks. Its purpose is to generate a good price to induce the desired purchase.

5.3.3 Affiliate Marketing

This digital marketing strategy refers to online platforms where advertisers are contacted who want to advertise their brand, their products, or their services with Web pages of all kinds and size that are intended to include advertising on the site for economic purposes (Peterson et al., 2010; Cook, 2014; Payne et al., 2017). The advertisers indicate the commission they offer to the media for using their advertising and, according to the commission, use an advertiser. Therefore the affiliate network rents the commission that the advertiser pays to the affiliate. Affiliate marketing is a form of Internet distribution based on commissions in which one company (advertiser) rewards economically another (publisher) for generating business through a series of links introduced in the website (Payne et al., 2017). These Hyperlinks can be routed to the direct sale of the product or sent to the user who has accessed the website to the selling company's page.

5.3.4 Search Engine Marketing

Search engine optimization consists of applying techniques and strategies to locate preferentially, via keywords or words key, websites in the search engines. The main motivation is that the companies want and need to be well-positioned in the network, which is equivalent to be among the first thirty results offered by different search engines (Munro & Richards, 2011). As per SLR, it is important to be positioned in the network; although "being" in the network means very little, the important thing is "to be visible to the user" (Gopalani & Shick, 2011; Munro & Richards, 2011).

5.3.5 Social Media Marketing (SMM)

The Internet makes it possible to have conversations between individual customers and companies that were simply impossible in the era of traditional media (Almazroue et al., 2020; Khasawneh et al., 2021b). Through Social Media, a company can converse individually with all of its customers in a unique and personalized way because the SMM are tools of communication that allow listening and speaking with the client (Heller and Parasnis, 2011; Rawson et al., 2013; Edelman and Heller, 2015). From a technical approach, SMMs can be described as platforms such as Facebook, Twitter, Instagram, MySpace, YouTube, LinkedIn, and others to publish content where the users make decisions about the published content (Al-Maroof et al., 2021). These users become people influencers whose thematic agenda is subject to personal circumstances and/or professionals of each.

One of the most significant SMM approaches is Facebook marketing (Edelman and Heller, 2015). Facebook, with more than 800 million users, works equivalently to a meeting place between customers or users with similar needs and interests who share, both professionally and personally. So it is considered the most important social network worldwide. From the perspective of marketing, it is especially interesting for products and services aimed at markets having huge consumption potential (Edelman and Heller, 2015). Many companies have their profile on Facebook and constitute one of the best tools when communicating and interacting with customers.

5.3.6 Corporate Blogs

According to Goodman (2019), the term blog corresponds to a Web frequently updated, chronologically structured offering information about one or several topics in an informal and accessible way, written by one or more authors and where the most recent one appears first and with the author retaining the freedom to publish what he/she thinks relevant. These are weblogs of personal, journalistic, business or corporate, technological, etc. Corporate blogs are published to achieve business objectives such as the positioning of the brand, the firm, and the internal communication where blogs act as a management tool for the knowledge and external communication to strengthen relationships with key groups of clients (Bughin et al., 2018).

5.3.7 Other Tools

Mobile Apps: Mobile applications are essential within an e-marketing strategy. The customer experience is fundamental, and the reputation of the brand is at stake. Some studies such as (Mitic & Kapoulas, 2012) and (Dennis et al., 2014) recommended including mobile applications within a company's mobile marketing strategy and taking advantage of the capabilities of mobile phones, making them simple and useful, integrating them into a marketing strategy of the company, using external Apps, choosing the right platform, including social networks, attracting the client and measuring the result.

Electronic product configurators and verifiers: An online product configurator is a special software that is inserted inside a site and allows the potential buyer to customize their product according to all the options available to them (Lemon & Verhoef, 2016). A product verifier is a system (software + hardware) that focuses on choosing options to show the product from all possible points of view. The navigator has experience as close as possible to that of a "physical test" of the product. Lemon and Verhoef (2016) highlighted that it is especially useful in products such as clothing and increases the customer perception of the brand and promotes trust.

Electronic coupons: e-coupons seek that the buyer prefers a brand, product/store in exchange for a discount on the purchase, a sample, or a free product (Jackson & Ahuja, 2016). The traditional coupon required going to the establishment to benefit from the discount; the new technologies reduced the process by just clicking on the computer or pressing a button on the phone to capture coupons to be exchanged electronically.

e-Survey: This tool facilitates marketing research by evaluating processes to determine the loyalty ratios, level of satisfaction with the purchase and customer service, etc. (Jackson & Ahuja, 2016). This tool also facilitates the client's relationship to ensure that the products, services, and prices satisfy them, know better the tastes, satisfactions and interests, and segment them to make a more personalized offer.

5.4 Market Segmentation in the Digital Era and Customer Experience

To adopt an effective digital marketing strategy in order to create a positive customer experience, it is essential to defines the profiles of the customers and places them in the appropriate segment. In the view of Hollebeek (2011), customer segmentation allows targeting particular groups of customers efficiently. Studies defined different market segments such as Backuppers, Bluetooths, Browsers, Hackers, Trojans, and Virals (Flavián et al., 2019; Iankova et al., 2018; Ryan, 2014; Verma et al., 2012). Examining these segments would also help in understanding how negative customer experience would hamper the company/business.

Backuppers: Corresponds to those satisfied customers who do not pronounce or exercise any kind of influence in the market (Verma et al., 2012). The business achieved through these clients is what they generate personally, i.e., the Medium Business Potential (MPN) or the VEC. These clients allow the company to survive without risk. The MPN is the weighted sum of the client's successful experience and the value that can generate the weighting coefficients depending on the strategic objectives of the company.

Bluetooth: The customers are satisfied enough to boast of their belonging to the company/organization (Ryan, 2014). They exert a positive influence on the market. The business will be the sum of the personal business plus the amount generated by its influence, which corresponds to the Net Positive Influence Index (NPII). The volume of the business that the clients of this segment suppose is: (NPII + 1) x MPN.

Browsers: The customers who, not being satisfied, do not keep the relationship, but they also have no influence on the market, so they do not destroy extra value, but customers lose only the MPN in this segment (Iankova et al., 2018).

Hackers: They are those customers who, in addition to not being satisfied with the products and/or services of the company, not only do not maintain the relationship but rather exert a strong negative influence among potential customers (Flavián et al., 2019) which corresponds to the Net Negative Influence Index (NNII). Therefore, the destruction of value due to this type of client is: (NNII + 1) x MPN.

Trojans: These are those customers that are maintained due to the different obstacles that condition their exit and that the company has established, for example, permanency contracts that under the assumption of loyalty establish some companies with which the user has no other option but to stay and continue paying, but the negative influence they exert on the market due to the mixture of dissatisfaction and frustration is very dangerous for the company (Iankova et al., 2018; Schneider, 2014). It corresponds to a specific destruction segment that can be evaluated with (NNII – 1) x MPN. The value (−1) is maintained during the period of time in which the obstacle for the exit of the client, that is, the time during which he continues to pay the fee for the service rendered.

Virals: These are those customers who are necessary to conquer more and more. If their experience is satisfactory with the acquisition of product/service, they communicate it to others, that is, exert a strong positive influence and attract undecided

Table 3 Digital marketing activities and related digital marketing strategies and tools

Marketing activities	Digital marketing strategies and tools	References
Market research	Electronic surveys; Online Focus Groups	Edelman (2010), Smith (2012)
Brand	Corporate blogs; Search engine marketing; The electronic word of mouth (eWOM)	Cook (2014)
Product	Digital product verifiers or the online product testing	Peterson et al. (2010)
Price	E-Auctions	Gill and VanBoskirk (2016)
Communication	RSS Marketing; Virtual communities	Homburg et al. (2017)
Promotion	Digital coupons or E-Merchandising	Dennis et al. (2014), Ryan (2016)
Advertising	Buzz marketing; Contextual advertising	Homburg et al. (2017)
Distribution	Infomediaries or affiliate marketing	Bolton et al. (2018), Mitic and Kapoulas (2012)
Marketing	E-Commerce portals; Social media, Blogs, Content creation	Kane et al. (2015)
Control	CRM and the electronic GRP	Heller and Parasnis (2011), Bolton et al. (2018)

customers or others who had not even considered it; therefore, the volume of business that they contribute, is the derivative of their influence, constituting a new continuity (Bilgihan, 2016; Flavián et al., 2019). Its evaluation corresponds to the expression: (NPII – 1) x MPN.

The following Table 3 shows that summary for the different market segments and indexes to measure customer experience (Heller and Parasnis, 2011; Rawson et al., 2013; Edelman and Heller, 2015) (Table 4).

6 Discussion

In contemporary years, it has become clear that all effective marketing strategies intended to improve the customer experience. Their voices are the ones that establish brands, form the market, and directly affect sales, so it appears rational that marketing determinations must work towards enhancing their experience (Tiago et al., 2014).

The SLR presented several strategies and tools which can enhance customer experience and generate a positive perception of the brand. One of the most important strategies is the eWOM campaign, which, as per the scholars, can effectively maximize customer experience (Kane et al., 2015; Ryan, 2016). This tool does not require

Table 4 Different market segments and indexes to measure customer experience

Different market segments	Indexes to measure customer experience	References
Backuppers	MPN	Verma et al. (2012)
Bluetooths	(NPII + 1) x MPN	Ryan (2014)
Browsers	-MPN	Iankova et al. (2018)
Hackers	(NNII + 1) x MPN	Flavián et al. (2019)
Trojans	(NNII – 1) x MPN	Iankova et al. (2018), Schneider (2014)
Virals	(NNII + 1) x MPN	Bilgihan (2016), Flavián et al. (2019)

much effort since the action of the users themselves transmits the message; the key is to find the key attraction that drives the multiplier diffusion. Likewise, it offers a quick brand promotion, a small investment volume, a very high response rate, and generates pre-existing social networks offering great efficiency and interactivity. However, Gill and VanBoskirk (2016) argued that customer experience could be affected badly if the loss of the message occurred or the interruption of the file by the antivirus preventing the reception of the message or the distortion of the message by the receiving audience; thus, creating a negative customer experience.

Likewise, the SLR highlighted e-mailing as the most powerful and effective e-marketing channel with the potential to maximize customer experience. In the view of Bilgihan (2016), the perception and experience of a customer concerning a certain product improve when the advertising message is personal and interactive. E-mailing, being a multidirectional channel, allows knowing the clients' opinions and allows real-time tracking of the action's effectiveness (Hudson et al., 2012). Likewise, the recipient of any communication via e-mail must have authorized and consented to send messages (permission marketing), what otherwise constitutes reportable illegality in several countries like the US and Spain. In addition, Homburg et al. (2017) maintained that to maximize customer experience. An e-mail must be short, no more than 30 or 40 characters, or a maximum of seven or eight words and include words such as "offer," "gift," "urgent," or questions that generate interest of the customers (Nash et al., 2013). Customization is the key when maximizing customer experience via e-mail marketing. However, similar to eWOM, email marketing can create a negative customer experience if messages are sent with false and misleading content. Customers can also become a Spam victim, which would also damage the customer experience (Hudson et al., 2012).

Another strategy is affiliate marketing, which aims to transform the customers into affiliates. Affiliation agreements between advertisers and sellers are operationalized through the so-called affiliate programs to which you can Access by click and by sale (Chou, 2019). Affiliate marketing can do a lot to construct trust, particularly when the affiliate has an enormous following. Instead of asking consumers to take the company's word for the quality of its goods and services, it asks someone else to do this for it while growing the experience and reach of the brand. Bolton et al. (2018) argued that the usage of referral programs enhances the impact of affiliate programs on customer experience. DropBox, Netflix, Uber are all incredibly successful examples of how "refer a friend" can boom a business. Some companies like Dell incorporate affiliate marketing with Electronic product configurators, a very complex tool from the technological point of view. A classic example of this type of tool is Dell computers' brand that allows customers to configure computers that they buy according to their online preferences without limiting those already configured with fixed characteristics (Bolton et al., 2018).

Furthermore, search engine marketing or SEO aims to create a positive customer experience by showing the customers the products on the first page of the engine. Studies show that users do not usually go beyond the third page of results. To achieve, it is necessary to optimize the Web page, that is, adapt it so that seekers understand it better and value it more (Dumitrescu et al., 2012). This is the most economical way to position the company on the Internet. The first thing for a positive customer experience is to choose the keywords for the activity to be published. Petit et al. (2019) maintained that it is necessary to perform comparative analyses with the competition through dashboards that allow companies to monitor certain words or combinations of words.

As mentioned, for the generation of positive customer experience, personalization, and customization, as well as interaction, are fundamental aspects of the digital era. Marketing via social media offers the customers the ease of access, participation, personalization to the user's taste and conversation, among others (Nash et al., 2013). The most striking feature of Social Media is the viral capacity that its contents acquire and from a business point of view. This relevance can occur between activities such as purchase, consumption or capture of attention. Numerous social networks aim to enhance the customer experience in a different approach and are suggested by several authors. The following table shows the types of these networks (Table 5).

As far as corporate blogging is concerned, similar to SMM, it tends to create decentralized information engaging the customers on a large scale (Tiago et al., 2014). Some of the studies regarded blogs as one of the most effective approaches to enhance the customer experience owing to faithfulness, loyalty, and trust among the customers. Faithfulness is based on the relationship of trust that is established with the blog editor, hence the importance of respecting their principles, and although they may become corporate vehicles or commercial the rules have to be maintained. Studies such as (Dörner & Edelman, 2015; Solis et al., 2014) argued that similar to e-mailing, if used properly, a blog allows the company's positioning as an organization of experts least in one suitable place. Because of its social environment, blogs allow the customers to participate publicly, free, and in real-time. Any reader can know the opinion of all the participants, which is a great advantage over the traditional written press. This is one of the characteristics that has made blogs the main drivers of the Web 2.0 phenomenon. However, they entail the danger that certain comments made without foundation affect the image and the positioning of some company, a risk it shares with viral/buzz marketing or eWOM (Petit et al., 2019).

Table 5 Types of social networks and respective approach to enhance customer experience

Types of social networks	Approach to enhance customer experience	Examples	References
Content, social networks	Builds relationships by joining profiles through published content; files, and documents shared	Scribd, Flickr, Bebo, Friendster, Dipity, StumbleUpon, and FileRide	Heller and Parasnis (2011)
Social networks of inert objects	Unites brands and places; make up a novel sector between social networks	Respectance	Nash et al. (2013), Rawson et al. (2013)
Sedentary social networks	By creating flexible and dynamic relations between people, the shared contents or the events created	Rejaw, Blogger, Kwippy, Plaxo, Bitacoras.com, Plurk	Edelman and Heller (2015)
Nomadic social networks	networks are composed and recomposed to the tenor of the subjects who are geographically close to the place where they find the user	Foursquare, Gowalla, Latitude, Brightkite, Fire Eagle, and Skout	Bolton et al. (2018)
Social Web Network	Based on a typical structure of Web and enhances customer experience via eWOM	Facebook, Twitter, Instagram, MySpace, YouTube, LinkedIn	Bolton et al. (2018)

7 Threats to Internal and External Validity

For the SLR findings, the key threats can be those associated with external and construct validity and reliability. This is because the chosen research method, i.e., SLR, mainly focused on the selection and categorization of previous studies.

The threat to external validity is when the competence to generalize the publication's findings is restricted (Klink & Smith, 2001). For the present study, a likely threat would be whether the chosen publications characterize a substantial portion of the past studies on current digital marketing strategies and customer experience. The researcher intended to reduce this threat by employing four online databases for the systematic search and employing the citation research method. Still, the researcher overlooked several digital databases and did not carry out manual research. Consequently, there are likely a few papers that could not be retrieved from the libraries and were not referred to by any of the publications carefully chosen.

Construct validity is the extent to which a test measures what it purports or claims to be measuring (Heale & Twycross, 2015). In the present study, the key threat was whether the search strings reflected the original intentions. The researcher aimed to reduce this threat by authenticating the strings from a marketing professional. When validating the strings, the researcher confirmed if publications were incorporated in the sample.

Reliability is related to what degree the data and the analysis are reliant on the particular researchers (Heale & Twycross, 2015). In the present study, the extraction of data was achieved by three experts within the field of marketing. Thus, this threat was minimized by selecting the maximum number of researchers for the task.

8 Conclusion

In the digital era, customer experience becomes positive if the information reached in the first instant is distributive, horizontal, and backed by many references such as peers and colleagues. This can be achieved most effectively via social media marketing, especially Facebook marketing. Digital marketing strategies such as social media marketing allows the development of products of the client's preference since the social networking platforms like Facebook and Twitter facilitate research in a more personalized way in order to develop products that necessarily enhance the customer experience. In addition, using social media does not generate high costs.

The customers who connect to social networks do so because they are interested in the information that is published by the contact list; while generating content that their friends follow, becoming a platform that allows society to access information valid outside the traditional media. When the customer experience is positive, these users become social media influencers, thus improving brand sales and productivity.

At present, social networks have consolidated as platforms oriented to the business world, and without having an active presence in it is practically impossible to achieve online reputation and image. Even corporate bogging can be done via social media platforms, and debates and discussions can be carried out on any subject in an agile way. The best way to increase the customer experience is to complement social media marketing strategy with tools such as electronic product configurators and verifiers or e-coupons. This would enhance customer perception of the brand.

Overall, this review is important for marketing professionals as it emphasizes the significance of selecting appropriate digital marketing strategies as per marketing activities to maximize customer experience. The Internet is the most powerful tool for businesses. Marketing managers who fail to utilize the Internet's importance in their business marketing strategy will be disadvantaged because the Internet is changing the brand, pricing, distribution, and promotion strategy. Future research should focus on selecting a particular industry or organization to conduct a qualitative or quantitative study to determine the effectiveness of social media marketing strategy.

References

Al Kurdi, B. H., & Alshurideh, M. T. (2021). Facebook advertising as a marketing tool: Examining the influence on female cosmetic purchasing behaviour. *International Journal of Online Marketing (IJOM), 11*(2), 52–74.

Alameeri, K., Alshurideh, M., Al Kurdi, B., & Salloum, S. A. (2020). The effect of work environment happiness on employee leadership. In *International conference on advanced intelligent systems and informatics* (pp. 668–680). Springer.

Al-Dmour, H., Alshuraideh, M., & Salehih, S. (2014). A study of Jordanians' television viewers habits. *Life Science Journal, 11*(6), 161–171.

Al-Dmour, R., AlShaar, F., Al-Dmour, H., Masa'deh, R., & Alshurideh, M. T. (2021). The effect of service recovery justices strategies on online customer engagement via the role of "Customer Satisfaction" during the Covid-19 pandemic: An empirical study. *The Effect of Coronavirus Disease (COVID-19) on Business Intelligence, 334*, 325.

Aljumah, A., Nuseir, M. T., & Alshurideh, M. T. (2021). The impact of social media marketing communications on consumer response during the COVID-19: Does the brand equity of a university matter. *The Effect of Coronavirus Disease (COVID-19) on Business Intelligence*, 367.

Al-Khayyal, A., Alshurideh, M., Al Kurdi, B., & Salloum, S. A. (2021). Factors influencing electronic service quality on electronic loyalty in online shopping context: Data analysis approach. In *Enabling AI applications in data science* (pp. 367–378). Springer.

Alkıtbı, S. S., Alshurıdeh, M., Al Kurdı, B., & Salloum, S. A. (2020). Factors affect customer retention: A systematic review. In *International conference on advanced intelligent systems and informatics* (pp. 656–667). Springer.

Almaazmi, J., Alshurideh, M., Al Kurdi, B., & Salloum, S. A. (2020). The effect of digital transformation on product innovation: A critical review. In *International conference on advanced intelligent systems and informatics* (pp. 731–741). Springer.

Al-Maroof, R., Ayoubi, K., Alhumaid, K., Aburayya, A., Alshurideh, M., Alfaisal, R., & Salloum, S. (2021). The acceptance of social media video for knowledge acquisition, sharing and application: A comparative study among YouYube users and TikTok users' for medical purposes. *International Journal of Data and Network Science, 5*(3), 197.

Almazrouei, F. A., Alshurideh, M., Al Kurdi, B., & Salloum, S. A. (2020). Social media impact on business: A systematic review. In *International conference on advanced intelligent systems and informatics* (pp. 697–707). Springer.

Alsharari, N. M., & Alshurideh, M. T. (2020). Student retention in higher education: the role of creativity, emotional intelligence and learner autonomy. *International Journal of Educational Management. International Journal of Educational Management, 35*(1), 233–247.

Alshurideh, M. (2022). Does electronic customer relationship management (E-CRM) affect service quality at private hospitals in Jordan? *Uncertain Supply Chain Management, 10*(2), 1–8.

Alshurideh, M., Nicholson, M., & Xiao, S. (2012). The effect of previous experience on mobile subscribers' repeat purchase behaviour. *European Journal of Social Sciences, 30*(3), 366–376.

Alshurideh, M., Al Kurdi, B. H., Vij, A., Obiedat, Z., & Naser, A. (2016). Marketing ethics and relationship marketing-An empirical study that measure the effect of ethics practices application on maintaining relationships with customers. *International Business Research, 9*(9), 78–90.

Alshurideh, M., Al Kurdi, B., Abu Hussien, A., & Alshaar, H. (2017). Determining the main factors affecting consumers' acceptance of ethical advertising: A review of the Jordanian market. *Journal of Marketing Communications, 23*(5), 513–532.

Alshurideh, M., Salloum, S. A., Al Kurdi, B., & Al-Emran, M. (2019). Factors affecting the social networks acceptance: An empirical study using PLS-SEM approach. In *Proceedings of the 2019 8th International Conference on Software and Computer Applications* (pp. 414–418).

Alshurideh, M. T., Al Kurdi, B., & Salloum, S. A. (2021). The moderation effect of gender on accepting electronic payment technology: A study on United Arab Emirates consumers. *Review of International Business and Strategy, 31*(3), 375–396.

Alzoubi, H. M., Alshurideh, M., & Ghazal, T. M. (2021). Integrating BLE Beacon technology with intelligent information systems IIS for operations' performance: A managerial perspective. In *The international conference on artificial intelligence and computer vision* (pp. 527–538). Springer.

Alzoubi, H., Alshurideh, M., Al Kurdi, B., Akour., I., & Azize, R. (2022) Does BLE technology contribute towards improving marketing strategies, customers' satisfaction and loyalty? The role of open innovation. *International Journal of Data and Network Science, 6*, 1–12.

Alzoubi, H. M., & Inairat, M. (2020). Do perceived service value, quality, price fairness and service recovery shape customer satisfaction and delight? A practical study in the service telecommunication context. *Uncertain Supply Chain Management, 8*(3), 579–588.

Bennett, D. R., & El Azhari, J. (2015). Omni-channel customer experience: An investigation into the use of digital technology in physical stores and its impact on the consumer's decision-making process.

Bilgihan, A. (2016). Gen Y customer loyalty in online shopping: An integrated model of trust, user experience and branding. *Computers in Human Behavior, 61*, 103–113.

Bolton, R. N., McColl-Kennedy, J. R., Cheung, L., Gallan, A., Orsingher, C., Witell, L., & Zaki, M. (2018). Customer experience challenges: Bringing together digital, physical and social realms. *Journal of Service Management, 29*(5), 776–808.

Bughin, J., Catlin, T., Hirt, M., & Willmott, P. (2018). *Why digital strategies fail*. McKinsey Quarterly.

Chaffey, D. (2010). Applying organisational capability models to assess the maturity of digital-marketing governance. *Journal of Marketing Management, 26*(3–4), 187–196.

Chaffey, D., & Ellis-Chadwick, F. (2019). *Digital marketing*. Pearson UK.

Chou, C. M. (2019). Social media characteristics, customer relationship and brand equity. *American Journal of Business, 10*(1).

Cook, G. (2014). Customer experience in the omni-channel world and the challenges and opportunities this presents. *Journal of Direct, Data and Digital Marketing Practice, 15*(4), 262–266.

Dennis, C., Brakus, J. J., Gupta, S., & Alamanos, E. (2014). The effect of digital signage on shoppers' behavior: The role of the evoked experience. *Journal of Business Research, 67*(11), 2250–2257.

Dörner, K., & Edelman, D. (2015). *What 'digital' really means*. McKinsey & Company Article.

Dumitrescu, L., Stanciu, O., Ţichindelean, M., & Vinerean, S. (2012). The importance of establishing customer experiences. *Studies in Business and Economics, 7*(1), 56–61.

Edelman, D. C. (2010). Branding in the digital age. *Harvard Business Review, 88*(12), 62–69.

Edelman, D., & Heller, J. (2015). *How digital marketing operations can transform business*. McKinsey Digital Report.

Flavián, C., Ibáñez-Sánchez, S., & Orús, C. (2019). The impact of virtual, augmented and mixed reality technologies on the customer experience. *Journal of Business Research, 100*, 547–560.

Gill, M., & VanBoskirk, S. (2016). *The digital maturity model 4.0*. Digital Transformation Playbook.

Goodman, J., 2019. Strategic customer service: Managing the customer experience to increase positive word of mouth, build loyalty, and maximize profits. Amazon.

Gopalani, A., & Shick, K. (2011). The service-enabled customer experience: A jump-start to competitive advantage. *Journal of Business Strategy, 32*(3), 4–12.

Hair, J. F., Celsi, M., Ortinau, D. J., & Bush, R. P. (2008). *Essentials of marketing research*. McGraw-Hill/Higher Education.

Heale, R., & Twycross, A. (2015). Validity and reliability in quantitative studies. *Evidence-Based Nursing, 18*(3), 66–67.

Heller Baird, C., & Parasnis, G. (2011). From social media to social customer relationship management. *Strategy and Leadership, 39*(5), 30–37.

Hollebeek, L. D. (2011). Demystifying customer brand engagement: Exploring the loyalty nexus. *Journal of Marketing Management, 27*(7–8), 785–807.

Homburg, C., Jozić, D., & Kuehnl, C. (2017). Customer experience management: Toward implementing an evolving marketing concept. *Journal of the Academy of Marketing Science, 45*(3), 377–401.

Hudson, S., Roth, M. S., & Madden, T. J. (2012). Customer communications management in the new digital era. *Center for marketing studies*. Darla Moore school of business.

Iankova, S., Davies, I., Archer-Brown, C., Marder, B., & Yau, A. (2018). A comparison of social media marketing between B2B, B2C and mixed business model. *Industrial Marketing Management*.

Jackson, G., & Ahuja, V. (2016). Dawn of the digital age and the evolution of the marketing mix. *Journal of Direct, Data and Digital Marketing Practice, 17*(3), 170–186.

Kane, G. C., Palmer, D., Phillips, A. N., & Kiron, D. (2015). Is your business ready for a digital future? *MIT Sloan Management Review, 56*(4), 37.

Kannan, P. K. (2017). Digital marketing: A framework, review and research agenda. *International Journal of Research in Marketing, 34*(1), 22–45.

Khasawneh, M. A., Abuhashesh, M., Ahmad, A., Masa'deh, R., & Alshurideh, M. T. (2021a). Customers online engagement with social media influencers' content related to COVID 19. In *The effect of coronavirus disease (COVID-19) on business intelligence* (pp. 385–404). Springer.

Khasawneh, M. A., Abuhashesh, M., Ahmad, A., Alshurideh, M. T., & Masa'deh, R. (2021b). Determinants of e-word of mouth on social media during COVID-19 outbreaks: An empirical study. In *The effect of coronavirus disease (COVID-19) on business intelligence* (pp. 347–366). Springer.

Klink, R. R., & Smith, D. C. (2001). Threats to the external validity of brand extension research. *Journal of Marketing Research, 38*(3), 326–335.

Kurdi, B., Alshurideh, M., & Al afaishata, T. (2020a). Employee retention and organizational performance: Evidence from banking industry. *Management Science Letters, 10*(16), 3981–3990.

Kurdi, B., Alshurideh, M., & Alnaser, A. (2020b). The impact of employee satisfaction on customer satisfaction: Theoretical and empirical underpinning. *Management Science Letters, 10*(15), 3561–3570.

Kurdi, B. A., Alshurideh, M., Nuseir, M., Aburayya, A., & Salloum, S. A. (2021). The effects of subjective norm on the intention to use social media networks: An exploratory study using PLS-SEM and machine learning approach. In *International conference on advanced machine learning technologies and applications* (pp. 581–592). Springer.

Lee, K., Azmi, N., Hanaysha, J., Alshurideh, M., & Alzoubi, H. (2022). The effect of digital supply chain on organizational performance: An empirical study in Malaysia manufacturing industry. *Uncertain Supply Chain Management, 10*, 1–16.

Lemon, K. N., & Verhoef, P. C. (2016). Understanding customer experience throughout the customer journey. *Journal of Marketing, 80*(6), 69–96.

Meyer, C., & Schwager, A. (2007). Understanding customer experience. *Harvard Business Review, 85*(2), 116.

Mitic, M., & Kapoulas, A. (2012). Understanding the role of social media in bank marketing. *Marketing Intelligence and Planning, 30*(7), 668–686.

Munro, J., & Richards, B. (2011). The digital challenge. *Destination Brands: Managing Place Reputation*, 141–154.

Nash, D., Armstrong, D., & Robertson, M. (2013). Customer experience 2.0: How data, technology, and advanced analytics are taking an integrated, seamless customer experience to the next frontier. *Journal of Integrated Marketing Communications, 1*(1), 32–39.

Nuseir, M. T., Al Kurdi, B. H., Alshurideh, M. T., & Alzoubi, H. M. (2021). Gender discrimination at workplace: Do Artificial Intelligence (AI) and Machine Learning (ML) have opinions about It. In *The international conference on artificial intelligence and computer vision* (pp. 301–316). Springer.

Parise, S., Guinan, P. J., & Kafka, R. (2016). Solving the crisis of immediacy: How digital technology can transform the customer experience. *Business Horizons, 59*(4), 411–420.

Payne, E. M, Peltier, J. W., & Barger, V. A. (2017). Omni-channel marketing, integrated marketing communications and consumer engagement: A research agenda. *Journal of Research in Interactive Marketing, 11*(2), 185–197.

Peterson, M., Gröne, F., Kammer, K., & Kirscheneder, J. (2010). Multi-channel customer management: Delighting consumers, driving efficiency. *Journal of Direct, Data and Digital Marketing Practice, 12*(1), 10–15.
Petit, O., Velasco, C., & Spence, C. (2019). Digital sensory marketing: Integrating new technologies into multisensory online experience. *Journal of Interactive Marketing, 45*, 42–61.
Rawson, A., Duncan, E., & Jones, C. (2013). The truth about customer experience. *Harvard Business Review, 91*(9), 90–98.
Ryan, D. (2014). *The best digital marketing campaigns in the World II*. Kogan Page Publishers.
Ryan, D. (2016). *Understanding digital marketing: marketing strategies for engaging the digital generation.* Kogan Page Publishers.
Schneider, J. (2014). *Higher satisfaction at lower costs: Digitizing customer care.* Mc Kinsey.
Smith, K. T. (2012). Longitudinal study of digital marketing strategies targeting Millennials. *Journal of Consumer Marketing, 29*(2), 86–92.
Solis, B., Li, C., & Szymanski, J. (2014). *The 2014 state of digital transformation.* Altimeter Group.
Sweiss, N., Obeidat, Z. M., Al-Dweeri, R. M., Mohammad Khalaf Ahmad, A., M. Obeidat, A., & Alshurideh, M. (2021). The moderating role of perceived company effort in mitigating customer misconduct within Online Brand Communities (OBC). *Journal of Marketing Communications*, 1–24.
Tariq, E., Alshurideh, M., Akour, I., & Al-Hawary, S. (2022) The effect of digital marketing capabilities on organizational ambidexterity of the information technology sector. *International Journal of Data and Network Science, 6*, 1–8.
Tiago, M. T. P. M. B., & Veríssimo, J. M. C. (2014). Digital marketing and social media: Why bother? *Business Horizons, 57*(6), 703–708.
Verhoef, P. C., Lemon, K. N., Parasuraman, A., Roggeveen, A., Tsiros, M., & Schlesinger, L. A. (2009). Customer experience creation: Determinants, dynamics and management strategies. *Journal of Retailing, 85*(1), 31–41.
Verma, R., Teixeira, J., Patrício, L., Nunes, N. J., Nóbrega, L., Fisk, R. P. & Constantine, L. (2012). Customer experience modeling: From customer experience to service design. *Journal of Service Management*.

Effects of Social Media Marketing on Consumer Perception in Liverpool, UK

Mohammed T. Nuseir, Ahmad Aljumah, Sarah Urabi, Barween Al Kurdi, and Muhammad Alshurideh

Abstract The purpose of this research is to analyze the effects of social media marketing on consumer perception for the people in Liverpool, UK. The descriptive research approach is used for this research. The research collected quantitative data and smart PLS analysis is done to analyze those collected data. It is found out from this research that the effect of social media marketing on consumer perception is positive and is increasing day by day because of the increasing use of the internet. The main limitation of this research was that the researcher was only able to work on the people of Liverpool. If the researcher would get more time and more financial support, the research could be done with vast information collected from a bigger sample size from other areas in the UK too. This research will help the companies to understand the power of social media platforms towards increasing consumer

M. T. Nuseir (✉)
Department of Business Administration, College of Business, Al Ain University Abu Dhabi Campus, Abu Dhabi, UAE
e-mail: mohammed.nuseir@aau.ac.ae

A. Aljumah
College of Communication and Media, Al Ain University Abu Dhabi Campus, P.O. Box 112612, Abu Dhabi, United Arab Emirates
e-mail: Ahmad.aljumah@aau.ac.ae

S. Urabi
Abu Dhabi Vocational Education & Training Institute (ADVETI), Abu Dhabi, UAE
e-mail: sarah.urabi@adveti.ac.ae

B. A. Kurdi
Department of Marketing, Faculty of Economics and Administrative Sciences, The Hashemite University, Zarqa, Jordan
e-mail: barween@hu.edu.jo

M. Alshurideh
Department of Management, College of Business Administration, University of Sharjah, Sharjah, United Arab Emirates
e-mail: m.alshurideh@ju.edu.jo

Department of Marketing, School of Business, The University of Jordan, Amman, Jordan

M. Alshurideh et al. (eds.), *The Effect of Information Technology on Business and Marketing Intelligence Systems*, Studies in Computational Intelligence 1056,
https://doi.org/10.1007/978-3-031-12382-5_3

perception and it will also show them the ways to properly use the social media platforms to retain current customers and get new customers.

Keywords Consumer perception · Social media marketing · Promotional marketing · Door to door marketing · Buying decision · Social media · Trust · Brand value · Smart-PLS analysis · Marketing tool

1 Introduction

Digital marketing or online marketing has helped to open a new era in the marketing field with different ways of marketing. Social media is one of the most powerful marketing tools used by organizations nowadays (Al Kurdi et al., 1339; Aljumah et al., 2021a; Almazrouei et al., 1261). Sajid (2016) Stated that easy access to the internet and social media by the people helps the organizations to deliver their unique marketing message towards their target customers. Social media helps the organizations to increase sales, share business information and also provide social support towards their customers (Al Khasawneh et al., 2021a, 2021b; Al-Maroof et al., 2021). Aljumah et al. (2021b), Sajid (2016) added that from the year 2004, the popularity of using social media as a marketing tool has increased as it delivers information about the products towards the users (Ahmad et al., 2021; Al Kurdi & Alshurideh, 2021; Nuseir et al., 2021a). Social media helps the organizations to contact their target customers without facing any local boundaries. This research is done based on the consumers located in Liverpool, UK.

1.1 Problem Statement

In the past few years, social media marketing has become a new marketing tool for organizations. As per the opinion of Stephen (2016), Nuseir et al. (2021b), Al-Dmour et al. (2021), Alshurideh (2022), the main aim of every organization to fulfill the desires of their customers and to make the highest profits through using the most effective marketing strategy as it helps to decrease the production cost. Different organizations use different social media as their marketing tool including Facebook, Twitter, Linkedin and many more. Nowadays, people have easy access to the internet in their smart mobile phones. Sajid (2016) stated that the firms use this rapid growth of using the internet as they now start using the internet as their marketing tool to affect consumer perception through social media marketing.

1.2 Purpose

The main purpose of this research is to understand the effects of social media marketing on consumer perception (Abuhashesh et al., 2021). In the recent past, easy access to the internet and social media has increased rapidly and people access social media sites regularly. The perception of people has changed by social media. Door to door selling was the most popular marketing strategy of the organizations before this social media era. However, now social media is helping the firms towards increasing their sales and making maximum profits (Al Dmour et al., 2014; Alshurideh et al., 2021; Suleman et al., 2021). Social media gave people many buying choices. People are now well known to many different brands due to the advertisements spread by organizations through social media sites. So, it can be said that social media has greater effects on consumer perception compared to another marketing tool.

1.3 Research Objective

The objectives of the research are–

- To understand the impacts of social media marketing on consumer perception.
- To assess the effects of the door to door marketing on consumer perception.
- To evaluate the roles of social media marketing as a promising marketing tool.

1.4 Research Questions

The research questions for this research are–

- What are the impacts of social media marketing on consumer perception?
- What are the effects of the door to door marketing on consumer perception?
- What are the roles of social media marketing as a promising marketing tool?

2 Conceptual Framework and Hypotheses Development

This section will elaborate on the existing theories about this topic. Research hypotheses will also be developed in this section.

2.1 Social Media Marketing

Social media marketing has become a very popular marketing tool for most of the organizations of all sizes as it helps them to reach their target customers. As per the opinion of Sajid (2016), effective marketing through social media may help the organizations to achieve remarkable business success through creating brand awareness among the customers. It also helps in maximizing sales of the organizations. Aissani and Awad ElRefae (2021), Zhu and Chen (2015) stated that social media marketing is a type of internet marketing that helps organizations to create and share business content in social media sites such as Facebook, Twitter, Pinterest, Instagram etc. towards achieving marketing goals. The social media marketing contents may include image updates, videos, text posting etc. It also includes paid social media advertising.

According to Stephen (2016), search marketers and social media are very closely related to each other. People often do searches on social media sites towards finding their desire social media content and discover new content. This discovery is considered a search activity. Zhu and Chen (2015) added that social media marketing is growing fast even faster than the internet. However, before creating social media marketing campaigning, one organization should understand their business goals clearly. There should be a proper social strategy that will help the organizations towards achieving their business goal through social media sites. Sajid (2016) stated that social media marketing helps the organization by creating conversations with its target customers, rising website traffic, increasing brand awareness, generating improved communication with their main audiences, and finally creating brand identity towards their target customers. Proper planning is necessary for any organization before starting marketing campaigns through social media sites.

2.2 Promotional Marketing

Chong et al. (2017) informed that promotional marketing helps the organizations to offer any special offers on their products such as special discounts, coupons, incentives, etc. towards attracting their target customers to make an immediate purchasing decision or future purchasing decision through making brand interaction. Promotion is more effective in changing the immediate purchasing behavior of the customers (Al-Dmour & Al-Shraideh, 2008; Al-Dmour et al., 2014; Alshurideh et al., 2017, 2018). Elsaadani (2020), West and McAllister (2013) opined that promotion marketing is often misused when it is handled with inexperienced people. The marketing people of any organization should keep it in mind that they should create positive brand awareness through promotional marketing and it should follow the rules of the internet protocol (Abu Zayyad et al., 2021; Alshurideh et al., 2015; Hamadneh et al., 2021; Sweiss et al., 2021). Hettche et al. (2017) added that promotional marketing could be used by organizations both in online and offline ways. One of the effective ways to draw the customers' attention is sweepstakes because people love to win prizes.

As per the opinion of West and McAllister (2013), promotional marketing is performed by many organizations, but it targets only the particular targeted customers of a particular organization. Most of the common people cannot know about the promotional marketing campaigns performed by multinational companies as their target audiences are highly profiled people (Al-Duhaish et al., 2014; Alshurideh & Shaltoni, 2014; Alshurideh et al., 2016). Chong et al. (2017) noted that promotional marketing activity towards the target customers is the best possible way for the organizations as it ensures a high positive return on their investments.

2.3 Door to Door Marketing

Door to door marketing has become a history now. The new generation of people who did not even hear of this kind of marketing has ever existed. On the contrary, Fergurson and Fergurson (2017) opined that door to door marketing was the best method of direct selling to the target customers as in this marketing, the salesperson makes direct real human contact with the customers. Nowadays, mostly marketing is performed through the internet or television or FM channels on the radio, where the organization cannot make direct human contact with their target customers. Fergurson and Fergurson (2017) contradicted that nowadays, the crime rate has increased so much that it has become natural that the prospects do not want to give entry to an unknown person to their house.

Fergurson and Fergurson (2017) informed that door to door marketing was a tough job as most of the housekeepers generally say a big NO to the salesperson when they knock the door. Direct marketing like a door to door marketing was an attractive selling method that an organization could use towards achieving the business deal at any position. Fergurson and Fergurson (2017) stated that direct marketing like a door to door marketing is the best way for the organization to achieve their marketing goals by using the information gathered from the customers. Door to door marketing is based on the good communication skills of the salesperson to attract the target customers.

2.4 Consumer Perception

According to Bilgihan et al. (2016), consumer perception is a concept of marketing that helps to understand the awareness, impressions, and consciousness of the customers about the organization and its products. Any organization can influence the perception of their target customers through reviews, advertising, personal experiences, social media, public relations and other methods (Alaali et al., 2021; Alshurideh & Xiao, 2012; Alzoubi et al., 2020; Madi Odeh et al., 2021). Hettche et al.

(2017) added that everything might affect the perception of the consumers including the position of the product in the store, its shape, colors etc. Sometimes, consumer perception cannot be controlled by the organization such as the time and place when the consumer is making interaction with their desired products as it also affects their perception (Alhamad et al., 2021; Al-Hamad et al., 2021; Joghee et al., 2021; Shah et al., 2021).

Vidal et al. (2016) stated that as perception theory, perception shows the moment when the customers become aware of some particular products through their senses. Every organization should use this theory of perception if they want to attract their customers. The viewpoint of the organization towards attracting customers from different income classes should be different. Bilgihan et al.(2016) added that in case of attracting high-end customers, the organizations should take care of quality, hygiene, packaging, lighting, cleanliness etc. in presenting their products or services. Customer segmentation should be done into different groups as these groups will help the organizations to understand the important things for each group of customers. Hettche et al. (2017) observed that it is important for the organizations to make their customers feel as if they are part of the organization as it will help to increase customer loyalty towards a brand.

2.5 *Research Hypothesis*

H1: Social media marketing has major effects on consumer perception.
H2: Social media marketing positively affects promotional marketing.
H3: Promotional marketing has positive effects on consumer perception.
H4: Door to door marketing has positive effects on consumer perception.

2.6 *Conceptual Framework*

Figure 1 is showing the research model followed by the conceptual framework.

3 Instrument Development and Data Collection

Instrument Development: This research is performed based on the quantitative questionnaire. 5-point Likert Scale is used by the researcher from 1 = Strongly Disagree to 5 = Strongly Agree. The questionnaire is formed from previous similar researches as it helped the researcher to enhance research validity. Customer perception is the most important construct for this research. This research helped to understand the impacts of marketing through social media on consumer perception. Other constructs

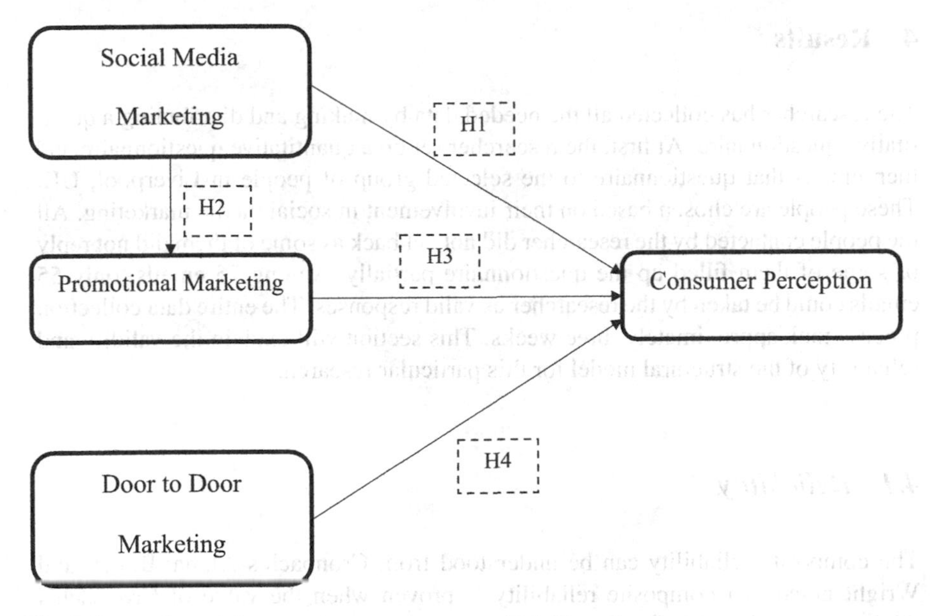

Fig. 1 Research model

for this research are promotional marketing and door to door marketing. The dependent variable for this research is social media marketing, which is another construct for this research.

Data Collection: The quantitative questionnaire is formed and distributed by the researcher among the people of Liverpool, UK. The researcher distributed an online questionnaire to the chosen group of people who are mostly involved in social media marketing. 75 emails with the questionnaire are sent by the researcher to 75 different people. Among them, 55 responses are received by the researcher after three weeks.

Research Method: Structural Equation Modelling (SEM) is a popular research method used by the researcher towards performing this research. Partial Least Square (PLS) method is also used in this research. As per the opinion of Alshurideh et al. (2019a), it is important to use the PLS method as it helps to understand the complex data within any low-structured situation. PLS method is very much helpful where it is hard to gather theoretical information such as the social media area. Ringle and Sarstedt (2016) added that the researcher could assess the reliability and validity of the research through the PLS method. A re-sampling method is also used by the researcher to perform significance testing.

4 Results

The researcher has collected all the needed data by making and distributing a quantitative questionnaire. At first, the researcher set up a quantitative questionnaire and then emails that questionnaire to the selected group of people in Liverpool, UK. These people are chosen based on their involvement in social media marketing. All the people contacted by the researcher did not get back as some of them did not reply or some of them filled up the questionnaire partially. Among 75 emails, only 55 emails could be taken by the researcher as valid responses. The entire data collection process took approximately three weeks. This section will explain the validity and reliability of the structural model for this particular research.

4.1 Reliability

The composite reliability can be understood from Cronbach's Alpha. Bonett and Wright noted that composite reliability is proven when the value of Cronbach's Alpha is more than 0.7. So, the reliability of any research survey can be tested with the help of the value of Cronbach's Alpha. The internal consistency of any research can also be computed by the value of Cronbach's Alpha. From Table 1, it is seen that all the values of Cronbach's Alpha for this particular research are more than 0.7. So, it can be said from those values that the reliability of this research is accepted. The quality criteria overview is seen from the below table.

4.2 Validity

As per the opinion of Heale and Twycross, Construct Validity and Content Validity help to test the validity for any research. The constructs for this particular research are social media marketing, promotional marketing, and door to door marketing. The researcher got help in the information system from these constructs.

On the other hand, face validity is tested by the researcher with the help of this research within the content validity. To perform the validity, the researcher prepared

Table 1 Quality criteria overview

	AVE	Composite reliability	R square	Cronbach's alpha
Social media marketing	0.668342	0.785658		0.775421
Promotional marketing	0.676823	0.89612	0.398744	0.793576
Door to door marketing	0.601308	0.737575		0.731564
Consumer perception	0.697112	0.726885	0.373425	0.735442

the quantitative questionnaire and sent them through emails to the chosen group of people in Liverpool, the UK, who are often involved in social media marketing. The substantial literature review is taken seriously by the researcher. Previous researches done on similar research topics has helped the researcher in setting up quantitative questionnaires. The researcher came up with the research content from the information gathered through these quantitative questionnaires. Egan et al. stated that the assessment of convergent and discriminant validity confirms convergent validity. The full form of AVE is the Average Variance Extracted. Ab Hamid et al. observed that AVE acts as a standard. AVE helps to analyze and calculate the convergent validity. Table 1 shows that the value of AVE for all the constructs is more than 0.5 which is anticipated. Again, the value of composite reliability should always be more than 0.7 and it should also be more than the value of AVE. Table 1 shows that both of these criteria are fulfilled for this research. So, it can be said that the validity of this research is accepted.

The testing of divergent validity can be performed in Tables 2 and 3 presented below.

The value of AVE should be more than the value of MSV (Maximum Shared Squared Variance), the value of ASV (Average Shared Squared Variance), and divergent validity.

AVE > MSV.
AVE > ASV.

Table 2 shows that–

0.817522 > {0.647913, 0.533122, 0.456482} and
0.822693 > {0.647913, 0.594372, 0.440076} and
0.775441 > {0.533122, 0.594372, 0.474514} and
0.834932 > {0.456482, 0.440076, 0.474514}.

Alarcón et al. noted that a researcher should assess the discrimination validity towards comparing the square of correlations of different variables with the value of AVE. Table 2 helps to verify the discrimination validity. A two-fold result can be

Table 2 Square of correlation between latent variables

	Social media marketing	Promotional marketing	Door to door marketing	Consumer perception
Social media marketing	0.817522			
Promotional marketing	0.647913	0.822693		
Door to door marketing	0.533122	0.594372	0.775441	
Consumer perception	0.456482	0.440076	0.474514	0.834932

Table 3 Cross loadings

	Social media marketing	Promotional marketing	Door to door marketing	Consumer perception
SMM1	0.630231	0.324166	0.165503	0.180532
SMM2	0.490231	0.298766	0.421044	0.297703
SMM3	0.49089	0.157767	0.232201	0.334209
SMM4	0.468091	0.195321	0.365054	0.234033
SMM5	0.732654	0.636022	0.450998	0.432011
PM1	0.665099	0.790332	0.609865	0.464401
PM2	0.415527	0.870489	0.632088	0.376608
DTDM1	0.304431	0.376655	0.790234	0.415433
DTDM2	0.56086	0.654409	0.75123	0.412229
CP1	0.430445	0.315033	0.322091	0.763421
CP2	0.397551	0.460224	0.36711	0.845712
CP3	0.198606	0.242331	0.36698	0.64322

Note SMM = Social Media Marketing; PM = Promotional Marketing; DTDM = Door to Door Marketing; CP = Consumer Perception

achieved for convergent and discrimination validity by testing the factor loadings for all the indicators. Ab Hamid et al. informed that the factor loading for every indicator should always be more compared to its constructs based on the value of the other factors.

Table 3 shows the factor loadings of every indicator that are more than the value of their constructs compared to other factors. This condition only can verify the divergent validity of this research.

For SMM 1, 0.630231 > {0.324166, 0.165503, 0.180532} and
For SMM 2, 0.490231 > {0.298766, 0.421044, 0.297703} and
For SMM 3, 0.49089 > {0.157767, 0.232201, 0.334209} and
For SMM 4, 0.468091 > {0.195321, 0.365054, 0.234033} and
For SMM 5, 0.732654 > {0.636022, 0.450998, 0.432011} and
For PM 1, 0.790332 > {0.665099, 0.609865, 0.464401} and
For PM 2, 0.870489 > {0.415527, 0.632088, 0.376608} and
For DTDM 1, 0.790234 > {0.304431, 0.376655, 0.415433} and
For DTDM 2, 0.75123 > {0.56086, 0.654409, 0.412229} and
For CP1, 0.763421 > {0.430445, 0.315033, 0.322091} and
For CP 2, 0.845712 > {0.397551, 0.460224, 0.36711} and
For CP 3, 0.64322 > {0.198606, 0.242331, 0.36698}.

Validity of this research is accepted that is confirmed from Table 3. Table 2 and 3 both helps to prove that the divergent validity for this research is also in acceptable level.

The test of total effects is seen from Tables 4, 5 and 6. Lakens noted that T-statistics should be greater than |1.96| if all the effects need to be accepted.

T-statistics = |Original Sample/Standard Error|

From Table 4, it can be said that the values of all T-statistics are more than |1.96| for this research, and that verifies that all the effects for this research are accepted.

O'Rourke and MacKinnon (2015) informed that the R^2 values for all three constructs for the research should be 0.6, 0.33, or 0.19 which are comparatively judged as substantial, moderate and weak.

Table 4 Test of total effects

	Original Sample (O)	Sample Mean (M)	Standard Deviation (STDEV)	Standard Error (STERR)	T Statistics (|O/STERR|)
Social media marketing – > Consumer perception	0.343254	0.321333	0.171163	0.171163	2.934220
Social media marketing – > Promotional marketing	0.607754	0.664334	0.055374	0.055374	12.71043
Promotional marketing – > Consumer perception	0.113177	0.129901	0.17566	0.17566	5.323210
Door to door marketing – > Consumer perception	0.255654	0.287601	0.179099	0.179099	4.765655

Table 5 Test of total effects

	R square
Social media marketing	
Promotional marketing	0.398744
Door to door marketing	
Consumer perception	0.373425

Table 6 Test of total effects

Total	SSO	SSE	1-SSE/SSO
Consumer perception	1284	1012.784132	0.211227
Promotional marketing	856	551.221037	0.355926

Table 7 Test of Goodness of Fit (GOF)

	Communality
Social media marketing	0.668342
Promotional marketing	0.676823
Door to door marketing	0.601308
Consumer perception	0.667112

$$R^2 \begin{cases} 0.19 \text{ Weak} \\ 0.33 \text{ Moderate} \\ 0.67 \text{ Substantial} \end{cases}$$

$$Q^2 = 1 - \text{SSE}/\text{SSO}$$

$$Q^2 = \begin{cases} 0.02 \text{ Weak} \\ 0.15 \text{ Moderate} \\ 0.35 \text{ Substantial} \end{cases}$$

It is seen that the R^2 values for promotional marketing and consumer perception are moderate, where the R^2 value for the door to door marketing is weak. On the other hand, all the value of Q^2 for consumer perception is moderate and for promotional marketing, it is substantial (Table 7).

$$\text{GOF} \begin{cases} 0.01 \text{ Weak} \\ 0.25 \text{ Moderate} \\ 0.36 \text{ Substantial} \end{cases}$$

$$\begin{aligned} \text{GOF} &= \sqrt{[\{\text{Average } (R^2) \times \text{ Average (Community)}\}]} \\ &= \sqrt{[0.39 \times 0.65]} \\ &= 0.5035 > 0.36 \end{aligned}$$

So, GOF is also in acceptance level for this research.

4.3 Structural Model

Smart PLS analysis software is used for this research as it helps to assess the validity of the research through structural paths. The path coefficients for this research need to be at 0.05.

Figure 2 shows that the R^2 value of promotional marketing is approximately 0.42% of the variance of promotional marketing, and it is accounted for social media

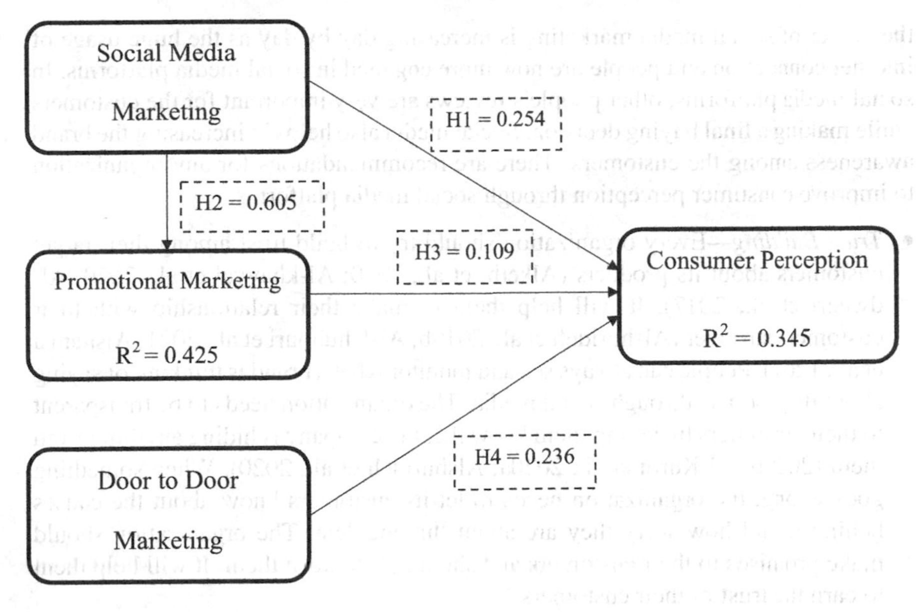

Fig. 2 Smart PLS analysis

marketing. Again, the R^2 value of consumer perception is approximately 0.34% of the variance of consumer perception and also accounted for social media marketing, promotional marketing and door to door marketing. From this PLS analysis, it can be said that consumer perception is, directly and indirectly, dependent upon social media marketing more than door to door marketing.

5 Recommendations and Implications

Customer perception is all about the opinion of the Customers about the products or businesses. It shows the feelings of the customers of any specific brand. It includes the direct and indirect experience of the customers with the product or the company. The company can monitor and analyze this customer perception to understand its drawbacks and improve its relationships with its customers. Customer perception helps the organization to retain their current customers and also increase the number of new customers. Customer perception helps to build a link between the interest of the customers and their actual purchase. This research shows various marketing tools that affect consumer perception. In today's world, both door to door marketing, and social media marketing positively affect customer perception through the effect of social media marketing is little more than door to door marketing. On the other hand, promotional marketing hugely affected by social media marketing and it also positively affect consumer perception. The result from this research is showing that the power of social media marketing is increasing day by day as the huge usage of internet connection and people are now more engaged in social media platforms. In social media platforms, other people's reviews are very important for the customers while making a final buying decision. Social media also helps in increasing the brand awareness among the customers. There are recommendations for any organization to improve consumer perception through social media platforms.

- *Trust Building*—Every organization should try to build trust among their target customers about its products (Alketbi et al., 2020; Al-khayyal et al., 2020; Al-dweeri et al., 2017). It will help them to make their relationship with their customers stronger (Alshurideh et al., 2019b; Al-Dhuhouri et al., 2021; Alshamsi et al., 1261). People can always see and monitor what a brand is thinking or saying about its products through social media. The organization needs to be transparent to their customers by letting them know that the company is hiding anything from them (2020b; Al Kurdi et al., 2020a; Alshurideh et al., 2020). When something goes wrong, the organization needs to let its customers know about the causes behind it and how sorry they are about this incident. The organization should make promises to their customers and should try to keep them. It will help them to earn the trust of their customers.
- *Showing Brand Values*—Generating brand values is very important for any organization as it helps them to successfully and effectively communicate through social media platforms. It is seen that most of the companies skip this step. However, the companies need to inform their stakeholders through social media platforms how they are related to the values created by their brands.
- *Showing interest in customer*—Through social media platforms, the organization can inform the customers about what they are thinking and regularly doing. The organization should spend more time building strong relations with the customers by responding to their queries and comments about the products. It will help the customers to think that someone cares about them and they will also feel interested in knowing more about the organization and their products.
- *Promoting product images and videos*—Sharing more images and videos about the product helps to generate more audience. Videos in social media help to generate interaction among the customers and the organization. The visual effect creates a clear image of the product and its usages in the minds of the customers. It helps the customers to make their final buying decisions.
- *Trying to be consistent*—Being consistent is one of the important factors in social media marketing. Reliable and constant messages help to improve the identity of the organization, and it also helps in promoting brand perception among customers. The key messages of the organization should be clear and specific so that the customers can understand the brand and the products.

6 Conclusion and Limitations

From this research, it is seen that all the independent variables including social media marketing, promotional marketing and door to door marketing, have effects on the dependent variable consumer perception. Hypotheses Analysis is done through PLS analysis too. From this PLS analysis, it can be said that social media marketing has a direct positive impact on consumer perception which is 0.254. So, H1 is accepted. Social media marketing also has a positive impact on promotional marketing which is 0.605. So, H2 is accepted. Again, promotional marketing has a positive impact on consumer perception which is 0.109 and H3 is supported. So, it is proved that social media marketing has a direct and indirect impact on consumer perception. The PLS analysis also sees that door to door marketing also has an impact on consumer perception which is 0.236 and H4 is accepted though this impact is less powerful compared to the impact of social media marketing on consumer perception.

Limitations of any research show that there are future possibilities for further research. This research helps to assess the impacts of social media marketing on consumer perception. The research shows that there is a positive but not very strong impact of social media marketing on consumer perception. Again, it is seen than door to door marketing also has positive but weak impacts on consumer perception. On the other hand, social media marketing has a comparatively strong positive impact on promotional marketing, and promotional marketing has weak but positive impacts on consumer perception.

The data for the research purpose are gathered from the people that are involved in social media marketing and whose marketing decision mar dependent on social media. This research is exploratory. So, different responses may be collected from different people. If the data were collected from the group of people whose belief in the door to door marketing, the researcher would get the different result of the research. It is the fact that the preferences or choices of people vary from place to place. Different people may act differently in case of seeing an advertisement for a product on social media. All the people do not love to buy online products and they believe in the authenticity of the products sold through the door to door marketing. If the research were performed in the locality where social media is not considered as very popular technology towards buying products, then the research result would be different. Another limitation was the limited time to perform this research. If the data were collected from a larger group of people who are comfortable in social media marketing as well as door to door marketing, then the research result would be more accurate. However, more time would have been needed to collect more data. So, more time in the future will help the researcher to produce a more accurate research result.

References

Abu Zayyad, H. M., Obeidat, Z. M., Alshurideh, M. T., Abuhashesh, M., Maqableh, M., & Masa'deh, R. (2021). Corporate social responsibility and patronage intentions: The mediating effect of brand credibility. *Journal of Marketing Communications, 27*(5). https://doi.org/10.1080/13527266.2020.1728565.

Abuhashesh, M. Y., Alshurideh, M. T., Ahmed, A., Sumadi, M., & Masa'deh, R. (2021). The effect of culture on customers' attitudes toward Facebook advertising: the moderating role of gender. *Review of International Business and Strategy, 31*(3). https://doi.org/10.1108/RIBS-04-2020-0045.

Ahmad, A., Alshurideh, M. T., Al Kurdi, B. H., & Salloum, S. A. (2021). *Factors impacts organization digital transformation and organization decision making during Covid19 pandemic* (Vol. 334).

Aissani, R., & Awad ElRefae, G., Aissani, R., & ElRefae, G. A. (2021). Trends of university students towards the use of YouTube and the benefits achieved; Field study on a sample of Al Ain University students. AAU *Journal of Business and Law* مجلة جامعة العين للأعمال والقانون, *5*(1), 1.

Al Dmour, H., Alshurideh, M., & Shishan, F. (2014). The influence of mobile application quality and attributes on the continuance intention of mobile shopping. *Life Science Journal, 11*(10).

Al Khasawneh, M., Abuhashesh, M., Ahmad, A., Masa'deh, R., & Alshurideh, M. T. (2021a). Customers online engagement with social media influencers' content related to COVID 19 (Vol. 334).

Al Khasawneh, M., Abuhashesh, M., Ahmad, A., Alshurideh, M. T., & Masa'deh, R. (2021b). Determinants of E-Word of mouth on social media during COVID-19 outbreaks: An empirical study (Vol. 334).

Al Kurdi, B. H., & Alshurideh, M. T. (2021). Facebook advertising as a marketing tool: Examining the influence on female cosmetic purchasing behaviour. *International Journal of Online Marketing, 11*(2), 52–74.

Al Kurdi, B., Alshurideh, M., Nuseir, M., Aburayya, A., Salloum, S. A. (2021). *The effects of subjective norm on the intention to use social media networks: An exploratory study using PLS-SEM and machine learning approach* (Vol. 1339).

Al Kurdi, B., Alshurideh, M., & Al afaishat, T. (2020a). Employee retention and organizational performance: Evidence from banking industry. *Management Science Letters, 10*(16). https://doi.org/10.5267/j.msl.2020a.7.011.

Al Kurdi, B., Alshurideh, M., & Alnaser, A. (2020b). The impact of employee satisfaction on customer satisfaction: Theoretical and empirical underpinning. *Management Science Letters, 10*(15). https://doi.org/10.5267/j.msl.2020b.

Alaali, N., et al. (2021). The impact of adopting corporate governance strategic performance in the tourism sector: A case study in the Kingdom of Bahrain. *Journal of Legal, Ethical and Regulatory Issues, 24*(Special Issue 1).

Al-Dhuhouri, F. S., Alshurideh, M., Al Kurdi, B., & Salloum, S. A. (2021). *Enhancing our understanding of the relationship between leadership, team characteristics, emotional intelligence and their effect on team performance: A critical review* (Vol. 1261). AISC.

Al-Dmour, H., & Al-Shraideh, M. T. (2008). The influence of the promotional mix elements on Jordanian consumer's decisions in cell phone service usage: an analytical study. *Jordan Journal of Business Administration, 4*(4), 375–392.

Al-Dmour, H. H., Alshurideh, M., & Salehih, S. (2014). A study of Jordanians' television viewers habits. *Life Science Journal, 11*(6).

Al-Dmour, R., AlShaar, F., Al-Dmour, H., Masa'deh, R., & Alshurideh, M. T. (2021). The effect of service recovery justices strategies on online customer engagement via the role of "Customer Satisfaction" during the Covid-19 pandemic: An empirical study (Vol. 334).

Al-Duhaish, A., Alshurideh, M., & Al-Zu'bi, Z. (2014). The impact of the basic reference group usage on the purchasing decision of clothes (A field study of Saudi youth in Riyadh city). *Dirasat: Administrative, 41*(2), 205–221.

Al-dweeri, R. M., Obeidat, Z. M., Al-dwiry, M. A., Alshurideh, M. T., & Alhorani, A. M. (2017). The impact of e-service quality and e-loyalty on online shopping: moderating effect of e-satisfaction and e-trust *International Journal of Marketing Studies, 9*(2), 92–103.
Alhamad, A. Q. M., Akour, I., Alshurideh, M., Al-Hamad, A. Q., Kurdi, B. A., & Alzoubi, H. (2021a). Predicting the intention to use google glass: A comparative approach using machine learning models and PLS-SEM. *International Journal of Data and Network Science, 5*(3). https://doi.org/10.5267/j.ijdns.2021.6.002.
Al-Hamad, M. Q., Mbaidin, H. O., Alhamad, A. Q. M., Alshurideh, M. T., Kurdi, B. H. A., & Al-Hamad, N. Q. (2021b). Investigating students' behavioral intention to use mobile learning in higher education in UAE during Coronavirus-19 pandemic. *International Journal of Data and Network Science, 5*(3). https://doi.org/10.5267/j.ijdns.2021.6.001.
Aljumah, A., Nuseir, M. T., & Alshurideh, M. T. (2021a). The impact of social media marketing communications on consumer response during the COVID-19: does the brand equity of a university matter? (Vol. 334).
Aljumah, A. I., Nuseir, M. T., & Alam, M. M. (2021b). Organizational performance and capabilities to analyze big data: do the ambidexterity and business value of big data analytics matter? *Business Process Management Journal*.
S. Alketbi, M. Alshurideh, and B. Al Kurdi, "The influence of service quality on customers'retention and loyalty in the UAE hotel sector with respect to the impact of customer'satisfaction, trust, and commitment: A qualitative study. *PalArch's Journal of Archaeology of Egypt/Egyptology 17*(4), pp. 541–561
Al-khayyal, A., Alshurideh, M., & Al, B. (2020). The impact of electronic service quality dimensions on customers E- shopping and E-loyalty via the impact of E-satisfaction and E-trust: A qualitative approach. *The International Journal of Innovation, Creativity and Change, 14*(9).
Al-Maroof, R., et al. (2021). The acceptance of social media video for knowledge acquisition, sharing and application: A comparative study among YouYube users and TikTok users' for medical purposes. *International Journal of Data and Network Science, 5*(3). https://doi.org/10.5267/j.ijdns.2021.6.013.
Almazrouei, F. A., Alshurideh, M., Al Kurdi, B., & Salloum, S. A. (2021). *Social media impact on business: A systematic review* (vol. 1261). AISC.
Alshamsi, A., Alshurideh, M., Kurdi, B. A., & Salloum, S. A. (2021). *The influence of service quality on customer retention: A systematic review in the higher education* (Vol. 1261). AISC.
Alshurideh, M. (2022). Does electronic customer relationship management (E-CRM) affect service quality at private hospitals in Jordan? *Uncertain Supply Chain Managements, 10*(2), 1–8.
Alshurideh, M., Nicholson, M., & Xiao, S. (2012). The effect of previous experience on mobile subscribers' repeat purchase behaviour *European Journal of Social Sciences, 30*(3).
Alshurideh, M., Bataineh, A., Alkurdi, B., & Alasmr, N. (2015). Factors affect mobile phone brand choices–studying the case of Jordan Universities students. *International Business Research, 8*(3). https://doi.org/10.5539/ibr.v8n3p141.
Alshurideh, M., Al Kurdi, B. H., Vij, A., Obiedat, Z., & Naser, A. (2016). Marketing ethics and relationship marketing-An empirical study that measure the effect of ethics practices application on maintaining relationships with customers. *International Business Research, 9*(9), 78–90.
Alshurideh, M., Al Kurdi, B., Abu Hussien, A., & Alshaar, H. (2017). Determining the main factors affecting consumers' acceptance of ethical advertising: A review of the Jordanian market. *Journal of Marketing Communications, 23*(5). https://doi.org/10.1080/13527266.2017.1322126.
Alshurideh, M., Al Kurdi, B., Abumari, A., & Salloum, S. (2018). Pharmaceutical promotion tools effect on physician's adoption of medicine prescribing: evidence from Jordan. *Modern Applied Science, 12*(11), 210–222.
Alshurideh, M., Salloum, S. A., Al Kurdi, B., & Al-Emran, M. (2019a). Factors affecting the social networks acceptance: an empirical study using PLS-SEM approach. In *Proceedings of the 2019a 8th International Conference on Software and Computer Applications* (pp. 414–418).
Alshurideh, M., Alsharari, N. M., & Al Kurdi, B. (2019b). Supply chain integration and customer relationship management in the airline logistics. *Theoretical Economics Letters, 9*(02), 392–414.

Alshurideh, M., Gasaymeh, A., Ahmed, G., Alzoubi, H., & Kurd, B. A. (2020). Loyalty program effectiveness: Theoretical reviews and practical proofs. *Uncertain Supply Chain Management, 8*(3). https://doi.org/10.5267/j.uscm.2020.2.003.

Alshurideh, M. T., & Shaltoni, A. M. (2014). Marketing communications role in shaping consumer awareness of cause-related marketing campaigns. *International Journal of Marketing Studies, 6*(2), 163.

Alshurideh M. T., et al. (2021). Factors affecting the use of smart mobile examination platforms by universities' postgraduate students during the COVID-19 pandemic: An empirical study. *Informatics, 8*(2). https://doi.org/10.3390/informatics8020032.

Alzoubi, H., Alshurideh, M., Kurdi, B. A., & Inairat, M. (2020). Do perceived service value, quality, price fairness and service recovery shape customer satisfaction and delight? A practical study in the service telecommunication context. *Uncertain Supply Chain Managements, 8*(3). https://doi.org/10.5267/j.uscm.2020.2.005.

Bilgihan, A., Barreda, A., Okumus, F., & Nusair, K. (2016). Consumer perception of knowledge-sharing in travel-related online social networks. *Tourism Management, 52*, 287–296.

Chong, A. Y. L., Ch'ng, E., Liu, M. J., & Li, B. (2017). Predicting consumer product demands via Big Data: the roles of online promotional marketing and online reviews. *International Journal of Production Researc, 55*(17), 5142–5156.

Elsaadani, M. A. (2020). Investigating the effect of E-management on customer service performance. *AAU Journal of Business Law مجلة جامعة العين للأعمال والقانون*, *3*(2), 1.

Fergurson, J. R., & Fergurson, J. (2017). Self-image congruence: An empirical look at consumer behavior in door-to-door sales! *Journal of Managerial Issues*, 262–277.

Hamadneh, S., Hassan, J., Alshurideh, M., Al Kurdi, B., & Aburayya, A. (2021). The effect of brand personality on consumer self-identity: the moderation effect of cultural orientations among British and Chinese consumers. *Journal of Legal, Ethical and Regulatory Issues, 24*, 1–14.

Hettche, M., Tupper, C., & Rooney, C. (2017). *Using a social machine for promotional marketing on campus: A case study.*

Joghee, S., et al. (2021). Expats impulse buying behaviour in UAE: A customer perspective. *Journal of Management Information and Decision Sciences, 24*(1), 1–24.

Madi Odeh, R. B. S., Obeidat, B. Y., Jaradat, M. O., Masa'deh, R., & Alshurideh, M. T. (2021) The transformational leadership role in achieving organizational resilience through adaptive cultures: the case of Dubai service sector. *The International Journal of Productivity and Performance Management.* https://doi.org/10.1108/IJPPM-02-2021-0093.

Nuseir, M. T., Aljumah, A., & Alshurideh, M. T. (2021a). *How the business intelligence in the new startup performance in UAE during COVID-19: The mediating role of innovativeness* (Vol. 334).

Nuseir, M. T., Ghaleb, A., & Aljumah, A. (2021b). The e-Learning of students and university's brand image (Post COVID-19): How successfully Al-Ain University have embraced the paradigm shift in digital learning. In *The effect of coronavirus disease (COVID-19) on business intelligence* (Vol. 334, p. 171).

Ringle, C. M., & Sarstedt, M. (2016). Gain more insight from your PLS-SEM results: The importance-performance map analysis. Industrial Management and Data Systems

Sajid, S. I. (2016) *Social media and its role in marketing.*

Shah, S. F., Alshurideh, M. T., Al-Dmour, A., & Al-Dmour, R. (2021). *Understanding the influences of cognitive biases on financial decision making during normal and COVID-19 pandemic situation in the United Arab Emirates* (Vol. 334).

Stephen, A. T. (2016). The role of digital and social media marketing in consumer behavior. *Current Opinion in Psychology, 10*, 17–21.

Suleman, M., Soomro, T. R., Ghazal, T. M., & Alshurideh, M. (2021). Combating against potentially harmful mobile apps. In *The international conference on artificial intelligence and computer vision* (pp. 154–173).

Sweiss, N., Obeidat, Z. M., Al-Dweeri, R. M., Mohammad Khalaf Ahmad, A., Obeidat, A. M., & Alshurideh, M. (2021). The moderating role of perceived company effort in mitigating customer

misconduct within Online Brand Communities (OBC). *Journal of Marketing Communications*. https://doi.org/10.1080/13527266.2021.1931942.
Vidal, L., Antúnez, L., Giménez, A., Varela, P., Deliza, R., & Ares, G. (2016). Can consumer segmentation in projective mapping contribute to a better understanding of consumer perception? *Food Quality and Preference, 47*, 64–72.
West, E., & McAllister, M. P. (2013). *The Routledge companion to advertising and promotional culture*. Routledge.
Zhu, Y.-Q., & Chen, H.-G. (2015). Social media and human need satisfaction: Implications for social media marketing. *Business Horizons, 58*(3), 335–345.

The Impacts of Social Media on Managing Customer Relationships with Brands in the UK

Mohammed T. Nuseir, Ahmad I. Aljumah, Sarah Urabi, Muhammad Alshurideh, and Barween Al Kurdi

Abstract This research is done to understand the impacts of social media in managing customer relationships for the selling business organizations within the b2b business environment. In today's world, the use of social media has increased rapidly. The selling business organizations use social media sites as their marketing tool. However, now, it is time for them to understand that social media sites can be used towards improving customer relationship performance. SEM-PLS model is used in this research to understand the effects of social media as the relationship tool. The data collected for the research shows that the use of social media has a direct positive impact on social CRM capabilities, which has a direct positive impact on customer relationship performance. The researcher has discussed the research result elaborately. In the end, proper implications, limitations, and recommendations are discussed by the researcher.

M. T. Nuseir (✉)
Department of Business Administration, College of Business, Al Ain University, Abu Dhabi Campus, Abu Dhabi, UAE
e-mail: mohammed.nuseir@aau.ac.ae

A. I. Aljumah
College of Communication and Media, Al Ain University, Abu Dhabi Campus, P.O. Box 112612, Abu Dhabi, United Arab Emirates

S. Urabi
Abu Dhabi Vocational Education and Training Institute (ADVETI), Abu Dhabi, UAE

M. Alshurideh
Department of Marketing, School of Business, The University of Jordan, Amman, Jordan

Department of Management, College of Business Administration, University of Sharjah, Sharjah, United Arab Emirates

B. Al Kurdi
Department of Marketing, Faculty of Economics and Administrative Sciences, The Hashemite University, Zarqa, Jordan
e-mail: barween@hu.edu.jo

M. Alshurideh et al. (eds.), *The Effect of Information Technology on Business and Marketing Intelligence Systems*, Studies in Computational Intelligence 1056,
https://doi.org/10.1007/978-3-031-12382-5_4

Keywords Customer relationship orientation · Business-to-business · Social media · CRM · Social CRM · Customer relationship performance · CRM in b2b · Social media in b2b · Social CRM in b2b · PLS-SEM model

1 Introduction

With the passing time, social media is becoming more powerful in maintaining customer relationships as people are becoming more active in communicating, sharing information, and creating social media contents day by day (Al-Maroof et al., 2021a; Nuseir et al., 2021b; El Refae et al., 2021). As per the opinion of Felix et al. (2017), Al-Hamad et al. (2021), and Suleman et al. (2021), the use of smart devices and social media will even increase more in the coming years. Different companies now spend more of their money on the social media platform rather than spending them on the traditional media (Akour et al., 2021; Alshurideh et al., 2021a; Kurdi et al., 2020b). Della Corte et al. (2015) stated that approximately 92% of the marketing divisions of different companies agreed that social media is essential for their business as it helps to create high value towards the business. Nowadays, the business organizations utilize social media sites properly to enhance and maintain customer relationships, to market through word of mouth, to generate community-based customer support and to co-creation (Abu Zayyad et al., 2021; Al Khasawneh et al., 2021a; Alshurideh et al., 2019b; Turki Alshurideh, 2014). Noe et al. (2017) added that organizations now use their existing customer database for improving their Customer Relationship Management (CRM), and in this process, they adopt the involvement of social media (Al Khasawneh et al., 2021b; Al Kurdi et al., 2021a; Aljumah et al., 2021a; Alshurideh et al., 2019a), which is known as Social Customer Relationship Management (SCRM).

The previous researches show the effectiveness of social media in the b2c (Business to Customer) context, including improving customer relationships towards increasing brand awareness, loyalty, and company sales (Alameeri et al., 2021; Almazrouei et al., 2020; Joghee et al., 2021b; Kabrilyants et al., 2021; Nuseir et al., 2021a). Kumar et al. (2016) opined that similar benefits could be achieved in the context of b2b (business to business). However, there were few numbers of researches done on the benefits of social media in b2b context. As per the statement of Felix et al. (2017), b2b marketers understood the potentiality of social media towards improving customer relationships and also enhancing brand value and how it can be achieved by following the path followed by the b2c marketers (Alshurideh et al., 2019d; Hamadneh & Al Kurdi, 2021; Hamadneh et al., 2021). Customers are a valuable financial asset for b2b companies. Noe et al. (2017) expected that there should be an increase in the significance of the SCRM in b2b organizations. It will help to improve the company's business performance through engaging more customers and creating more values through these customer engagements.

This research aims to understand the importance of Social Media on Customer Relationship Management activities in the b2b context. Kumar et al. (2016) and

Alshurideh (2022) stated that previous researches were focused either on the CRM in the b2b context or Social Media in a b2b context. The primary purpose of this research is to test the process through which the business organizations in the UK properly utilize Social Media as their CRM tool in the b2b context. The below section shows the research questions which are drawn for this particular research study.

Research Question 1: What are the impacts of Social Media as the Customer Relationship Management tool on b2b customer relationships?

The answer to this question will help the researcher to understand how organizations improve their b2b customer relationships through their social media activities by using it as the customer relationship management tool. It will provide the company's perspective.

Research Question 2: How social media as a CRM tool create a match between the expectations of the company and the experience of its customers?

By answering this research question, the researcher will assess if the expectations of the companies match the real situation. It will provide the customer's perspective.

2 Conceptual Framework and Hypotheses Development

In this section, the previous related theories about the research topic will be discussed. The development of a conceptual framework and research hypotheses will also be done in this section.

2.1 *Concept of Business-To-Business (b2b)*

Business to business (b2b) marketing is entirely different from business to customer (b2c) marketing. Stott et al. (2016) observed that before any product comes to the b2c market, it needs to pass different stages involved in the b2b market. Before selling any product to its final consumer, the producer organization needs to sell that product to the wholesalers. So, the wholesalers or the retailers at first buy different products from the producer companies, and then they sell those products to the final consumers (Alshurideh et al., 2017, 2019c, 2021b; Obeidat et al., 2019). As per the opinion of Wirtz et al. (2016), unlike the b2c market, in the b2b market, there are a smaller number of buyers who buy in a more substantial amount. The demand in the b2b market comes from the demand made by the end customers. It is a complex process, so many professional people are involved in evaluating different criteria. These professional people need information about the products and also the demands of the end consumers (Al Kurdi & Alshurideh, 2021; Al-Dmour & Al-Shraideh, 2008; Alshurideh, 2016). According to Stott et al. (2016), a business relationship performs an important part that needs to be researched more within the b2b context.

2.2 Concept of Social Media in b2b Context

As per the opinion of Agnihotri et al. (2016), social media plays an essential part in the process of promoting different products towards customers. It also helps in making and maintaining a strong relationship between the selling companies and the customers (Al-Dhuhouri et al., 2020; Alshurideh, 2017; Alshurideh et al., 2016; Hayajneh et al., 2021; Svoboda et al., 2021). Several social media channels are there to help in performing social media marketing. Siamagka et al. (2015) noted that the selling organizations share the messages, photos, product descriptions, video, and audio files about their products through social media channels. These organizations make their online presence through different social media channels (Ahmad et al., 2021; Alaali et al., 2021; Joghee et al., 2021a). Cawsey and Rowley (2016) added that there are mainly five types of social media are blogs, microblogs, social networks, collaboration community, and content sharing community. It is hard to categorize these different applications of SM as the obstructions among different categories overlie each other.

B2c companies mostly use SM instead of b2b companies. Most of the uses of SM by the b2b companies are done for their internal and external processes through internal use is more than external use. As per the statement made by Wang et al. (2016), social networking sites, blogs, microblogs, and discussion forums are the most effective and popular tools for the external use of SM towards their customers. Siamagka et al. (2015) stated that project communication, communication, and collaboration, and network management are three essential functions of partner-related use of SM. SM provides sales support and helps in making communication with partner organizations and customers. Some barriers come when SM is used in the b2b context, which includes inappropriate measurement of the efficiency of SM, a high priority of the other projects, a lack of proper understanding of the available options, a smaller number of related case studies, etc.

Wang et al. (2016) showed the advantages of using SM in b2b context, including increasing involvement of the customers, increased profitability through increased sales, increased customer loyalty, brand awareness, improved relationship with the customers, helping in providing improved customer services. Cawsey and Rowley (2016) added that one of the important advantages of SM is the low-cost coverage and high impact on the target customers at the same time. SM helps the b2b organizations to build personal interactions with their vendors and stakeholders, which ultimately helps in making a strong relationship that helps to improve their corporate credibility (Abuhashesh et al., 2021; Al-Dmour et al., 2021a; Harahsheh et al., 2021; Leo et al., 2021). Agnihotri et al. (2016) stated that proper and effective b2b sales decisions might generate higher sales for the organizations in the long term. SM helps in the process of transaction that helps towards increasing sales for the current customers and also helps in the promotional process for the new customers (Alhamad et al., 2021; Ghazal et al., 2021; Madi Odeh et al., 2021). The organizations that use SM as their marketing tool use SM towards setting up their companies as the leaders, increasing brand awareness, and making contacts with their stakeholders. Siamagka et al. (2015) informed that SM also helps the organizations towards optimizing the search engine results and also helps in increasing their website homepage traffic.

2.3 Concept of Customer Relationship Management (CRM)

As per the opinion of Terziev and Banabakova (2017), customer relationship management (CRM) is a strategic management approach that includes the strategies, practices, and technologies used by the organizations towards analyzing and managing their interaction with their customers. It helps the companies to use the customer lifecycle data towards achieving the goal towards enhancing the relationships with the customers, retaining their existing customers, and improving the sales (Al-Dmour et al., 2014; Alshurideh, 2012; Zu'bi et al., 2012). Hassan et al. (2015) stated that CRM helps the organizations to collect and save the data about their customers across various points of contact or channels between the organization and their customers (Al Kurdi et al., 2021b; Al-Dmour et al., 2021b; Shannak et al., 2012; Sweiss et al., 2021). These points of contact or channels include the company's website, live chat, telephone, direct mail, and of course, social media (Al-Khayyal et al., 2021; Alsharari & Alshurideh, 2021; Alshurideh et al., 2020; Kurdi et al., 2020a; Obeidat et al., 2021). Terziev and Banabakova (2017) added that effective CRM provides the detailed needed personal information of the customers to the customer care employees of the organization. It also provides the buying preferences of the customers, their purchase history, and also their anxiety.

2.4 CRM in the Context of b2b

Lipiäinen (2015) observed that as in the b2b context, there are a smaller number of customers and a higher volume of purchase compared to the b2c context, CRM is more important for b2b compared to b2c context. The organizations are forced to generate and maintain a closer relationship with their lower number of customers within the b2b context. It helps the organizations towards minimizing the risk and also generating maximum value. As per the opinion of Kumar and Reinartz (2018), there are three different perspectives in managing customer asset, which includes relationship, sales, and network. Among these perspectives, the network helps the organizations in maintaining the highest depth in managing their customer assets. It also provides maximum scope for the customer care managers of the organizations to generate value within the organization-customer environment. Aissani and Awad ElRefae (2021) and Lipiäinen (2015) added that customers are one of the important assets of the organization, and without the customers, the organizations cannot achieve a competitive advantage.

2.5 *Social CRM in the Context of b2b*

According to Nunan et al. (2018), social CRM is the customer relationship management that can be achieved by organizations through making communication with their customers through social media sites such as Facebook or Twitter. Social CRM helps organizations to interact with their customers with a particular brand, and also it helps to improve the relationship with their customers. Agnihotri et al. (2016) and Nuseir et al. (2021b) defined the uses of social CRM in the b2b context as it helps in making social sales, internal collaboration, and building sales genius. The organization can increase social sales by using the social information of their customers. It is provided by social CRM, which helps to maintain the relationship with the current customers and also build a new relationship with the new customers within the b2b context. Guesalaga (2016) informed that social CRM helps organizations in making internal or partner collaboration through consumer's social networks. Again, Social CRM helps to build sales genius, which helps the organizations to find similar types of customers that might desire to buy similar products or services.

2.6 *Hypotheses Development*

2.6.1 Customer Relationship Orientation

Lee and Hwang (2016) observed that it is very much important to understand market orientation to understand customer relationship orientation (CRO) as these two are interrelated. Market orientation helps to understand the demands and needs of the customers and also helps to fulfill those needs. On the other hand, Aljumah et al. (2021b) and Terho et al. (2015) opined that CRO helps organizations to concentrate on customer loyalty, retention, and the mutual relationship between the customers and the organizations during the business acts. The relationship marketing principle is also related to CRO, which shows that a competitive advantage can be gained and maintained by satisfying the needs of the customers. It can be done by maintaining continuous mutual beneficial relationships.

2.6.2 Social CRM Capabilities

As per the information provided by Elsaadani (2020), Kamboj et al. (2016), and Al Batayneh et al. (2021), social CRM capabilities show the competencies of the organizations towards developing, incorporating, and responding to the information collected from the interactions with their customers through social media. There are three dimensions of the social CRM capabilities, including generating information, disseminating information, and responsiveness towards this information. Sigala (2016) noted that the organizations that use high-level social media technology use

SM technology along with the customer-centric management system, which helps to generate more SCRM capabilities compared to the competitor companies that use a comparatively low level of SM technology. So, it can be said that social CRM capabilities create a relationship between the use of social media and customer relationship performance.

H_1: Customer relation orientation is positively related to social CRM capabilities.

2.6.3 Social Media Use

Four functions of social media are essential for CRM, including relationships, conversations, groups, and sharing. As per the opinion of Boulianne (2015), relationships include a set of technologies with the help of which the business organizations and their customers generate association networks. The organizations gather and use the information from these networks. Again, the conversation is used as applications that help in accelerating the activities of the organizations with their customers and also gather information about the customers from the conversation. Aichner and Jacob (2015) added that groups include the technological means which helps in the development of the communities and sub-communities for the online users, which are focused on particular products, topics, or brands. In Sharing, interacting, and sharing information with the customers are performed by the organization with different processes and activities. Boulianne (2015) stated that sharing helps organizations to get a positive effect in managing relationships with their customers.

H_2: Social media use has a positive relationship with social CRM capabilities.

H_3: Customer relationship orientation is positively related to social media use.

2.6.4 Customer Relationship Performance

Hassan et al. (2015) stated that Customer relationship performance could be achieved when CRM is successfully implemented as it helps in increasing the profits and values for the organizations and as well as for the customers. Effective CRM helps organizations in providing improved customer service, increasing marketing efficiency, improving competitiveness, reducing costs, increasing customer loyalty and market awareness, and increasing profitability. For this research, CRP is represented as the degree for measuring customer satisfaction, retention, and loyalty.

H_4: Social CRM capabilities have a positive relationship with customer relationship performance.

H_5: Customer relation orientation is positively related to customer relationship performance.

Figure 1 represents the research model, followed by the conceptual framework.

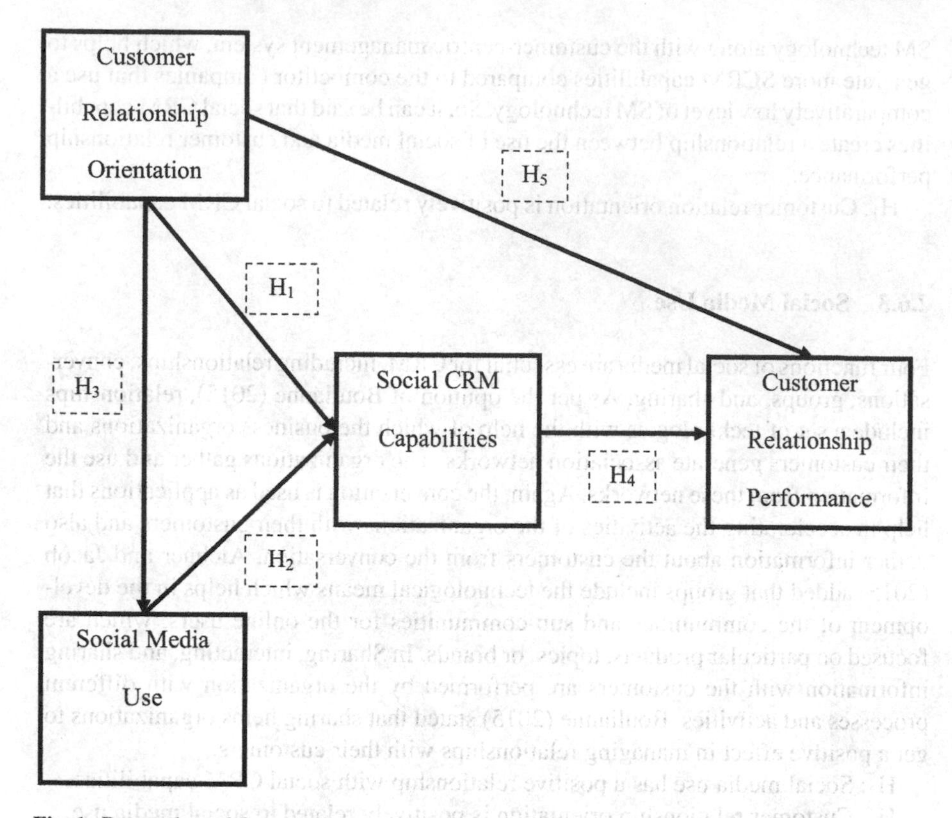

Fig. 1 Research model

3 Instrument Development and Data Collection

Instrument Development: A questionnaire will help to perform this research the researcher has used Likert scale from 5 = strongly agree to 1 = strongly disagree. Previous researchers have helped the researcher to develop this questionnaire, which has helped in enhancing the validity of the research. Customer relationship is one of the essential constructs for this research. In this research, the impacts of social media on customer relationships with the organizations are understood. Another construct for this research is social CRM capabilities. It is measured by customer relationship orientation. The dependent variable for this research is the customer relationship orientation, which is also a construct for this research.

Data Collection: The quantitative questionnaire is distributed among small business organizations in Birmingham, UK. The researcher sent the online questionnaire to the management of the chosen small business organizations through email. The chosen participants for this research are only the small business organization that buys products from the more prominent business organization and perform in the b2b context by using social media. The researcher sent 50 soft copies of questionnaires to the small business organizations through emails. The researcher got a total of 35 responses after four weeks of sending the emails.

Research Method: The researcher has used Structural Equation Modelling (SEM) for this research. The researcher has also used the Partial Least Square (PLS) method. Al-Maroof et al. (2021b) observed that the PLS method helps the researcher to understand complex data within the low-structured situation. Hult et al. (2018) informed that PLS is very much useful in the areas of social media as it is tough to get any theoretical information in this area. PLS helps the researcher to assess the validity and reliability of the research. The researcher does significance testing by using the resampling method.

4 Results

The researcher gathers all the needed information through the quantitative questionnaire. The researcher prepared the quantitative questionnaire and emailed them to the target small business organizations that are involved in the b2b business with the more prominent business organizations. The researcher sent the email with the questionnaire to 50 small business organizations. Among them, only 35 organizations replied with all the needed information for this research. This whole process of data collection took approximately 4 weeks. The reliability and validity of the structural model for this research will be explained in this section in a detailed manner.

Reliability

Cronbach's Alpha shows composite reliability. The value of Cronbach's Alpha should be greater than 0.7. Cronbach's Alpha helps in testing the reliability of the survey of any research. Trizano-Hermosilla and Alvarado (2017) added that it is also used in measuring internal consistency for any research. The value of Cronbach's Alpha for this testing is more than 0.7. From the below table, it can be said that the reliability is accepted for this particular research. The below table shows the overview of quality criteria.

Validity

The validity of this research is tested with the help of content validity and construct validity. Within the content validity, face validity is tested through this research. Face validity testing is performed by sending the quantitative questionnaire to the small business organizations involved in the b2b marketing through emails and asking the company marketing personals to fill up those forms. The researcher has taken into account the substantial literature review. The constructs for this research are Customer Relationship Orientation, social media use, and Social Customer Relationship Management Capabilities. These constructs have helped the researcher in the information system. Currently, available research works on the similar topic have helped the researcher to prepare the quantitative questionnaire. The information gathers through these quantitative questionnaires have helped the researcher enrich the research contents. As per the opinion of Yamato et al. (2017), convergent validity can be ensured through assessing discriminant and convergent validity. AVE shows the Average Variance Extracted, and it performs as a criterion. Alarcón et al. (2015) added that AVE is used to measure and test convergent validity. The value of AVE is always expected to be more than 0.5. Table 1 shows the values of AVE and all the AVE for all the constructs are more than 0.5. Cegarra-Navarro et al. (2016) stated that in the case of discriminant validity, PLS helps to measure the extent up to the level where there is a difference between a particular construct with the other constructs. The value of composite reliability should always be more than AVE, and also it should be more than 0.7. It is seen from Table 1 that the value of composite reliability is more than 0.7 and more than the value of AVE for all constructs. So, from Table 1, it can be understood that the validity of this research is accepted.

Square of correlation among the latent variables is shown in Table 2 where cross loadings are shown in Table 3. Divergent validity testing for this research can be done with the help of Tables 2 and 3.

As per the opinion of Alarcón et al. (2015), it is needed to confirm that the value of Average Variance Extracted (AVE) is more than the divergent validity, the value of MSV (Maximum Shared Squared Variance) and value of ASV (Average Shared Squared Variance).

AVE > MSV

AVE > ASV

It is seen from Table 2 that

0.986671 > {0.02772, 0.042641, 0.173419} and

0.714738 > {0.02772, 0.142834, 0.119818} and

0.749631 > {0.042641, 0.142834, 0.703512} and

0.855889 > {0.173419, 0.119818, 0.703512}.

Cegarra-Navarro et al. (2016) observed that it is important to assess the discrimination validity to compare the square of the correlations among the variables with the AVE. Table 2 shows the square of correlations among different latent variables. It will help to confirm the discriminant validity. Yamato et al. (2017) added that testing the factor loadings for every indicator helps to get a two-fold result for the convergent

Table 1 Overview of quality criteria

	AVE	Composite reliability	R square	Cronbach's alpha
Customer relationship orientation	0.97352	0.986582		0.9728
Social media use	0.510851	0.764277	0.340768	0.947628
Social customer relationship management capabilities	0.561947	0.752086	0.421899	0.85566
Customer relationship performance	0.732546	0.890322	0.7816	0.810019

Table 2 Square of correlation between latent variables

	Customer relationship orientation	Social media use	Social customer relationship management capabilities	Customer relationship performance
Customer relationship orientation	**0.986671**			
Social media use	0.02772	**0.714738**		
Social customer relationship management capabilities	0.042641	0.142834	**0.749631**	
Customer relationship performance	0.173419	0.119818	0.703512	**0.855889**

and discriminant validity both. The factor loading for any indicator should always be more than its constructs when compared with other factors.

Cross loadings are shown in Table 3. It includes factor loadings of the indicators, which are more than their constructs comparing with other factors. The value of the red cells in every row of Table 3 should be more than the other cells of the same row. This condition confirms the divergent validity.

For CRO1, 0.9866 > {0.030976, 0.046624, 0.168913} and
For CRO2, 0.986742 > {0.023745, 0.037545, 0.17329} and
For SMU1, 0.53156 > {0.001578, 0.077049, 0.015338} and
For SMU2, 0.859736 > {0.031766, 0.122122, −0.150665} and
For SCRMC1, 0.570522 > {0.050194, 0.145343, 0.540354} and
For SCRMC2, 0.870945 > {−0.012473, 0.110302, 0.628633} and
For SCRMC3, 0.121996 > {−0.094954, −0.159483, −0.71053} and
For SCRMC4, 0.776416 > {0.130899, 0.125381, 0.750261} and
For SCRMC5, 0.886804 > {−0.019179, −0.096823, 0.64445} and
For SCRMC6, 0.810901 > {−0.00188, 0.077145, 0.70523} and
For SCRMC7, 0.662697 > {0.053638, 0.003076, 0.658247} and
For SCRMC8, 0.722868 > {0.003081, −0.146486, 0.474074} and
For CRP1, 0.726161 > {0.194897, 0.11079, 0.656614} and
For CRP2, 0.872359 > {0.035171, −0.271159, 0.731392} and
For CRP3, 0.953581 > {0.209034, −0.129291, 0.841247} and
For CRP3, 0.953114 > {−0.20903, 0.175248, 0.821993}.

Table 3 confirms that the validity of this research is obtained to an acceptable level. Tables 2 and 3 show that the divergent validity is also accepted for this research.

Tables 4, 5, and 6 help to perform a test of total effects. It is needed for all effects to be accepted; the T-statistics must be greater than |1.96| (Aissani & Awad ElRefae, 2021). The formula for calculating T-statistics is |Original Sample/Standard Error|.

Table 3 Cross loadings

	Customer relationship orientation	Social media use	Social customer relationship management capabilities	Customer relationship performance
CRO1	0.9866	0.030976	0.046624	0.168913
CRO2	0.986742	0.023745	0.037545	0.17329
SMU1	0.001578	0.53156	0.077049	0.015338
SMU2	0.031766	0.859736	0.122122	−0.150665
SCRMC1	0.050194	0.145343	0.570522	0.540354
SCRMC2	−0.012473	0.110302	0.870945	0.628633
SCRMC3	−0.094954	−0.159483	0.121996	−0.71053
SCRMC4	0.130899	0.125381	0.776416	0.750261
SCRMC5	−0.019179	−0.096823	0.886804	0.64445
SCRMC6	−0.00188	0.077145	0.810901	0.70523
SCRMC7	0.053638	0.003076	0.662697	0.658247
SCRMC8	0.003081	−0.146486	0.722868	0.474074
CRP1	0.194897	0.11079	0.656614	0.726161
CRP2	0.035171	−0.271159	0.731392	0.872359
CRP3	0.209034	−0.129291	0.841247	0.953581

Note CRO = customer relationship orientation; SMU = social media use; SCRMC = social customer relationship management capabilities; CRP = customer relationship performance

Table 4 shows that all values of T-statistics for this research are more than |1.96|, which confirms that all the effects are accepted for this particular research.

The test of total effects is shown in Tables 5 and 6. Ulrich et al. (2018) observed that values of R^2 for the three constructs for the research are 0.67, 0.33, and 0.19, which are considered comparatively as substantial, moderate, and weak.

Table 4 Test of total effects

	Original sample (O)	Sample mean (M)	Standard deviation (STDEV)	Standard ERROR (STERR)	T statistics (|O/STERR|)
Customer relationship orientation -> Social CRM capabilities	0.038711	0.026131	0.109734	0.109734	4.352775
Social media use -> Social CRM capabilities	0.141761	0.053139	0.194107	0.194107	8.730325
Customer relationship orientation -> Social media use	0.02772	0.009774	0.136158	0.136158	9.203589
Social CRM capabilities -> Social relationship performance	0.867695	0.430055	0.755501	0.755501	11.1485
Customer relationship orientation -> Social relationship performance	0.13642	0.133173	0.047738	0.047738	2.857642

Table 5 Test of total effects

	R square
Customer relationship orientation	
Social media use	0.340768
Social customer relationship management capabilities	0.421899
Customer relationship performance	0.7816

Table 6 Test of total effects

Total	SSO	SSE	$Q^2 = 1 - SSE/SSO$
Social media use	472	270.899207	0.426061
Social customer relationship management capabilities	1888	1069.851936	0.433341
Customer relationship performance	708	291.033761	0.588935

$$R^2 \begin{cases} 0.19 \text{ Weak} \\ 0.33 \text{ Moderate} \\ 0.67 \text{ Substantial} \end{cases}$$

$$Q^2 \begin{cases} 0.02 \text{ Weak} \\ 0.15 \text{ Moderate} \\ 0.35 \text{ Substantial} \end{cases}$$

$$Q^2 = 1 - SSE/SSO$$

From Table 5, it can be said that the values of R^2 for Social Media Use and Customer Relationship Management Capabilities are moderate, and the value of R^2 for Customer Relationship Performance is Substantial. From Table 6, it is seen that all the values of Q^2 are substantial.

The test of Goodness of Fit (GOF) is shown in Table 7.

$$GOF \begin{cases} 0.01 \text{ Weak} \\ 0.25 \text{ Moderate} \\ 0.36 \text{ Substantial} \end{cases}$$

$$\begin{aligned} GOF &= \surd\left[\{\text{Average}\left(R^2\right) \times \text{Average (Community)}\}\right] \\ &= \surd[0.51 \times 0.69] = 0.5952 > 0.36 \end{aligned}$$

GOF is accepted for this particular research.

Table 7 Test of goodness of fit (GOF)

	Communality
Customer relationship orientation	0.97352
Social media use	0.732546
Social customer relationship management capabilities	0.561947
Customer relationship performance	0.510851

Structural Model

The researcher has used Smart PLS Analysis software to analyze the collected data for this research. Al-Maroof et al. (2021b) observed that the validity of the research could be assessed with the help of R^2s and structural paths. The path coefficients for this research should be positively important to be at the level of 0.05.

From Fig. 2, it can be stated that the R^2 of social CRM capabilities is approximately 0.33% of the variance in social CRM capabilities, which is accounted for the customer relationship orientation and social media use. Again, the R^2 of social media use is approximately 0.015% of the variance of social media use, which is accounted for the customer relationship orientation. The R^2 of the customer relationship performance is approximately 0.75% of the variance of customer relationship performance and accounted for the customer relationship orientation and social CRM capabilities. So, PLS analysis helps to understand the use of social media influences that customer relationship performance.

PLS analysis helps in performing hypotheses analysis. It is seen from the PLS analysis that customer relationship orientation has positive impacts on social CRM capabilities, which is 0.327. It can be said that H_1 is supported, and the customer relationship orientation has positive impacts on social CRM capabilities. PLS analysis also shows that customer relationship orientation has positive effects on social media use, which is 0.124. This value ensures that H_3 is supported and customer relationship

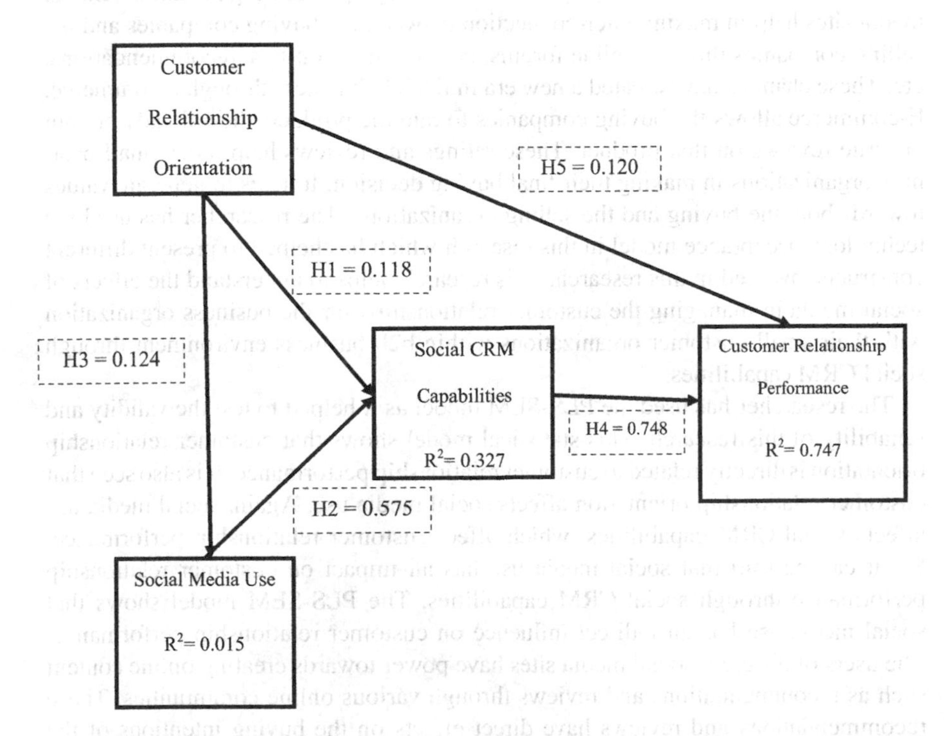

Fig. 2 Smart PLS analysis

orientation has a positive effect on social media use. Again, it is also seen that the use of social media sites has a positive effect on social CRM capabilities, which is 0.575. This value indicates that the use of social media has a very strong positive effect on social CRM capabilities. So, it ensures that H_2 is supported, and increasing the use of social media helps to enhance the social CRM capabilities. Social CRM capabilities have a strong positive effect on customer relationship performance, which is 0.748. So, it is seen that H_4 is supported. So, it can be said that ultimately, customer relationship orientation has a positive effect on customer relationship performance. Again, PLS analysis shows that customer relationship orientation has direct positive effects on customer relationship performance, which is 0.120. It proves that H_5 is a supporter, and customer relationship performance is positively affected by customer relationship orientation.

5 Discussion and Implications

The availability of the internet has increased in recent years, which have a significant impact on the buying decision of the customer companies and the selling companies within a b2b business environment. Social media sites help to provide needed information about the products towards buying small organizations. Social media sites help in making inter-connection between the buying companies and the selling companies through online forums, communities, ratings, recommendations, etc. These changes have created a new era in the b2b business through e-commerce. E-commerce allows the buying companies to rate the products they already bought or write reviews on that product. These ratings and reviews help other small business organizations in making their final buying decision. It helps to generate values towards both the buying and the selling organizations. The researcher has used the technology acceptance model in this research which has helped to present different constructs involved in this research. This research helps to understand the effects of social media in managing the customer relationships for the business organization with their small customer organizations within b2b business environment through social CRM capabilities.

The researcher has used the PLS-SEM model as it helped to test the validity and reliability of this research. This statistical model shows that customer relationship orientation is directly related to customer relationship performance. It is also seen that customer relationship orientation affects social media use. Again, social media use affects social CRM capabilities, which affect customer relationship performance. So, it can be said that social media use has an impact on customer relationship performance through social CRM capabilities. The PLS-SEM model shows that social media use has an indirect influence on customer relationship performance. The users of different social media sites have power towards creating online content such as recommendations and reviews through various online communities. These recommendations and reviews have direct effects on the buying intentions of the other customers, which are the other small business organizations in the case of b2b business. The result of this research shows that the increased use of social media has a direct effect on social customer relationship management capabilities, and it affects the customer relationship performance of the big selling business organizations to maintain their small customer organizations. Customer relationship orientation has positive effects on social CRM capabilities. Social media use also has positive effects on social CRM capabilities. Customer relationship orientation has positive impacts on social media use where social CRM capabilities have positive effects on customer relationship performance. It is seen from the PLS-SEM model that the positive effects of social media use on social CRM capabilities are very strong, which is 0.575, and the positive impacts of social CRM capabilities on customer relationship performance is also very strong which is 0.748. This proves that social media use has strong indirect positive impacts on customer relationship performance through social CRM capabilities.

The implications drawn by the researcher shows that it is needed to continue developing further social CRM capabilities through using the main two strong social media sites such as YouTube and Twitter. It will help not only to collect information but will also help to respond to this information. The selling organizations should focus on social analytics, live feeds, photo sharing, etc. as it will help to increase social CRM capabilities. The selling organizations should make more investment on social media sites as a CRM tool as it has positive impacts on customer relationship performance. The researcher has suggested for performing further research on the reason why YouTube plays the role of intermediary among social CRM capabilities and customer relationship orientation and their impacts. Future research should test the conceptual model with a larger sample size towards minimizing the research limitations.

Proper research ethics is maintained by the researcher while performing this research. The researcher promised the sample group of companies that the primary data collected from them will be used only for this research and will not be published anywhere else. The researcher maintained the privacy of these companies in an ethical manner. The quantitative questionnaire set by the researcher in this way so that they do not create any harmful effects on these companies. The researcher has maintained the code of ethics while performing this research.

6 Conclusion and Limitations

It is seen from this research that most of the business organizations involved in the b2b business do not understand the full value of using social media. Most of them still use social media as their marketing tool, and they do not realize that social media can be used for improving customer relationships. So, their customer relationship performances do not improve. It is also seen that some of the companies realized the full value of social media as a relationship tool, but they are in very early stages towards developing their social media strategies. These companies did not even develop the needed social customer relationship management capabilities so that they can attain a higher customer relationship performance.

In conclusion, it can be said that the companies that are customer relationship-oriented, should use social media towards improving their customer relationship performance through developing social CRM capabilities. More and useful use of YouTube and Twitter may help the selling companies involved in the b2b business environment to develop more improved social CRM capabilities, and it may help these organizations to obtain a higher level of customer relationship performance.

Limitations of the research indicate that there are future scopes of the study. This researcher helps to understand the effects of social media in managing customer relationships for the selling business organizations within the b2b business environment. The research result shows that the use of social media has a strong positive impact on customer relationship performance. Though it is also seen that most of the companies do not realize that Social media can be used as a relationship tool, and they only utilize social media as their marketing tool. This research highlights that social media is a tool to make a quick response to their customers. Social media allows companies to engage with their customers. It helps to make a conversation with the customers. It helps the companies to ask their customer companies to provide feedback and also thank them for their valuable time to write the feedback. Social media helps the selling companies to provide personal touch towards their customer companies. Providing rewards towards customer companies through social media sites can be a way towards making an active community. Different contests and competitions through social media sites help to sell organizations to know more about their customer companies. Social media sites can be used by the selling organizations to provide their customer companies various promotional offers, sales, or coupons. Social media sites also help to sell organizations to find out their unhappy customers and the reasons for their unhappiness. It will help them to improve their services. The selling organizations should follow User-generated content (UGC) as it will inform the followers of that company that the ideas and views of the followers are taken into account towards building the brand. It will help to customer relationship performance of the selling organizations.

Like every other research, there are many limitations faced by the researcher while performing this research. One of the most critical limitations faced by the researcher at the time of performing quantitative research was that the sample size was too small. So that it was hard to find any particular conclusion as the research findings could be taken in any direction. The sample size should be increased to get the perfect research result. The small sample size influenced the validity, quality, and reliability of this research. More data should be collected to address the validity of the research. Another limitation faced by the researcher was that some customers were contacted partially by the companies on behalf of the researcher. This may provide incomplete information as in this case; the companies selected the customers whom they were comfortable to contact with. It may create a systematic error in the sample. Another limitation was that the companies always do not provide the correct information for maintaining the privacy of their companies. This may generate changes in the research result. The limited-time period was another major limitation of this research. Again, another critical limitation was limited financial support. If the researcher would get more time and needed financial support, then the research would be more productive and specific with more required data.

References

Abu Zayyad, H. M., Obeidat, Z. M., Alshurideh, M. T., Abuhashesh, M., Maqableh, M., & Masa'deh, R. (2021). Corporate social responsibility and patronage intentions: The mediating effect of brand credibility. *Journal of Marketing Communications, 27*(5). https://doi.org/10.1080/13527266.2020.1728565.

Abuhashesh, M. Y., Alshurideh, M. T., & Sumadi, M. (2021). The effect of culture on customers' attitudes toward Facebook advertising: The moderating role of gender. *Review of International Business and Strategy*.

Agnihotri, R., Dingus, R., Hu, M. Y., & Krush, M. T. (2016). Social media: Influencing customer satisfaction in B2B sales. *Industrial Marketing Management, 53*, 172–180.

Ahmad, A., Alshurideh, M., Al Kurdi, B., Aburayya, A., & Hamadneh, S. (2021). Digital transformation metrics: A conceptual view. *Journal of Management Information and Decision Sciences, 24*(7), 1–18.

Aichner, T., & Jacob, F. (2015). Measuring the degree of corporate social media use. *International Journal of Market Research, 57*(2), 257–276.

Aissani, R., & Awad ElRefae, G. (2021). Trends of university students towards the use of YouTube and the benefits achieved—Field study on a sample of Al Ain University students. *AAU Journal of Business and Law* مجلة جامعة العين للأعمال والقانون, 5(1), 1.

Akour, I., Alshurideh, M., Al Kurdi, B., Al Ali, A., & Salloum, S. (2021). Using machine learning algorithms to predict people's intention to use mobile learning platforms during the COVID-19 pandemic: Machine learning approach. *JMIR Medical Education, 7*(1), 1–17.

Al Batayneh, R. M., Taleb, N., Said, R. A., Alshurideh, M. T., Ghazal, T. M., & Alzoubi, H. M. (2021). IT governance framework and smart services integration for future development of Dubai infrastructure utilizing AI and big data, its reflection on the citizens standard of living. In *The International Conference on Artificial Intelligence and Computer Vision* (pp. 235–247).

Al Khasawneh, M., Abuhashesh, M., Ahmad, A., Alshurideh, M. T., & Masa'deh, R. (2021a). Determinants of e-word of mouth on social media during COVID-19 outbreaks: An empirical study. In *The effect of coronavirus disease (COVID-19) on business intelligence* (Vol. 334).

Al Khasawneh, M., Abuhashesh, M., Ahmad, A., Masa'deh, R., & Alshurideh, M. T. (2021b). *Customers online engagement with social media influencers' content related to COVID 19* (Vol. 334).

Al Kurdi, B. H., & Alshurideh, M. T. (2021). Facebook advertising as a marketing tool: Examining the influence on female cosmetic purchasing behavior. *International Journal Online Marketing, 11*(2), 52–74.

Al Kurdi, B., Alshurideh, M., Nuseir, M., Aburayya, A., & Salloum, S. A. (2021a). The effects of subjective norm on the intention to use social media networks: An exploratory study using PLS-SEM and machine learning approach. In *Advanced Machine Learning Technologies and Applications: Proceedings of AMLTA 2021* (pp. 581–592).

Al Kurdi, B., Elrehail, H., Alzoubi, H. M., Alshurideh, M., & Al-adaileh, R. (2021b). The interplay among HRM practices, job satisfaction and intention to leave: An empirical investigation. *24*(1), 1–14.

Alaali, N., et al. (2021). The impact of adopting corporate governance strategic performance in the tourism sector: A case study in the Kingdom of Bahrain. *Journal of Legal, Ethical and Regulatory Issues, 24*(Special Issue 1).

Alameeri, K. A., Alshurideh, M. T., & Al Kurdi, B. (2021). *The effect of Covid-19 pandemic on business systems' innovation and entrepreneurship and how to cope with it: A theatrical view* (Vol. 334).

Alarcón, D., Sánchez, J. A., & De Olavide, U. (2015). Assessing convergent and discriminant validity in the ADHD-R IV rating scale: User-written commands for average variance extracted (AVE), composite reliability (CR), and heterotrait-monotrait ratio of correlations (HTMT). In *Spanish STATA meeting* (Vol. 39).

Al-Dhuhouri, F. S., Alshurideh, M., Al Kurdi, B., & Salloum, S. A. (2020). Enhancing our understanding of the relationship between leadership, team characteristics, emotional intelligence and their effect on team performance: A critical review. In *International Conference on Advanced Intelligent Systems and Informatics* (pp. 644–655).

Al-Dmour, H., & Al-Shraideh, M. T. (2008). The influence of the promotional mix elements on Jordanian consumer's decisions in cell phone service usage: An analytical study. *Jordan Journal of Business Administration, 4*(4), 375–392.

Al-Dmour, H., Alshuraideh, M., & Salehih, S. (2014). A study of Jordanians' television viewers habits. *Life Science Journal, 11*(6), 161–171.

Al-Dmour, R., AlShaar, F., Al-Dmour, H., Masa'deh, R., & Alshurideh, M. T. (2021a). The effect of service recovery justices strategies on online customer engagement via the role of "customer satisfaction" during the Covid-19 pandemic: An empirical study. In *The effect of coronavirus disease (COVID-19) on business intelligence* (Vol. 334).

Al-Dmour, A., Al-Dmour, H., Al-Barghuthi, R., Al-Dmour, R., & Alshurideh, M. T. (2021b). Factors influencing the adoption of e-payment during pandemic outbreak (COVID-19): Empirical evidence. In *The effect of coronavirus disease (COVID-19) on business intelligence* (Vol. 334).

Alhamad, A. Q. M., Akour, I., Alshurideh, M., Al-Hamad, A. Q., Kurdi, B. A., & Alzoubi, H. (2021). Predicting the intention to use google glass: A comparative approach using machine learning models and PLS-SEM. *International Journal of Data and Network Science, 5*(3). https://doi.org/10.5267/j.ijdns.2021.6.002.

Al-Hamad, M. Q., Mbaidin, H. O., Alhamad, A. Q. M., Alshurideh, M. T., Kurdi, B. H. A., & Al-Hamad, N. Q. (2021). Investigating students' behavioral intention to use mobile learning in higher education in UAE during Coronavirus-19 pandemic. *International Journal of Data and Network Science, 5*(3). https://doi.org/10.5267/j.ijdns.2021.6.001.

Aljumah, A., Nuseir, M. T., & Alshurideh, M. T. (2021a). The impact of social media marketing communications on consumer response during the COVID-19: Does the brand equity of a university matter. In *The effect of coronavirus disease (COVID-19) on business intelligence* (Vol. 334, pp. 384–367).

Aljumah, A. I., Nuseir, M. T., & Alam, M. M. (2021b). Organizational performance and capabilities to analyze big data: Do the ambidexterity and business value of big data analytics matter? *Business Process Management Journal*.

Al-Khayyal, A., Alshurideh, M., Al Kurdi, B., & Salloum, S. A. (2021). Factors influencing electronic service quality on electronic loyalty in online shopping context: Data analysis approach. In *Enabling AI applications in data science* (pp. 367–378). Springer.

Almazrouei, F. A., Alshurideh, M., Al Kurdi, B., & Salloum, S. A. (2020). Social media impact on business: A systematic review. In *International Conference on Advanced Intelligent Systems and Informatics* (pp. 697–707).

Al-Maroof, R., et al. (2021a). The acceptance of social media video for knowledge acquisition, sharing and application: A comparative study among YouTube users and TikTok users' for medical purposes. *International Journal of Data and Network Science, 5*(3). https://doi.org/10.5267/j.ijdns.2021.6.013.

Al-Maroof, R. S., Alshurideh, M. T., Salloum, S. A., AlHamad, A. Q. M., & Gaber T. (2021b). Acceptance of google meet during the spread of coronavirus by Arab university students. *Informatics, 8*(2). https://doi.org/10.3390/informatics8020024.

Alsharari, N. M., & Alshurideh, M. T. (2021). Student retention in higher education: The role of creativity, emotional intelligence and learner autonomy. *International Journal of Educational Management, 35*(1). https://doi.org/10.1108/IJEM-12-2019-0421.

Alshurideh. (2012). The effect of previous experience on mobile subscribers' repeat purchase behaviour.

Alshurideh, M. (2016). Exploring the main factors affecting consumer choice of mobile phone service provider contracts. *International Journal of Communications, Network and System Sciences, 9*(12), 563–581.

Alshurideh, M., Al Kurdi, B., Vij, A., Obiedat, Z., & Naser, A. (2016). Marketing ethics and relationship marketing—An empirical study that measure the effect of ethics practices application on maintaining relationships with customers. *International Business Research, 9*(9), 78–90.

Alshurideh, M. (2017). A theoretical perspective of contract and contractual customer-supplier relationship in the mobile phone service sector. *International Journal of Business and Management, 12*(7), 201–210.

Alshurideh, M., Al Kurdi, B., Abu Hussien, A., & Alshaar, H. (2017). Determining the main factors affecting consumers' acceptance of ethical advertising: A review of the Jordanian market. *Journal of Marketing Communications, 23*(5). https://doi.org/10.1080/13527266.2017.1322126.

Alshurideh, et al. (2019a). Understanding the quality determinants that influence the intention to use the mobile learning platforms: A practical study. *International Journal of Interactive Mobile Technologies, 13*(11), 157–183.

Alshurideh, M., Al Kurdi, B., & Salloum, S. A. (2019b). Examining the main mobile learning system drivers' effects: A mix empirical examination of both the expectation-confirmation model (ECM) and the technology acceptance model (TAM). In *International Conference on Advanced Intelligent Systems and Informatics* (pp. 406–417).

Alshurideh, M., Alsharari, N. M., & Al Kurdi, B. (2019c). Supply chain integration and customer relationship management in the airline logistics. *Theoretical Economics Letters, 9*(02), 392–414.

Alshurideh, M., Salloum, S. A., Al Kurdi, B., & Al-Emran, M. (2019d). Factors affecting the social networks acceptance: An empirical study using PLS-SEM approach. In *PervasiveHealth: Pervasive computing technologies for healthcare* (Vol. Part F1479). https://doi.org/10.1145/3316615.3316720.

Alshurideh, M., Al Kurdi, B., Salloum, S. A., Arpaci, I., & Al-Emran, M. (2020). Predicting the actual use of m-learning systems: A comparative approach using PLS-SEM and machine learning algorithms. *Interactive Learning Environments*, 1–15.

Alshurideh, M. T., et al. (2021a). Factors affecting the use of smart mobile examination platforms by universities' postgraduate students during the COVID 19 pandemic: An empirical study. *Informatics, 8*(2), 32.

Alshurideh, M. T., Al Kurdi, B., & Salloum, S. A. (2021b). The moderation effect of gender on accepting electronic payment technology: A study on United Arab Emirates consumers. *Review of International Business and Strategy*.

Alshurideh, M. (2022). Does electronic customer relationship management (E-CRM) affect service quality at private hospitals in Jordan? *Uncertain Supply Chain Management, 10*(2), 1–8.

Boulianne, S. (2015). Social media use and participation: A meta-analysis of current research. *Information Communications Society, 18*(5), 524–538.

Cawsey, T., & Rowley, J. (2016). Social media brand building strategies in B2B companies. *Marketing Intelligence and Planning*.

Cegarra-Navarro, J.-G., Soto-Acosta, P., & Wensley, A. K. P. (2016). Structured knowledge processes and firm performance: The role of organizational agility. *Journal of Business Research, 69*(5), 1544–1549.

Della Corte, V., Iavazzi, A., & D'Andrea, C. (2015). Customer involvement through social media: The cases of some telecommunication firms. *Journal of Open Innovation: Technology, Market, and Complexity, 1*(1), 10.

El Refae, G. A., Nuseir, M., & Alshurideh, M. (2021). The influence of social media regulations boundary on marketing and commerce of industries in UAE. *Journal of Legal, Ethical and Regulatory Issues, 24*(Special Issue 6), 1–14.

Elsaadani, M. A. (2020). Investigating the effect of e-management on customer service performance. *AAU Journal of Business and Law* مجلة جامعة العين للأعمال والقانون, *3*(2), 1.

Felix, R., Rauschnabel, P. A., & Hinsch, C. (2017). Elements of strategic social media marketing: A holistic framework. *Journal of Business Research, 70*, 118–126.

Ghazal, T. M., et al. (2021). IoT for smart cities: Machine learning approaches in smart healthcare—A review. *Future Internet, 13*(8). https://doi.org/10.3390/fi13080218.

Guesalaga, R. (2016). The use of social media in sales: Individual and organizational antecedents, and the role of customer engagement in social media. *Industrial Marketing Management, 54*, 71–79.

Hamadneh, S., & Al Kurdi, B. (2021). The effect of brand personality on consumer self-identity: The moderation effect of cultural orientations among British and Chinese consumers. *Journal of Legal, Ethical and Regulatory Issues, 24*(Special Issue 1), 1–14.

Hamadneh, S., Pedersen, O., & Al Kurdi, B. (2021). An investigation of the role of supply chain visibility into the Scottish blood supply chain. *Journal of Legal, Ethical and Regulatory Issues, 24*(Special Issue 1), 1–12.

Harahsheh, A. A., Houssien, A. M. A., & Alshurideh, M. T. (2021). The effect of transformational leadership on achieving effective decisions in the presence of psychological capital as an intermediate variable in private Jordanian. In *The effect of coronavirus disease (COVID-19) on business intelligence* (pp. 243–221). Springer Nature.

Hassan, R. S., Nawaz, A., Lashari, M. N., & Zafar, F. (2015). Effect of customer relationship management on customer satisfaction. *Procedia Economics and Finance, 23*, 563–567.

Hayajneh, N., Suifan, T., Obeidat, B., Abuhashesh, M., Alshurideh, M., & Masa'deh, R. (2021). The relationship between organizational changes and job satisfaction through the mediating role of job stress in the Jordanian telecommunication sector. *Management Science Letters, 11*(1), 315–326.

Hult, G. T. M., Hair, J. F., Jr., Proksch, D., Sarstedt, M., Pinkwart, A., & Ringle, C. M. (2018). Addressing endogeneity in international marketing applications of partial least squares structural equation modeling. *Journal of International Marketing, 26*(3), 1–21.

Joghee, S., et al. (2021a). Expats impulse buying behaviour in UAE: A customer perspective. *Journal of Management Information and Decision Sciences, 24*(1), 1–24.

Joghee, S., Alzoubi, H. M., Alshurideh, M., & Al Kurdi, B. (2021b). The role of business intelligence systems on green supply chain management: Empirical analysis of FMCG in the UAE. In *The International Conference on Artificial Intelligence and Computer Vision* (pp. 539–552).

Kabrilyants, R., Obeidat, B. Y., Alshurideh, M., & Masa'deh, R. (2021). The role of organizational capabilities on e-business successful implementation. *International Journal of Data and Network Science, 5*(3). https://doi.org/10.5267/j.ijdns.2021.5.002.

Kamboj, S., Yadav, M., Rahman, Z., & Goyal, P. (2016). Impact of social CRM capabilities on firm performance: Examining the mediating role of co-created customer experience. *International Journal of Information Systems in the Service Sector (IJISSS), 8*(4), 1–16.

Kumar, V., & Reinartz, W. (2018). CRM issues in the business-to-business context. In *Customer relationship management* (pp. 265–283). Springer.

Kumar, A., Bezawada, R., Rishika, R., Janakiraman, R., & Kannan, P. K. (2016). From social to sale: The effects of firm-generated content in social media on customer behavior. *Journal of Marketing, 80*(1), 7–25.

Kurdi, B. A., Alshurideh, M., & Salloum, S. A. (2020a). Investigating a theoretical framework for e-learning technology acceptance. *International Journal of Electrical and Computer Engineering, 10*(6). https://doi.org/10.11591/IJECE.V10I6.PP6484-6496.

Kurdi, B. A., Alshurideh, M., Salloum, S. A., Obeidat, Z. M., & Al-dweeri, R. M. (2020b). An empirical investigation into examination of factors influencing university students' behavior

towards elearning acceptance using SEM approach. *International Journal of Interactive Mobile Technologies, 14*(2). https://doi.org/10.3991/ijim.v14i02.11115.
Lee, J. J., & Hwang, J. (2016). An emotional labor perspective on the relationship between customer orientation and job satisfaction. *International Journal of Hospitality Management, 54*, 139–150.
Leo, S., Alsharari, N. M., Abbas, J., & Alshurideh, M. T. (2021). From offline to online learning: A qualitative study of challenges and opportunities as a response to the COVID-19 pandemic in the UAE higher education context. In *The effect of coronavirus disease (COVID-19) on business intelligence* (Vol. 334, pp. 203–217).
Lipiäinen, H. S. M. (2015). CRM in the digital age: Implementation of CRM in three contemporary B2B firms. *Journal of Systems and Information Technology*.
Madi Odeh, R. B. S., Obeidat, B. Y., Jaradat, M. O., Masa'deh, R., & Alshurideh, M. T. (2021). The transformational leadership role in achieving organizational resilience through adaptive cultures: The case of Dubai service sector. *International Journal of Productivity and Performance Management*. https://doi.org/10.1108/IJPPM-02-2021-0093.
Noe, R. A., Hollenbeck, J. R., Gerhart, B., & Wright, P. M. (2017). *Human resource management: Gaining a competitive advantage*. McGraw-Hill Education.
Nunan, D., Sibai, O., Schivinski, B., & Christodoulides, G. (2018). Reflections on 'social media: Influencing customer satisfaction in B2B sales' and a research agenda. *Industrial Marketing Management, 75*, 31–36.
Nuseir, M. T., Aljumah, A., & Alshurideh, M. T. (2021a). How the business intelligence in the new startup performance in UAE during COVID 19: The mediating role of innovativeness. In *The effect of coronavirus disease (COVID-19) on business intelligence* (Vol. 334).
Nuseir, M., El Refae, G. A., & Alshurideh, M. (2021b). The impact of social media power on the social commerce intentions: Double mediating role of economic and social satisfaction. *Journal of Legal, Ethical and Regulatory Issues, 24*(Special Issue 6), 1–15.
Nuseir, M. T., Ghaleb, A., & Aljumah, A. (2021c). The e-learning of students and university's brand image (Post COVID-19): How successfully Al-Ain University have embraced the paradigm shift in digital learning. In *The effect of coronavirus disease (COVID-19) on business intelligence* (Vol. 334, p. 171).
Obeidat, Z. M., Alshurideh, M. T., Al Dweeri, R., & Masa'deh, R. (2019). The influence of online revenge acts on consumers psychological and emotional states: Does revenge taste sweet? In *Proceedings of the 33rd International Business Information Management Association Conference, IBIMA 2019: Education Excellence and Innovation Management through Vision 2020* (pp. 4797–4815).
Obeidat, U., Obeidat, B., Alrowwad, A., Alshurideh, M., Masadeh, R., & Abuhashesh, M. (2021). The effect of intellectual capital on competitive advantage: The mediating role of innovation. *Management Science Letters, 11*(4), 1331–1344.
Shannak, R. O., Masa'deh, R. M. T., Al-Zu'bi, Z. M. F., Obeidat, B. Y., Alshurideh, M., & Altamony, H. (2012). A theoretical perspective on the relationship between knowledge management systems, customer knowledge management, and firm competitive advantage. *European Journal of Social Sciences, 32*(4).
Siamagka, N.-T., Christodoulides, G., Michaelidou, N., & Valvi, A. (2015). Determinants of social media adoption by B2B organizations. *Industrial Marketing Management, 51*, 89–99.
Sigala, M. (2016). Social CRM capabilities and readiness: Findings from Greek tourism firms. *Information and Communication Technologies in Tourism, 2016*, 309–322.
Stott, R. N., Stone, M., & Fae, J. (2016). Business models in the business-to-business and business-to-consumer worlds—What can each world learn from the other? *Journal of Business and Industrial Marketing*.
Suleman, M., Soomro, T. R., Ghazal, T. M., & Alshurideh, M. (2021). Combating against potentially harmful mobile apps. In *The International Conference on Artificial Intelligence and Computer Vision* (pp. 154–173).
Svoboda, P., Ghazal, T. M., Afifi, M. A. M., Kalra, D., Alshurideh, M. T., & Alzoubi, H. M. (2021). Information systems integration to enhance operational customer relationship management in the

pharmaceutical industry. In *The International Conference on Artificial Intelligence and Computer Vision* (pp. 553–572).

Sweiss, N., Obeidat, Z. M., Al-Dweeri, R. M., Mohammad Khalaf Ahmad, A., Obeidat, A. M., & Alshurideh, M. (2021). The moderating role of perceived company effort in mitigating customer misconduct within online brand communities (OBC). *Journal of Marketing Communications*, 1–24. https://doi.org/10.1080/13527266.2021.1931942.

Terho, H., Eggert, A., Haas, A., & Ulaga, W. (2015). How sales strategy translates into performance: The role of salesperson customer orientation and value-based selling. *Industrial Marketing Management, 45*, 12–21.

Terziev, V., & Banabakova, V. (2017). Customer relationship management (CRM) as base for organization's behavior. *Psychological Review*.

Turki Alshurideh, M. (2014). The influence of mobile application quality and attributes on the continuance intention of mobile shopping.

Ulrich, R., Miller, J., & Erdfelder, E. (2018). Effect size estimation from t-statistics in the presence of publication bias. *Zeitschrift für Psychologie*.

Wang, W. Y. C., Pauleen, D. J., & Zhang, T. (2016). How social media applications affect B2B communication and improve business performance in SMEs. *Industrial Marketing Management, 54*, 4–14.

Wirtz, B. W., Pistoia, A., Ullrich, S., & Göttel, V. (2016). Business models: Origin, development and future research perspectives. *Long Range Planning, 49*(1), 36–54.

Yamato, T. P., Maher, C., Koes, B., & Moseley, A. (2017). The PEDro scale had acceptably high convergent validity, construct validity, and interrater reliability in evaluating methodological quality of pharmaceutical trials. *Journal of Clinical Epidemiology, 86*, 176–181.

Zu'bi, Z., Al-Lozi, M., Dahiyat, S., Alshurideh, M., & Al Majali, A. (2012). Examining the effects of quality management practices on product variety. *European Journal of Economics, Finance and Administrative Sciences, 51*(1), 123–139.

Impacts of Social Media on Managing Customer Relationships in b2b Business Environment in Birmingham, UK

Mohammed T. Nuseir, Ahmad I. Aljumah, Muhammad Alshurideh, Sarah Urabi, and Barween Al Kurdi

Abstract This research aims to understand the impacts of social media in managing customer relationships for the seller business organizations within the Business-to-Business (b2b) business environment. This research will help the seller organizations involved in b2b businesses to understand the full advantages of social media they can take towards achieving better customer relationship performance. Accordingly, the seller business organizations can use social media sites to improve social CRM capabilities and retaining more customers. The researcher has followed a descriptive research approach. Quantitative data were collected and used for this research. SEM-PLS model is used in this research. It can be said from the data analysis that social media has strong positive effects on building customer relationships within the b2b business environment though it is also seen that most of the organizations use social

M. T. Nuseir (✉)
Department of Business Administration, College of Business, Al Ain University, Abu Dhabi Campus, Abu Dhabi, UAE
e-mail: mohammed.nuseir@aau.ac.ae

A. I. Aljumah
College of Communication and Media, Al Ain University, P.O. Box 112612, Abu Dhabi CampusAbu Dhabi, United Arab Emirates
e-mail: Ahmad.aljumah@aau.ac.ae

M. Alshurideh
Department of Marketing, School of Business, The University of Jordan, Amman, Jordan
e-mail: malshurideh@sharjah.ac.ae

Department of Management, College of Business Administration, University of Sharjah, Sharjah, United Arab Emirates

S. Urabi
Abu Dhabi Vocational Education & Training Institute (ADVETI), Abu Dhabi, UAE
e-mail: sarah.urabi@adveti.ac.ae

B. A. Kurdi
Department of Marketing, Faculty of Economics and Administrative Sciences, The Hashemite University, Zarqa, Jordan
e-mail: barween@hu.edu.jo

M. Alshurideh et al. (eds.), *The Effect of Information Technology on Business and Marketing Intelligence Systems*, Studies in Computational Intelligence 1056,
https://doi.org/10.1007/978-3-031-12382-5_5

media as only their marketing tool, and they do not use it as a relationship tool. The main limitations of this research were the small sample size, limited time, and limited financial support. More time, financial support, and bigger sample size will help to perform comprehensive research in the future.

Keywords Customer relationship orientation · Business-to-business · Social media · CRM · Social CRM · Customer relationship performance · CRM in b2b · Social media in b2b · Social CRM in b2b

1 Introduction

With the passing time, social media is becoming more powerful in maintaining customer relationships as people are becoming more active in communicating, sharing information, and creating social media contents day by day (Al Khasawneh et al., 2021a; Nuseir et al., 2021; Refae et al., 2021). As per the opinion of Felix et al. (2017), Al-Maroof et al. (2021), Aburayya and Salloum (2021), the use of smart devices and social media will even increase more in the coming years. Different companies now spend more of their money on the social media platform rather than spending them on the traditional media (Al Khasawneh et al., 2021b; Aljumah et al., 2021a; Almazrouei et al., 2020). Della Corte et al. (2015) stated that approximately 92% of the marketing divisions of different companies agreed that social media is significant for their business as it helps to create high value towards the business. Nowadays, the business organizations utilize social media sites properly to enhance and maintain customer relationships, to market through word of mouth, to generate community-based customer support and to co-creation (Abu Zayyad et al., 2021; Ahmad et al., 2021; Alshurideh et al., 2019a). Alshurideh (2012), Noe et al. (2017), Svoboda et al. (2021), Sweiss et al. (2021) added that organizations now use their existing customer database for improving their Customer Relationship Management (CRM), and in this process, they adopt the involvement of social media, which is known as Social Customer Relationship Management (SCRM).

The previous researches show the effectiveness of social media in the b2c (Business to Customer) context, including improving customer relationships towards increasing brand awareness, loyalty, and company sales (Hamadneh & Al Kurdi, 2021). Al-khayyal et al. (2020), Alshurideh et al. (2022), Kumar et al. (2016) opined that similar benefits could be achieved in the context of b2b (business to business). However, there were few numbers of researches done on the benefits of social media in b2b context. As per the statement of Felix et al. (2017), b2b marketers understood the potentiality of social media towards improving customer relationships and also enhancing brand value and how it can be achieved by following the path followed by the b2c marketers. Customers are a valuable financial asset for b2b companies. Noe et al. (2017) expected that there should be an increase in the significance of the

SCRM in b2b organizations. It will help to improve the company's business performance through engaging more customers and creating more values through these customer engagements.

This research aims to understand the importance of Social Media on Customer Relationship Management activities in the b2b context. Kumar et al. (2016) stated that previous researches were focused either on the CRM in the b2b context or Social Media in a b2b context. The main purpose of this research is to test the process through which the business organizations in the UK properly utilize Social Media as their CRM tool in the b2b context. The below section shows the research questions which are drawn for this particular research study.

Research Question 1: What are the impacts of Social Media as the Customer Relationship Management tool on b2b customer relationships?

The answer to this question will help the researcher to understand how organizations improve their b2b customer relationships through their social media activities by using it as the customer relationship management tool. It will provide the company's perspective.

Research Question 2: How social media as a CRM tool create a match between the expectations of the company and the experience of its customers?

By answering this research question, the researcher will assess if the expectations of the companies match the real situation. It will provide the customer's perspective.

2 Conceptual Background

In this section, the previous related theories about the research topic will be discussed. The development of a conceptual framework and research hypotheses will also be done in this section.

3 Concept of Business-To-Business (B2B)

Business to business (b2b) marketing is entirely different from business to customer (b2c) marketing. Stott et al. (2016) observed that before any product comes to the b2c market, it needs to pass different stages involved in the b2b market. Before selling any product to its final consumer, the producer organization needs to sell that product to the wholesalers. So, the wholesalers or the retailers at first buy different products from the producer companies, and then they sell those products to the final consumers (Alketbi et al., 2020; Al-Khayyala et al., 2020; Alkitbi et al., 1261; Alshamsi et al., 2020; Alshurideh et al., 2020a; Alzoubi et al., 2020; Kurdi et al., 2020a). As per the opinion of Wirtz et al. (2016), unlike the b2c market, in the b2b market, there are a smaller number of buyers who buy in a more substantial amount. The demand in the b2b market comes from the demand made by the end customers. It is a complex process; so many professional people are involved in evaluating different criteria.

These professional people need information about the products and also the demands of the end consumers (Abuhashesh et al., 2021; Alshurideh, 2016, 2019; Alshurideh et al., 2012, 2016; Joghee et al., 2021). According to Stott et al. (2016), a business relationship performs an important part that needs to be researched more within the b2b context.

4 Concept of Social Media in B2B Context

As per the opinion of Nunan et al. (2018) social media plays an essential part in the process of promoting different products towards customers. It also helps in making and maintaining a strong relationship between the seller companies and the customers. Several social media channels are there to help in performing social media marketing. Siamagka et al. (2015) noted that the seller organizations share the messages, photos, product descriptions, video, and audio files about their products through social media channels. These organizations make their online presence through different social media channels (Abuhashesh et al., 2021; Alhamad et al., 2021; Al-Hamad et al., 2021; Alshurideh et al., 2021b; Nunan et al., 2018; Siamagka et al., 2015). Cawsey and Rowley (2016) added that there are mainly five types of social media are blogs, microblogs, social networks, collaboration community, and content sharing community. It is hard to categorize these different applications of SM as the obstructions among different categories overlie each other.

B2c companies mostly use SM instead of b2b companies. Most of the uses of SM by the b2b companies are done for their internal and external processes through internal use is more than external use. As per the statement made by Wang et al. (2016), social networking sites, blogs, microblogs, and discussion forums are the most effective and popular tools for the external use of SM towards their customers. Siamagka et al. (2015) stated that project communication, communication, and collaboration, and network management are three essential functions of partner-related use of SM. SM provides sales support and helps in making communication with partner organizations and customers. Some barriers come when SM is used in the b2b context, which includes inappropriate measurement of the efficiency of SM, a high priority of the other projects, a lack of proper understanding of the available options, a smaller number of related case studies, etc.

Wang et al. (2016) showed the advantages of using SM in b2b context, including increasing involvement of the customers, increased profitability through increased sales, increased customer loyalty, brand awareness, improved relationship with the customers, helping in providing improved customer services. Cawsey and Rowley, (2016), Elsaadani (2020) added that one of the important advantages of SM is the low-cost coverage and high impact on the target customers at the same time. SM helps the b2b organizations to build personal interactions with their vendors and stakeholders, which ultimately helps in making a strong relationship that helps to improve their corporate credibility. Nunan et al. (2018) stated that proper and effective b2b sales decisions might generate higher sales for organizations in the long term.

SM helps in the process of transaction that helps towards increasing sales for the current customers and also helps in the promotional process for the new customers (Al Kurdi et al., 2020; Al-Dmour et al., 2014; Alshurideh et al., 2019b, 2019c; Kurdi et al., 2020b). The organizations that use SM as their marketing tool use SM towards setting up their companies as the leaders, increasing brand awareness, and making contacts with their stakeholders. Siamagka et al. (2015) informed that SM also helps the organizations towards optimizing the search engine results and also helps in increasing their website homepage traffic.

5 Concept of Customer Relationship Management (CRM)

As per the opinion of Terziev and Banabakova (2017), customer relationship management (CRM) is a strategic management approach that includes the strategies, practices, and technologies used by the organizations towards analyzing and managing their interaction with their customers. It helps the companies to use the customer lifecycle data towards achieving the goal towards enhancing the relationships with the customers, retaining their existing customers, and improving the sales. Hassan et al. (2015) stated that CRM helps the organizations to collect and save the data about their customers across various points of contact or channels between the organization and their customers. These points of contact or channels include the company's website, live chat, telephone, direct mail, and of course, social media. Terziev and Banabakova (2017) added that effective CRM provides the detailed needed personal information of the customers to the customer care employees of the organization. It also provides the buying preferences of the customers, their purchase history, and also their anxiety.

6 CRM in the Context of B2B

Lipiäinen (2015) observed that as in the b2b context, there are a smaller number of customers and a higher volume of purchase compared to the b2c context, CRM is more critical for b2b compared to b2c context. The organizations are forced to generate and maintain a closer relationship with their lower number of customers within the b2b context. It helps the organizations towards minimizing the risk and also generating maximum value. As per the opinion of Aljumah et al. (2021b), Kumar and Reinartz (2018), there are three different perspectives in managing customer asset, which includes relationship, sales, and network. Among these perspectives, the network helps the organizations in maintaining the highest depth in managing their customer assets (Alshurideh, 2012; Hamadneh et al., 2021; Turki Alshurideh, 2014). It also provides maximum scope for the customer care managers of the organizations to generate value within the organization-customer environment. Lipiäinen

(2015), Nuseir et al. (2021) added that customers are one of the essential assets of the organization, and without the customers, the organizations cannot achieve a competitive advantage.

7 Social CRM in the Context of B2B

According to Nunan et al. (2018), social CRM is the customer relationship management that can be achieved by organizations through making communication with their customers through social media sites such as Facebook or Twitter. Social CRM helps organizations to interact with their customers with a particular brand, and also it helps to improve the relationship with their customers. Nunan et al. (2018) defined the uses of social CRM in the b2b context as it helps in making social sales, internal collaboration, and building sales genius. The organization can increase social sales by using the social information of their customers. It is provided by social CRM, which helps to maintain the relationship with the current customers and also build a new relationship with the new customers within the b2b context. Aissani and Awad ElRefae (2021), Guesalaga (2016) informed that social CRM helps organizations in making internal or partner collaboration through consumer's social networks. Again, Social CRM helps to build sales genius, which helps the organizations to find similar types of customers that might desire to buy similar products or services.

8 Hypotheses Development

8.1 Customer Relationship Orientation (CRO)

Lee and Hwang (2016) observed that it is very much important to understand market orientation to understand customer relationship orientation (CRO) as these two are interrelated. Market orientation helps to understand the demands and needs of the customers and also helps to fulfill those needs. On the other hand, Terho et al. (2015) opined that CRO helps organizations to concentrate on customer loyalty, retention, and the mutual relationship between the customers and the organizations during the business acts. The relationship marketing principle is also related to CRO, which shows that a competitive advantage can be gained and maintained by satisfying the needs of the customers. It can be done by maintaining continuous mutual beneficial relationships.

8.2 Social CRM Capabilities

As per the information provided by Kamboj et al. (2016), social CRM capabilities show the competencies of the organizations towards developing, incorporating, and responding to the information collected from the interactions with their customers through social media. There are three dimensions of the social CRM capabilities, including generating information, disseminating information, and responsiveness towards this information. Sigala (2016) noted that the organizations that use high-level social media technology use the SM technology along with the customer-centric management system, which helps to generate more SCRM capabilities compared to the competitor companies that use a comparatively low level of SM technology. So, it can be said that social CRM capabilities create a relationship between the use of social media and customer relationship performance.

H_1: Customer relation orientation is positively related to social CRM capabilities.

8.3 Social Media Use

Four functions of social media are essential for CRM, including relationships, conversations, groups, and sharing. As per the opinion of Boulianne (2015), relationships include a set of technologies with the help of which the business organizations and their customers generate association networks. The organizations gather and use the information from these networks. Again, a conversation is used as an application that helps in accelerating the activities of the organizations with their customers and also gathers information about the customers from the conversation. Aichner and Jacob (2015) added that groups include the technological means which helps in the development of the communities and sub-communities for the online users, which are focused on particular products, topics, or brands. In Sharing, interacting, and sharing information with the customers are performed by the organization with different processes and activities. Boulianne (2015) stated that sharing helps organizations to get a positive effect in managing relationships with their customers.

H_2: Social media use has a positive relationship with social CRM capabilities.

H_3: Customer relationship orientation is positively related to social media use.

9 Customer Relationship Performance

Hassan et al. (2015) stated that Customer relationship performance could be achieved when CRM is successfully implemented as it helps in increasing the profits and values for the organizations and as well as for the customers. Effective CRM helps organizations in providing improved customer service, increasing marketing efficiency, improving competitiveness, reducing costs, increasing customer loyalty and market

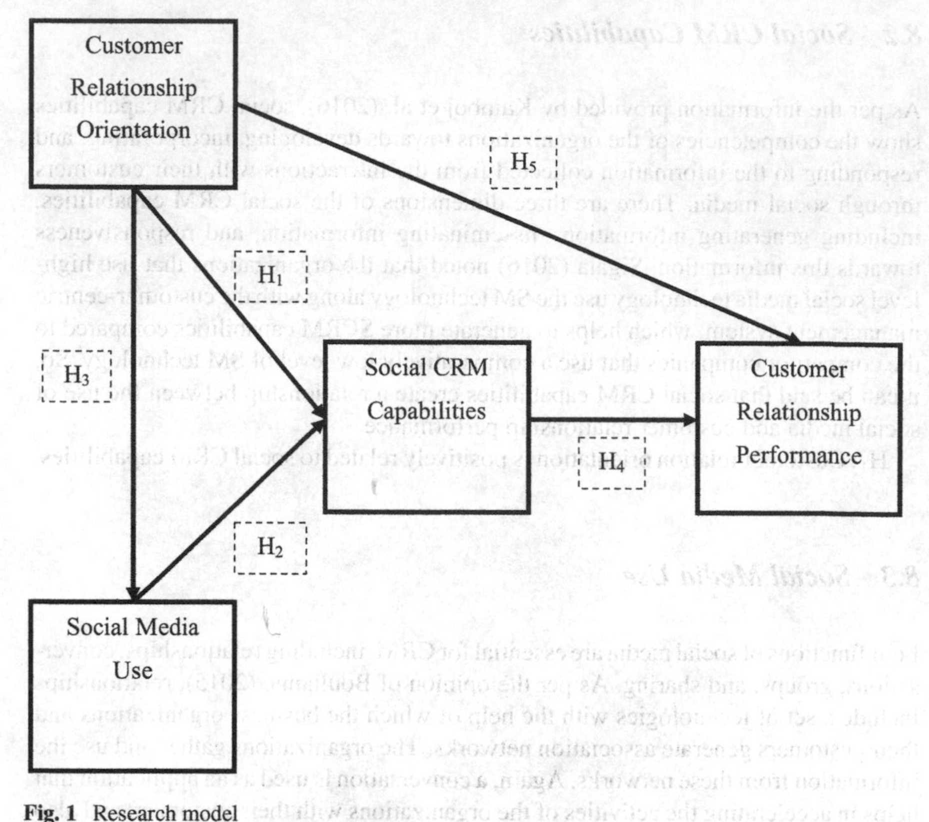

Fig. 1 Research model

awareness, and increasing profitability. For this research, CRP is represented as the degree for measuring customer satisfaction, retention, and loyalty.

H_4: Social CRM capabilities have a positive relationship with customer relationship performance.

H_5: Customer relation orientation is positively related to customer relationship performance.

Figure 1 represents the research model, followed by the conceptual framework.

10 Method

Instrument Development: A questionnaire will help to perform this research the researcher has used Likert scale from 5 = strongly agree to 1 = strongly disagree. Previous researchers have helped the researcher to develop this questionnaire, which has helped in enhancing the validity of the research. Customer relationship is one of the essential constructs for this research. In this research, the impacts of social media on customer relationships with the organizations are understood. Another construct for this research is social CRM capabilities. It is measured by customer relationship orientation. The dependent variable for this research is the customer relationship orientation, which is also a construct for this research.

Data Collection: The quantitative questionnaire is distributed among small business organizations in Birmingham, UK. The researcher sent the online questionnaire to the management of the chosen small business organizations through email. The chosen participants for this research are only the small business organization that buys products from the more significant business organization and perform in the b2b context by using social media. The researcher sent 50 soft copies of questionnaires to the small business organizations through emails. The researcher got a total of 35 responses after four weeks of sending the emails.

Research Method: The researcher has used Structural Equation Modelling (SEM) for this research. The researcher has also used the Partial Least Square (PLS) method. Al-Maroof et al. (2021) observed that the PLS method helps the researcher to understand complex data within the low-structured situation. Hult et al. (2018) informed that PLS is very much useful in the areas of social media as it is tough to get any theoretical information in this area. PLS helps the researcher to assess the validity and reliability of the research. The researcher does significance testing by using the resampling method.

11 Results and Discussion

The researcher gathers all the needed information through the quantitative questionnaire. The researcher prepared the quantitative questionnaire and emailed them to the target small business organizations that are involved in the b2b business with the more prominent business organizations. The researcher sent the email with the questionnaire to 50 small business organizations. Among them, only 35 organizations replied with all the needed information for this research. This whole process of data collection took approximately 4 weeks. The reliability and validity of the structural model for this research will be explained in this section in a detailed manner.

11.1 Reliability

Cronbach's Alpha shows composite reliability. The value of Cronbach's Alpha should be greater than 0.7. Cronbach's Alpha helps in testing the reliability of the survey of any research. Trizano-Hermosilla and Alvarado (2016) added that it is also used in measuring internal consistency for any research. The value of Cronbach's Alpha for this testing is more than 0.7. From the below table, it can be said that the reliability is accepted for this particular research. The below table shows the overview of quality criteria.

11.2 Validity

The validity of this research is tested with the help of content validity and construct validity. Within the content validity, face validity is tested through this research. Face validity testing is performed by sending the quantitative questionnaire to the small business organizations involved in the b2b marketing through emails and asking the company marketing personals to fill up those forms. The researcher has taken into account the substantial literature review. The constructs for this research are Customer Relationship Orientation, social media use, and Social Customer Relationship Management Capabilities. These constructs have helped the researcher in the information system. Currently, available research works on the similar topic have helped the researcher to prepare the quantitative questionnaire. The information gathers through these quantitative questionnaires have helped the researcher enrich the research contents. As per the opinion of Wirtz et al. (2016), convergent validity can be ensured through assessing discriminant and convergent validity. AVE shows the Average Variance Extracted, and it performs as a criterion. Alarcón et al. (2015) added that AVE is used to measure and test convergent validity. The value of AVE is always expected to be more than 0.5. Table 1 shows the values of AVE and all the AVE for all the constructs are more than 0.5. Cegarra-Navarro et al. (2016) stated that in the case of discriminant validity, PLS helps to measure the extent up to the level where there is a difference between a particular construct with the other constructs. The value of composite reliability should always be more than AVE, and also it should be more than 0.7. It is seen from Table 1 that the value of composite reliability is more than 0.7 and more than the value of AVE for all constructs. So, from Table 1, it can be understood that the validity of this research is accepted.

Square of correlation among the latent variables is shown in Table 2, where cress leadings are shown in Table 3. Divergent validity testing for this research can be done with the help of Tables 2 and 3.

As per the opinion of Alarcón et al. (2015), it is needed to confirm that the value of Average Variance Extracted (AVE) is more than the divergent validity, the value of MSV (Maximum Shared Squared Variance) and value of ASV (Average Shared Squared Variance).

Table 1 Overview of quality criteria

	AVE	Composite reliability	R square	Cronbach's alpha
Customer relationship orientation	0.97352	0.986582		0.9728
Social media use	0.510851	0.764277	0.340768	0.947628
Social customer relationship management capabilities	0.561947	0.752086	0.421899	0.85566
Customer relationship performance	0.732546	0.890322	0.7816	0.810019

Table 2 Square of correlation between latent variables

	Customer relationship orientation	Social media use	Social customer relationship management capabilities	Customer relationship performance
Customer relationship orientation	0.986671			
Social media use	0.02772	0.714738		
Social customer relationship management capabilities	0.042641	0.142834	0.749631	
Customer relationship performance	0.173419	0.119818	0.703512	0.855889

Table 3 Cross loadings

	Customer relationship orientation	Social media use	Social customer relationship management capabilities	Customer relationship performance
CRO1	0.9866	0.030976	0.046624	0.168913
CRO2	0.986742	0.023745	0.037545	0.17329
SMU1	0.001578	0.53156	0.077049	0.015338
SMU2	0.031766	0.859736	0.122122	−0.150665
SCRMC1	0.050194	0.145343	0.570522	0.540354
SCRMC2	−0.012473	0.110302	0.870945	0.628633
SCRMC3	−0.094954	−0.159483	0.121996	−0.71053
SCRMC4	0.130899	0.125381	0.776416	0.750261
SCRMC5	−0.019179	−0.096823	0.886804	0.64445
SCRMC6	−0.00188	0.077145	0.810901	0.70523
SCRMC7	0.053638	0.003076	0.662697	0.658247
SCRMC8	0.003081	−0.146486	0.722868	0.474074
CRP1	0.194897	0.11079	0.656614	0.726161
CRP2	0.035171	−0.271159	0.731392	0.872359
CRP3	0.209034	−0.129291	0.841247	0.953581

Note CRO = Customer Relationship Orientation; SMU = Social Media Use; SCRMC = Social Customer Relationship Management Capabilities; CRP = Customer Relationship Performance

AVE > MSV.
AVE > ASV.
It is seen from Table 2 that.
0.986671 > {0.02772, 0.042641, 0.173419} and
0.714738 > {0.02772, 0.142834, 0.119818} and
0.749631 > {0.042641, 0.142834, 0.703512} and
0.855889 > {0.173419, 0.119818, 0.703512}.

Cegarra-Navarro et al. (2016) observed that it is important to assess the discrimination validity to compare the square of the correlations among the variables with the AVE. Table 2 shows the square of correlations among different latent variables. It will help to confirm the discriminant validity. Wirtz et al. (2016) added that testing the factor loadings for each indicator helps to get a two-fold result for the convergent and discriminant validity both. The factor loading for any indicator should always be more than its constructs when compared with other factors.

Cross loadings are shown in Table 3. It includes factor loadings of the indicators, which are more than their constructs comparing with other factors. The value of the red cells in every row of Table 3 should be more than the other cells of the same row. This condition confirms the divergent validity.

For CRO1, 0.9866 > {0.030976, 0.046624, 0.168913} and
For CRO2, 0.986742 > {0.023745, 0.037545, 0.17329} and
For SMU1, 0.53156 > {0.001578, 0.077049, 0.015338} and
For SMU2, 0.859736 > {0.031766, 0.122122, -0.150665} and
For SCRMC1, 0.570522 > {0.050194, 0.145343, 0.540354} and
For SCRMC2, 0.870945 > {-0.012473, 0.110302, 0.628633} and
For SCRMC3, 0.121996 > {-0.094954, -0.159483, -0.71053} and
For SCRMC4, 0.776416 > {0.130899, 0.125381, 0.750261} and
For SCRMC5, 0.886804 > {-0.019179, -0.096823, 0.64445} and
For SCRMC6, 0.810901 > {-0.00188, 0.077145, 0.70523} and
For SCRMC7, 0.662697 > {0.053638, 0.003076, 0.658247} and
For SCRMC8, 0.722868 > {0.003081, -0.146486, 0.474074} and
For CRP1, 0.726161 > {0.194897, 0.11079, 0.656614} and
For CRP2, 0.872359 > {0.035171, -0.271159, 0.731392} and
For CRP3, 0.953581 > {0.209034, -0.129291, 0.841247} and
For CRP3, 0.953114 > {-0.20903, 0.175248, 0.821993}.

Table 3 confirms that the validity of this research is obtained to an acceptable level. Tables 2 and 3 shows that the divergent validity is also accepted for this research.

Tables 4, 5, and 6 help to perform a test of total effects. It is needed for all effects to be accepted, and the T-statistics must be greater than |1.96| (Salkind & Frey, 2021). The formula for calculating T-statistics is |Original Sample/Standard Error|. Table 4 shows that all values of T-statistics for this research are more than |1.96|, which confirms that all the effects are accepted for this particular research.

The test of total effects is shown in Tables 5 and 6. Ulrich et al. (2018) observed that values of R^2 for the three constructs for the research are 0.67, 0.33, and 0.19, which are considered comparatively as substantial, moderate, and weak.

Table 4 Test of total effects

	Original Sample (O)	Sample Mean (M)	Standard Deviation (STDEV)	Standard Error (STERR)	T Statistics (|O/STERR|)
Customer Relationship Orientation – > Social CRM Capabilities	0.038711	0.026131	0.109734	0.109734	4.352775
Social media use – > Social CRM Capabilities	0.141761	0.053139	0.194107	0.194107	8.730325
Customer Relationship Orientation – > Social media use	0.02772	0.009774	0.136158	0.136158	9.203589
Social CRM Capabilities – > Social Relationship Performance	0.867695	0.430055	0.755501	0.755501	11.1485
Customer Relationship Orientation – > Social Relationship Performance	0.13642	0.133173	0.047738	0.047738	2.857642

Table 5 Test of total effects

	R square
Customer relationship orientation	
Social media use	0.340768
Social customer relationship management capabilities	0.421899
Customer relationship performance	0.7816

Table 6 Test of total effects

Total	SSO	SSE	Q^2 = 1-SSE/SSO
Social media use	472	270.899207	0.426061
Social customer relationship management capabilities	1888	1069.851936	0.433341
Customer relationship performance	708	291.033761	0.588935

Table 7 Test of Goodness of Fit (GOF)

	Communality
Customer relationship orientation	0.97352
Social media use	0.732546
Social customer relationship management capabilities	0.561947
Customer relationship performance	0.510851

$$R^2 \begin{cases} 0.19 \text{ Weak} \\ 0.33 \text{ Moderate} \\ 0.67 \text{ Substantial} \end{cases}$$

$$Q^2 \begin{cases} 0.02 \text{ Weak} \\ 0.15 \text{ Moderate} \\ 0.35 \text{ Substantial} \end{cases}$$

$$Q^2 = 1 - SSE/SSO$$

From Table 5, it can be said that the values of R^2 for Social Media Use and Customer Relationship Management Capabilities are moderate, and the value of R^2 for Customer Relationship Performance is Substantial. From Table 6, it is seen that all the values of Q^2 are substantial.

The test of Goodness of Fit (GOF) is shown in Table 7.

$$GOF \begin{cases} 0.01 \text{ Weak} \\ 0.25 \text{ Moderate} \\ 0.36 \text{ Substantial} \end{cases}$$

$$\begin{aligned} GOF &= \sqrt{[\{\text{Average}\ (R^2) \times \text{Average (Community)}\}]} \\ &= \sqrt{[0.51 \times 0.69]} \\ &= 0.5952 > 0.36 \end{aligned}$$

GOF is accepted for this particular research.

11.3 *Structural Model*

The researcher has used Smart PLS Analysis software to analyze the collected data for this research. Al-Maroof et al. (2021) observed that the validity of the research could be assessed with the help of R^2s and structural paths. The path coefficients for this research should be positively essential to be at the level of 0.05.

From Fig. 2, it can be stated that the R^2 of social CRM capabilities is approximately 0.33% of the variance in social CRM capabilities, which is accounted for customer relationship orientation and social media use. Again, the R^2 of social media use is approximately 0.015% of the variance of social media use, which is accounted for the customer relationship orientation. The R^2 of the customer relationship performance is approximately 0.75% of the variance of customer relationship performance and accounted for the customer relationship orientation and social CRM capabilities. So, PLS analysis helps to understand the use of social media influences that customer relationship performance.

PLS analysis helps in performing hypotheses analysis. It is seen from the PLS analysis that customer relationship orientation has positive impacts on social CRM capabilities, which is 0.327. It can be said that H1 is supported, and the customer relationship orientation has positive impacts on social CRM capabilities. PLS analysis also shows that customer relationship orientation has positive effects on social media use, which is 0.124. This value ensures that H3 is supported and customer relationship orientation has a positive effect on social media use. Again, it is also seen that the use of social media sites has a positive effect on social CRM capabilities, which is 0.575. This value indicates that the use of social media has a powerful positive effect on social CRM capabilities. So, it ensures that H2 is supported, and increasing the

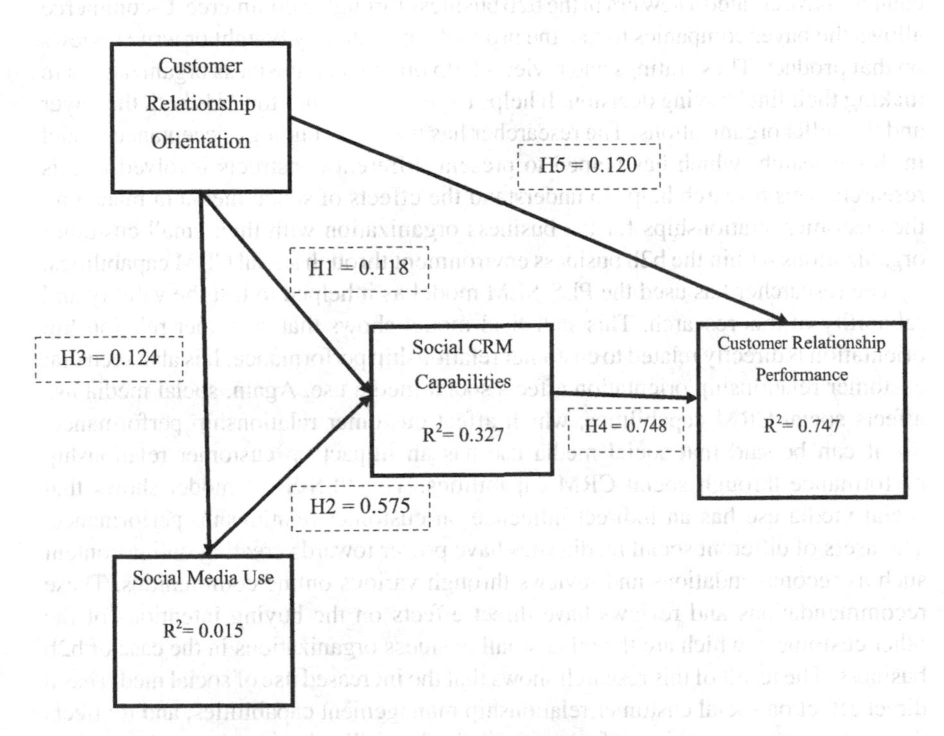

Fig. 2 Smart PLS analysis

use of social media helps to enhance the social CRM capabilities. Social CRM capabilities have a robust positive effect on customer relationship performance, which is 0.748. So, it is seen that H4 is supported. So, it can be said that ultimately, customer relationship orientation has a positive effect on customer relationship performance. Again, PLS analysis shows that customer relationship orientation has direct positive effects on customer relationship performance, which is 0.120. It proves that H5 is a supporter, and customer relationship performance is positively affected by customer relationship orientation.

12 Practical Implications

The availability of the internet has increased in recent years, which have a significant impact on the buying decision of the customer companies and the seller companies within a b2b business environment. Social media sites help to provide needed information about the products towards the buyer's small organizations. Social media sites help in making inter-connection between the buyer companies and the seller companies through online forums, communities, ratings, recommendations, etc. These changes have created a new era in the b2b business through e-commerce. E-commerce allows the buyer companies to rate the products they already bought or write reviews on that product. These ratings and reviews help other small business organizations in making their final buying decision. It helps to generate values towards both the buyer and the seller organizations. The researcher has used a technology acceptance model in this research, which has helped to present different constructs involved in this research. This research helps to understand the effects of social media in managing the customer relationships for the business organization with their small customer organizations within the b2b business environment through social CRM capabilities.

The researcher has used the PLS-SEM model as it helped to test the validity and reliability of this research. This statistical model shows that customer relationship orientation is directly related to customer relationship performance. It is also seen that customer relationship orientation affects social media use. Again, social media use affects social CRM capabilities, which affect customer relationship performance. So, it can be said that social media use has an impact on customer relationship performance through social CRM capabilities. The PLS-SEM model shows that social media use has an indirect influence on customer relationship performance. The users of different social media sites have power towards creating online content such as recommendations and reviews through various online communities. These recommendations and reviews have direct effects on the buying intentions of the other customers, which are the other small business organizations in the case of b2b business. The result of this research shows that the increased use of social media has a direct effect on social customer relationship management capabilities, and it affects the customer relationship performance of the big seller business organizations to maintain their small customer organizations. Customer relationship orientation has positive effects on social CRM capabilities. Social media use also has positive effects on social CRM capabilities. Customer relationship orientation has positive impacts on social media use where social CRM capabilities have positive effects on customer relationship performance. It is seen from the PLS-SEM model that the positive effects of social media use on social CRM capabilities are convincing, which is 0.575, and the positive impacts of social CRM capabilities on customer relationship performance is also very strong which is 0.748. This proves that social media use has strong indirect positive impacts on customer relationship performance through social CRM capabilities.

The implications drawn by the researcher shows that it is needed to continue developing further social CRM capabilities by using the main two influential social media sites such as YouTube and Twitter. It will help not only to collect information but will also help to respond to this information. The seller organizations should focus on social analytics, live feeds, photo sharing, etc. as it will help to increase social CRM capabilities. The seller organizations should make more investment on social media sites as a CRM tool as it has positive impacts on customer relationship performance. The researcher has suggested for performing further research on the reason why YouTube plays the role of intermediary among social CRM capabilities and customer relationship orientation and their impacts. Future research should test the conceptual model with a larger sample size towards minimizing the research limitations.

Proper research ethics is maintained by the researcher while performing this research. The researcher promised the sample group of companies that the primary data collected from them will be used only for this research and will not be published anywhere else. The researcher maintained the privacy of these companies in an ethical manner. The quantitative questionnaire set by the researcher in this way so that they do not create any harmful effects on these companies. The researcher has maintained the code of ethics while performing this research.

13 Conclusion

It is seen from this research that most of the business organizations involved in the b2b business do not understand the full value of using social media. Most of them still use social media as their marketing tool, and they do not realize that social media can be used for improving customer relationships. So, their customer relationship performances do not improve. It is also seen that some of the companies realized the full value of social media as a relationship tool, but they are in very early stages towards developing their social media strategies. These companies did not even develop the needed social customer relationship management capabilities so that they can attain a higher customer relationship performance.

In conclusion, it can be said that the companies that are customer relationship-oriented, should use social media towards improving their customer relationship performance through developing social CRM capabilities. More and useful use of YouTube and Twitter may help the seller companies involved in the b2b business environment to develop more improved social CRM capabilities, and it may help these organizations to obtain a higher level of customer relationship performance.

Limitation and recommendations for future studies

Limitations of the research indicate that there are future scopes of the study. This researcher helps to understand the effects of social media in managing customer relationships for the seller business organizations within the b2b business environment. The research result shows that the use of social media has a strong positive impact on customer relationship performance. Though it is also seen that most of the companies do not realize that Social media can be used as a relationship tool, and they only utilize social media as their marketing tool. This research highlights that social media is a tool to make a quick response to their customers. Social media allows companies to engage with their customers. It helps to make a conversation with the customers. It helps the companies to ask their customer companies to provide feedback and also thank them for their valuable time to write the feedback. Social media helps the seller companies to provide personal touch towards their customer companies. Providing rewards towards customer companies through social media sites can be a way towards making an active community. Different contests and competitions through social media sites help the seller organizations to know more about their customer companies. Social media sites can be used by the seller organizations to provide their customer companies various promotional offers, sales, or coupons. Social media sites also help the seller organizations to find out their unhappy customers and the reasons for their unhappiness. It will help them to improve their services. The seller organizations should follow User-generated content (UGC) as it will inform the followers of that company that the ideas and views of the followers are taken into account towards building the brand. It will help to customer relationship performance of the seller organizations.

Like every other research, there are many limitations faced by the researcher while performing this research. One of the most critical limitations faced by the researcher at the time of performing quantitative research was that the sample size was too small. So that it was hard to find any particular conclusion as the research findings could be taken in any direction. The sample size should be increased to get a perfect research result. The small sample size influenced the validity, quality, and reliability of this research. More data should be collected to address the validity of the research. Another limitation faced by the researcher was that some customers were contacted partially by the companies on behalf of the researcher. This may provide incomplete information as in this case, and the companies selected the customers whom they were comfortable to contact with. It may create a systematic error in the sample. Another limitation was that the companies always do not provide the correct information for maintaining the privacy of their companies. This may generate changes in the research result. The limited-time period was another major limitation of this research. Again, another critical limitation was limited financial support. If the researcher would get more time and needed financial support, then the research would be more productive and specific with more required data.

References

Abu Zayyad, H. M., Obeidat, Z. M., Alshurideh, M. T., Abuhashesh, M., Maqableh, M., & Masa'deh, R. (2021). Corporate social responsibility and patronage intentions: The mediating effect of brand credibility. *Journal of Marketing Communications, 27*(5). https://doi.org/10.1080/13527266.2020.1728565.

Abuhashesh, M. Y., Alshurideh, M. T., & Sumadi, M. (2021). The effect of culture on customers' attitudes toward Facebook advertising: the moderating role of gender. *Review of International Business and StrategyReview of International Business and Strategy.*

Aburayya, A., & Salloum, S. A. (2021). The effects of subjective norm on the intention to use social media networks: An exploratory study using PLS-SEM and machine learning approach.

Ahmad, A., Alshurideh, M., Al Kurdi, B., Aburayya, A., & Hamadneh, S. (2021). Digital transformation metrics: A conceptual view. *Journal of Management Information and Decision Sciences, 24*(7), 1–18.

Aichner, T., & Jacob, F. (2015). Measuring the degree of corporate social media use. *International Journal of Market Research, 57*(2), 257–276.

Aissani, R., & Awad ElRefae, G. (2021). Trends of university students towards the use of youtube and the benefits achieved-field study on a sample of Al Ain University students. *AAU Journal of Business Law مجلة جامعة العين للأعمال والقانون*, 5(1), 1.

Al Khasawneh, M., Abuhashesh, M., Ahmad, A., Masa'deh, R., & Alshurideh, M. T. (2021a). *Customers online engagement with social media influencers' content related to COVID 19* (Vol. 334).

Al Khasawneh, M., Abuhashesh, M., Ahmad, A., Alshurideh, M. T., & Masa'deh, R. (2021b). *Determinants of E-word of mouth on social media during COVID-19 outbreaks: An empirical Study* (Vol. 334).

Al Kurdi, B. H., & Alshurideh, M. T. (2021). Facebook advertising as a marketing tool: examining the influence on female cosmetic purchasing behaviour. *International Journal of Online Marketing, 11*(2), 52–74.

Al Kurdi, B., Alshurideh, M., & Al afaishata, T. (2020). Employee retention and organizational performance: Evidence from banking industry. *Management Science Letters, 10*(16), 3981–3990.

Alarcón, D., Sánchez, J. A., & De Olavide, U. (2015). Assessing convergent and discriminant validity in the ADHD-R IV rating scale: User-written commands for Average Variance Extracted (AVE), Composite Reliability (CR), and Heterotrait-Monotrait ratio of correlations (HTMT). In *Spanish STATA meeting* (Vol. 39).

Al-Dmour, H., Alshuraideh, M., & Salehih, S. (2014). A study of Jordanians' television viewers habits. *Life Science Journal, 11*(6), 161–171.

Alhamad, A. Q. M., Akour, I., Alshurideh, M., Al-Hamad, A. Q., Kurdi, B. A., & Alzoubi, H. (2021). Predicting the intention to use google glass: A comparative approach using machine learning models and PLS-SEM. *Int. J. Data Netw. Sci.*, vol. 5, no. 3, 2021, doi: https://doi.org/10.5267/j.ijdns.2021.6.002.

Al-Hamad, M. Q., Mbaidin, H. O., Alhamad, A. Q. M., Alshurideh, M. T., Kurdi, B. H. A., & Al-Hamad, N. Q. (2021). Investigating students' behavioral intention to use mobile learning in higher education in UAE during Coronavirus-19 pandemic. *International Journal of Data and Network Science,* 5(3). https://doi.org/10.5267/j.ijdns.2021.6.001.

Aljumah, A., Nuseir, M. T., & Alshurideh, M. T. (2021a). The impact of social media marketing communications on consumer response during the COVID-19: Does the brand equity of a university matter. In *The effect of coronavirus disease (COVID-19) on business intelligence* (Vol. 334, pp. 384–367)

Aljumah, A. I., Nuseir, M. T., Alam, M.M. (2021b). Organizational performance and capabilities to analyze big data: do the ambidexterity and business value of big data analytics matter? *Business Process Management Journal.*

Alketbi, S., Alshurideh, M. & Al Kurdi, B. (2020). The influence of service quality on customers'retention and loyalty in the uae hotel sector with respect to the impact of customer'satisfaction, trust, and commitment: A qualitative study. *PalArch's Journal of Archaeology of Egypt/Egyptology, 17*(4), 541–561.

Al-khayyal, A., Alshurideh, M., & Al, B. (2020). The impact of electronic service quality dimensions on customers ' E- shopping and E-loyalty via the impact of E-satisfaction and E-trust : A qualitative approach. *The International Journal of Innovation, Creativity and Change, 14*(9).

Al-Khayyala, A., Alshuridehb, M., Al Kurdic, B., & Aburayyad, A. (2020). The impact of electronic service quality dimensions on customers' E-shopping and E-loyalty via the Impact of E-satisfaction and E-trust: A qualitative approach. *The International Journal of Innovation, Creativity and Change, 14*(9), 257–281.

Alkitbi, S. S., Alshurideh, M., Al Kurdi, B., & Salloum, S. A. (2021). *Factors affect customer retention: A systematic review* (Vol. 1261). AISC.

Al-Maroof, R. S., Alshurideh, M. T., Salloum, S. A., AlHamad, A. Q. M., & Gaber T. (2021) Acceptance of google meet during the spread of coronavirus by Arab university students. *Informatics, 8*(2). https://doi.org/10.3390/informatics8020024.

Al-Maroof, R. et al. (2021), The acceptance of social media video for knowledge acquisition, sharing and application: A comparative study among YouYube users and TikTok users' for medical purposes. *International Journal of Data and Network Science, 5*(3). https://doi.org/10.5267/j.ijdns.2021.6.013.

Almazrouei, F. A., Alshurideh, M., Al Kurdi, B., & Salloum, S. A. (2020). Social media impact on business: A systematic review. In *International conference on advanced intelligent systems and informatics* (pp. 697–707).

Alshamsi, A., Alshurideh, M., Al Kurdi, B., & Salloum, S. A. (2020). The influence of service quality on customer retention: A systematic review in the higher education. In *International conference on advanced intelligent systems and informatics* (pp. 404–416).

Alshurideh. (2012). The effect of previous experience on mobile subscribers' repeat purchase behaviour.

Alshurideh, M. (2016). Scope of Customer Retention Problem in the Mobile Phone Sector: A Theoretical Perspective. *J. Mark. Consum. Res., 20*, 64–69.

Alshurideh, M. T., et al. (2021b). Factors affecting the use of smart mobile examination platforms by universities' postgraduate students during the COVID 19 pandemic: An empirical study. *Informatics, 8*(2), 32.

Alshurideh. (2019). Do electronic loyalty programs still drive customer choice and repeat purchase behaviour?," *International Journal of Electronic Customer Relationship Management, 12*(1), 40–57.

Alshurideh, M. (2022). Does electronic customer relationship management (E-CRM) affect service quality at private hospitals in Jordan? *Uncertain Supply Chain Managements, 10*(2), 1–8.

M. Alshurideh, R. M. T. Masa'deh, and B. Alkurdi, "The effect of customer satisfaction upon customer retention in the Jordanian mobile market: An empirical investigation," *Eur. J. Econ. Financ. Adm. Sci.*, no. 47, 2012.

Alshurideh, M., Bataineh, A., Al kurdi, B., & Alasmr, N. (2015). Factors affect mobile phone brand choices–studying the case of Jordan universities students. *International Business Research,* 8(3), 141–155.

Alshurideh, M., Al Kurdi, B., Vij, A., Obiedat, Z., & Naser, A. (2016). Marketing ethics and relationship marketing-An empirical study that measure the effect of ethics practices application on maintaining relationships with customers. *Journal of International Business Research, 9*(9), 78–90

Alshurideh, M., Salloum, S. A., Al Kurdi, B., & Al-Emran, M. (2019a). Factors affecting the social networks acceptance: An empirical study using PLS-SEM approach. In *PervasiveHealth: pervasive computing technologies for healthcare* (Vol. Part F1479). https://doi.org/10.1145/3316615.3316720.

Alshurideh et al. (2019b). Understanding the quality determinants that influence the intention to use the mobile learning platforms: A practical study. *International Journal of Interactive Mobile Technologies,* 13(11), 157–183.

Alshurideh, M., Alsharari, N. M., & Al Kurdi, B. (2019c). Supply chain integration and customer relationship management in the airline logistics. *Theoretical Economics Letters, 9*(02), 392–414.

Alshurideh, M., Gasaymeh, A., Ahmed, G., Alzoubi, H., & Al Kurd, B. (2020a). Loyalty program effectiveness: Theoretical reviews and practical proofs. *Uncertain Supply Chain Managements, 8*(3), 599–612. https://doi.org/10.5267/j.uscm.2020a.2.003.

Alshurideh, M., Al Kurdi, B., Salloum, S. A., Arpaci, I., & Al-Emran, M. (2020b). Predicting the actual use of m-learning systems: a comparative approach using PLS-SEM and machine learning algorithms. *Interactive Learning Environments,* 1–15.

Alshurideh, M. T., Al Kurdi, B., & Salloum, S. A. (2021a). The moderation effect of gender on accepting electronic payment technology: A study on United Arab Emirates consumers. *Review of International Business and Strategy.*

Alshurideh, M. T., et al. (2022). Fuzzy assisted human resource management for supply chain management issues. *Annals of Operations Research,* 1–19.

Alzoubi, H., Alshurideh, M., Al Kurdi, B., & Inairat, M. (2020). Do perceived service value, quality, price fairness and service recovery shape customer satisfaction and delight? A practical study in the service telecommunication context. *Uncertain Supply Chain Managements, 8*(3), 579–588. https://doi.org/10.5267/j.uscm.2020.2.005.

Boulianne, S. (2015). Social media use and participation: A meta-analysis of current research. *Information, Communication and Society,* 18(5), 524–538.

Cawsey, T., & Rowley, J. (2016) Social media brand building strategies in B2B companies. *Marketing Intelligence and Planning.*

Cegarra-Navarro, J.-G., Soto-Acosta, P., & Wensley, A. K. P. (2016). Structured knowledge processes and firm performance: The role of organizational agility. *Journal of Business Research,* 69(5), 1544–1549

Della Corte, V., Iavazzi, A., & D'Andrea, C. (2015). Customer involvement through social media: the cases of some telecommunication firms. *Journal of Open Innovation: Technology, Market, and Complexity, 1*(1), 10.

Elsaadani, M. A. (2020). Investigating the effect of e-management on customer service performance. *AAU Journal of Business Law مجلة جامعة العين للأعمال و القانون*, *3*(2), 1.

Felix, R., Rauschnabel, P. A., & Hinsch, C. (2017). Elements of strategic social media marketing: A holistic framework. *Journal of Business Research, 70,* 118–126.

Guesalaga, R. (2016). The use of social media in sales: Individual and organizational antecedents, and the role of customer engagement in social media. *Industrial Marketing Management, 54,* 71–79.

Hamadneh, S., & Al Kurdi, B. (2021). The effect of brand personality on consumer self-identity: The moderation effect of cultural orientations among British and Chinese consumers. *Journal of Legal, Ethical and Regulatory Issues, 24*(Special Issue 1), 1–14.

Hamadneh, S., Pedersen, O., & Al Kurdi, B. (2021). An investigation of the role of supply chain visibility into the scottish bood supply chain. *Journal of Legal, Ethical and Regulatory Issues, 24*(Special Issue 1), 1–12.

Hassan, R. S., Nawaz, A., Lashari, M. N., & Zafar, F. (2015). Effect of customer relationship management on customer satisfaction. *Procedia Economics and Finance, 23,* 563–567.

Hult, G. T. M., Hair, J. F., Jr., Proksch, D., Sarstedt, M., Pinkwart, A., & Ringle, C. M. (2018). Addressing endogeneity in international marketing applications of partial least squares structural equation modeling. *Journal of International Marketing, 26*(3), 1–21.

Joghee, S., et al. (2021). Expats impulse buying behaviour in UAE: A Customer Perspective. *Journal of Management Information and Decision Sciences, 24*(1), 1–24.

Kamboj, S., Yadav, M., Rahman, Z., & Goyal, P. (2016). Impact of social CRM capabilities on firm performance: Examining the mediating role of co-created customer experience. International *Journal of Information Systems in the Service Sector (IJISSS), 8*(4), 1–16.

Kumar, A., Bezawada, R., Rishika, R., Janakiraman, R., & Kannan, P. K. (2016). From social to sale: The effects of firm-generated content in social media on customer behavior. *Journal of Marketing, 80*(1), 7–25.

Kumar, V., & Reinartz, W. (2018). CRM issues in the Business-To-Business context. In *Customer relationship management* (pp. 265–283). Springer.
Kurdi, B. A., Alshurideh, M., & Alnaser, A. (2020a). The impact of employee satisfaction on customer satisfaction: Theoretical and empirical underpinning. *Management Science Letters, 10*(15). https://doi.org/10.5267/j.msl.2020a.6.038.
Kurdi, B. A., Alshurideh, M., Salloum, S. A., Obeidat, Z. M., & Al-dweeri, R. M. (2020b). An empirical investigation into examination of factors influencing university students' behavior towards elearning acceptance using SEM approach. *International Journal of Interactive Mobile Technologies, 14*(2). https://doi.org/10.3991/ijim.v14i02.11115.
Lee, J. J., & Hwang, J. (2016). An emotional labor perspective on the relationship between customer orientation and job satisfaction. *International Journal of Hospitality Management, 54*, 139–150.
Lipiäinen, H. S. M. (2015). CRM in the digital age: implementation of CRM in three contemporary B2B firms. *Journal of Systems and Information Technology*.
Noe, R. A., Hollenbeck, J. R., Gerhart, B., & Wright, P. M. (2017). *Human resource management: Gaining a competitive advantage*. McGraw-Hill Education.
Nunan, D., Sibai, O., Schivinski, B., & Christodoulides, G. (2018). Reflections on 'social media: Influencing customer satisfaction in B2B sales' and a research agenda. *Industrial Marketing Management, 75*, 31–36.
Nuseir, M., El Refae, M., & Alshurideh, G. (2021). The impact of social media power on the social commerce intentions: Double mediating role of economic and social satisfaction. *Journal of Legal, Ethical and Regulatory Issues, 24*(Special Issue 6), 1–15.
Nuseir, M. T., Aljumah, A., & Alshurideh, M. T. (2021). *How the business intelligence in the new startup performance in UAE during COVID-19: The mediating role of innovativeness* (vol. 334).
El Refae, M., Nuseir, G., & Alshurideh, M. (2021). The influence of social media regulations boundary on marketing and commerce of industries in UAE. *J Journal of Legal, Ethical and Regulatory Issues, 24*(Special Issue 6), 1–14.
Salkind, N. J., & Frey, B. B. (2021). *Statistics for people who (think they) hate statistics: Using microsoft excel*. Sage Publications.
Siamagka, N.-T., Christodoulides, G., Michaelidou, N., & Valvi, A. (2015). Determinants of social media adoption by B2B organizations. *Industrial Marketing Management, 51*, 89–99.
Sigala, M. (2016). Social CRM capabilities and readiness: Findings from Greek tourism firms. *Information and Communication Technologies in Tourism, 2016*, 309–322.
Stott, R. N., Stone, M., & Fae, J. (2016). Business models in the business-to-business and business-to-consumer worlds–what can each world learn from the other? *Journal of Business and Industrial Marketing*.
Svoboda, P., Ghazal, T. M., Afifi, M. A. M., Kalra, D., Alshurideh, M. T., & Alzoubi, H. M.: Information systems integration to enhance operational customer relationship management in the pharmaceutical industry. In *The international conference on artificial intelligence and computer vision* (pp. 553–572).
Sweiss, N., Obeidat, Z. M., Al-Dweeri, R. M., Mohammad Khalaf Ahmad, A., Obeidat, A. M., & Alshurideh, M. (2021). The moderating role of perceived company effort in mitigating customer misconduct within Online Brand Communities (OBC). *Journal of Marketing Communications*, 1–24. https://doi.org/10.1080/13527266.2021.1931942.
Terho, H., Eggert, A., Haas, A., & Ulaga, W. (2015). How sales strategy translates into performance: The role of salesperson customer orientation and value-based selling. *Industrial Marketing Management, 45*, 12–21.
Terziev, V., & Banabakova, V. (2017). Customer Relationship Management (CRM) As base for organization's behavior. *Psychological Review*.
Turki Alshurideh, M. (2014). *The influence of mobile application quality and attributes on the continuance intention of mobile shopping*.
Ulrich, R., Miller, J., & Erdfelder, E. (2018). Effect size estimation from t-statistics in the presence of publication bias. *Zeitschrift für Psychologie*.

Wang, W. Y. C., Pauleen, D. J., & Zhang, T. (2016). How social media applications affect B2B communication and improve business performance in SMEs. *Industrial Marketing Management, 54*, 4–14.

Wirtz, B. W., Pistoia, A., Ullrich, S., & Göttel, V. (2016). Business models: Origin, development and future research perspectives. *Long Range Planning, 49*(1), 36–54.

An Empirical Study Investigating the Role of Team Support in Digital Platforms and Social Media Marketing Towards Consumer Brand Awareness: A Case of the United Arab Emirates

Mohammed T. Nuseir, Abu Reza Mohammad Islam, Sarah Urabi, Muhammad Alshurideh, and Barween Al Kurdi

Abstract The advent of social media marketing and team support for innovation is considered simultaneously as the two main antecedents of consumer brand awareness. To support this assertion, this study explored the consequences of support by marketing team members toward innovative approaches to social media marketing. The study randomly selected 247 participants from the United Arab Emirates. Data were analyzed using multiple linear regression. Results showed that both social media marketing and team support for innovation were positively associated with consumer brand awareness in that social media marketing and team support played as catalysts reinforcing brand awareness. In addition, the team support for innovation was positively associated with social media marketing; alternatively, they complement each other towards brand awareness.

M. T. Nuseir
Department of Business Administration, College of Business, Al Ain University, Abu Dhabi Campus, Abu Dhabi, UAE

A. R. M. Islam
Department of Business Administration, College of Business, Al Ain University, Abu Dhabi, UAE

S. Urabi
Abu Dhabi Vocational Education & Training Institute (ADVETI), Abu Dhabi, UAE
e-mail: sarah.urabi@adveti.ac.ae

M. Alshurideh (✉)
Department of Marketing, School of Business, The University of Jordan, Amman, Jordan
e-mail: m.alshurideh@ju.edu.jo

Department of Management, College of Business Administration, University of Sharjah, Sharjah, United Arab Emirates

B. A. Kurdi
Department of Marketing, Faculty of Economics and Administrative Sciences, The Hashemite University, Zarqa, Jordan
e-mail: barween@hu.edu.jo

M. Alshurideh et al. (eds.), *The Effect of Information Technology on Business and Marketing Intelligence Systems*, Studies in Computational Intelligence 1056,
https://doi.org/10.1007/978-3-031-12382-5_6

Keywords Brand awareness · Social media platforms · Digital marketing · Social media marketing · Innovation · Team support for innovation

1 Introduction

There is a high prevalence that consumers now a day spend extraordinarily a higher amount of time on digital platforms, an unprecedented rate of engagement with the brands that they usually support and advocate for (Alshurideh et al., 2019a, 2019b; Aljumah et al., 2021a, 2021; Kurdi et al., 2021). In the past, any brand wanting to connect with prospects of finding an avenue, the usual tendency was either to make an advertisement in the paper or invest in billboard (Alshurideh et al., 2017; Abuhashesh et al., 2021a, 2021b; Al Kurdi et al., 2021). Nowadays, customers make an intensive usbe of digital media to find brands (Abu Sweiss et al., 2021; Zayyad et al., 2021). Consequently, there has been on a plethora of dynamic change in momentum shifting the exclusive control from the brand itself to the consumers. Further, the advent of the internet and explosion of digital technology has brought about major changes in the world of consumer behavior due to the massive application of websites, blogs, reviews, testimonials, and other forms of online content. All these avenues have strengthened consumers to conduct research on anything about any product, service, or any brand enabling them to decide upon their purchases. Consequently, consumers expect the brands to have their digital presence and communication with brands through the digital channels. This necessitates the cultivation of a digital brand through a digital-first approach. Then consumers would expect the brands to be responsive on digital platforms; a bad customer experience, however, means when the majority do not get any response after reaching out to a brand on a social network. In some instances, customers when using social platforms like Facebook to reach out to them for satisfaction or resolution with a problem, and do not get responses as quickly as they would like to, half of them would hardly conduct business with the company again.

Social media as platforms for marketing are gaining substantial attention from the public as well as from researchers. Marketing practitioners have been continuously and consistently amazed by the power of social media (Lipsman et al., 2012). Research has suggested that the major benefits of using social media platforms for marketing is that they provide a very convenient vehicle for developing consumer brand awareness in the target market (Hanna et al., 2011; Berthon et al., 2012).

Social media platforms have become a major part of the daily lives of many people now a day, and this has been increasingly so over the years. A single post by a brand on these platforms can, therefore, reach to a much larger proportion of the population compared to other marketing channels (Michaelidou et al., 2011). Many previous studies focused on the variables of brand awareness and social media marketing (Thackeray et al., 2008; Lipsman et al., 2012; Berthon et al., 2012), but those studies did not examine the concept of innovation and the crucial need for team members supporting the innovative approaches to marketing, and the social media marketing per se.

Given the above backdrop, the present study has identified and aimed at filling this gap by investigating the role of team member supports for engendering the innovative approaches when using the social media platforms to create the consumer brand awareness in society (Nuseir et al., 2021a, 2021b; Al-Dhuhouri et al., 2020; Alshurideh et al., 2022).

In the following three sections, we discuss relevant important variables and issues considered in this study and their relationships with the study objectives.

1.1 Consumer Brand Awareness

The concept of consumer brand awareness is one of the most crucial dimensions of branding (Hamadneh et al., 2021; Keller, 2001). As explained by Keller (1993), brand awareness is the ability of a consumer to recall a brand and to center their buying decisions based on this memory. Considering the possible impact of brand awareness on brand choice (Elsaadani, 2020; Hays et al., 2013) and the intentions of consumers making the final purchases (Alshurideh et al., 2015; Hutter et al., 2013), this particular concept and issue have been treated with much importance in the marketing domain as well as in the literature.

Importance of a brand: Research has shown that consumers are relatively more confident about purchasing the brands readily that they are aware of (Low & Lamb Jr., 2000; Homburg et al., 2010). When consumers deal with unknown brands, they resort to making comparisons with known brands in terms of characteristics such as, product packaging, product quality, and price (Dick et al., 1997). Considering the kind of competition that exists in the present-day business sector, brands are competing with each other intensely than ever before. It is therefore, imperative and very important for companies to use the most effective technique (s) to maintain their brand portfolios in order to retain their respective positions in the industry (Cobb-Walgren et al., 1995; Malik et al., 2013; Nguyen et al., 2018). Other factors such as population demographics (Matherly et al., 2018) and the design and content of advertising efforts, can also have a great impact on brand equity (Aljumah et al., 2021a, 2021b, 2021c; Guitart et al., 2018).

1.2 The Concept of Digital Marketing

Within the field of marketing, digital marketing has increased in popularity lately, which led to have increased its importance (Hutter et al., 2013; Khasawneh et al., 2021a, 2021b) in social media. As a result, other marketing media such as print advertising, television advertising and outdoor advertising, all have lost their popularity and importance across industries and business sectors alike (Al-Dmour et al., 2014; Al-Maroof et al., 2021; Bruhn et al., 2012).

Another important factor or determinant that has created an impact on the shift in marketing media is economic pressure (Kirtiş & Karahan, 2011). It is no wonder that the increase in advertising costs has put limitations on many companies on the classical style of marketing ventures. This has led most companies with a large majority of brands developing their preference for digital/online media marketing, as they are found to be less expensive potentially and thus, providing the managers across companies with the ability and control to manage their advertising budgets rather more effectively and efficiently (Hoffman & Fodor, 2010; Kirtiş & Karahan, 2011; Almazrouei et al., 2020).

1.3 *The Concept of Team Support for Innovation*

In the business world, innovation refers to the introduction of novel and improved ways of carrying out an activity in the workplace (Alameeri et al., 2021; Martins & Terblanche, 2003). Research has shown that an open approach to innovation can lead to more creative solutions (Al Suwaidi et al., 2020; Ghannajeh et al., 2015; Hakansson & Waluszewski, 2007) and better work outcomes (Autio et al., 2014; Obeidat et al., 2021). Team members who are supportive of such work behaviors are open to change, and are, therefore, more likely to succeed in (Enkel et al., 2009; Almaazmi et al., 2021).

With each passing day, social media platforms are improving their functionality, with new and improved features added at a regular interval (Shankar et al., 2011). Professionals working in the marketing fields are therefore, encouraged to be open in creative ideas, thinking and innovation. In addition, it is advised that the marketing departments must demonstrate the attitude towards innovative ideas and actions specifically, by their team members (Al Naqbia et al., 2020; Dodgson, Gann, & Phillips, 2013; Enkel et al., 2009; Georgellis et al., 2000).

2 Literature Review

Now a day, marketing activities through platforms like, Twitter, Facebook, and Instagram have become immensely popular among the public. This has brought about a significant shift in marketing practices (Shankar et al., 2011, Nuseir et al., 2021a, 2021b). Marketing professionals have new opportunities in the form of digital marketing (Mangold & Faulds, 2009). These social media platforms not only help in the marketing of products, but also help companies to gain necessary insights into the needs of consumers through their feedback more often than otherwise (Hansen et al., 2018; Rapp et al., 2013). Eventually, this new insight is capitalized to help assist in designing effective marketing and advertising strategies in the modern-day marketing. Consequently, such information is used further to improve products at regular intervals, helping the product development and diversification across industries (Mount & Martinez, 2014; Rapp et al., 2013).

However, within the classical form of marketing and advertising efforts, any rapid changes usually take some time to be operative. For instance, the in-stream advertisements and Facebook stories are a relatively new phenomenon. Because of severe competition among different social media platforms, once a feature is introduced by one platform, it is often adopted later by other platforms (Mangold & Faulds, 2009). This requires the team members of the marketing department (especially, in digital platforms) with high readiness and alert to be flexible enough in approaching the pragmatic and competitive marketing technique (s) that they must apply or adhere to. However, the failure to follow suit the commensurate marketing technique (s) in time required to sustain the current dynamic environment or otherwise rigidity can create hindrances that may lead to negative consequences for the businesses (Alshurideh et al., 2016; Antikainen et al., 2010). Therefore, one would expect the benefits of using a flexible approach with better results, as it always encourages positive changes than resisting them (; Enkel et al., 2009).

Team innovation is considered therefore, the backbone of every successful company helping it to grow and prosper, setting the business apart from the competition for a while. In this process, the team members feel themselves connected to the company with the motivation and incentives to be creative and innovative. The members, therefore, consider themselves to be in the loop on a firm's strategies and challenges inviting their input. Team members involved early in the processes and plans of the company business would like to see them through to completion—their active participation helps hitherto in fueling more innovative ideas and ventures towards branding a product or service.

Research has shown that teams that are supportive of each other's novel ideas and creativity are more open to learning new things (Martins & Terblanche, 2003). Such teams are better able to produce innovative solutions, which in long run perspective, would help assist the brand or company in gaining a competitive edge in the market making of the brand to be more successful (Alameeri et al., 2020; Evans, 2010; Mount & Martinez, 2014).

2.1 Conceptual Framework

Figure 1 depicts the conceptual framework used for this study. This framework shows two antecedents of consumer brand awareness. It also reflects the consequences of a supportive approach of team members for innovation in the practice of social media marketing.

Fig. 1 Conceptual framework

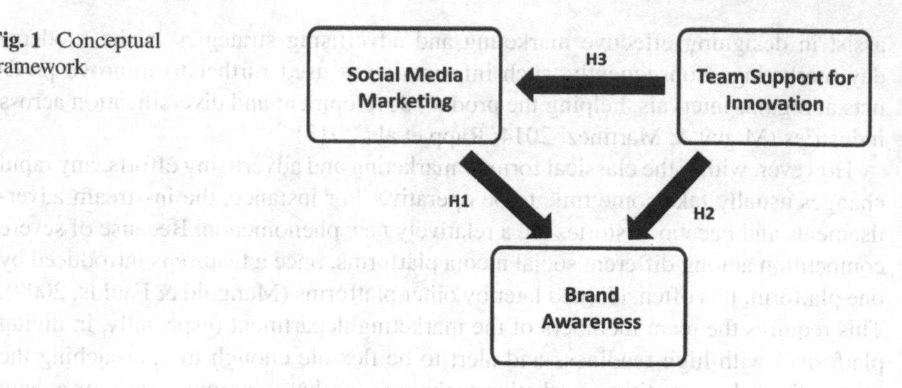

3 Research Hypotheses

3.1 *Hypotheses Related to the Antecedents of Consumer Brand Awareness*

3.1.1 Social Media Marketing

The online platforms of social media were developed according to the basic principles of communication. These channels of communication encourage the flow of information among the users, which in turn, leads to the development of relationships (Alshurideh, 2022; Homburg et al., 2010) and collaboration among the people living in different locations all over the world (Michaelidou et al., 2011). Therefore, when marketing professionals use these platforms for marketing their products, they create healthy and unrestricted channels of communication.

The posts and the marketing content created and shared by companies often go viral. Consequently, this leads to millions of views in a very short span of time (Hoffman & Fodor, 2010; Scott, 2015). In the end, this high exposure ultimately helps to create brand awareness even beyond the target market at which the advertisement was initially meant and aimed for (Marchand et al., 2017; Mount & Martinez, 2014). Social media platforms have not only shown their power at the business-to-consumer (B2C) level (Moore et al., 2013), but in some cases, they have also helped to increase the level of profits at the business-to-business (B2B) level (Michaelidou et al., 2011). Such objectives are mainly achieved by delivering information about the products and services to the general public, which eventually, helps to create a higher level of brand awareness (Hays et al., 2013; Klostermann et al., 2018). In this regard, research has also shown that the loyal consumers also contribute to brand awareness by engaging themselves in electronic word-of-mouth (e-WOM) (Eelen et al., 2017).

Given the discussions above, we investigated the association between the social media marketing and consumer brand awareness; the first research hypothesis (H1) was proposed as follows.

H1. *Social media marketing is positively associated with consumer brand awareness.*

3.1.2 Team Support for Innovation

Once innovations and brands that companies adopted to continue in the market, the usage of Internet and Digital marketing have been very common (Alzoubi et al., 2022; Mount & Martinez, 2014). Companies that usually do not adapt to the changing trends or fail to adhere are found to be to lose out of the competition (Roberts & Piller, 2016). Management's support for creativity and innovation is regarded to be a highly essential component of marketing success (Bassett-Jones, 2005; Dodgson, Gann, & Phillips, 2013; Roberts & Piller, 2016). According to Wu and Ho (2014), even the perception of brand innovation among consumers can have a stronger impact on the level of brand awareness, perceived product quality, and purchase intentions among them.

Given the above backdrop, we investigated the association between team support for innovation and consumer brand awareness, and as such, a second hypothesis (H2) was proposed as below.

H2. *Team support for innovation is positively associated with consumer brand awareness.*

3.2 Hypothesis Related to the Consequences of Team Support for Innovation

Social media marketing has become a common practice, with almost all companies, large and small, invariably; use these media for marketing and advertising purposes (Gallaugher & Ransbotham, 2010). In a time of severe competition, brands and companies need to be differentiated from one another (Dodgson et al., 2013). One can achieve this objective only by ensuring that the social media marketing campaigns developed by brands are innovative (Shankar et al., 2011).

One major benefit of social media marketing is that people from outside the workforce such as, consumers can also be involved in the process, allowing the workforce to gain required insights into more creative marketing options (Scott, 2015). However, the team members of the marketing department must be capable enough of identifying such creative and innovative opportunities from the marketing team members as well as from outside the workforce (Antikainen et al., 2010).

Therefore, we studied the consequences of team support for innovation in terms of the use of social media marketing tools with the third study hypothesis (H3) proposed as follows.

H3. *Team support for innovation is positively associated with social media marketing.*

4 Data Collection and Analysis

4.1 Participants

Data were collected from a sample of 247 participants who were randomly selected from the United Arab Emirates. To qualify for the study, a basic understanding of social media platforms was required. To simplify data collection, the researcher used online media to send the questionnaires to participants. This also enabled the automatic recording of information.

Since social media, platforms are very popular and are commonly used by people irrespective of ages, ethnicities, and nationalities, no candidates were disqualified based on age and above selection criterion. It is worth pointing that more than 50% of participants were between 21 and 26 years old. The sample had relatively more male than female participants did (males: 138, 55.9%, females: 109, 44.1%).

4.2 Measures and Procedures

A survey approach was used to collect quantitative data. To measure the variable "social media marketing" (SMM), a 5-item scale was adapted from Bambauer-Sachse & Mangold. To measure the level of "consumer brand awareness" (CBA), the scale developed by Yoo et al. (2000) was adapted according to the requirements of this study. For the variable "team support for innovation", (TSI) and this scale also had five items. The scale developed by Anderson and West (1998) was adapted to measure "team support for innovation" (TSI), which included eight items in total.

The first section of the questionnaire (Appendix A) was related to demographic information; the second section included the scales designed to measure the three variables of interests. In answering the questions, respondents used a 5-point Likert scale; response options ranged from 1 = Strongly Disagree to 5 = Strongly Agree.

To check the reliability of the study constructs, the statistical technique of Chronbach's alpha was used. The alpha values for the scales of SMM, CBA, and TSI were 0.781, 0.794, and 0.860, respectively (Table 1). All of these values were higher than 0.70, which was believed to be the cut-off point for the reliability of any scale (Nunnally & Bernstein, 1994).

Table 1 Reliability testing of scales

Scale	N	Chronbach's Alpha	Source
Social media marketing (SMM)	5	0.781	Bambauer-Sachse and Mangold
Consumer brand awareness (CBA)	5	0.794	Yoo et al. (2000)
Team support for innovation (TSI)	8	0.860	Anderson and West (1998)

5 Findings and Discussion

To test the three proposed hypotheses of this study and to find out the impact of three determinants precisely on consumer brand awareness (CBA), we performed a regression analysis (both simple linear and multiple regressions) by using SPSS. Using the multiple linear regression analysis, a strongly positive correlation was found between the dependent variable, CBA, with each of the independent variables, TSI, and SMM ($R = 0.733$). A significant, positive impact was also found of the independent variables (TSI and SMM) on CBA ($R^2 = 0.537$, $p < 0.05$) (Table 2).

Our findings support the first and second hypotheses of the study, in that the study results supported the assertion that SMM and TSI were positively associated with CBA. The estimated regression equation was found to be as follows.

CBA = 0.298 + 0.343 (TSI) + 0.516 (SMM).

Previous studies also showed a positive impact of the use of digital and social media marketing on consumer brand awareness (Hanna et al., 2011; Berthon et al., 2012; Bruhn et al., 2012; Hutter et al., 2013). Lipsman et al. (2012) were of the view that the social media tools held lots of power, and therefore, be considered as essential media for marketing purposes.

Support for the second hypothesis of the study was also in line with those of previous studies, demonstrating also a positive impact of team support for innovation on consumer brand awareness (Antikainen et al., 2010; Enkel et al., 2009; Evans, 2010; Mount & Martinez, 2014). Marketing teams were more likely to bring more creative solutions with a high rate of success, provided they encouraged innovative ideas offered by other team members (Autio et al., 2014; Enkel et al., 2009; Hakansson & Waluszewski, 2007).

Using simple linear regression analysis, a medium positive correlation was observed between TSI and SMM ($R = 0.601$). In addition, a statistically significant positive impact of TSI was also observed for SMM ($R^2 = 0.362$; $p < 0.05$). (See Table 3).

H3 of the study was, therefore, supported. The estimated regression equation drawn from the study is:

SMM = 1.257 + 0.669 (TSI).

Table 2 Multiple regression analysis for Hypothesis 1 and Hypothesis 2

Variable	B	SE	t	Significance
Constant	0.298	0.235	1.269	0.206
Team support for innovation	0.343	0.054	6.368	0.000
Social media marketing	0.516	0.060	8.607	0.000
Dependent variable	CBA			
R	0.733			
R^2	0.537			
F (Significance)	141.420 (0.000)			

Table 3 Linear regression analysis for Hypothesis 3

Variable	B	SE	t	Significance
Constant	1.257	0.266	4.722	0.000
Team support for innovation	0.669	0.057	11.782	0.000
Dependent variable	SMM			
R	0.601			
R^2	0.362			
F (Significance)	138.806 (0.000)			

The study result is in line with the previous research work in the literature suggesting that the team support for innovation was positively associated with the effectiveness of the tools and techniques of social media marketing (Dodgson et al., 2013; Enkel et al., 2009; Georgellis et al., 2000). Further, using innovative techniques and content ideas for social media marketing, we were able to observe a precipitative impact of generating competitive advantage. The use of such creative ideas, however, might also lead to a higher chance of media posts "going viral," which could lead to greater exposure and hence, a higher level of brand awareness among the potential, as well as existing consumers.

6 Limitations

Even though this study provided some interesting results, it has one main limitation. The inclusion criteria for this study were very broad, as people from different functional backgrounds were included. Considering the nature of variables examined in this study, more relevant findings could have been obtained, had the research focused only on marketing professionals working in the United Arab Emirates. The current participants might have responded to the questionnaire statements, based on their assumptions only, rather than on strong academic knowledge or some other practical experiences in the field of marketing.

7 Conclusion

The purpose of this study was to investigate the role of social media marketing, and team support for innovation, as the two main antecedents of consumer brand awareness. The study also explored the impact of a supportive approach of team members on innovation concerning the techniques of social media marketing. After the analysis of the data, we found that social media marketing and team support for innovation were positively associated with consumer brand awareness. In addition, we found that team support for innovation was positively associated with the effectiveness of social media marketing. All the three hypotheses proposed in this study were verified and supported in line with those of the previous research studies in the field.

8 Implications and Directions for Future Research

8.1 Practical Implications

We recommend that marketing professionals should make use of the different tools and techniques of social media marketing in order to create greater brand awareness. Social media platforms were widely used and had large audiences, and that led the messages posted on these media, reach more likely relatively to a larger section of population. It is asserted that, an open culture needs to be nourished and cherished within the companies' providing supports and encouragement, especially for the entrepreneurial and innovative approaches of marketing team members. Social media platforms require an upgrade at regular intervals; marketing professionals should also keep abreast of new developments when designing the marketing strategies and applying them.

8.2 Opportunities for Future Research

In order to improve the quality of data when studying the variables of social media marketing, the team support for innovation, and consumer brand awareness, we recommend the recruitment of only the professionals from the marketing domain. In addition, we suggest a case study approach on a particular brand type to be adopted for a clearer vision. Besides, comparisons between two or more companies developing different brands of a product could be made concerning their respective organizational structure (s) and other cultural characteristics.

Appendix A: Questionnaire

Demographic information

Gender

- Male
- Female

Age

- 20 years or under
- 21–26 years
- 27–33 years
- 34–40 years
- 41–46 years
- 47–53 years
- 54 years or above

Education

- Below high school
- High school
- Diploma
- Associate degree
- Bachelor's degree
- Master's degree
- Postgraduate

Monthly income

- Less than 1,000 AED
- 1,000 AED–8,999 AED
- 9,000 AED–16,999 AED
- 17,000 AED–24,999 AED
- 25,000 AED–32,999 AED
- 33,000 AED–40,999 AED
- 41,000 AED–48,999 AED
- 49,000 AED–56,999 AED
- More than 57,000 AED

Do you use social media for marking and advertising?

- Yes
- No

If yes, which of the following social media do you use for marking and advertising? (Select as many as apply.)

- Facebook
- Twitter
- Instagram
- YouTube
- All of the above

Which ONE of the following is your most preferred medium of social media for marking and advertising? (Please select one only.)

- Facebook

- Instagram
- Snapchat
- YouTube
- Other (Specify): ______________________________

Instructions: For each of the statements given below, please signify your level of agreement or disagreement by marking any one of the options provided: Strongly Disagree (SD), Disagree (D), Neutral (N), Agree (A), or Strongly Agree (SA).

Level of agreement		SD	D	N	A	SA
Social media marketing statements (*Adapted from* Bambauer-Sachse & Mangold, 2011)						
1	To make sure that they are buying the right products and/or services, consumers often read other consumers' online product reviews					
2	Our consumers can read others' online product reviews to know what products and/or services our company/brand offers					
3	Consumers often consult other consumers' online product reviews to help choose the right product/brand					
4	Consumers frequently gather information from online consumers' product reviews before they buy a certain product and/or service					
5	If consumers do not read others' online reviews before buying a product/brand, they worry about their decisions					
Consumer brand awareness (*Adapted from* Yoo et al., 2000)						
6	Consumers are aware of most of the products and/or services offered by our company/brand					
7	Consumers can easily recognize the products and/or services offered by our company/brand					
8	Consumers can easily recognize the logo of our company/brand					
9	Consumers are aware of the quality of products and/or services offered by our company/brand					
10	Consumers can tell the difference between products and/or services of our company/brand and those of others					
Team support for innovation (*Adapted from* Anderson & West, 1998)						
11	My team is always moving toward the development of new answers					
12	Assistance in developing new ideas is readily available					
13	My team is open and responsive to change					
14	People on my team are always searching for fresh and new ways of looking at problems					
15	On my team, we have the time needed to develop new ideas					
16	People on my team cooperate in order to help develop and apply new ideas					

(continued)

(continued)

Level of agreement		SD	D	N	A	SA
17	Members of my team provide and share resources to help in the application of new ideas					
18	My team members provide practical support for new ideas and their applications					

References

Aaker, D. A. (1991). *Managing brand equity*. Free Press.

Abu Zayyad, H. M., Obeidat, Z. M., Alshurideh, M. T., Abuhashesh, M., Maqableh, M., & Masa'deh, R. E. (2021). Corporate social responsibility and patronage intentions: the mediating effect of brand credibility. *Journal of Marketing Communications*, 1–24.

Abuhashesh, M. Y., Alshurideh, M. T., & Sumadi, M. (2021). The effect of culture on customers' attitudes toward Facebook advertising: the moderating role of gender. *Review of International Business and Strategy, 31*(3), 416–437.

Aissani, R., Awad ElRefae, G. (2021). Trends of university students towards the use of YouTube and the benefits achieved; Field study on a sample of Al Ain University students. *AAU Journal of Business and Law, 5*(1).

Al Kurdi, B. H., & Alshurideh, M. T. (2021). Facebook advertising as a marketing tool: Examining the influence on female cosmetic purchasing behaviour. *International Journal of Online Marketing (IJOM), 11*(2), 52–74.

Al Naqbia, E., Alshuridehb, M., AlHamadc, A., & Al, B. (2020). The impact of innovation on firm performance: a systematic review. *International Journal of Innovation, Creativity and Change 14*(5), 31–58.

Al Suwaidi, F., Alshurideh, M., Al Kurdi, B., & Salloum, S. A. (2020). The impact of innovation management in SMEs performance: a systematic review. In *International conference on advanced intelligent systems and informatics* (pp. 720–730). Springer.

Alameeri, K., Alshurideh, M., Al Kurdi, B., & Salloum, S. A. (2020). The effect of work environment happiness on employee leadership. In *International conference on advanced intelligent systems and informatics* (pp. 668–680). Springer.

Alameeri, K. A., Alshurideh, M. T., & Al Kurdi, B. (2021). The effect of Covid-19 pandemic on business systems' innovation and entrepreneurship and how to cope with it: A theatrical view. *The Effect of Coronavirus Disease (COVID-19) on Business Intelligence, 334*, 275.

Al-Dhuhouri, F. S., Alshurideh, M., Al Kurdi, B., & Salloum, S. A. (2020). Enhancing our understanding of the relationship between leadership, team characteristics, emotional intelligence and their effect on team performance: A critical review. In *International conference on advanced intelligent systems and informatics* (pp. 644–655). Springer.

Al-Dmour, H., Alshuraideh, M., & Salehih, S. (2014). A study of Jordanians' television viewers habits. *Life Science Journal, 11*(6), 161–171.

Aljumah, A., Nuseir, M. T., & Alshurideh, M. T. (2021a). The impact of social media marketing communications on consumer response during the COVID-19: Does the brand equity of a university matter. *The Effect of Coronavirus Disease (COVID-19) on Business Intelligence*, 384–367.

Aljumah, A. I., Nuseir, M. T., & Alam, M. M. (2021b), Organizational performance and capabilities to analyze big data: do the ambidexterity and business value of big data analytics matter? *Business Process Management Journal, 27*(4), 1088–1107. https://doi.org/10.1108/BPMJ-07-2020-0335.

Aljumah, A. I., Nuseir, M. T., & Alam, M. M. (2021c). Traditional marketing analytics, big data analytics and big data system quality and the success of new product development. *Business Process Management Journal, 27*(4), 1108–1125. https://doi.org/10.1108/BPMJ-11-2020-0527

Almaazmi, J., Alshurideh, M., Al Kurdi, B., & Salloum, S. A. (2020). The effect of digital transformation on product innovation: a critical review. In *International conference on advanced intelligent systems and informatics* (pp. 731–741). Springer.

Al-Maroof, R., Ayoubi, K., Alhumaid, K., Aburayya, A., Alshurideh, M., Alfaisal, R., & Salloum, S. (2021). The acceptance of social media video for knowledge acquisition, sharing and application: A comparative study among YouYube users and TikTok users' for medical purposes. *International Journal of Data and Network Science, 5*(3), 197–214

Almazrouei, F. A., Alshurideh, M., Al Kurdi, B., & Salloum, S. A. (2020, October). Social media impact on business: a systematic review. In *International conference on advanced intelligent systems and informatics* (pp. 697–707). Springer.

Alshurideh, M. (2022). Does electronic customer relationship management (E-CRM) affect service quality at private hospitals in Jordan? *Uncertain Supply Chain Management, 10*(2), 1–8.

Alshurideh, M., Bataineh, A., Alkurdi, B., & Alasmr, N. (2015). Factors affect mobile phone brand choices–studying the case of Jordan universities students. *International Business Research, 8*(3), 141–155.

Alshurideh, M., Al Kurdi, B. H., Vij, A., Obiedat, Z., & Naser, A. (2016). Marketing ethics and relationship marketing-an empirical study that measure the effect of ethics practices application on maintaining relationships with customers. *International Business Research, 9*(9), 78–90.

Alshurideh, M., Al Kurdi, B., Abu Hussien, A., & Alshaar, H. (2017). Determining the main factors affecting consumers' acceptance of ethical advertising: A review of the Jordanian market. *Journal of Marketing Communications, 23*(5), 513–532.

Alshurideh, M., Salloum, S. A., Al Kurdi, B., & Al-Emran, M. (2019a). Factors affecting the social networks acceptance: an empirical study using PLS-SEM approach. In *Proceedings of the 2019a 8th International Conference on Software and Computer Applications* (pp. 414–418).

Alshurideh, M., Kurdi, B. A., Shaltoni, A. M., & Ghuff, S. S. (2019b). Determinants of pro-environmental behaviour in the context of emerging economies. *International Journal of Sustainable Society, 11*(4), 257–277.

Alshurideh, M. T., Al Kurdi, B., Alzoubi, H. M., Ghazal, T. M., Said, R. A., AlHamad, A. Q., & Al-kassem, A. H. (2022). Fuzzy assisted human resource management for supply chain management issues. *Annals of Operations Research*, 1–19.

Alzoubi, H., Alshurideh, M., Al Kurdi, B., Akour., I, & Azize, R. (2022) Does BLE technology contribute towards improving marketing strategies, customers' satisfaction and loyalty? The role of open innovation. *International Journal of Data and Network Science, 6*, 1–12.

Anderson, N. R., & West, M. A. (1998). Measuring climate for work group innovation: Development and validation of the team climate inventory. *Journal of Organizational Behavior: THe International Journal of Industrial, Occupational and Organizational Psychology and Behavior, 19*(3), 235–258.

Antikainen, M., Makipaa, M., & Ahonen, M. (2010). Motivating and supporting collaboration in open innovation. *European Journal of Innovation Management, 13*(1), 100–119.

Autio, E., Kenney, M., Mustar, P., Siegel, D., & Wright, M. (2014). Entrepreneurial innovation: The importance of context. *Research Policy, 43*(7), 1097–1108.

Bassett-Jones, N. (2005). The paradox of diversity management, creativity and innovation. *Creativity and Innovation Management, 14*(2), 169–175.

Berthon, P. R., Pitt, L. F., Plangger, K., & Shapiro, D. (2012). Marketing meets Web 2.0, social media, and creative consumers: Implications for international marketing strategy. *Business Horizons, 55*(3), 261–271.

Bruhn, M., Schoenmueller, V., & Schäfer, D. B. (2012). Are social media replacing traditional media in terms of brand equity creation? *Management Research Review, 35*(9), 770–790.

Cobb-Walgren, C. J., Ruble, C. A., & Donthu, N. (1995). Brand equity, brand preference, and purchase intent. *Journal of Advertising, 24*(3), 25–40

Dick, A., Jain, A., & Richardson, P. (1997). How consumers evaluate store brands. *Pricing Strategy and Practice, 5*(1), 18–24.

Dodgson, M., Gann, D. M., & Phillips, N. (Eds.). (2013). *The Oxford handbook of innovation management*. Oxford University Press (OUP.com).

Eelen, J., Özturan, P., & Verlegh, P. W. (2017). The differential impact of brand loyalty on traditional and online word of mouth: The moderating roles of self-brand connection and the desire to help the brand. *International Journal of Research in Marketing, 34*(4), 872–891.

Elsaadani, M. A. (2020). Investigating the effect of E-management on customer service performance. *AAU Journal of Business and Law., 3*(2), 6–24.

Enkel, E., Gassmann, O., & Chesbrough, H. (2009). Open R&D and open innovation: Exploring the phenomenon. *R&D Management, 39*(4), 311–316.

Evans, D. (2010). *Social media marketing: The next generation of business engagement*. Wiley.

Gallaugher, J., & Ransbotham, S. (2010). Social media and customer dialog management at Starbucks. *MIS Quarterly Executive, 9*(4), 197–212.

Georgellis, Y., Joyce, P., & Woods, A. (2000). Entrepreneurial action, innovation and business performance: The small independent business. *Journal of Small Business and Enterprise Development, 7*(1), 7–17.

Ghannajeh, A. M., AlShurideh, M., Zu'bi, M. F., Abuhamad, A., Rumman, G. A., Suifan, T., & Akhorshaideh, A. H. O. (2015). A qualitative analysis of product innovation in Jordan's pharmaceutical sector. *European Scientific Journal, 11*(4), 474–503.

Guitart, I. A., Gonzalez, J., & Stremersch, S. (2018). Advertising non-premium products as if they were premium: The impact of advertising up on advertising elasticity and brand equity. *International Journal of Research in Marketing., 35*(1), 471–489.

Hakansson, H., & Waluszewski, A. (Eds.). (2007). *Knowledge and innovation in business and industry: The importance of using others*. Routledge.

Hamadneh, S., Hassan, J., Alshurideh, M., Al Kurdi, B., & Aburayya, A. (2021). The effect of brand personality on consumer self-identity: The moderation effect of cultural orientations among British and Chinese consumers. *Journal of Legal, Ethical and Regulatory, 24*, 1–14.

Hanna, R., Rohm, A., & Crittenden, V. L. (2011). We're all connected: The power of the social media ecosystem. *Business Horizons, 54*(3), 265–273.

Hansen, N., Kupfer, A. K., & Hennig-Thurau, T. (2018). Brand crises in the digital age: The short- and long-term effects of social media firestorms on consumers and brands. *International Journal of Research in Marketing, 35*(4), 557–574

Hays, S., Page, S. J., & Buhalis, D. (2013). Social media as a destination marketing tool: Its use by national tourism organizations. *Current Issues in Tourism, 16*(3), 211–239.

Hoffman, D. L., & Fodor, M. (2010). Can you measure the ROI of your social media marketing? *MIT Sloan Management Review, 52*(1), 41.

Homburg, C., Klarmann, M., & Schmitt, J. (2010). Brand awareness in business markets: When is it related to firm performance? *International Journal of Research in Marketing, 27*(3), 201–212.

Hutter, K., Hautz, J., Dennhardt, S., & Füller, J. (2013). The impact of user interactions in social media on brand awareness and purchase intention: The case of MINI on Facebook. *Journal of Product & Brand Management, 22*(5/6), 342–351.

Keller, K. L. (1993). Conceptualizing, measuring, and managing customer-based brand equity. *Journal of Marketing, 57*(1), 1–22.

Keller, K. L. (2001). *Building customer-based brand equity: A blueprint for creating strong brands* (pp. 3–27). Marketing Science Institute.

Khasawneh, M. A., Abuhashesh, M., Ahmad, A., Alshurideh, M. T., & Masa'deh, R. (2021a). Determinants of e-word of mouth on social media during COVID-19 outbreaks: An empirical study. In *The effect of coronavirus disease (COVID-19) on business intelligence* (pp. 347–366). Springer.

Khasawneh, M. A., Abuhashesh, M., Ahmad, A., Masa'deh, R., & Alshurideh, M. T. (2021b). Customers online engagement with social media influencers' content related to COVID 19. In *The effect of coronavirus disease (COVID-19) on business intelligence* (pp. 385–404). Springer.

Kirtiş, A. K., & Karahan, F. (2011). To be or not to be in social media arena as the most cost-efficient marketing strategy after the global recession. *Procedia-Social and Behavioral Sciences, 24*, 260–268.
Klostermann, J., Plumeyer, A., Böger, D., & Decker, R. (2018). Extracting brand information from social networks: Integrating image, text, and social tagging data. *International Journal of Research in Marketing, 35*(4), 538–556.
Kurdi, B. A., Alshurideh, M., Nuseir, M., Aburayya, A., & Salloum, S. A. (2021). The effects of subjective norm on the intention to use social media networks: an exploratory study using PLS-SEM and machine learning approach. In *International conference on advanced machine learning technologies and applications* (pp. 581–592). Springer.
Lipsman, A., Mudd, G., Rich, M., & Bruich, S. (2012). The power of "like": How brands reach (and influence) fans through social-media marketing. *Journal of Advertising Research, 52*(1), 40–52.
Low, G. S., & Lamb, C. W., Jr. (2000). The measurement and dimensionality of brand associations. *Journal of Product and Brand Management, 9*(6), 350–370.
Malik, M. E., Ghafoor, M. M., Hafiz, K. I., Riaz, U., Hassan, N. U., Mustafa, M., & Shahbaz, S. (2013). Importance of brand awareness and brand loyalty in assessing purchase intentions of consumer. *International Journal of Business and Social Science, 4*(5).
Mangold, W. G., & Faulds, D. J. (2009). Social media: The new hybrid element of the promotion mix. *Business Horizons, 52*(4), 357–365.
Marchand, A., Hennig-Thurau, T., & Wiertz, C. (2017). Not all digital word of mouth is created equal: Understanding the respective impact of consumer reviews and micro-blogs on new product success. *International Journal of Research in Marketing, 34*(2), 336–354
Martins, E. C., & Terblanche, F. (2003). Building organizational culture that stimulates creativity and innovation. *European Journal of Innovation Management, 6*(1), 64–74.
Matherly, T., Arens, Z. G., & Arnold, T. J. (2018). Big brands, big cities: How the population penalty affects common, identity relevant brands in densely populated areas. *International Journal of Research in Marketing, 35*(1), 15–33.
Michaelidou, N., Siamagka, N. T., & Christodoulides, G. (2011). Usage, barriers and measurement of social media marketing: An exploratory investigation of small and medium B2B brands. *Industrial Marketing Management, 40*(7), 1153–1159.
Moore, J. N., Hopkins, C. D., & Raymond, M. A. (2013). Utilization of relationship-oriented social media in the selling process: A comparison of consumer (B2C) and industrial (B2B) salespeople. *Journal of Internet Commerce, 12*(1), 48–75.
Mount, M., & Martinez, M. G. (2014). Social media: A tool for open innovation. *California Management Review, 56*(4), 124–143.
Nguyen, H. T., Zhang, Y., & Calantone, R. J. (2018). Brand portfolio coherence: Scale development and empirical demonstration. *International Journal of Research in Marketing, 35*(1), 60–80.
Nunnally, J. C., & Bernstein, I. H. (1994). *Psychometric theory* (3rd ed.). McGraw-Hill.
Nuseir, M. T., & Aljumah, A. (2021). Digital marketing adoption influenced by relative advantage and competitive industry: A UAE tourism case study International Journal of Innovation. *Creativity and Change, 11*(2), 617–631.
Nuseir, M. T., Al Kurdi, B. H., Alshurideh, M. T., & Alzoubi, H. M. (2021a). Gender discrimination at workplace: Do Artificial Intelligence (AI) and Machine Learning (ML) have opinions about it. In *The international conference on artificial intelligence and computer vision* (pp. 301–316). Springer.
Nuseir, M. T., El-Refae, G. A., Aljumah, A. (2021b). The e-learning of students and university's brand image (Post COVID-19): How successfully Al-Ain university have embraced the paradigm shift in digital learning. *Studies in Systems, Decision and Control, 334*, 171–187.
Obeidat, U., Obeidat, B., Alrowwad, A., Alshurideh, M., Masadeh, R., & Abuhashesh, M. (2021). The effect of intellectual capital on competitive advantage: The mediating role of innovation. *Management Science Letters, 11*(4), 1331–1344.
Owen, L., & Othman, A. (2001). Developing an instrument to measure customer service quality (SQ) in Islamic banking. *International Journal of Islamic Financial Services, 3*(1), 1–26.

Rapp, A., Beitelspacher, L. S., Grewal, D., & Hughes, D. E. (2013). Understanding social media effects across seller, retailer, and consumer interactions. *Journal of the Academy of Marketing Science, 41*(5), 547–566.

Roberts, D. L., & Piller, F. T. (2016). Finding the right role for social media in innovation. *MIT Sloan Management Review, 57*(3), 41.

Scott, D. M. (2015). *The new rules of marketing and PR: How to use social media, online video, mobile applications, blogs, news releases, and viral marketing to reach buyers directly*. Wiley.

Shankar, V., Inman, J. J., Mantrala, M., Kelley, E., & Rizley, R. (2011). Innovations in shopper marketing: Current insights and future research issues. *Journal of Retailing, 87*, S29–S42.

Sweiss, N., Obeidat, Z. M., Al-Dweeri, R. M., Mohammad Khalaf Ahmad, A., M. Obeidat, A., & Alshurideh, M. (2021). The moderating role of perceived company effort in mitigating customer misconduct within Online Brand Communities (OBC). *Journal of Marketing Communications*, 1–24.

Thackeray, R., Neiger, B. L., Hanson, C. L., & McKenzie, J. F. (2008). Enhancing promotional strategies within social marketing programs: Use of Web 2.0 social media. *Health Promotion Practice*, *9*(4), 338–343.

Wu, S. I., & Ho, L. P. (2014). The influence of perceived innovation and brand awareness on purchase intention of innovation product—an example of iPhone. *International Journal of Innovation and Technology Management, 11*(04), 1450026.

Yoo, B., Donthu, N., & Lee, S. (2000). An examination of selected marketing mix elements and brand equity. *Journal of the Academy of Marketing Science, 28*(2), 195–211.

The Influence of Sharing Fake News, Self-Regulation, Cyber Bullying on Social Media Fatigue During COVID-19 Work Technology Conflict as Mediator Role

Mohammed T. Nuseir, Ghaleb A. El Refae, Muhammad Alshurideh, Sarah Urabi, and Barween Al Kurdi

Abstract The purpose of this paper is to analyze the impact of Sharing Fake news, self-regulation, Cyber Bullying on social media fatigue during COVID-19 work technology conflict as mediator role. The current study uses quantitative with cross sectional design to examine the effect of Sharing Fake news, self-regulation, Cyber Bullying on social media fatigue during COVID-19 and suing work technology conflict as mediator role. The respondents were situation from different top sites, such as twitter, Facebook and Instagram sample of 132, and population of this study is 200 users were selected for this study, sample size is calculated through ROA soft. The dissemination of unverified information has been showcased as a significant challenge during the COVID-19 pandemic. The role of social media in this process is exemplified by its increased use during COVID-19, as, for example, a recent report

M. T. Nuseir (✉)
Department of Business Administration, College of Business, Al Ain University, Abu Dhabi Campus, Abu Dhabi, UAE
e-mail: mohammed.nuseir@aau.ac.ae

G. A. E. Refae
Department of Business Administration, College of Business, Al Ain University, Al Ain Campus, Abu Dhabi, UAE
e-mail: ghalebelrefae@aau.ac.ae

M. Alshurideh
Department of Management, College of Business Administration, University of Sharjah, Sharjah, United Arab Emirates
e-mail: m.alshurideh@ju.edu.jo; malshurideh@sharjah.ac.ae

Department of Marketing, School of Business, The University of Jordan, Amman, Jordan

S. Urabi
Abu Dhabi Vocational Education & Training Institute (ADVETI), Abu Dhabi, UAE
e-mail: sarah.urabi@adveti.ac.ae

B. A. Kurdi
Department of Marketing, Faculty of Economics and Administrative Sciences, The Hashemite University, Zarqa, Jordan
e-mail: barween@hu.edu.jo

M. Alshurideh et al. (eds.), *The Effect of Information Technology on Business and Marketing Intelligence Systems*, Studies in Computational Intelligence 1056,
https://doi.org/10.1007/978-3-031-12382-5_7

shows that the use of Facebook hit record levels during the pandemic. This study revealed that potential mechanisms for counteracting fake news creating Facebook pages of real news and using this advertising to disseminate accurate information this paper will enhanced the understanding the effects of SMA, cyberbullying and self-regulation on mental health of individual through the use of social cognitive perspectives. To enhance efficacy in the role of social media in this process to reduce the gap between theory and practice, social marketers should include messages that empower recipients. Campaigns should show recommended behaviors and highlight their usefulness and effectiveness. This paper has been methodological as well as theoretical limitations, first using CLT might be regarded limiting even through it has been adopted in past studies that are based on social media. The primary concern is the cognitive load theory is still essential an instructional science theory even through it has been adopted, use widely in HCI & also shown to explain not only learning but also acquiring knowledge from new theories, articles might be more useful for conceptualizing fake news sharing.

Keywords Sharing fake news · Self-regulation · Cyber bullying · Social media · Social media fatigue · COVID-19

1 Introduction

Internet has powerful changed the type of news that customers receive (Alshurideh et al., 2016; Obeidat et al., 2012). In past times, consumers were basically relied on traditional type of media such as television and radio, which cover comparatively more and fever established sources of news (Alshurideh et al., 2019a; Obeidat et al., 2019; Zayyad et al., 2021), but currently consumers are reveal to online information sources through social networking sites, which allow any person to share content without "fact-checking or editorial judgement (Al-Dmour et al., 2014; Liu et al., 2021). Whereas social media enables firms to introduce creative and useful practices of management by introducing new models of business, communication reshaping, collaboration enhancing and seeking improve and shared knowledge (Almazrouei et al., 2020; Khan et al., 2021; Kurdi et al., 2021). The circulation of fake news sharing was significantly posing challenge for organization and brands fake news promote a specific option about brand, product or firm which may not be true can be intentionally designed to mislead customers. Social media is the term that enables users to connect with others (Hamadneh et al., 2021; Khasawneh et al., 2021a, 2021b). Past studies revealed that it can provide platforms to individuals that overcome the obstacles of time and distance to connect and reconnect with others and thereby increase the strength their offline networks and exchanges (Alshurideh et al., 2019b; Bekalu et al., 2019).

Social media platforms such as Facebook describe as several driven motivators such as wish for entertainments, stay informed and the desire to know the social activities of mutual users (friends) (Abuhashesh et al., 2021; Al Kurdi et al., 2021;

Laato et al., 2020). Social media basically plays a significant role during COVID as it enables individuals to share news as well as personal viewpoints and experiences with one another globally and in real-time (Aljumah et al., 2021a, 2021b; Al-Maroof et al., 2021). The issue of social media regulation platform has always been a controversial one for many in large society. The traditional method of media regulation is that there are relatively small number of users who produce the media coupled with a large number of those who consume it and powerless to directly influence the content (Zabedah & Kosmin, 2012). Regulatory actions is a social media is often focused upon disclosure of interest, children protection, code practices and offensive material prohibition (Ranchordás & Goanta, 2019). Almost every country in the world facing the COVID-19 situation during this time social media become the main information sources about pandemic (Akour et al., 2021; Al-Dmour et al., 2021a; Alshurideh et al., 2021; Leo et al., 2021).

2 Problem Statement

The abundance of uncertain, inaccurate information during existing situation led to overload information and accelerated health cyberchondria as well as misinformation sharing as social media intensity the spread of news and people read that news through shared link on social websites (Allcott & Gentzkow, 2017). The specific concern of fake news effects 2016 US election and about 62% US adults get news on social media and mostly fake news were widely shared on Facebook and become most popular mainstream of fake stories fortunately many people who see those fake news reported that they believe fake news arises because it is cheaper to provide precise signals as the consumer cannot causelessly infer accuracy and also because consumers enjoy partisan news (Chiou, 2018). However, internet has significant modification type of news that consumer acquired. Nowadays, consumers are revealing to online information sources for example social media sites which allow any person to share content without "fact-checking or editorial judgement" (Allcott & Gentzkow, 2017).

About two third adult's use Facebook, 16% adult's uses twitter and 20% users uses Instagram critics allege that most of fake new traffic were originates from these websites and that referrals account for a huge fraction of fake news sites than real news sites (Meinert et al., 2018) on the other hand cyberbullying is characterized as online communication it has negative effect on mental health with 32% victims reporting symptoms of stress and 38% were experiencing emotional distress (Das, 2020). This study examines the impact of fake news, self-regulation and cyberbullying factors on social media during the current situation of pandemic particularly people driven by entertainment and do not seem equally concerned about the information reliability they share on social media and investigate strategies with aim to reduce these factors on social media.

3 Literature Review

For the existing study, we adopt the affordance lens for understanding how users of social media connects with different platforms during pandemic situation. In this section, the author will present both theoretical & empirical approaches in order to connect them to the topic of this study.

3.1 Cognitive Load Theory

Cognitive load theory was developed in the late 1980s, it refers to use the amount of work memory resources in which information is stored in long-term after being attended to the experience of cognitive load is not same in everyone this theory differentiates three types of cognitive load i.e. germane, extraneous and intrinsic (Chong, 2005). The extraneous load has been investigated more often and in the boarder approach of human-capital interaction HCI, indicate to the environmental stimuli in which human brain reacts. Intrinsic cognitive load on the other hand is the load resulting from processing this information and is affected by individual's psychological state of mind as well as their prior knowledge (Sweller, 2010). Accordingly, well-structured information and earlier expertise of the learner can both decreases intrinsic cognitive load. The evolutionary still affects the behavior of human and it play especially acquiring new knowledge also referred as the human comfort zone. Vygotsky proximal zone development and CLT overload information can be conceptualized to occur when individual are overcome with further novel information or are taken too far away from their comfort zone. Consequently, overload information leads to impulses to step from new knowledge, back to the zone of proximal establishment.

4 Conceptual Framework

The proposed conceptual framework of this particular research study based on three independent variables which include sharing of fake news, cyberbullying, self-regulation. On the other hand, this study contains one dependent variable which is social media fatigue (SMF) COVID-19. The given below Fig. 1 shows the conceptual framework of this research study:

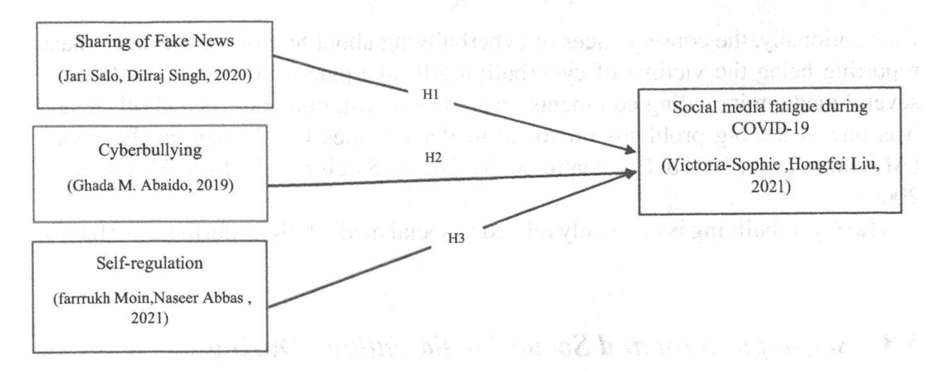

Fig. 1 Conceptual framework

5 Hypothesis Development

5.1 *Sharing of Fake News and Social Media Fatigue During COVID-19*

According to Nuseir et al. (2021), Social media fatigue (SMF) refers the tendency of users to pull back from social media when they become overwhelmed with too many social media sites. The term fatigue relates to overload cognitive however it reduces the concept of fatigue to the two components information and communication overload scholars argues that social media fatigue has significant negative implications for both service operators and the businesses (Yoo & Zhang, 2020). Social media fatigue can be detrimental for both business and service operator because fatigue results in withdrawal from service use, which translates into lower profits for companies and service operators (Aljumah et al., 2021a, 2021b, 2021c) (Dhir et al., 2018).

SMF is described as "social media users" aptitude to slow down social media usage when they feel overwhelmed with too many content pieces, friends and so much time consumed keeping these contacts (Cao, 2018).The supporting theory has been used to look how social media characteristics give rises to stressors, such as privacy invasion and life invasion, which then lead to SMF (Adhikari & Panda, 2020).

H1: Sharing of fake news is positively related to social media fatigue during COVID-19.

5.2 *Cyberbullying and Social Media Fatigue During COVID-19*

Cyber bullying take place over digital devices such as computers, cell phones and tablets it includes sharing, posting or sending negative or false mean content about someone else. The concept of cyber bullying that separate it from traditional bullying.

Conventionally, the consequences of cyberbullying about half of the teenagers have reporting being the victims of cyberbullying It take places in cyberspace through several medium including comments on someone post, online chats and others and it is one of the big problems on social media websites like Instagram, Facebook (Al-Dmour et al., 2021b; Elsaadani, 2020a, 2020b; Sweiss et al., 2021; Vartak et al., 2000).

H2: Cyberbullying is positively related to social media fatigue during COVID-19.

5.3 *Self-regulation and Social Media Fatigue During COVID-19*

Self-regulation is referred to as the process of turning goals into actions it defined the process of self-generated thoughts, feelings and behaviors are periodically and planned. Individuals suffer from DS-R trouble regulating their actions such as they are more susceptible to acting based on impulses rather than planned cognition and behavior it is connected to sub-optimal and irresponsible behavior and reduced psychological well-being and internet addiction (Al-Khayyal et al., 2021; Matzat & Vrieling, 2016; Sultan et al., 2021).

H3: Self-regulation is positively related to social media fatigue during COVID-19.

The knowledge gap of this study is that these variables are not study in this sequence with social media fatigue during COVID-19 with mediating variable of work technology conflict, in future authors can also examine mediating variable like fear of mission out (FOMO). However, the variables are adopted from different papers furthermore, this study is not previous done in Pakistani context thus this paper fill the gap to study these variables in Pakistan using different social media websites to examine how these variables effects human behaviors and social media.

6 Methodology

This paper is quantitive in nature, and data is collected from a survey questionnaire. The questionnaire is divided into two parts one part is demographics such as income, age, gender, monthly income, and next part include variables questions as seen in the appendix Table 1. The research design will used in study is cross-sectional & deductive approach will be used as the following study is based on preexisting theories and hypothesis in order to identify the association between independent and dependent variable. We used different website such as twitter, Instagram, Facebook to determine how fake news, cyberbullying and regulating predicator's effects social media during COVID-19 Non-Probability will be used in further chapters and examine the regression, descriptive through SMARTPLS.

Table 1 Cronbach Alpha values for the study factors

Constructs	Sources	Items	Cronbach Alpha
Sharing of fake news	Jari et al.	4	0.857
Self-regulation	Moin and Abbas	4	0.785
Cyberbullying	Abaido (2019)	4	0.866
Social Media Fatigue	Victoria-Sophie and Liu	4	0.879

6.1 Respondents Profile

The survey method helped to determine the impact of social media fatigue during current situation from different top sites, such as twitter, Facebook and Instagram sample of 132, and population of this study is 200 users were selected for this study, sample size is calculated through ROA soft. Convenient sampling technique was selected for the respondents. Survey questionnaire was randomly conducted among Instagram, Facebook, and twitter by sending online questionnaire to collect the individual opinion from respondents.

6.2 Measurement and Scales

The survey questionnaire divided into two sections. Section one has demographic-related questions i.e. (age, gender) and section two has questions on 4 constructs and 16 indicators adopted from literature as seen in the appendix Table 1. These indicators were measured on the 5-point likert scale where 1 represents strongly agree and 5 represents strongly disagree. Moreover, the measurement scale summary is presented in Table 1.

In the following table it demonstrates the outcomes of pilot testing which show variables reliability, that are calculated through selected software SPSS. Subsequently, Chiou (2018) illustrates that the acceptable reliability value is 0.70 or greater than 0.70. Thus, this test was applied on the data collected from 15 respondents which was 5% of actual sample size of this research study.

7 Results

7.1 Descriptive Analysis

As seen the above table 2, the Skewness of all variables ranged between −0.446 and −1.614 that means the distribution is moderately to highly Skewed. The Kurtosis data

Table 2 Means, Std. Dev., Variance, Skewness and Kurtosis for the study factors

	Mean	Std. Dev	Variance	Skewness	Kurtosis
Sharing of fake news	3.95	0.605	0.366	−1.614	5.077
Cyberbullying	3.61	0.766	0.587	−1.116	1.716
Self-regulation	3.41	0.720	0.519	−0.446	0.249
Social media fatigue	3.64	0.723	0.522	−1.252	−0.775

Table 3 Mean, Std. deviation, composite reliability and AVE for the study factors

Constructs	Mean	Std. deviation	Composite reliability	(AVE)
Sharing of fake news	3.56	0.72	0.821	0.606
Cyberbullying	3.42	0.62	0.834	0.569
Self-regulation	3.67	0.88	0.902	0.650
Work technology conflict	3.88	0.89	0.798	0.774
Social media fatigue	3.88	0.89	0.798	0.774

range between 5.077 and −0.775 which means most variables shows the population very likely has positive excess kurtosis and only one variable shows the population very likely has negative excess kurtosis. The value with positive kurtosis value follows t-distribution whereas negative values follow the beta distribution (Barret & Morgan, 2015).

7.2 *Convergent Validity*

The results in Table 3 show that all composite reliability values are at least 0.70; and the values of AVE are at least 0.60 which confirm that the respective variables indicators have adequately convergent validity.

7.3 *Discriminant Validity*

The results in Table 4 show that the square of each pair of the correlation is lesser that the square of the explained variance. These results confirmed that all constructs are distinct and unique.

Table 4 The study factors correlations

	Sharing of fake news	Cyberbullying	Self-regulation	Social media fatigue
Sharing of fake news	0.754			
Cyberbullying	0.638	0.755		
Self-regulation	0.706	0.641	0.778	
Social media fatigue	0.561	0.623	0.867	0.812

Table 5 The study hypotheses testing

	Beta	T-stats	P-val	Results
Sharing of fake news – > Social Media Fatigue (H1)	−0.174	3.092	0.0021	Accepted
Cyberbullying – > Social Media Fatigue (H2)	0.144	2.448	0.0147	Accepted
Self-regulation – > Social Media Fatigue (H3)	0.863	20.322	0.000	Accepted

7.4 Path Coefficients

The results in Table 5 and Figs. 2 and 3 show that all four direct hypotheses were accepted, and three independent hypotheses are also accepted.

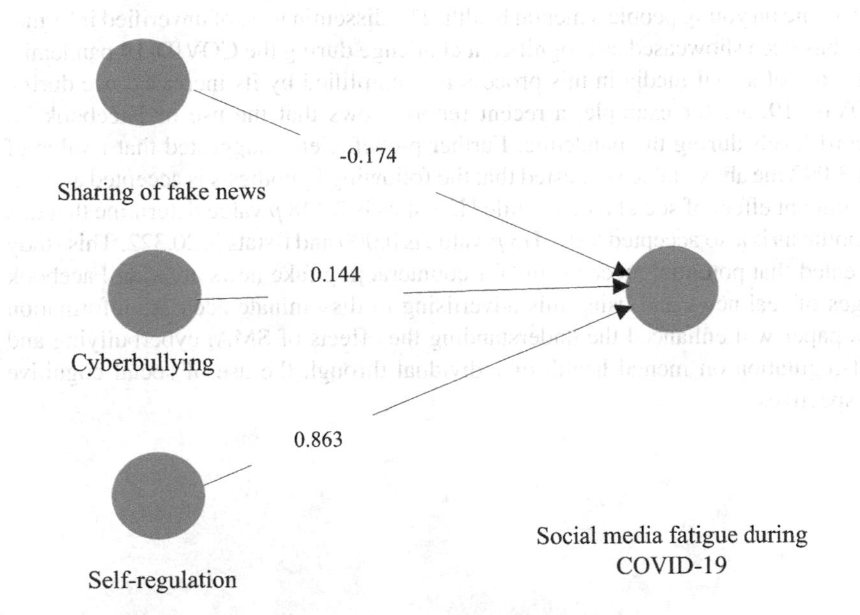

Fig. 2 Measurement model

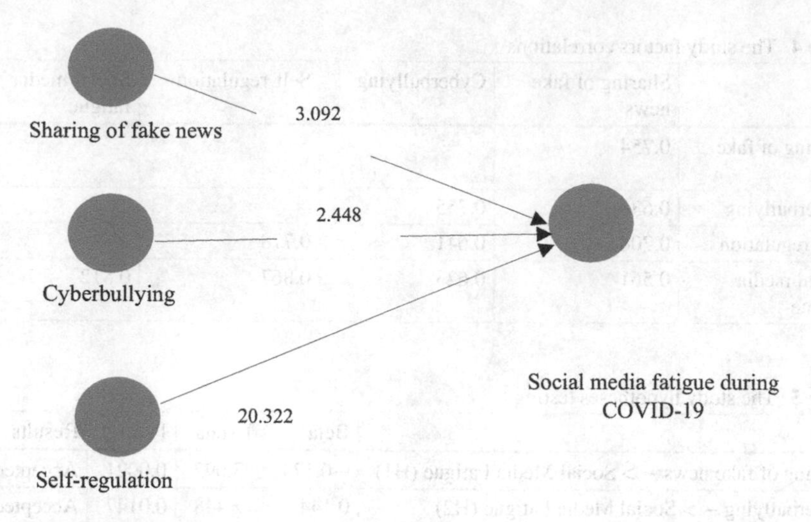

Fig. 3 Structural model

8 Conclusion

Technology is an integral part of our lives an important aspect which effects youth this research explored the dark side of social media during the first pandemic lockdown in the UK and adds to empirical evidence of the devastating effect of the COVID-19 pandemic on young people's mental health. The dissemination of unverified information has been showcased as a significant challenge during the COVID-19 pandemic. The role of social media in this process is exemplified by its increased use during COVID-19, as, for example, a recent report shows that the use of Facebook hit record levels during the pandemic. Further p-coefficients suggested that t-value of H1 3.092 the above table suggested that the following hypothesis is accepted and has significant effect of social media while H2 t-stats is 2.448 *p* value determine that this hypothesis is also accepted lastly H3 *p*-value is 0.000 and t-stats is 20.322. This study revealed that potential mechanisms for counteracting fake news creating Facebook pages of real news and using this advertising to disseminate accurate information this paper will enhanced the understanding the effects of SMA, cyberbullying and self-regulation on mental health of individual through the use of social cognitive perspectives.

9 Limitations

This paper has been methodological as well as theoretical limitations, first using CLT might be regarded limiting even through it has been adopted in past studies that are based on social media. The primary concern is the cognitive load theory is still essential an instructional science theory even through it has been adopted, use widely in HCI & also shown to explain not only learning but also acquiring knowledge from new theories, articles might be more useful for conceptualizing fake news sharing. Additionally, we chose factors with the help of previous studies and noticing gaps in prior literature. With regards to data collection, we collected cross-sectional responses from Pakistani social media users during the COVID-19 pandemic while the author also ensured the reliability and validity of our data, cultural and contextual specifically may be introduced in the outcome. Due to collecting data among adults our results may contain some bias. Furthermore, our sample consisted of younger adults and calculated through ROA soft, while in future other researchers may use different sample size technique such as rule of 5 and 10.

10 Recommendations

This research proposed sharing of fake news, cyberbullying and self-regulation on social media fatigue during COVID-19. Based on the findings of the study following are recommendations:

- Results demonstrates that there are several purposes of social media have an impact on sharing fake news, cyberbullying particularly people driven by entertainment, for example do not seem to be equally concerned about the information reliability they share on social media.
- We encourage future research into different purpose of social media (entertainment, sharing information and self-promotion) and their impact on the sharing misinformation.
- Future study needs to characterize the factors associated with the mechanisms through which emotional connection to social media and health-related outcomes are associated.

Appendix A

See Table A.1.

References

Abaido, G. M. (2019). Cyberbullying on social media platforms among university students in the United Arab Emirates. *United Arab Emirates: International Journal of Adolescence and Youth, 25*(1), 407–420.

Zayyad, A., Obeidat, H. M., Alshurideh, Z. M., Abuhashesh, M. T., Maqableh, M., & Masa'deh, R. E. (2021). Corporate social responsibility and patronage intentions: the mediating effect of brand credibility. *Journal of Marketing Communications*, 1–24.

Abuhashesh, M. Y., Alshurideh, M. T., & Sumadi, M. (2021). The effect of culture on customers' attitudes toward Facebook advertising: the moderating role of gender. *Review of International Business and Strategy, 31*(3), 416–437.

Table A.1 Constructs and items in questionnaire

Sharing of fake News
Based on your experience which social media platforms offers a more fertile environment for the spread of a fake story
I share fake news because I don't have time to check facts through trusted sources
I share information or news on COVID-19 without checking facts through trusted sources
I share information or news on COVID-19 even if sometimes I feel the information may not be correct
Cyberbullying
Cyberbullying is normal in the world of social media
I would like to witness more kindness and respect on social media
Cyberbullying is a crime like any other crime
If someone is being cyberbullied, it is important to inform an adult
Self-regulation
My behavior is not that different from other people's
I have so many plans that it's hard for me to focus on any one of them
I don't seem to learn from my mistakes
I enjoy a routine, and like things to stay the same
Social Media fatigue
I find it difficult to relax after continually using social media
After a session of using social media, I feel really fatigued
After using social media, it takes effort to concentrate in my spare time
After using social media, it takes effort to concentrate in my spare time

Adhikari, K., & Panda, R. K. (2020). Examining the role of social networking fatigue toward discontinuance intention: The multigroup effects of gender and age. *Journal of Internet Commerce, 19*(2), 125–152.

Akour, I., Alshurideh, M., Al Kurdi, B., Al Ali, A., & Salloum, S. (2021). Using machine learning algorithms to predict people's intention to use mobile learning platforms during the COVID-19 pandemic: Machine learning approach. *JMIR Medical Education, 7*(1), 1–17.

Al Kurdi, B. H., & Alshurideh, M. T. (2021). Facebook advertising as a marketing tool: Examining the influence on female cosmetic purchasing behaviour. *International Journal of Online Marketing (IJOM), 11*(2), 52–74.

Al-Dmour, H., Alshuraideh, M., & Salehih, S. (2014). A study of Jordanians' television viewers habits. *Life Science Journal, 11*(6), 161–171

Al-Dmour, A., Al-Dmour, H., Al-Barghuthi, R., Al-Dmour, R., & Alshurideh, M. T. (2021a). Factors influencing the adoption of E-payment during pandemic outbreak (COVID-19): Empirical evidence. *The Effect of Coronavirus Disease (COVID-19) on Business Intelligence, 334*, 133–154.

Al-Dmour, R., AlShaar, F., Al-Dmour, H., Masa'deh, R., & Alshurideh, M. T. (2021b). The effect of service recovery justices strategies on online customer engagement via the role of "Customer Satisfaction" during the Covid-19 pandemic: An empirical study. *The Effect of Coronavirus Disease (COVID-19) on Business Intelligence, 334*, 325–346.

Aljumah, A., Nuseir, M. T., & Alshurideh, M. T. (2021a). The Impact of Social Media Marketing Communications on Consumer Response During the COVID-19: Does the Brand Equity of a University Matter. *The Effect of Coronavirus Disease (COVID-19) on Business Intelligence*, 367–384.

Aljumah, A. I., Nuseir, M. T., & Alam, M. M. (2021b). Organizational performance and capabilities to analyze big data: do the ambidexterity and business value of big data analytics matter? *Business Process Management Journal, 27*(4), 1088–1107. https://doi.org/10.1108/BPMJ-07-2020-0335.

Aljumah, A. I., Nuseir, M. T., & Alam, M. M. (2021c). Traditional marketing analytics, big data analytics and big data system quality and the success of new product development. *Business Process Management Journal, 27*(4), 1108–1125. https://doi.org/10.1108/BPMJ-11-2020-0527.

Al-Khayyal, A., Alshurideh, M., Al Kurdi, B., & Salloum, S. A. (2021). Factors influencing electronic service quality on electronic loyalty in online shopping context: data analysis approach. In *Enabling AI applications in data science* (pp. 367–378). Springer.

Allcott, H., & Gentzkow. (2017). Social media and fake news in the 2016 election. *New York: Journal of Economic Perspectives., 31*(2), 211–236.

Al-Maroof, R., Ayoubi, K., Alhumaid, K., Aburayya, A., Alshurideh, M., Alfaisal, R., & Salloum, S. (2021). The acceptance of social media video for knowledge acquisition, sharing and application: A comparative study among YouYube users and TikTok users' for medical purposes. *International Journal of Data and Network Science, 5*(3), 197–214.

Almazrouei, F. A., Alshurideh, M., Al Kurdi, B., & Salloum, S. A. (2020). Social media impact on business: a systematic review. In *International conference on advanced intelligent systems and informatics* (pp. 697–707). Springer.

Alshurideh, M., Al Kurdi, B. H., Vij, A., Obiedat, Z., & Naser, A. (2016). Marketing ethics and relationship marketing-An empirical study that measure the effect of ethics practices application on maintaining relationships with customers. *International Business Research, 9*(9), 78–90.

Alshurideh, M., Salloum, S. A., Al Kurdi, B., & Al-Emran, M. (2019a). Factors affecting the social networks acceptance: an empirical study using PLS-SEM approach. In *Proceedings of the 2019a 8th International Conference on Software and Computer Applications* (pp. 414–418).

Alshurideh, M., Salloum, S. A., Al Kurdi, B., Monem, A. A., & Shaalan, K. (2019b). Understanding the quality determinants that influence the intention to use the mobile learning platforms: A practical study. *International Journal of Interactive Mobile Technologies, 13*(11), 183–157.

Alshurideh, M. T., Kurdi, B. A., AlHamad, A. Q., Salloum, S. A., Alkurdi, S., Dehghan, A., & Masa'deh, R. E. (2021). Factors affecting the use of smart mobile examination platforms by universities' postgraduate students during the COVID 19 pandemic: an empirical study. In *Informatics* (Vol. 8, No. 2, p. 32). Multidisciplinary Digital Publishing Institute.

Barret & Morgan. (2015). *College of management sciences*. Market Forces.

Bekalu, M. A., McCloud, R. F., & Viswanath, K. (2019). Association of social media use with social well-being, positive mental health, and self-rated health: disentangling routine use from emotional connection to use. *Health Education and Behavior, 46*(2_suppl), 69S–80S.

Cao, X. (2018). The stimulators of social media fatigue among students: role of moral disengagement. *China: Journal of Educational Computing Research. Journal of Educational Computing Research, 57*(5), 1083–1107.

Chiou, L., & Tucker, C. (2018). *Fake news and advertising on social media: A study of the anti-vaccination movement (No. w25223).* (National Bureau of Economic Research. Working paper, 1–35). Available at: https://www.nber.org/system/files/working_papers/w25223/w25223.pdf

Chiou, L. T. (2018). *Fake news and advertising on social media: A study of the anti-vaccination movement*. USA.

Chong, S. T. (2005). Recent advances in cognitive load theory research: Implications for instructional designers. *Malaysia : Malaysian Online Journal of Instructional Technology (MOJIT)., 2*(3), 106–117.

Dhir, A., Chen S. (2018). Online social media fatigue and psychological wellbeing—A study of compulsive use, fear of missing out, fatigue, anxiety and depression. *Vietnam: International Journal of Information Management, 24*(June), 141–152.

Elsaadani, M. A. (2020b). Investigating the effect of E-management on customer service performance. *AAU Journal of Business and Law., 3*(2), 1–24.

Elsaadani, M. A. (2020a). Investigating the effect of E-management on customer service performance. *AAU Journal of Business and Law, 3*(2), 1–24.

Hamadneh, S., Hassan, J., Alshurideh, M., Al Kurdi, B., & Aburayya, A. (2021). The effect of brand personality on consumer self-identity: The moderation effect of cultural orientations among British and Chinese consumers. *Journal of Legal, Ethical and Regulatory, 24*, 1–14.

Islam, A. N., Laato, S., Talukder, S., & Sutinen, E. (2020). Misinformation sharing and social media fatigue during COVID-19: An affordance and cognitive load perspective. *Technological Forecasting and Social Change, 159*, 120201.

Karmakar, S. D. (2020). *Evaluating the impact of COVID-19 on cyberbullying through bayesian trend analysis*. Florida.

Keles, B., McCrae, N., & Grealish, A. (2020). A systematic review: The influence of social media on depression, anxiety and psychological distress in adolescents. *International Journal of Adolescence and Youth, 25*(1), 79–93.

Khan, N. A., Khan, A. N., & Moin, M. F. (2021). Self-regulation and social media addiction: A multi-wave data analysis in China. *Technology in Society, 64*, 101527.

Khasawneh, M. A., Abuhashesh, M., Ahmad, A., Masa'deh, R., & Alshurideh, M. T. (2021a). Customers online engagement with social media influencers' content related to COVID 19. In *The effect of coronavirus disease (COVID-19) on business intelligence* (pp. 385–404). Springer.

Khasawneh, M. A., Abuhashesh, M., Ahmad, A., Alshurideh, M. T., & Masa'deh, R. (2021b). Determinants of e-word of mouth on social media during COVID-19 outbreaks: An empirical study. In *The effect of coronavirus disease (COVID-19) on business intelligence* (pp. 347–366). Springer.

Kurdi, B. A., Alshurideh, M., Nuseir, M., Aburayya, A., & Salloum, S. A. (2021). The effects of subjective norm on the intention to use social media networks: an exploratory study using PLS-SEM and machine learning approach. In *International Conference on advanced machine learning technologies and applications* (pp. 581–592). Springer.

Laato, S., Islam, A. N., & Laine, T. H. (2020). Did location-based games motivate players to socialize during COVID-19? *Telematics and Informatics, 54*, 101458.

Leo, S., Alsharari, N. M., Abbas, J., & Alshurideh, M. T. (2021). From offline to online learning: A qualitative study of challenges and opportunities as a response to the COVID-19 pandemic in the UAE higher education context. *The Effect of Coronavirus Disease (COVID-19) on Business Intelligence, 334*, 203–217.

Liu, H., Liu, W., Yoganathan, V., & Osburg, V. S. (2021). COVID-19 information overload and generation Z's social media discontinuance intention during the pandemic lockdown. *Technological Forecasting and Social Change, 166*, 120600.

Matzat, U., & Vrieling, E. M. (2016). Self-regulated learning and social media–a 'natural alliance'? Evidence on students' self-regulation of learning, social media use, and student–teacher relationship. *Learning, Media and Technology, 41*(1), 73–99.

Meinert, J., Mirbabaie, M., Dungs, S., & Aker, A. (2018). Is it really fake?–towards an understanding of fake news in social media communication. In International conference on social computing and social media (pp. 484–497). Springer.

Nuseir, M. T., Aljumah, A. (2021). Digital marketing adoption influenced by relative advantage and competitive industry: A UAE tourism case study. *International Journal of Innovation, Creativity and Change, 11*(2), 617–631.

Nuseir, M.T., El-Refae, G.A., Aljumah, A.(2021). The e-learning of students and university's brand image (Post COVID-19): How successfully Al-Ain university have embraced the paradigm shift in digital learning. *Studies in Systems, Decision and Control, 334*, 171–187.

Obeidat, B., Sweis, R., Zyod, D., & Alshurideh, M. (2012). The effect of perceived service quality on customer loyalty in internet service providers in Jordan. *Journal of Management Research, 4*(4), 224–242.

Obeidat, R., Alshurideh, Z., Al Dweeri, M., & Masa'deh, R. (2019). The influence of online revenge acts on consumers psychological and emotional states: does revenge taste sweet? In *33 IBIMA Conference Proceedings-Granada* (pp. 10–11). Spain.

Ranchordás, S., & Goanta, C. (2019). The Regulation of social media influencers: An introduction. *Netherland*. https://doi.org/10.4337/9781788978286

Salo, J., Singh, D. (2020). Sharing of fake news on social media: Application of the honeycomb framework and the third-person effect hypothesis. *Norway: Journal of Retailing and Consumer Services. 57*, 102197.

Sultan, R. A., Alqallaf, A. K., Alzarooni, S. A., Alrahma, N. H., AlAli, M. A., & Alshurideh, M. T. (2021). How students influence faculty satisfaction with online courses and do the age of faculty matter. In *The international conference on artificial intelligence and computer vision* (pp. 823–837). Springer.

Sweiss, N., Obeidat, Z. M., Al-Dweeri, R. M., Mohammad Khalaf Ahmad, A., M. Obeidat, A., & Alshurideh, M. (2021). The moderating role of perceived company effort in mitigating customer misconduct within Online Brand Communities (OBC). *Journal of Marketing Communications*, 1–24.

Sweller, J. (2010). Element interactivity and intrinsic, extraneous, and germane cognitive load. *Educational Psychology Review, 22*(2), 123–138.

Vartak, S., Vaydande, A., Varule, J. (2000) Prediction of cyberbullying incident on social media network. *India: International Research Journal of Engineering and Technology (IRJET), 7*(4), 3334–3337.

Yoo, S., Zhang, Y. (2020). Elaboration of social media performance measures: From the perspective of social media discontinuance behavior. *China: Sustainabilitym, 12*, 7962. https://doi.org/10.3390/su12197962.

Zabedah, S., & Kosmin, R. (2012). Regulating content in broadcasting, media and the Internet: A case study on public understanding of their role on self regulation. *International Journal of Humanities and Social Sciences, 2*(23), 140–150.

The Impact of Social Media Usage on Companies' Customer Relationship Management (CRM)

Abdallah AlShawabkeh, Mohammed T. Nuseir, and Sarah Urabi

Abstract This study examines the possibility of enhancing Customer Relationship Management through Social Media. As Social Media is a trending topic in companies' strategies, the aim of this research is to find a possible link between the use of Social Media, and its impact on Fashion retails' Customer Relationship Management in both UK and Germany. The use of Social Media platforms was explored by using a semi-structured survey. Significant differences were found in the usage of social media by age group and sex. The results revealed that using Social Media in marketing strategies for fashion retailors would be most effective when targeting younger demographics, offering companies the benefit of longer customer lifetime value when enhancing their Customer Relationship Management through Social Media.

Keywords Social media · Social media networking platforms · Social media marketing · Customer relationship management · CRM tools

1 Introduction

For the recent times, it is an undeniable fact that the evolution of social media has revolutionized the way customer relationship is managed (Malthouse et al., 2013; Almazrouei et al., 2020; Kurdi et al., 2021). Social media has changed the

---

A. AlShawabkeh · M. T. Nuseir (✉)
Department of Business Administration, College of Business, Al Ain University Abu Dhabi Campus, Abu Dhabi, UAE
e-mail: mohammed.nuseir@aau.ac.ae

A. AlShawabkeh
e-mail: Abdallah.alshawabkeh@aau.ac.ae

S. Urabi
Abu Dhabi Vocational Education & Training Institute (ADVETI), Abu Dhabi, UAE
e-mail: sarah.urabi@adveti.ac.ae

M. Alshurideh et al. (eds.), *The Effect of Information Technology on Business and Marketing Intelligence Systems*, Studies in Computational Intelligence 1056,
https://doi.org/10.1007/978-3-031-12382-5_8

way by which individuals communicate, interact, create content and share information including the effective and competent customer relationship management (Khasawneh et al., 2021a, b). The excessive and increasing usage of social media has affected every walk of life including the effective maintenance of the customer relationship management (Aljumah et al., 2021; Cretu & Brodie, 2007).

The use of technology and specifically the use of social media applications have made a significant change in the field of business (Al-Maroof et al., 2021; Alshurideh et al., 2019a). Firms and organizations of the current era have transformed and reallocated their budgets from traditional media of marketing and advertising to online platforms (Al Dmour et al., 2014; Brekke, 2009). Research has shown that the use of social media has much positive effect on the customer relationship management and the effective utilization of social media services can help organizations to achieve priority and preference of consumers over the offerings of their competitors (Alemán & Wartman, 2009; Palmer & Koenig-Lewis, 2009; Rodriguez, et al., 2015; Swarts, 2014).

In this regard, the most prominent social media applications include Facebook, YouTube and Twitter (Abuhashesh et al., 2021; Al Kurdi & Alshurideh, 2021). The customers can receive regular updates about the upcoming products of their preferred companies which explain how the use of these platforms play a leading role in helping the companies retain their current consumers as well as to attract new consumes across different walks of life (Soares et al., 2012; Alshurideh et al., 2019b, 2021). Moreover, the use of social media services helps in generating a positive word of mouth among the customers which additionally helps the organization to maintain a good reputation regarding their products and services (Abu Zayyad et al., 2021; Alshurideh, 2022; Buss and Begorgis, 2015).

Although many of the researched articles have indicated that there is a connection between Social Media and the positive outcomes of the implemented marketing strategies (Alshurideh et al., 2016; Katona et al., 2011), this research will investigate this apparent connection further. Moreover, the possibility of this connection positively influencing the companies' acquisition of new customers and the maintenance of the existing customers has also been examined (Alkitbi et al., 2020; Alsharari & Alshurideh, 2020; Alshamsi et al., 2020). Through finding a possible connection between these different factors, the research problem of 'why companies should be using Social Media to enhance their Customer Relationship Management' has been examined.

Therefore, the aim of this research is to detect if the use of social media can positively influence the companies' Customer Relationship Management (CRM). As there are still companies refusing to exploit new media technologies within their strategies, this research aims at looking at social media from a different perspective to highlight their importance on the marketing outcomes of the company. Therefore, the research question that this study aims to answer is; can companies enhance their CRM by using Social Media?

In order to reach the stated aim and to answer the mentioned research question, the objectives are as follows:

1. Analysing the medium of social media as a means of communication, by assessing literature, journals and articles dated no earlier than 2000, concentrating more intently on current literature.
2. Design and evaluate a questionnaire based on the framework laid out in this research. The questionnaire will be issued to people ranging from the age of 17 to 40+. The survey collected from this target group will offer a wide range of data to investigate the connection of social media to the marketing outcomes of the company as well as the possible relations of social media to specific demographics and therefore the possible effects that this may have on the companies' CRM.
3. Compare the findings from the current literature with the survey results and gain insight into the possible links between the two main factors, that is: social media and Customer Relationship Management (CRM).
4. Draw conclusions as to how CRM may be enhanced through the companies' usage of social media.

2 Literature Review

The concept of social media can be understood by assessing a statement by Sterne (1955, p. 51): "Social media is the connective tissue of social networks". The rise of social media as a new communication platform has enabled customers to become active participants within markets rather than passive recipients of marketing campaigns (Heckadon, 2010; Hennig-Thurau et al., 2010; Kleinberg, 2008). This has led to companies having to rethink their marketing campaigns, as the traditional marketing models have become more complicated due to the on-going changes within communication channels (Alshurideh et al., 2014; Füller et al., 2007).

Existing research by Heckadon (2010) has investigated critical success factors when implementing marketing strategies on social media platforms. In his Master Thesis, he defines ten critical success factors, namely "1. Having a social media marketing strategy, 2. Integrating the social media marketing strategy into the broader marketing strategy, 3. Social media optimization, 4. Creating a community, 5. Encouraging user content creation and feedback, 6. Being open and honest, 7. Keeping content fresh and relevant, 8. Making the user feel special, 9. Identifying with a cause, 10. Measuring social media marketing performance." (Heckadon, 2010, pp. 12–13).

Moving on, with regards to the communication strategies of companies, with the goals of increasing sales, recruiting new customers or even staff, strengthening the company's brand and image or improving Customer Relationship Management (CRM), communication between companies and customers has become one of the most important processes a company needs to concentrate on (Alshurideh et al., 2017; Alyammahi et al., 2020; Wallace et al., 2014).

Means of communication present a wide range of possibilities for a company (AlHamad et al., 2022; Alshurideh et al., 2022; Ammari et al., 2017; Ashal et al., 2021). Communication, be it from business to business (B2B) or from business to consumers (B2C), can take place in the form of public relations, fairs or exhibitions, direct sales, classic advertisements, sales promotion, sponsoring, event-marketing or multimedia communication (Alzoubi et al., 2020; Felix, et al., 2017; Wang & Kim, 2017). As the more traditional communication channels are beginning to lose their effectiveness due to the rise of Social Media (Hennig-Thurau et al., 2010), it is important to understand why customers are so attracted to this new communication platform.

Having understood that social media can play an important part in interacting with customers and assessing new ideas (Hennig-Thurau et al., 2010; Maklan et al., 2008), the question at hand is, whether this can help companies enhancing their Customer Relationship Management? Hennig-Thurau et al. (2010) introduced the idea of a 'Pinball-Framework' to highlight the impact of these new social media platforms on customer relationships. They state, that companies using new media channels are merely playing Pinball. By throwing the 'marketing ball' into the web, companies supply customers with a message, however in what way this message is received or passed along is no longer in the company's hands, as even though they have 'flippers' to guide the message, the ball may not move in the anticipated direction, which could lead to catastrophic results. So does the possibility exist of enhancing Customer Relationship Management by using social media, or would it rather have the opposite effect?

When reviewing the literature, it seems that there is definite potential for social media to enhance companies' Customer Relationship Management (Heckadon, 2010; Hennig-Thurau et al., 2010; Katona et al., 2011). And so the following research hypothesis has been formulated:

H1 Social media has a positive impact on companies' Customer Relationship Management.

3 Method

To assess the developed research hypotheses, a quantitative approach were adopted that made use of a self-administered questionnaire. This questionnaire was distributed in 2016 to people from ages 17–40+ living in the United Kingdom or Germany. This target group was selected for this study to enable consideration of the possible influences of age, occupation status and culture on the survey results. The questionnaire had a very good response rate which is mainly because of the close contact to University lecturers and reliable contacts throughout the communities.

For this study, a total of 325 questionnaires were distributed. From these completed and returned questionnaires, 57 questionnaires were excluded from this study, because of the missing data in them. Therefore, the residual 268 responses denote an effective response rate of 82.46%, which is a fairly high rate, ensuring that the collected data is valid and the depicted results justified. The respondents were fairly equally divided in terms of gender, with 53.4% being female and 46.6% being male. In addition to this, 40.3% of all respondents were over 30 years of age, while 59.7%

were under 30 years old, which enabled the author to further distinguish social media usage in correlation to age.

3.1 Measures

The developed scales for this study were based on a 5—point rating scale (changed from four-point rating scale to five-point rating scale after the pilot-study), using the values (1) non-applicable, (2) strongly disagree, (3) disagree, (4) agree and (5) strongly agree. A pilot study was undertaken to test the understandability of the questionnaire and was handed out to ensure that nothing had been overlooked in the design of the questionnaire. After having conducted the pilot study, the following changes were made to the questionnaire, based on the feedback from the respondents:

- In Sect. 1 of the questionnaire, a 'Foundation Degree', 'Master Degree' and 'PhD' option were added to the existing options of first year, second year or third-year student in question 5
- In addition to the options of 'strongly agree', 'agree', 'disagree' and 'strongly disagree' within Sect. 2 of the questionnaire, a fifth option ('non-applicable') was added in order to allow respondents to state honestly if the stated question didn't apply to them
- The numbering on the questions was changed from discontinuous to continuous numbering in order to make the later analysis easier and give each question a definite identifier

Even though the changes made to the questionnaire were of minor nature, the analysis having more options in breaking down the respondents' demographics, and further decreasing the complexity that came from the discontinuous numbering, especially when cross-referencing the individual questions to the framework and indicators for all factors. In order to conduct the analysis, the IBM Statistical Package for Social Sciences (SPSS) version 20 was used.

4 Results and Discussion

In order to test the impact of social media on the Customer Relationship Management (CRM), the Pearson correlation was calculated (Table 1).

Table 1 Correlation of social media with CRM (Pearson coefficient)

Factor	Measurement	Social media	CRM
Social media	Pearson correlation	1	0.992
	Sig. (2-tailed)		0.000
	N	268	268

Table 2 Means of social media criteria

Factor	Criterion	Mean (in %)
Social media		54
	Information technology management	56
	Communication channel	52
	Collaboration (customer engagement)	54
	Building virtual communities/Networks	56

The results show a significant relationship between Social Media and Customer Relationship Management, which indicates that CRM can be enhanced through Social Media. Based on these insights it is important to determine which criterion of Social Media poses the most important, so that companies could specifically address these criteria in order to enhance their customer performance. Table 2 depicts the results from the conducted survey for each criterion of Social Media.

The survey results showed that the criterion 'Information Technology Management' as well as 'Building virtual communities' scored the highest at 2.8 (56%). Even though this poses only an average outcome, the results indicate that the strongest link between Social Media and CRM lies within the creation of virtual communities through Information Technology Management. This further indicates that companies' use of these criteria would promise the best results for enhancing their Customer Relationship Management.

4.1 *How Important Is the Criterion Information Technology Management for CRM?*

The frequency analysis for the questions linked to Information Technology Management is shown in Table 3. The most significant results are those on item 6, 9, 12, 13 and 16. Due to the choice of using a Fashion Retailers' case study, the questions were asked in accordance to fashion retailer interests. According to Hennig-Thurau et al. (2010), Kleinberg (2008) and Sterne (1955) Social Media offers companies the opportunity to perform market research at low costs, with the further ability to detect trends and attitudinal shifts within the market. However, when asking the survey respondents, if they discussed fashion trends with their friends on Social media platforms (item 6) 42.7% strongly disagreed. Yet 12.7% agreed to the statement and a further 1.9% strongly agreed. These numbers indicate, that fashion retailers would be able to discover trends based on sample groups of Social Media users. In addition to this it has been said in the literature that Social Media offers companies access to customer data (Heckadon, 2010; Katona et al., 2011). The results to item 9, 12

Table 3 Frequency analysis for criterion information technology management

Item no.	Statement	Strongly agree (%)	Agree (%)	Disagree (%)	Strongly disagree (%)	Non-applicable (%)
6	I always discuss fashion trends with my friends on social networking sites	1.9	12.7	30.7	42.7	12
7	I always take part in questionnaires issued by companies even if they are not rewarded	1.1	18.4	36.8	35.7	7.9
8	I always accept friend requests made by companies on social networking sites	1.1	14.2	34.3	40.7	9.7
9	I am friends with more than one company on social networking sites (e.g. Primark)	6.7	21.6	25.7	34.3	11.6
10	I always 'like' companies social networking sites	3.4	22.4	32.5	31.7	10.1
11	I frequently comment on posts made by companies on social networking sites	1.5	5.6	33.8	48.9	10.2
12	On my social networking site profile, I always share personal information (e.g. date of birth, job etc.)	7.5	36.2	26.5	21.3	8.6

(continued)

Table 3 (continued)

Item no.	Statement	Strongly agree (%)	Agree (%)	Disagree (%)	Strongly disagree (%)	Non-applicable (%)
13	I don't mind at all letting companies see my personal profile	1.9	24.0	31.8	35.2	7.1
16	I strongly prefer sending direct messages through social networking sites to friends than posting something on my wall	28.8	39.3	12.7	10.1	9
Average		5.99	21.6	29.42	33.4	9.6
Mean (in %)		56				

and 13 verify this fact, as for each of these questions more than 20% of respondents agreed.

According to Wallace et al. (2014) Social Media users prefer more personalized advertising efforts by companies, which is emphasized by the fact, that the majority of respondents (68.1% in total) strongly preferred sending personal messages to their contacts on these networking platforms than openly displaying their interactions with others.

The comparison of respondent's answers with their age (see Fig. 1) showed a significant difference to the overall outcome for the criterion Information Technology Management. The respondents of under 30 years of age were far more willing to interact with companies as well as display their personal information on Social Media sites. Furthermore the respondents' answers to items 9 and 10 demonstrated that younger Social Media users are very interested in interacting with companies on Social Media sites, which emphasises companies' ability to engage customers as co-creators of value through using Information Technology (Füller et al., 2010).

The results of the survey furthermore depict that the possibilities for companies to use Social Media as a market research system (Kleinberg, 2008) are given, which means Information Technology Management should play an important role in Customer Relationship Management (Alshurideh et al., 2019a, b; Svoboda et al., 2021), as the amount of data available to companies can help improve customer services and company offerings (Aburayya et al., 2020; Al-Khayyal et al., 2021; Odeh et al., 2021).

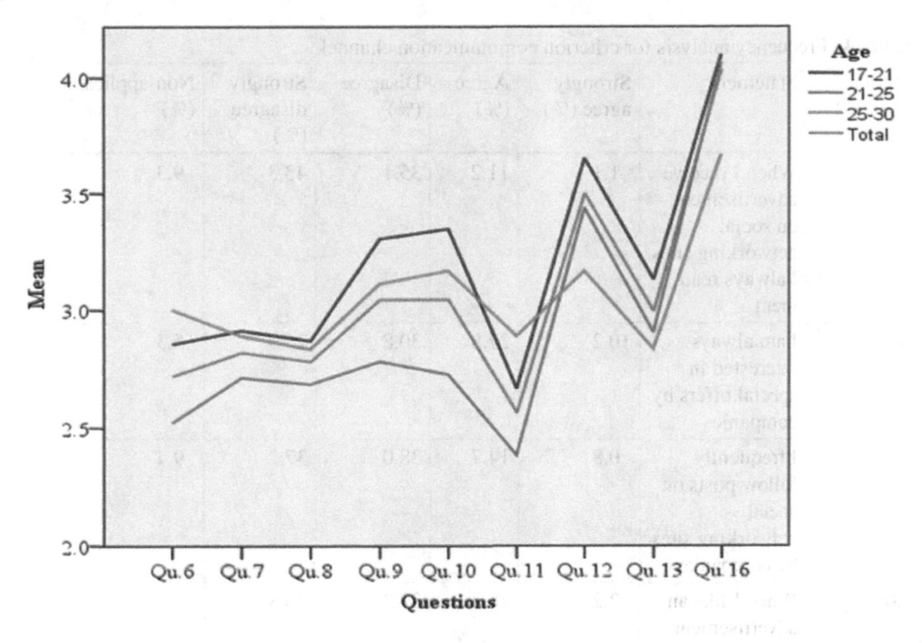

Fig. 1 Comparison of means between age and ITM

4.2 How Important Is the Criterion Communication Channel for CRM?

Companies using Social Media generally have the aim of communicating the companies' offerings and benefits to a broad audience, in order to reach a large crowd through the same mean. Within the survey, respondents were questioned on their behaviour when encountering advertising on social media platforms (Table 4).

This was done in order to determine the significance of Social Media as a communication channel. Interestingly enough, 43.3% of respondents strongly disagreed with reading advertisements on social media platforms. This confirms what Soares et al. (2012) have stated within their research. They say that Social Media users pose a new type of cyberspace consumer, who perceives advertisements on Social Media sites as intrusive and furthermore disruptive to their online activities. According to Sterne (1955) however, Social Media as a platform for communication is more important in terms of customers spreading the advertisements or messages of companies. In regard to this, respondents' answers to items 20 through 24 showed that on average 20%[1] had forwarded advertisements to friends or received advertisements from friends. This means that customers only find advertisements on Social Media sites interesting, when they appeal to their own individual needs in addition to delivering an important message.

[1] (2.2 + 18.7 + 2.6 + 15.4 + 3.4 + 13.8 + 2.2 + 21.7)/5 to evaluate the percentage of respondents that agreed with the statements.

Table 4 Frequency analysis for criterion communication channel

Item no.	Statement	Strongly agree (%)	Agree (%)	Disagree (%)	Strongly disagree (%)	Non-applicable (%)
17	When I receive advertisement on social networking sites I always read them	1.1	11.2	35.1	43.3	9.3
18	I am always interested in special offers by companies	10.2	27.8	30.8	25.9	5.3
19	I frequently follow posts on social networking sites by companies	0.8	14.7	38.0	37.2	9.4
20	When I like an advertisement on social networking sites I always respond (e.g. by 'liking' the comment)	2.2	18.7	29.2	40.8	9.0
22	When I receive a post or message I like I always forward it to my friends or post it on my wall	2.6	15.4	40.4	32.6	9.0
23	When I find an interesting advertisement I always post it on my wall or forward it to my friends on social networking sites	3.4	13.8	29.9	42.5	10.4
24	I always receive posts by my friends when they find interesting advertisements	2.2	21.7	32.6	33.0	10.5

(continued)

Table 4 (continued)

Item no.	Statement	Strongly agree (%)	Agree (%)	Disagree (%)	Strongly disagree (%)	Non-applicable (%)
46	Advertisement on social networking sites always makes me purchase merchandise	1.1	4.5	36.6	47.8	10.1
Average		2.95	15.98	34.08	37.89	9.13
Mean (in %)		52				

Therefore the average outcome of 52% for Social Media as a communication channel indicates that this criterion is less important than the other criteria for Social Media usage in companies' marketing strategies in order to enhance Customer Relationship Management.

4.3 How Important Is the Criterion Collaboration for CRM?

The results for the criterion 'collaboration' showed only an average outcome with an overall mean of 2.7 (see Table 5). This means that the respondents viewed this criterion as less important than 'Information Technology Management' and 'Building virtual communities/networks', answering the research question 'What role does collaboration on Social Media play within CRM?'. When looking at the results, there were significant opinions expressed. The majority of respondents (41.2%) agreed to always commenting back on comments made by their friends on social media platforms, which correlates to Herrmann, et al. (2000) research, where it was stated that companies would be able to use customers' need to share to their advantage.

However, when asking the respondents whether they had used social media platforms to praise or complain to companies 50.6% strongly disagreed to writing complaints and 49.6% strongly disagreed to writing praise. This diminishes companies' ability to improve their services by using customers as co-creators of value, which was stated as critical factor in the literature (Hennig-Thurau et al, 2010; Maklan et al., 2008). In order to distinguish whether this was true for all users of Social Media, a mean comparison was undertaken, to correlate the importance of collaboration with age (see Table 6).

The results show that younger users found collaboration on social media platforms more important (62%) than older users (46%). Even though this outcome is still fairly low, it indicates that companies using Social Media would be able to engage more effectively with younger customers than older customers on these platforms,

Table 5 Frequency analysis for criterion collaboration

Item no.	Statement	Strongly agree (%)	Agree (%)	Disagree (%)	Strongly disagree (%)	Non-applicable (%)
15	I always share everything that happens to me with my friends on social networking sites	3	11.2	38.2	39.7	7.9
25	When a friend posts something on my wall I always comment back	13.9	41.2	23.6	11.6	9.7
26	When I have ideas on how to improve products or services I always contact companies through their social networking sites	1.1	7.1	35.1	46.3	10.4
27	I have written many complaints to companies through social networking sites	1.5	3.7	32.6	50.6	11.6
28	I have written many praise to companies through social networking sites	0.4	6.8	30.8	49.6	12.4
29	I am definitely active in more than one group on social networking sites	11.7	29.7	24.4	25.2	9.0
Average		5.27	16.62	30.78	37.17	10.17
Mean (in %)		54				

Table 6 Mean comparison between age and criterion collaboration

Criterion	Age					
	17–21	21–25	25–30	30–40	>40	Total
Collaboration (Customer engagement) (%)	62	54	60	56	46	54

by integrating them into their strategies and encouraging them to interact with the company.

4.4 How Important Is the Criterion Virtual Communities/networks for CRM?

Based on Heckadon's (2010) research, companies using Social Media should have the main aim of creating online communities in order to benefit from customers' social capital (Alemán & Wartman, 2009; Herrmann, et al., 2000). Therefore respondents were asked questions on their behaviours towards their online Social Media to come to conclusions of companies' ability to benefit from these.

The majority of respondents (35.6%) agreed to have friends from all over the world on their Social Media platforms. This indicates that companies could address large customer bases in order to acquire new customers even on an international basis. Surprisingly only 28% of respondents agreed to appreciating companies responding to inquiries, which indicates that the majority of Social Media users prefers to keep their platforms in a private environment. The high number of respondents disagreeing with question 33 (40.7%) shows that peer pressure, in this case to always have the same interests as friends, was not essential for the respondents' behaviour on these platforms, depicting the differences in social structures on Social Media sites (Ansari et al., 2011; Kleinberg, 2008). Based on this, companies would be able to access data on customers with more diverse interests, increasing the companies' ability to assess a broader range of perceptions (Table 7).

As the only average outcome (56%) was unexpected, the results were analysed separately for the different age groups (see Table 8). The analysis clearly shows that younger respondents' answers were very high. This means that companies targeting this demographic would have access to large and diverse communities when using Social Media.

4.5 Testing of Hypotheses

After having analysed each criterion for the different factors thoroughly, it can be said that there is a definite influence of age on companies' ability to utilize Social Media in order to enhance their Customer Relationship Management. Furthermore, the extraordinary significance of the different factors on each other indicates a positive confirmation of the proposed hypothesis (see Table 9).

The correlation between all factors is more than significant (as the correlation becomes significant at the 0.01 level (2-tailed)). Each factor being significantly higher than 0.01 it is indicated that Customer Relationship Management could assuredly be

Table 7 Frequency analysis for criterion virtual communities

Item no.	Statement	Strongly agree (%)	Agree (%)	Disagree (%)	Strongly disagree (%)	Non-applicable (%)
21	I currently have friends from all over the world on social networking sites	21.0	35.6	20.6	14.6	8.2
30	I have always used social networking sites to find new friends	4.9	15.7	32.8	36.9	9.7
31	I feel very sad when my friends on social networking sites do not comment on my posts	3.4	10.1	36.3	38.6	11.6
33	I always join groups on social networking sites if my friends are in it	1.1	11.9	40.7	37.3	9.0
47	I feel very happy, when companies respond to my enquiries	7.8	28.0	15.7	26.1	22.4
Average		7.64	20.26	29.22	30.7	12.18
Mean (in %)		56				

Table 8 Survey results by age for the criterion virtual communities

Independent variable	Criterion	Corresponding questionnaire questions				
Age	Virtual communities (%)	Item 21 (%)	Item 30 (%)	Item 31 (%)	Item 33 (%)	Item 47 (%)
17–21	64.28	77.2	60.6	58	60.6	65.2
21–25	55.68	68	55.6	50.4	52	52.4
25–30	64.88	78.8	64.4	54.4	62.2	64.4
Total	60.42	73.2	58.8	54.2	56.8	59.2

Table 9 Correlation of different framework factors (Pearson correlation coefficient)

Factor	Measurement	SM	CRM	Acquisition	Maintenance
Social Media	Pearson correlation	1	0.992	0.965	0.968
	Sig. (2-tailed)		0.000	0.000	0.000
	N	268	268	268	268

Table 10 One-way Anova for hypothesis

Factor	Measurement	Sum of squares	df	Mean square	F	Sig
Social media	Between groups	121.830	102	1.194	166.912	0.000
	Within groups	1.181	165	0.007		

enhanced through Social Media. However to fully confirm this, a one-way analysis of variance (one-way ANOVA) has been conducted (see Table 10).

In order to test the hypotheses for this research a further one-way Anova test was run to depict the level of significance of Social Media on CRM (see Table 10).

The results clearly show that there is a strong significance Social Media to Customer Relationship Management. This is due to the significance for Social Media is less than 0.001, which is the set-out level in the analysis of variance. This means that the hypothesis of this research is supported:

H1 Social media has a positive impact on companies' Customer Relationship Management.

5 Discussion

This research analysed, whether the use of Social Media has a positive impact on companies' ability to enhance their Customer Relationship Management. Overall the analysis showed that there is a significant relationship between the two factors and that all four criterion factors of Social Media (information technology management, communication channel, collaboration (customer engagement), and building virtual communities/networks) are also important in this regards. These findings enabled the verification of the proposed hypotheses.

The analysis of the positive impact that Social Media has on Customer Relationship Management was verified through strong correlations between the two factors. Also, the main criteria for Social Media to be effective in companies' strategies were depicted in this research as Information Technology Management, Communication Channel, collaboration and building virtual communities. The analysis showed that the criteria Information Technology Management as well as building virtual communities and collaboration were perceived as most significant to respondents. The results of the research further revealed that the younger generations were the definite target audience when using Social Media in communication strategies, which was in attunement with what was stated in the literature (e.g., Heckadon, 2010; Hennig-Thurau et al., 2010; Katona et al., 2011; Rodriguez et al., 2015; Wallace et al., 2014).

On the whole, these findings therefore highlight that using Social Media in company strategies can effectively enhance company performance, even during difficult times as the recession. In order to investigate possibilities on how best to do so, Sect. 3 of the questionnaire examined the respondents' knowledge on companies already using Social Media and to further gain insights into their interests in companies exploiting it. The literature had offered several activities of companies on Social Media and so the respondents were asked to choose the activities they would like companies to pursue most (see Fig. 2). Respondents were asked this question simply to derive an entry gate into the medium of Social Media in order to enhance Customer Relationship Management effectively.

The results clearly showed that Social Media users were most interested in discounts, special offers and new products, which relates to what has been stated within the literature in terms of increasing customers' awareness towards company offerings using Social Media (Heckadon, 2010; Sterne, 1955). The answers furthermore revealed that the respondents were very interested in receiving information on companies, which they later elaborated on by requesting information on companies charity work, Corporate Social responsibility and policies. If companies were to make this information available to customers on Social Media platforms, they would be able to improve their reputation and therefore enhance the acquisition of new customers and the maintenance of existing customers. The greatest possibility to enhance Customer Relationship Management through Social Media however, was

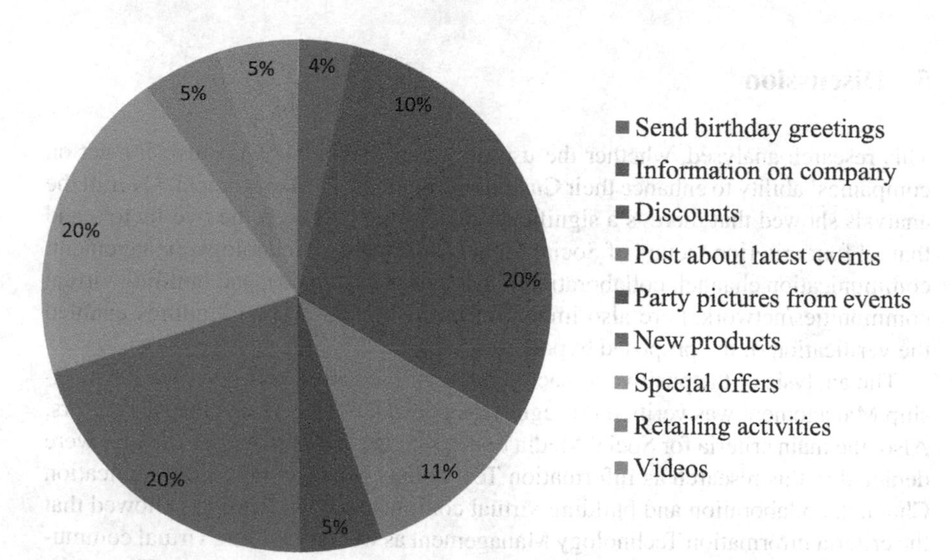

Fig. 2 Activities that social media users are most likely to pursue

through being honest with customers. Survey respondents requested that companies openly display any critique or complaints they encountered by other customers, so that they could make informed decisions. This reflected what Heckadon (2010) had listed as a Critical Success Factor for implementing Social Media Marketing campaigns.

Interestingly enough, 31.12% of respondents were already following companies' Facebook pages which suggest that the relevant companies therefore have access to all of their followers' Social Media platforms, which increases the effective number of people they can address by far. Furthermore 64% of all respondents were able to name companies that were already active on Facebook, which confirms what Sterne (1955) and Heckadon (2010) have stated on increasing customer awareness using Social Media. Even if all the respondents were not following these companies on Facebook, their knowledge of the companies' sites displays the fact that they were reached. This is most likely due to their friends on Facebook following these companies and therefore coming to their attention.

Companies' using Social Media platforms also benefit from the phenomenon of word-of-mouth, which has come to be known online as a part of viral marketing, which is also known as word-of-mouth marketing (Felix, et al., 2017; Wang & Kim, 2017). Social Media enables customers to act pro-actively when engaging with companies (Hennig-Thurau et al., 2010), which gives companies the opportunity to integrate customers in their strategies and use their contributions to create greater value in their offerings. By incorporating Social Media into companies' marketing campaigns, companies are therefore able to incorporate customer insight into their strategies by engaging customers in conversations rather than simply addressing campaigns at them (Hennig-Thurau et al., 2010; Maklan et al., 2008).

However, the fact that Social Media as a communication channel (advertising) was not well received by respondents lies within the fact that most Social Media users feel that advertisements on Social Media site are disruptive to their personal aims when using this medium (Soares et al., 2012). This means that advertisement efforts on Social Media platform need to be constructive and informative rather than overpowering.

6 Conclusions

Social Media has become one of the most talked-about platforms in today's' interconnected world. It offers different opportunities for people to engage with each other and share their thoughts and feelings as well as photos, videos and music. Even though research has shown the profitability of using social media in marketing strategies, some companies are still hesitant to enter this new communication platform. With respect to this fact, the aim of this research was to determine whether social media has a positive influence on companies' Customer Relationship Management (CRM). By doing so, it was hoped to eliminate any remaining doubts of Social Media's potential in companies marketing strategies especially during a time when recession is persistent in the economic climate. By linking their marketing strategies to social media and accordingly to Customer Relationship Management, this research aimed at providing evidence, that social media can effectively enhance Customer Relationship Management and that companies should therefore not dismiss this new medium as part of their strategy.

6.1 Practical Implications

The main aims within Customer Relationship Management are to acquire new and maintain existing customers. Due to the boom in Internet technologies, companies' need to enter into this platform becomes evident. Social Media as a current hot topic in company strategies poses an effective way for companies' to connect with their customers with low investment costs and several opportunities to interact effectively. In such a competitive environment as we live in today, companies seek every opportunity to differentiate themselves from their competitors. The recession that has been presiding worldwide has further led to customers switching brands, which has made Customer Relationship Management an essential tool for companies in order to maintain customers and avoid loss of customer loyalty.

By using Information Technology Management as part of Social Media, companies will be able to develop customer-oriented products and services based on their close interactions with customers and the insight into customer data. The market research that is enabled through Social Media will enable companies to benefit further from lower costs in developing new products and services. In addition to this, it is necessary for companies to continually update their Social Media sites in order to show customers that they care about them and further anticipate customers' to respond even when companies post about challenges they may be facing. This will enable companies to increase their customer loyalty resulting in longer lifetime value and effectively enhance their Customer Relationship Management. Therefore companies need to effectively make use of the fact that Social Media enhances customer relationships and can be used to improve their customer experience.

6.2 Limitations and Suggestions for Future Research

Like most research studies, this study also had some limitations. Because of the limitations of time, data that was collected in 2016 was used for this study. There is a likely chance that the opinions of the people have changed from back then particularly about the marketing approaches of companies on social media platforms since frequent changes are made on these mediums and new features are rapidly added over time. Also, the different techniques of marketing on social media platforms also need to be assessed to gain an insight into the particular kind of techniques that can have the most positive influence on companies' CRM.

Appendix

Appendix 1 Research Questionnaire

Section 1 - Personal Profile

Please tick the appropriate box

1. **What is your sex?**
   o Female o Male

2. **How old are you?**
   o<17 o 17-21 o 21-25 o 25-30 o30-40 o>40

3. **What country do you live in?**
   ____

4. **What is your current occupation (job description/job title)?**
   ____

5. **If you are a student, what year are you in at University?**
   o 1st o 2nd o 3rd o Foundation Degree o Master Degree o PhD

Section 2 - Project-Questions

Below you will find different statements to the different areas that are discussed within my dissertation. Please tick the box, which most expresses your opinion. Note that there are no multiple answers possible unless stated otherwise.

Q (No)	Statement	Strongly agree	Agree	Disagree	Strongly disagree	Non Applicable
6	I always discuss fashion trends with my friends on social networking sites					
7	I always take part in questionnaires issued by companies even if they are not rewarded					
8	I always accept friend requests made by companies on social networking sites					
9	I am friends with more than one company on social networking sites (e.g. Primark)					
10	I always 'like' companies social networking sites					
11	I frequently comment on posts made by companies on social networking sites					

Q (No)	Statement	Strongly agree	Agree	Disagree	Strongly disagree	Non Applicable
12	On my social networking site profile, I always share personal information (e.g. date of birth, job etc.)					
13	I don't mind at all letting companies see my personal profile					
14	I comment frequently on friends posts on social networking sites					
15	I always share everything that happens to me with my friends on social networking sites					
16	I strongly prefer sending direct messages through social networking sites to friends than posting something on my wall					
17	When I receive advertisement on social networking sites I always read them					
18	I am always interested in special offers by companies					
19	I frequently follow posts on social networking sites by companies					
20	When I like an advertisement on social networking sites I always respond (e.g. by 'liking' the comment)					
21	I currently have friends from all over the world on social networking sites					
22	When I receive a post or message I like I always forward it to my friends or post it on my wall					
23	When I find an interesting advertisement I always post it on my wall or forward it to my friends on social networking sites					
24	I always receive posts by my friends when they find interesting advertisements					
25	When a friend posts something on my wall I always comment back					
26	When I have ideas on how to improve products or services I always contact companies through their social networking sites					
27	I have written many complaints to companies through social networking sites					
28	I have written many praise to companies through social networking sites					
29	I am definitely active in more than one group on social networking sites					
30	I have always used social networking sites to find new friends					

Q (No)	Statement	Strongly agree	Agree	Disagree	Strongly disagree	Non Applicable
31	I feel very sad when my friends on social networking sites do not comment on my posts					
32	All of my friends on social networking sites currently have the same friends on social networking sites as me					
33	I always join groups on social networking sites if my friends are in it					
34	I always trust the judgment of my friends on social networking sites concerning purchasing experiences					
35	There are some friends on social networking sites whom I rely on more than others					
36	I frequently communicate with friends via social networking sites					
37	I always accept friend requests from people I don't know					
38	If I receive a friend request from someone who knows one of my friends I always accept					
39	Without my friends on social networking sites I would not know what to do all day					
40	I currently have many friends on social networking sites whom I don't know personally					
41	*All* of my friends on social networking sites are also friends offline					
42	I am always interested in reading about others' status on social networking sites					
43	I am strongly influenced by others experiences posted on social networking sites					
44	My friends frequently purchase merchandise from fashion retailers such as Primark, Peacocks or New Look					
45	I frequently purchase merchandise from certain fashion retailers due to my friends on social networking sites recommending them					
46	Advertisement on social networking sites always makes me purchase merchandise					
47	I feel very happy, when companies respond to my enquiries					

Appendix 2 Criteria and indicators for social media

Criteria/Value	Information technology management	Communication channel	Collaboration (customer engagement)	Building virtual communities/Networks
Customers	(ITM1) Discover trends and shifts in the market; SM as market research system (Sterne, 1955)	(A1) Send message from company to customer; build awareness towards company (Heckadon, 2010; Sterne, 1955)	(C1) Use customers' need to share to company's advantage (e.g., use critique to improve service) (Ang, 2011)	
	(ITM2) Involve customers in the business; Research on customers is conducted in their own interest rather than just being 'researched' (Ang, 2011; Maklan et al., 2008)	(A2) Reaching a large number of customers (Sterne, 1955)	(C2) Interacting with customers to assess new ideas (Füller et al., 2010; Hennig-Thurau et al, 2010; Maklan et al., 2008)	
	(ITM3) Integrate the personal information gained from SM into strategies; SNS as sources for member data, friendship data, communication data etc.; collection & comparison to enhance the customer experience (Ang, 2011; Ansari et al., 2011; Payne & Frow, 2006)		(C3) Interactive collaboration; customer as co-creator of value (Füller et al., 2010; Hennig-Thurau et al, 2010; Maklan et al., 2008)	

(continued)

(continued)

Criteria/Value	Information technology management	Communication channel	Collaboration (customer engagement)	Building virtual communities/Networks
Company	(ITMC1) Path of communication: direct messaging or comment (Ansari et al., 2011; Zsolt et al., 2011)	(AC1) Customer reaction (e.g., passed message along) (Sterne, 1955)	(C4) Online relationships are formed to forge collaborative relationships (Ansari et al., 2011:725)	(CBC1) SM helps to build social network; Increase social capital (Ang, 2011; Martínez Alemán & Lynk Wartman, 2009)

Appendix 3 Criteria and indicators for customer relationship management

Criteria/Value	Social structures	Relationships
Success factors depicted from the company's perspective	(ACQ1) Increase revenue (Sterne, 1955)	(MAI1) Improve customer satisfaction (Sterne, 1955)
		(MAI2) Involve customers in the business (Sterne, 1955)
	(ACQ3) Getting potential customers' attention (Sterne, 1955)	(MAI3) Identify customer needs (Payne & Frow, 2006)
		(MAI4) Reduce costs of gaining new customers by satisfying existing customers (Sterne, 1955)
	(ACQ5) Convince them of the company's benefits (leading them into purchase) (Sterne, 1955)	(MAI5) Response to queries (Sterne, 1955)
	(ACQ6) Interaction with potential customers (Sterne, 1955)	

References

Abuhashesh, M. Y., Alshurideh, M. T., & Sumadi, M. (2021). The effect of culture on customers' attitudes toward Facebook advertising: The moderating role of gender. *Review of International Business and Strategy, 31*(3), 416–437.

Aburayya, A., Alshurideh, M., Al Marzouqi, A., Al Diabat, O., Alfarsi, A., Suson, R., Salloum, S. A., et al. (2020). An empirical examination of the effect of TQM practices on hospital service quality: An assessment study in UAE hospitals. *Systematic Reviews in Pharmacy, 11*(9), 347–362.

Abu Zayyad, H. M., Obeidat, Z. M., Alshurideh, M. T., Abuhashesh, M., Maqableh, M., & Masa'deh, R. E. (2021). Corporate social responsibility and patronage intentions: the mediating effect of brand credibility. *Journal of Marketing Communications*, 1–24.

Al Dmour, H., Alshurideh, M., & Shishan, F. (2014). The influence of mobile application quality and attributes on the continuance intention of mobile shopping. *Life Science Journal, 11*(10), 172–181.

Alemán, M. A. M., & Wartman, K. (2009). *Online social networking on campus: Understanding what matters in student culture*. Routledge: Francis Taylor Group.

AlHamad, A., Alshurideh, B., Alomari, K., Al Kurdi, B., & Alzoubi, H. (2022). The effect of electronic human resources management on organizational health of telecommunications companies in Jordan. *International Journal of Data and Network Science, 6*, 1–10.

Aljumah, A., Nuseir, M. T., & Alshurideh, M. T. (2021). The impact of social media marketing communications on consumer response during the COVID-19: Does the Brand Equity of a University Matter. *The Effect of Coronavirus Disease (COVID-19) on Business Intelligence*, 367.

Al-Khayyal, A., Alshurideh, M., Al Kurdi, B., & Salloum, S. A. (2021). Factors influencing electronic service quality on electronic loyalty in online shopping context: Data analysis approach. In *Enabling AI Applications in Data Science* (pp. 367–378). Cham: Springer.

Alkitbi, S. S., Alshurideh, M., Al Kurdi, B., & Salloum, S. A. (2020). Factors affect customer retention: A systematic review. In *International Conference on Advanced Intelligent Systems and Informatics* (pp. 656–667). Cham: Springer.

Al Kurdi, B. H., & Alshurideh, M. T. (2021). Facebook advertising as a marketing tool: Examining the influence on female cosmetic purchasing behaviour. *International Journal of Online Marketing (IJOM), 11*(2), 52–74.

Al-Maroof, R., Ayoubi, K., Alhumaid, K., Aburayya, A., Alshurideh, M., Alfaisal, R., & Salloum, S. (2021). The acceptance of social media video for knowledge acquisition, sharing and application: A comparative study among YouYube users and TikTok users' for medical purposes. *International Journal of Data and Network Science, 5*(3), 197.

Almazrouei, F. A., Alshurideh, M., Al Kurdi, B., & Salloum, S. A. (2020). Social media impact on business: A systematic review. In *International Conference on Advanced Intelligent Systems and Informatics* (pp. 697–707). Cham: Springer.

Alshamsi, A., Alshurideh, M., Al Kurdi, B., & Salloum, S. A. (2020). The influence of service quality on customer retention: A systematic review in the higher education. In *International Conference on Advanced Intelligent Systems and Informatics* (pp. 404–416). Cham: Springer.

Alsharari, N. M., & Alshurideh, M. T. (2020). Student retention in higher education: the role of creativity, emotional intelligence and learner autonomy. *International Journal of Educational Management*.

Alshurideh, M., Shaltoni, A., & Hijawi, D. (2014). Marketing communications role in shaping consumer awareness of cause-related marketing campaigns. *International Journal of Marketing Studies, 6*(2), 163–168.

Alshurideh, M., Al Kurdi, B. H., Vij, A., Obiedat, Z., & Naser, A. (2016). Marketing ethics and relationship marketing-An empirical study that measure the effect of ethics practices application on maintaining relationships with customers. *International Business Research, 9*(9), 78–90.

Alshurideh, M., Al Kurdi, B., Abu Hussien, A., & Alshaar, H. (2017). Determining the main factors affecting consumers' acceptance of ethical advertising: A review of the Jordanian market. *Journal of Marketing Communications, 23*(5), 513–532.

Alshurideh, M., Alsharari, N. M., & Al Kurdi, B. (2019a). Supply chain integration and customer relationship management in the airline logistics. *Theoretical Economics Letters, 9*(02), 392.

Alshurideh, M., Salloum, S. A., Al Kurdi, B., & Al-Emran, M. (2019b). Factors affecting the social networks acceptance: An empirical study using PLS-SEM approach. In *Proceedings of the 2019b 8th International Conference on Software and Computer Applications* (pp. 414–418).

Alshurideh, M. T., Kurdi, B. A., AlHamad, A. Q., Salloum, S. A., Alkurdi, S., Dehghan, A., Masa'deh, R. E., et al. (2021). Factors affecting the use of smart mobile examination platforms

by universities' postgraduate students during the COVID 19 pandemic: an empirical study. In *Informatics* (Vol. 8, No. 2, p. 32). Multidisciplinary Digital Publishing Institute.
Alshurideh, M. (2022). Does electronic customer relationship management (E-CRM) affect service quality at private hospitals in Jordan? *Uncertain Supply Chain Management, 10*(2), 1–8.
Alshurideh, M. T., Al Kurdi, B., Alzoubi, H. M., Ghazal, T. M., Said, R. A., AlHamad, A. Q., ... & Al-kassem, A. H. (2022). Fuzzy assisted human resource management for supply chain management issues. *Annals of Operations Research,* 1–19.
Alzoubi, H. M., & Inairat, M. (2020). Do perceived service value, quality, price fairness and service recovery shape customer satisfaction and delight? A practical study in the service telecommunication context. *Uncertain Supply Chain Management, 8*(3), 579–588.
Ammari, G., Alkurdi, B., Alshurideh, A., & Alrowwad, A. (2017). Investigating the impact of communication satisfaction on organizational commitment: A practical approach to increase employees' loyalty. *International Journal of Marketing Studies, 9*(2), 113–133.
Ang, L. (2011). Community relationship management and social media. *Journal of Database Marketing & Customer Strategy Management, 18,* 31–38.
Ansari, A., Koenigsberg, O., & Stahl, F. (2011). Modeling multiple relationships in social networks. *Journal of Marketing Research, 48*(4), 713–728.
Ashal, N., Alshurideh, M., Obeidat, B., & Masa'deh, R. (2021) The impact of strategic orientation on organizational performance: Examining the mediating role of learning culture in Jordanian telecommunication companies. *Academy of Strategic Management Journal, 21*(Special Issue 6), 1–29.
Alyammahi, A., Alshurideh, M., Al Kurdi, B., & Salloum, S. A. (2020). The impacts of communication ethics on workplace decision making and productivity. In *International Conference on Advanced Intelligent Systems and Informatics* (pp. 488–500). Cham: Springer.
Brekke, V. (2009). *Getting Started With Social Media Technologies.* [Online] Retrieved Jan 4, 2012, from http://www.slideshare.net/vidarbrekke/getting-started-with-social-media-technologies.
Buss, O., & Begorgis, G. (2015). *The impact of social media as a customer relationship management tool: A B2B perspective.* Karlstad University.
Cretu, A. E., & Brodie, R. J. (2007). The influence of brand image and company reputation where manufacturers market to small firms: A customer value perspective. *Industrial Marketing Management, 36*(2), 230–240.
Felix, R., Rauschnabel, P. A., & Hinsch, C. (2017). Elements of strategic social media marketing: A holistic framework. *Journal of Business Research, 70,* 118–126.
Füller, J., Jawecki, G., & Mühlbacher, H. (2007). Innovation creation by online basketball communities. *Journal of Business Research, 60*(1), 60–71.
Heckadon, D. (2010). *Critical success factors for creating and implementing effective social media marketing campaigns.* [Online] Retrieved Nov 3, 2011 from https://doi.org/10.2139/ssrn.1734586.
Hennig-Thurau, T., Malthouse, E. C., Friege, C., Gensler, S., Lobschat, L., Rangaswamy, A., & Skiera, B. (2010). The impact of new media on customer relationships. *Journal of Service Research, 13*(3), 311–330.
Herrmann, A., Huber, F., & Braunstein, C. (2000). Market-driven product and service design: Bridging the gap between customer needs, quality management, and customer satisfaction. *International Journal of Production Economics, 66*(1), 77–96.
Katona, Z., Zubcsek, P. P., & Sarvary, M. (2011). Network effects and personal influences: The diffusion of an online social network. *Journal of Marketing Research, 48*(3), 425–443.
Khasawneh, M. A., Abuhashesh, M., Ahmad, A., Masa'deh, R., & Alshurideh, M. T. (2021a). Customers online engagement with social media influencers' content related to COVID 19. In *The Effect of Coronavirus Disease (COVID-19) on Business Intelligence* (pp. 385–404). Springer.
Khasawneh, M. A., Abuhashesh, M., Ahmad, A., Alshurideh, M. T., & Masa'deh, R. (2021b). Determinants of e-word of mouth on social media during COVID-19 outbreaks: An empirical study. In *The Effect of Coronavirus Disease (COVID-19) on Business Intelligence* (pp. 347–366). Springer, Cham.

Kleinberg, J. (2008). The convergence of social and technological networks. *Communications of the ACM, 51*(11), 66–72.
Kurdi, B. A., Alshurideh, M., Nuseir, M., Aburayya, A., & Salloum, S. A. (2021). The effects of subjective norm on the intention to use social media networks: An exploratory study using PLS-SEM and machine learning approach. In *International Conference on Advanced Machine Learning Technologies and Applications* (pp. 581–592). Springer, Cham.
Maklan, S., Knox, S., & Ryals, L. (2008). New trends in innovation and customer relationship management: A challenge for market researchers. *International Journal of Market Research, 50*(2), 221–240.
Malthouse, E. C., Haenlein, M., Skiera, B., Wege, E., & Zhang, M. (2013). Managing customer relationships in the social media era: Introducing the social CRM house. *Journal of Interactive Marketing, 27*(4), 270–280.
Martínez Alemán, A., & Lynk Wartman, K. (2009). *Online Social Networking on Campus—Understanding What Matters in Student Culture*. Routledge Taylor & Francis Group.
Odeh, R. B. M., Obeidat, B. Y., Jaradat, M. O., & Alshurideh, M. T. (2021). The transformational leadership role in achieving organizational resilience through adaptive cultures: the case of Dubai service sector. *International Journal of Productivity and Performance Management*.
Palmer, A., & Koenig-Lewis, N. (2009). *Direct Marketing: An International Journal., 3*(3), 162–176.
Payne, A., & Frow, P. (2006). Customer relationship management: From strategy to implementation. *Journal of Marketing Management, 22*, 135–168.
Rodriguez, M., Peterson, R. M., & Ajjan, H. (2015). CRM/social media technology: Impact on customer orientation process and organizational sales performance. In *Ideas in Marketing: Finding the New and Polishing the Old* (pp. 636–638). Springer, Cham.
Soares, A. M., Pinho, J. C., & Nobre, H. (2012). From social to marketing interactions: The role of social networks. *Journal of Transnational Management, 17*(1), 45–62.
Sterne, J. (1955). *Social Media Metrics*. Wiley and Sons Inc.
Svoboda, P., Ghazal, T. M., Afifi, M. A., Kalra, D., Alshurideh, M. T., & Alzoubi, H. M. (2021). Information systems integration to enhance operational customer relationship management in the pharmaceutical industry. In *The International Conference on Artificial Intelligence and Computer Vision* (pp. 553–572). Springer, Cham.
Swarts, K.M. (2014). *Social media as a customer relationship management tool within the building and construction industry*. Doctoral dissertation, University of Tasmania.
Wallace, E., Buil, I., & de Chernatony, L. (2014). Consumer engagement with self-expressive brands: Brand love and WOM outcomes. *Journal of Product and Brand Management, 23*(1), 33–42.
Wang, Z., & Kim, H. G. (2017). Can social media marketing improve customer relationship capabilities and firm performance? Dynamic capability perspective. *Journal of Interactive Marketing, 39*, 15–26.

Regulating Social Media and Its Effects on Digital Marketing: The Case of UAE

Mohammed T. Nuseir, Ghaleb A. El Refae, and Sarah Urabi

Abstract Digital Marketing is the most efficient way to market a product/service in the modern era. The most crucial constituent of digital marketing is social media, and it is used by half of the world. People use it to share content in text, images, videos, and live streams. Major social media platforms are Facebook, Youtube, Instagram, Twitter, and LinkedIn. In recent times, social media platforms put regulations on using their platforms regarding sharing content related to violence-inciting, spreading nudity, sexual harassment, etc. This research has discussed these social media regulations and their effects on digital marketing by performing a questionnaire-based survey by leading digital marketing firms. Results are showing that social media regulations have positive effects on digital marketing.

Keywords Digital marketing · Social media regulation · Violence inciting · Social media marketing

M. T. Nuseir (✉)
Department of Business Administration, College of Business, Al Ain University, Abu Dhabi Campus, Abu Dhabi, UAE
e-mail: mohammed.nuseir@aau.ac.ae

G. A. E. Refae
Department of Business Administration, College of Business, Al Ain University, Al Ain Campus, Al Ain, UAE
e-mail: ghalebelrefae@aau.ac.ae

S. Urabi
Abu Dhabi Vocational Education & Training Institute (ADVETI), Abu Dhabi, UAE
e-mail: sarah.urabi@adveti.ac.ae

M. Alshurideh et al. (eds.), *The Effect of Information Technology on Business and Marketing Intelligence Systems*, Studies in Computational Intelligence 1056,
https://doi.org/10.1007/978-3-031-12382-5_9

1 Introduction

Social Media is an interactive modern-day digital technology that allows the users to create and/or share content like information, videos, pictures, etc. with the public by using different virtual networks and communities such as Twitter, Youtube, Facebook, Instagram, Tiktok, etc. (Khasawneh et al., 2021a, b; Kietzmann et al., 2011; Obar & Wildman, 2015). Twitter is a social media platform where people create and share content in the form of short messages. Youtube is a platform where people share their content in the form of videos. Tiktok and Instagram are used to share short videos, while Facebook is used for all types of content to share with the public, including videos, photos, and text (Storm, 2020). According to a Global Digital Overview report, many people use social media worldwide; there are 4.5 billion people who use the internet, and 3.8 million use social media. It means around 60% population of the planet uses social media (Dwivedi et al., 2021; Kemp, 2020). Several features attract social media, remaining in touch with family members, friends and reading their favorite celebrities and politicians (Aljumah et al., 2021a, b; Kurdi et al., 2021; Smith, 2011).

As more than half of the world uses social media, it attracts business. In the modern era, especially during the pandemic, businesses are going on the online mode (Akour et al., 2021; Al-Dmour et al., 2021; Alshurideh et al., 2021a, b; Leo et al., 2021). People were bringing online sales/purchase features to their businesses because it was the only way to survive (Al-Khayyal et al., 2021; Obeidat et al., 2019; Sweiss et al., 2021). According to UNCTAD, there was an increase of 10% regarding online shopping during pandemic (UNCTAD, 2020). As the business world is moving online, the marketing industry also went online (Mashaqi et al., 2020; Sultan et al., 2021). Social media platforms use social media platforms to market a product is termed social media marketing (Al-Maroof et al., 2021; Nuseir et al., 2021; Nuseir & Aljumah, 2021; Tuten, 2020). Social media marketing is performed by sharing content, videos, and/or photos with the public; all the marketing is performed using digital technology known as Digital Marketing (Aljumah et al., 2021c; Almazrouei et al., 2020; Alshurideh et al., 2022; Lee et al., 2022a, b; Tariq et al., 2022). Digital marketing is a broad term related to marketing on the internet, either on social media or outside social media like google, games-ads, etc. (Almaazmi et al., 2020; Chaffey & Ellis-Chadwick, 2019; Aissani & ElRefae, 2021). Digital marketing is a great way to market a product, 4.5 billion people using the internet is the market where you can promote your product. Most of the public uses social media; the main constituent of digital marketing is social media marketing, and it is such a powerful technique to promote a business by reaching the customers via social media (Ahmed et al., 2020, 2021; Alshurideh, 2022; Anjum et al., 2012). The marketing objects of different companies' products and services have been obtained by digital marketing at relatively low rates to conventional marketing (Ajina, 2019). This can provide great success to one's business by raising brand awareness, increasing traffic on the website, developing the brand's identity, the positive association of brand, increasing the leads, and improving communication with customers by direct contact on social media. Conclusively, digital media is mandatory to develop a business in this modern era (Abu Zayyad et al., 2021; Alshurideh et al., 2015, 2019a, b; Hamadneh et al., 2021; Morris, 2009).

The topmost famous and worthy social media platforms are Facebook, Twitter, Instagram, Youtube, and Linkedin. According to stats, more than two billion users, men, and women, between having ages of 18–65 years on Facebook; here, the users can share images, videos, text notes, live streams, develop a community, and initiate a business page. Twitter has the reach of 330 million users of age range 18–49, and users can write a concise, engaging, and informative short message having 280 characters. Instagram has reached more than one billion people ages 18–64; here, users mostly come with photos, short videos, and live streams. It is the best platform for marketing the young urban area public. Youtube has two billion-plus users within the age of 18–65 years, and here users can share videos of all kinds. At last, there is LinkedIn, a platform with 303 million users of ages 25–64 years. It makes the user able to share professional content (WebFX, 2021).

One of the main reasons for social media's success as compared to mainstream media was uncensored content (Abuhasheshv et al., 2021; Al Kurdi & Alshurideh, 2021). People use it with the freedom to share, discuss everything they want, which is not possible on the mainstream media. On the other hand, it is noted that social media is used for cyber-bullying, gang-related crimes, and harassment. Besides, it is noted that social media has been a reason for self-harm, mainly cyber-suicide (Cash et al., 2013; Hinduja & Patchin, 2010; Ruder et al., 2011). The bullying performed using the computer and internet-related communication platforms are termed cyberbullying (Menesini et al., 2012; Smith et al., 2008).

Furthermore, humans were facing purposeful misinformation and dehumanization of a particular group and false allegations; but in the era of social media, the process of purposeful misinformation has been accelerated. Violent racists have gained access to a public platform to incite hate and violence even across the borders. On social media, the extremist views and conflict narratives are spread very quickly and grab public attention (Miro-Llinares & Rodriguez-Sala, 2016; Mutahi & Kimari, 2017).

Technologies related to digital and social media are widely implemented to awareness creating of political promotions (Grover et al., 2019; Hossain et al., 2020; Kapoor & Dwivedi, 2015; Shareef et al., 2018). Politicians use social media to campaign against the opposite party; social media was used for political misperception. It was used to blame the opposite parties, spread fake news, and reduce a person or party (Elsaadani, 2020; Garrett, 2019). We have an example of Myanmar, where outright incitement and dehumanizing language were amplified to mass murder through Twitter and Facebook, mainly targeting the Rohingya Muslim minority (Lee, 2019). Facebook released a report related to the Rohingya Muslim killing and accept that "Facebook has become a means of those seeking to spread hate and cause hate." There is another political violence incitement example via social media; in Syria, people popularized a hashtag on Twitter, #SyriaHoax, where they discredit the chemical attack evidence on civilians like Douma and Khan Dheikhoun (MRGI, 2019). On the other hand, the killing of 40 Indian soldiers in Kashmir was stoked by social media between Pakistan and India (Gettleman & Yasir, 2019).

After an intensive misuse of social media for Violence incitement, Sexual Harassment, Mass Killing, Misperception, etc., the social media platforms decided to bring "Social Media" regulations against those who misuse these platforms. In this connection, all the social media platforms developed their regulations and refined the content based on these regulations. As we talk about Facebook, they developed regulations against violence incitement, either political or public; they do not allow any individual or organization on their platform to perform or originate terroristic activities, violence, mass murder, etc. On Facebook, the sale of non-medical drugs and ammunition is a ban. Further, they do not allow any content under the boundaries of suicide, sexual exploitation of children and adults, nudity, harassment, bullying, and privacy violation (Facebook, 2022).

On the other hand, Twitter also developed a list of regulations; no one can threaten any other individual or group of people, no one can incite violence. They cannot be the promotion of terrorist activities. Harassment and child sexual exploitation has no room on Twitter. Self-harm, suicide, and hate based on race and religion can't be promoted. Illegal goods, including drugs and ammunition, cannot be discussed, purchased, and sell on Twitter (Twitter, n.d.). Similarly, Youtube, Instagram, and LinkedIn have regulations to share and spread the platforms' content (Instagram, 2013; LinkedIn, 2020; YouTube, n.d.).

Recently, a political complication has noticed during the presidential election in the United States. Where the thousands of Ex-President Trump's followers breached and stormed Capitol Hill. These followers were incited for violence by Ex-President Trump through his social media accounts, resulting in four civilians and one police officer. Facebook and Twitter get into action and perform a detailed analysis of Trump's social media posts and tweets and found Ex-President Trump guilty for violating the regulations of "Violence Incitement" Facebook and Twitter blocked his social media accounts. Facebook blocked permanently, and Twitter blocked his account temporarily with the warning. In reply, Ex-President Trump smashed Twitter with the allegations of keeping him away from the right of Freedom of Speech (BBC, 2021), as shown in Fig. 1.

This has become a red-hot debatable topic; either the regulations of Facebook and Twitter are against the freedom of speech, and their act of terminating Ex-President trump's social media accounts is an act of keeping Trump from his right of freedom of speech. Different journalists and researchers address different perceptions. Some are saying the banning of Trump's accounts is not against the freedom of speech. Trump violated private social media platforms' regulations, and they have taken actions against him as per their regulations (Romano, 2021). Some say the termination of Trump's account is double standards of social media platforms (Mehmood, 2021). However, German Chancellor 'AngelMerkel' said that Donal Trump's Twitter account's termination is problematic. In addition to this, Navalny, a Kremlin critic, said Trump's ban is against freedom of speech; this precedent will be exploited by the enemies of freedom of speech around the world, in Russia as well. Every time when they need to silence someone, they will say: this is just standard practice; even Trump got blocked on Twitter (Campbell, 2021). So, there are diversified perceptions regarding the ban of Trump's social media account and social media regulations. In

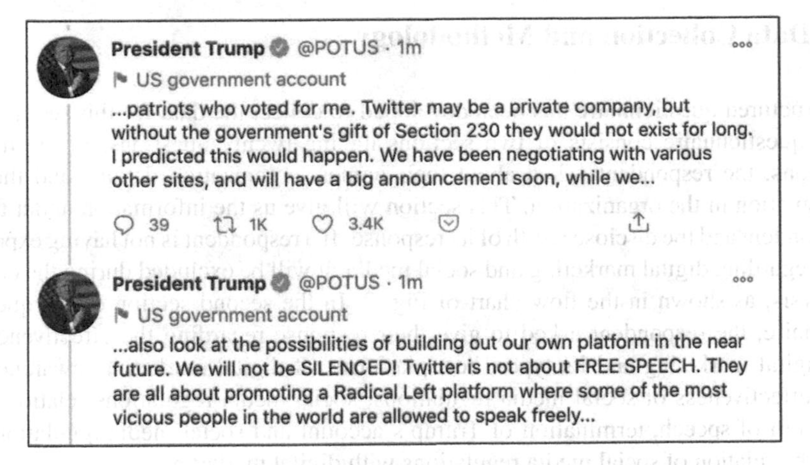

Fig. 1 Trump's response on the termination of his social media accounts

this paper, this question has been asked by the social media experts and tried to find out the 'point of view' of social media experts regarding the termination of Trump's account and freedom of speech.

2 Problem Statement

As the effective implementation of regulations by social media platforms, including Trump's account suspension, leaves a perception of an attack on users' freedom of speech. It might be possible that the users quit using leading social media platforms where the business advertisement through digital marketing occurs. It is possible the social media regulations put an adverse effect on the reach of leading social media platforms which ultimately affects the reach of digital marketing.

3 Purpose of Research

After Trump's ban by social media platforms, it is noticed that the social media platforms are strictly following their regulations even after intense opposition. This makes people cautious regarding digital marketing. To find out the effects of social media regulations on digital marketing, this study has been made. Because digital marketing is the best marketing way in the modern era, businesses' development is based on digital marketing. In this regard, this research paper is developed by performing actual market-based research where the social media regulations and their effects on digital marketing are discussed with the digital marketing experts, including Trump's social media accounts suspension.

4 Data Collection and Methodology

A structured questionnaire has been developed to collect the data for this research. The questionnaire consists of two sections having twenty questions; in the first sections, the respondent asked about their names, organization details, and their designation in the organization. This section will give us the information about the respondent and the disclose worth of its response. If a respondent is not having expertise regarding digital marketing and social media, it will be excluded during the data analysis, as shown in the flow chart of Fig. 2. In the second section of the questionnaire, the respondent asked to give their response regarding the effectiveness of digital marketing and its types, best platforms of digital marketing, awareness and effectiveness of social media regulations, social media regulations relation to freedom of speech, termination of Trump's account and social media regulations, and the relation of social media regulations with digital marketing.

This structured questionnaire has been developed on Google® Forms to be easily shared with the participants and record the responses efficiently. These research participants are the digital marketing experts from twelve leading digital marketing firms working in the United Arab Emirates. The questionnaire was shared with these organizations and get the fifteen responses from the participants within ten days. After this duration, the responses stopped to receive and start analyzing the data. The initial data analysis has been performed in Microsoft® Excel, and for the final data analyzing the collected data from respondents has been added in IBM® SPSS.

Data analysis has been performed and finds out fractions of the responses as per their nature. Moreover, the data of essential responses have been visualizing, including the best social media platform among respondents, one of the most beneficial regulations, and Trump's account termination responses. After the data visualization, the results have been explained and discussed.

5 Findings

The survey questionnaire has been responded to by Digital Marketing Experts, and their responses are analyzed using SPSS software. All participants confirm that digital marketing is better than conventional marketing, and 66.7% of the participants believe that digital marketing is highly effective while remaining marked digital marketing as effective only. On the other hand, all of the participants said that the best type of digital marketing is social media marketing; furthermore, 60% of the participants believe that Facebook is the best platform for marketing, 20% marked for Instagram, 13% said youtube is best for marketing and remaining said Twitter is better for marketing, as shown in Fig. 3.

Most of the participants believe that the social media regulations are effective for the social media users; 42% of them marked that social media regulations are highly effective for the users, while 35.7% believe it is just effective for the users. More

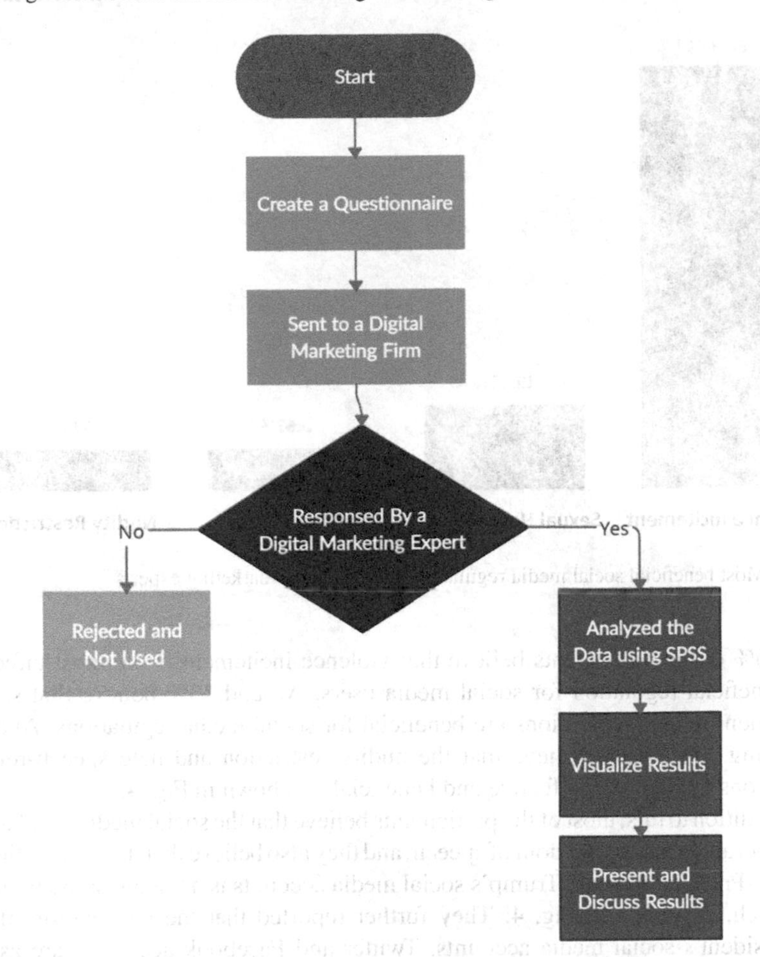

Fig. 2 Reserach flow chart

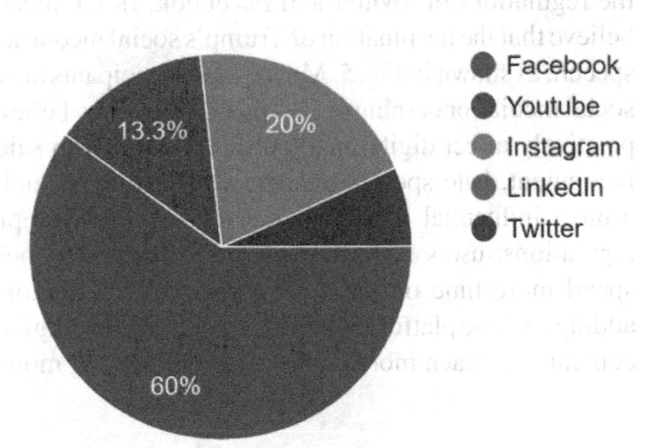

Fig. 3 Best platform for social media marketing among digital marketing experts

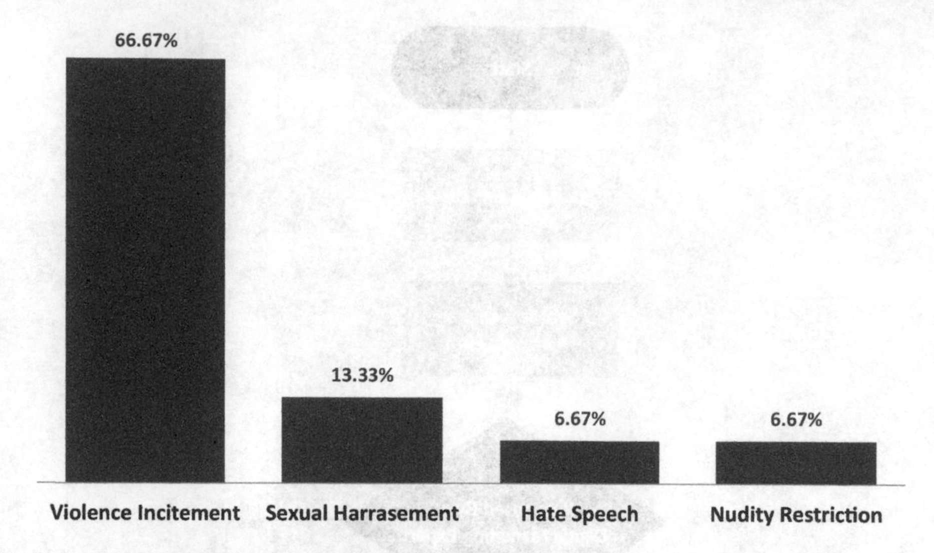

Fig. 4 Most beneficial social media regulation among digital marketing experts

than 66% of the participants believe that violence incitement is the most effective and beneficial regulation for social media users. Around 13% believe that sexual harassment-related regulations are beneficial for social media regulations. And the remaining participants believe that the nudity restriction and hate speech-related regulations are the most effective and beneficial, as shown in Fig. 4.

In addition to this, most of the participants believe that the social media regulations are not an attack on the freedom of speech, and they also believe that the termination of U.S. Ex-President Donald Trump's social media accounts is not against the freedom of speech, as shown in Fig. 4. They further reported that the termination of the Ex-President's social media accounts, Twitter and Facebook accounts, are as per the regulations of Twitter and Facebook. In the meantime, 13% of the participants believe that the termination of Trump's social media account is against the freedom of speech, as shown in Fig. 5. Most of the participants are also favoring the regulations of social media concerning digital marketing. They believe social media regulations will positively affect digital marketing. Most of them said the regulations like violence incitement, hate speech, data privacy, nudity, sexual harassment, etc., are playing a most influential role in digital marketing development. By implementing these regulations, users feel more secure to use these social media platforms, and they spend more time on these platforms and share more content; new users are also adding to these platforms. This is a good sign for digital marketing, and the marketing content will reach more people make marketing more effective.

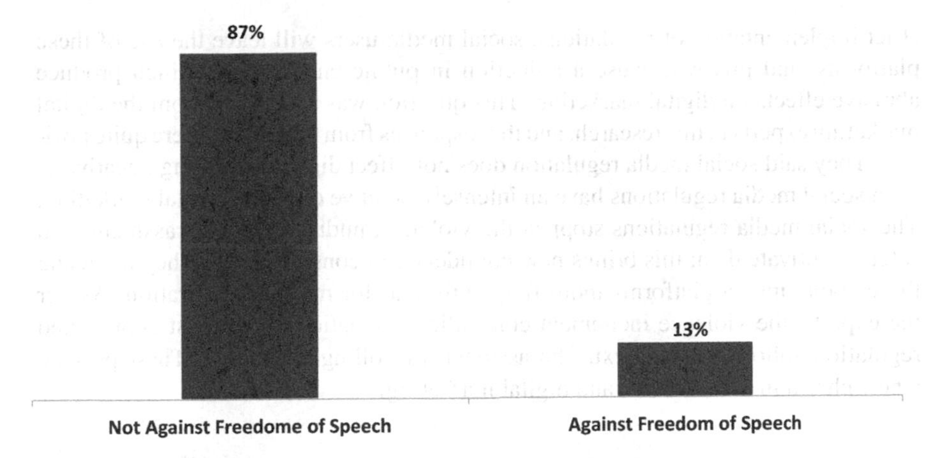

Fig. 5 Participants responses on the termination of Trump's Twitter and Facebook accounts

6 Discussions

In this research, digital marketing and the effects of social media regulations on digital marketing have been studied. It has been noticed that digital marketing is the mandatory way to market a business in the modern era (Abed et al., 2015, 2016; Felix et al., 2017; Shareef et al., 2019; Shiau et al., 2017). However, it has some problems; the most important way of digital marketing is social media marketing. In recent times, social media get subsequent fame among the public, and most people use that is why people social media marketing is the best way to promote a product/services. Moreover, the digital marketers also confirmed the social media marketing is way more beneficial than conventional media and other digital marketing ways. The digital marketers also found that the most beneficial platform for social media marketing is Facebook. There are more than a hundred million business-related pages, while 88% of the global businesses utilize Twitter for marketing. (Lister, 2018).

On the other hand, it is also found that social media platforms are used for cyber-bullying, violence incitement, nudity, sexual harassment, and much other non-ethical content. In this connection, social media platforms developed a series of regulations, and they decided not to allow anyone to use their platform for violence incitement, cyber-bullying, etc. In recent times, Ex-President Trump was found to incite violence in the USA's people to attack Capitol Hill and the social media giants, Facebook and Twitter, terminated his accounts. In response, Trump questioned them with the allegation to keep him away from his right to freedom of speech. This becomes a debate in the world, and in this research, we also asked about the experts of social marketing, and most of them confirmed that Donald Trump had shaken the regulations of social media. As a result, they terminated their accounts. There is nothing against freedom of speech, both of the platforms are private social media companies, and they have taken action against their user who violated their regulations. However, this raised a question on the survival of digital marketing; it might be possible that with the strict implementation of regulations, social media users will leave the use of these platforms, and this will cause a reduction in public engagement, which produce abrasive effects on digital marketing. This question was also asked from the digital marketing experts in this research, and the responses from the experts were quite positive. They said social media regulation does not affect digital marketing negatively; even social media regulations have an intensely positive effect on digital marketing. The social media regulations stopped the violence, nudity, sexual harassment, and safety of private data; this brings new confidence in consumers, and they are using these social media platforms more frequently and for more time duration. As per the experts, the violence incitement controlling regulation is the most appreciated regulation, followed by the sexual harassment controlling regulations. These provide a new charm in social media and digital marketing.

7 Conclusion

Digital Marketing and its relationship with the social media regulations implementation has been studied in this research. Experts asked about the effects of social media regulations on digital marketing and found that social media regulations positively affect digital marketing. People are appreciating the social media regulations and using these platforms more frequently. While the experts are also asked regarding the regulations and freedom of speech with the example of Ex-President of USA, they responded that there is not any violation of the right of freedom of speech. These regulations of social media bring a new shine towards the rise of digital marketing. In this research, the original research purpose has been achieved; it is found that Trump's account termination is not an act of keeping him away from freedom of speech; on the other hand, it is also determined that there are beneficial relationships among the new social media regulations and digital marketing. These regulations will enhance the reach of digital marketing and make it more effective.

8 Practical Application

The outcome of this paper can assist digital marketers, business owners, and future research regarding the impact of social media regulations on digital marketing. Business owners and digital markers will understand these regulations will enhance digital marketing reach by attracting more users. There was also a research gap regarding this topic, and this research is also going to fill that research gap and assist future researchers.

9 Future Recommendations

The use of social media and e-commerce is increasing day by day, and this will also increase the values of digital marketing, and social media marketing will be used exponentially to reach the marketing limit. However, another crucial factor regarding social media regulations is 'Local Government Social Media Regulations.' In the future, the effects of local government social media regulations should be involved in the research and find that either the local social media regulations can affect them adversely or effectively to digital marketing.

10 Limitations

- This research is limited to the five social media platforms, namely, Facebook, YouTube, LinkedIn, Twitter, and Instagram.
- It is just limited to the point of view of the United Arab Emirates digital marketing firms.
- It has another limitation that there is a limited number of organizations participated and with their limited responses.

References

Abed, S. S., Dwivedi, Y. K., & Williams, M. D. (2016). Social commerce as a business tool in Saudi Arabia's SMEs. *International Journal of Indian Culture and Business Management, 13*, 1–19. https://doi.org/10.1504/IJICBM.2016.077634

Abed, S. S., Dwivedi, Y. K., & Williams, M. D. (2015). SMEs' adoption of e-commerce using social media in a Saudi Arabian context: A systematic literature review. *International Journal of Business Information Systems, 19*, 159–179. https://doi.org/10.1504/IJBIS.2015.069429

Abuhashesh, M. Y., Alshurideh, M. T., & Sumadi, M. (2021). The effect of culture on customers' attitudes toward Facebook advertising: The moderating role of gender. *Review of International Business and Strategy, 31*(3), 416–437.

Abu Zayyad, H. M., Obeidat, Z. M., Alshurideh, M. T., Abuhashesh, M., Maqableh, M., & Masa'deh, R. E. (2021). Corporate social responsibility and patronage intentions: The mediating effect of brand credibility. *Journal of Marketing Communications*, 1–24.

Ahmed, A., Alshurideh, M., Al Kurdi, B., & Salloum, S. A. (2020). Digital transformation and organizational operational decision making: A systematic review. In *International Conference on Advanced Intelligent Systems and Informatics* (pp. 708–719). Springer, Cham.

Ahmad, A., Alshurideh, M. T., Al Kurdi, B. H., & Salloum, S. A. (2021). Factors impacts organization digital transformation and organization decision making during covid19 pandemic. *The Effect of Coronavirus Disease (COVID-19) on Business Intelligence, 334*, 95–106.

Aissani, R., ElRefae, G. A. (2021). Trends of university students towards the use of YouTube and the benefits achieved; Field study on a sample of Al Ain University students. *AAU Journal of Business and Law, 5*(1).

Ajina, A. S. (2019). The perceived value of social media marketing: An empirical study of online word-of-mouth in Saudi Arabian context. *Entrepreneurship and Sustainability Issues, 6*(3), 1512–1527.

Anjum, A., More, V., Ghouri, A. M. (2012). Social media marketing: A paradigm shift in business (SSRN Scholarly Paper No. ID 2149910). Social Science Research Network, Rochester, NY.

Akour, I., Alshurideh, M., Al Kurdi, B., Al Ali, A., & Salloum, S. (2021). Using machine learning algorithms to predict people's intention to use mobile learning platforms during the covid-19 pandemic: Machine learning approach. *JMIR Medical Education, 7*(1), 1–17.

Al-Dmour, R., AlShaar, F., Al-Dmour, H., Masa'deh, R., & Alshurideh, M. T. (2021). The effect of service recovery justices strategies on online customer engagement via the role of "customer satisfaction" during the covid-19 pandemic: An empirical study. *The Effect of Coronavirus Disease (COVID-19) on Business Intelligence, 334*, 325–346.

Aljumah, A. I., Nuseir, M. T., & Alam, M. M. (2021a). Organizational performance and capabilities to analyze big data: Do the ambidexterity and business value of big data analytics matter? *Business Process Management Journal, 27*(4), 1088–1107. https://doi.org/10.1108/BPMJ-07-2020-0335

Aljumah, A. I., Nuseir, M. T., & Alam, M. M. (2021b). Traditional marketing analytics, big data analytics and big data system quality and the success of new product development. *Business Process Management Journal, 27*(4), 1108–1125. https://doi.org/10.1108/BPMJ-11-2020-0527

Aljumah, A., Nuseir, M. T., & Alshurideh, M. T. (2021c). The impact of social media marketing communications on consumer response during the covid-19: Does the brand equity of a university matter. *The Effect of Coronavirus Disease (COVID-19) on Business Intelligence*, 367–384.

Al-Khayyal, A., Alshurideh, M., Al Kurdi, B., & Salloum, S. A. (2021). Factors influencing electronic service quality on electronic loyalty in online shopping context: Data analysis approach. In *Enabling AI Applications in Data Science* (pp. 367–378). Springer, Cham.

Almaazmi, J., Alshurideh, M., Al Kurdi, B., & Salloum, S. A. (2020). The effect of digital transformation on product innovation: A critical review. In *International Conference on Advanced Intelligent Systems and Informatics* (pp. 731–741). Springer, Cham.

Al Kurdi, B. H., & Alshurideh, M. T. (2021). Facebook advertising as a marketing tool: Examining the influence on female cosmetic purchasing behaviour. *International Journal of Online Marketing (IJOM), 11*(2), 52–74.

Al-Maroof, R., Ayoubi, K., Alhumaid, K., Aburayya, A., Alshurideh, M., Alfaisal, R., & Salloum, S. (2021). The acceptance of social media video for knowledge acquisition, sharing and application: A comparative study among YouYube users and TikTok users' for medical purposes. *International Journal of Data and Network Science, 5*(3), 197.

Almazrouei, F. A., Alshurideh, M., Al Kurdi, B., & Salloum, S. A. (2020). Social media impact on business: a systematic review. In *International Conference on Advanced Intelligent Systems and Informatics* (pp. 697–707). Springer, Cham.

Alshurideh, M. (2022). Does electronic customer relationship management (E-CRM) affect service quality at private hospitals in Jordan? *Uncertain Supply Chain Management, 10*(2), 1–8.

Alshurideh, M., Bataineh, A., Alkurdi, B., & Alasmr, N. (2015). Factors affect mobile phone brand choices–Studying the case of Jordan universities students. *International Business Research, 8*(3), 141–155.

Alshurideh, M., Salloum, S. A., Al Kurdi, B., & Al-Emran, M. (2019a). Factors affecting the social networks acceptance: An empirical study using PLS-SEM approach. In *Proceedings of the 2019a 8th International Conference on Software and Computer Applications* (pp. 414–418).

Alshurideh, M., Salloum, S. A., Al Kurdi, B., Monem, A. A., & Shaalan, K. (2019b). Understanding the quality determinants that influence the intention to use the mobile learning platforms: A practical study. *International Journal of Interactive Mobile Technologies, 13*(11), 183–157.

Alshurideh, M. T., Kurdi, B. A., AlHamad, A. Q., Salloum, S. A., Alkurdi, S., Dehghan, A., Masa'deh, R. E., et al. (2021a). Factors affecting the use of smart mobile examination platforms by universities' postgraduate students during the COVID 19 pandemic: An empirical study. In *Informatics* (Vol. 8, No. 2, p. 32). Multidisciplinary Digital Publishing Institute.

Alshurideh, M. T., Kurdi, B. A., AlHamad, A. Q., Salloum, S. A., Alkurdi, S., Dehghan, A., Masa'deh, R. E., et al. (2021b). Factors affecting the use of smart mobile examination platforms by universities' postgraduate students during the COVID 19 pandemic: An empirical study. *Informatics*, 8 (2), 21–31.

BBC (2021). Twitter "permanently suspends" Trump's account. BBC News.

Campbell, H. (2021). Donald Trump's bans from social media fire freedom of speech debate [WWW Document]. Euronews. Retrieved Jan 3, 2021, from https://www.euronews.com/2021/01/11/president-trumps-platform-bans-opens-debate-about-freedom-of-speech-thecube.

Cash, S. J., Thelwall, M., Peck, S. N., Ferrell, J. Z., & Bridge, J. A. (2013). Adolescent suicide statements on MySpace. *Cyberpsychology, Behavior, and Social Networking, 16*(3), 166–174.

Chaffey, D., & Ellis-Chadwick, F. (2019). *Digital Marketing*. Pearson.

Dwivedi, Y. K., Ismagilova, E., Hughes, D. L., Carlson, J., Filieri, R., Jacobson, J., Wang, Y., et al. (2021). Setting the future of digital and social media marketing research: Perspectives and research propositions. *International Journal of Information Management, 59*, 1–37.

Elsaadani, M. A. (2020). Investigating the effect of e-management on customer service performance. *AAU Journal of Business and Law, 3*(2), 1–24.

Facebook (2022). Community Standards [WWW Document]. Retrieved Jan 3, 2021, from https://www.facebook.com/communitystandards/safety.

Felix, R., Rauschnabel, P. A., & Hinsch, C. (2017). Elements of strategic social media marketing: A holistic framework. *Journal of Business Research, 70*, 118–126. https://doi.org/10.1016/j.jbusres.2016.05.001

Garrett, R. K. (2019). Social media's contribution to political misperceptions in U.S. Presidential elections. *PLOS ONE, 14*, e0213500. https://doi.org/10.1371/journal.pone.0213500.

Gettleman, J., Yasir, S. (2019). *Kashmir Militants Kill Again as Trouble Grows Between India and Pakistan*. N. Y. Times.

Grover, P., Kar, A. K., Dwivedi, Y. K., & Janssen, M. (2019). Polarization and acculturation in U.S. Election 2016 outcomes—Can Twitter analytics predict changes in voting preferences. *Technological Forecasting and Social Change, 145*, 438–460. https://doi.org/10.1016/j.techfore.2018.09.009

Hamadneh, S., Hassan, J., Alshurideh, M., Al Kurdi, B., & Aburayya, A. (2021). The effect of brand personality on consumer self-identity: The moderation effect of cultural orientations among British and Chinese consumers. *Journal of Legal, Ethical and Regulatory, 24*, 1–14.

Hinduja, S., & Patchin, J. W. (2010). Bullying, cyberbullying, and suicide. *Archives of Suicide Research, 14*, 206–221. https://doi.org/10.1080/13811118.2010.494133

Hossain, T. M. T., Akter, S., Kattiyapornpong, U., & Dwivedi, Y. (2020). They are reconceptualizing integration quality dynamics for omnichannel marketing. *Industrial Marketing Management, 87*, 225–241. https://doi.org/10.1016/j.indmarman.2019.12.006

Instagram. (2013). Terms of Use • Instagram [WWW Document]. Retrieved Mar 13, 2021, from https://www.instagram.com/about/legal/terms/before-january-19-2013/.

Kapoor, K. K., & Dwivedi, Y. K. (2015). Metamorphosis of Indian electoral campaigns: Modi's social media experiment. *International Journal of Indian Culture and Business Management, 11*(4), 496–516. https://doi.org/10.1504/IJICBM.2015.072430

Khasawneh, M. A., Abuhashesh, M., Ahmad, A., Masa'deh, R., & Alshurideh, M. T. (2021a). Customers online engagement with social media influencers' content related to COVID 19. In *The Effect of Coronavirus Disease (COVID-19) on Business Intelligence* (pp. 385–404). Springer.

Khasawneh, M. A., Abuhashesh, M., Ahmad, A., Alshurideh, M. T., & Masa'deh, R. (2021b). Determinants of e-word of mouth on social media during COVID-19 outbreaks: An empirical study. In *The Effect of Coronavirus Disease (COVID-19) on Business Intelligence* (pp. 347–366). Springer.

Kemp, S. (2020). Digital 2020: 3.8 billion people use social media [WWW Document]. We Are Soc. Retrieved Feb 27, 2021, from https://wearesocial.com/blog/2020/01/digital-2020-3-8-billion-people-use-social-media.

Kietzmann, J. H., Hermkens, K., McCarthy, I. P., & Silvestre, B. S. (2011). Social media? Get serious! Understanding the functional building blocks of social media. *Business Horizons, 54*(3), 241–251. https://doi.org/10.1016/j.bushor.2011.01.005

Kurdi, B. A., Alshurideh, M., Nuseir, M., Aburayya, A., & Salloum, S. A. (2021). The effects of subjective norm on the intention to use social media networks: An exploratory study using PLS-SEM and machine learning approach. In *International Conference on Advanced Machine Learning Technologies and Applications* (pp. 581–592). Springer, Cham.

Lee, R. (2019). Extreme speech| extreme speech in Myanmar: The role of state media in the Rohingya forced migration crisis. *International Journal of Communication, 13*, 3203–3224.

Lee, K., Azmi, N., Hanaysha, J., Alshurideh, M., & Alzoubi, H. (2022a). The effect of digital supply chain on organizational performance: An empirical study in Malaysia manufacturing industry. *Uncertain Supply Chain Management, 10*, 1–16.

Lee, K., Ramiz, P., Hanaysha, J., Alzoubi, H., & Alshurideh, M. (2022b). Investigating the impact of benefits and challenges of IOT adoption on supply chain performance and organizational performance: An empirical study in Malaysia. *Uncertain Supply Chain Management, 10*(2), 1–14.

Leo, S., Alsharari, N. M., Abbas, J., & Alshurideh, M. T. (2021). From offline to online learning: A qualitative study of challenges and opportunities as a response to the covid-19 pandemic in the UAE higher education context. *The Effect of Coronavirus Disease (COVID-19) on Business Intelligence, 334*, 203–217.

LinkedIn. (2020). User Agreement | LinkedIn [WWW Document]. Retrieved Mar 13, 2021, from https://www.linkedin.com/legal/user-agreement.

Lister, M. (2018). 40 Essential Social Media Marketing Statistics for WordStream [WWW Document]. Retrieved Mar 13, 2021, from https://www.wordstream.com/blog/ws/2017/01/05/social-media-marketing-statistics.

Mashaqi, E., Al-Hajri, S., Alshurideh, M., & Al Kurdi, B. (2020). The impact of e-service quality, e-recovery services on e-loyalty in online shopping: Theoretical foundation and qualitative proof. *PalArch's Journal of Archaeology of Egypt/egyptology, 17*(10), 2291–2316.

Mehmood, R. (2021) The double standards in Facebook and Twitter's Trump ban [WWW Document]. Retrieved Mar 13, 2021, from https://www.aljazeera.com/opinions/2021/1/29/the-double-standards-in-facebook-and-twitters-trump-ban.

Menesini, E., Nocentini, A., Palladino, B. E., Frisén, A., Berne, S., Ortega-Ruiz, R., Smith, P. K., et al. (2012). Cyberbullying definition among adolescents: A comparison across six European countries. *Cyberpsychology, Behavior, and Social Networking, 15*(9), 455–463. https://doi.org/10.1089/cyber.2012.0040

Miro-Llinares, F., & Rodriguez-Sala, J. J. (2016). Cyber hate speech on Twitter: Analyzing disruptive social media events to build a violent communication and hate speech taxonomy. *International Journal of Design, Nature and Ecodynamics, 11*, 406–415. https://doi.org/10.2495/DNE-V11-N3-406-415

Morris, N. (2009). Understanding digital marketing: Marketing strategies for engaging the digital generation. *Journal of Direct, Data and Digital Marketing Practice, 10*, 384–387. https://doi.org/10.1057/dddmp.2009.7

MRGI. (2019). Peoples under Threat 2019: Social media's role in exacerbating violence—World [WWW Document]. ReliefWeb. Retrieved Mar 13, 2021, from https://reliefweb.int/report/world/peoples-under-threat-2019-role-social-media-exacerbating-violence.

Mutahi, P., & Kimari, B. (2017). The impact of social media and digital technology on electoral violence in Kenya. IDS.

Nuseir, M.T., El-Refae, G.A., Aljumah, A. (2021). The e-learning of students and university's brand image (Post COVID-19): How successfully Al-Ain University have embraced the paradigm shift in digital learning. *Studies in Systems, Decision and Control, 334*, 171–187.

Nuseir, M. T., & Aljumah, A. (2021). Digital marketing adoption influenced by relative advantage and competitive industry: A UAE tourism case study. *International Journal of Innovation Creativity and Change, 11*(2), 617–631.

Obar, J. A., & Wildman, S. (2015). Social media definition and the governance challenge: An introduction to the special issue. *Telecommunication Policy, Special Issue on the Governance of Social Media, 39*, 745–750. https://doi.org/10.1016/j.telpol.2015.07.014

Obeidat, R., Alshurideh, Z., Al Dweeri, M., & Masa'deh, R. (2019). The influence of online revenge acts on consumers psychological and emotional states: Does revenge taste sweet? *33 IBIMA Conference Proceedings-Granada* (pp. 10–11). Granada.

Romano, A. (2021). Kicking people off social media is not about free speech [WWW Document]. Vox. Retrieved Jan 1, 2021, from https://www.vox.com/culture/22230847/deplatforming-free-speech-controversy-trump.

Ruder, T. D., Hatch, G. M., Ampanozi, G., Thali, M. J., & Fischer, N. (2011). Suicide announcement on Facebook. *Crisis, 32*, 280–282. https://doi.org/10.1027/0227-5910/a000086

Shareef, M. A., Mukerji, B., Alryalat, M. A. A., Wright, A., & Dwivedi, Y. K. (2018). Advertisements on Facebook: Identifying the persuasive elements in the development of positive attitudes in consumers. *Journal of Retailing and Consumer Services, 43*, 258–268. https://doi.org/10.1016/j.jretconser.2018.04.006

Shareef, M. A., Mukerji, B., Dwivedi, Y. K., Rana, N. P., & Islam, R. (2019). Social media marketing: Comparative effect of advertisement sources. *Journal of Retailing and Consumer Services, 46*(58–69), 58–69. https://doi.org/10.1016/j.jretconser.2017.11.001

Shiau, W. L., Dwivedi, Y. K., & Yang, H. S. (2017). Co-citation and cluster analyses of extant literature on social networks. *International Journal of Information Management, 37*(5), 390–399. https://doi.org/10.1016/j.ijinfomgt.2017.04.007

Smith, Aa. (2011). Why Americans use social media. Pew Research Center. Internet Science and Technology. Retrieved Feb 27, 2021, from https://www.pewresearch.org/internet/2011/11/15/why-americans-use-social-media/.

Smith, P. K., Mahdavi, J., Carvalho, M., Fisher, S., Russell, S., & Tippett, N. (2008). Cyberbullying: Its nature and impact in secondary school pupils. *Journal of Child Psychology and Psychiatry, 49*(4), 376–385. https://doi.org/10.1111/j.1469-7610.2007.01846.x

Storm, M. (2020). 5 Types of Social Media and Examples of Each. WebFX Blog. Retrieved Feb 27, 2021, from https://www.webfx.com/blog/social-media/types-of-social-media/.

Sultan, R. A., Alqallaf, A. K., Alzarooni, S. A., Alrahma, N. H., AlAli, M. A., & Alshurideh, M. T. (2021). how students influence faculty satisfaction with online courses and do the age of faculty matter. In *The International Conference on Artificial Intelligence and Computer Vision* (pp. 823–837). Springer, Cham.

Sweiss, N., Obeidat, Z. M., Al-Dweeri, R. M., Mohammad Khalaf Ahmad, A., M. Obeidat, A., & Alshurideh, M. (2021). The moderating role of perceived company effort in mitigating customer misconduct within Online Brand Communities (OBC). *Journal of Marketing Communications*, 1–24.

Tariq, E., Alshurideh, M., Akour, I., & Al-Hawary, S. (2022). The effect of digital marketing capabilities on organizational ambidexterity of the information technology sector. *International Journal of Data and Network Science, 6*, 1–8.

Tuten, T. L. (2020). *Social Media Marketing*. Sage.

Twitter. (n.d.). The Twitter Rules [WWW Document]. Retrieved Jan 3, 2021, from https://help.twitter.com/en/rules-and-policies/twitter-rules.

UNCTAD. (2020). COVID-19 has changed online shopping forever, survey shows | UNCTAD [WWW Document]. Retrieved Feb 28, 2021, from https://unctad.org/news/covid-19-has-changed-online-shopping-forever-survey-shows.

WebFX. (2021). Best Social Media Platforms for Businesses in 2021 [WWW Document]. Retrieved Feb 28, 2021, from https://www.webfx.com/social-media/which-social-media-platforms-are-right-for-your-business.html.

YouTube. (n.d.). YouTube Community Guidelines & Policies—How YouTube Works [WWW Document]. YouTube Community Guidel. Policies—YouTube Works. Retrieved Mar 13, 2021, from https://www.youtube.com/howyoutubeworks/policies/community-guidelines/.

The Impact of Facebook Advertisements on Customer Attentions of Jordanian Female Young Users

Younes Megdadi, Mohammad Hammouri, and Zaid Megdadi

Abstract This research aimed to investigate the impact of facebook advertisements on customer attentions of Jordanian female young users. Analytical descriptive approach was used. A suitable sample used which includes of (400) young female customer users. The researchers used the questionnaire as a tool for collecting primary data. The results found that there is an impact of the facebook advertisements by facebook on customer attentions of Jordanian young female users. The researchers set up a number of recommendations and the most is acquired more advertisements with different design and entertainments to influence customer's attentions positively toward their personal products.

Keywords Facebook · Advertisements · Customer attentions · Jordanian female young

1 Introduction

Due to the technological development and the diversity of means of communication with customers, electronic advertisements for the marketing of companies' products have become dependent on social media networks, as they are networks that work on the interaction of members with each other, and at the same time they are considered a means of concern with the requirements of their personal lives. Hence, companies

Y. Megdadi (✉)
Faculty of Business, Amman Arab University, Amman, Jordan
e-mail: megdadi@aau.edu.jo

M. Hammouri
Assistance Professor of Business Management, Open Arab University, Kuwait City, Kuwait
e-mail: Mohamed.hammouri@mcbs.edu.om; M.hamouri@arabou.edu.kw

Z. Megdadi
Ph.D. Candidate of Business Management, Faculty of Business, Girne American University, Girne, Cyprus
e-mail: zaidmgdad23@yahoo.com

M. Alshurideh et al. (eds.), *The Effect of Information Technology on Business and Marketing Intelligence Systems*, Studies in Computational Intelligence 1056,
https://doi.org/10.1007/978-3-031-12382-5_10

have noted the importance of social media networks in marketing work, including the Facebook network because of the benefits it enjoys on the personal and social level for customers for the purposes of advertising their products, not to mention what Facebook enjoys such as speed of spread and flexibility in dealing with advertising and its content, and controlling the number Advertising times, ease of directing the advertisement to the target groups, ease of customer opinion poll taking into account all the comments, opinions and suggestions of customers towards the advertised products, the flexibility of designing the content of the advertisement in terms of images, videos and sounds, and the ease of identifying the details of the advertised products, which makes it easier to influence customers attentions positively, in addition to that, customer's will be able to compare with competing goods. Companies realized the importance of Facebook advertisements benefits for customers for different age groups, and ease to identify and allocating target customers to influence them positively.

2 Research Problem

Facebook ads have become one of the main and important means for companies to market their products to influence the generation of customers' attention towards the advertised products. This research seeks to examine the impact of Facebook ads on the attention of young female users who are users.

2.1 *Research Conceptual Framework*

The use of Facebook as one of the most common tool of social networks, which widespread use through electronic advertisements which aim to give young customers an attension by the marketed products of business firms and as one of the major network used by young users in Jordan, which noticed that there is a real problem in the marketing process of its various products, which has become directed towards electronic advertisements messages through Facebook network as one of the modern technological means in giving customers attension towards business firms products to improve their marketing performance.

2.2 *Research Questions*

The main question: Is there an impact of Facebook advertisements by its sub-dimensions (advertisement content, advertisement design, advertisement artistic production, and advertisement entertainments) on the attention in its combined dimensions (perception, interest, desire, and preference) among young female's users? It is divided into the following sub-questions:

1. Is there the impact of Facebook advertisement by its sub-dimensions (advertisement content, advertisement design, advertisement artistic production, and advertisement entertainments) on the perception of young female's users?
2. Is there the impact of Facebook advertisements by its sub-dimensions (advertisement content, advertisement design, advertisement artistic production, and advertisement entertainments) on interest of female's users?
3. Is there the impact of Facebook advertisements by its sub-dimensions (advertisement content, advertisement design, advertisement artistic production, and advertisement entertainments) on the desire of female's users?
4. Is there the impact of Facebook advertisements by its sub-dimensions (advertisement content, advertisement design, advertisement artistic production, and advertisement entertainments) on the preference of female's users?

3 Research Hypotheses

Main hypothesis (HO): There is no statistically significant impact at the significance level (0.05 $\geq \alpha$) of Facebook advertisements by its sub-dimensions (advertisement content, advertisement design, advertisement artistic production, and advertisement entertainments) on attention by its dimensions (perception, interest, desire, and preference) of young females? The following sub-hypotheses stem from it:

(H01): There is no statistically significant impact at the significance level (0.05 $\geq \alpha$) of Facebook advertisements by its sub-dimensions (advertisement content, advertisement design, advertisement artistic production, and advertisement entertainments) on the perception of young females?

(H02): There is no statistically significant impact at the significance level (0.05 $\geq \alpha$) of Facebook advertisements by its sub-dimensions (advertisement content, advertisement design, advertisement artistic production, and advertisement entertainments) on the interest of young females.

(H03): There is no statistically significant impact at the significance level (0.05 $\geq \alpha$) of Facebook advertisement by its sub-dimensions (advertisement content, advertisement design, advertisement artistic production, and advertisement entertainments) on the desire of young females?

(H04): There is no statistically significant impact at the significance level (0.05 $\geq \alpha$) of Facebook advertisement by its sub-dimensions (advertisement content, advertisement design, advertisement artistic production, and advertisement entertainments) on the preference of young females.

4 Research Model

Based on previous studies, the researchers formulated a hypothetical model for the research and based on previous studies related to this research.

4.1 Discussion of the Proposed Model

As showed in Fig. 1, the researchers proposed the Facebook advertisements messages by a number of sub-dimensions were as (advertisement content, advertisement design, advertisement artistic production, and advertisement entertainments) on young customers attention by a number of sub-dimensions were as (perception, interest, desire, and preference) to help business firms of influencing customers attention toward business firms products and performing marketing activities efficiently and with less cost, by enhancing the efficiency of the used social media tools to communicate with potential customers as far as the young customers. It also used in data analytics which helps the operations as well; research data and results can be powerfully used by decisions makers of business firms for a better future of performing marketing activities by using social media network as far as Facebook.

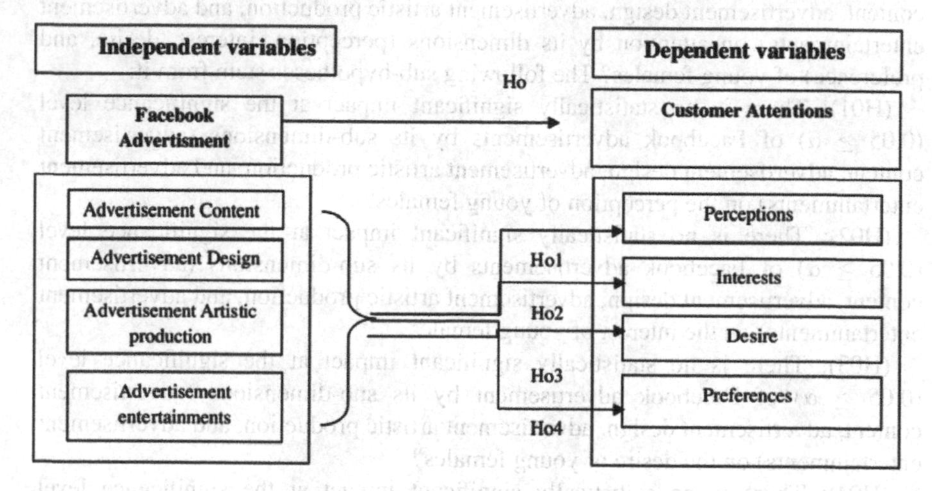

Fig. 1 Research model

5 Theoretical Background and Literature Review

5.1 The Concept of Electronic Advertising

There are several advantages to electronic advertising, the most important of which are: delivering the company's message to millions of people, establishing permanent and continuous relationships with customers, lowering the cost of electronic advertising due to the absence of printing and production costs, easy access to information and data in detail about the product, and the possibility of direct purchase from the site. The actions of customers and obtaining continuous feedback. Using technical methods and various technical media to draw the viewer's attention. The possibility of knowing the extent to which consumers accept new products and services. Knowing the impact of the advertising campaign first-hand and knowing the efficiency and effectiveness of advertising and linking it to the level of costs. And making any adjustments directly and quickly, and the ability to control the way the ad appears, as you can specify its appearance for its target group, a specific time, or a specific day. Increasing product recall rates by customers, mental enhancement of the brand name and trademark among consumers, and the ability to enter the company's website with just one click (Boyd & Ellison, 2008).

Electronic advertising takes many forms on the Internet, the most important of which is ads on social networks, which are ads that appear to users of social networking sites while they browse them, for example, the funded ads that appear on Facebook, which the user often does not follow these pages, or Twitter ads or YouTube ads that are Mandatory in some sections. The advertisement through the account of a famous person on social networking sites (influencers): which is for a company to agree with some personalities that have a large number of followers on its accounts on social networking sites to make paid ads to promote its products and take advantage of the large number of followers it has (Scott et al., 2007).

Facebook is a social network that communicates and interacts between members to share their thoughts, feelings, pictures, and videos that belong to them, and this makes it one of the most important marketing tools for companies (Fazalur et al., 2014).

The artistic output of the electronic advertisement has been known as an integrated technical unit that both the advertisement designer and director contribute to its preparation and manufacture, in order to reach the advertising idea to be presented to consumers in order to attract attention and draw attention to these electronic advertisements. This depends on the use of all components of digital images, drawings, lines and colors in an optimal and homogeneous way to come out to the consumer audience in a wonderful way because the artistic output of electronic advertisements is not a goal but rather a means aimed at highlighting the contents of the advertisement in a way that helps draw consumer attention and arouse interest, and then reach the purchase decision-making (Al-Junaid, 2017).

The text of the advertisement has been known as composing digital texts with the aim of promoting a company, person, product, opinion or idea, and the main goal of writing this text is to persuade the consumer to buy the product or subscribe to the service or to pay for a specific idea, behavior or belief, and includes the texts of the electronic advertisement The various pillars of advertising, from the structure of the literary text and the title to the logo, in addition to direct e-mail messages or advertisement songs, and the text of digital advertisements, and the writers of electronic advertising texts use methods to choose words that are entitled to the best ranks in the search engines. This is done by choosing strategic words and repeating key words and phrases on the pages of the sites, it has become the duty of the writer of electronic ads to attract the words of the search engines, and hold them with an advanced rank, not to mention that it captivates the website visitors from the consumers (Heath, 2012).

The characteristics, type, and duration of the electronic advertisement display and the nature of the sites on which the advertisement appears is able to perceived quality of the commodities appearing in this advertisement (Abu Kharma et al., 2011). The electronic advertising has a positive impact in terms of (electronic advertising characteristics, the content of the e-mail, the mental image of the electronic advertisement, and the electronic advertising medium) combined in influencing the purchasing behavior of the Jordanian youth consumer of cars (Hassan, 2015).The electronic advertising has an effect on consumer behavior, but this effect is in the behavioral stages, attention, interest and desire, but there is no effect on the buying process (Ibrahim, 2017). High percentage of respondents stated that the quality of information provided by electronic advertising plays a role in influencing the attitudes of individuals, and electronic advertising works on appealing to the individual, and inciting feelings of experimenting with the services of the institution (Belakahleh, 2017).

A positive relationship between the AIDA model and the effectiveness of advertising in private insurance companies, and as the study recommended and one of the most important of them is that private insurance companies should add the AIDA model to their advertising strategies (Gharibi, 2012). The majority of individuals access ads on Facebook and see that they target youth, and the results also showed that (54.8%) interact with the ads they receive on Facebook, and that (45.2%) do not interact with them (Ago, 2015). The information and entertainment in advertisements through social media have a significant impact on awareness of the products offered (Abuhashesh et al., 2021; Al Kurdi & Alshurideh, 2021; Rahim, 2015).

5.2 Research Approach

This research relied on the descriptive approach. This approach was used to review the most important literature related to the impact of Facebook ads on the attention of female users of social networks.

5.3 Research Population and Study Sample

The population of the research consisted of the Jordanian youth group whose ages ranged between (18–40 years) and their total number was (54,662,3) young of both genders in according to the report of the Department of Statistics for the year (2020). A random sample was selected from all governorates of the Kingdom with total (400) female.

5.4 Research Tool and Validation

The research tool was developed that covered all variables with its independent and dependent dimensions, and it was judged by specialists in scientific research according to the aim and purpose of the research. The statistical standard and its percentage (80%) were adopted to indicate the validity of the paragraph, based on the opinions of the arbitrators. The researchers considered the opinions of young females and their amendments as an indication of the validity of the content of the research tool.

Table 1 shows the results of Cronbach alpha coefficient values to all sub-dimensions of the independent and sub-study variables ranged, and the scale as a whole were higher than (0.60), which is acceptable ratios based on the current research purposes.

Table 1 Alpha Cronbach test

Ind. & Dep. variables	Cronbach's alpha coefficient result
Facebook Advertisments	
Advertisement content	0.821
Advertisement design	0.744
Advertisement artistic production	0.771
Advertisement entertainments	0.841
Customer attentions	
Perception	0.811
Interest	0.845
Desire	0.799
Preference	0.778
Overall	0.802

5.5 Data Analysis: Descriptive Statistics Results

First: The arithmetic averages and standard deviations were extracted to identify the responses of respondents about the level of Facebook ads dimensions.

Table 2 shows that the arithmetic averages of the level of Facebook ads. dimensions ranged between (4.20–4.45), with an average of (4.29) which indicates a high degree. Advertisement entertainments it was in the first rank, with arithmetic mean of (4.45) and standard deviation of (0.69) which indicates a high degree and in the least rank is advertisement design with arithmetic mean of (4.20), and standard deviation (0.58), which indicates a high degree. The overall arithmetic mean of Facebook advertisements dimensions indicate (4.29) with standard deviation of (0.64) which indicates a high degree.

Second: The arithmetic averages and standard deviations were extracted to identify the responses of respondents about the level of customer's attentions dimensions.

Table 3 shows that the overall arithmetic averages of the level of customers' attention ranged between (4.28–4.44), with average of (4.36) with a high degree. Interest dimension was ranked no. 1, with an arithmetic mean of (4.44), with standard deviation of (0.77) which indicates a high degree. The least dimension was preference with an arithmetic mean of (4.28), with standard deviation of (0.63), which indicates a high degree. The overall arithmetic mean of (4.36), with standard deviation of (0.69) which indicates a high degree. This indicates that the customer's attentions is high.

Table 2 The arithmetic averages and standard deviations of Facebook ads. dimensions

No	Facebook advertisements	Mean	S.D	Rank	Degree
1	Advertisement content	4.28	0.67	2	High
2	Advertisement design	4.20	0.58	4	High
3	Advertisement artistic production	4.22	0.61	3	High
4	Advertisement entertainments	4.45	0.71	1	High
	Overall	4.29	0.64		High

Table 3 The arithmetic averages and standard deviations of customer's attentions dimensions

No	Customers attentions	Mean	S.D	Rank	Degree
1	Perception	4.36	0.69	2	High
2	Interest	4.44	0.77	1	High
3	Desire	4.34	0.67	3	High
4	Preference	4.28	0.63	4	High
	Overall	4.36	0.69		High

Table 4

Facebook advertisements	B	S.D	Beta	t. value	Sig
Advertisement content	0.452	0.045	0.489	10.675	0.000*
Advertisement design	0.134	0.048	0.145	2.776	0.000*
Advertisement artistic production	0.192	0.032	0.234	4.766	0.000*
Advertisement entertainments	0.145	0.036	0.196	4.664	0.000*

* Statistically significant at $(0.05 \geq \alpha)$ level (t) tabular value = (± 1.96)

5.6 Research Hypotheses Tests

(Ho): Multiple regression analysis test was used in order to identify the impact of Facebook advertisements by its dimensions (advertisement content, advertisement design, advertisement artistic production, and advertisement entertainments) on customers attentions by its dimensions (perception, interest, desire, and preference) of young females users. Table 4 illustrates this:

Table 4, following the (t) test values that the sub-variables related to Facebook ads with their dimensions (advertisement content, advertisement design, advertisement artistic production, and advertisement entertainments) had an impact on customers attention of female youth customers, as the values of (t) calculated for the three dimensions reached (10.675, 2.776, 4.766 and 4.664) respectively, which are significant values at $(0.05 \geq \alpha)$ significance level, and it was found that all dimensions had an impact on the attention of young females customers using the Facebook network (Table 5).

Stepwise Multiple Regression analysis used to determine the importance of each independent variable separately in contributing to the model that represents the effect of the characteristics of Facebook ads on the attention of young female customers users of Facebook, as it is clear that the order of entry of the independent variables into the regression equation, The advertisement entertainment dimension came in the first place and explained of (72.2%) of the variance in the dependent variable, and the second dimension is advertisement content, and explained of (73.4%) of the variance in the dependent variable, and the third dimension is advertisement artistic production

Table 5 Results of stepwise multiple regression analysis to predict the attention span of females using Facebook

Model	The of entry of the independent sub-variables into the prediction equation	Value R^2	Value (F)	Sig
1	Advertisement entertainments	0.722[a]	388.210	0.000*
1 + 2	Advertisement content	0.734[b]	245.258	0.000*
1 + 2 + 3	Advertisement artistic production	0.742[c]	162.224	0.000*
1 + 2 + 3 + 4	Advertisement design	0.748 [d]	144.376	0.000*

*Statistically significant at $(0.05 \geq \alpha)$ level (t) tabular value = (± 1.96)

with the previous variables and explained of (74.2%) of the variance in the dependent variable, and the fourth dimension is advertisement design with the previous variables and explained of (74.8%) of the variance in the dependent variable. Accordingly, the null hypothesis is rejected, and the alternative hypothesis is accepted, meaning that there is a significant impact. Statistical significance at the level of significance (0.05 ≥ α) for Facebook ads by (advertisement content, advertisement design, advertisement artistic production, and advertisement entertainments) on customers' attention in its dimensions (perception, interest, desire, and preference) among young females who use the Facebook network.

(H01): There is no statistically significant impact at the level of significance (0.05 ≥ α) for Facebook ads on young female's perception of Facebook users. Multiple regression tests were used for knowing the impact of Facebook ads in their dimensions (advertisement content, advertisement design, advertisement artistic production, and advertisement entertainments) on young female customer's perception, and Table 6 illustrates this:

Table 6, shows that the (t) test values is related to Facebook ads with its dimensions (advertisement content, advertisement design, advertisement artistic production, and advertisement entertainments) had an impact on young female perception, as the values of (t) calculated for the three dimensions reached (16.234, −1.229, 4.344 and 4.677) respectively, which are significant values at (0.05 ≥ α) significance level, and it was found that all dimensions had an impact on females customers perceptions who use the Facebook network.

(H02): There is no statistically significant impact at the significance level (0.05 ≥ α) of the Facebook ads on the interest of young female users. Multiple regression tests were used in order to identify the impact of Facebook advertisements by its dimensions (advertisement content, advertisement design, advertisement artistic production, and advertisement entertainments) on young female customers interest, and Table 7 illustrates this:

Table 7, shows that the (t) test values is related to Facebook ads with their dimensions (advertisement content, advertisement design, advertisement artistic production, and advertisement entertainments) had an impact on young female interest, as the values of (t) calculated for the three dimensions reached (5.568, 2.213, 4.567 and 4.544) respectively, which are significant values at (0.05 ≥ α) significance level, and it was found that all dimensions had an impact on young females interest who use the Facebook.

Table 6 .

Facebook advertisements	B	S.D	Beta	t. value	Sig
Advertisement content	0.812	0.048	0.712	16.234	0.000*
Advertisement design	−0.069	0.058	−0.066	−1.229	0.191
Advertisement artistic production	0.166	0.039	0.171	4.344	0.000*
Advertisement entertainments	0.148	0.038	0.196	4.677	0.000*

*Statistically significant at (0.05 ≥ α) level (t) tabular value = (±1.96)

Table 7 .

Facebook advertisements	B	S.D	Beta	t. value	Sig
Advertisement content	0.254	0.051	0.289	5.568	0.000*
Advertisement design	0.112	0.048	0.131	2.213	0.028*
Advertisement artistic production	0.154	0.036	0.221	4.567	0.000*
Advertisement entertainments	0.148	0.036	0.196	4.544	0.000*

*Statistically significant at (0.05 ≥ α) level (t) tabular value = (± 1.96)

(H03): There is no statistically significant effect at the significance level (0.05 ≥ α) of the Facebook ads on young female's desire. Multiple regression tests were used in order to identify the impact of Facebook advertisements by its dimensions (advertisement content, advertisement design, advertisement artistic production, and advertisement entertainments) on young female desire, and Table 8 illustrates this:

Table 8, shows that the (t) test values is related to Facebook ads with its dimensions (advertisement content, advertisement design, advertisement artistic production, and advertisement entertainments) had an impact on young female desire, as the values of (t) calculated for the three dimensions reached (5.775, 2.453, −1.498 and 4.632) respectively, which are significant values at (0.05 ≥ α) significance level, and it was found that all dimensions had an impact on young females desire who use Facebook.

(H04): There is no statistically significant effect at the significance level (0.05 ≥ α) of Facebook ads on young females preferences. Multiple regression tests were used in order to identify the impact of Facebook advertisements by its dimensions (advertisement content, advertisement design, advertisement artistic production, and advertisement entertainments) on young female preferences, and Table 9 illustrates this:

Table 9, shows that the (t) test values that is related to Facebook ads with their dimensions (advertisement content, advertisement design, advertisement artistic production, and advertisement entertainments) had an impact on customers preferences of female youth customers, as the values of (t) calculated for the three dimensions reached (6.441, 2.545, 4.672 and 4.559) respectively, which are significant values at (0.05 ≥ α) significance level, and it was found that all dimensions had an impact on females customers preferences who use the Facebook.

Table 8 .

Facebook advertisements	B	S.D	Beta	t. value	Sig
Advertisement content	0.254	0.051	0.289	5.775	0.000*
Advertisement design	0.112	0.048	0.131	2.453	0.000*
Advertisement artistic production	−0.144	0.095	−0.092	−1.498	0.129
Advertisement entertainments	0.148	0.036	0.046	4.632	0.000*

*Statistically significant at (0.05 ≥ α) level (t) tabular value = (± 1.96)

Table 9 .

Facebook advertisements	B	S.D	Beta	t. value	Sig
Advertisement content	0.266	0.054	0.299	6.441	0.000*
Advertisement design	0.121	0.046	0.142	2.545	0.000*
Advertisement artistic production	0.157	0.089	0.066	4.672	0.000*
Advertisement entertainments	0.151	0.041	0.052	4.559	0.000*

*Statistically significant at ($0.05 \geq \alpha$) level (t) tabular value = (± 1.96)

6 Findings

- The arithmetic averages of the level of Facebook advertisements dimensions it was high, that the arithmetic averages of the level of Facebook ads dimensions ranged between (4.20–4.45), with an average of (4.29) which indicates a high degree. The overall arithmetic mean of Facebook advertisements dimensions indicate (4.29) with standard deviation of (0.64) which indicates a high degree.
- The arithmetic averages of the level of customers' attention dimensions ranged between (4.28–4.44), with average of (4.36) which indicates a high degree. The overall arithmetic mean of (4.36), with standard deviation of (0.69) with a high degree. This indicates that the customer's attentions is high.
- Facebook ads by its dimensions (advertisement content, advertisement design, advertisement artistic production, and advertisement entertainments) had an impact on young female's attention, as the values of (t) calculated for the three dimensions reached (10.675, 2.776, 4.766 and 4.664) respectively, which are significant values at ($0.05 \geq \alpha$) significance level, and it was found that all dimensions had an impact on the attention of young females using Facebook.
- Advertisement entertainments it was ranked number one with a high degree by Facebook were characterized as harmony between sound and image, which increases the attentions of young female users by Facebook depends on the reality of the markets and the characteristics of young females. Facebook advertisements enjoys modernity in terms of choosing the appropriate colors, shapes and sounds according to the characteristics and characteristics of the customers to have a clear impact on the customers' response (Aljumah et al., 2021; Al-Maroof et al., 2021; Almazrouei et al., 2020; Alshurideh et al., 2019)
- There is an impact of Facebook advertisements on the perceptions of young female users, and it has also been found that the ads content contains all information about the products trademark, trade name, values, quality, and specifications in a way that is striking in attentions (Khasawneh et al. 2021a, b; Kurdi 2021).
- There is an impact of Facebook advertisements on young females attentions towards the advertisement artistic production due to its ability to build and change the convictions of young female positively towards the advertising products to help them of perceiving, interesting, desirable, and selecting the favorable products they prefer.

- There is an impact of Facebook advertisements on young female's attentions towards the advertisement design based on young female characteristics in terms of age, attitudes, behaviours, and culture to ensure the attention ability of them towards the advertisements, and to help them to become more interested of following up Facebook advertisements as one of the famous social media used recently by business and marketers for promoting products with ease of customer's reachability at any time.

7 Conclusion

Business firms and marketing managers require understanding carefully the young customers attention by Facebook advertisements messages on the bases of messages by a number of sub-dimensions were as (advertisement content, advertisement design, advertisement artistic production, and advertisement entertainments) on young customers attention by a number of sub-dimensions were as (perception, ınterest, desire, and preference).

Business firms and marketing managers require to follow up young customers to design a number of advertising strategies to match young customers as one of biggest target segment by several business firms' products to persuade and changing their perceptions positively to help them of purchasing advertised products.

8 Research Limitations

Based on the results that have been reached, the researcher provides a number of recommendations were as: Marketing management required to pay more attention for young customers' females users of Facebook advertisements to influence and to attract them positively to become a regular followers for products advertisements concerning young female's customers, and marketing management required to pay more attention to young female's customers' perceptions, interests, desires, and preferences toward products advertisements to influence their attitudes and behaviours positively, and marketing management required to pay more efforts for developing advertisements content, design, artistic production, and entertainments to match young customers' female personal characteristics in bases of cultures, values, habits, personal interest, and life styles, and marketing management required to update the advertisements content in regard to products information in terms of the trade name and trade mark, products values and benefits, using products informations to keep young female customers aware of product knowledge which is a basic for advertisements credibility and preferences, and marketing management required to ensure young female customer's privacy and response to adopt Facebook advertisements easily.

9 Future Implications

Conducting several future studies that will address other dimensions of Facebook Advertisements as far as advertisement applications, online shopping, online in different environments with different population and samples.

References

Abu Kharma, Thaer, Al-Qaisi, & Qutaiba (2011). *The impact of electronic advertising on product quality perceptions*. Unpublished MA thesis, An-Najah National University, Nablus, Palestine.

Abuhashesh, M. Y., Alshurideh, M. T., & Sumadi, M. (2021). The effect of culture on customers' attitudes toward Facebook advertising: The moderating role of gender. *Review of International Business and Strategy, 31*(3), 416–437.

Ago, V. (2015). Influence of Facebook advertisement on the buying behaviors of students of a Nigerian University. *International Journal of Humanities and Social Science., 5*(7), 135–148.

Al Kurdi, B. H., & Alshurideh, M. T. (2021). Facebook advertising as a marketing tool: Examining the influence on female cosmetic purchasing behaviour. *International Journal of Online Marketing (IJOM), 11*(2), 52–74.

Al-Junaid, B. (2017). *Civil Liability for Electronic Advertising via the Internet* (1st ed.). Arab Studies Center for Publishing and Distribution.

Al-Maroof, R., Ayoubi, K., Alhumaid, K., Aburayya, A., Alshurideh, M., Alfaisal, R., & Salloum, S. (2021). The acceptance of social media video for knowledge acquisition, sharing and application: A comparative study among YouYube users and TikTok users' for medical purposes. *International Journal of Data and Network Science, 5*(3), 197–214.

Aljumah, A., Nuseir, M. T., & Alshurideh, M. T. (2021). The impact of social media marketing communications on consumer response during the COVID-19: Does the brand equity of a university matter. *The Effect of Coronavirus Disease (COVID-19) on Business Intelligence, 367*.

Almazrouei, F. A., Alshurideh, M., Al Kurdi, B., & Salloum, S. A. (2020). Social media impact on business: A systematic review. In *International Conference on Advanced Intelligent Systems and Informatics* (pp. 697–707). Springer, Cham.

Alshurideh, M., Salloum, S. A., Al Kurdi, B., & Al-Emran, M. (2019). Factors affecting the social networks acceptance: An empirical study using PLS-SEM approach. In *Proceedings of the 2019 8th International Conference on Software and Computer Applications* (pp. 414–418).

Belakahleh, Sh. (2017). *The impact of electronic advertising on individuals attitudes: A field study of the "Mobilis" communications corporation*. Unpublished MA thesis, Elaraby Ben Mehidi University, Oum El Bouaghi, Algeria.

Boyd, D., & Ellison, N. (2008). Social network sites: Definition, history, and scholarship. *Journal of Computer-Mediated Communication, 13*, 210–230.

Fazalur, R., Muhammad, I., Tariq, N., & Shabir, H. (2014). How Facebook advertising affects buying behavior of young consumers: The moderating role of gender. *Academic Research International, 5*(4); Department of Management Sciences, COMSATS Institute of Information Technology, Pakistan.

Gharibi, S. (2012). Explain the effectiveness of advertising using the ADIA model. *Interdisciplinary Journal of Contemporary Research in Business., 4*(2), 926–940.

Hassan, A. (2015). *The impact of electronic advertising on the purchasing behavior of the Jordanian car consumer: An empirical study on youth in Amman, Jordan*. Unpublished Master's thesis, Zarqa University, Zarqa, Jordan.

Heath, R. (2012). *Seducing the Subconscious: The Psychology of Emotional Influence in Advertising* (1st ed.). Wiley Blackwell.

Ibrahim, I. (2017). The role of the electronic promotional mix in influencing the Algerian consumer behavior. *Journal of Financial Economic Studies,* (10), Part (1), 47–64.

Khasawneh, M. A., Abuhashesh, M., Ahmad, A., Alshurideh, M. T., & Masa'deh, R. (2021a). Determinants of e-word of mouth on social media during COVID-19 outbreaks: An empirical study. In *The Effect of Coronavirus Disease (COVID-19) on Business Intelligence* (pp. 347–366). Springer, Cham.

Khasawneh, M. A., Abuhashesh, M., Ahmad, A., Masa'deh, R., & Alshurideh, M. T. (2021b). Customers online engagement with social media influencers' content related to COVID 19. In *The Effect of Coronavirus Disease (COVID-19) on Business Intelligence* (pp. 385–404). Springer, Cham.

Kurdi, B. A., Alshurideh, M., Nuseir, M., Aburayya, A., & Salloum, S. A. (2021). The effects of subjective norm on the intention to use social media networks: An exploratory study using PLS-SEM and machine learning approach. In *International Conference on Advanced Machine Learning Technologies and Applications* (pp. 581–592). Springer, Cham.

Rahim, H. (2015). Social media advertising value: A study on consumer's perception. *International Academic Research Journal of Business and Technology., 3*(1), 1–5.

Scott, M., Andrea, E., Peter, P., & Dennis, F. (2007). The effects of online advertising. *Communications of the ACM, 50*(3), 84–88.

The Impact of Social Media on Purchase Intention at Jordanian Women Clothing Sector

Maram Amer Alkhlifat, Sanaà Nawaf Al-Nsour, Faraj Mazyed Faraj Aldaihani, Raed Ismael Ababneh, Mohammad Issa Ghafel Alkhawaldeh, Muhammad Turki Alshurideh, and Sulieman Ibraheem Shelash Al-Hawary

Abstract The aim of the study was to examine the impact of social media on purchase intention. Therefore, it focused on Jordan-based companies operating in the women clothing sector. Data were primarily gathered through self-reported questionnaires created by Google Forms which were distributed to a random sample of (384) responses. In total, (370) responses were received including (6) invalid to statistical analysis due to uncompleted or inaccurate. Hence, the final sample contained (364) responses suitable to analysis requirements that were formed a response rate of (94.80%). structural equation modeling (SEM) was conducted to test hypotheses. The results showed that social media dimensions had a positive impact on purchase intention. Moreover, the results indicated that the highest impact was for Sharing Contents. Based on the previous results, the current study recommends organizations in the clothing sector to pay attention to social media and take advantage of

M. A. Alkhlifat · S. N. Al-Nsour
Business and Finance Faculty., World Islamic Science and Education University (WISE), Postal Code 11947, P.O. Box 1101, Amman, Jordan

F. M. F. Aldaihani
Kuwait Civil Aviation, Ishbiliyah Bloch 1, Street 122, Home 1, Kuwait City, Kuwait

R. I. Ababneh
Policy, Planning, and Development Program, Department of International Affairs, College of Arts and Sciences, Qatar University, 2713 Doha, Qatar

M. I. G. Alkhawaldeh
Directorates of Building and Land Tax, Ministry of Local Administration, Zarqa, Jordan

M. T. Alshurideh
Department of Marketing, School of Business, The University of Jordan, Amman 11942, Jordan
e-mail: m.alshurideh@ju.edu.jo; malshurideh@sharjah.ac.ae

Department of Management, College of Business, University of Sharjah, 27272 Sharjah, United Arab Emirates

S. I. S. Al-Hawary (✉)
Department of Business Administration, School of Business, Al Al-Bayt University, P.O. Box 130040, Mafraq 25113, Jordan
e-mail: dr_sliman73@aabu.edu.jo

M. Alshurideh et al. (eds.), *The Effect of Information Technology on Business and Marketing Intelligence Systems*, Studies in Computational Intelligence 1056,
https://doi.org/10.1007/978-3-031-12382-5_11

it in advertising and marketing for the women's clothing sector in Jordan, due to the impact of these sites on the purchase intention of consumers, emphasizing the importance of clarifying all information for products.

Keywords Social media · Purchasing intention · Clothes sector · Jordan

1 Introduction

Social media has become a contemporary phenomenon; As companies began to expand the operations of offering their products and marketing operations based on these media as one of the tributaries of their revenues, the great development in the Internet has provided these media with wide opportunities for the spread and promotion of advertisements for companies of all kinds, and thus helped their users to know the contents of advertisements for different companies, and organizations that provides detailed information for each of its products and services (Al-Hawary & Alhajri, 2020; Al-Hawary & Obiadat, 2021; Altarifi et al., 2015). Sanad (2019) believes that for the maximum benefit of these media, it is necessary to understand what is meant by social media, and how to apply it in many ways. Social media is distinguished from the various sites on the Internet in its ability to enable customers to produce, publish and criticize and help them classify content and interact with it online.

By studying the intent to purchase, social media has played an important role in searches, discovering new customers, and identifying the desires and needs of existing customers, and it also constitutes an opportunity to discover as much information as possible about what interests companies or customers, such as: buying a product, or Searching for job opportunities, or learning about content for the purpose of browsing only to learn about their content (Al-Hawary & Harahsheh, 2014; Al-Hawar & Al-Menhaly, 2016; Al-Hawary & Al-Smeran, 2017; Al-Hawary & Hussien, 2017; Alolayyan et al., 2018;). The use of social media as a marketing channel has spread, as organizations have found several ways to send advertising messages through social media. Advertising is used through private user content in social media, through the advertiser finding a link between the advertisement in social media and the company's website (Abuhashesh et al., 2021; Al Kurdi et al., 2021; Al-Nesour et al, 2016; Alshurideh et al., 2017a, b).

Reviewing the managerial literature related to the intention of purchasing, and social media, finds that there is ambiguity in the knowledge of the role of social media in determining the intent to purchase in the women's clothing sector in Jordan (Al-Hawary, 2013; Al-Hawary et al., 2013). It emerges through conducting research and presentation of previous studies, that there are paradoxes about the interpretation of the content of these variables, which indicates that there is a cognitive problem that requires preparation and setting scientific and cognitive determinants for it; therefore, this study contributes to the preparation of an intellectual and theoretical framework for the topics of the study, and works to provide a scientific contribution to its variables, and to contribute to the enrichment and strengthening of Arab libraries that lack research and studies on the subject of intention to purchase, and social media, which may open new ways for researchers in this topics.

The problem of the study lies in knowing the motives behind the purchase intent of customers in the women's clothing sector in Jordan, by using social media, and identifying the factors affecting purchase intent; Social media has evolved to become effective media that allows users to interact and communicate with acquaintances and friends (Al-Alwan, 2018; Aljumah et al., 2021; Khasawneh et al., 2021a, b; Kurdi et al., 2021). The number of social media users has become very large, and thus advertisers and marketers have alerted to the need to use these media to market their products, and to introduce their customers to their various products, and thus benefit by increasing their sales, and gaining larger numbers of customers by revealing their image in front of customers and stimulating customers' intention to buy and in the end maximizing its profits (Al-Hadithi, 2017; Almazrouei et al., 2020; Al-Maroof et al., 2021).

This study attempts to contribute to spreading the importance of social media in the work environment, which leads to an increase in the awareness of the competent parties and becomes more knowledgeable about their work. It also provides assistance to managers and workers in the Jordanian women's clothing sector in the field of detecting strengths and weaknesses and identifying the elements that affect the intention of e-purchasing, and social media. This study seeks to present a simple effort and a qualitative addition to the practical side of the strategic thinking of organizations, by revealing the impact of social media on purchasing intention in the women's clothing sector in Jordan.

2 Theoretical Framework and Hypothesis Building

2.1 Social Media

The emergence of social media has completely changed the world and the way it works and has brought people of all cultures and nationalities closer to each other, and the previous year's witnessed a remarkable investment by companies in Internet marketing operations through social media to advertise their services and products such as the following sites: Instagram, Twitter, Facebook as modern advertising platforms that allow individuals to communicate with each other. Social media is distinguished from other sites in its ability to enable users to produce, criticize, categorize, publish and interact with content over the internet (Khasawneh et al., 2021b; Tariq et al., 2022a, b). Thus enabling them to interact directly and exchange information, which makes these media a very popular and desirable feature, as their widespread spread made institutions seek to maintain their sites in social media by promoting their products, and working to find various modern methods to interact with their customers, and build relationships positive interaction with social media users, and the interaction of individuals via social media sites provided a new behavior that affected the purchasing intentions of consumers, as a new and useful platform for providing information. (Al-Nusour et al., 2016; Alshurideh, 2022; Mohammad et al., 2020).

There are many definitions of the concept of social media, and it is defined as a set of electronic systems that enable Internet users to create their own pages, and their ability to link them with other members through electronic social systems; to share products, information, and services as well as a variety of issues (Qaseem et al., 2018). Thabet (2017) defined social media as a number of web pages that facilitate interactions between members actively participating in social networks located on the Internet and that provide all the media that enables individuals to interact with each other and help them to do so. Sounia and Youssef (2017) defined social media as: a term that expresses a set of global networks that enable individuals to communicate with each other in virtual community environments; Where they are brought together by different links, such as interest or belonging to a class or country and studied or a specific group of the global system for the transmission of information and data. The researchers define social media as a group of websites on the Internet that allow individuals to create pages and groups and disseminate information on a large scale at the level of countries all over the world, through which it is possible to establish social and economic relations in addition to exchanging information and making purchases and sales through its various media which these sites provide easily.

Social media marketing is a new trend and a fast-growing way in which businesses reach their target customers easily, social media marketing can be defined simply as the use of social media channels to promote a company and its products, this type of marketing can be considered as a subset of online marketing activities that complement traditional web-based promotion strategies, such as email newsletters, online advertising campaigns, and encourage users to disseminate messages to personal contacts (Alwan & Alshurideh, 2022; Nadaraja & Yazdanifard, 2013).

There are many dimensions resulting from marketing through social media, the researcher believes that the dimensions of marketing through social media are summarized as follows; **Informatics**: Information is the primary material for making any decision, and the informatics system is meant as a homogeneous and interconnected group of business, and the elements that work to collect and control data, with the aim of producing and communicating useful information to decision makers through a network, lines and communication channels (Al-Sahn, 2007). **Interaction**: It is one of the most important concepts in the contemporary business world to ensure the survival of companies in the market, because marketers nowadays face many challenges due to technological efficiency, high competition, and the spread of market segmentation. The importance of interaction in the marketing process lies in the processes of education, educating the individual and giving him information to obtain services or goods that help change lifestyles and levels, and achieve various social benefits for the individual and society (Ahmad et al., 2020, 2021; Al-Hamad et al., 2021; Almaazmi et al., 2020; Alshurideh et al., 2019a, b). Reliability: reliability is one of the most important factors on which the quality of service depends in marketing, and fulfilling the credibility of services is the most important way that facilitates the provision of services by organizations with their customers, that is, organizations that seek to achieve success are the ones who put the customer in mind first, and then provide all the necessary processes and activities to meet all the needs and desires of customers in the target markets. **Content sharing**: Creating and sharing valuable and informative content helps develop the brand, thus increasing the likelihood of success of the marketing campaign capabilities, especially those of the company and what it offers as well, to be published on the Internet afterwards and on various social media sites, bringing the content to the largest number of people interested by searching.

2.2 Purchasing Intent

Social media has contributed to adding a new concept to the intention of purchasing by directing users to social media to search for products and information about the product, and make purchasing decisions, as the main objective of marketing operations through social media sites is to achieve and satisfy the desires and needs of the consumer, and develop all products to meet the consumers need and achieve their satisfaction, and the success and prosperity of organizations is not limited to discovering the consumer requirements, but rather work with increasing effort to identify requirements and needs or other motives of intentions for making purchasing decisions (Akour et al., 2021; Al Dmour et al., 2014; Al-Nesour et al., 2016; Shishan et al., 2022).

This process requires careful analysis and a deep study of the motives and precise factors that drive the purchasing intention of consumers, where the concept of purchase intention can be defined as the presence of a need among consumers to be satisfied, and these needs differ either to be basic or non-essential. Services and goods are one of the basic means used to satisfy the needs and desires of prospective consumers, and the need for these products is discovered through the interaction of internal motives and external stimuli that generate a person's feeling of need (Abu Qaaud et al., 2011; Al-Hawary & Al-Khazaleh, 2020; Al-Hawary et al., 2017; Alshurideh et al., 2017a, b; Metabis & Al-Hawary, 2013; Qaseem et al., 2018). Purchasing intent is defined as the pattern or means by which the individual consumer in his behavior to evaluate, search for, purchase or use goods, ideas and services that are expected to satisfy his desires and satisfy his needs, and it can also be defined as a set of behaviors that include the purchase and use of services and goods. It also includes those that precede these behaviors and determine the movements that the individual consumer makes while making the purchase decision (Al-Nesour et al., 2016). The researchers define it as a set of behaviors and patterns followed by individual consumers to search for, evaluate, purchase or use goods that he believes may meet their needs and desires, through which he determines the behavior and movements of the individual when making a purchasing decision.

Many factors and variables play an important role in the purchasing intention of individual consumers, whether they are internal or external, that directly affect the purchasing decision related to them. The most important of these factors that affect the purchasing intention of consumers are demographic factors (such as marital status, gender, race, and age)., where the purchasing behavior of females towards services and goods differs from the behavior of males and the behavior of the elderly consumer individuals differs from the behavior of the young consumers. The social status of individuals affects their purchasing behavior, as the marketer can identify the desires and needs of individual consumers and respond to them effectively according to demographic factors, which motivates the consumer to be more loyal and loyal to the marketing organization and the products issued by it, and this loyalty can be strengthened through promotional programs Directed and properly planned towards the needs and motives of the consumer, which is based on persuasion and influence on the intent of purchase, and the adoption of appropriate promotional methods that serve the objectives that the organization is trying to achieve (Al-Dmour et al., 2008; Kuester, 2012).

The changes that occurred in the behavior of consumers and the developments that resulted from the external environment led to the difference and multiplicity of factors affecting the purchasing decision, which made it difficult to study this behavior on consumers, and made institutions work to try to adapt to these factors and variables by distinguishing and diversifying their variables, whether in terms of prices or quality through Relying on distribution and advertising methods, in accordance with the capabilities of the organization and in a manner that suits the consumer, in response to the foregoing, it is clear that making the purchasing decision is not an easy process, but rather a complex process that passes through many stages, where the individual works after making the intention to purchase a particular product or service as a result of the urgency of the need or the necessity of its use, as he determines the possible alternatives to it that are presented to the Many sites and determine the specifications that he needs and the characteristics that distinguish each good or service from others, and many factors can be based in choosing the appropriate alternative such as price, quality and previous experiences of the consumer himself or the individuals surrounding him, and then he makes the purchase decision, and then the extent of customer satisfaction is determined about the service or product or not, it is worth noting that e-marketing has a very important role in making the purchasing decision, which works to attract consumers towards a specific commodity through the methods of contacts that work to entice them with the product and its specifications.

2.3 *Social Media and Purchasing Intent*

Social media has added a new concept to the intention of purchasing through the continuous increase in the quest to direct users towards social media in order to search for information and make purchasing decisions. Marketing through various activities has played a key role in achieving the satisfaction of the customer's needs and desires, and working to develop products to achieve customer satisfaction, so that the success of institutions is not limited only to discovering what the consumer needs, but rather work doubly to determine the reasons for finding that need or other intentions and motives for decision-making, and this in turn requires an in-depth study and careful analysis of the internal factors, and motives that motivate the purchase (Abu Zayyad et al., 2021; Al-Nesour et al., 2016).

Social media helps increase customer satisfaction through retailers realizing the competitive advantage of using the internet, and many of them have adopted social media as a way to interact, engage and create value with customers. Many well-known electronic retail stores such as Amazon have gained experience to deal with customers and have become fully aware of the customer and what he needs. Many studies have addressed the importance of the role of these media in increasing customer and customer satisfaction and influencing their purchase intention. In light of the cause-and-effect relationship of online customer satisfaction and their intention to buy, and repeating purchases from the same retailer several times means the desire to get the same product, and this indicates a high satisfaction. In other words, the more satisfied customer is more likely to buy repeatedly in the same manner and method (Al-Adayleh, 2015). Accordingly, the study hypothesis can be formulated as follows:

There is a statistically significant effect of social media on purchasing intention in the Jordanian women's clothing sector.

3 Study model

See Fig. 1.

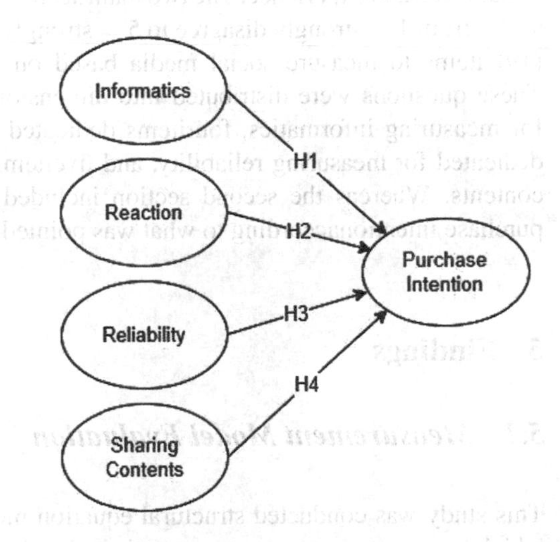

Fig. 1 Research model

4 Methodology

4.1 Population and Sample Selection

A qualitative method based on a questionnaire was used in this study for data collection and sample selection. The major aim of the study was to examine the impact of social media on purchase intention. Therefore, it focused on Jordan-based companies operating in the women clothing sector. Data were primarily gathered through self-reported questionnaires created by Google Forms which were distributed to a random sample of (384) responses. In total, (370) responses were received including (6) invalid to statistical analysis due to uncompleted or inaccurate. Hence, the final sample contained (364) responses suitable to analysis requirements that were formed a response rate of (94.80%), where it proved to be sufficient to the extent that was predictable and allowed for a presumption of data saturation (Sekaran & Bougie, 2016).

4.2 Measurement Instrument

A self-reported questionnaire that consists of two main sections along with a section regarding control variables was used as the measurement instrument. Control variables considered as categorical measures were composed of gender, age group, educational level, and experience. The two main sections were dealt with a five-point Likert scale (from 1 = strongly disagree to 5 = strongly agree). The first section contained (19) items to measure social media based on (Nadaraja & Yazdanifard, 2013). These questions were distributed into dimensions as follows: fiveitems dedicated for measuring informatics, fouritems dedicated for measuring reaction, fiveitems dedicated for measuring reliability, and fiveitems dedicated for measuring sharing contents. Whereas the second section included nineitems developed to measure purchase intentionaccording to what was pointed by (Kuester, 2012).

5 Findings

5.1 Measurement Model Evaluation

This study was conducted structural equation modeling (SEM) to test hypotheses, which represents a contemporary statistical technique for testing and estimating the relationship between factors and variables (Al-Adamat et al., 2020; Al-Gasawneh and Al-Adamat, 2020; Al-Hawary & Al-Syasneh, 2020; Al-Nady et al., 2016;

Table 1 Results of validity and reliability tests

Constructs	1	2	3	4	5
1. Informatics	**0.764**				
2. Reaction	0.415	**0.746**			
3. Reliability	0.366	0.364	**0.747**		
4. Sharing contents	0.391	0.458	0.381	**0.768**	
5. Purchase intention	0.625	0.597	0.604	0.622	**0.758**
VIF	1.834	1.552	2.341	1.784	–
Loadings range	0.694–0.864	0.612–0.802	0.671–861	0.733–0.816	0.581–0.841
AVE	0.584	0.557	0.559	0.590	0.574
MSV	0.441	0.512	0.419	0.381	0.451
Internal consistency	0.873	0.830	0.861	0.875	0.921
Composite reliability	0.875	0.833	0.863	0.878	0.923

Note Bold fonts indicate to square root of AVE

AlTaweel & Al-Hawary, 2021; Wang & Rhemtulla, 2021). Accordingly, the reliability and validity of the constructs were tested using confirmatory factor analysis (CFA) through the statistical program AMOSv24. Table 1 summarizes the results of convergent and discriminant validity, as well the indicators of reliability.

Table 1 shows that the standard loading values for the individual items were within the domain (0.581–0.864), these values greater than the minimum retention of the elements based on their standard loads (Al-Lozi et al., 2018; Sung et al., 2019). Average variance extracted (AVE) is a summary indicator of the convergent validity of constructs that must be above 0.50 (Howard, 2018). The results indicate that the AVE values were greater than 0.50 for all constructs, thus the used measurement model has an appropriate convergent validity. Rimkeviciene et al. (2017) suggested the comparison approach as a way to deal with discriminant validity assessment in covariance-based SEM. This approaches based on comparing the values of maximum shared variance (MSV) with the values of AVE, as well as comparing the values of square root of AVE ($\sqrt{AVE}$) with the correlation between the rest of the structures. The results show that the values of MSV were smaller than the values of AVE, and that the values of $\sqrt{AVE}$ were higher than the correlation values among the rest of the constructs. Therefore, the measurement model used is characterized by discriminative validity. The internal consistency measured through Cronbach's Alpha coefficient (α) and compound reliability by McDonald's Omega coefficient (ω) was conducted as indicators to evaluate measurement model. The results listed in Table 1 demonstrated that both values of Cronbach's Alpha coefficient and McDonald's Omega coefficient were greater than 0.70, which is the lowest limit for judging on measurement reliability (De Leeuw et al., 2019).

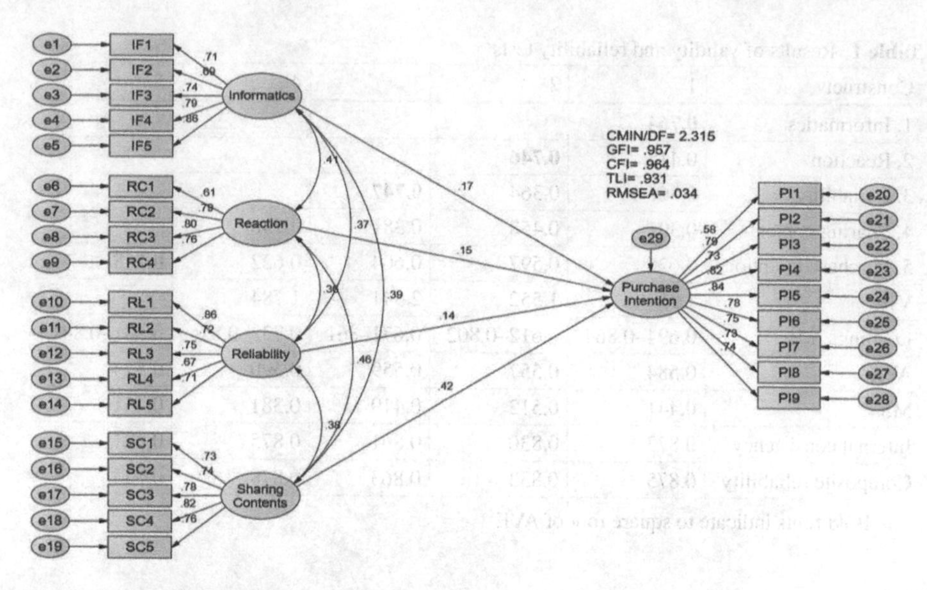

Fig. 2 SEM results of the social media effect on purchase intention

5.2 *Structural Model*

The structural model illustrated no multicollinearity issue among predictor constructs because variance inflation factor (VIF) values are below the threshold of 5, as shown in Table 1 (Hair et al., 2017). This result is supported by the values of model fit indices shown in Fig. 1 (Fig. 2).

The results in Fig. 1 indicated that the chi-square to degrees of freedom (CMIN/DF) was 2.315, which is less than 3 the upper limit of this indicator. The values of the goodness of fit index (GFI), the comparative fit index (CFI), and the Tucker-Lewis index (TLI) were upper than the minimum accepted threshold of 0.90. Moreover, the result of root mean square error of approximation (RMSEA) indicated to value 0.034, this value is a reasonable error of approximation because it is less than the higher limit of 0.08. Consequently, the structural model used in this study was recognized as a fit model for predicting the purchase intentionand generalization of its result (Ahmad et al., 2016; Shi et al., 2019). To verify the results of testing the study hypotheses, structural equation modeling (SEM) was used, the results of which are listed in Table 2.

The results demonstrated in Table 2 show that social media dimensions had a positive impact on purchase intention. Moreover, the results indicated that the highest impact was for Sharing Contents ($\beta = 0.423$, $t = 8.510$, $p = 0.000$), followed by informatics ($\beta = 0.172$, $t = 3.997$, $p = 0.002$), then reaction ($\beta = 0.152$, $t = 3.737$, $p = 0.007$), and finally the lowest impact was for reliability ($\beta = 0.136$, $t = 2.792$, $p = 0.000$).

Table 2 Hypothesis testing

Hypothesis	Relation	Standard beta	*t* value	*p* value
H1	Informatics→Purchase intention	0.172	3.997**	0.002
H2	Reaction→Purchase intention	0.152	3.737**	0.007
H3	Reliability→Purchase intention	0.136	2.792**	0.000
H4	Sharing contents→Purchase intention	0.423	8.510***	0.000

Note $*p < 0.05$, $**p < 0.01$, $***p < 0.001$

6 Discussion

The results indicated that the respondents use social media to a high degree, as it is an important source for obtaining information, especially in this sector, and that the world is witnessing various transformations and rapid leaps towards the digital economy because of the easy access to service, in record time and with less effort, and that women have a desire to interact, and this means that social media is an effective means that encourages customers to interact in terms of expressing their comments about products, which provides service providers with information about products that are useful in improving and developing them in order to increase customer satisfaction and retention. Accordingly, after the interaction between customers and companies producing the Jordanian women's clothing sector, the two sides of the process can achieve their interests, at the customers level, it is possible to obtain good products according to their desires, and at the providers level of these products, they are able to improve the quality of products and achieve their goals in increasing their sales numbers, achieving profits and at the same time attracting new customers and maintaining old customers. Social media is a good source of information related to the Jordanian women's clothing sector, as advertisements through social media provide new, sufficient and adequate information about products and everything related to them in terms of new and modern, as well as giving an idea of prices to customers, meaning that customers can obtain information easily, and without effort and time.

Women have the ability and support to participate in the content, and this result means that social media provides the advantage of the customer sharing information about the brand with other users, regarding products in the Jordanian women's clothing sector, and this is a benefit and a cumulative advertising method known as the snowball; That is, the user shares information via social media with other users, and other users can share the same information with new users; Thus, the information reaches a wider number of users, which brings benefit to users in knowing new products, as well as benefits product providers to the women's clothing sector in advertising and marketing their products electronically without effort, and in record time. The respondents have the desire and intention to buy through social media because of the information these media provide about products. Through these media, it is possible to learn about the experiences and experiments of old customers who made the purchase through these media, and these media provide information about trusted and preferred brands, or other brands, which encourage customers to buy.

The results of the study concluded that there is a statistically significant effect of social media on purchasing intention in the women's clothing sector in Jordan. This result means that social media of all kinds provides an important advantage for the consumer, which is information about products, whether those related to prices, quality, quality, offers, etc., which encourages a change in consumers' tendency to purchase products in the women's clothing sector in Jordan. Reliable, honest and real information about the product attracts the person's attention and may consider sharing it with others and interacting with the producers to express opinions and observations that may be useful in improving, developing or updating the products, which benefits all parties to the purchasing process. This result is consistent with the finding of Al Saifi study (2018) which found that perceived interactivity, credibility, privacy, and advertiser reputation are among the influencing factors influencing consumer attitudes towards social network advertising. The result is also consistent with Alwan study (2018), which showed that social media provides features, including: interactive motivation, interactiveness and information, and it has a role in purchasing intentions. The result is also consistent with Al-Yassin study (2017), which showed a statistically significant effect of the dimensions of social networking sites through the electronic spoken word in the fashion sector. The result was also consistent with Al-Nesour et al. (2016), which showed an effect of marketing using social media on purchase intention. The result is consistent with Al-Kilani and Al-Qurashi (2015), which showed the existence of an impact of the social network on the consumer purchasing decision-making process in the city of Amman. The result coincides with Ismail, which found a significant effect of marketing through social media on brand loyalty and awareness. The result is consistent with Murtiningsihi & Murad (2016), which found that there is an effect of social media marketing on brand loyalty. The result met with the result of Tatar & Eren-Erdogmus (2016), which showed a positive impact of marketing through social media on trust and loyalty to the hotel brand.

7 Recommendations

Based on the previous results, the current study recommends organizations in the clothing sector to pay attention to social media and take advantage of it in advertising and marketing for the women's clothing sector in Jordan, due to the impact of these sites on the purchase intention of consumers, emphasizing the importance of clarifying all information for products, and the importance of interaction, credibility and content sharing, and to intensify the process of promoting their products through social media, in order to ensure that they reach the largest possible number of users of social networking sites. It also recommends adopting social media in its business and focusing its efforts in increasing the effectiveness of its content on social networks. And update, develop and improve their commercial advertisements in a convincing and effective way to provide customers with all that is new about those products. Finally, the producers should update the information on these products so that customers can obtain this information easily and without effort and time to obtain it, and provide incentives and prizes to customers and users who share their promotional ads for their products with other users to reach the largest possible number of people.

References

Abu Qaaud, F., Al-Shoura, M., & Al-Hawary, S. I. (2011). The impact of the service marketing mix in the service quality of health services from the viewpoint of patients in government hospitals in Amman "A field study." *Abhath Al-Yarmouk, 27*(1B), 417–441.

Abuhashesh, M. Y., Alshurideh, M. T., & Sumadi, M. (2021). The effect of culture on customers' attitudes toward Facebook advertising: the moderating role of gender. *Review of International Business and Strategy, 31*(3), 416–437.

Abu Zayyad, H. M., Obeidat, Z. M., Alshurideh, M. T., Abuhashesh, M., Maqableh, M., & Masa'deh, R. E. (2021). Corporate social responsibility and patronage intentions: the mediating effect of brand credibility. *Journal of Marketing Communications, 27*(5), 533–510.

Ahmad, S., Zulkurnain, N., & Khairushalimi, F. (2016). Assessing the validity and reliability of a measurement model in structural equation modeling (SEM). *British Journal of Mathematics & Computer Science, 15*(3), 1–8. https://doi.org/10.9734/BJMCS/2016/25183

Ahmed, A., Alshurideh, M., Al Kurdi, B., & Salloum, S. A. (2020). Digital transformation and organizational operational decision making: a systematic review. In *International Conference on Advanced Intelligent Systems and Informatics* (pp. 708–719). Springer, Cham.

Ahmad, A., Alshurideh, M. T., Al Kurdi, B. H., & Alzoubi, H. M. (2021). Digital strategies: A systematic literature review. In *The International Conference on Artificial Intelligence and Computer Vision* (pp. 807–822). Springer, Cham.

Akour, I., Alshurideh, M., Al Kurdi, B., Al Ali, A., & Salloum, S. (2021). Using machine learning algorithms to predict people's intention to use mobile learning platforms during the COVID-19 pandemic: Machine learning approach. *JMIR Medical Education, 7*(1), 1–17.

Al-Adamat, A., Al-Gasawneh, J., & Al-Adamat, O. (2020). The impact of moral intelligence on green purchase intention. *Management Science Letters, 10*(9), 2063–2070.

Al-Adayleh, M. (2015). The role of social networks in influencing the online consumer purchasing decision. *The Jordanian Journal of Business Administration, 11*(1), 154–159.

Al-Dmour, H., & Al-Shraideh, M. T. (2008). The influence of the promotional mix elements on Jordanian consumer's decisions in cell phone service usage: An analytical study. *Jordan Journal of Business Administration, 4*(4), 375–392.

Al Dmour, H., Alshurideh, M., & Shishan, F. (2014). The influence of mobile application quality and attributes on the continuance intention of mobile shopping. *Life Science Journal, 11*(10), 172–181.

Al Kurdi, B. H., & Alshurideh, M. T. (2021). Facebook advertising as a marketing tool: Examining the influence on female cosmetic purchasing behaviour. *International Journal of Online Marketing (IJOM), 11*(2), 52–74.

Al-Gasawneh, J. A., & Al-Adamat, A. M. (2020). The relationship between perceived destination image, social media interaction and travel intentions relating to Neom city. *Academy of Strategic Management Journal, 19*(2), 1–12.

Al-Hadithi, A. L. (2017). The marketing effects of social media in the Saudi society. *King Khalid University Journal of Human Sciences, 25*(2), 195–238.

Al-Hamad, M., Mbaidin, H., AlHamad, A., Alshurideh, M., Kurdi, B., & Al-Hamad, N. (2021). Investigating students' behavioral intention to use mobile learning in higher education in UAE during Coronavirus-19 pandemic. *International Journal of Data and Network Science, 5*(3), 321–330.

Al-Hawary, S. I. (2013). The role of perceived quality and satisfaction in explaining customer brand loyalty: Mobile phone service in Jordan. *International Journal of Business Innovation and Research, 7*(4), 393–413.

Al-Hawary, S. I. S., & Alhajri, T. M. S. (2020). Effect of Electronic customer relationship management on customers' electronic satisfaction of communication companies in Kuwait. *Calitatea, 21*(175), 97–102.

Al-Hawary, S. I., & Al-Khazaleh A, M. (2020). The mediating role of corporate image on the relationship between corporate social responsibility and customer retention. *Test Engineering and Management, 83*(516), 29976–29993.

Al-Hawary, S. I., & Al-Menhaly, S. (2016). The Quality of e-government services and its role on achieving beneficiaries satisfaction. *Global Journal of Management and Business Research: A Administration and Management, 16*(11), 1–11.

Al-Hawary, S. I., & Al-Smeran, W. (2017). Impact of electronic service quality on customers satisfaction of Islamic banks in Jordan. *International Journal of Academic Research in Accounting, Finance and Management Sciences, 7*(1), 170–188.

Al-Hawary, S. I., & Al-Syasneh, M. S. (2020). Impact of dynamic strategic capabilities on strategic entrepreneurship in presence of outsourcing of five stars hotels in Jordan. *Business: Theory and Practice, 21*(2), 578–587.

Al-Hawary, S. I., & Harahsheh, S. (2014). Factors affecting Jordanian consumer loyalty toward cellular phone brand. *International Journal of Economics and Business Research, 7*(3), 349–375.

Al-Hawary, S. I., & Hussien, A. J. (2017). The impact of electronic banking services on the customers loyalty of commercial banks in Jordan. *International Journal of Academic Research in Accounting, Finance and Management Sciences, 7*(1), 50–63.

Al-Hawary, S. I. S., & Obiadat, A. A. (2021). Does mobile marketing affect customer loyalty in Jordan? *International Journal of Business Excellence, 23*(2), 226–250.

Al-Hawary, S. I., Al-Hawajreh, K., AL-Zeaud, H., & Mohammad, A. (2013). The impact of market orientation strategy on performance of commercial banks in Jordan. *International Journal of Business Information Systems, 14*(3), 261–279.

Al-Hawary, S. I., Batayneh, A. M., Mohammad, A. A., & Alsarahni, A. H. (2017). Supply chain flexibility aspects and their impact on customers satisfaction of pharmaceutical industry in Jordan. *International Journal of Business Performance and Supply Chain Modelling, 9*(4), 326–343.

Aljumah, A., Nuseir, M. T., & Alshurideh, M. T. (2021). The impact of social media marketing communications on consumer response during the COVID-19: Does the brand equity of a university matter. In *The effect of coronavirus disease (COVID-19) on business intelligence* (pp. 367–384).

Al-Kilani, Y., & Al-Qurashi, Z. (2015). The impact of the social network on the purchasing decision-making process for the consumer in the city of Amman. *An-Najah University Journal of Research, 29*(12), 2410–2444.

Al-Lozi, M. S., Almomani, R. Z. Q., & Al-Hawary, S. I. S. (2018). Talent Management strategies as a critical success factor for effectiveness of human resources information systems in commercial banks working in Jordan. *Global Journal of Management and Business Research: A Administration and Management, 18*(1), 30–43.

Almaazmi, J., Alshurideh, M., Al Kurdi, B., & Salloum, S. A. (2020). The effect of digital transformation on product innovation: A critical review. In *International conference on advanced intelligent systems and informatics* (pp. 731–741). Springer, Cham.

Al-Maroof, R., Ayoubi, K., Alhumaid, K., Aburayya, A., Alshurideh, M., Alfaisal, R., & Salloum, S. (2021). The acceptance of social media video for knowledge acquisition, sharing and application: A comparative study among YouTube users and TikTok users' for medical purposes. *International Journal of Data and Network Science, 5*(3), 197–214.

Almazrouei, F. A., Alshurideh, M., Kurdi, B. A., & Salloum, S. A. (2020). Social media impact on business: A systematic review. In *International Conference on Advanced Intelligent Systems and Informatics* (pp. 697–707). Springer, Cham.

Al-Nady, B. A., Al-Hawary, S. I., & Alolayyan, M. (2016). the role of time, communication, and cost management on project management success: An empirical study on sample of construction projects customers in Makkah City, Kingdom of Saudi Arabia. *International Journal of Services and Operations Management, 23*(1), 76–112.

Al-Nesour, H., Al-Manasra, A., & Al-Zayyat, M. (2016). The impact of marketing using social media on purchase intention in Jordan. *The Jordanian Journal of Business Administration, 12*(3), 519–530.

Alolayyan, M., Al-Hawary, S. I., Mohammad, A. A., & Al-Nady, B. A. (2018). Banking service quality provided by commercial banks and customer satisfaction. A structural equation modelling approaches. *International Journal of Productivity and Quality Management, 24*(4), 543–565.

Al Saifi, N. (2018). Factors affecting consumers' attitudes towards social network advertisements and their relationship to their behavioral response. *The Arab Journal of Media and Communication, 19*(1), 97–142.

Al-Sahn, A. (2007). *Marketing management in the environment of globalization and the Internet*. Dar Al-Fikr University.

Alshurideh, M. (2022). Does electronic customer relationship management (E-CRM) affect service quality at private hospitals in Jordan? *Uncertain Supply Chain Management, 10*(2), 325–332.

Alshurideh, M., Al Kurdi, B., Abu Hussien, A., & Alshaar, H. (2017a). Determining the main factors affecting consumers' acceptance of ethical advertising: A review of the Jordanian market. *Journal of Marketing Communications, 23*(5), 513–532.

Alshurideh, M., Salloum, S. A., Al Kurdi, B., Monem, A. A., & Shaalan, K. (2019b). Understanding the quality determinants that influence the intention to use the mobile learning platforms: A practical study. *International Journal of Interactive Mobile Technologies, 13*(11), 183–157.

Alshurideh, M., Al-Hawary, S. I., Batayneh, A. M., Mohammad, A., & Al-Kurdi, B.(2017b). The impact of Islamic banks' service quality perception on Jordanian customers loyalty. *Journal of Management Research, 9*(2), 139–159.

Alshurideh, M., Salloum, S. A., Al Kurdi, B., & Al-Emran, M. (2019a). Factors affecting the social networks acceptance: an empirical study using PLS-SEM approach. In *Proceedings of the 2019a 8th international conference on software and computer applications* (pp. 414–418).

Altarifi, S., Al-Hawary, S. I. S., & Al Sakkal, M. E. E. (2015). Determinants of e-shopping and its effect on consumer purchasing decision in Jordan. *International Journal of Business and Social Science, 6*(1), 81–92.

AlTaweel, I. R., & Al-Hawary, S. I. (2021). The mediating role of innovation capability on the relationship between strategic agility and organizational performance. *Sustainability, 13*(14), 7564.

Alwan, M., & Alshurideh, M. (2022). The effect of digital marketing on purchase intention: Moderating effect of brand equity. *International Journal of Data and Network Science, 10*(3), 1–12.

Alwan. (2018). Investigating the impact of advertising features on social media on customer buying intent. *International Journal of Information Management, 42*, 56–77.

Al-Yassin, L. (2017). *The effect of using social networking sites on purchasing behavior through the electronic spoken word: An applied study of the clothing sector in Jordan*. Master's Thesis, Middle East University, Jordan.

De Leeuw, E., Hox, J., Silber, H., Struminskaya, B., & Vis, C. (2019). Development of an international survey attitude scale: Measurement equivalence, reliability, and predictive validity. *Measurement Instruments for the Social Sciences, 1*(1), 9. https://doi.org/10.1186/s42409-019-0012-x

Hair, J. F., Babin, B. J., & Krey, N. (2017). Covariance-based structural equation modeling in the journal of advertising: Review and recommendations. *Journal of Advertising, 46*(1), 163–177. https://doi.org/10.1080/00913367.2017.1281777

Howard, M. C. (2018). The convergent validity and nomological net of two methods to measure retroactive influences. *Psychology of Consciousness: Theory, Research, and Practice, 5*(3), 324–337. https://doi.org/10.1037/cns0000149

Khasawneh, M. A., Abuhashesh, M., Ahmad, A., Alshurideh, M. T., & Masa'deh, R. (2021a). Determinants of e-word of mouth on social media during COVID-19 outbreaks: An empirical study. In *The effect of coronavirus disease (COVID-19) on business intelligence* (pp. 347–366). Springer.

Khasawneh, M. A., Abuhashesh, M., Ahmad, A., Masa'deh, R., & Alshurideh, M. T. (2021b). Customers online engagement with social media influencers' content related to COVID 19. In *The effect of coronavirus disease (COVID-19) on business intelligence* (pp. 385–404). Springer.

Kuester, S. (2012). MKT 301: Strategic marketing & marketing in specific industry contexts. *University of Mannheim, 110*, 393–404.

Kurdi, B. A., Alshurideh, M., Nuseir, M., Aburayya, A., & Salloum, S. A. (2021, March). The effects of subjective norm on the intention to use social media networks: An exploratory study using PLS-SEM and machine learning approach. In *International conference on advanced machine learning technologies and applications* (pp. 581–592). Springer, Cham.

Metabis, A., & Al-Hawary, S. I. (2013). The impact of internal marketing practices on services quality of commercial banks in Jordan. *International Journal of Services and Operations Management, 15*(3), 313–337.

Mohammad, A. A., Alshura, M. S., Al-Hawary, S. I. S., Al-Syasneh, M. S., & Alhajri, T. M. (2020). The influence of internal marketing practices on the employees' intention to leave: A study of the private hospitals in Jordan. *International Journal of Advanced Science and Technology, 29*(5), 1174–1189.

Murtiningsih, D., & Murad, A. (2016). The effect of social media marketing to brand loyalty. *International Journal of Business and Management Invention., 5*(5), 50–63.

Nadaraja, R., & Yazdanifard, R. (2013). Social media marketing: advantages and disadvantages. *Center of Southern New Hempshire University*, 1–10.

Qassem, R., Latif, R., Taher, V., & Habib, J. (2018). *The impact of social media as advertising channels on purchase intention: A field study in Al-Moussawi Lalizocam Company*. University of Maysan.

Rimkeviciene, J., Hawgood, J., O'Gorman, J., & De Leo, D. (2017). Construct validity of the acquired capability for suicide scale: Factor structure, convergent and discriminant validity. *Journal of Psychopathology and Behavioral Assessment, 39*(2), 291–302. https://doi.org/10.1007/s10862-016-9576-4

Sanad, N. (2019). *The impact of marketing using social media on purchase intention: A field study on the fashion sector in Jordan*. Unpublished Master's Thesis, Al al-Bayt University, Amman, Jordan.

Sekaran, U., & Bougie, R. (2016). *Research methods for business: A skill-building approach* (Seventh edition). Wiley.

Shi, D., Lee, T., & Maydeu-Olivares, A. (2019). Understanding the model size effect on SEM fit indices. *Educational and Psychological Measurement, 79*(2), 310–334. https://doi.org/10.1177/0013164418783530

Shishan, F., Mahshi, R., Al Kurdi, B., Alotoum, F. J., & Alshurideh, M. T. (2022). Does the past affect the future? An analysis of consumers' dining intentions towards green restaurants in the UK. *Sustainability, 14*(1), 1–14.

Sounia, C., & Youssef, B. (2017). The impact of social media on consumer behavior towards Islamic fashion—A study on a sample of veiled female students at Jijel University. *Academic Journal of Social and Human Studies, 37*(1), 29–37.

Sung, K.-S., Yi, Y. G., & Shin, H.-I. (2019). Reliability and validity of knee extensor strength measurements using a portable dynamometer anchoring system in a supine position. *BMC Musculoskeletal Disorders, 20*(1), 1–8. https://doi.org/10.1186/s12891-019-2703-0

Tariq, E., Alshurideh, M., Akour, E., Al-Hawaryd, S., & Al Kurdi, B. (2022a). The role of digital marketing, CSR policy and green marketing in brand development at UK. *International Journal of Data and Network Science, 6*(3), 1–10.

Tariq, E., Alshurideh, M., Akour, I., & Al-Hawary, S. (2022b). The effect of digital marketing capabilities on organizational ambidexterity of the information technology sector. *International Journal of Data and Network Science, 6*(2), 401–408.

Tatar, ŞB., & Eren-Erdoğmuş, İ. (2016). The effect of social media marketing on brand trust and brand loyalty for hotels. *Information Technology & Tourism, 16*(3), 249–263.

Thabet, H. (2017). *Viral marketing and its impact on purchasing decision-making among consumers of social networking sites for the student segment at the Islamic University in the Gaza Strip.* Master's Thesis, The Islamic University of Gaza, Palestine.

Wang, Y. A., & Rhemtulla, M. (2021). Power analysis for parameter estimation in structural equation modeling: A discussion and tutorial. *Advances in Methods and Practices in Psychological Science, 4*(1), 1–17. https://doi.org/10.1177/2515245920918253

The Impact of the Digital Marketing for Education Services on the Mental Image for Students in Private Universities in Jordan

Abdullah Matar Al-Adamat, Nisreen Ahmad Fares Falaki, Majed Kamel Ali Al-Azzam, Faraj Mazyed Faraj Aldaihani, Reham Zuhier Qasim Almomani, Anber Abraheem Shlash Mohammad, Mohammed Saleem Khlif Alshura, Sulieman Ibraheem Shelash Al-Hawary, D. Barween Al Kurdi, and Muhammad Turki Alshurideh

Abstract This study aims to identifying the impact of digital marketing on the mental image of students at private universities in Jordan. The population of the study consisted of all the higher education students in the private Jordanian universities in (18) universities numbering (6381). The study sample was chosen randomly,

A. M. Al-Adamat · N. A. F. Falaki
Department of Business Administration & Public Administration, School of Business, Al Al-Bayt University Jordan, P.O. Box 130040, Mafraq 25113, Jordan
e-mail: aaladamat@aabu.edu.jo

M. K. A. Al-Azzam
Department of Business Administration, Faculty of Economics and Administrative Sciences, Yarmouk University, P.O. Box 566, Irbid 21163, Jordan

F. M. F. Aldaihani
Kuwait Civil Aviation, Ishbiliyah bloch 1, street 122, home 1, Kuwait City, Kuwait

R. Z. Q. Almomani
Business Administration, Amman, Jordan

A. A. S. Mohammad
Marketing Department, Faculty of Administrative and Financial Sciences, Petra University, P.O. Box 961343, Amman 11196, Jordan

M. S. K. Alshura
Management Department, Faculty of Money and Management, The World Islamic Science University, P.O. Box 1101, Amman 11947, Jordan

S. I. S. Al-Hawary (✉)
Department of Business Administration, School of Business, Al Al-Bayt University, P.O.Box 30040, Mafraq 25113, Jordan
e-mail: dr_sliman73@aabu.edu.jo

D. B. Al Kurdi
Department of Marketing, Faculty of Economics and Administrative Sciences, The Hashemite University, Zarqa, Jordan
e-mail: barween@hu.edu.jo

M. Alshurideh et al. (eds.), *The Effect of Information Technology on Business and Marketing Intelligence Systems*, Studies in Computational Intelligence 1056,
https://doi.org/10.1007/978-3-031-12382-5_12

and it consisted of (359) male and female students. The distribution of the questionnaire was done through an electronic link by (Google Drive) and was sent to the students with the collaboration of the deanship of students' affairs. To conduct this study, the analytical and descriptive method was used. The results of the study showed a statistically significant effect of digital marketing on the mental image of students at private universities in Jordan. Based on the study results, the researchers recommend managers and decision makers of the private universities in Jordan to enhance their interactive advertisements on the pages of others, enhance the pricing of their educational services, and make updates on their marketing channels to achieve a positive image.

Keywords Digital marketing · Education services · Mental image · Private universities · Jordan

1 Introduction

The rapid developments in the business environment have led organizations to apply digital marketing to their products and services, and to shift from traditional marketing methods into digital methods (Al-Hawary & Al-Smeran, 2017; Al-Hawary & Hussien, 2017; Al-Hawary & Obiadat, 2021). The nature of the tasks carried out by the organization and the external surrounding factors should be taken into account. Hence, it has become necessary for modern organizations to take advantage of strategies, scientific methods and applied experiences in digital marketing in order to increase the number of their clients and enhance their confidence through studying modern trends for the development of digital marketing (Al-Hawary & Alhajri, 2020; Altarifi et al., 2015; AlTaweel & Al-Hawary, 2021). The development of information and communication technology since (1990) until now has led to many changes in organizations' use for information technology. A number of digital platforms have been gradually integrated into marketing plans, which has led to the significant spread of digital marketing, even though it was associated with customer relationship management applications (Al-Nady et al., 2016; Al-Hawary & Alwan, 2016; Al-Hawary & Al-Syasneh, 2020; Allahow et al., 2018; Al-Hawary & Ismael, 2010). Recently, it has developed in conjunction with the development of electronic applications, which made it easy to apply digital marketing strategies. Undoubtedly, it contributes to improve the effectiveness of the organization's marketing plans and achieving its future goals (Ahmad et al., 2021a, b, c; Alhalalmeh et al., 2020; Tariq et al., 2022a, b).

M. T. Alshurideh
Department of Marketing, School of Business, The University of Jordan, Amman 11942, Jordan
e-mail: m.alshurideh@ju.edu.jo; malshurideh@sharjah.ac.ae

Department of Management, College of Business, University of Sharjah, 27272 Sharjah, United Arab Emirates

Many organizations tended to take advantage of modern digital applications that provided them the opportunity to market their services through the information network (Ahmed et al., 2020; Almaazmi et al., 2020; Alwan & Alshurideh, 2022; Lee et al., 2022a, b). Private universities are among the most prominent organizations affected by the transition to the area of information, and aware to the importance of benefiting from modern technological concepts by using them in their business environment (Al Kurdi et al., 2020a; Alshurideh et al., 2019a, b; Alshurideh, 2022). Given that perspective, education is one of the most affected fields by the transition into the information area (Al-Hawary & Batayneh, 2010). Many universities in various countries of the world have tended to market their educational services through electronic applications, which include a virtual environment of electronic applications through which universities can market their educational services (Al-Hawary, 2010; Al Kurdi et al., 2020b; Alshurideh et al., 2020; Alshurideh et al., 2021a, b; Akour et al., 2021; Al-Hamad et al., 2021; Leo et al., 2021). In light of the intense competition between private universities towards marketing their educational services, using strategies that contribute to the formation of a mental image to their students, universities tend to achieve the competitive advantage through the formation of a positive mental image for the public (Al-Hawary & Al-Khazaleh, 2020, Alsharari & Alshurideh, 2020).

Accordingly, it has become important for universities to understand the needs of their students and society to understand the needs of their students and society to make them able to market their educational services according to modern and previously studied strategies, in order to form a perceptive mental image about the educational services provided by universities, which lead to affect their perception, motivation and desire to study in these universities, especially in light of the escalating competition between private universities to reach the competitive advantage. since the mental image of university students and the public reflects the mental representation that students can feel or hear about universities, the digital marketing of educational services, if it is built on modern marketing foundations and strategies, may have its effects on society's perception on the image of educational services provided by universities; this influence may raise the motivation and desires of individuals to study in private universities in Jordan.

The digital marketing of educational services is one of the issues related to modernity in the field of marketing technology; therefore it sheds light on the extent to which private universities in Jordan keep pace with modern managerial systems, models and strategies, as well as the extent to which private universities in Jordan use modern digital applications to contribute in dealing with all marketing relationship parties. In addition to clarifying the role played by digital marketing for educational services in shaping the mental image for its students and society as a whole. Accordingly, the results of this study emerged from a set of recommendations that help decision-makers in universities to reconsider their digital marketing strategies, which reflects a positive mental image for students in private universities in Jordan, and their prospective audience at the local and international levels. Here, this study came to reveal the impact of digital marketing for educational services on the mental image of students in private universities in Jordan.

2 Theoretical Framework and Hypothesis Development

2.1 Digital Marketing

The concept of digital marketing has become concerning about the marketing of commercial services and products, but this concept has developed to be more comprehensive, as this digital transformation coincided with the development of universities. Great trends have emerged to use digital marketing tools in marketing educational services to universities by providing a set of scientific foundations and rules that contribute to keep pace with technological developments to bridge the gap between the requirements of marketing educational services and modern technology (Ahmed et al., 2017). Technological developments in the field of digital marketing have made it a marketing equivalent to regular marketing, with the addition of a new character, which is adapting to the requirements of modern marketing, and the tends of public. Digital marketing has become more advanced than conventional marketing because of its speed in transferring and downloading data easily and accurately. Moreover, digital marketing support interaction and communication between individuals and organizations implement marketing activities, contribute to adopt with modern developments as well (Al-Hawary & Obiadat, 2021).

Many universities have started to market their educational services by employing marketing activities and practices in a scientific manner, believing in the importance of marketing these services as one of the most important entrances to development and keeping pace with modernity. Digital marketing for educational services helps improve their competitive capabilities and raise the level of their marketing performance. From this point of view, digital marketing has become one of the scientific trends that universities have adopted through designing strategies and meaningful marketing plans to meet the needs of the public from these educational services (Muhammad, 2021).

Al Yamani indicated that digital marketing has many benefits, including: reducing costs by reducing the need for usual marketing uses such as printing and paper publishing, increasing marketing power by entering global markets, and contributing to increase competition in the provision of services or products. This leads to enable organizations to compete, improve communication between the public and the organization, and have control by following up individuals and knowing their tendencies.

Badous (2017: 442) indicated that the marketing of university educational services is: "analysis, planning, implementation, and control on university services they provide. It is a preparation that achieves the voluntary exchange of things that have value in the targeted markets to achieve the goals of the university based on the needs and desires of students, by using effective methods." for pricing, promotion, and distribution. On the other hand, Kotler explained that marketing educational services contributes to help students understand the wide changing context that is occurring at the university, by providing them with mechanisms through which they can extrapolate the future. Moreover, it provides them with important feedback as a mechanism for determining the difference in specializations and building exchanges in needs and demand forms for university clients. Through this, the university knows about the quality of educational services that achieve student satisfaction over time.

Basha study (2019) indicated that it highlights the importance of digital marketing for educational services in the power of technological applications and marketing social media which contribute to share marketing information with a global audience; This marketing information for educational services, that is shared in the form of videos, photos or information, can have a significant role in influencing the decision-making process for students. It should be noted that marketing technology means, especially social websites, are followed-up by the majority of students. The adoption of popular marketing channels in marketing educational services contributes to reach the right audience, and also enhances the university's competitiveness and marketing performance. In order to let the university meet the needs of students, obtain a market share and achieve financial resources, it must have its own marketing mixture. Perhaps the most prominent digital marketing tools used by organizations are the following:

E-mail: E-mail is one of the digital media at which digital messages can be exchanged. Marketing through e-mail is called direct marketing, at which marketer sends his commercial messages to a group of people through digital messages, and these messages are newsletters sent by the organization to its clients to build a participatory relationship with them. Promotional messages can also be sent to the public that include specific advertisements and offers for services or products.

Social Media: Social media is one of the digital media that can be used as a type of digital marketing. These sites allow the organization to reach the largest number of clients by digital messaging and digital publishing that allows sharing commercial messages through these sites, whether these messages are written, audio, images or videos. These sites have contributed to transform the power of digital marketing from marketers to customers such as (Facebook, YouTube, Twitter, Instagram, WhatsApp) (AlHamad et al., 2021; Al-Maroof et al., 2021; Alshamsi et al., 2020). These sites are the most common, famous, used and least expensive in marketing, and it is used to promote and advertise digital marketing elements (Salloum et al., 2020).

Website: this is used by designing a digital page to be accessed by anyone through search engines on the Internet. Creating the organization's website requires images and information about the nature of the products or services it provides. The website plays an important role in marketing if the design is comprehensive, integrated, and attractive to users, and this contributes to supporting the online marketing process. The organization can show digital content on the site such as advertising, promotion, pricing, and the products or services it offers; in addition it can make it possible for the user to order directly (Al-Adili, 2015; Al-Hawary, 2013; Al-Hawary & Al-Menhaly, 2016; Al-Hawary & Harahsheh, 2014; Mohammad et al., 2020).

Smartphone Apps: Marketing applications available on mobile phones is an important tool of digital marketing, buying, and selling, where organizations can market their services or products through digital applications. Studies have proven that these applications have opened great horizons in the digital marketing process, as users can download these applications on their mobile devices and see the most important products and services offered by organizations. These applications are characterized by giving the user the ability to try online shopping and follow up transactions easily (Al Dmour et al., 2014; Kurdi et al., 2021; Suleman et al., 2021).

2.2 Mental Image

The mental image became one of the important topics for researchers in the fields of media and management. Measuring and achieving it attracted the organizations by forming a mental image of the organization through planning and implementing its decisions, due to its importance in the social behavior on which the organizations are based. The formation of a positive mental image of the public is a goal that all organizations seek at all levels. The organization has always realized the importance of the emotional and mental connection between the organization and the public. The positive mental image of the public plays an important role in attracting dealers, gaining their satisfaction and loyalty to the organization. This contributes to increase its sales and enhance its reputation for the public opinion; thus, obtaining a competitive market portion for other organizations (Abu Qaaud et al., 2011; Al-Hawary et al., 2013, 2017; Alolayyan et al., 2018; Alshurideh et al., 2017; Metabis & Al-Hawary, 2013).

The definition of the mental image depends on the nature of the organizations' work, and the perspective from which this concept is seen. The concept of the mental image from the media point of view differs in some matters with the concept of the mental image from the point of view of management and marketing scholars. One of the most prominent definitions of the managerial mental image is: "impressions and values that are formed in the minds of others, from services and products provided by the organization to gain the client satisfaction" (Qaseem & Al-Aksha, 2020: 11). It is also defined as: the process of designing a set of mental differences through which the organization can distinguish its services and products from other competitors and their products in the similar market.

The researcher defines the mental image of educational services as: the impressions and perceptions of the public and university students about the level of quality of services and educational programs provided by a university, by realizing their importance, the extent of their impact on their motivation, and their desire to benefit from it if they have a positive mental image. The mental image has many dimensions at which the level of this image is measured for individuals; these dimensions are: **Perception** is the individual's reception of information related to the things surrounding him, and the ability of the individual to sort and organize the information he receives. In order to reach the perception, there should be a coherent and effective image that makes the individual see things in the same way seen by others (Al-Shammari, 2017). **Influence** refers to the result of the interaction conducted by the individual and the organization, which is summarized in the extent to which the organization convinces individuals with its services or products, through its moral commitment towards them, as well as their efforts to achieve success which makes individuals ready to deal with them. **Motivation** refers to the internal self-power that encourages the individuals to deal with the organization, as a result of the organization's fulfillment of their desires and needs, through their feelings, emotions, behavior and motives towards it (Gallab, 2011). **Desire** refers to the extent to which the organization satisfy and meet the needs and desires of individuals, according to their future aspirations and ambitions as a result of dealing with an organization. Regarding university students, the desire is fulfilled when they find that universities make them special and responsible people to achieve success, as well as respecting them by the society (Al-Amri, 2015).

2.3 The Relationship Between Digital Marketing and Mental Image

Abu Naser and El Talla, explained the reality of the effectiveness of electronic marketing in technical colleges in Palestine; he found that there is a correlation between smart dimensions and sustainability. While Nizar and Janathanan sees that customers of online stores can choose the suitable product through social networking sites, and this kind of shopping is safe and effective, as well as digital marketing through social media affects the buying behavior of the consumer. Aljawarneh et al. showed that the e-marketing elements represented in the dimensions of (nature of banking service, pricing of banking services, and promotion of banking services) have a positive impact on forming the mental image of customers.

Oluwasola, indicated that private universities mostly use digital marketing tools such as university website, social media marketing and digital marketing. Nurbait et al., explained that there are differences in digital marketing applied in universities listed in the TopTwenty category) with unlisted universities, at which listed universities implemented digital marketing in (8) types, namely (Search engine optimization, content marketing, influencing marketing, marketing automation, and social media marketing) which made listed universities gain a consuming community of unlisted universities during the COVID-19 pandemic. Thus, the hypothesis of the study can be formulated as follows:

There is a statistically significant effect at the significance level ($\alpha > 0.05$) of digital marketing for educational services on the mental image of private university students in Jordan.

3 Study model

See Fig. 1.

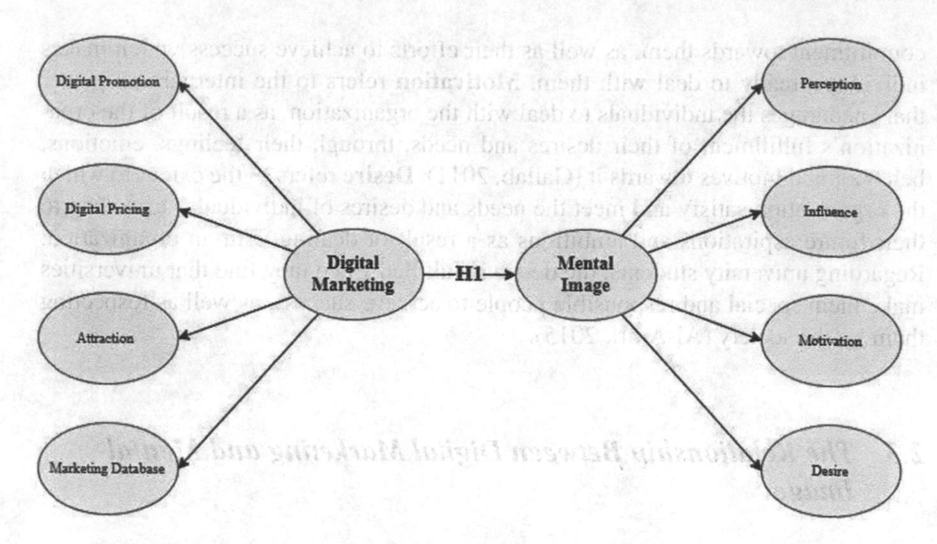

Fig. 1 Research model

4 Methodology

4.1 Population and Sample Selection

A qualitative method based on a questionnaire was used in this study for data collection and sample selection. The major aim of the study was to examine the impact of digital marketing on a mental image. Therefore, it was focused on private universities in Jordan. Data were primarily gathered through self-reported questionnaires created by Google Forms which were distributed to a random sample of (430) postgraduate students. In total, (359) responses were received including (13) invalid to statistical analysis due to uncompleted or inaccurate. Hence, the final sample contained (346) responses suitable to analysis requirements that were formed a response rate of (80.46%), where it proved to be sufficient to the extent that was predictable and allowed for a presumption of data saturation (Sekaran & Bougie, 2016).

4.2 Measurement Instrument

A self-reported questionnaire that consists of two main sections along with a section regarding control variables was used as the measurement instrument. Control variables considered as categorical measures were composed of gender, age group, educational level, and experience. The two main sections were dealt with a five-point Likert scale (from 1 = strongly disagree to 5 = strongly agree). The first section contained (23) items to measure digital marketing based on (Kusumawati, 2019; Warokka,

2020). These questions were distributed into dimensions as follows: six items dedicated for measuring digital promotion, five items dedicated for measuring digital pricing, five items dedicated for measuring attraction, and seven items dedicated for measuring marketing database. Whereas the second section included (21) items developed to measure mental image according to what was pointed by (Larson et al., 2009). This variable was divided into four dimensions: perception that was measured through five items, influence which measured by six items, motivation was measured using five items, and desire that were measured by five items.

5 Findings

5.1 *Measurement Model Evaluation*

This study was conducted structural equation modeling (SEM) to test hypotheses, which represents a contemporary statistical technique for testing and estimating the relationship between factors and variables (Wang & Rhemtulla, 2021). Accordingly, the reliability and validity of the constructs were tested using confirmatory factor analysis (CFA) through the statistical program AMOSv24. Table 1 summarizes the results of convergent and discriminant validity, as well the indicators of reliability.

Table 1 shows that the standard loading values for the individual items were within the domain (0.641–0.894), these values greater than the minimum retention of the elements based on their standard loads (Al-Lozi et al., 2018; Sung et al., 2019). Average variance extracted (AVE) is a summary indicator of the convergent validity of constructs that must be above 0.50 (Howard, 2018). The results indicate that the AVE values were greater than 0.50 for all constructs, thus the used measurement model has an appropriate convergent validity. Rimkeviciene et al. (2017) suggested the comparison approach as a way to deal with discriminant validity assessment in covariance-based SEM. This approaches based on comparing the values of maximum shared variance (MSV) with the values of AVE, as well as comparing the values of square root of AVE ($\sqrt{AVE}$) with the correlation between the rest of the structures. The results show that the values of MSV were smaller than the values of AVE, and that the values of $\sqrt{AVE}$ were higher than the correlation values among the rest of the constructs. Therefore, the measurement model used is characterized by discriminative validity. The internal consistency measured through Cronbach's Alpha coefficient (α) and compound reliability by McDonald's Omega coefficient (ω) was conducted as indicators to evaluate measurement model. The results listed in Table 1 demonstrated that both values of Cronbach's Alpha coefficient and McDonald's Omega coefficient were greater than 0.70, which is the lowest limit for judging on measurement reliability (De Leeuw et al., 2019).

Table 1 Results of validity and reliability tests

Constructs	1	2	3	4	5	6	7	8
1. DPRO	**0.778**							
2. DPRI	0.517	**0.796**						
3. ATTR	0.497	0.538	**0.793**					
4. MADB	0.435	0.511	0.492	**0.780**				
5. PERC	0.409	0.446	0.537	0.563	**0.789**			
6. INFL	0.628	0.638	0.624	0.597	0.628	**0.785**		
7. MOTI	0.579	0.599	0.662	0.681	0.571	0.601	**0.754**	
8. DESI	0.638	0.658	0.674	0.671	0.552	0.668	0.647	**0.797**
VIF	3.008	3.466	3.189	3.415	–	–	–	–
Loadings range	0.684–0.883	0.692–0.874	0.704–0.891	0.657–0.894	0.691–0.884	0.641–0.877	0.704–0.855	0.695–0.864
AVE	0.605	0.633	0.628	0.609	0.623	0.615	0.569	0.635
MSV	0.503	0.497	0.516	0.467	0.514	0.503	0.497	0.506
Internal consistency	0.897	0.893	0.892	0.911	0.889	0.903	0.866	0.894
Composite reliability	0.901	0.895	0.894	0.915	0.891	0.905	0.868	0.896

Note DPRO: digital promotion, DPRI: digital pricing, ATTR: attraction, MADB: marketing database, PERC: perception, INFL: influence, MOTI: motivation, DESI: desire, bold fonts in the table indicate to root square of AVE

5.2 Structural Model

The structural model illustrated no multicollinearity issue among predictor constructs because variance inflation factor (VIF) values are below the threshold of 5, as shown in Table 1 (Hair et al., 2017). This result is supported by the values of model fit indices shown in Fig. 1 (Fig. 2).

The results in Fig. 1 indicated that the chi-square to degrees of freedom (CMIN/DF) was 2.064, which is less than 3 the upper limit of this indicator. The values of the goodness of fit index (GFI), the comparative fit index (CFI), and the Tucker-Lewis index (TLI) were upper than the minimum accepted threshold of 0.90. Moreover, the result of root mean square error of approximation (RMSEA) indicated to value 0.049, this value is a reasonable error of approximation because it is less than the higher limit of 0.08. Consequently, the structural model used in this study was recognized as a fit model for predicting the DEP and generalization of its result (Ahmad et al., 2016; Shi et al., 2019). To verify the results of testing the study hypotheses, structural equation modeling (SEM) was used, the results of which are listed in Table 2.

The results demonstrated in Table 2 show that digital marketing had a positive impact relationship on the mental image. Moreover, the results indicated that the highest impact was for marketing database ($\beta = 0.464$, $t = 29.905$, $p = 0.000$), followed by digital pricing ($\beta = 0.245$, $t = 19.241$, $p = 0.000$), then attraction ($\beta = 0.153$, $t = 18.273$, $p = 0.032$), and finally the lowest impact was for digital promotion ($\beta = 0.122$, $t = 13.522$, $p = 0.040$).

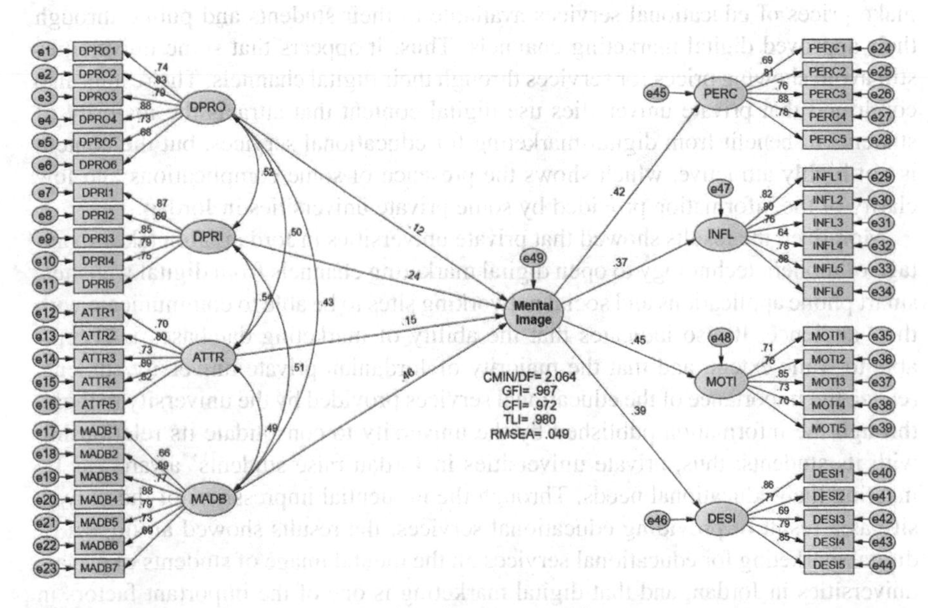

Fig. 2 SEM results of the digital marketing effect on mental image

Table 2 Hypothesis testing

Relation	Unstandardized coefficient (B)	Standardized error (S.E.)	Standardized coefficient (β)	*t* value	*p* value
Digital promotion → Mental image	0.311	0.023	0.122	13.522	0.040
Digital pricing → Mental image	0.558	0.029	0.245	19.241	0.000
Attraction → Mental image	0.402	0.022	0.153	18.273	0.032
Marketing database → Mental image	0.628	0.021	0.464	29.905	0.000

Note * $p < 0.05$, ** $p < 0.01$, *** $p < 0.001$

6 Discussion

The study results showed that private universities keep pace with technological developments in the field of digital marketing for educational services, and this indicates the presence of strengths in most cases, and there are some matters that should be improved in order to promote digital marketing of educational services in these universities. The results indicate that private universities in Jordan often make prices of educational services available to their students and public through their approved digital marketing channels. Thus, it appears that some universities still avoid showing prices for services through their digital channels. The results also confirmed that private universities use digital content that attracts the majority of students to benefit from digital marketing for educational services, but the content is not highly attractive, which shows the presence of some complications and low clarity of the information provided by some private universities in Jordan.

Moreover, the results showed that private universities in Jordan often take advantage of modern technology to open digital marketing channels from digital websites, smart phone applications and social networking sites to be able to communicate with their audience. It also indicates that the ability of marketing databases is acceptable to some extent, and that the majority of Jordanian private university students realize the importance of the educational services provided by the university to them, through the information published by the university to consolidate its relationship with its students; thus, private universities in Jordan raise students' awareness by meeting their educational needs. Through the influential impressions of the university as a result of providing educational services, the results showed an impact of digital marketing for educational services on the mental image of students of private universities in Jordan, and that digital marketing is one of the important factors in forming the mental image of students. This was supported by Stefco et al., whose results showed the existence of a correlation and impact of electronic marketing tools on the image of the university.

It can be concluded that whenever private universities in Jordan increase their interest in marketing educational services, this will reflect positively on the mental image of students, which will lead to transfer the mental image to the public. This also makes students partners of the university in marketing educational services. The results indicate a positive impact of digital promotion on the mental image of graduate students, but this effect is small, due to the presence of weakness in the planning of marketing strategies at private universities, as private universities were not successful in choosing methods and promotional strategies that enhance their mental image among students.

In addition, the results showed that the digital content in universities does not meet the aspirations to some extent, and this may be due to the digital marketing channels of the universities that still need to be updated and developed in a way that makes their characteristics, advantages and design in line with the aspirations of students.

7 Recommendations

Based on the study results, it recommends the university administrations to enhance interactive advertisements on other pages, funded digital advertisements through social networking sites to contribute to attracting targeted students should be at the internal and external levels to benefit from their enrollment in the study programs they offer, as well as promote the pricing of its educational services, by setting attractive price offers for students, which works to achieve the comparative advantage of universities through price differentiation strategies.

The study also recommends universities to develop their marketing channels, especially their official websites, to be in line with the current developments and within modern designs that are attractive to students. Moreover, they need to conduct periodic surveys to reveal the strengths and weaknesses in their digital marketing strategies for educational services, to find a kind of continuous improvement on their digital marketing.

References

Abu Qaaud, F., Al-Shoura, M., & Al-Hawary, S. I. (2011). The impact of the service marketing mix in the service quality of health services from the viewpoint of patients in government hospitals in Amman "A field study." *Abhath Al-Yarmouk, 27*(1B), 417–441.

Ahmad, S., Zulkurnain, N., & Khairushalimi, F. (2016). Assessing the validity and reliability of a measurement model in Structural Equation Modeling (SEM). *British Journal of Mathematics & Computer Science, 15*(3), 1–8. https://doi.org/10.9734/BJMCS/2016/25183

Ahmed, M., Abdel Rahman, S., & Al-Hajj, R. (2017). The impact of marketing educational services on achieving the goals of Shaqra University: A field study. *Journal of the American Arab Academy for Science and Technology, 8*(24), 73–86.

Ahmed, A., Alshurideh, M., Al Kurdi, B., & Salloum, S. A. (2020). Digital transformation and organizational operational decision making: A systematic review. In *International Conference on Advanced Intelligent Systems and Informatics* (pp. 708–719). Springer, Cham.

Ahmad, A., Alshurideh, M. T., Al Kurdi, B. H., & Alzoubi, H. M. (2021a). Digital strategies: A systematic literature review. In *The International Conference on Artificial Intelligence and Computer Vision* (pp. 807–822). Springer, Cham.

Ahmad, A., Alshurideh, M. T., Al Kurdi, B. H., & Salloum, S. A. (2021b). Factors impacts organization digital transformation and organization decision making during covid19 pandemic. In *The Effect of Coronavirus Disease (COVID-19) on Business Intelligence* (pp. 95–106). Springer, Cham.

Ahmad, A., Alshurideh, M., Al Kurdi, B., Aburayya, A., & Hamadneh, S. (2021c). Digital transformation metrics: A conceptual view. *Journal of Management Information & Decision Sciences, 24*(7), 1–18.

Akour, I., Alshurideh, M., Al Kurdi, B., Al Ali, A., & Salloum, S. (2021). Using machine learning algorithms to predict people's intention to use mobile learning platforms during the COVID-19 pandemic: Machine learning approach. *JMIR Medical Education, 7*(1), 1–17.

Alhalalmeh, M. I., Almomani, H. M., Altarifi, S., Al- Quran, A. Z., Mohammad, A. A., & Al-Hawary, S. I. (2020). The nexus between corporate social responsibility and organizational performance in Jordan: The mediating role of organizational commitment and organizational citizenship behavior. *Test Engineering and Management, 83*, 6391–6410.

AlHamad, M., Akour, I., Alshurideh, M., Al-Hamad, A., Kurdi, B., & Alzoubi, H. (2021). Predicting the intention to use Google glass: A comparative approach using machine learning models and PLS-SEM. *International Journal of Data and Network Science, 5*(3), 311–320.

Almaazmi, J., Alshurideh, M., Al Kurdi, B., & Salloum, S. A. (2020). The effect of digital transformation on product innovation: A critical review. In *International Conference on Advanced Intelligent Systems and Informatics* (pp. 731–741). Springer, Cham.

Alolayyan, M., Al-Hawary, S. I., Mohammad, A. A., & Al-Nady, B. A. (2018). Banking service quality provided by commercial banks and customer satisfaction. A structural equation modelling approaches. *International Journal of Productivity and Quality Management, 24*(4), 543–565.

Allahow, T. J. A. A., Al-Hawary, S. I. S., & Aldaihani, F. M. F. (2018). Information technology and administrative innovation of the central agency for information technology in Kuwait. *Global Journal of Management and Business, 18*(11-A), 1–16.

Al Dmour, H., Alshurideh, M., & Shishan, F. (2014). The influence of mobile application quality and attributes on the continuance intention of mobile shopping. *Life Science Journal, 11*(10), 172–181.

Al Kurdi, B., Alshurideh, M., & Salloum, S. A. (2020a). Investigating a theoretical framework for e-learning technology acceptance. *International Journal of Electrical and Computer Engineering (IJECE), 10*(6), 6484–6496.

Al Kurdi, B., Alshurideh, M., Salloum, S., Obeidat, Z., & Al-dweeri, R. (2020b). An empirical investigation into examination of factors influencing university students' behavior towards elearning acceptance using SEM approach. *International Journal of Interactive Mobile Technologies, 14*(2), 19–24.

Al-Adili, M. (2015). *E-Marketing*. Dar Amjad for Publishing and Distribution.

Al-Amri, A. (2015). *The role of visual presentation in strengthening the mental status of the organization—An analytical study of the opinions of some workers in commercial complexes in the center of the province of Babylon*, Unpublished Master's thesis, University of Karbala, Karbala, Iraq.

Al-Hamad, M., Mbaidin, H., AlHamad, A., Alshurideh, M., Kurdi, B., & Al-Hamad, N. (2021). Investigating students' behavioral intention to use mobile learning in higher education in UAE during Coronavirus-19 pandemic. *International Journal of Data and Network Science, 5*(3), 321–330.

Al-Hawary, S. I. (2010). Marketing public higher education: A social perspective. *Al Manara for Research and Studies, 16*(4), 9–32.

Al-Hawary, S. I. (2013). The role of perceived quality and satisfaction in explaining customer brand loyalty: Mobile phone service in Jordan. *International Journal of Business Innovation and Research, 7*(4), 393–413.

Al-Hawary, S. I. S., & Alwan, A. M. (2016). Knowledge management and its effect on strategic decisions of Jordanian public universities. *Journal of Accounting-Business & Management, 23*(2), 24–44.

Al-Hawary, S. I. S., & Alhajri, T. M. S. (2020). Effect of electronic customer relationship management on customers' electronic satisfaction of communication companies in Kuwait. *Calitatea, 21*(175), 97–102.

Al-Hawary, S. I., & Al-Khazaleh A. M. (2020). The mediating role of corporate image on the relationship between corporate social responsibility and customer retention. *Test Engineering and Management, 83*(516), 29976–29993.

Al-Hawary, S. I., & Al-Menhaly, S. (2016). The quality of e-government services and its role on achieving beneficiaries satisfaction. *Global Journal of Management and Business Research: A Administration and Management, 16*(11), 1–11.

Al-Hawary, S. I., & Al-Smeran, W. (2017). Impact of electronic service quality on customers satisfaction of Islamic Banks in Jordan. *International Journal of Academic Research in Accounting, Finance and Management Sciences, 7*(1), 170–188.

Al-Hawary, S. I., & Al-Syasneh, M. S. (2020). Impact of dynamic strategic capabilities on strategic entrepreneurship in presence of outsourcing of five stars hotels in Jordan. *Business: Theory and Practice, 21*(2), 578–587.

Al-Hawary, S. I., & Batayneh, A. M. (2010). The effect of marketing communication tools on non-Jordanian students' choice of Jordanian public universities: A field study. *International Management Review, 6*(2), 90–99.

Al-Hawary, S. I., & Harahsheh, S. (2014). Factors affecting Jordanian consumer loyalty toward cellular phone brand. *International Journal of Economics and Business Research, 7*(3), 349–375.

Al-Hawary, S. I., & Hussien, A. J. (2017). The impact of electronic banking services on the customers loyalty of commercial banks in Jordan. *International Journal of Academic Research in Accounting, Finance and Management Sciences, 7*(1), 50–63.

Al-Hawary, S. I., & Ismael, M. (2010). The effect of using information technology in achieving competitive advantage strategies: A field study on the Jordanian pharmaceutical companies. *Al Manara for Research and Studies, 16*(4), 196–203.

Al-Hawary, S. I. S., & Obiadat, A. A. (2021). Does mobile marketing affect customer loyalty in Jordan? *International Journal of Business Excellence, 23*(2), 226–250.

Al-Hawary, S. I., Al-Qudah, K., Abutayeh, P., Abutayeh, S., & Al-Zyadat, D. (2013). The impact of internal marketing on employee's job satisfaction of commercial banks in Jordan. *Interdisciplinary Journal of Contemporary Research in Business, 4*(9), 811–826.

Al-Hawary, S. I., Batayneh, A. M., Mohammad, A. A., & Alsarahni, A. H. (2017). Supply chain flexibility aspects and their impact on customers satisfaction of pharmaceutical industry in Jordan. *International Journal of Business Performance and Supply Chain Modelling, 9*(4), 326–343.

Al-Lozi, M. S., Almomani, R. Z. Q., & Al-Hawary, S. I. S. (2018). Talent management strategies as a critical success factor for effectiveness of human resources information systems in commercial banks working in Jordan. *Global Journal of Management and Business Research: A Administration and Management, 18*(1), 30–43.

Al-Maroof, R., Ayoubi, K., Alhumaid, K., Aburayya, A., Alshurideh, M., Alfaisal, R., & Salloum, S. (2021). The acceptance of social media video for knowledge acquisition, sharing and application: A comparative study among YouYube users and TikTok users' for medical purposes. *International Journal of Data and Network Science, 5*(3), 197–214.

Al-Nady, B. A., Al-Hawary, S. I., & Alolayyan, M. (2016). The role of time, communication, and cost management on project management success: An empirical study on sample of construction projects customers in Makkah City, Kingdom of Saudi Arabia. *International Journal of Services and Operations Management, 23*(1), 76–112.

Al-Shammari, M. (2017). *Digital marketing and its role in enhancing the organization's mental image: A pilot study for the opinions of employees of the Iraqi Asiacell Telecom Company—Holy Karbala Branch.* Unpublished Master's thesis, Karbala University, Karbala, Iraq.

Alshamsi, A., Alshurideh, M., Al Kurdi, B., & Salloum, S. A. (2020). The influence of service quality on customer retention: A systematic review in the higher education. In *International Conference on Advanced Intelligent Systems and Informatics* (pp. 404–416). Springer, Cham.

Alsharari, N. M., & Alshurideh, M. T. (2020). Student retention in higher education: The role of creativity, emotional intelligence and learner autonomy. *International Journal of Educational Management, 35*(1), 233–247.

Alshurideh, M. (2022). Does electronic customer relationship management (E-CRM) affect service quality at private hospitals in Jordan? *Uncertain Supply Chain Management, 10*(2), 325–332.

Alshurideh, M., Al-Hawary, S. I., Batayneh, A. M., Mohammad, A., & Al-Kurdi, B. (2017). The impact of Islamic Banks' service quality perception on Jordanian customers loyalty. *Journal of Management Research, 9*(2), 139–159.

Alshurideh, M., Al Kurdi, B., & Salloum, S. A. (2019a). Examining the main mobile learning system drivers' effects: A mix empirical examination of both the Expectation-Confirmation Model (ECM) and the Technology Acceptance Model (TAM). In *International Conference on Advanced Intelligent Systems and Informatics* (pp. 406–417). Springer, Cham.

Alshurideh, M., Salloum, S. A., Al Kurdi, B., Monem, A. A., & Shaalan, K. (2019b). Understanding the quality determinants that influence the intention to use the mobile learning platforms: A practical study. *International Journal of Interactive Mobile Technologies, 13*(11), 183–157.

Alshurideh, M., Al Kurdi, B., Salloum, S. A., Arpaci, I., & Al-Emran, M. (2020). Predicting the actual use of m-learning systems: A comparative approach using PLS-SEM and machine learning algorithms. *Interactive Learning Environments,* 1–15.

Alshurideh, M. T., Al Kurdi, B., & Salloum, S. A. (2021a). The moderation effect of gender on accepting electronic payment technology: A study on United Arab Emirates consumers. *Review of International Business and Strategy, 31*(3), 375–396.

Alshurideh, M. T., Kurdi, B. A., AlHamad, A. Q., Salloum, S. A., Alkurdi, S., Dehghan, A., Masa'deh, R. E., et al. (2021b). Factors affecting the use of smart mobile examination platforms by universities' postgraduate students during the COVID 19 pandemic: An empirical study. In *Informatics* (vol. 8, no. 2, pp. 1–21). Multidisciplinary Digital Publishing Institute.

Altarifi, S., Al-Hawary, S. I. S., & Al Sakkal, M. E. E. (2015). Determinants of e-shopping and its effect on consumer purchasing decision in Jordan. *International Journal of Business and Social Science, 6*(1), 81–92.

AlTaweel, I. R., & Al-Hawary, S. I. (2021). The mediating role of innovation capability on the relationship between strategic agility and organizational performance. *Sustainability, 13*(14), 7564.

Alwan, M., & Alshurideh, M. (2022). The effect of digital marketing on purchase intention: Moderating effect of brand equity. *International Journal of Data and Network Science, 10*(3), 1–12.

Badous, W. (2017). Suggested scenarios for the future of marketing educational services in Egyptian universities in light of the models of some foreign universities. *Educational Journal, 37*(1), 430–525.

De Leeuw, E., Hox, J., Silber, H., Struminskaya, B., & Vis, C. (2019). Development of an international survey attitude scale: Measurement equivalence, reliability, and predictive validity. *Measurement Instruments for the Social Sciences, 1*(1), 9. https://doi.org/10.1186/s42409-019-0012-x

Gallab, E. (2011). *Management of bge.* Dar Safaa for Publishing and Distribution.

Hair, J. F., Babin, B. J., & Krey, N. (2017). Covariance-based structural equation modeling in the journal of advertising: Review and recommendations. *Journal of Advertising, 46*(1), 163–177. https://doi.org/10.1080/00913367.2017.1281777

Howard, M. C. (2018). The convergent validity and nomological net of two methods to measure retroactive influences. *Psychology of Consciousness: Theory, Research, and Practice, 5*(3), 324–337. https://doi.org/10.1037/cns0000149

Kurdi, B. A., Alshurideh, M., Nuseir, M., Aburayya, A., & Salloum, S. A. (2021). The effects of subjective norm on the intention to use social media networks: an exploratory study using PLS-SEM and machine learning approach. In *International Conference on Advanced Machine Learning Technologies and Applications* (pp. 581–592). Springer, Cham.

Kusumawati, A. (2019). Impact of digital marketing on student decision-making process of higher education institution: A case of Indonesia. *Journal of E-Learning and Higher Education, 2019*(2019), 1–11. https://doi.org/10.5171/2019.267057

Lee, K., Azmi, N., Hanaysha, J., Alshurideh, M., & Alzoubi, H. (2022a). The effect of digital supply chain on organizational performance: An empirical study in Malaysia manufacturing industry. *Uncertain Supply Chain Management, 10*(2), 1–16.

Lee, K., Ramiz, P., Hanaysha, J., Alzoubi, H., & Alshurideh, M. (2022b). Investigating the impact of benefits and challenges of IOT adoption on supply chain performance and organizational performance: An empirical study in Malaysia. *Uncertain Supply Chain Management, 10*(2), 1–14.

Leo, S., Alsharari, N. M., Abbas, J., & Alshurideh, M. T. (2021). From offline to online learning: A qualitative study of challenges and opportunities as a response to the COVID-19 pandemic in the UAE higher education context. In *The Effect of Coronavirus Disease (COVID-19) on Business Intelligence* (pp. 203–217). Springer, Cham.

Metabis, A., & Al-Hawary, S. I. (2013). The impact of internal marketing practices on services quality of commercial banks in Jordan. *International Journal of Services and Operations Management, 15*(3), 313–337.

Mohammad, A. A., Alshura, M. S., Al-Hawary, S. I. S., Al-Syasneh, M. S., & Alhajri, T. M. (2020). The influence of Internal Marketing Practices on the employees' intention to leave: A study of the private hospitals in Jordan. *International Journal of Advanced Science and Technology, 29*(5), 1174–1189.

Muhammad, S. (2021). The importance of university marketing in achieving the goals of universities: A survey study for students of the College of Administration and Economics at the Iraqi University. *Journal of Media Studies and Research, 1*(1), 167–187.

Qassem, I., & Aksha, A. (2020). The role of marketing ethics in enhancing the mental image: A field study for users of mobile companies in Gaza. *Palestine Technical College Journal, 7*(1), 1–38.

Rimkeviciene, J., Hawgood, J., O'Gorman, J., & De Leo, D. (2017). Construct validity of the acquired capability for suicide scale: Factor structure, convergent and discriminant validity. *Journal of Psychopathology and Behavioral Assessment, 39*(2), 291–302. https://doi.org/10.1007/s10862-016-9576-4

Salloum, S. A., Alshurideh, M., Elnagar, A., & Shaalan, K. (2020). Mining in educational data: Review and future directions. In *The International Conference on Artificial Intelligence and Computer Vision* (pp. 92–102). Springer, Cham.

Sekaran, U., & Bougie, R. (2016). *Research methods for business: A skill-building approach* (Seventh edition). Wiley.

Shi, D., Lee, T., & Maydeu-Olivares, A. (2019). Understanding the model size effect on SEM fit indices. *Educational and Psychological Measurement, 79*(2), 310–334. https://doi.org/10.1177/0013164418783530

Suleman, M., Soomro, T. R., Ghazal, T. M., & Alshurideh, M. (2021). Combating against potentially harmful mobile apps. In *The International Conference on Artificial Intelligence and Computer Vision* (pp. 154–173). Springer, Cham.

Sung, K.-S., Yi, Y. G., & Shin, H.-I. (2019). Reliability and validity of knee extensor strength measurements using a portable dynamometer anchoring system in a supine position. *BMC Musculoskeletal Disorders, 20*(1), 1–8. https://doi.org/10.1186/s12891-019-2703-0

Tariq, E., Alshurideh, M., Akour, E., Al-Hawaryd, S., & Al Kurdi, B. (2022a). The role of digital marketing, CSR policy and green marketing in brand development at UK. *International Journal of Data and Network Science, 6*(3), 1–10.

Tariq, E., Alshurideh, M., Akour, I., & Al-Hawary, S. (2022b). The effect of digital marketing capabilities on organizational ambidexterity of the information technology sector. *International Journal of Data and Network Science, 6*(2), 401–408.

Wang, Y. A., & Rhemtulla, M. (2021). Power analysis for parameter estimation in structural equation modeling: A discussion and tutorial. *Advances in Methods and Practices in Psychological Science, 4*(1), 1–17. https://doi.org/10.1177/2515245920918253

Warokka, A. (2020). Digital marketing support and business development using online marketing tools: An experimental analysis. *International Journal of Psychosocial Rehabilitation, 24*(1), 1181–1188. https://doi.org/10.37200/IJPR/V24I1/PR200219.

Impact of Social Media Marketing on Creating Brand Responsiveness

Nancy Abdullah Shamaileh, Mohammed Saleem Khlif Alshura, Enas Ahmad Alshuqairat, Anber Abraheem Shlash Mohammad, Zaki Abdellateef Khalaf Khalaylah, Barween Al Kurdi, Sulieman Ibraheem Shelash Al-Hawary, Muhammad Turki Alshurideh, and Maali M. Al-mzary

Abstract The aim of the study to examine the Impact of social media marketing on creating brand responsiveness. It focused on public universities in Jordan. Data were primarily gathered through self-reported questionnaires creating by Google Forms which were distributed to a sample (785) student via email. In total, (564) responses were received including (38) invalid to statistical analysis due to uncompleted or inaccurate. Hence, the final sample contained (526) responses suitable to analysis requirements. Structural equation modeling (SEM) was conducted to test hypotheses. The results showed that social media marketing dimensions had impact on creating the response to the brand. However, the results indicated that the highest impact was for e-word of mouth. Based on the results of the study, the researchers recommend

---

N. A. Shamaileh · S. I. S. Al-Hawary (✉)
Department of Business Administration, School of Business, Al Al-Bayt University, P.O.Box 130040, Mafraq 25113, Jordan
e-mail: dr_sliman73@aabu.edu.jo

M. S. K. Alshura · E. A. Alshuqairat · Z. A. K. Khalaylah
Management Department, Faculty of Money and Management, The World Islamic Science University, P.O. Box 1101, Amman 11947, Jordan

A. A. S. Mohammad
Marketing Department, Faculty of Administrative and Financial Sciences, Petra University, P.O. Box 961343, Amman 11196, Jordan

B. Al Kurdi
Department of Marketing, Faculty of Economics and Administrative Sciences, The Hashemite University, Zarqa, Jordan
e-mail: barween@hu.edu.jo

M. T. Alshurideh
Department of Marketing, School of Business, The University of Jordan, Amman 11942, Jordan
e-mail: m.alshurideh@ju.edu.jo; malshurideh@sharjah.ac.ae

Department of Management, College of Business, University of Sharjah, Sharjah 27272, United Arab Emirates

M. M. Al-mzary
Department of Applied Science. Irbid College, Al-Balqa Applied University, Salt, Jordan

M. Alshurideh et al. (eds.), *The Effect of Information Technology on Business and Marketing Intelligence Systems*, Studies in Computational Intelligence 1056,
https://doi.org/10.1007/978-3-031-12382-5_13

educating young people about the concepts associated with the use of marketing social media from breach of privacy and addiction to its use.

Keywords Social media marketing · Brand responsiveness · Public Universities · Jordan

1 Introduction

Social media is primarily a source of communication, information, entertainment, and for commercial transactions; As one of its main uses is marketing, communication, and branding through social sites (such as Facebook, Twitter) in addition to communicating with customers to know their response and making the necessary improvements to offered products (Al-Hawary & Hussien, 2017; Al-Hawary & Mohammad, 2011; Al-Maroof et al., 2021a, b). Social media is primarily a source of communication, information, entertainment and a medium social media is one of the effective marketing tools that is concerned with applying knowledge and techniques to enhance economic and social exchanges, and satisfy users' needs and desires based on the marketing products and services implications analysis, and maintaining relationships with users and making friends and social interaction (Al-Hawary & Al-Fassed, 2022; Al-Hawary & Alhajri, 2020; Al-Qudah et al., 2012; Altarifi, 2015; Khasawneh et al., 2021a, b).

Study importance highlights social media marketing role in marketing, interaction and information exchange between customers, which helps companies and brand owners in promoting their products and reaching all customers in a faster and less costly way than traditional (Al-Hawary & Al-Khazaleh, 2020; Alshurideh et al., 2019). The data published on the Social Packers website, which specializes in social networking statistics, stated that Jordan ranked seventh in the list of Arab countries that use the social networking site Facebook, with an average of one million subscribers, and ranked 60th globally out of 213 countries surveyed. Social media importance increasing its popularity among Arab youth, as a social media report issued by the Mohammed bin Rashid School of Government shows that the number of Facebook users in 22 Arab countries, including Jordan, reached 81.3 million in May 2014, 67% of whom are under 30 As for Twitter, the number of users reached 57 million, generating 17.2 million tweets per day, while the average use of LinkedIn was 8.4 million, of whom 68% were between the ages of 18–35 years (Abdel-Fattah, 2016). Brand awareness and identity can be built through social media, people-to-people communication and interaction, and information free exchange; especially the youth group, as they are the group that uses social media the most (Aljumah et al., 2021; Kurdi et al., 2021). Social media spread such as Facebook and Twitter via the Internet among a large segment of people, especially the youth group, has forced many organizations to resort to these means as promotional and marketing tools for their products and an important way to raise the brand reputation level, because the user of social media via the Internet interacts with emerged information and brands a way that is expected to have a greater impact than traditional media (Almazrouei et al., 2020; Alwan & Alshurideh, 2022). Hence, this study comes to show the impact that social media had on the youth category (a sample of Jordanian public university students) in building a brand.

2 Literature Review and Hypotheses Development

2.1 Social Media Marketing

The term social media derives from two words: media, which refers to advertising and communication, which includes ideas and information through marketing social media, and the word social: means individuals interaction within a group (Al-Hawary & Abu-Laimon, 2013; Al-Hawary & Metabis, 2012; Alshurideh, 2022). These media refer to the information democratization and the transformation of users from content readers to content publishers, and take several forms, such as: Internet forums, messages, blogs, commercial images and videos (Al-Hawary & Ismael, 2010; Allahow et al., 2018; Neti, 2011; Lee et al., 2022a, b). Social media is defined as a group of Internet-based applications built on ideological and technological foundations from the Web (2.0) that allow the creation and exchange of user-generated content), the exchange of content and the ability to interact with other users and brand supporters; makes it an ideal management in marketing and consumer engagement in the media as a tool for enabling decision-making and obtaining reliable information (Al-Hawary & Batayneh, 2010; Koshy, 2013; Tariq et al., 2022a, b). Baruah (2012) defined it as the media for social interaction as a comprehensive that goes beyond marketing social communication and includes several types, namely: collaborative projects (Wikidia), blogs, micro-blogs (Twitter), content communities (YouTube), and networking sites such as (FB). The researchers assure that media marketing contributes to convincing consumers that the company's products and services are worthwhile, through the online communities use and knowledge, concepts and techniques necessary application to advance economic and social ends, and is concerned with analyzing marketing policies implications, activities and decisions. It is a strategic and systematic process to establish the company's influence and enhance the brand's reputation (Abu Zayyad et al., 2021; Al-Hawary & Harahsheh, 2014; Neti, 2011). Social media provides two-way communication, as there is no marketing tool that provides effective two-way communication, through which reactions can be known, customer service and care can be provided to obtain real business results. It contributes to building trust and credibility (Al-Hawary & Al-Menhaly, 2016; Al-Hawary & Al-Namlan, 2018; Almaazmi et al., 2020; Divya & Bulomine, 2014). One of the most important dimensions of social media is the **information quality**: information quality refers to the information content that can attract the user to visit the site again on a regular basis for e-commerce, information and media content quality on the Facebook page and their response to customers' needs, and also refers to the information quality about products and services, and the quality of system outputs to make sound decisions in the purchasing process, as the high quality of information increases the chance of maintaining customers to return to the site by providing accurate, detailed and timely information (Ahmed et al., 2020; ; Thumsamisorn & Rittippant, 2011). **Electronic service quality**: electronic service quality is defined as the customer's experience with the service provider, through a specific electronic channel, without human intervention, as social media is the main channel for providing electronic services (Abbad & Al-Hawary, 2011; Abu Qaaud et al., 2011; Al-Hawary et al., 2012; Alshurideh, 2019; Alshurideh et al., 2021; Metabis & Al-Hawary, 2013). **Social Interaction**: the interaction with the brand that these media provide is based on the message and perceived risk of the buying process, prompting the user to think more about product quality and usefulness (Al-Hawary et al., 2011; Alolayyan et al., 2018; Alshurideh et al., 2017; Loanas & Stoica, 2014). **Social impact:** Social networking sites are seen as a means of advertising, which every brand must adopt, as social media marketing—as a means of promotion—contributes to a positive relationship between the brand and the audience in the Internet environment for social impact and maintaining stronger attitudes towards the brand and use of social media and platforms around the industry, competitors, products and companies that increasingly reach customers through social networks (Abuhashesh et al., 2021; Baruah, 2012). **Friendship Dimension:** Social media is effective for connecting with friends, family, and entertainment. These media allow individuals to generate content, engage in conversations, collaborate, participate, and help (Hasan et al., 2022; Parsons & College, 2011). **Social Engagement:** Sharing a long-term relationship with customers is essential for businesses and is used by many marketers to form the bond between the company and customers. Knowing the customers reaction to a particular product or brand, as obtaining the best customer participation is useful for companies in encouraging customers to make decisions (Barhemmati & Ahmed, 2015; Al-Dmour et al., 2021). **Entertainment**: refers to the advertising media ability to meet consumer needs to enjoy them, and studies have shown that entertainment advertising can meet the needs of the audience for entertainment, and that it has a positive impact on the attitude of consumers towards advertising and communication with the brand. **Wom** is also defined as the consumer's participation in the brand related to the electronic word through marketing social media, and this participation is positive or negative by potential, actual or former customers about the product or company that is available through Internet companies and websites Electronic Communication (Al-Maroof et al., 2021a, b; Thurau et al., 2004).

2.2 *Brand Responsiveness*

The way to build a distinctive brand is the ability to choose a name, logo, symbol, packaging design, and other attributes that define the product and distinguish it (Keller, 2013; Al-Hawary, 2013). A brand is defined as a set of assets and liabilities associated with the brand, its name and symbol, that add or subtract from the value it provides to the company's product or service (Al-Hawary et al., 2013a, b; George et al., 2015; Kayaman & Arasli, 2008). Marketing tools directed towards social media marketing are increasing steadily, which indicates the interest of marketers in establishing the brand through social media marketing and interaction with users, which helps in shaping their experiences and benefiting from their opinions.

We can build a brand by creating a mark in memory, creating a brand image and increasing sales (Al-Hawary & Al-Smeran, 2017; Al-Hawary & Obiadat, 2021). Marketing communications include advertising and sales promotion, direct marketing, personal selling, word of mouth, and interactive marketing that includes activities and programs aimed at engaging customers on the Internet and raising awareness. In the context of social media marketing, companies can keep consumers close to the brand and its name, by taking advantage of the positive and direct communication provided by the marketing social media, the activities that they offer and giving them discounts and promotions, or using the contact page as a channel to direct sales, increase sales by spreading positive word of mouth, this gives an opportunity to interact with customers, and influence others in social networks through the experiences of others with the brand, so that it contributes to supporting if the experience is positive. Positive emotional connections with the brand help users become advocates of the brand. Companies must encourage commencers to like the brand, and they must add content that adds value to the brand's page users (Tsimonis & Dimitriadis, 2014). Creating the response towards the brand is through knowing the appropriate customers response towards the brand, and it is known through the customers' judgments towards the brand and their feelings towards it, and the following is an explanation for each of them: Brand judgments: Brand judgments: are customers' personal opinions of the brand and brand performance evaluation. Judgments can be about quality, credibility, and superiority. Brand position and brand quality are the brand overall divisions. Brand position depends on specific brand attributes; it includes the brand's credibility, brand characteristics (innovative or pioneering), the entertainment availability, brand pleasure, as well as the brand superiority to build an active relationship with customers (Hamadneh et al., 2021; Sweiss et al., 2021). As for the brand's feelings, it represents the brand emotional reaction, and how the brand affects the feelings of customers and their relationship with others, these feelings can be positive or negative, They include: warmth feelings, fun (optimistic), excitement, and security: comfort, confidence, social situations (looking at them positively and acknowledging them), Self-esteem: a sense of pride, accomplishment, and fulfillment (Keller, 2013). The brand is a long-term commitment to live up with brand promise made, focus on consistency, clarity, and convergence to build brand content, and manage the brand by delivering on promises and cultivating brand trust.

2.3 *Social Media and Brand Responsiveness*

Social networks dimensions have an impact on the purchasing decision (Adaileh, 2015), and social networks play an important role in marketing and in managing the customer relationship due to the characteristics of these means of communication, social interaction and dialogue between companies and the customer without spatial and temporal restrictions (Noureddine, 2014). Al-Khatib (2014) added that there is an impact of the electronically transmitted word dimensions on electronically spoken word confidence, and the impact of confidence in the electronically transmitted word on the trend towards hotel service, and there is an effect of electronically spoken word confidence in making the purchasing decision. Customer engagement via social media leads to an emotional connection between customers and the brand according to Barhemmati & Ahmed (2015). Customers have positive attitudes toward companies and brands that use social media platforms, and customer engagement via social media leads to an emotional connection between customers and the brand, and to promote customer loyalty to the company, and customer communications via social media led to generating favorable word of mouth towards the brand, where positive word of mouth via social media influences consumer purchasing behavior. Ramnarain and Govender (2013) emphasized that companies with social media activities have a long-lasting and strong brand because of their nurturing and concern for customers, and that credibility has a positive impact on the brand because providing information on networks to help customers know their responses and share content has favorable brand impact, as well as the positive brand impact of quick customer access. Consumers in Pakistan are interested in using social media platforms, blogs, Twitter, and YouTube, and prefer social networking sites Facebook and Google, according to Bilal et al. (2014), because these means provide reliable information about brands, products, and services, as well as the impact of the word transmitted electronically on decisions. Consumers buy about the brand through comments and shares provided by social networks, which contribute to the impact of the word transmitted electronically on decisions. Kim and Ko (2010) show that entertainment has a positive and significant impact on the relationship with customers, that entertainment, personalization and direction positively affect the relationship with customers, and that the luxury industry aims to provide added value to customers, and it remains for the brand to focus on providing Free services, customized services, and additional services, which contribute to strengthening relationships with customers. It also indicated that by using social media marketing; It can retain old customers, attract new ones, and encourage competition among luxury fashion brands. The results also found that the dimensions of entertainment, personalization, and trend positively affect customer relationship, brand loyalty, and frequency of purchase. Accordingly, the study hypothesis can be formulated as follows:

There is a statistically significant effect of the use of social media marketing in creating brand responsiveness.

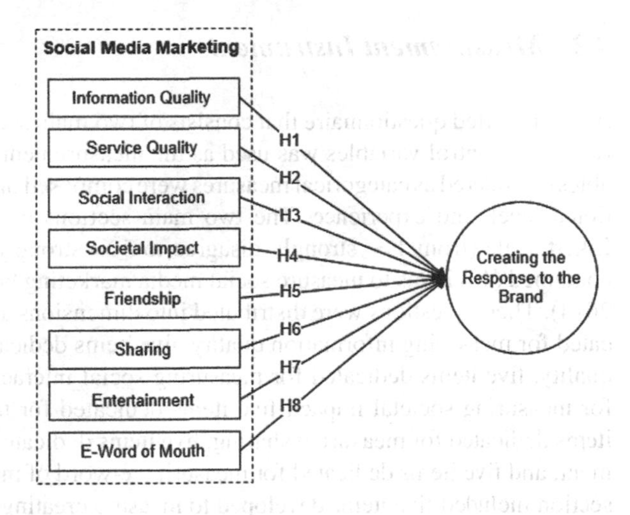

Fig. 1 Research model

3 Study Model

See Fig. 1.

4 Methodology

4.1 Population and Sample Selection

A qualitative method based on a questionnaire was used in this study for data collection and sample selection. The major aim of the study was to examine the impact of social media marketing on creating the response to the brand. Therefore, it focused on public universities in Jordan. Data were primarily gathered through self-reported questionnaires creating by Google Forms which were distributed to a sample (785) students via email. In total, (564) responses were received including (38) invalid to statistical analysis due to uncompleted or inaccurate. Hence, the final sample contained (526) responses suitable to analysis requirements that were formed a response rate of (67%), where it proved to be sufficient to the extent that was predictable and allowed for a presumption of data saturation (Sekaran & Bougie, 2016).

4.2 Measurement Instrument

A self-reported questionnaire that consists of two main sections along with a section regarding control variables was used as the measurement instrument. Control variables considered as categorical measures were composed of gender, age group, educational level, and experience. The two main sections were dealt with a five-point Likert scale (from 1 = strongly disagree to 5 = strongly agree). The first section contained (40) items to measure social media marketing based on (Loanas & Stoica, 2014). These questions were distributed into dimensions as follows: five items dedicated for measuring information quality, five items dedicated for measuring service quality, five items dedicated for measuring social interaction, five items dedicated for measuring societal impact, five items dedicated for measuring friendship, five items dedicated for measuring sharing, five items dedicated for measuring entertainment, and five items dedicated for measuring e-word of mouth. Whereas the second section included five items developed to measure creating the response to the brand according to what was pointed by (Barhemmati & Ahmed, 2015).

5 Findings

5.1 Measurement Model Evaluation

This study was conducted structural equation modeling (SEM) to test hypotheses, which represents a contemporary statistical technique for testing and estimating the relationship between factors and variables (Wang & Rhemtulla, 2021). Accordingly, the reliability and validity of the constructs were tested using confirmatory factor analysis (CFA) through the statistical program AMOSv24. Table 1 summarizes the results of convergent and discriminant validity, as well the indicators of reliability.

Table 1 shows that the standard loading values for the individual items were within the domain (0.641–0.903), these values greater than the minimum retention of the elements based on their standard loads (Al-Lozi et al., 2018; Sung et al., 2019). Average variance extracted (AVE) is a summary indicator of the convergent validity of constructs that must be above 0.50 (Howard, 2018). The results indicate that the AVE values were greater than 0.50 for all constructs, thus the used measurement model has an appropriate convergent validity. Rimkeviciene et al. (2017) suggested the comparison approach as a way to deal with discriminant validity assessment in covariance-based SEM. This approach is based on comparing the values of maximum shared variance (MSV) with the values of AVE, as well as comparing the values of square root of AVE ($\sqrt{AVE}$) with the correlation between the rest of the structures. The results show that the values of MSV were smaller than the values of AVE, and that the values of $\sqrt{AVE}$ were higher than the correlation values among the rest of the constructs. Therefore, the measurement model used is characterized by discriminative validity. The internal consistency measured through Cronbach's Alpha

Table 1 Results of validity and reliability tests

Constructs	1	2	3	4	5	6	7	8	9
1. INQ	**0.751**								
2. SEQ	0.264	**0.781**							
3. SIN	0.337	0.258	**0.766**						
4. SIM	0.274	0.325	0.311	**0.727**					
5. FRS	0.294	0.354	0.346	0.364	**0.760**				
6. SHA	0.351	0.395	0.285	0.332	0.285	**0.767**			
7. ENT	0.341	0.226	0.297	0.316	0.384	0.221	**0.742**		
8. EWM	0.335	0.274	0.349	0.357	0.264	0.325	0.332	**0.778**	
9. CRB	0.415	0.428	0.485	0.439	0.406	0.459	0.394	0.408	**0.789**
VIF	1.251	1.034	2.415	1.335	1.371	1.954	1.285	1.574	–
L.R	0.684–0.835	0.562–0.903	0.641–0.871	0.682–0.764	0.725–0.801	0.719–0.835	0.722–0.764	0.729–0.864	0.715–0.848
AVE	0.565	0.610	0.587	0.528	0.577	0.588	0.551	0.605	0.622
MSV	0.351	0.441	0.385	0.497	0.451	0.462	0.495	0.482	0.473
I.C	0.864	0.882	0.873	0.846	0.870	0.875	0.857	0.881	0.890
C.R	0.866	0.885	0.875	0.848	0.872	0.877	0.860	0.884	0.891

Note INQ: information quality, SEQ: service quality, SIN: social interaction, SIM: societal impact, FRS: friendship, SHA: sharing, ENT: entertainment, EWM: e-word of mouth, CRB: creating the response to the brand, L.R: loadings range, I.C: internal consistency, C.R: composite reliability coefficient (α) and compound reliability by McDonald's Omega coefficient (ω) was conducted as indicators to evaluate measurement model. The results listed in Table 1 demonstrated that both values of Cronbach's Alpha coefficient and McDonald's Omega coefficient were greater than 0.70, which is the lowest limit for judging on measurement reliability (De Leeuw et al., 2019).

5.2 *Structural Model*

The structural model illustrated no multicollinearity issue among predictor constructs because variance inflation factor (VIF) values are below the threshold of 5, as shown in Table 1 (Hair et al., 2017). This result is supported by the values of model fit indices shown in Fig. 1 (Fig. 2)

The results in Fig. 1 indicated that the chi-square to degrees of freedom (CMIN/DF) was 2.617, which is less than 3 the upper limit of this indicator. The values of the goodness of fit index (GFI), the comparative fit index (CFI), and the Tucker-Lewis index (TLI) were upper than the minimum accepted threshold of 0.90. Moreover, the result of root mean square error of approximation (RMSEA) indicated to value 0.061, this value is a reasonable error of approximation because it is less than the higher limit of 0.08. Consequently, the structural model used in this study was recognized as a fit model for predicting the creating the response to the brand and generalization of its result (Ahmad et al., 2016; Shi et al., 2019). To verify the results of testing the study hypotheses, structural equation modeling (SEM) was used, the results of which are listed in Table 2.

The results demonstrated in Table 2 show that most of social media marketing dimensions had no impact on creating the response to the brand. However, the results indicated that the highest impact was for e-word of mouth ($\beta = 0.411$, $t = 9.866$, $p = 0.000$), followed by sharing ($\beta = 0.388$, $t = 6.678$, $p = 0.000$), and finally the lowest impact was for societal impact ($\beta = 0.294$, $t = 4.400$, $p = 0.000$).

6 Discussion

The results of the study showed that social media enables consumers to exchange information with their peers about the brand and allows them to exchange conversations. Consumers prefer social media marketing sites to keep up with brand products and promotional campaigns, and their desire to see updated content about the brand, the popularity of content among friends increases brand loyalty, and the effect of electronically transmitted word positively on brand loyalty and this is consistent with. This emphasizes the possibility of building brand commitment and loyalty through social media marketing that helps improve brand visibility away from barriers that prevent information flow, in addition to the ease of use of the Internet by marketers to communicate with consumers and the target audience. The results of the study showed

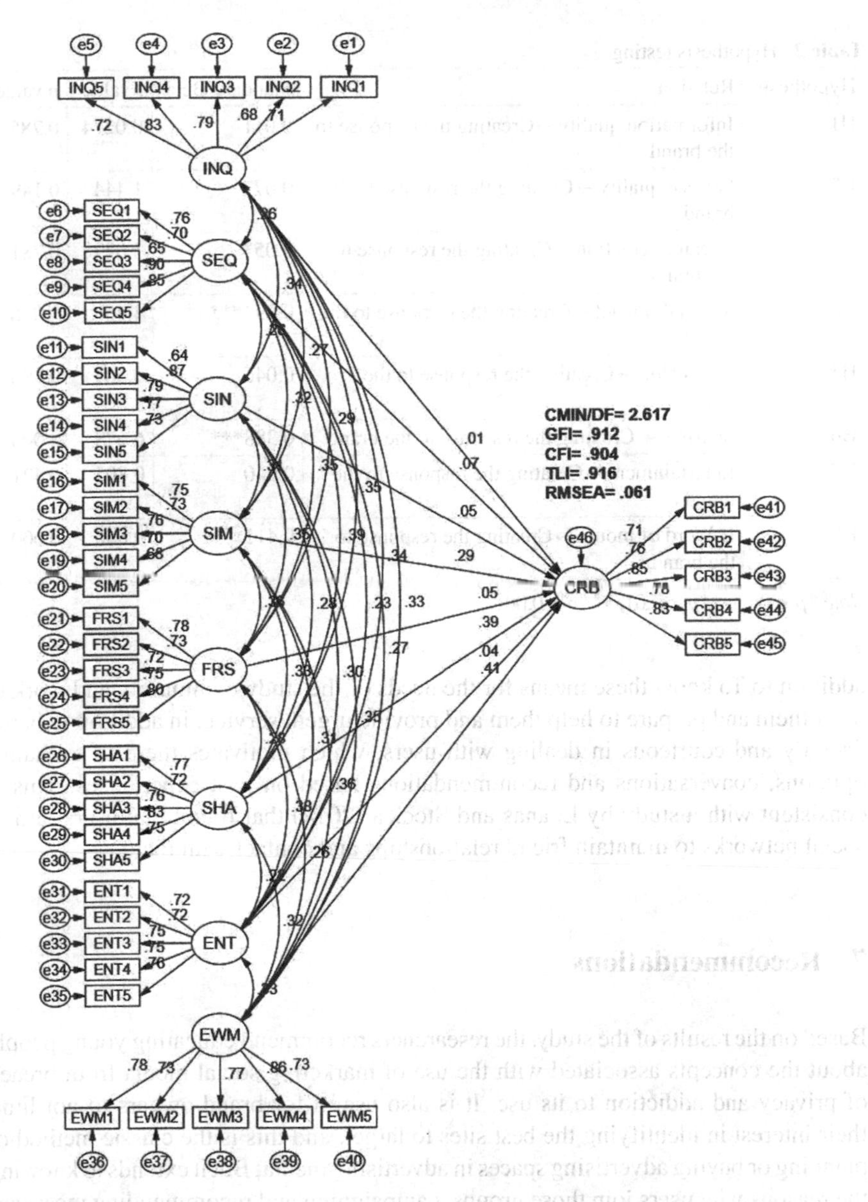

Fig. 2 SEM results of the social media marketing effect on creating the response to the brand

that there is a statistically significant effect towards the use of marketing social media in creating brand responsiveness. We conclude from this that the marketing social media is an ideal and positive environment for promoting social marketing communication and an ideal and positive environment to enhance social interaction and dissemination of knowledge and information and communication with friends in

Table 2 Hypothesis testing

Hypothesis	Relation	Standard Beta	*t* value	*p* value
H1	Information quality→Creating the response to the brand	0.011	0.0274	0.785
H2	Service quality→Creating the response to the brand	0.077	1.444	0.149
H3	Social interaction→Creating the response to the brand	0.053	1.079	0.281
H4	Societal impact→Creating the response to the brand	0.294***	4.400	0.000
H5	Friendship→Creating the response to the brand	0.046	0.867	0.386
H6	Sharing→ Creating the response to the brand	0.388***	6.678	0.000
H7	Entertainment→Creating the response to the brand	0.040	0.805	0.421
H8	E-Word of mouth→Creating the response to the brand	0.411***	9.866	0.000

Note $*p < 0.05$, $**p < 0.01$, $***p < 0.001$

addition to To know these means for the needs of the study community and work to meet them and prepare to help them and provide urgent service, in addition to being friendly and courteous in dealing with users, which motivates them to exchange opinions, conversations and recommendations based on past experiences. This is consistent with a study by Loanas and Stocica (2014) that found that users turn to social networks to maintain friend relationships and contact with them.

7 Recommendations

Based on the results of the study, the researchers recommend educating young people about the concepts associated with the use of marketing social media from breach of privacy and addiction to its use. It is also useful for brand owners to not limit their interest in identifying the best sites to target, and this is the classic method of planning or buying advertising spaces in advertising media; But it extends to knowing the reasons why users join those groups, campaigning and recommending messages to create groups that consumers want to join, so that its informational message covers more consumers more effectively. Modern marketing social media has transformed from a textual media tool to a comprehensive audio-visual media tool, which indicates the need for those in charge of managing social media marketing communications and marketers to know how students use marketing social networks and their impact on them, to enhance the share of companies in competitive markets Develop brand loyalty and enhance its reputation.

References

Abbad, J., & Al-Hawary, S. I. (2011). Measuring banking service quality in Jordan: A case study of Arab Bank. *Abhath Al-Yarmouk, 27*(3), 2179–2196.

Abdel Fattah, F. (2016). Towards activating the Arab League communication between youth, the strategic network.

Abu Qaaud, F., Al-Shoura, M., & Al-Hawary, S. I. (2011). The impact of the service marketing mix in the service quality of health services from the viewpoint of patients in government hospitals in Amman "A field study." *Abhath Al-Yarmouk, 27*(1B), 417–441.

Abu Zayyad, H. M., Obeidat, Z. M., Alshurideh, M. T., Abuhashesh, M., Maqableh, M., & Masa'deh, R. E. (2021). Corporate social responsibility and patronage intentions: The mediating effect of brand credibility. *Journal of Marketing Communications, 27*(5), 533–510.

Abuhashesh, M. Y., Alshurideh, M. T., & Sumadi, M. (2021). The effect of culture on customers' attitudes toward Facebook advertising: The moderating role of gender. *Review of International Business and Strategy, 31*(3), 416–437.

Adaileh, M. J. (2015). The role of social networks in influencing the consumer's purchasing decision via the Internet: An analytical study at Qassim University—Saudi Arabia. *The Jordanian Journal of Business Administration, 11*(1), 153–170, University of Jordan, Amman, Jordan.

Ahmad, S., Zulkurnain, N., & Khairushalimi, F. (2016). Assessing the validity and reliability of a measurement model in Structural Equation Modeling (SEM). *British Journal of Mathematics & Computer Science, 15*(3), 1–8. https://doi.org/10.9734/BJMCS/2016/25183

Ahmed, A., Alshurideh, M., Al Kurdi, B., & Salloum, S. A. (2020). Digital transformation and organizational operational decision making: A systematic review. In *International Conference on Advanced Intelligent Systems and Informatics* (pp. 708–719). Springer, Cham.

Al-Dmour, R., AlShaar, F., Al-Dmour, H., Masa'deh, R., & Alshurideh, M. T. (2021). The effect of service recovery justices strategies on online customer engagement via the role of "Customer Satisfaction" during the covid-19 pandemic: An empirical study. *The Effect of Coronavirus Disease (COVID-19) on Business Intelligence, 334*, 325–346.

Al-Hawary, S. I., & Al-Fassed, K. J. (2022). The impact of social media marketing on building brand loyalty through customer engagement in Jordan. *International Journal of Business Innovation and Research* (In Press)

Al-Hawary, S. I. (2013). The role of perceived quality and satisfaction in explaining customer brand loyalty: Mobile phone service in Jordan. *International Journal of Business Innovation and Research, 7*(4), 393–413.

Al-Hawary, S. I. S., & Alhajri, T. M. S. (2020). Effect of electronic customer relationship management on customers' electronic satisfaction of communication companies in Kuwait. *Calitatea, 21*(175), 97–102.

Al-Hawary, S. I. S., & Obiadat, A. A. (2021). Does mobile marketing affect customer loyalty in Jordan? *International Journal of Business Excellence, 23*(2), 226–250.

Al-Hawary, S. I. S., Alhamali, R. M., & Alghanim, S. A. (2011). Banking service quality provided by commercial banks and customer satisfaction. *American Journal of Scientific Research, 27*, 68–83.

Al-Hawary, S. I., & Abu-Laimon, A. A. (2013). The impact of TQM practices on service quality in cellular communication companies in Jordan. *International Journal of Productivity and Quality Management, 11*(4), 446–474.

Al-Hawary, S. I., & Al-Khazaleh A. M. (2020). The mediating role of corporate image on the relationship between corporate social responsibility and customer retention. *Test Engineering and Management, 83*(516), 29976–29993.

Al-Hawary, S. I., & Al-Menhaly, S. (2016). The Quality of e-government services and its role on achieving beneficiaries satisfaction. *Global Journal of Management and Business Research: A Administration and Management, 16*(11), 1–11.

Al-Hawary, S. I., & Al-Namlan, A. (2018). Impact of electronic human resources management on the organizational learning at the private hospitals in the State of Qatar. *Global Journal of Management and Business Research: A Administration and Management, 18*(7), 1–11.
Al-Hawary, S. I., & Al-Smeran, W. (2017). Impact of electronic service quality on customers satisfaction of Islamic Banks in Jordan. *International Journal of Academic Research in Accounting, Finance and Management Sciences, 7*(1), 170–188.
Al-Hawary, S. I., & Batayneh, A. M. (2010). The effect of marketing communication tools on Non-Jordanian students' choice of Jordanian public universities: A field study. *International Management Review, 6*(2), 90–99.
Al-Hawary, S. I., & Harahsheh, S. (2014). Factors affecting Jordanian consumer loyalty toward cellular phone brand. *International Journal of Economics and Business Research, 7*(3), 349–375.
Al-Hawary, S. I., & Hussien, A. J. (2017). The impact of electronic banking services on the customers loyalty of commercial banks in Jordan. *International Journal of Academic Research in Accounting, Finance and Management Sciences, 7*(1), 50–63.
Al-Hawary, S. I., & Ismael, M. (2010). The effect of using information technology in achieving competitive advantage strategies: A field study on the Jordanian pharmaceutical companies. *Al Manara for Research and Studies, 16*(4), 196–203.
Al-Hawary, S. I., & Metabis, A. (2012). Service quality at Jordanian commercial banks: What do their customers say? *International Journal of Productivity and Quality Management, 10*(3), 307–334.
Al-Hawary, S. I., & Mohammad, A. A. (2011). The role of the internet in marketing the services of travel and tourism agencies in Jordan. *Abhath Al-Yarmouk, 27*(2B), 1339–1359. Retrieved from http://journals.yu.edu.jo/ayhss/Issues/Vol272B2011.pdf.
Al-Hawary, S. I., Al-Hawajreh, K., AL-Zeaud, H., & Mohammad, A. (2013a). The impact of market orientation strategy on performance of commercial banks in Jordan. *International Journal of Business Information Systems, 14*(3), 261–279.
Al-Hawary, S. I., Al-Nady, B. A., & Alolayyan, M. (2013b). Effect of brand name and price on business to business (B2B) success: An empirical study on sample of food hypermarket retailers in Amman City. *International Journal of Information and Coding Theory, 2*(2/3), 115–139.
Al-Hawary, S. I., AL-Zeaud, H., & Matabes, A. (2012). Measuring the quality of educational services offered to postgraduate students at the Faculty of Business and Finance: A field study on the Universities of the North Region. *Al Manara for Research and Studies, 18*(1), 241–278.
Aljumah, A., Nuseir, M. T., & Alshurideh, M. T. (2021). The impact of social media marketing communications on consumer response during the covid-19: Does the brand equity of a university matter. *The Effect of Coronavirus Disease (COVID-19) on Business Intelligence*, 367–384.
Al-Khatib, H. (2014) The *impact of the electronically transmitted word on determining the purchasing of hotel service for customers of five-star hotels in Amman.* Master's Thesis, Middle East University, Amman, Jordan.
Allahow, T. J. A. A., Al-Hawary, S. I. S., & Aldaihani, F. M. F. (2018). Information technology and administrative innovation of the central agency for information technology in Kuwait. *Global Journal of Management and Business, 18*(11-A), 1–16.
Al-Lozi, M. S., Almomani, R. Z. Q., & Al-Hawary, S. I. S. (2018). Talent Management strategies as a critical success factor for effectiveness of human resources information systems in commercial banks working in Jordan. *Global Journal of Management and Business Research: A Administration and Management, 18*(1), 30–43.
Almaazmi, J., Alshurideh, M., Al Kurdi, B., & Salloum, S. A. (2020). The effect of digital transformation on product innovation: A critical review. In *International Conference on Advanced Intelligent Systems and Informatics* (pp. 731–741). Springer, Cham.
Al-Maroof, R. S., Alshurideh, M. T., Salloum, S. A., AlHamad, A. Q. M., & Gaber, T. (2021a). Acceptance of Google Meet during the spread of Coronavirus by Arab university students. *Informatics, 8*(2), 1–17.
Al-Maroof, R., Ayoubi, K., Alhumaid, K., Aburayya, A., Alshurideh, M., Alfaisal, R., & Salloum, S. (2021b). The acceptance of social media video for knowledge acquisition, sharing and application:

A comparative study among YouYube users and TikTok users' for medical purposes. *International Journal of Data and Network Science, 5*(3), 197–214.

Almazrouei, F. A., Alshurideh, M., Kurdi, B. A., & Salloum, S. A. (2020). Social media impact on business: A systematic review. In *International conference on advanced intelligent systems and informatics* (pp. 697–707). Springer, Cham.

Alolayyan, M., Al-Hawary, S. I., Mohammad, A. A., & Al-Nady, B. A. (2018). Banking service quality provided by commercial banks and customer satisfaction. A structural equation modelling approaches. *International Journal of Productivity and Quality Management, 24*(4), 543–565.

Al-Qudah, K. A., Al-Hawary, S. I., & Al-Mehsen, M. A. (2012). Electronic credit cards usage and their impact on bank's profitability: The rate of return on owners equity model "An empirical study." *Interdisciplinary Journal of Contemporary Research in Business, 4*(7), 828–841.

Alshurideh, D. M. (2019). Do electronic loyalty programs still drive customer choice and repeat purchase behaviour? *International Journal of Electronic Customer Relationship Management, 12*(1), 40–57.

Alshurideh, M. (2022). Does electronic customer relationship management (E-CRM) affect service quality at private hospitals in Jordan? *Uncertain Supply Chain Management, 10*(2), 325–332.

Alshurideh, M. T., Al Kurdi, B., & Salloum, S. A. (2021). The moderation effect of gender on accepting electronic payment technology: A study on United Arab Emirates consumers. *Review of International Business and Strategy, 31*(3), 375–396.

Alshurideh, M., Al-Hawary, S. I., Batayneh, A. M., Mohammad, A., & Al-Kurdi, B. (2017). The impact of Islamic Banks' service quality perception on Jordanian customers loyalty. *Journal of Management Research, 9*(2), 139–159.

Alshurideh, M., Salloum, S. A., Al Kurdi, B., & Al-Emran, M. (2019). Factors affecting the social networks acceptance: an empirical study using PLS-SEM approach. In *Proceedings of the 2019 8th International Conference on Software and Computer Applications* (pp. 414–418).

Altarifi, S., Al-Hawary, S. I. S., & Al Sakkal, M. E. E. (2015). Determinants of e-shopping and its effect on consumer purchasing decision in Jordan. *International Journal of Business and Social Science, 6*(1), 81–92.

Alwan, M., & Alshurideh, M. (2022). The effect of digital marketing on purchase intention: Moderating effect of brand equity. *International Journal of Data and Network Science, 10*(3), 1–12.

Barhemmati, N., & Ahmad, A. (2015). Effects of social network marketing (SNM) on consumer purchase behavior through customer engagement. *Journal of Advanced Management Science, 3*(4), 307–311.

Baruah, T., (2012). Effectiveness of social media as a tool of communication and it's potential for technology enabled connection's: Amicro—Level study. *International Journal of Scientific and Research Publications, 2*(5), ISSN 2250-3153.

Bilal, C., Ahmed, M., & Shahzad, N. (2014). Role of Social media networks in consumer decision marking: A case of the garment sector. *International Journal of Multi-Disciplinary Sciences and Engineering, 5*(3), 1–9.

de Leeuw, E., Hox, J., Silber, H., Struminskaya, B., & Vis, C. (2019). Development of an international survey attitude scale: Measurement equivalence, reliability, and predictive validity. *Measurement Instruments for the Social Sciences, 1*(1), 9. https://doi.org/10.1186/s42409-019-0012-x

Divya, S., & Bulomine, R. (2014). An empirical study of effectiveness of social media as a marketing tool. *International Journal of Current Research and Academic Review, 2*(3), 163–168.

Christodoulides, G., Cadogan, J. W., & Veloutsou, C. (2015). Consumer based brand equity measurement: Lessons Learned from an international study. *International Marketing Review, 32*(3/4), 307–328.

Hair, J. F., Babin, B. J., & Krey, N. (2017). Covariance-based structural equation modeling in the journal of advertising: Review and recommendations. *Journal of Advertising, 46*(1), 163–177. https://doi.org/10.1080/00913367.2017.1281777

Hamadneh, S., Hassan, J., Alshurideh, M., Al Kurdi, B., & Aburayya, A. (2021). The effect of brand personality on consumer self-identity: The moderation effect of cultural orientations among British and Chinese consumers. *Journal of Legal, Ethical and Regulatory Issues, 24*, 1–14.

Hasan, O., McColl, J., Pfefferkorn, T., Hamadneh, S., Alshurideh, M., & Kurdi, B. (2022). Consumer attitudes towards the use of autonomous vehicles: Evidence from United Kingdom taxi services. *International Journal of Data and Network Science, 6*(2), 537–550.

Howard, M. C. (2018). The convergent validity and nomological net of two methods to measure retroactive influences. *Psychology of Consciousness: Theory, Research, and Practice, 5*(3), 324–337. https://doi.org/10.1037/cns0000149

Kayaman, R., & Arasli, H. (2007). Customer based brand equity: Evidence from the hotel industry. *Managing Service Quality, 17*(1), 92–109.

Keller, K., (2013). *Strategic brand management building, measuring; and managing brand equity* (Fourth Edition). ISBN 0-13-266425-7:29 590.

Khasawneh, M. A., Abuhashesh, M., Ahmad, A., Alshurideh, M. T., & Masa'deh, R. (2021a). Determinants of e-word of mouth on social media during COVID-19 outbreaks: An empirical study. In *The Effect of Coronavirus Disease (COVID-19) on Business Intelligence* (pp. 347–366). Springer, Cham.

Khasawneh, M. A., Abuhashesh, M., Ahmad, A., Masa'deh, R., & Alshurideh, M. T. (2021b). Customers online engagement with social media influencers' content related to COVID 19. In *The Effect of Coronavirus Disease (COVID-19) on Business Intelligence* (pp. 385–404). Springer, Cham.

Kim, A., & Ko, E. (2010). Impacts of Luxury Fashion brand's social media marketing on customer relationship and purchase intention. *Journal of Global Fashion Marketing, 1*(3), 71–164.

Koshy, S. (2013). Factors that affect the use of Facebook and Twitter as marketing tools in the UAE. In *The 18th Annual International Systems, United Kingdom* (pp. 1–7).

Kurdi, B. A., Alshurideh, M., Nuseir, M., Aburayya, A., & Salloum, S. A. (2021). The effects of subjective norm on the intention to use social media networks: An exploratory study using PLS-SEM and machine learning approach. In *International Conference on Advanced Machine Learning Technologies and Applications* (pp. 581–592). Springer, Cham.

Lee, K., Azmi, N., Hanaysha, J., Alshurideh, M., & Alzoubi, H. (2022a). The effect of digital supply chain on organizational performance: An empirical study in Malaysia manufacturing industry. *Uncertain Supply Chain Management, 10*(2), 1–16.

Lee, K., Ramiz, P., Hanaysha, J., Alzoubi, H., & Alshurideh, M. (2022b). Investigating the impact of benefits and challenges of IOT adoption on supply chain performance and organizational performance: An empirical study in Malaysia. *Uncertain Supply Chain Management, 10*(2), 1–14.

Loanas, E., & Stoica, I. (2014). Social media and it's impact on consumers behavior. *International Journal of Economic Practices and Theories, 4*(2), 295–303.

Metabis, A., & Al-Hawary, S. I. (2013). The impact of internal marketing practices on services quality of commercial banks in Jordan. *International Journal of Services and Operations Management, 15*(3), 313–337.

Neti, S. (2011). Social media and it's role in marketing. *International Journal of Enterprise Computing and Business Systems, 1*(2), 1–16.

Noureddine, M. (2014). *The role of marketing through social networks in managing the relationship with the customer*. Master's Thesis, Kasdi Merbah University, Ouargla, Algeria.

Parsons, A., & College, K. (2011). Social media from a corporate perspective: A content analysis of official Facebook pages. *Proceedings of the Academy of Marketing Studies, 16*(2), 11–14.

Ramnarain, Y., & Govender, K. (2013). Social media browsing and consumer behaviour: Exploring the youth market. *African Journal of Business Management, 7*(18), 1885–1893.

Rimkeviciene, J., Hawgood, J., O'Gorman, J., & De Leo, D. (2017). Construct validity of the acquired capability for suicide scale: Factor structure, convergent and discriminant validity. *Journal of Psychopathology and Behavioral Assessment, 39*(2), 291–302. https://doi.org/10.1007/s10862-016-9576-4

Sekaran, U., & Bougie, R. (2016). *Research methods for business: A skill-building approach* (Seventh edition). Wiley.

Shi, D., Lee, T., & Maydeu-Olivares, A. (2019). Understanding the model size effect on SEM fit indices. *Educational and Psychological Measurement, 79*(2), 310–334. https://doi.org/10.1177/0013164418783530

Sung, K.-S., Yi, Y. G., & Shin, H.-I. (2019). Reliability and validity of knee extensor strength measurements using a portable dynamometer anchoring system in a supine position. *BMC Musculoskeletal Disorders, 20*(1), 1–8. https://doi.org/10.1186/s12891-019-2703-0

Sweiss, N., Obeidat, Z. M., Al-Dweeri, R. M., Mohammad Khalaf Ahmad, A., M. Obeidat, A., & Alshurideh, M. (2021). The moderating role of perceived company effort in mitigating customer misconduct within Online Brand Communities (OBC). *Journal of Marketing Communications,* 1–24.

Tariq, E., Alshurideh, M., Akour, E., Al-Hawaryd, S., & Al Kurdi, B. (2022a). The role of digital marketing, CSR policy and green marketing in brand development at UK. *International Journal of Data and Network Science, 6*(3), 1–10.

Tariq, E., Alshurideh, M., Akour, I., & Al-Hawary, S. (2022b). The effect of digital marketing capabilities on organizational ambidexterity of the information technology sector. *International Journal of Data and Network Science, 6*(2), 401–408.

Thumsamisorn, A., & Rittippant, N. (2011). The engagement of social media in Facebook: the case of college student in Thailand. *EPPM, Singapore, 20*(21), 227–238.

Thurau, T., Gwinner, K., Walsh, G., & Gremler, D. (2004). Electronic Word-of-Mouth via Consumer-opinion platforms: What Motivates consumers to articulate themselves on the internet. *Journal of interactive Marketing, 18*(1), 38–52.

Tsimonis, G., & Dimitriadis, S. (2014). Brand strategies in social media. *Marketing Intelligence & Planning, 32*(3), 328–344.

Wang, Y. A., & Rhemtulla, M. (2021). Power analysis for parameter estimation in structural equation modeling: A discussion and tutorial. *Advances in Methods and Practices in Psychological Science, 4*(1), 1–17. https://doi.org/10.1177/2515245920918253

Evaluation of the Digital Divide Status and Its Impact on the Use of TikTok Platform Through E-Commerce Activities

Dmaithan Abdelkarim Almajali, Ala'aldin Alrowwad, and Ra'ed Masa'deh

Abstract The current paper was looking at how the factors of content richness, trust, perceived ease of use, perceived usefulness and user satisfaction impact TikTok acceptance. Participants of this study were 145 Telecenter users from three regions in Jordan. Their use of TikTok platform in e-commerce activities was examined. The data from the survey were analyzed and the obtained results led to the discussion on the policy implications in guiding both government agencies and customers in their efforts in improving their e-commerce activities via TikTok platform. Avenues for forthcoming research were discussed as well.

Keywords Perceived ease of use · Perceived usefulness · TikTok · User satisfaction · Trust

1 Introduction

The e-commerce adoption has been expansively studied, as demonstrated by Kiwanuka (2015), Williams et al. (2015), and it has been countlessly reported, for instance, by Venkatesh et al. (2003) that the behavior of individuals during their online purchase is majorly influenced by intrinsic and extrinsic variables and also by variables associated with the features of the new technology no matter the form of

D. A. Almajali (✉)
Department of Management Information System, Faculty of Business, Applied Science Private University, Amman, Jordan
e-mail: d_almajali@asu.edu.jo

A. Alrowwad
Department of Business Management, The University of Jordan, Aqaba, Jordan
e-mail: a.alrowwad@ju.edu.jo

R. Masa'deh
Department Management Information Systems, School of Business, University of Jordan, Amman, Jordan
e-mail: r.masadeh@ju.edu.jo

M. Alshurideh et al. (eds.), *The Effect of Information Technology on Business and Marketing Intelligence Systems*, Studies in Computational Intelligence 1056,
https://doi.org/10.1007/978-3-031-12382-5_14

electronic commerce (e-commerce), product or service. Equally, as stated in Sánchez-Torres et al. (2017), Torres and Cañada (2016), notable differences associated with the digital divide appear to exist in this technology diffusion within various world regions. This is particularly salient when looking at issues relating to the infrastructures, access to information and communication technologies or ICTs, high access tariffs, the internet network, and training in the technology's usage, which adversely impact the adoption of user of e-commerce (Landau, 2012).

The actual impact of the digital divide on virtual channel usage has been examined in several studies. For instance, in a study carried out by Muñoz and Amaral (2011), it was found that elderlies face more difficulties in technology usage owing to the lack of training, as they had no experience in the use of the technology for the most part of their lives. Additionally, the imposed regulation, infrastructure, and the market are among the factors found to impact the access of people to e-commerce. Many macroeconomic studies carried out by international organizations have been examining the aforementioned form of digital divide, in relation to the growth and development (Deursen & Dijk, 2013; Hilbert, 2016; Obeidat et al., 2017; Peral-Peral et al., 2015; Pick & Nishida, 2015; Tarhini et al., 2015; Wei et al., 2011; White et al., 2011). Somehow, in relation to the models of e-commerce adoption, this phenomenon has not been closely examined.

Among developing nations especially, the integral role of information and communication technologies (ICT) in socio-economic development, has led governments, non-government bodies as well as International Telecommunication Union (ITU) to fund Telecentres (Sey & Fellows, 2009). For the facilitation of local community development, Telecentres make available to the general public, the access to computers as well as the associated ICTs (Gurstein, 2007). As mentioned by the International Telecommunication Union, all over the world in total, only 26.6% of the citizens have homes internet access. Somehow, these people were mostly those living in developed countries (Clark & Gomez, 2011). Notably, the past twenty years have demonstrated the commonness of public computers and the Internet accessibility by way of Telecentres platform (Clark & Gomez, 2011). The use allows the sharing of network resources among various users, for instance, the sharing of communication lines, and this has allowed ICTs to be expanded to many other users. Meanwhile, the term "Telecentres" that is used in this study for the context of Jordan encompasses a wide concept, and the term is referred differently in different countries.

Various Telecentres which provide training in information and communication technology are available in Jordan. These centers can be found in various cities in the country including Irbid, Balqa, Zarqa, Tafelah, and Amman as well (Amman is the capital city of Jordan). Essentially, Telecentres are for promoting digital literacy, easing digital divide and poverty, providing prospects of employment to the people, in addition to being part of the e-governance and the e-commerce initiatives (Andersson & Grönlund, 2009). There are indeed various advantages of Telecentres, and yet, these Telecentres have not been receiving the appropriate academic interest especially in terms of its user acceptance and use (Sey & Fellows, 2009).

Notably, with the application of a modified form of unified theory of acceptance (UTAUT) and technology acceptance model (TAM), several determinants of potential users' acceptance of Telecentres have been mentioned in several past studies within the context of developing nations (Al Kurdi et al., 2020a, b; Almajali, 2021a; Alshurideh et al., 2019a, b; Davis, 1989; Wang & Shih, 2008). Considering the above discussion, an exploratory study should be carried out, to ascertain whether or not the digital divide impact on the adoption of e-commerce in a negative manner. Relevantly, Almajali et al. (2021) in their study have mentioned the need to explore the applications of social media that appeal to millennials, and TikTok is one of these applications. As with certain other applications, TikTok also allows user to engage in e-commerce activities.

Sensor Tower accordingly reported that TikTok was downloaded over 188 million times in the first quarter of 2019. Indeed, TikTok is social media App that has been downloaded the most frequently all over the world (Al-Maroof et al., 2021; SensorTower, 2019), and its solid technical strength and good product experience makes TikTok appear usable to various users. TikTok is a free platform that allows user to express them and establish new relationships (Omar & Dequan, 2020). Among the major elements of TikTok found to be highly attractive among its users include its "music dubbing," "funny story," and "mind relaxing" (Ma et al., 2019). In China, Provincial Health Committees and other authorities have registered official accounts on TikTok (Zhu et al., 2021). TikTok has indeed rapidly expanded its user base.

TikTok encompasses Social Network Sites (SNSs) encompassing common platforms of social media (Almazrouei et al., 2021; Alyammahi et al., 2021). The internet realm and communication methods have been transformed by social sites. Videos and the experiences shared affect the life of people (Alqudah et al., 2021; Chatzoglou et al., 2020). TikTok has recently introduced a micro-video feature in its technology, and during the pandemic of COVID-19, governments all over the world employed short videos in their efforts of disseminating the news associated with the disease. TikTok in this regard is viewed as a fitting application that makes available to user the short video socialization, and as can be observed in China, TikTok has been a major application in the dissemination of countless of topics to the public (Al-Maroof et al., 2020; Wang & Fu, 2020; Zhu et al., 2021).

Notwithstanding the great opportunities offered by TikTok in data sharing and exchanging as highlighted in Mhamdi et al. (2018), Salloum et al. (2018), the effectiveness of TikTok as a key source within the domain of e-commerce that compels its users to accept the technology has not been sufficiently explored, as indicated in several studies (Salloum et al., 2017, 2018; Aburayya et al., 2020; Abu Zayyad et al., 2020; Al-Dmour et al., 2017, 2020; Alhashmi et al., 2019, 2020; Alshurideh, 2018). Utilizing a conceptual model for evaluating users' acceptance, the present study looks into the factors that can inspire users to gain from this platform.

In past studies on TikTok, the examined variables include travel intention (Bian & Zhu, 2020), purchase intention (Wang, 2019), usage motivation (Omar & Dequan, 2020), information service quality involving the Chinese government accounts on TikTok (Chen & Zhang, 2020), and intention to continue using TikTok (Susilo, 2020). On the other hand, the impact of TikTok usage on the positive emotion and relationships among e-commerce customers has not been examined especially in the setting of Jordan, and this has created a significant gap. As such, the present study will explore how TikTok usage would influence positive emotion and relationship among e-commerce customers in Jordan.

A comprehensive model was constructed involving several external factors associated with the platform of TikTok, and among the external factors included in the model are content richness, readiness to trust new technology, and user satisfaction. Content richness relates to the wealth of the content, time restriction as well as the latest information. Readiness to trust new technology and users' satisfaction are the other two factors included, and these factors are associated with the customers themselves (Al-Dmour et al., 2021; Alshurideh, 2022; Kurdi et al., 2020). Essentially, the main purpose of carrying out the present study was to examine the acceptance of customers towards TikTok usage, and thus far, the literature has not provided any evidence on past studies that examined TikTok as an e-commerce platform. Accordingly, a conceptual framework that measures customers' acceptance towards TikTok as a social media platform was constructed. This framework specifically examined the status of digital divide in Jordan, alongside its impact on the platforms of e-commerce including TikTok.

2 Literature Review

TikTok have previously been examined but past studies on this subject were mostly looking at different important factors impacting TikTok acceptance in both the educational (Al-Maroof et al., 2020) and non-educational contexts. In education, Kurdi et al. (2020), Habes et al. (2019) reported that this application has been playing the role of a learning tool. The impact of different factors such as motives and benefits on the acceptance of users especially educators and students, towards TikTok were accordingly investigated. Comparatively, TikTok has been examined in non-educational settings, specifically to ascertain its impact on the settings outside of universities and schools. The present study therefore measured the effectiveness of Tiktok social media platforms in understanding the acceptance of customers in the domain of e-commerce.

A conceptual framework was accordingly constructed in the present study. The framework measures users' acceptance of TikTok social platform. The extrinsic and intrinsic tasks were currently distinguished by scholars, whereby the former is associated with the information system that supports in-built goal accomplishment, while the latter encompasses tasks that are applied in extracting the various types of information, delivered in video format, that customers can view and share at will (Gefen & Straub, 2000; Ryan & Deci, 2000). Accordingly, TAM was applied here in this study in the illustration of the process of customer acceptance (Davis, 1989). Here, the constructs of perceived usefulness and perceived ease of use are considered as part of extrinsic type of motivation. The constructs of customer satisfaction, Trust, and content richness were added to TAM as they could add to the acceptance of TikTok, as has been demonstrated in Ong and Lai (2006), Pituch and Lee (2006), Sánchez and Hueros (2010).

2.1 Development of Research Hypotheses and the Conceptual Model

Content richness encompasses resources of learning which ease knowledge acquirement and use. In view of that, adequate amount of content richness will increase the readiness of customers to accept the technology. The three key constructs of content richness are relevance, sufficiency, and timeliness. Specifically, relevance entails the manner in which the shared content has value, while sufficiency means the manner in which the shared content demonstrates comprehensiveness and illustrates a cohesive concept, whereby timeliness relates to the competebailable the most recent development information within the required content to its users (Jung et al., 2009; Sánchez & Hueros, 2010; Tung & Chang, 2008; Wulf et al., 2006).

Content richness has a clear direction with technology acceptance. It was reported by De Wulf et al. (2006) that content positively affects customer's acceptance of technology. Also, it was reported that higher content richness leads to the increase in usefulness and users' satisfaction, and thus, technology acceptance. As mentioned in related studies including Jung et al. (2009) high level of content richness will make the technology very useful. As for the construct of timeliness, it was found that outdated content is less useful, and thus, elements that are time-sensitive will impact the perception of user towards the technology. The following hypotheses are thus formed:

H1: Content richness has a positive impact on user satisfaction.

H2: Content richness has a positive impact on TikTok acceptance.

Trust

As a concept, trust has been described in various perspectives (McKnight et al., 2002). In Mayer et al. (1995), the notion of trust was described as the manifestation of persistent views from the actions of rivals, while Doney and Cannon (1997) described the concept as the confidence felt by someone towards others. Within the context of online business, the impact of trust has been examined as in Doney and Cannon (1997), Mayer et al. (1995), Jones and Leonard (2008). Gaining trust of consumer is ab challenging task for companies (Kim & Park, 2013), and it has been reported that the online purchase decisions of consumer is impacted by trust (Almajali, 2021b; Hajli, 2012) Equally, purchase intentions of consumer have been shown to be greatly affected by trust (Kim, 2011; Kim et al., 2008). Relevantly, in online businesses, Kim and Park (2013), Kim et al. (2008) and Mikalef et al., (2013) concluded that word of mouth (WOM) and low prices are among the two factors that affect trust.

A solid link between trust and satisfaction was concluded in Bauer et al. (2002), and as highlighted in Kennedy et al. (2001), the company is obliged to fulfill the needs and wants of its consumers. Notably, when consumer experiences desired expectations, he or she expresses satisfaction (Anderson & Sullivan, 1993). Satisfaction is dictated by the initial expectation of consumer (Anderson & Sullivan, 1993; Beyari & Abareshi, 2018). and for this reason, satisfaction is quantifiable through what is wanted by consumer and by whether (or not) the consumer attains the tangible aspects, including security of web system, delivery time, in addition to the intangible aspects, for instance, feelings of happiness and joy, or anger after the completion of purchase of a given product or service (Ashraf et al., 2017; Flavián et al., 2006). As indicated in Flavián et al. (2006), consumer will feel a sense of satisfaction when the sense of trust is high, when there is honesty, and when the website is viewed as competent. Trust thus becomes an element that significantly impacts consumer satisfaction in their intention to make purchase from certain website (Kim & Park, 2013).

In the adoption of a given technology, Gefen et al. (2003) stated trust as a key driver considering that trust is integral in dealing with two important provisos of digital means namely vulnerability risk and uncertainty (Doney & Cannon, 1997; Gambetta & Gambetta, 2000). Within the context of e-commerce, trust is crucial (Harris & Goode, 2004). This owes to some issues associated with conveyance of electronic service including the dynamicity of the environment, inexpert user (Ruyter et al., 2001), in addition to the non-existence of direct interaction. Furthermore, within the context of electronic services, trust has been explored in the context of e-commerce for instance in McKnight et al. (2002), Gefen et al. (2003). The notion of trust in the context of online behavior is important in understanding user acceptance towards electronic services (Carter & Weerakkody, 2008). The hypotheses below are therefore proposed:

H3: Trust has a positive impact on user satisfaction.

H4: Trust has a positive impact on TikTok acceptance.

User Satisfaction

User satisfaction is the level to which user feels pleased and satisfied in technology usage. In general, scholars perceive user satisfaction as an attained result that is manifested through the emotion towards the use of a given technology. In other words, it is possible that user would perceive a given technology as satisfying even when they have no preference towards it (Liao et al., 2009; Szymanski & Hise, 2000; Wixom & Todd, 2005). Additionally, Alshurideh et al. (2012) found positive influence of user satisfaction positively on e-service usage. Reporting similar result, Ashraf et al. (2017) added that such positive impact has direct linkage to both perceived usefulness and perceived ease of use. Relevantly, Al-Hawari and Mouakket (2010), George and Kumar (2013) and Teo (2011) mentioned user satisfaction as a prominent factor impacting the acceptance of technology (Liaw, 2008 (notably mentioned that higher level satisfaction will increase the inclination of user to use the technology. On the other hand, low satisfaction level will reduce the inclination of user towards technology acceptance or usage. As such, the hypothesis to be tested is:

H5: User satisfaction (US) has a positive impact on TikTok acceptance.

Technology Acceptance Model

Technology Acceptance Model (TAM) is an acceptance model that has been commonly applied in scholarly works relating to acceptance and adoption of technology in countless of domains. TAM, as mentioned in some studies (Agarwal & Karahanna, 2000; Niederhauser & Perkmen, 2010), has been effective particularly in measuring the successfulness of a given system and the outcomes of an experience in question. In this regard, two leading constructs of TAM, namely perceived ease of use (PEOU) and perceived usefulness (PU) have been shown to impact the socio-technical aspects, and these constructs have been applied in examining users' behavioral intention in utilizing a given system. These constructs greatly facilitate researcher in the comprehension of users' attitudes or beliefs toward a given information system. Both constructs relate to the level to which user is sure that the system is not only easy to use, but useful as well (Davis, 1989), and past studies (Alshurideh et al., 2020; Krejcie & Morgan, 1970; Limayem et al., 2001; Teo, 2011) reported that these constructs greatly impact the acceptance of technology and the behavioral intention towards its usage. The hypotheses below are therefore presented:

H6: Perceived ease of use (PEOU) positively impacts the acceptance of TikTok.

H7: Perceived usefulness (PU) has a positive impact on the acceptance of TikTok.

3 Research Methodology

The research methodology is discussed in this section, specifically covering the discussion on the research model proposed, the variables involved in this study, the research hypotheses, the instrument used in gathering the study data, the research population and sample, as well as the data analysis and result.

3.1 Research Model

The study model as illustrated in Fig. 1 displays the used variables and the proposed linkage between them.

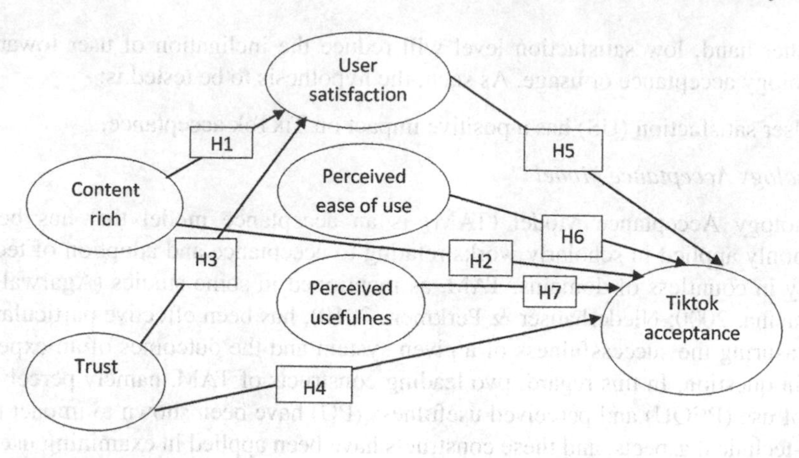

Fig. 1 Research model

3.2 Data Collection

This descriptive-analytical study adopted a deductive technique involving a cross-sectional design. Meanwhile, the data were obtained from the study questionnaires which were dispersed to the respondents online. The respondents in this study were users of Telecenters in Jordan. The study duration began on 15th March 2021 and ended on 20th June 2021. The questionnaire link was sent to the respondents through official emails and social media channels, including WhatsApp. The samples of the study were chosen using stratified random sampling methods. Jordan has various Telecentres (www.ks.gov.jo), and 196 were involved in this study. From these 196 Telecentres, 51 were in rural areas, 68 were operating in a semi-urban area, while 77 Telecentres were operating in an urban area.

This study distributed a total of 200 questionnaires, 145 of which, or 72%, were usable for analysis. Structural equation modeling techniques run with AMOS 22 were used in testing the proposed research model. The individuals made up the study's unit of analysis, and they were chosen for this research as they were the end-customers of ecommerce. The use of individuals as key respondents was also demonstrated in past studies including. In terms of sample size, the current study followed the suggestion of Chuan and Penyelidikan (2006). Here, the appropriate sample size was based on the collective number of 145 valid responses as 132 respondents are the expected sample size for a population size of 196. Structural Equation Simulation (SEM) was used in the evaluation of the sample size (Bagozzi & Yi, 1988), as the hypothesis validation. Table 1 accordingly presents the characteristics of the respondents, particularly in terms of their gender, age, level of education and experience of social media usage.

Table 1 Demographic data of the respondents

Demographic	Factor	Frequency	Percentage (%)
Gender	Female	90	62
	Male	55	38
Age	18–29	70	48
	30–39	40	28
	40–49	20	14
	50–59	15	10
Education level	Diploma	100	69
	Bachelor degree	10	7
	Others	35	24
TikTok usage frequency	Less than 10	15	10
	10–30	40	28
	More than 30	90	62
	Total	145	100

3.3 Demographic Information

Table 1 presents the demographic data of the respondents. As can be viewed, females made up 62% of the overall respondents, while males were the remaining 38%. In terms of age, 48% were between 18 and 29 years, while the remaining 52% were above 29 years. As for the level of education of the respondents, the table shows that: 7% obtained a bachelor's degree, 69% had diploma, and 24% were holders to other types of paper qualification. In the use of TikTok in a day, the majority stated that they used it more than 30 times, while 28% stated that they used the application between 10 and 30 times, whereas 10% indicated the use of Tiktok of less than 10 times in a day. The IBM SPSS Statistics version 23 was employed in the demographic data measurement.

3.4 Study Instrument

The study's proposed hypotheses were validated through the use of data obtained through a survey instrument, namely questionnaire. There were six constructs included in the questionnaire represented by 22 items (see Table 2). The items were adopted from past studies, and each item employs 5-point Likert scale, from the score of 1 to denote "strongly disagree" to the score of 5 to denote "strongly agree." A pre-test was carried out on the initial instrument. For the purpose, three semi-structured interviews involving TikTok users and five Department of Management Information

Table 2 Measurement items

Constructs	Items	Instruments
TikTok acceptance	TA1	Using TikTok is highly preferred for e-commerce activities
	TA2	Using TikTok is highly preferred for customers in the e-commerce environment
	TA3	Using TikTok is more preferred than other social media platforms
User satisfaction	US1	In general, my experience with TikTok as a customer was satisfactory
	US2	In general, my experience with TikTok satisfied all my needs as a customer
	US3	In general, my experience with TikTok is more satisfactory than on other social media platforms
Perceived ease of use	PEOU1	I think that TikTok is easy to use compared to other e-commerce activities
	PEOU2	I can easily find the information I need with TikTok
	PEOU3	I think that TikTok can replace other social media platforms due to its ease of use
Perceived usefulness	PU1	I think that TikTok helps in the acquirement and sharing of e-commerce information among customers
	PU2	I think that TikTok improves my desire to get new e-commerce information regularly
	PU3	I think that TikTok is more useful than other social media in the application of the e-commerce
Content richness	CR1	TikTok provides customers with sufficient content for e-commerce activities
	CR2	TikTok is rich with e-commerce content as compared with other social media platforms
	CR3	TikTok has the latest e-commerce information that can be shared and applied in e-commerce environment
	CR4	TikTok's short videos are more satisfactory than those presented on other social media platforms
	CR5	TikTok has provided customers with sufficient e-commerce information whenever they need it
	CR6	TikTok has provided customers with sufficient e-commerce activities whenever they need it
	CR7	TikTok has more sufficient e-commerce information than other social media platforms
Trust	TR1	I trust this TikTok platform
	TR2	This TikTok is reliable
	TR3	This TikTok is trustworthy

Systems (MIS) professors from the applied science private university. The purpose of this pre-test was to ascertain the completeness, relevance, clarity, and length of the questionnaire. Several necessary adjustments were made after the pre-test.

4 Data Analysis and Results

4.1 Measurement Model

AMOS software version 22 was applied in the hypotheses examination, involving the use of Confirmatory Factor Analysis (CFA). CFA determines whether the data fits the conjectured measurement model. Next, the proposed structural model was examined using Structural Equation Modelling (SEM) involving the use of path analysis with latent variables as demonstrated in Hair (2007), Kline (2011) and Newkirk and Lederer (2006). Accordingly, the model fit was accomplished through the execution of a robust statistics test for instance, x^2/ degrees of freedom (df), the Incremental Fit Index (IFI), Tucker-Lewis Index (TLI), Comparative Fit Index (CFI), Goodness-of-Fit Index (GFI), Adjusted Goodness-of-Fit Index (AGFI), and the Root Mean Square Error of Approximation (RMSEA). In this regard, Table 3 can be referred. In this study, the preliminary CFA model failed to show sufficient fit, resulting in the removal of items US3, CR3 and CR6, to improve the fit for the measurement model, with the following: the chi-square (x^2/df) value of the model = 1.86, IFI = 0.82, TLI = 0.83, GFI = 0.80, CFI = 0. 81 and RMSEA = 0.075. The form of measurement consequently demonstrated adequate fit to the data, as was also demonstrated in Hair et al. (2014), Kline (2011) and Newkirk and Lederer (2006).

The internal consistency of multi-item constructs was determined through Cronbach's Alpha evaluation. Creswell (2009) accordingly recommended that the value should be greater than 0.6, while Larcker et al. (1981) proposed by a factor loading of greater than 0.6 to the research items. Hair (2007) stated that all constructs should have composite reliabilities that are greater than the threshold value of 0.6. Additionally, Hair (2007) and Larcker et al. (1981) mentioned that the Average Variance Extracted (AVE) from a set of measurements of a latent variable should be greater than the value of 0.5. Table 4 accordingly displays the outcomes of Cronbach's Alpha, composite reliability, factor loadings, and AVE for the present study's research variables.

For all constructs in this study, based on Hair (2007), Larcker et al. (1981) and Ryu et al., (2009,(the attained Cronbach's Alpha was greater than 0.7, while the

Table 3 Measurement model fit indices

Model	x^2	Df	P	x^2/df	IFI	TLI	GFI	CFI	RMSEA
Initial Estimation	1321.114	521	0.00	2.53	0.71	0.70	0.77	0.74	0.121
Final model	641.122	343	0.00	1.86	0.82	0.83	0.80	0.81	0.075

Table 4 Properties of the measurement model

Construct	Standard loading	Standard error	Square multiple correlation	Error Variance	Cronbach's Alpha	Composite reliability	AVE
TikToK acceptance					0.86	0.96	0 0.91
TA1	0.751	0.083	0.075	0.043			
TA2	0.611	0.106	0.220	0.061			
TA3	0.559	0.084	0.470	0.031			
User satisfaction					0.83	0 0.96	0 0.86
US1	0.525	0.088	0.239	0.050			
US2	0.587	0.112	0.441	0.034			
Perceived ease of use					0 0.70	0 0.87	0.77
PEOU1	0.572	0.095	0.078	0.045			
PEOU2	0.504	0.115	0.085	0.066			
PEOU3	0.536	0.142	0.103	0.027			
Perceived usefulness					0.90	0.94	0.93
PU1	0.617	0.90	0.117	0.022			
PU2	0.573	0.360	0.371	0.057			
PU3	0.547	0.107	0.145	0.65			
Content richness					0.78	0.80	0.84
CR1	0.738	0.055	0.491	0.701			
CR2	0.693	0.073	0.394	0.627			
CR4	0.680	0.052	0.562	0.751			
CR5	0.544	0.050	0.612	0.780			
CR7	0.502	0.068	0.617	0.783			
Trust					0.82	0.94	0.81
TR1	0.622	0.068	0617	0.786			
TR2	0.589	0.069	0.711	0.843			
TR3	0.561	0.215	0.419	0.027			

*Applying Larcker et al.'s (Larcker et al., 1981) formula, the computation of composite reliability is based on the following: Composite Reliability = $(\Sigma \text{ Li})^2/((\Sigma \text{ Li})^2 + \Sigma \text{ Var (Ei)})$, where Li denotes the standardised factor loadings for each indicator, and Var (Ei) denotes the error variance related to the individual indicator variables

**The formula for the variance extracted is: Average Variance Extracted = $\Sigma \text{ Li}^2/(\Sigma\text{Li}^2 + \Sigma\text{Var(Ei)})$ where Li denotes the standardised factor loadings for each indicator, while Var (Ei) denotes the error variance related to the individual indicator variables.

indicators of the factor loadings were all greater than 0.50, which means that the items had convergent validity. Furthermore, the fact that all AVE values obtained in this study were greater than 0.50 was denoting convergent validity, as proposed in Creswell (2009) and Hair (2007). Table 4 accordingly displays the attained means, standard deviations, AVEs, and the square of correlations together with the respective constructs. As demonstrated by the findings, the relationships between construct pairs were smaller compared to the square root of the estimates of the AVE of the two constructs. Based on Creswell (2009), there is discriminant validity.

Minimum recommended value x^2/df = 1; IFI = 0.80; TFI = 0.80; CFI = 0.80; GFI = 0.80; AGFI = 0.80; RMSEA = 0.05.

4.2 Structural Model

As shown by the following Table 5 on the outcomes of SEM analysis, content richness imparts a significant positive and direct impact on user satisfaction (β = 0.206, C.R = 2.420, p = 0.006), denoting support to H1. The results also support H2 by demonstrating the significant positive and direct impact of content richness on TikTok acceptance (β = 0.513, C.R = 7.211, p = ***). Trust imparts a positive significant effect on user satisfaction (β = 0.175, C.R = 3.042, p = 0.002). Hence, H3 was supported. For H4, the results affirm the positive and significant direct impact of trust on TikTok acceptance (β = 0.110, C.R = 2.138, p = 0.032), and therefore, H4 was supported. Further, the results demonstrate a positive direct impact of user satisfaction on TikTok acceptance (β = 0.114, C.R = 2.312, p = 0.041), which means that H5 was supported. In examining H6, the results show significant positive direct impact of perceive ease of use on TikTok acceptance (β = 0.141, C.R = 2.411, p = 0.018), denoting support to H6. A positive direct impact of perceived usefulness on TikTok acceptance was shown by the results (β = 0.111, C.R = 2.444, p = 0.035). As such, H7 was supported.

Table 5 Summary of the proposed results for the theoretical model

Research proposed paths	*t*-value (CR)	Coefficient value (std. estim)	*P*-value	Decision
CR ⇨ US	2.420	0.206	0.006	Supported
CR ⇨ TA	7.211	0.513	***	Supported
TR ⇨ US	3.042	0.175	0.002	Supported
TR ⇨ TA	2.138	0.110	0.032	Supported

(continued)

Table 5 (continued)

Research proposed paths	*t*-value (CR)	Coefficient value (std. estim)	*P*-value	Decision
US ⟹ TA	2.312	0.114	0.041	Supported
PEOU ⟹ TA	2.411	0.141	0.018	Supported
PU ⟹ TA	2.444	0.111	0.035	Supported

(***$P \leq 0.005$, **$P \leq 0.01$, *$P \leq 0.05$)
Notes Path = Relationship between independent variable on dependent variable; C.R = Critical ratio; S.E = Standard error; P = Level of significance.
Note CR: Content Richness, TA: TikToK acceptance, TR: Trust, US: user satisfaction, PEOU: Perceived ease of use, PU: Perceived usefulness

5 Discussion

The literature review evidences the significance of social media within the learning environment, and some factors have been found to significantly impact social media acceptance (Oum & Han, 2011; Wang, 2020). Notably, the major determinants in TikTok acceptance for e-commerce activities were highlighted in this study. In addition, the status of digital divide among users was ascertained, especially within the context of Jordan. The vision of TikTok was expanded in this study by delving into e-commerce activities via this application. The results obtained demonstrate the positive impact of content richness on user satisfaction, showing support to H1.

In a related study, Yang and Zilberg (2020) looked into the impacts of features of short-form videos on the psychological and acceptance of users towards TikTok, and reported the ability of TikTok's short videos in greatly improving the acceptance of users. Such acceptance was attributed to the wide-spread humor feature. Contrariwise, the present study found that TikTok contains more general information generated by non-expert users, and these users created the information just to pass the time. Initially, TikTok was created for just socialization and self-expression, allowing users to share funny and merry emotions, with no regard to the content type being (Boyinbode et al., 2017). Content creators do exist on TikTok but they appear to be focusing on the general topics only. In other words, there is no specialized type of content for e-commerce users.

In addition, content creators in TikTok upload their content merely to pass time. This is different from their counterparts on other social media platform who are more specific in their content type. Another point worth noting is that TikTok video creators are generally uninterested in the issues associated with e-commerce. Notably, TikTok has been known as a platform that hosts short inspirational unspecialized videos that inspire and bring joy to people. In other words, TikTok videos are able to create a positive atmosphere. Additionally, as highlighted in Boyinbode et al., (2017), influencers in TikTok were mostly individuals in the age group of 13–64 years old, meaning that these influencers encompass teenagers, young or middle-aged adults.

Referring to H2, this study found positive impact of content richness on Tiktok acceptance. H2 was therefore supported. Additionally, past related studies have also reported the important role of content richness on TikTok acceptance, as it impacts the keenness of user to consistently utilize social media. As such, governments and companies should create superior social media content to entice more users. Kuan and Bock (2007) relevantly reported content richness as a significant determinant of WhatsApp acceptance, as this application offers its users up-to-date information.

In discussing H3, this study found positive impact of trust on user satisfaction, and thus, H3 was supported. Relevantly, past studies have found that trust of high level among consumers would cause them to have more intention to purchase the goods and services, in this study context, purchase them online. In their study, Paulov (2003) reported trust as a major factor in testing the elements affecting consumer satisfaction.

The obtained results suggested a positive impact of trust on Tiktok acceptance, and thus, H4 was supported. As shown in past studies, trust impacts user behavior in technology adoption (Abuhashesh et al., 2019; Gyampah, 2004; Masa'deh et al., 2014, 2018; Stoel, 2009; Wang et al., 2007). In other words, untrusted service provider will prevent customer from accepting the provided service. Meanwhile, trust towards the acceptance of TikTok will decrease the effort in verifying, monitoring and controlling the service interaction. Contrariwise, low level of trust will cause customer to focus more on evaluating the service as a way to prevent opportunism on TikTok acceptance.

Results on H5 show the positive effect of user satisfaction on Tiktok acceptance, and thus, H5 was supported. Studies have reported user satisfaction as the pivotal factor underpinning the intention to use technology. For this reason, user satisfaction has been linked to long term usage, rather than the initial acceptance. Also, this factor has been regarded as a critical factor in the measurement of the satisfaction level within the domain of e-commerce, whereby motivation and commitment were found to impact the effectiveness of the information. Pertinently, past studies have shown that high level of satisfaction contributes to the constant acceptance of technology (Deci, 1985; Masa'deh et al., 2014; Shee & Wang, 2008; Wu et al., 2010).

The results demonstrate a significant influence of perceived ease of use and perceived usefulness on Tiktok acceptance, which means that both H6 and H7 were supported. The acceptance of TikTok is significantly justified through TAM. Further, perceived usefulness is regarded as a direct predictor of behavioral intention to accept both TikTok, and this finding contradicts the proposed hypotheses. Apparently, PEOU and PU in the facilitation of acceptance are significant, and technology that is easy and useful is likely to be more accepted. TikTok has been deemed a beneficial tool in stimulating extrinsic motivation in the determination of technology usage behaviour. Acceptance might hence be directly impacted by goal-oriented behaviour efforts (Lee & Lehto, 2013; Shang et al., 2005). In this study, user satisfaction was added to TAM as it has been found to have a direct linkage to the actual and practical experience of users.

5.1 Theoretical and Practical Implications

At present, TikTok is the most popular social networks among users globally, and it has been deemed the most entertaining short video-sharing app. Its brief time and content, as highlighted in Lee and Lehto (2013) and Yang and Zilberg (2020), has captured the interest of millions of viewers. In theory, e-commerce customers would enjoy the use of TikTok as a source of information. For this reason, the prospect of providing more specialized and current videos that will improve the activities and experience of customers should be increased. As videos on TikTok are short, they add knowledge to users but in restricted manner, and according to Boyinbode et al. (2017), its ability to provide enjoyable experience to user is the only striking feature of TikTok. It is thus important that the content is evaluated in order that creators of content of TikTok could produce information that is richer, up-to-date and specialized.

In practice, website developers should be concentrating on the generation of features that could facilitate the classification of the short video in order that the target users could be reached more easily. It has been recently reported that as a platform of social media, TikTok alongside its mobile short video features can indeed enrich the knowledge and experience of users. In this regard, developers of TikTok need to consider that videos with time duration that is too short may limit knowledge acquirement and experience of user, and thus, more features should be added in order that users in e-commerce domain could continue using TikTok.

5.2 Limitations and Suggestions for Future Studies

Several limitations have been identified in this study. Firstly, only one type of social network was addressed in this study namely TikTok. On the other hand, there are various other platforms with comparable functions like Reels. As such, these other platforms could be examined in future studies, particularly in terms of their effectiveness in the e-commerce context. Secondly, as this study was addressing the present conception and perception customer of TikTok, next studies may consider looking into the continuous intention to use such platform on long-term basis. Thirdly, considering that Jordanian customers were the sole respondents of this study, the results may not be generalizable to customers in other countries. As such, future studies should include customers in other countries in order that comparison could be made in terms of the use of TikTok as a social network.

References

Abu Zayyad, H. M., Obeidat, Z. M., Alshurideh, M. T., Abuhashesh, M., Maqableh, M., & Masa'deh, R. (2020). Corporate social responsibility and patronage intentions: The mediating effect of brand credibility. *Journal of Marketing Communications*. https://doi.org/10.1080/13527266.2020.1728565.

Abuhashesh, M., Al-Khasawneh, M., Al-Dmour, R., & Masa'deh, R. (2019). The impact of facebook on Jordanian consumers' decision process in the hotel selection. *IBIMA Business Review*. https://doi.org/10.5171/2019.928418.

Aburayya, A., Alshurideh, M., Al Marzouqi, A., Al Diabat, O., Alfarsi, A., Suson, R., Bash, M., & Salloum, S. A. (2020). An empirical examination of the effect of TQM practices on hospital service quality: An assessment study in UAE hospitals. *Systematic Reviews in Pharmacy, 11*(9), 347–362.

Agarwal, R., & Karahanna, E. (2000). Cognitive absorption and beliefs about information technology usage. *MIS Quarterly, 24*(4), 665–694.

Al Kurdi, B., Alshurideh, M., & Salloum, S. A. (2020a). Investigating a theoretical framework for e-learning technology acceptance. *International Journal of Electrical and Computer Engineering (IJECE), 10*(6), 6484–6496.

Al Kurdi, B., Alshurideh, M., Salloum, S., Obeidat, Z., & Al-dweeri, R. (2020b). An empirical investigation into examination of factors influencing university students' behavior towards elearning acceptance using SEM approach. *International Journal of Interactive Mobile Technologies, 14*(2), 19–24.

Al-Dmour, R. H., Masa'deh, R., & Obeidat, B. Y. (2017). Factors influencing the adoption and implementation of HRIS applications: Are they similar? *International Journal of Business Innovation and Research, 14*(2), 139–167. https://doi.org/10.1504/IJBIR.2017.086276

Al-Dmour, H., Masa'deh, R., Salman, A., Abuhashesh, M., & Al-Dmour, R. (2020). Influence of social media platforms on public health protection against the COVID-19 pandemic via the mediating effects of public health awareness and behavioral changes: Integrated model. *Journal of Medical Internet Research, 22*(8). https://doi.org/10.2196/19996.

Al-Dmour, R., AlShaar, F., Al-Dmour, H., Masa'deh, R., & Alshurideh, M. T. (2021). The effect of service recovery justices strategies on online customer engagement via the role of "customer satisfaction" during the covid-19 pandemic: An empirical study. *The Effect of Coronavirus Disease (COVID-19) on Business Intelligence, 334*, 325–346.

Alhashmi, S. F., Salloum, S. A., & Mhamdi, C. (2019). Implementing artificial intelligence in the United Arab Emirates healthcare sector: an extended technology acceptance model. *International Journal of Information Technology and Language Studies, 3*(3).

Alhashmi, S., Salloum, S., & Abdallah, S. (2020). Critical success factors for implementing Artificial Intelligence (AI) projects in Dubai Government United Arab Emirates (UAE) health sector: Applying the extended Technology Acceptance Model (TAM). *Advances in Intelligent Systems and Computing* (vol. 1058). https://doi.org/10.1007/978-3-030-31129-2_36.

Al-Hawari, M. A., & Mouakket, S. (2010). The influence of technology acceptance model (TAM) factors on students' e-satisfaction and e-retention within the context of UAE e-learning. *Education, Business and Society: Contemporary Middle Eastern Issues, 3*(4), 299–314.

Almajali, D. (2021a). Antecedents of ecommerce on actual use of international trade center: Literature review. *Academy of Strategic Management Journal, 20*(2), 1–8.

Almajali, D. (2021b). Diagnosing the effect of green supply chain management on firm performance: An experiment study among Jordan industrial estates companies. *Uncertain Supply Chain Management, 9*(4), 1–8.

Almajali, D., Hammouri, Q., Majali., T., AL-Gasawneh., J., & Dahalin, Z. (2021). Antecedents of consumers' adoption of electronic commerce in developing countries. *International Journal of Data and Network Science, 5*(4), 1–10.

Al-Maroof, R., Salloum, S., Hassanien, A., & Shaalan, K. (2020). Fear from COVID-19 and technology adoption: The impact of Google Meet during Coronavirus pandemic. *Interactive Learning Environments*. https://doi.org/10.1080/10494820.2020.1830121

Al-Maroof, R., Ayoubi, K., Alhumaid, K., Aburayya, A., Alshurideh, M., Alfaisal, R., & Salloum, S. (2021). The acceptance of social media video for knowledge acquisition, sharing and application: A comparative study among YouYube users and TikTok users' for medical purposes. *International Journal of Data and Network Science, 5*(3), 197–214.

Almazrouei, F.A., Alshurideh, M., Al Kurdi, B., & Salloum, S. A. (2021). Social media impact on business: a systematic review. *Advances in Intelligent Systems and Computing* (AISC) (vol. 1261). https://doi.org/10.1007/978-3-030-58669-0_62.

Alqudah, A., Salloum, S., & Shaalan, K. (2021). The role of technology acceptance in healthcare to mitigate covid-19 outbreak. *During the Era of COVID-19 Pandemic, 348*, 223.

Alshurideh, M. (2018). Pharmaceutical promotion tools effect on physician's adoption of medicine prescribing: Evidence from Jordan. *Modern Applied Science, 12*(11), 210–222.

Alshurideh, M. (2022). Does electronic customer relationship management (E-CRM) affect service quality at private hospitals in Jordan? *Uncertain Supply Chain Management, 10*(2), 1–8.

Alshurideh, M., Masa'deh, R., & Alkurdi, B. (2012). The effect of customer satisfaction upon customer retention in the Jordanian mobile market: An empirical investigation. *European Journal of Economics, Finance and Administrative Sciences, 47*(12), 69–78.

Alshurideh, M., Al Kurdi, B., & Salloum, S. A. (2019a). Examining the main mobile learning system drivers' effects: A mix empirical examination of both the Expectation-Confirmation Model (ECM) and the Technology Acceptance Model (TAM). In *International Conference on Advanced Intelligent Systems and Informatics* (pp. 406–417). Springer, Cham.

Alshurideh, M., Salloum, S. A., Al Kurdi, B., Monem, A. A., & Shaalan, K. (2019b). Understanding the quality determinants that influence the intention to use the mobile learning platforms: A practical study. *International Journal of Interactive Mobile Technologies, 13*(11), 183–157.

Alshurideh, M., Al Kurdi, B., Salloum S. A. (2020) Examining the main mobile learning system drivers' effects: A mix empirical examination of both the Expectation-Confirmation Model (ECM) and the Technology Acceptance Model (TAM). In A. Hassanien, K. Shaalan, M. Tolba (eds.), *Proceedings of the International Conference on Advanced Intelligent Systems and Informatics 2019. AISI 2019. Advances in Intelligent Systems and Computing*, (vol. 1058). Springer, Cham. https://doi.org/10.1007/978-3-030-31129-2_37.

Alyammahi, A., Alshurideh, M., Kurdi, B. A., & Salloum, S. A. (2021). The impacts of communication ethics on workplace decision making and productivity. *Advances in Intelligent Systems and Computing* (AISC) (vol. 1261). https://doi.org/10.1007/978-3-030-58669-0_44.

Anderson, E. W., & Sullivan, M. W. (1993). The antecedents and consequences of customer satisfaction for firms. *Marketing Science, 12*(2), 125–143.

Andersson, A., & Grönlund, A. (2009). A conceptual framework for e-learning in developing countries: A critical review of research challenges. *The Electronic Journal on Information Systems in Developing Countries, 38*, 1–16.

Ashraf, M., Jaafar, N. I., & Sulaiman, A. (2017). The mediation effect of trusting beliefs on the relationship between expectation-confirmation and satisfaction with the usage of online product recommendation. *The South East Asian Journal of Management, 10*(1), 75–94.

Bagozzi, R., & Yi, Y. (1988). On the evaluation of structural equation models. *Journal of the Academy of Marketing Sciences, 16*(1), 74–94.

Bauer, H. H., Grether, M., & Leach, M. (2002). Building customer relations over the Internet. *Industrial Marketing Management, 31*(2), 155–163.

Bavarsad, B., & Mennatyan, M. A. (2013). A study of the effects of technology acceptance factors on users' satisfaction of e-government services. *World Applied Programming, 3*(5), 190–199.

Beyari, H., & Abareshi, A. (2018). Consumer satisfaction in social commerce: An exploration of its antecedents and consequences. *The Journal of Developing Areas, 52*(2), 55–72.

Bian, R., & Zhu, M. (2020). The relationship between ritual, personal involvement and travel intention: A study of check-in-travel on DouYin. *American Journal of Industrial Business Management, 10*(02), 451–467.

Boyinbode, O. K., Agbonifo, O. C., & Ogundare, A. (2017). Supporting mobile learning with WhatsApp based on media richness. *Circulation in Computer Science, 2*(3), 37–46.

Carter, L., & Weerakkody, V. (2008). E-government adoption: A cultural comparison. *Information Systems Frontiers, 10*, 473–482.

Chatzoglou, P., Chatzoudes, D., Ioakeimidou, D., & Tokoutsi, A. (2020). Generation Z: Factors affecting the use of Social Networking Sites (SNSs). In *2020 15th International Workshop on Semantic and Social Media Adaptation and Personalization (SMA)* (pp. 1–6). IEEE.

Chen, P., & Zhang, X. (2020). Evaluation and empirical study on the information service quality of TikTok government accounts. *Eurasian Journal of Social Sciences, 8*(2), 53–69.

Chuan, C. L., & Penyelidikan, J. (2006). Sample size estimation using Krejcie and Morgan and Cohen statistical power analysis: A comparison. *Jurnal Penyelidikan IPBL, 7*, 78–86.

Clark, M., & Gomez, R. (2011). The negligible role of fees as a barrier to public access computing in developing countries. *Electronic Journal of Information Systems in Developing Countries, 46*, 1–14.

Creswell, J. W. (2009). *Research design: Qualitative, quantitative, and mixed methods approaches* (3rd ed.). Thousand Oaks.

Davis, F. D. (1989). Perceived usefulness, perceived ease of use, and user acceptance of information technology. *MIS Quarterly*, 319–340.

De Ruyter, K., Wetzels, M., & Kleijnen, M. (2001). Customer adoption of e-service: An experimental study. *International Journal of Service Industry Management, 12*, 184–207.

De Wulf, K., Schillewaert, N., Muylle, S., & Rangarajan, D. (2006). The role of pleasure in web site success. *Information & Management, 43*(4), 434–446.

Deci, E. (1985). L Intrinsic motivation and self-determination in human behavior/EL Deci, RM Ryan.

Deursen, A. J., & Dijk, J. A. (2013). The digital divide shifts to differences in usage. *New Media & Society, 16*(3), 507–526.

Doney, P., & Cannon, J. (1997). An examination of the nature of trust in buyer-seller relationships. *The Journal of Marketing, 61*, 35–51.

Flavián, C., Guinalíu, M., & Gurrea, R. (2006). The role played by perceived usability, satisfaction and consumer trust on website loyalty. *Information & Management, 43*(1), 1–14.

Gambetta, D. (2000). Can we trust trust? In D. Gambetta (Ed.), *Trust: Making and breaking cooperative relations* (Electronic, pp. 213–237). University of Oxford.

Gefen, D., & Straub, D. (2000). The relative importance of perceived ease of use in IS adoption: A study of e-commerce adoption. *Journal of the Association for Information Systems, 1*, 1–28.

Gefen, D., Karahanna, E., & Straub, D. W. (2003). Trust and TAM in online shopping: An integrated model. *MIS Quarterly, 27*(1), 51–90.

George, A., & Kumar, G. S. (2013). Antecedents of customer satisfaction in internet banking: Technology acceptance model (TAM) redefined. *Global Business Review, 14*(4), 627–638.

Gurstein, M. (2007). What is community informatics: (And why does it matter)? Polimetrica, Milano.

Gyampah, A. (2004). An extension of the technology acceptance model in an ERP implementation environment. *Information & Management, 41*(6), 731–745.

Habes, M., Salloum, S. A., Alghizzawi, M., & Mhamdi, C. (2019). The relation between social media and students' academic performance in Jordan: YouTube perspective. In *International Conference on Advanced Intelligent Systems and Informatics* (pp. 382–392). Springer.

Hair, J. (2007). Research methods for business. In *Education + Training* (4th ed., vol. 49, Issue 4). Wiley. ISBN0-470-03404-0. https://doi.org/10.1108/et.2007.49.4.336.2.

Hair, J., Black, W., Babin, B., & Anderson, R. (2014). Multivariate data analysis. *Pearson Custom Library*. https://doi.org/10.1038/259433b0

Hajli, M. (2012). An integrated model for e-commerce adoption at the customer level with the impact of social commerce. *International Journal of Information Science and Management (IJISM)*, 77–97.

Harris, L., & Goode, M. (2004). The four levels of loyalty and the pivotal role of trust: A study of online service dynamics. *Journal of Retailing, 80*, 139–158.

Hilbert, M. (2016). The bad news is that the digital access divide is here to stay: Domestically installed bandwidths among 172 countries for 1986–2014. *Telecommunications Policy, 40*(6), 1–23.

Jones, K., & Leonard, L. N. (2008). Trust in consumer-to-consumer electronic commerce. *Information & Management, 45*(2), 88–95.

Jung, Y., Perez-Mira, B., & Wiley-Patton, S. (2009). Consumer adoption of mobile TV: Examining psychological flow and media content. *Computers in Human Behavior, 25*(1), 123–129.

Kennedy, M. S., Ferrell, L. K., & LeClair, D. T. (2001). Consumers' trust of salesperson and manufacturer: An empirical study. *Journal of Business Research, 51*(1), 73–86.

Kim, Y. (2011). Market analysis and issues of social commerce in Korea. *KISDI, 23*(11), 41–63.

Kim, S., & Park, H. (2013). Effects of various characteristics of social commerce (s-commerce) on consumers' trust and trust performance. *International Journal of Information Management, 33*(2), 318–332.

Kim, D. J., Ferrin, D. L., & Rao, H. R. (2008). A trust-based consumer decision-making model in electronic commerce: The role of trust, perceived risk, and their antecedents. *Decision Support Systems, 44*(2), 544–564.

Kiwanuka, A. (2015). Acceptance process: The missing link between UTAUT and diffusion of innovation theory. *American Journal of Information Systems, 3*(2), 40–44.

Kline, R. B. (2011). *Principles and practice of structural equation modeling* (2nd ed.). The Guilford Press.

Krejcie, R. V., & Morgan, D. W. (1970). Determining sample size for research activities. *Educational and Psychological Measurement, 30*(3), 607–610.

Kuan, H.-H., & Bock, G. W. (2007). Trust transference in brick and click retailers: An investigation of the before-online-visit phase. *Information & Management, 44*(2), 175–187.

Kurdi, B., Alshurideh, M., & Alnaser, A. (2020). The impact of employee satisfaction on customer satisfaction: Theoretical and empirical underpinning. *Management Science Letters, 10*(15), 3561–3570.

Landau, L. (2012). Estado de La Banda Ancha En América Latina y El Caribe, 2012. Informe Del Obervatorio Regional de Banda Ancha (ORBA), Zhurnal Eksperimental'noi i Teoreticheskoi Fiziki. Retrieved Nov 26, 2014, from http://scholar.google.com/scholar?hl=en&btnG=Search&q=intitle:No+Title#0.

Larcker, D. F., Fornell, C., & Larcker, D. F. (1981). Evaluating structural equation models with unobservable variables and measurement error. *Journal of Marketing Research, 18*(1), 456–464.

Lee, D. Y., & Lehto, M. R. (2013). User acceptance of YouTube for procedural learning: An extension of the technology acceptance model. *Computers and Education, 61*(1), 193–208.

Liao, C., Palvia, P., & Chen, J.-L. (2009). Information technology adoption behavior life cycle: Toward a Technology Continuance Theory (TCT). *International Journal of Information Management, 29*(4), 309–320.

Liaw, S.-S. (2008). Investigating students' perceived satisfaction, behavioral intention, and effectiveness of e-learning: A case study of the Blackboard system. *Computers & Education, 51*(2), 864–873.

Limayem, M., Hirt, S. G., & Chin, W. W. (2001). Intention does not always matter: the contingent role of habit on IT usage behavior. *ECIS 2001 Proceedings,* 56.

Ma, L., Feng, J., Feng, Z., & Wang, L. (2019). Research on user loyalty of short video app based on perceived value—Take Tik Tok as an example. *Paper presented at the 2019 16th International Conference on Service Systems and Service Management (ICSSSM)*, Shenzhen, China.

Masa'deh, R., Alananzeh, O., Tarhini, A., & Algudah, O. (2018). The effect of promotional mix on hotel performance during the political crisis in the Middle East. *Journal of Hospitality and Tourism Technology, 9*(1), 32–47. https://doi.org/10.1108/JHTT-02-2017-0010.

Masa'deh, R. M. T., Maqableh, M. M., & Karajeh, H. (2014). A theoretical perspective on the relationship between leadership development, knowledge management capability, and firm performance. *Asian Social Science, 10*(6), 128–137. https://doi.org/10.5539/ass.v10n6p128

Mayer, R., Davis, J., & Schoorman, F. (1995). An integrative model of organizational trust. *The Academy of Management Review, 20*, 709–734.

McKnight, D., Choudhury, V., & Kacmar, C. (2002). The impact of initial consumer trust on intentions to transact with a web site: A trust building model. *Journal of Strategic Information Systems, 11*, 297–323.

Mhamdi, C., Al-Emran, M., & Salloum, S. A. (2018). Text mining and analytics: A case study from news channels posts on Facebook. *Studies in Computational Intelligence* (vol. 740). https://doi.org/10.1007/978-3-319-67056-0_19.

Mikalef, P., Giannakos, M., & Pateli, A. (2013). Shopping and word-of-mouth intentions on social media. *Journal of Theoretical and Applied Electronic Commerce Research, 8*(1), 17–34.

Muñoz, T., & Amaral, T. (2011). Factores determinantes del comercio electrónico en España. *Boletín Económico De ICE, 30*(16), 51–65.

Newkirk, H. E., & Lederer, A. L. (2006). The effectiveness of strategic information systems planning under environmental uncertainty. *Information and Management, 43*(4), 481–501.

Niederhauser, D. S., & Perkmen, S. (2010). Beyond self efficacy: Measuring pre-service teachers' instructional technology outcome expectations. *Computers in Human Behavior, 26*(3), 436–442.

Obeidat, B. Y., Al-Hadidi, A., Tarhini, A., & Masa'deh, R. (2017). Factors affecting strategy implementation: A case study of pharmaceutical companies in the Middle East. *Review of International Business and Strategy, 27*(3), 386–408. https://doi.org/10.1108/RIBS-10-2016-0065

Omar, B., & Dequan, W. (2020). Watch, share or create: The influence of personality traits and user motivation on TikTok mobile video usage. *International Journal of Interactive Mobile Technologies, 14*(4).

Ong, C.-S., & Lai, J.-Y. (2006). Gender differences in perceptions and relationships among dominants of e-learning acceptance. *Computers in Human Behavior, 22*(5), 816–829.

Oum, S., & Han, D. (2011). An empirical study of the determinants of the intention to participate in user-created contents (UCC) services. *Expert Systems with Applications, 38*(12), 15110–15121.

Park, Y., Son, H., & Kim, C. (2012). Investigating the determinants of construction professionals' acceptance of web-based training: An extension of the technology acceptance model. *Automation in Construction, 22*, 377–386.

Paulov, P. (2003). Consumer acceptance of electronic commerce: Integrating trust and risk with the technology acceptance model. *International Journal of Electronic Commerce, 7*(3), 101–134.

Peral-Peral, B., Arenas-Gaitán, J., & Villarejo-Ramos, Á. -F. (2015). From digital divide to psycho-digital divide: Elders and online social networks. *Comunicar, 23*(45), 10–11.

Pick, J. B., & Nishida, T. (2015). Digital divides in the world and its regions: A spatial and multivariate analysis of technological utilization. *Technological Forecasting and Social Change, 91*(1), 1–17.

Pituch, K. A., & Lee, Y. (2006). The influence of system characteristics on e-learning use. *Computers & Education, 47*(2), 222–244.

Ryan, R. M., & Deci, E. L. (2000). Intrinsic and extrinsic motivations: Classic definitions and new directions. *Contemporary Educational Psychology, 25*(1), 54–67.

Ryu, M.-H., Kim, S., & Lee, E. (2009). Understanding the factors affecting online elderly user's participation in video UCC services. *Computers in Human Behavior, 25*(3), 619–632.

Salloum, S., Al-Emran, M., Abdallah, S., & Shaalan, K. (2017). Analyzing the Arab Gulf newspapers using text mining techniques. In *International Conference on Advanced Intelligent Systems and Informatics* (pp. 396–405). Springer. https://doi.org/10.1007/978-3-319-64861-3_37.

Salloum, S., Maqableh, W., Mhamdi, C., Al Kurdi, B., & Shaalan, K. (2018). Studying the social media adoption by university students in the United Arab Emirates. *International Journal of Information Technology and Language Studies, 2*(3).
Sánchez, R. A., & Hueros, A. D. (2010). Motivational factors that influence the acceptance of Moodle using TAM. *Computers in Human Behavior, 26*(6), 1632–1640.
Sánchez-Torres, J. A., Arroyo-Cañada, F., Varon-Sandobal, A., & Sánchez-Alzate, J. A. (2017). Differences between e-commerce buyers and non-buyers in Colombia: The moderating effect of educational level and socioeconomic status on electronic purchase intention. *DYNA, 84*(202), 175–189.
SensorTower. (2019). Top social networking apps worldwide for Q1 2019 by downloads. https://sensortower.com/blog/top-social-networking-apps-worldwide-q1-2019.
Sey, A., & Fellows, M. (2009). *Literature review on the impact of public access to information and communication technologies*. CIS Working Paper No. 6. http://www.globalimpactstudy.org/wp-content/uploads/2010/12/TASCHA_Public-Access-Review_2009.pdf.
Shang, R.-A., Chen, Y.-C., & Shen, L. (2005). Extrinsic versus intrinsic motivations for consumers to shop on-line. *Information & Management, 42*(3), 401–413.
Shee, D. Y., & Wang, Y.-S. (2008). Multi-criteria evaluation of the web-based e-learning system: A methodology based on learner satisfaction and its applications. *Computers & Education, 50*(3), 894–905.
Stoel, S. (2009). Consumer e-shopping acceptance: Antecedents in a technology acceptance model. *Journal of Business Research, 62*(5), 565–571.
Susilo, D. (2020). Unlocking the secret of e-loyalty: A study from Tiktok users in China. *International Journal of Economics, Business, Entrepreneurship, 3*(1), 37–49.
Szymanski, D. M., & Hise, R. T. (2000). E-satisfaction: An initial examination. *Journal of Retailing, 76*(3), 309–322.
Tarhini, A., Mgbemena, C., Trab, M. S. A., & Masa'deh, R. (2015). User adoption of online banking in Nigeria: A qualitative study. *Journal of Internet Banking and Commerce, 20*(3). https://doi.org/10.4172/1204-5357.1000132.
Teo, T. (2011). Technology acceptance research in education. In *Technology acceptance in education* (pp. 1–5). Springer.
Torres, J. A., & Cañada, F. X. (2016). Diferencias de la adopción del comercio electrónico entre países. *Suma De Negocios, 7*(16), 1–10.
Tung, F.-C., & Chang, S.-C. (2008). Nursing students' behavioral intention to use online courses: A questionnaire survey. *International Journal of Nursing Studies, 45*(9), 1299–1309.
Venkatesh, V., Morris, M. G., Davis, G. B., & Davis, F. D. (2003). User acceptance of information technology: Toward a unified view. *MIS Quarterly, 27*(3), 425–478.
Venkatesh, V., Thong, J. Y., & Xu, X. (2012). Consumer acceptance and use of information technology: Extending the unified theory of acceptance and use of technology. *MIS Quarterly, 36*(1), 157–178.
Wang, Y. (2020). Humor and camera view on mobile short-form video apps influence user experience and technology adoption intent, an example of TikTok (DouYin). *Computers in Human Behavior, 110*, 106373.
Wang, Y. S., & Shih, Y. W. (2008). A validation of the unified theory of acceptance and use technology. *Government Information Quarterly, 36*, 158–165.
Wang, Y. S., Wang, H.-Y., & Shee, D. Y. (2007). Measuring e-learning systems success in an organizational context: Scale development and validation. *Computers in Human Behavior, 23*(4), 1792–1808.
Wang, R. (2019). *Effect of e-wom message of opinion leaders on purchase intention of female consumers in China: Case of Ddouyin (Tik Tok)*. [Doctoral dissertation], Universidade De Lisboa, Lisbon, Portugal. https://www.repository.utl.pt/bitstream/10400.5/18315/1/DM-WR-2019.pdf.
Wang, S., & Fu, R. (2020). Research on the influencing factors of the communication effect of Tik Tok short videos about intangible cultural heritage. In *International Conference on Applied Human Factors and Ergonomics* (pp. 275–282).

Wei, K., Teo, H.-H., Chan, H., & Tan, B. (2011). Conceptualizing and testing a social cognitive model of the digital divide. *Information Systems Research, 22*(1), 170–187.

White, D. S., Gunasekaran, A., Shea, T. P., & Ariguzo, G. C. (2011). Mapping the global digital divide. *International Journal of Business Information Systems, 7*(2), 207–219.

Williams, M.-D., Rana, N.-P., & Dwivedi, Y.-K. (2015). The unified theory of acceptance and use of technology (UTAUT): A literature review. *Journal of Enterprise Information Management, 28*(3), 433–488.

Wixom, B. H., & Todd, P. A. (2005). A theoretical integration of user satisfaction and technology acceptance. *Information Systems Research, 16*(1), 85–102.

Wu, J.-H., Tennyson, R. D., & Hsia, T.-L. (2010). A study of student satisfaction in a blended e-learning system environment. *Computers & Education, 55*(1), 155–164.

Yang, Y., & Zilberg, I. E. (2020). Understanding young adults' TikTok usage. https://communication.ucsd.edu/_files/undergrad/yang-yuxin-understanding-young-adults-tiktok-usage.pdf.

Zhu, Y., Dong, J., Qi, X., & Deng, J. (2021). Intention to use governmental micro-video in the pandemic of covid-19: An empirical study of governmental Tik Tok in China. In *2021 6th International Conference on Inventive Computation Technologies (ICICT)*.

The Impact of Social Media Usage on Customer Decision Making-Process in Holiday Travel Planning Context, Applied Study Among Petra Visitors

Bahaa Mohammad Alhamad, Naseem Mohammad Twaissi, Zaid Ahmad Alabaddi, and Ra'ed Masa'deh

Abstract The study aimed to explore the impact of social media usage on customer decision-making process among Petra visitors in holiday travel planning context. In order to achieve the objectives of the study, the analytical descriptive approach methodology and convenience sample technique were adopted, the researcher designed a questionnaire based on previous studies, which consisted of (33) items to gather the information from the study sample. The Statistical Package for Social Sciences (SPSS) program was used to analyze and examine the data and the hypotheses. The research hypotheses were tested using descriptive statistical methods and simple linear regression test. The main results were: There is a moderate level of using social media in customer purchase decision process among Petra visitors, where the average of using social media was (3.05) in search information phase, (3.15) in evaluation of alternatives phase, (3.54) in purchase decision phase and (3.64) in post-purchase evaluation phase. There is statistically significant impact of the using social media on customer decision making process in all phases at the significance level ($\alpha \leq 0.05$). Also, the study found that the information provided by friends, relatives or other travelers on social media tools is more trusted in comparison to other source of information. The study recommends for companies and agents engaged in tourism industry to adopt social media as a marketing strategy due to its great advantages.

B. M. Alhamad
Researcher in Business Management Issues, Ma'an, Jordan
e-mail: bahaa_alhamad@yahoo.com

N. M. Twaissi (✉)
Department of Business Administration, Al-Hussein Bin Talal University, Ma'an, Jordan
e-mail: n.twaissi@ahu.edu.jo

Z. A. Alabaddi
Department of Management Information System, Al-Hussein Bin Talal University, Ma'an, Jordan
e-mail: zaid.abaddi@yahoo.com

R. Masa'deh
Department of Management Information System, School of Business, The University of Jordan, Amman, Jordan
e-mail: r.masadeh@ju.edu.jo

M. Alshurideh et al. (eds.), *The Effect of Information Technology on Business and Marketing Intelligence Systems*, Studies in Computational Intelligence 1056,
https://doi.org/10.1007/978-3-031-12382-5_15

Specially, After the end of the COVID-19 pandemic, so companies operating in the field of tourism are in dire need of promotional and advertising campaigns in line with the modern technologies.

Keywords Social media using · Customer decision-making process · Social media information trust

1 Introduction

The advent of the Information and Communications Technology (ICT) revolution had a major impact on tourism, so the effects of this revolution will continue to change the nature of contemporary tourism. The globalization of information, innovation and improved access also contributed to the design of a new scientific model that led to the growth of social media to be one of the most effective means for tourists to obtain information and exchange travel experiences. The digital dealings also played an effective role in the past decades, which in turn paved the way for commercial interactions between individuals on a large scale and set new directions to reshape the market (Lee et al., 2022a, b; Tariq et al., 2022a, b). The channels of communication are being transformed from being a marginal engine to the main market enabler (Fotis, 2015; Öztamur & Karakadılar, 2014) Social media has achieved tremendous success over the past years: Facebook, a social media site, has announced that "our products empower more than 2 billion users around the world to share ideas, more than 50% of them log in every daily" (Facebook, 2019), and more than 330 million monthly active users on Twitter, by average 145 million of them daily log in Statista (2019). Whereas YouTube announced that "Over 2 Billion logged-in users visit YouTube each month and everyday people watch over a billion hours of video and generate billions of views and more than 70% of them comes from mobile devices" (YouTube, 2019). And many other social media sites that have become networks for the exchange of knowledge and experience that directly affect consumer behavior by encouraging and promoting social interaction and information creation and sharing (Abuhashesh et al., 2019; Masa'deh et al., 2018; Masa'deh et al., 2013; Vinerean, 2017; Wren & Power, 2011). With this advent and growing of using ICT specially through social media as statistics seen at official websites of some social media platforms, that use may have an effect at customer decision-making process in context of travel decision and choosing destination, so this study explores this effect in new model.

On the other side, Jordan ranked only 90th out of 136 countries on the smart tourism index for the past year, despite the high technical readiness of the Kingdom, which indicates that tourism decline from the goals of the strategic plan of the Ministry of Tourism and Antiquities (2018–2020) as published on its website. Badado attributed the reasons for the decline in smart tourism to the traditional marketing of Jordanian tourist sites, as the official Jordanian promotion of tourism lacks creativity and innovation in accordance with the concept of "digitizing tourism", as well as the traditional tourism media, which was reflected on the rank of the Jordanian tourism services on the international tourism map, due to the failure in communicating the tourism message about Jordan, with its cultural symbols and connotations (Alrowwad et al., 2020; Nairoukh, 2018). Therefore, it has become a necessary to work on research on the issue of tourism and technology to provide results and recommendations that could improve Jordan's global tourism position in line with the Ministry's strategic plans to adopt the principle of innovation in promoting Jordanian tourism.

Petra tourism city, witnessed during the past year 2019 an unprecedented tourism activity in the number of its visitors, where the number of visitors reached one million and 135 thousand and 300 visitors of all nationalities, which increased over 37 percent in comparison to the year 2018, in which the number of visitors reached 828,952 from different nationalities. The statistics issued by the Petra Development Tourism Region Authority stated that the number of foreigners was one million and 553 visitors, while the number of Jordanians and Arabs were 129,144 visitors, and student trips were 5,603 visitors, the increase in the number of foreigners is reaching to 49 percent. The active tourism movement during the past two years has made the Petra region authority work continuously to develop tourism services and improve infrastructure and organize the movement of visitors in the archaeological sites in a way that preserves the privacy of these archaeological sites and create sustainable development parallel to the overall tourism process and in various sectors (Petra Development and Tourism Region Authority annual report, 2019).

The current study results would hopefully provide a number of effective theoretical and practical implications and will attempt to Explore to what extent Petra city visitors using social media in their purchase decision making process in context of holiday travel planning; and exploring the impact of social media usage at customer decision-making process in holiday travel planning. And exploring the level of customer trust towards social media in relation to other sources of holiday related information. As a result, this study will be looking forward and try to answer the following main questions:

What is the extent of Petra city visitors using social media in their purchase decision making process in context of holiday travel planning?
How do social media impact customer decision-making process throughout the holiday travel planning?
What is the level of trust towards social media in relation to other sources of holiday travel related information?

2 Conceptual Model and Hypotheses Development

As presented in the following Fig. 1, four main aspects of (search information phase, evaluation of alternatives phase, purchase decision phase, post-purchase evaluation phase) impact of using social media on customer decision making process when choosing destination.

The following hypotheses were constructed, based on the study problem, and its various questions to achieve the objectives of the current study, they are as follows:

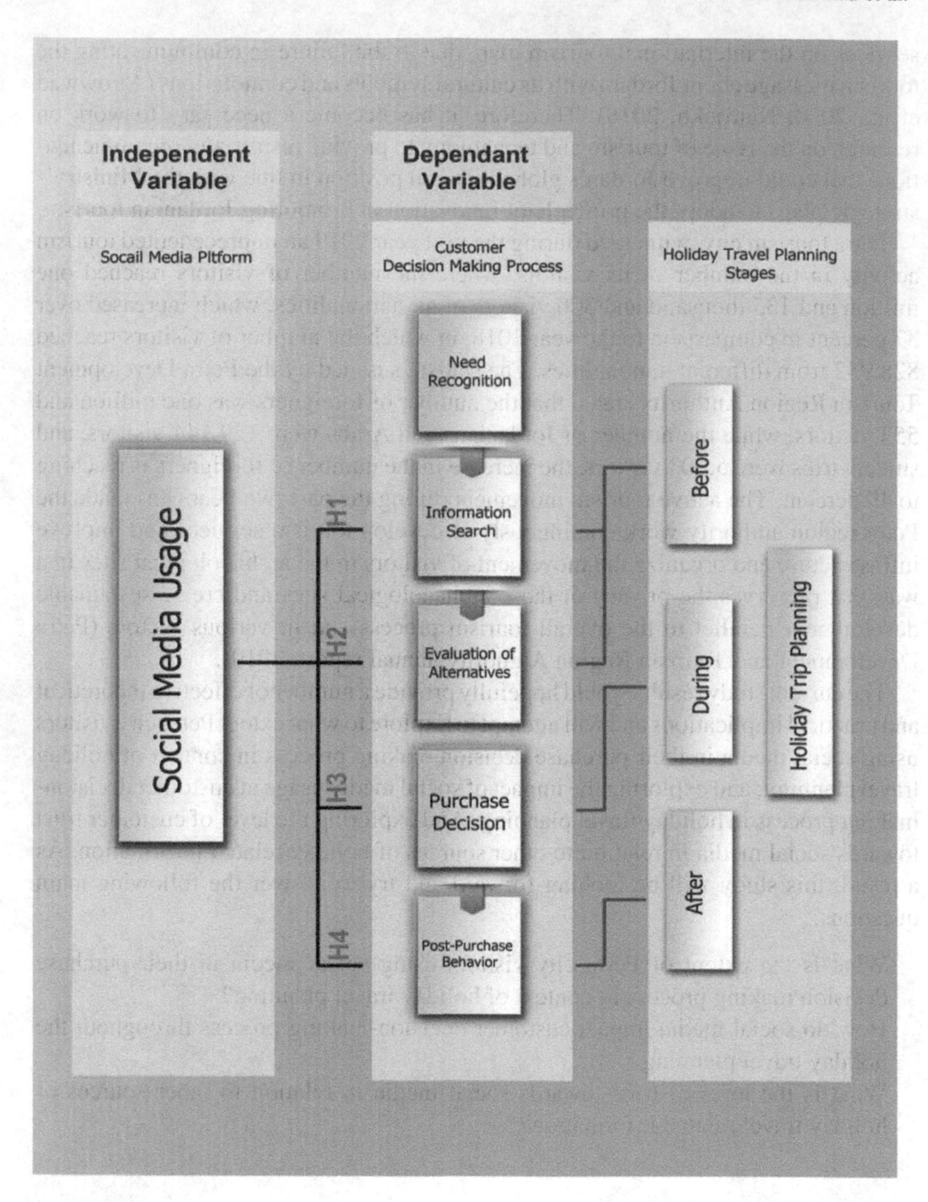

Fig. 1 Study model

H1. There is a statistically significant impact of using social media on customer decision making process in search information phase when choosing destination at the significance level ($\alpha \leq 0.05$).

H2. There is a statistically significant impact of using social media on customer decision making process in evaluation of alternatives phase when choosing destination at the significance level ($\alpha \leq 0.05$).

H3. There is a statistically significant impact of using social media on customer decision making process in purchase decision phase at the significance level ($\alpha \leq 0.05$).

H4. There is a statistically significant impact of using social media on customer decision making process in post-purchase evaluation phase at the significance level ($\alpha \leq 0.05$).

2.1 Customer Decision Making Process

2.1.1 Concept of Customer Decision Making Process (CDMP)

Author	Concept	Definition
Basil et al. (2013)	CDMP	The consumer decision-making process can be defined as the phases that consumers go through to take a final purchase decision. The task of a marketer is to focus on the whole series of the purchasing process instead of emphasizing solely on a purchase decision, because consumers experience different phases before reaching a conclusion
Kotler et al. (2005)	CDMP	A consumer decision making process includes of five stages to acquire a product or services. From the very beginning stage consumer recognize the need, gather the information and sources, evaluate alternatives and make the decision

2.1.2 Customer Decision Making Phases

The model of customer decision-making process consists of five stages/phases, these are connected tightly with each other as explained in the Fig. 2. Consumer uses all five phases during decision-making process in those products which are purchased rarely or occasionally with the high involvement. The consumer purchase goods or services with low involvement in decision making if the product is daily usage or low-cost products (Qazzafi, 2019). Every phase depends on the other one, and everyone described in details below for better understanding of the customer decision-making process:

Phase 1: Problem recognition

Every customer who wants to buy something, should go through the buying process which starts with the desire (problem recognition), in holiday travel context there

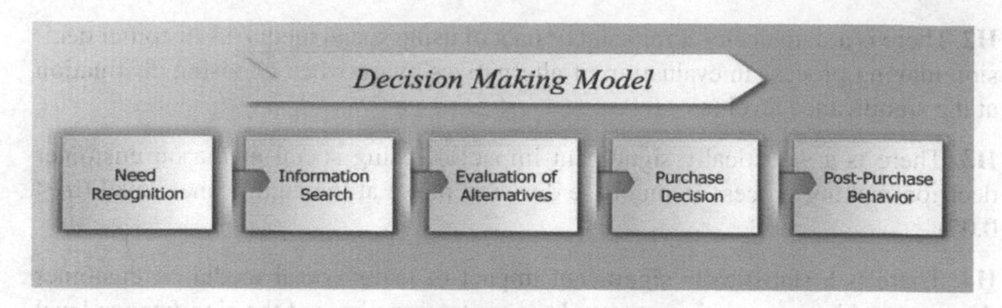

Fig. 2 Customer decision-making model (Kotler & Keller, 2012)

are some related stimuli such as pictures, videos or any other related information found on social media and these stimuli should be identified by tourism marketers to develop the right online marketing strategy and to move to the next phase (Aljumah et al., 2021; Al-Maroof et al., 2021; Alshurideh et al., 2019; Alwan & Alshurideh, 2022; ; Kotler & Keller, 2012; Kurdi et al., 2021; Obeidat et al., 2017, 2019; Tarhini et al., 2015)

Phase 2: Information search

It is the phase which seems to be the most crucial, so it provide the users/customers with the information about the available options. According to two categories which are internal and external, first one depends on the memory and previous experience of the customer his/herself. Normally, it concerns only low involvement products which are frequently purchased. When the second one is concerned with more complex product that need external search while there is a little or insufficient knowledge about them. According to Kotler and Keller (2012) there are three basic sources of external information search:

Personal, they are the people like friends and family represented by social network.

Companies, like the official websites and advertisements.

Public, represented by consumers rating platform such as forums, blogs and micro-blogs.

Phase 3: Evaluation of alternatives

During the phase of evaluation of alternatives, users interact with social media as an external source of information as well as his/her limited knowledge and skills to evaluate the right option. Social media platform could play an important role in this phase (Kotler & Keller, 2012).

Phase 4: Purchase decision

It is the phase of the actual purchase, when the customer accepts a specific option to buy after collecting information from several sources, evaluate it and decided where to purchase and what to purchase. Consumer purchase the brand or product which he/she gives the highest rank in the evaluation phase. The purchase decision also influenced by the surrounding environment who/when to buy and don't buy, and depend on someone else experiences, price, place and offers available (Abu Zayyad et al., 2020; Qazzafi, 2019).

Phase 5: Post-purchase evaluation

It is the last phase and may be the most important phase for the organization to receive positive or negative feedback, so it comes after the customer take his/her action. The companies work doesn't stop if the customer buys a product. The companies should know the behavior or view of the consumer towards the product. After the use of the product, the customer might be satisfied or dissatisfied. If the consumer satisfied, then the chances of retention are more of the same product and satisfied consumer can also influence the other people to buy the product and vice versa (Qazzafi, 2019). In this phase social media play an important role for the company reputation and brand, so companies should take care for the whole series not just specific phase.

2.1.3 Customer Decision Making in Holiday Travel Planning Context

Moreover, when defining a tourism concepts, there is an important point should take into consideration which is consumers activities. Tourism, particularly vacation and recreation as an activity decided by the consumer his/herself independently (Dwityas & Briandana, 2017; Zeng & Gerritsen, 2014). Therefore, there is a main focus on customer behavior when talk about tourism. Study on customer behavior grows nowadays when the marketing strategies affected by social media and have been started to consider in many companies. The field of customer behavior is the study of individuals, groups or organizations and processes they use to select, decide, use as well as post-use of a product, service, experience as well as idea to meet the demand and impact of the process to customers and society (Hawkins & Mothersbaugh, 2010; Senanayake & Anise, 2019). Customer decision-making process come as an extent to customer behavior science, it can identify through these steps: desire, information search, alternatives evaluation, selection and post-purchasing evaluation. Figure 3 shows these steps throughout the three main steps in holiday trip decision which are pre-Trip, during-Trip and post-Trip (Dwityas & Briandana, 2017).

Pre-Trip phase, the phase when someone has a desire to dose a travel, and it consists of some activities such as the desire to travel, search information that related to the destination and some tourism activity, and travel decision making about purchasing some products to do travel as an airline ticket.

During-Trip phase, the phase when someone take the decision and the destination already decided and consist of some activities such as consumption of products in the scope of the tourism, like accommodation, transportation, attraction and food. It also includes searching additional information required around the scope of the tourism products and services during the trip.

Post-Trip phase, the phase when that a series of activities accomplished and the travel already has done. In short, this phase is after the travelers have come back to home, social media is usually use in this phase to share experiences and memories,

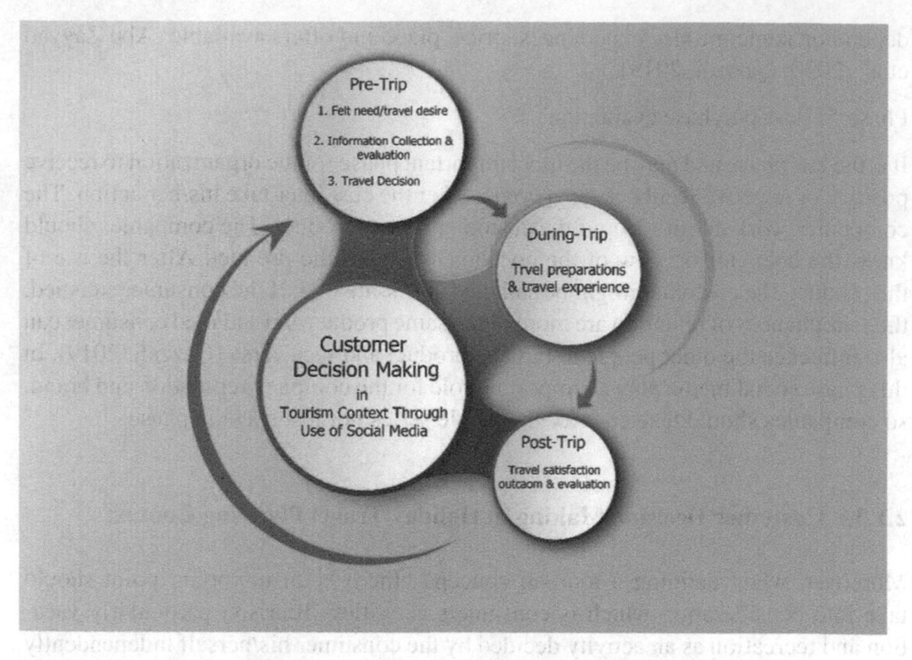

Fig. 3 Designed by researcher based on Dwityas and Briandana (2017) study

and to provide evaluation and recommendations about the destination which could be a satisfaction rate.

2.2 *Social Media*

2.2.1 Social Media as a Concept Definition

Author	Concept	Definition
Abd Al-samee (2012)	SC	Social media is considered as one of the most powerful tools of networking across the Internet that has connected the social with economic aspects in the real world
Smith (2009)	SC	Over the last decade. Social media has emerged as a new phenomenon, and the user generated content increased on (WWW) World Wide Web through this new technology such as social networks, blogs and media sharing platform. Collectively called social media

(continued)

(continued)

Author	Concept	Definition
Powers et al. (2012)	SC	Social media is a new platform that allows the users to connect more easily and more efficiently and it is provide the consumers with a new platform to search information and publish their opinions and experiences

2.2.2 Social Media Advantages and Disadvantages

According to Amy Looy in Social Media Management book (Looy, 2016):

Advantages

- The first advantage is speed. social media characterized by speed, it is faster in react or share content than traditional media. In addition, it is less limited to geographical distance. E.g., electronic message on social media is faster than postal letter.
- The second advantage is scalability, this means the organizations can reach more people in a lower budget. This advantage is useful to SMEs Small and Medium Enterprises.
- Third one, monitoring social media and analyzing data are less expensive than collecting information from customer through face-to-face interviews.
- Last one, interactivity between parties instead of traditional interactivity between organization and its customer through Ads or publish information on official websites.

Disadvantages

- First, social media are still new technology, many organizations are still learning how to create business value and take social media advantage in proper way.
- Second, social media content has some degree of undeletable due to the reason of the ability to electronic backup. Social media users could have a stored copy of the content deleted elsewhere.
- Lastly, privacy still the biggest concerns matter on social media. Due to ethical and legal issues. For instance, regarding to sell information about the customers to third parties.

2.2.3 Types of Social Media Platform

This study introduces the travel related social media as six types: blogs, microblogs, content community sites, consumer review sites, forums and social network sites.

Blogs

Blogs as a type of social media are considered as personal website (maybe stand alone or hosted at specific platform such as blogger/http://www.wordpress.com). It contains information about personal diary, personal experiences, thoughts, stores or ideas in a personal or informal style. Blogs provide a costumer with platform that enable publishing personal impression and communicate ideas (Fotis et al., 2011). In our tourism context, blogs are contributing to share visitors their experiences.

Microblogs

Microblogs are considered small blogs that contain limited characters, individual image, or video link. Best example of microblogs is twitter application and WhatsApp status. Twitter is an application that limits the users in 140 letters called tweet. According to Fotis (2015), all micro-blogging applications share three characteristics: (A) Short messages, limited to a specific number of characters, (B) instant message delivery, usually supported via multiple platforms, (C) users subscribe to users to receive posts.

Social Network Sites (SNS)

It is the most popular social media to share posts and experiences. It is a platform that provides the users/consumers to publish text, images and videos about their experiences, opinions and feelings through posts, reviews and comments. The best examples of social network sites are Facebook, Instagram, Snapchat, Pinterest and LinkedIn. In Fig. 4 some statistics on Ebizmba.com. (2020) about the most popularity social network sites (SNS).

Content Communities

Content communities are web-based applications that enable users to share media content such as videos, photos, documents and presentations, music and web links (Fotis, 2015). YouTube is the most popular content community website, with‖1,850,000,000—Unique Monthly Visitors ‖ according to Ebizmba.com. (2020).

Consumer Review Websites (CRW)

It is a platform where user can upload products/services related reviews and ratings by her/his experiences and opinions to an impressive range of features such as prices comparisons, wish-lists, multi variable ratings, advanced search, price comparisons, price history charts, buy/hold recommendations, deals ranking, merchants/retailers' evaluations, personalized shopping and more (Fotis, 2015). Those sites have obviously positive or negative influence on customer decision-making process and final purchase as many online users utterly rely on experiences of others (Hunaiti et al., 2009; Roučkova, 2015). In our context, TripAdvisor is the most popular one.

Forums

It is a platform consider as discussion site, a virtual session to exchange opinions and experiences. In comparison to social network, in forums there is no need to personally

Logo	SNS	Visitors	Per
facebook	Facebook	2,200,000,000	Unique Monthly Visitors
Instagram	Instagram	1,100,000,000	Unique Monthly Visitors
Pinterest	Pinterest	250,000,000	Unique Monthly Visitors
	Snapchat	110,000,000	Unique Monthly Visitors
LinkedIn	LinkedIn	85,000,000	Unique Monthly Visitors

Fig. 4 http://www.bizmba.com (2020)

connection between the members and the content can be read by any people. Figure 5 describe social media types and tools more specifically:

3 Research Methods

3.1 Study Approach

To achieve the objectives of the study and answer its questions, the analytical descriptive approach methodology was adopted as the study has a specific clear problem statement and specific hypotheses (Malhotra, 2004), so it is the most suitable approaches for the nature of the study and its goals in terms of conducting data analysis procedures and subjecting them to the statistical analysis and tests its hypotheses.

	Social media types	Examples of social media tools per type
1	Social communities	Facebook, LinkedIn, Google+, Yammer
2	Social text publishing tools	Blogs, Wikipedia, SlideShare, Quora
3	Microblogging tools	Twitter, Tumblr
4	Social photo publishing tools	Pinterest, Instagram, Flickr, Picasa
5	Social audio publishing tools	Spotify, iTunes, Podcast.com
6	Social video publishing tools	YouTube, Vimeo, Vine
7	Social gaming tools	World of Warcraft
8	Live casting tools	Live365, Justin.tv
9	Virtual worlds	Second Life, Kaneva
10	Mobile tools	Foursquare, Swarm
11	Productivity tools	SurveyMonkey, Google Docs, Doodle
12	Aggregators	MyYahoo!,iGoogle (until November 2013)

Fig. 5 Social media types and tools (Looy, 2016)

3.2 Study Population and Sample

The study adopted visitors of Petra tourism city as a population. It is not targeted to a specific industry like (shopping centers, restaurants or hotels …etc.), so the Convenience sample technique (Accidental sample) was chosen, so it is the most suitable sample technique as the nature of the study and its population. The questionnaire was randomly distributed to the members of sample in 10 days distribution program from 5 to 15 May 2020 simultaneously with governmental lockdown and unstable conditions due to COVID-19 epidemic, so researchers couldn't distribute more than 175 questionnaires. Table 1 describes the characteristics of the sample.

Table 1 Sample demographic characteristics

Sample characteristic			
Total	175	Valid for statistical analysis 148	
Gender		Age	
Male	71	18–29	58
Female	77	30–49	62
		+50	28

3.3 *Questionnaire Design Process*

A Questionnaire was designed to achieve the objectives of the study, and to answer its questions, as it is instrument to collect data from the sample to test study hypotheses and consists of three parts. The first part of the questionnaire provides a general information to describe the demographic characteristics of the respondents as shows in Table 1. Second one consists of 32 questions to measure the social media usage and the four phases of customer decision making process variables, to test hypotheses of the study, this part relies mainly on (Fotis et al., 2011; Malhotra, 2004; Roučkova, 2015), studies and restructured to correspond the goals and objectives of this study. Whereas, paragraph number 33 which consist of a question and 4 nominal options, to achieve the third objective based on (Fotis et al., 2011) study. A 5-point Likert scale anchored by "strongly disagree" (1) to "strongly disagree" (5) was used as the attitude measurement for the independent and dependent variables. The degree of approval was concluded through the following steps according to Chen and Tabari (2017):

- Extracting the scale range ($5 - 1 = 4$).
- Divide the average scale by the number of required levels, total of 3 (Low, Moderate, and High) $4 \div 3 = 1.33$, which the category length, and thus resulting in three levels:

  1.00–2.33 low level of approval.
  2.34–3.67 Moderate level of approval.
  3.68–5.00 High level of approval.

Whereas, question number 33, which consist of a question and 4 nominal options, tested by calculation of frequencies and percentage.

4 Data Analysis

4.1 *Reliability Test*

The reliability of a measure indicates the stability and consistency with which the instrument measures the concept and helps to assess the 'goodness' of a measure (Dancey & Reidy, 2004). The reliability of the questionnaire was measured by extracting the Cronbach Alpha coefficient to measure the coefficient of the scale paragraphs' consistency. According to each variable, note that all of them higher than the value of (0.7), which indicates the reliability of the study tool. Table 2 shows the study tool reliability of coefficients.

Table 2 Reliability of the study tool by Cronbach alpha

Variable	Dimension	Number of paragraphs	Coefficient of Cronbach Alpha
Social media	Social media usage	8	0.849
Customer decision-making process	Information search	7	0.884
	Evaluation of alternatives	5	0.889
	Purchase decision	6	0.797
	Post-purchase evaluation	6	0.873

Table 3 Validity of the study by KMO and Bartlett's test

KMO and Bartlett's test		
Kaiser-Meyer-Olkin measure of sampling adequacy		0.673
Bartlett's test of sphericity	Approx. Chi-square	16,753.023
	df	221
	Sig.	0.000

4.2 Validity Test

To verify the validity of the questionnaire items used in the study, it had been presented to academic arbitrators specialized in this field and are working in Jordanian Universities, in order to ensure the accuracy and validity of the paragraphs. To issue out their decision regarding the validity of the paragraphs, the soundness of their formulation and its relevance to the study subject. The questionnaire was assessed by four academic arbitrators specialized in business administration, marketing and strategic management. Furthermore, construct validity was adopted as validity measurement and factor analysis was used to measure the construct validity (Dancey & Reidy, 2004), the details of factor analysis as shows in Table 3 is appropriate because the KMO value is 0.673 is greater than 0.6 and approximate Chi-Square 16,753.023 with df 221 at the significant level 0.000 less than 0.05. Hence factor analysis indicated that it is appropriate for further data analysis.

4.3 Instrument Descriptive Statistics

To achieve the first objective of the study (Explore to what extent Petra city visitors using social media in their purchase decision making process in context of holiday travel planning), Mean and Standard deviations adopted to analysis of the data as shown in Table 4.

Table 4 Descriptive statistics of the study instrument

	Customer decision making-process	Mean	Standard deviation	The level of usage
1	Information search phase	3.05	1.147	Moderate
2	Evaluation of alternatives phase	3.15	1.210	Moderate
3	Purchase decision phase	3.54	1.082	Moderate
4	Post-purchase evaluation phase	3.64	1.093	Moderate

Data from Table 4 shows that, there is a moderate level of using social media among Petra city visitors in all phases of purchase decision-making process in holiday trip decision.

4.4 Hypotheses Test

For this study, five hypotheses were tested using Simple Regression Linear and as following in Table 5:

H1. There is a statistically significant impact of using social media on the customer decision making process in search information phase when choosing destination at the significance level ($\alpha \leq 0.05$).

Table 5 shows that, the result from this table is statistically accepted, where F-value (79.684) and it is statistically significant at the significance level ($\alpha \leq 0.05$), which confirms that, there is statistically significant of the impact of the social media usage on the customer decision making-process in information search phase when choosing destination, and the value of (T) calculated (6.377) statistically significant at the significance level ($\alpha \leq 0.05$). That is indicated by the value of the coefficient (β) (0.625), where the independent variable is interpreted according to the coefficient

Table 5 Regression factor for the study hypotheses

Hypotheses	Regression		(F) Test		(T) Test		R^2	R
	β	Std. error	F	Sig.	T	Sig.		
H1	0.625	0.060	79.684	0.000	6.377	0.000	0.657	0.562
H2	0.485	0.060	45.633	0.000	6.755	0.000	0.438	0.588
H3	0.455	0.060	73.633	0.000	8.071	0.000	0.410	0.640
H4	0.471	0.058	40.927	0.000	6.397	0.000	0.505	0.768

*Statistically significant at ($\alpha \leq 0.05$) level

of (R^2) estimated by (0.657) of the variance in the dependent variable, which mean (65.7%) of the changes in information search phase when choosing destination, caused by changes in the level of social media usage.

H2. There is a statistically significant impact of using social media on customer decision making process in evaluation of alternatives phase when choosing destination at the significance level ($\alpha \leq 0.05$).

Table 5 shows the results from this table is statistically accepted, where F-value (45.633) and it is statistically significant at the significance level ($\alpha \leq 0.05$), which confirms that, there is statistically significant of the impact of the social media usage on the customer decision making-process in evaluation of alternatives phase when choosing destination, and the value of (T) calculated (6.755) statistically significant at the significance level ($\alpha \leq 0.05$). That is indicated by the value of the coefficient (β) (0.485), where the independent variable is interpreted according to the coefficient of (R^2) estimated by (0.438) of the variance in the dependent variable, which mean (43.8%) of the changes in evaluation of alternatives phase when choosing destination, caused by changes in the level of social media usage.

H3. There is a statistically significant impact of using social media on customer decision making process in purchase decision phase at the significance level ($\alpha \leq$ 0.05).

Table 5 shows that, the result from this table statistically accepted, where F-value (73.633) and it is statistically significant at the significance level ($\alpha \leq 0.05$), which confirms that, there is statistically significant of the impact of the social media usage at the customer decision making-process in purchase decision phase, and the value of (T) calculated (8.071) statistically significant at the significance level ($\alpha \leq 0.05$). That is indicated by the value of the coefficient (β) (0.455), where the independent variable is interpreted according to the coefficient of (R^2) estimated by (0.410) of the variance in the dependent variable, which mean (41.0%) of the changes in purchase decision phase, caused by changes in the level of social media usage.

H4. There is a statistically significant impact of using social media on customer decision making process in post-purchase evaluation phase at the significance level ($\alpha \leq 0.05$).

Table 5 shows that, the result from this table statistically accepted, where F-value (40.927) and it is statistically significant at the significance level ($\alpha \leq 0.05$), which confirms that, there is statistically significant of the impact of the social media usage on the customer decision making-process in post-purchase decision phase, and the value of (T) calculated (6.397) statistically significant at the significance level (α $\leq$ 0.05). That is indicated by the value of the coefficient (β) (0.471), where the independent variable is interpreted according to the coefficient of (R^2) estimated by (0.505) of the variance in the dependent variable, which mean (50.5%) of the changes in post-purchase decision phase, caused by changes in the level of social media usage.

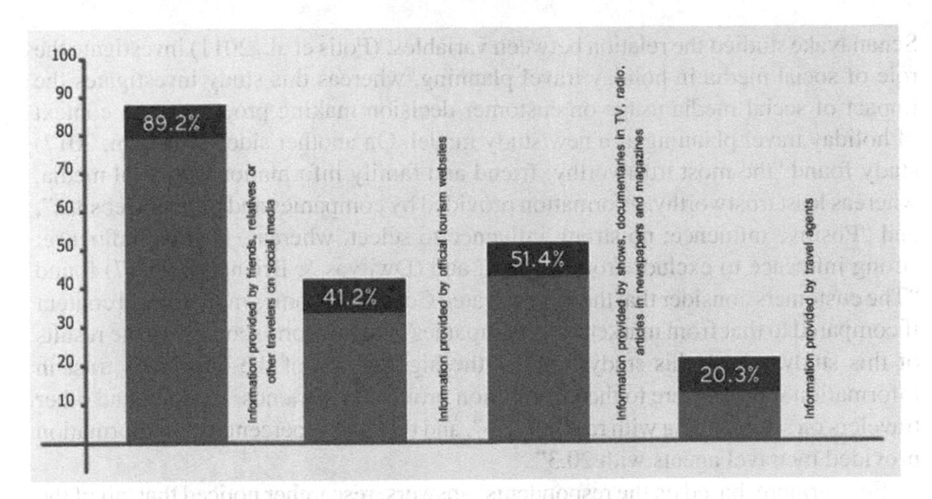

Fig. 6 The level of trust towards information provided by social media

4.5 Objective No. 3

To achieve the third objective of the study (Exploring the level of customer trust towards social media in relation to other sources of holiday related information), frequencies and percentage technique adopted as shows in Fig. 6.

Figure 6 shows that, 89.2 % of the sample trust the information provided by their friends, relatives or other travelers on social media tools in holiday trip decision, 41.2 % trust the information provided by official tourism websites, 51.4 % trust the information provided by shows, documentaries in TV, radio, and articles in newspapers and magazines, whereas the lowest ratio 20.3 % came of the information provided by travel agents. From the data in figure above can notices that, the highest percentage came to the first column on the left where 89.2% of the sample trust the information provided by their friends, relatives or other travelers on social media tools. In contrast to, other sources of information.

5 Discussion

This study found a significant impact of social media usage on customer decision making process in holiday travel planning, which in fact, corresponds with (Fotis, 2015; Fotis et al., 2011; Senanayake & Anise, 2019) studies, where Senanayake found a strong relation between social media usage and travel decision making process, a strong relation between travel decision making process and travel motives throughout social media usage. Notices that, this study focused at the customer decision making in context holiday travel planning, whereas Senanayake focused at the travel decision making process. This study studied the impact of variable on another one, whereas

Senanayake studied the relation between variables. (Fotis et al., 2011) investigate the role of social media in holiday travel planning, whereas this study investigates the impact of social media usage on customer decision making process in the context of holiday travel planning as a new study model. On another side, (Vinerean, 2017) study found "the most trustworthy: friend and family information on social media, whereas least trustworthy, information provided by companies and official websites", and "Positive influence; no strong influence to select, whereas, Positive influence; strong influence to exclude from choice", and (Dwityas & Briandana, 2017) found "The customers consider that the User Created Content becomes more trusted content if compared to that from marketers or companies", which correspond with the results of this study, where this study found " the highest rate of the customers trust in information sources were to the information provided by friends, relatives and other travelers on social media with ratio 89.2%", and the lowest percentage to information provided by travel agents with 20.3".

Furthermore, based on the respondents' answers, researcher noticed that, all of the paragraphs those talked about Advertisements on the social media in all dimensions of the study came with low-to-med level. First dimension Social Media Usage, paragraph (4) came with (2.80), which states "I browse the Ads that related to tourism services". In Information Search Phase, paragraph (3) came with (2.84), which states "The adverts on social media are important when you want to try a new place", and paragraph no (4) came with (2.77), which states "Ads published on social media are important when choosing a destination". In Purchase Decision Phase, paragraph (5) came with (2.70), which states "Ads on social media during my holiday is stimulates me to trying a new products/services". On the other side, in the results of the fourth objective, noticed that the lowest ratio of trust in information sources came to information provided by Travel Agent with ratio 20.3%, where this results in corresponding with founds of Fotis et al. (2011) "study found that information from other travelers in various websites is trusted more than official tourism websites, and travel agents". Therefore, could conclude that, there is a relation between decreasing the level of trust in travel agent's information and the decreasing the level of browsing tourism services adverts by customers.

6 Conclusion

6.1 Implications

In the light of this study, as it is discussed the impact of social media on customer decision making process in context of holiday travel planning, and proved in scientific methods that, there is a great effect of the social media on the customer decision process, and how it could be the success key for the companies engaged in tourism industry. Therefore, study recommends those engaged of such industry by the following recommendations:

- The companies and agents engaged in tourism industry to analyze current marketing strategies and development it to adapt this technology due to its great advantages.
- Design and develop a new training programs to improve the employee's skills to obtain the maximum benefits of this marketing strategies.
- Study showed an important information to those work in hospitality industry, when the paragraph "I use social media when seeking ideas and information on accommodation options)" came with level (3.36), which provide an opportunity for hotels to have a strong advantage of social media.
- A recommendation to a public sector, the lowest paragraph between the all, states that "The information and Ads published about Petra city on social media was the reason to choose it as a destination" came with lowest level (2.26) among all paragraphs, which drew attention to improve the marketing strategies through the social media.
- After the end of the COVID-19 pandemic, companies operating in the field of tourism are in dire need of promotional and advertising campaigns in line with the modern technologies, so they could take social media tools in consideration as marketing strategy.

6.2 Limitations and Future Recommendations

This research process provides some insights to the researchers, and these insights could be viewed in the shed light of limitations. This study, showed the impact of social media usage on customer-decision making process in all phases except the first one (problem recognition phase). Future studies to be done on the whole series. Also, this study adopted the quantitative research approach and questionnaire study tool. Future study could combine the two approaches quantitative and qualitative, and interview study tool and could adopt e-questionnaire to increase the number of respondents. Furthermore, this study, applied in tourism sector among the Petra city visitors. Future study could apply in another sector and exploring what is the impact of social media on customer decision making process in another industry field.

References

Abd Al-samee, M. (2012). Role of social media sites in marketing Egypt as an international touristic destination, Ministry of Tourism. Ministry of Tourism Contest for the Year 2012.

Abu Zayyad, H. M., Obeidat, Z. M., Alshurideh, M. T., Abuhashesh, M., & Maqableh, M. (2020). Corporate social responsibility and patronage intentions: The mediating effect of brand credibility. *Journal of Marketing Communications*. https://doi.org/10.1080/13527266.2020.1728565

Abuhashesh, M., Al-Khasawneh, M., & Al-Dmour, R. (2019). The impact of Facebook on Jordanian consumers' decision process in the hotel selection. *IBIMA Business Review, 2019*. https://doi.org/10.5171/2019.928418.

Aljumah, A., Nuseir, M. T., & Alshurideh, M. T. (2021). The impact of social media marketing communications on consumer response during the COVID-19: Does the brand equity of a university matter. In *The Effect of Coronavirus Disease (COVID-19) on Business Intelligence* (pp. 367–384).

Al-Maroof, R., Ayoubi, K., Alhumaid, K., Aburayya, A., Alshurideh, M., Alfaisal, R., & Salloum, S. (2021). The acceptance of social media video for knowledge acquisition, sharing and application: A comparative study among YouYube users and TikTokusers' for medical purposes. *International Journal of Data and Network Science, 5*(3), 197–214.

Alrowwad, A., Abualoush, S. H., & Masa'deh, R. (2020). Innovation and intellectual capital as intermediary variables among transformational leadership, transactional leadership, and organizational performance. *Journal of Management Development, 39*(2), 196–222. https://doi.org/10.1108/JMD-02-2019-0062

Alshurideh, M., Salloum, S. A., Al Kurdi, B., & Al-Emran, M. (2019). Factors affecting the social networks acceptance: an empirical study using PLS-SEM approach. In *Proceedings of the 2019 8th international conference on software and computer applications* (pp. 414–418).

Alwan, M., & Alshurideh, M. (2022). The effect of digital marketing on purchase intention: Moderating effect of brand equity. *International Journal of Data and Network Science, 10*(3), 1–12.

Basil, G., Etuk, E., & Ebitu, E. T. (2013). The marketing mix element as determinants of consumer's choice of made-in-Nigeria shoes in Cross River state. *European Journal of Business and Management, 5*(6), 141–147.

Chen, W., & Tabari, S. (2017). A study of negative customer online reviews and managerial responses on social media—Case study of the Marriott hotel group in Beijing. *Journal of Marketing and Consumer Research, 41*, 53–64.

Dancey, C., & Reidy, J. (2004). *Statistics without Math's for psychology using SPSS for windows.* Prentice Hall.

Dwityas, N., & Briandana, B. (2017). Social media in travel decision making process. *International Journal of Humanities and Social Science, 7*(7), 193–201.

Ebizmba.com. (2020). Top 15 Most Popular Social Networking Sites | Feb 2020. http://www.ebizmba.com/articles/social-networking-websites. Accessed 26 Feb 2020.

Facebook. (2019). Statistics. https://www.about.fb.com/company-info/. Accessed 16 Dec 2019.

Fotis, J., Buhalis, D., & Rossides, N. (2011). Social media impact on holiday travel planning. *International Journal of Online Marketing, 1*(4), 1–19.

Fotis, J. (2015). The use social media and its impacts on consumer behavior: The context of holiday travel. Bournemouth University. Unpublished Doctoral Thesis.

Hawkins, D., & Mothersbaugh, D. (2010). Consumer behavior: Building marketing strategy, 11th edn. McGraw-Hill International Edition.

Hunaiti, Z., Mansour, M., & Al-Nawafleh, A. (2009). Electronic commerce adoption barriers in small and medium-sized enterprises (SMEs) in developing countries: the case of Libya. In *Innovation and knowledge management in twin track economies challenges and solutions—Proceedings of the 11th international business information management association conference, IBIMA 2009*, 1–3 (pp. 1375–1383).

Khasawneh, M. A., Abuhashesh, M., Ahmad, A., Alshurideh, M. T., & Masa'deh, R. (2021b). Determinants of e-word of mouth on social media during COVID-19 outbreaks: An empirical study. In *The Effect of Coronavirus Disease (COVID-19) on Business Intelligence* (pp. 347–366). Cham: Springer.

Khasawneh, M. A., Abuhashesh, M., Ahmad, A., Masa'deh, R., & Alshurideh, M. T. (2021a). Customers online engagement with social media influencers' content related to COVID 19. In *The Effect of Coronavirus Disease (COVID-19) on Business Intelligence* (pp. 385–404). Cham: Springer.

Kotler, P., & Keller, K. L. (2012). *Marketing management* (14th ed.). Pearson Education Inc.

Kotler, P., Wong, V., Saunder, J., & Armstrong, G. (2005). *Principle of marketing*, 4th ed (European). Pearson Education Inc.

Kurdi, B. A., Alshurideh, M., Nuseir, M., Aburayya, A., &Salloum, S. A. (2021, March). The effects of subjective norm on the intention to use social media networks: an exploratory study using PLS-SEM and machine learning approach. In *International Conference on Advanced Machine Learning Technologies and Applications* (pp. 581–592). Cham: Springer.

Lee, K., Azmi, N., Hanaysha, J., Alshurideh, M., & Alzoubi, H. (2022a). The effect of digital supply chain on organizational performance: An empirical study in Malaysia manufacturing industry. *Uncertain Supply Chain Management, 10*(2), 1–16.

Lee, K., Ramiz, P., Hanaysha, J., Alzoubi, H., & Alshurideh, M. (2022b). Investigating the impact of benefits and challenges of IOT adoption on supply chain performance and organizational performance: An empirical study in Malaysia. *Uncertain Supply Chain Management, 10*(2), 1–14.

Looy, A. (2016). *Social media management* (pp. 7–9). Springer International Publishing Switzerland.

Malhotra, N. K. (2004). Marketing research: An applied orientation, 4th edn. New Jersey: Prenticall-Hall.

Masa'deh, R., Alananzeh, O., Tarhini, A., & Algudah, O. (2018). The effect of promotional mix on hotel performance during the political crisis in the Middle East. *Journal of Hospitality and Tourism Technology, 9*(1), 32–47. https://doi.org/10.1108/JHTT-02-2017-0010

Masa'deh, M. T., & R., Shannak, R.O., & Mohammad Maqableh, M. (2013). A structural equation modeling approach for determining antecedents and outcomes of students' attitude toward mobile commerce adoption. *Life Science Journal, 10*(4), 2321–2333.

Nairoukh. (2018). International indicators: Jordan Retreating on the Smart Tourism Index. http://www.alsaa.net/article-62318. Accessed 18 Dec 2019.

Obeidat, B. Y., Al-Hadidi, A., & Tarhini, A. (2017). Factors affecting strategy implementation: A case study of pharmaceutical companies in the Middle East. *Review of International Business and Strategy, 27*(3), 386–408. https://doi.org/10.1108/RIBS-10-2016-0065

Obeidat, Z. M., Alshurideh, M. T., Al Dweeri, R., & Masa'deh, R. (2019). The influence of online revenge acts on consumers psychological and emotional states: does revenge taste sweet? In *Proceedings of the 33rd international business information management association conference, IBIMA 2019: Education excellence and innovation management through vision 2020* (pp. 4797–4815).

Öztamur, D., & Karakadılar, I. (2014). Exploring the role of social media for SMEs: As a new marketing strategy tool for the firm performance perspective. In *Social and behavioral sciences, vol.150. 10th International strategic management conference* (pp. 511–520). https://doi.org/10.1016/j.sbspro.2014.09.067.

Powers, T., Advincula, D., Austin, M. S., Graiko, S., & Snyder, J. (2012). Digital and social media in the purchase decision process. *Journal of Advertising Research, 52*(4), 479–489.

Qazzafi, S. (2019). Consumer buying decision process toward products. *International Journal of Scientific Research and Engineering Development, 2*(5), 130–134.

Roučkova, V. (2015). Social media in customer decision-making process—The role of reviews. Copenhagen Business School. Unpublished Master Thesis.

Senanayake, S., & Anise, R. (2019). The influence of social media on millennial's travel decision making process. *Colombo Journal of Advanced Research, 1*(1), 192–205.

Smith, T. (2009). The social media revolution. *International Journal of Market Research, 51*(4), 559–561.

Statista. (2019). https://www.statista.com/statistics/282087/number-of-monthly-active-twitter-users/. Accessed 16 Dec 2019.

Tarhini, A., Mgbemena, C., &Trab, M. S. A. (2015). User adoption of online banking in Nigeria: A qualitative study. *Journal of Internet Banking and Commerce, 20*(3). https://doi.org/10.4172/1204-5357.1000132.

Tariq, E., Alshurideh, M., Akour, E., Al-Hawaryd, S., & Al Kurdi, B. (2022a). The role of digital marketing, CSR policy and green marketing in brand development at UK. *International Journal of Data and Network Science, 6*(3), 1–10.

Tariq, E., Alshurideh, M., Akour, I., & Al-Hawary, S. (2022b). The effect of digital marketing capabilities on organizational ambidexterity of the information technology sector. *International Journal of Data and Network Science, 6*(2), 401–408.

Vinerean, S. (2017). Importance of strategic social media marketing. *Expert Journal of Marketing, 5*(1), 28–35.

Wren, G., & Power, D. (2011). Impact of social media and web 2.0 on decision-making. *Journal of Decision Systems, 13*(1).

YouTube. (2019). YouTube by the numbers. https://www.youtube.com/about/press/. Accessed 16 Dec 2019.

Zeng, B., & Gerritsen, R. (2014). What do we know about social media in tourism? a review. *Tourism Management Perspectives, 10*(1), 27–36.

Factors Influencing Online Shopping During Fear of Covid-19 Pandemic in Jordan: A Conceptual Framework

Ra'ed Masa'deh, Dmaithan Abdelkarim Almajali, Mohmmad Reyad Almajali, Eman Reyad Almajali, and Muhammad Turki Alshurideh

Abstract Aside from causing fear worldwide, COVID-19 has significantly affected global marketing and the behavior and attitudes of consumer. In Jordan, COVID-19 has affected the lifestyle, buying, and consumption patterns of consumer whereby in-store purchases are becoming less popular, and consumers are increasingly favoring other product sources. With the increasing usability of the internet in the business world, the popularity of online shopping is increasing, evidenced by the increased number of online retailers globally. A psychometric Fear of COVID-19 Scale (FCV-19S) was developed to evaluate COVID-19 fear and online shopping and it was employed to gauge consumers in developing countries like Jordan. This descriptive study presented a theoretical framework comprising trust issues, perceived ease of use, perceived risk and fear of complexity in online shopping among Jordanian consumers. The findings show that trust, perceived ease of use, fear of COVID-19 and perceived risk affect online shopping behavior. The increase in confidence in a given online site increases the frequency of online shopping, and the decrease in the risk perceived increases the decision to make an online purchase.

R. Masa'deh (✉)
Department Management Information Systems, School of Business, University of Jordan, Amman, Jordan
e-mail: r.masadch@ju.edu.jo

D. A. Almajali
Department of Management Information System, Applied Science Private University, Amman, Jordan
e-mail: d_almajali@asu.edu.jo

M. R. Almajali · E. R. Almajali
An Independent Researcher, Al-Karak, Jordan

M. T. Alshurideh
Department of Marketing, School of Business, The University of Jordan, Amman, Jordan
e-mail: m.alshurideh@ju.edu.jo

University of Sharjah, Sharjah, UAE

M. Alshurideh et al. (eds.), *The Effect of Information Technology on Business and Marketing Intelligence Systems*, Studies in Computational Intelligence 1056,
https://doi.org/10.1007/978-3-031-12382-5_16

Keywords Perceive risk · Trust · Ease of use · Fear of Covid-19 · Online shopping behavior

1 Introduction

The novel coronavirus 2019 (COVID-19) is a deadly infectious disease that has shaken the world and managing this disease has been highly challenging. Symptoms of this disease usually would appear within 2–14 days, and among them are fatigue, fever, dry cough, dyspnea and myalgia (Wang et al., 2020). In China, the mortality rate of this disease was reported at 3.6% as of March 1, 2020, and the rate reported for other nation was 1.5% (Baud et al., 2020). By March 14, 2020, a total of 135 countries/territories had reported confirmed cases of COVID-19 (World Health Organisation (World Health Organization, 2020). COVID-19 is highly infectious and very deadly, and this has caused fear among people. Lin (2020) reported that COVID-19 patients are feared and this fear may indeed exacerbate the harm caused by the disease. Also, there has been worldwide fear towards COVID-19, resulting in stigma in certain situations (Alsyouf et al., 2021; Centers for Disease Control & Prevention, 2022; Guan et al., 2019; Huang et al., 2020; Lin, 2020; Lin et al., 2021).

Infectious disease often generates fear, as opposed to non-infectious ones, and this fear has direct linkage to its rate of transmission, the medium of infection, and its morbidity and mortality rate as well. Hence, the more rapid and the more undiscernible the disease is, the more fearful the disease becomes. Fear could result in psychosocial problems for instance, loss, discrimination, and stigmatization as well (Pappas et al., 2009). Extreme fear could impair rational thinking of individuals towards COVID-19. In combatting this disease, countries all over the world are currently focusing on control of infection, effective vaccine, and treatment cure rate (Abuhashesh et al., 2021; Almajali et al., 2021; Almajali, 2021a, b; Wang et al., 2020). However, in achieving a COVID-19-free society, the psychosocial aspect has to be taken into account as well as the social media impacts (Al Khasawneh et al., 2021a, b; Al-Dmour et al., 2020, 2022).

In Jordan, the impact of COVID-19 was rather late, but the severity was just as high especially on the country's economics and health, and globalization has contributed to Jordan falling victim of the pandemic. Several measures have been implemented by Jordan like social distancing, lockdowns, movement restrictions, as well as closure of some businesses, particularly those classed as non-essential. Relevant to that, the pandemic has drastically changed the way consumers in Jordan live their life and make purchases of goods and services. Like other consumers globally, those in Jordan had to search for other ways in their product purchase affairs due to restrictions of in-store purchases involving both essential and non-essential product retailers. Furthermore, because of the pandemic, many consumers in Jordan had to deal with the new personal and social situations, and many were facing changes in their income and leisure time. All the aforementioned have impact on their attitudes and behaviors as consumer.

New personal circumstances like different income level, different lifestyle (from free moving to home bound), different leisure time (from more time at work to more time at home, or full time homebound) have led to the change in values and priorities of consumers. Among the changes caused by the pandemic include less purchases on non-essential items like clothing, jewelries, accessories and transports (just to name a few) which has led to a plummet in the sales of the aforementioned items. In fact, the pandemic has severely hit the fashion and the apparel industry. The impact of the pandemic on the economy was severe, and many consumers were uncertain about their future financial situations.

In addition to reducing their purchases on non-essential items, consumers also have learned to make purchases online, on various items. Online purchasing saves consumer from having to visit the physical store which can put them at risk of catching the coronavirus from other infected individuals. The ease of online purchases has reduced their desire to visit the physical stores. However, there are issues associated with online shopping. In a study conducted by Almajali and Hammouri (2021), the authors found that online shopping stores are facing several issues namely trust, perceived risk, and perceived ease of use. The authors further mentioned that trust in the online store will decrease the perceived risk, resulting in the execution of online purchase. In Jordan, there has been an increase in online purchasing and such increase has been associated with the increasing number of online stores available in social media. It is clear that people in Jordan are interested in making purchases online.

In this study, the ineffectiveness of in-store purchases during COVID-19 pandemic and fear towards COVID-19 are discussed. The decisions of retail brands to sell their products online to provide convenience to their target customers are explored. The resultant new lifestyle of consumer from of the pandemic is elaborated, particularly in respect to online shopping.

2 Literature Review

Online shopping can be performed anywhere, so long that there is internet access. People engage in online shopping because it offers better convenience, selection, information, and price. In exploring the subject of online shopping in the context of Jordan, Theory of Planned Behavior by Ajzen (1991) and Technology Acceptance Model (TAM) by Davis (1989) were the two theoretical adoption models applied in this study. Both theories have been used in evaluating the readiness of people in utilizing technology (Almajali et al., 2021; Manaseer et al., 2019; Hsu et al., 2006; Wu & Chen, 2005; Gefen et al., 2003).

TAM has been altered in various studies to increase its fittingness to the study context (Venkatesh & Davis, 2000). In this study, TPB was examined strength-wise to increase the potency of TAM, specifically through the inclusion of external independent variables with significant effect on technology adoption decision-making process of consumer. TAM does not consider the impact of social and interpersonal variables on the decisions to adopt technology (Ukoha et al., 2011), and thus, through TPB, the constructs of perceived behavioral control and subjective norms were added to TAM. Both constructs describe the ease or difficulty in performing certain act amidst resource constraints, and as reported by Orapin (2009), both factors impact consumers in their online shopping.

In their study, Nachit and Belhcen (2020) reported that COVID-19 pandemic has dramatically altered the behavior of consumer in Morocco. According to the authors, the priorities of Moroccan consumers have shifted because of the pandemic. This was demonstrated by their high concerns towards the availability of some essential products in the market like the hygiene products as evidenced by panic buying of these products. The pandemic has indeed increased the willingness of Moroccans in spending more on hygiene and food items. Additionally, external information, alternative attractiveness, conformity, regret, and loyalty were reported by Liao et al. (2017) as the factors impacting online shoppers. In their study on digital transformation readiness of consumers in Jordan, Nachit and Belhcen (2020) found the impact of COVID-19 and government measures, on knowledge, attitude, and perception on digital transformation of public and private sector. Furthermore, certain digital tools were introduced, in order that consumers were able sufficiently access their sought-after products and services.

As described by Moling (2011), online shopping is a process of direct goods and services purchase made by customers from a seller over the internet with no intermediary service present. Internet shopping allows customers to review the goods and services that they want to purchase with ease. This type of shopping does not involve or require face-to-face communication between seller and consumer. In terms of trust in online shopping, it is considered as a long-term investment, and in this type of shopping, trust cannot be controlled, it can only be supported. Hence, trust must be earned. For many people, online shopping is challenging because it is a fairly new form of shopping and many people are not acclimatized to it (Monsuwé et al., 2004).

Kumar and Grisaffe (2004) mentioned Perceived Risk as an important element in buying behavior. Gefen et al., (2002) described perceived risk as an attribute of an alternate decision that reflects the difference of its potential outcomes. Perceived risk affects purchase decisions, especially when the outcomes of the purchase are unknown. Hong and Yi (2012) found that consumers who favor Internet transactions over the conventional ones appear to have low risk avoidance profiles. Hence, alternating, postponing and canceling a purchase demonstrate risk concerns.

Perceived ease of use relates to the level to which a person is sure that the use of certain technology will be easy (Almajali & Masa'deh, 2021; Venkatesh & Davis, 2000). The present study hence will examine the effect of perceived ease of use on the use of various online stores. Also, it was discovered that COVID-19 pandemic and restrictions imposed by government have affected consumer behavior. Armando, (2021) indicated that during the pandemic, consumers, irrespective of generation, are all into making purchases of goods and services online. In fact, a significant swing towards e-commerce spending in addition to the increase of shopping frequency, have been reported by Armando (2021). In examining online consumer behavior during

COVID-19 pandemic, Barbu et al., (2021) identified several factors that propel online consumer behavior. These factors are: strong and unremitting increase in the amount of Internet users, increased knowledge towards online shopping, increase in the number of online products being introduced, and low prices of products purchased in bulk. The pandemic itself, and the requirements of social distancing and staying at home are other factors contributing to the increase in online shopping activities.

The study at hand is of value to both scholars and retail marketers in increasing their knowledge on the perceptions of online shopping intentions of consumer during COVID-19 health safety crisis and also in managing COVID-19 fears. Based on the findings, appropriate marketing strategies could be developed in encouraging online shopping among consumers in Jordan particularly.

3 Research Model and Hypotheses Formulation

Based on the above discussion on online shopping behavior among consumers in Jordan, the hypotheses proposed in this study are, as seen in Fig. 1, as follows.

H1 Perceived risk will have a positive impact on online shopping behavior.
H2 Trust will have a positive impact on online shopping behavior.
H3 Perceived ease of use will have a positive impact on online shopping behavior.
H4 Fear of COVID-19 will have a positive impact on online shopping behavior.

The measurement items in this study were adapted from past related studies. Details on them are as presented in Table 1.

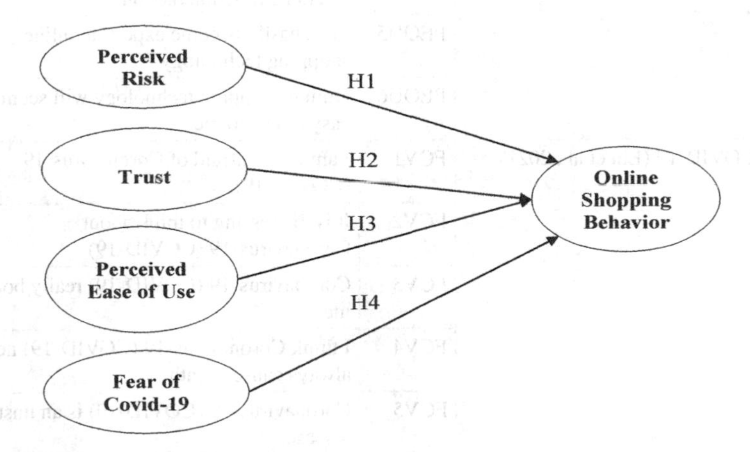

Fig. 1 The proposed model

Table 1 Measurements of factors

Factor	Code	Measurement dimensions of factors
Online purchasing behavior (Forsythe et al., 2006; George, 2004; Swinyard & Smith, 2003)	OPB1	It is easy to use the Internet when I shop online
	OPB2	The opinion of my family and friends is important when I make a purchase online
	OPB3	I think I will face no problems with online shopping if my family and friends also face no problems with online shopping
	OPB4	I will not shop online if the webpage is slow to download
Perceived risk (Miyazaki & Fernandez, 2000)	PR1	I think that online shopping today is less risky
	PR2	I think that online shopping today is safe
Trust (Miyazaki & Fernandez, 2000)	TR1	I am confident about buying products or services online
	TR2	I am anxious about buying products or services online
	TR3	Online shopping today is reliable
Perceived ease of use (Davis, 1989)	PEOU1	I think that learning the technology would not be easy for me
	PEOU2	I think that online shopping does not need much thinking
	PEOU3	I think that using technology in doing my online shopping tasks is easy
	PEOU4	I think that online shopping technology offers flexible interaction
	PEOU5	I can easily become expert in online shopping technology
	PEOU6	Online shopping technology will seem easy to use to me
Fear of COVID-19 (Lin et al., 2021)	FCV1	I am really afraid of Coronavirus-19 (COVID-19)
	FCV2	It is distressing to think about Coronavirus-19 (COVID-19)
	FCV3	Coronavirus-19 (COVID-19) really bothers me
	FCV4	I think Coronavirus-19 (COVID-19) nearly always causes death
	FCV5	Coronavirus-19 (COVID-19) is an unstable disease
	FCV6	The thought of Coronavirus-19 (COVID-19) makes my hands clammy

(continued)

Table 1 (continued)

Factor	Code	Measurement dimensions of factors
	FCV7	Dying from Coronavirus-19 (COVID-19) frightens me
	FCV8	I feel anxious when watching news and stories about Coronavirus-19 (COVID-19) on social media
	FCV9	The thought of Coronavirus-19 (COVID-19) makes me unable to sleep
	FCV10	The thought of being infected by Coronavirus-19 (COVID-19) gives me heart palpitations

4 Population and Sample Size

The population of this study comprised online shoppers in Jordan, and the participants selected in this study were recommended by informants or the study participants themselves. The recommended participants were those with the potential interest in taking part in the study through social networks. All participants were those who prefer to shop online.

5 Discussion and Conclusion

Trust refers to the readiness of customer in accepting flaws in an online transaction as a compensation for desirable anticipations in future online store behavior (Kimery & McCard, 2002). Online shopping can only occur when there is trust because shoppers or consumers will be trading with unknown parties, in this case, sellers. In terms of payment, it may be required prior to or after the product or service is delivered. If the payment is made before the product or service is received, then, the consumer would be transferring money without assurance if the seller will actually send the product or service. In order to reduce uncertainty and build trust, good communication between seller and consumer is important. Additionally, the provision of a detailed product description will increase the confidence of consumer in making the purchase of products or services online.

Online shopping is also affected by Perceived Risk which comes in two types according to Park and John (2010). The first type of perceived risk is behavioral risk, and this risk comes from online retailers that employ internet marketing techniques in monitor all transactions. The associated risks are psychological risks and product risks (i.e., time and convenience risks). Another risk is the environmental risk created by the internet, and this risk cannot be controlled by both the retailer and consumer. This could lead to financial and security risks. In general, respondents understand the risk associated with online purchasing. In order to reduce the associated risk, respondents will evaluate the site, and examine the products of interest, in all matters like price, deals, reviews, just to name a few.

Perceived ease of use is another factor affecting online shopping behavior. It relates to the perception of user concerning the difficulty of technology usage. In other words, perceived ease of use concerns the level of ease of using a given technology (Davis, 1989). A system that is viewed as complicated will not be used when alternative method which is perceived as easier to use, is available. In the context of online shopping, technologies that are difficult to utilize are considered as less useful. Also, uncertain consumer demand and supply chain issues could impact COVID-19 fear. As this coronavirus is a pandemic, many merchants have been affected. Filimonau et al., (2021) and Alshurideh et al., (2021) indicated that during the pandemic, these merchants were facing reduction in sales, disrupted supply chain, and increase in the purchases of certain products, especially hygiene related products.

Online shopping is greatly impacted by trust and perceived risk, and consumers will likely engage in online shopping if they are confident towards the online site. Perceived lower risk also increases the likelihood of online purchasing. Ease-of-use quality in online purchasing site entices consumers to make the purchase online as well. Additionally, an evaluation instrument of COVID-19 fear during pandemic needs further scrutiny, in order to be validated. To date, at least ten countries have verified the satisfactory psychometric properties of the instrument. Moreover, the relationship between variables, and the validated hypotheses have to be examined more deeply.

Consumers need assurance of security from online stores in order that they could stay loyal to the stores. Also, assurance of security will solidify the trust of consumer, while reducing the possibility of fraud. In increasing trust and hence online purchasing, online stores should provide details of items offered so that consumer could make accurate purchasing decisions, use established and trusted delivery service in order that the product purchased could reach the consumer safely and on time and provide fair and appropriate treaty agreements in order that the consumer will feel satisfied and will make repeat purchase or even promote the store to other potential consumers.

References

Abuhashesh, M., Al-Dmour, H., Masa'deh, R., Salman, A., Al-Dmour, R., Boguszwick, M., and AlAmaireh, Q. (2021). The role of social media in raising public health awareness during pandemic Covid-19: An international comparative study. *Informatics, 8*(4), 1–19.

Ajzen, I. (1991). The theory of planned behavior. *Organizational Behavior and Human Decision Process, 50*, 179–211.

Al Khasawneh, M., Abuhashesh, M., Ahmad, A., Masa'deh, R., & Alshurideh, M. (2021a). Determinants of e-word of mouth on social media during Covid 19 outbreak: An empirical study (347–366). In M. Alshurideh, A.E. Hassanien, & R. Masa'deh (Eds), *The Effect of Coronavirus Disease (COVID-19) on Business Intelligent Systems. Studies in Systems, Decision and Control*, vol. 334. Springer. https://doi.org/10.1007/978-3-030-67151-8_20.

Al Khasawneh, M., Abuhashesh, M., Ahmad, A., Masa'deh, R., & Alshurideh, M. (2021b). Customers online engagement with social media influencers' content related to COVID 19 (385–404). In M. Alshurideh, A.E. Hassanien, & R. Masa'deh (Eds.), *The Effect of Coronavirus Disease (COVID-19) on Business Intelligent Systems. Studies in Systems, Decision and Control*, vol. 334. Springer. https://doi.org/10.1007/978-3-030-67151-8_22.

Al-Dmour, H., Masa'deh, R., Salman, A., Abuhashesh, M., & Al-Dmour, R. (2022). The influence of mass media interventions on public health awareness and protection against COVID-19 pandemic: Empirical study. *SAGE Open*, in press.

Al-Dmour, H., Masa'deh, R., Salman, A., Abuhashesh, M., & Al-Dmour, R. (2020). Influence of social media platforms on public health protection against the COVID-19 pandemic via the mediating effects of public health awareness and behavioral changes: Integrated model. *Journal of Medical Internet Research, 22*(8), e19996.

Almajali, D. A. (2021a). Diagnosing the effect of green supply chain management on firm performance: An experiment study among Jordan industrial estates companies. *Uncertain Supply Chain Management, 9*(4), 897–904.

Almajali, D. A. (2021b). Antecedents of e-commerce on actual use of international trade center: Literature review. *Academy of Strategic Management Journal, 20*(2), 1–8.

Almajali, D. A., & Hammouri, Q. (2021). Predictors of online shopping during Covid-19 pandemic in developing country: Qualitative analysis. *Annals of the Romanian Society for Cell Biology, 25*(6), 12970–12977.

Almajali, D. A., Hammouri, Q., Majali, T., Al-Gasawneh, J. A., & Dahalin, Z. M. (2021). Antecedents of consumers' adoption of electronic commerce in developing countries. *International Journal of Data and Network Science, 5*(4), 681–690.

Almajali, D. A., & Masa'deh, R. (2021). Antecedents of students' perceptions of online learning through COVID-19 pandemic in Jordan. *International Journal of Data and Network Science, 5*(4), 587–592.

Alshurideh, M., Al Kurdi, B., AlHamad, A., Salloum, S., Alkurdi, S., Dehghan, A., Abuhashesh, M., & Masa'deh, R. (2021). Factors affecting the use of smart mobile exam platforms by universities' postgraduate students during the Covid 19 pandemic: An empirical study. *Informatics, 8*(2), 1–20.

Alsyouf, A., Masa'deh, R., Albugami, M., Al-Bsheish, M., Lutfi, A., & Alsubahi, N. (2021). Risk of fear and anxiety in utilizing health app surveillance due to COVID-19: Gender differences analysis. *Risks, 9*(10), 1–19.

Armando, R. L. C. (2021). Disruption in the consumer decision-making? critical analysis of the consumer's decision making and its possible change by the COVID-19. *Turkish Journal of Computer and Mathematics Education, 12*, 1468–1480.

Barbu, C. M., Florea, D. L., Dabija, D.-C., & Barbu, M. C. R. (2021). Customer experience in Fintech. *Journal of Theoretical and Applied Electronic Commerce Research, 16*, 1415–1433.

Baud, D., Qi, X., Nielsen-Saines, K., Musso, D., Pomar, L., & Favre, G. (2020). Real estimates of mortality following COVID-19 infection. *The Lancet Infectious Diseases*. https://doi.org/10.1016/S1473-3099(20)30195-X

Centers for Disease Control and Prevention. (2022). Coronavirus disease 2019 (COVID-19): Reducing stigma. https://www.cdc.gov/coronavirus/2019-ncov/about/related-stigma.html. Retrieved 11 Mar 2022.

Davis, F. D. (1989). Perceived usefulness, perceived ease of use, and user acceptance of information technology. *MIS Quarterly, 13*(1), 319–339.

Filimonau, V., Beer, S., & Ermolaev, V. A. (2021). The covid-19 pandemic and food consumption at home and away: An exploratory study of English households. *Socio-Economic Planning Sciences*, in press.

Forsythe, S., Liu, C., Shannon, D., & Gardner, L. C. (2006). Development of a scale to measure the perceived benefits and risks of online shopping. *Journal of Interactive Marketing, 20*(2), 55–75.

Gefen, D., Karahanna, E., & Straub, D. (2003). Inexperience and experience with online stores: The importance of TAM and trust. *IEEE Transactions on Engineering Management, 50*(3), 307–321.

Gefen, D., Rao, V.S., & Tractinsky, N. (2002). The conceptualization of trust, risk and their relationship in electronic commerce: The need for clarifications. In *Proceedings of the 36th Hawaii International Conference on System Sciences*.

George, J. F. (2004). The theory of planned behaviour and Internet purchasing. *Internet Research, 14*(3), 198–212.

Guan, W. J., Ni, Z. Y., Hu, Y., Liang, W. H., Ou, C. Q., He, J. X., & Du, B. (2019). Clinical characteristics of coronavirus disease. *International Journal of Mental Health and Addiction, 2020*,. https://doi.org/10.1056/NEJMoa2002032

Hong, Z., & Yi, L. (2012). Research on the influence of perceived risk in consumer on-line purchasing decision. *Physics Procedia, 24*(B), 1304–1310.

Hsu, M. H., Yen, C. H., & Chang, C. M. (2006). A longitudinal investigation of continued online shopping behavior: An extension of the theory of planned behavior. *International Journal of Human-Computer Studies, 64*, 889–904.

Huang, C., Wang, Y., Li, X., Ren, L., Zhao, J., Hu, Y., et al. (2020). Clinical features of patients infected with2019 novel coronavirus in Wuhan, China. *The Lancet, 395*(10223), 497–506.

Kimery, K. M., & McCard, M. (2002). Third-party assurance: Mapping the road to trust in e retailing. *Journal of Information Technology Theory and Application, 4*(2), 63–82.

Kumar, A., & Grisaffe, D. B. (2004). Effects of extrinsic attributes on perceived quality, customer value, and behavioral intentions in B2B settings: A comparison across goods and service industries. *Journal of Business-to-Business Marketing, 11*(4), 43–74.

Liao, C., Lin, H. N., Luo, M. M., & Chea, S. (2017). Factors influencing online shoppers' repurchase intentions: The roles of satisfaction and regret. *Information & Management, 54*(5), 651–668.

Lin, C.-Y. (2020). Social reaction toward the 2019 novel coronavirus (COVID- 19). *Social Health and Behavior, 3*, 1–2.

Lin, C. Y., Hou, W. L., Mamun, M. A., Aparecido da Silva, J., Broche-Pérez, Y., Ullah, I., & Pakpour, A. H. (2021). Fear of COVID-19 scale (FCV-19S) across countries: Measurement invariance issues. *Nursing Open, 8*(4), 1892–1908.

AL Manaseer, M., Maqableh, M., Alrowwad, A., & Masa'deh, R. (2019). Impact of information technology on organizational performance in Jordanian public government entities. *Jordan Journal of Business Administration, 15*(4), 489–516.

Miyazaki, A. D., & Fernandez, A. (2000). Internet privacy and security: An examination of online retailer disclosures. *Journal of Public Policy and Marketing, 19*(1), 54–61.

Moling, L. (2011). The analysis of strengths and weakness of online shopping. In M. Dai (Ed.), *Innovative Computing and Information. ICCIC 2011. Communications in Computer and Information Science*, vol. 231. Berlin, Heidelberg: Springer. https://doi.org/10.1007/978-3-642-23993-9_66.

Monsuwé, T., Dellaert, B., & Ruyter, K. (2004). What drives consumers to shop online? a literature review. *International Journal of Service Industry Management, 15*(1), 102–121.

Nachit, H., & Belhcen, L. (2020). Digital transformation in times of COVID-19 pandemic: The case of Morocco. https://www.ssrn.com/abstract=3645084 or https://doi.org/10.2139/ssrn.3645084.

Orapin, L. (2009). Factors influencing internet shopping behavior: A survey of consumers in Thailand. *Journal of Fashion Marketing and Management, 13*(4), 501–513.

Pappas, G., Kiriaze, I. J., Giannakis, P., & Falagas, M. E. (2009). Psychosocial consequences of infectious diseases. *Clinical Microbiology and Infection, 15*(8), 743–747.

Park, J., & John, D. R. (2010). Got to get into my life. Do brand personalities rub off on consumers? *Journal of Consumer Research, 37*(4), 655–669.

Swinyard, W. R., & Smith, S. M. (2003). Why people (don't) shop online: A lifestyle study of the internet consumer. *Psychology and Marketing, 20*(7), 567–597.

Ukoha, O., Awa, H., Nwuche, C., & Asiegbu, I. (2011). Analysis of explanatory and predictive architectures and the relevance in explaining the adoption of IT in SMEs. *Interdisciplinary Journal of Information, Knowledge, and Management, 6*, 217–230.

Venkatesh, V., & Davis, F. D. (2000). A theoretical extension of the technology acceptance model: Four longitudinal field studies. *Management Science, 46*(2), 186–204.

Wang, D., Hu, B., Hu, C., Zhu, F., Liu, X., Zhang, J., & Zhao, Y. (2020). Clinical characteristics of 138 hospitalized patients with 2019 novel coronavirus–infected pneumonia in Wuhan, China. *JAMA, 323*(11), 1061–1069.

World Health Organization. (2020). Coronavirus disease (COVID-2019): situation report-54. https://www.who.int/docs/default-source/coronaviruse/situation-reports/20200314-sitrep-54-covid-19.pdf?sfvrsn=dcd46351_2020. Retrieved 10 Apr 2020.

Wu, I. L., & Chen, J. L. (2005). An extension of trust and TAM model with TPB in the initial adoption of on-line tax: An empirical study. *International Journal of Human Computer Studies, 62*(6), 784–808.

Learning- E-learning and M-learning

Agility in Higher Education Institutions to Management of Covid-19 Disaster in UAE

Mohammed T. Nuseir, Amer Qasim, and Ghaleb A. El Refae

Abstract The covid-19 pandemic has severely affected the economy all around the globe and also dramatic effects on the education sector worldwide. This situation needs to be examined and capture the attention of recent studies and regulators. Thus, the present study aim is to examine the impact of student readiness, faculty readiness, and information technology readiness on the management of the Covid-19 disaster in higher education institutions in UAE. This article has followed the quantitative approach of data collection and taken the questionnaires for this purpose. The current study has executed the smart-PLS to analyze the nexus among the constructs. The results indicated that student readiness, faculty readiness, and information technology readiness have a positive linkage with the management of the Covid-19 disaster in higher education institutions in the UAE. This study has guided the policymakers that they should enhance their focus towards the management of the Covid-19 disaster in the education sector and develop effective policies that improve the education sector's ability to manage the effects of the Covid-19 disaster.

Keywords Covid-19 disaster · Student readiness · Faculty readiness · Information technology readiness

M. T. Nuseir (✉) · A. Qasim · G. A. E. Refae
Department of Business Administration, College of Business, Al Ain University, Abu Dhabi Campus, UAE
e-mail: mohammed.nuseir@aau.ac.ae

A. Qasim
e-mail: amer.qasim@aau.ac.ae

G. A. E. Refae
e-mail: ghalebelrefae@aau.ac.ae

M. Alshurideh et al. (eds.), *The Effect of Information Technology on Business and Marketing Intelligence Systems*, Studies in Computational Intelligence 1056,
https://doi.org/10.1007/978-3-031-12382-5_17

1 Introduction

The covid-19 pandemic is a contagious disease caused by severe acute respiratory syndrome coronavirus 2. First, it started in Wuhan, China, in December 2019. The disease has spread across the world, leading to an ongoing pandemic. Covid-19 transfers from one person to another when people are exposed to a virus which consists of respiratory droplets and airborne particles exhaled by an infected person. These particles are inhaled or reach the eyes, nose, or mouth of an individual through touching or being coughed on (Akour et al., 2021; Al-Dmour et al., 2021a; Alshurideh et al., 2021a; Daniel, 2020; Liguori & Winkler, 2020). The risk of covid-19 is high in crowded or poorly ventilated places. That is the reason the covid-19 epidemic has affected the educational system around the world (Al Khasawneh et al., 2021a; Alameeri et al., 2021; Al-Dmour et al., 2021b; Aljumah et al., 2021a; Amarneh et al., 2021). It causes the near-total closure of educational institutions (Ahmad et al., 2021; Al Khasawneh et al., 2021b; Alshurideh et al., 2021b; Shah et al., 2021). Most of the governments made the decision to close educational institutions in a struggle to reduce the spread of covid-19. As per the worldwide statistics of 2021, about 825 million learners are currently influenced by the closure of educational institutions as a result of the pandemic. As per UNICEF monitoring, 23 countries are presently imposing nationwide closures, and about 40 countries are found to be imposing local closure. This all affects about 47% of the world's student population. While at the beginning of this contagious disease, more countries were forced to close educational institutions (Alshamsi et al., 2021; Alsharari & Alshurideh, 2020; Chang et al., 2021; Leo et al., 2021; Salloum et al., 1153).

As education has become one of the basic needs of life and also an economic plateform which provides the nation builders, great producers, and businessmen, even in this severe situation it has been given special attention. The educational institutions have been trying their best to manage the issues arising from the universal contagious disease covid-19 (Al-Hamad et al., 2021; Alshurideh et al., 2019a; Kurdi et al., 2020a). The institutions have been applying the maximum possible ways to provide education to students and maintain continuity in education. Thus, the online classes are being carried on (Al Kurdi et al., 2020a; Alshurideh et al., 2019b; Theoret & Ming, 2020). Our study analyzes the agility in the higher educational institutions and examines how they help in overcoming the covid-19 disaster. Some educational institutions, especially higher educational institutions, maintains the quality of the resources, technology, techniques, teaching methods, human resources, and the production of talented students as they can respond to the sudden changes in the circumstances (Alshurideh, 2014; Alshurideh & Xiao, 2012; Alshurideh et al., 2019c). During covid-19, the educational institutions are forced to close traditional face-to-face classes; their agility in all the organizational processes enables them to continue the teaching classes online. The agility in educational institutions is derived by information technology readiness, the readiness of faculty at education institutions online teaching classes, and the student's readiness for attending online learning classes (Alomari et al., 2019; Chick et al., 2020; Elshamy et al., 2017).

This study analyzes the impacts of agility in higher educational institutions on the management of the covid-19 disaster in the United Arab Emirates (UAE). UAE has been declared by the reports of certain local and international organizations in 2014 as one of the most developed regions of the world. According to the reports, UAE lies among happy countries. Here, the citizens and other residents are happy having a high living standard. Most of the economic fields like investment, information, technology, industry, and service industry have high growth. Thus, UAE is a successful economic world. UAE has a good education system that produces skilled persons. UAE is regarded as one of the top 20 best education systems across the world (Ashour, 2020). Though UAE is a wealthiest country and has a successful education system throughout the world, it is not safe from the covid-19 pandemic. This universal contagious disease has affected the higher education institutions in UAE. UAE universities are moving typical face-to-face education to remove or distance education. Most educational institutions are moving towards online education. Students are acquiring education online having seated in front of their computer screens at their homes (AlHamad & Al Qawasmi, 2014; AlHamad et al., 2012; McMinn et al., 2020). The education sector in UAE growing rapidly and shows that both public and private enrolment will grow around 0.7%, resulting in more than 5,500 extra enrolments by 2021. These figures are shown in Fig. 1.

Our study aims at encouraging online education in higher educational institutions of UAE in case all the higher educational institutions, including universities, are bound to close while covid-19, the universal contagious disease, is spreading. This study elaborates how with the agility in higher education institutions in the form of information technology readiness, faculty readiness, and students' readiness, online education can be promoted, and thus, covid-19 disaster in higher educational institutions can be overcome. This study has chosen UAE as a center of analysis in this context because it can provide more valid consequences of the agility in higher education institutions and their role in overcoming the covid-19 disaster. UAE consists of seven emirates Abu Dhabi, Ajman, Dubai, Fujairah, Ras Al Khaimah, Sharjah,

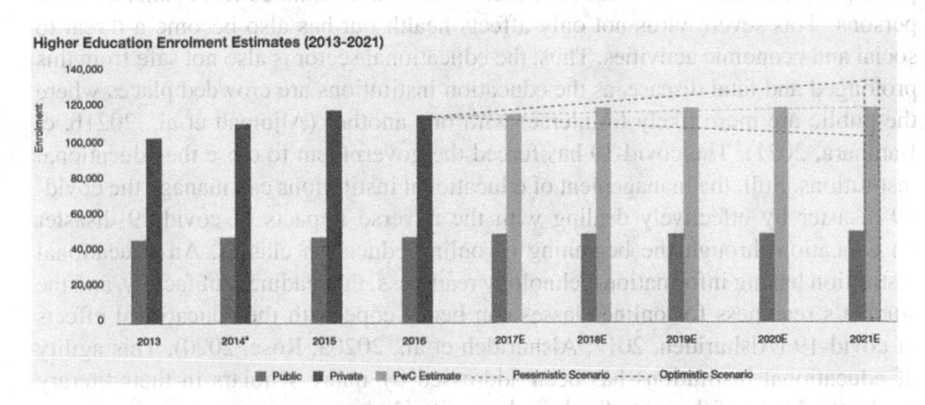

Fig. 1 Higher education enrollment in UAE

and Umm Al Quwain. And UAE education system is one of the top 20 education systems across the world (Alhashmi & Moussa-Inaty, 2021). Thus, the study conducted in UAE would be more general as compared to any study dealing with the aforementioned concept in any particular country.

The aim of our study is to examine the agility in higher educational institutions and check their influences on the management of the covid-19 pandemic. In this context, it examines information technology readiness, the readiness of faculty at education institutions' online teaching classes, and the student's readiness for attending online learning classes. From the beginning of 2021 till now, many authors have given attention to the impacts of covid-19 disasters on education and the factors helping in overcoming the covid-19 disaster. These authors have discussed the efforts of institutional management as a whole while analyzing the management of the covid-19 disaster, or they have discussed the role of information technology readiness, the readiness of faculty at education institutions online teaching classes, and the student's readiness for attending online learning classes separately. This study secures a distinctive place in the literature as it initiates to categorize the agility in higher education institutions into information technology readiness, faculty readiness, and students' readiness and discuss them separately in detail in relation to the management of covid-19 disaster.

2 Literature Review

The spread of the covid-19 pandemic has been a serious threat to the health of people, as it fast spreads among the people through touching and reaching the virus to the nose, eyes, or mouth when the affected shared something either knowingly or unknowingly or coughs (Akour et al., 2020; AlHamad & AlHammadi, 2018; Qasim AlHamad et al., 2014). It is more likely to affect the health of the people in some places where the people are in close contact with one another or in some crowded place places where people cannot be safe from viruses exhaled from some affected persons. This severe virus not only affects health but has also become a threat to social and economic activities. Thus, the educational sector is also not safe from this prolonged and fatal disease, as the education institutions are crowded places where the public are more likely to interact with one another (Aljumah et al., 2021b, c; Batubara, 2021). The covid-19 has forced the government to close the educational institutions. Still, the management of educational institutions can manage the covid-19 disaster by effectively dealing with the adverse impacts of covid-19 disaster on education through the beginning of online education classes. An educational institution having information technology readiness, the readiness of faculty, and the student's readiness for online classes can better cope with the educational effects of covid-19 (Alshurideh, 2017; Alshurideh et al., 2020a; Rose, 2020). This agility of educational institutions has been addressed by many scholars in their literary workouts. Many of these studies have been cited below.

Information technology is the utilization of computing hardware, application software, storage devices, networking (internet and digital devices), and other devices, infrastructure, and processes to acquire, process, store, save and transfer information (Aissani & Awad ElRefae, 2021; AlShurideh et al., 2019; Sun et al., 2021). Information technology readiness in an educational institution refers to the availability of good quality computer hardware components, required software applications, storage devices, effective networking systems, smart devices, and infrastructure to respond to the changes in circumstances (Nawaz et al., 2021; Nuseir et al., 2021a, b). Informational technology readiness enables the organization to manage the contagious disease covid-19 as it helps to mitigate the adverse influences of a covid-19 pandemic on the continuity of education through the start of online teaching classes. A study conducted by He et al. (2021) on the management of covid-19 disaster elaborates that the educational institutions which have an effective information technology system consisting of all the essential devices, good quality infrastructure, and fast & smooth internet have the ability to proceed to begin online classes (Al Kurdi et al., 2020b; Alshurideh et al., 2020b; Kurdi et al., 2020b). This study elaborates that it is the ready information system that enables the institution to collect essential information about the provision of online study, keep in contact with the student even after the close of traditional face-to-face classes, and make them go through the course of study online, and prepare them for final exams. Though the covid-19 pandemic places a barrier around the educational institutions in the form of closure of face-to-face educational services, the information technology readiness brings continuity in the education. Similarly, the literary workout of Sittig and Singh (2020) also supports this view by telling that the renowned higher education institutions manage to maintain the quality of all the components of information technology in order to immediately respond to the change in circumstances, have been analyzed to start online teaching classes and overcome the disastrous impacts of covid-19 on the education. The online classes require the computer or other digital devices run on android, speaker, camera, specific apps to start a conversation with the students and to design and show effective presentations, and good quality infrastructure to deliver lecture without any barrier or difficulty. All this is possible under the availability and maintenance of good quality information technology, and fluent online classes are a powerful tool to handle the covid-19 disaster (Her, 2020; Nuseir et al., 2021c; Sultan et al., 2021).

H1: Information technology readiness has a positive association with the management of covid-19 disaster.

The survival, performance, and success of any educational institution depend on the abilities, skills, and responsiveness of the teaching staff. It is the member's staff with the help of which an education institution faces different problems and overcome difficulties. When any disaster comes and disturbs the education system, it is the teachers who come to the ground and fight against the disaster and saves the education. When the Covid-19 the contagious disease which affects the health of people without any discrimination of creed, religion, or region, social or economic group, somehow come into existence, the skilled teachers have shown active performance in maintaining continuity in the education (Bettayeb et al., 2020; Rafique et al., 2021; Taryam et al., 2020). The more effective way to cope with the disastrous influences of a covid-19 pandemic on education is to keep online contact with the students and provide them online education. The provision of online education is dependent on the readiness of students to take online classes. The study by Moralista and Oducado (2020) has investigated the covid-19 disastrous effects on education and their management to save the education. This study posits that though the spreading of the covid-19 pandemic is a barrier to the way education, if the faculty at the institutions have the willingness to provide online education, is interested in the online classes, motivation to provide education to the students to the best of their abilities, they can continue the education. Similarly, the faculty who have the capacity to create interest and motivation in the students to get online education and perform effectively is a powerful tool against the covid-19 disaster whose victim is not only the health but also the education. The research conducted by the renowned scholar elaborates that when the students have prior knowledge of online education, the essentials of online classes, the ability to design online course study and make them implemented, and the self-efficacy to deal with the students online, it is more likely for the educational institutions to carry online classes and overcome the issue of closure of education institutions created by the spread of covid-19 (Al-Maroof et al., 2021a; Alshurideh et al., 2021c; Elsaadani, 2020). Hence, it can be hypothesized.

H2: Faculty readiness for the online teaching classes has a positive association with the management of the covid-19 pandemic.

When any sort of disaster attacks education, it is the ability and capacity of students to be steadfast and ready to fight against the problems and issues to which they are exposed because of disaster. When the covid-19 attacked the health of the people across the world, it became a serious threat to the people to lose their health or life and thereby, has caused the closure the educational institutions; the students can fight against the disaster and carry their study through online learning classes. But these online classes and their success depend on the readiness of students (Pokhrel & Chhetri, 2021). The students are ready for the online classes when they have sufficient knowledge of digital devices, relevant software apps, and other related devices, the ability to use these devices and apps, have these facilities in actuality, understand the subject being discussed, and respond to this effectively determine the possibility and effectiveness of online classes and the ability of educational institutions to manage covid-19 disaster (Leo et al., 2021). The students have complete knowledge about digital devices like computer or smart mobile devices, speaker or camera, monitor with a large and clear screen, apps to build a virtual conversation with the tutor, and to note down the key points discussed in the ongoing online class, and to respond to them in an effective manner can participate in the online classes. Contrary to this, a student having no knowledge or support in this regard is unable to benefit from online classes and overcome the covid-19 disaster. In case the spreading of covid-19 pandemic, only those students continue their study fighting against the covid-19 disaster who have interest in the online education, knowledge, experience, and financial resources which are helpful in getting an online education (AlHamad et al., 2021; Al-Maroof et al., 2021b; Baticulon et al., 2021).

H3: Student readiness for the online learning classes has a positive association with the management of the covid-19 disaster.

3 Research Methods

The methodology used in this study required direct information from students. Therefore, a data collection instrument was designed to examine the impact of student readiness, faculty readiness, and information technology readiness on managing the Covid-19 disaster in higher education institutions in the UAE. The quantitative approach of data collection designed a questionnaire for this purpose. primary data were collected based on a survey of students at two public universities–United Arab Emirates University and Zayed University, as well as one private university-Al Ain University.

The respondents that are selected based on simple random sampling. Approximately, 400 students are enrolled in the last semester of each institution in business administration departments. Thus, the total population of the study is 1200 students. The researchers have used online forms to distribute the surveys to the respondents and a total of 455 surveys have been forwarded to the respondents, and after twenty days, only 292 were received which about 64.18 percent response rate.

The current study has executed the smart-PLS to analyze the nexus among the constructs. The researchers have used this effective tool due to the large sample size along with the hypotheses testing purpose of the research (Hair et al., 2017). This research has used three predictors such as student readiness (SR) with fifteen items taken from the study of Dray et al. (2011), faculty readiness (FR) with six items that are taken from the study of Petko et al. (2018) and information technology readiness (ITR) with eight items extracted from the article of Blayone et al. (2020). This study has also taken the management of the Covid-19 disaster (MCD) with five items that are taken from the article of Martono et al. (2019). These constructs are given in Fig. 2.

4 Findings

This study has examined the assessment of the measurement model by using the convergent validity that is about the links among items. The statistics have been shown that the factor loadings and AVE are larger than 0.50 while Alpha and composite reliability (CR) values are cross the limits of 0.70. These figures indicated the high linkage among items. These values are shown in Table 1.

This study has examined the discriminant validity that is about the links among variables. The statistics about Heterotrait Monotrait (HTMT) ratios have not larger than 0.85. These figures indicated the low linkage among variables. These values are shown in Table 2 (Fig. 3).

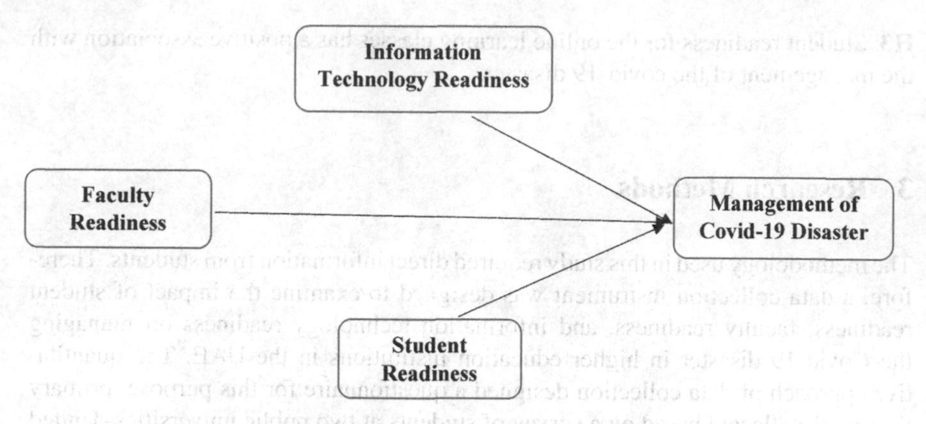

Fig. 2 Theoretical model

The structural model assessment has been shown the nexus among the constructs, and the results indicated that student readiness, faculty readiness, and information technology readiness have a positive linkage with the management of the Covid-19 disaster in higher education institutions in UAE and accept H1 H2 and H3. The results among the variables are significant because the t-statistics are larger than 1.64. These links have been mentioned in Table 3 (Fig. 4).

5 Discussions and Implications

The study results have indicated that the readiness of information technology in an educational institution has a positive relationship with the management of the covid-19 pandemic. The study implies that when the education institution has an effective information technology system encompassing computers (hardware & software), storage, networking, physical devices, infrastructure, and processes to create, process, store, save and exchange knowledge and information, it is in a better position to cope with the disasters of a covid-19 pandemic on education. These results are in line with the recent study of Whitelaw et al. (2020), which has been conducted to check the effects of a covid-19 pandemic on students' education and the performance of the educational institution and examines the management of disaster of covid-19 in this regard. This study suggests that the educational institutions which have the information technology which already has the capacity to meet the online education requirements in case the institutions are not allowed to provide education through physical interaction with the students can mitigate the effects of covid-19 on education. These results are also in line with the previous study of Urbaczewski and Lee (2020).

This study elaborates those good educational institutions, whether they provide school education or higher education (colleges or universities), always take care of

Table 1 Convergent validity

Constructs	Items	Loadings	Alpha	CR	AVE
Faculty readiness	FR1	0.845	0.927	0.943	0.733
	FR2	0.868			
	FR3	0.859			
	FR4	0.821			
	FR5	0.884			
	FR6	0.858			
Information technology readiness	ITR1	0.773	0.906	0.922	0.597
	ITR2	0.752			
	ITR3	0.823			
	ITR4	0.762			
	ITR5	0.723			
	ITR6	0.728			
	ITR7	0.794			
	ITR8	0.820			
Management of Covid-19 disaster	MCD1	0.654	0.848	0.892	0.625
	MCD2	0.782			
	MCD3	0.882			
	MCD4	0.730			
	MCD5	0.881			
Student readiness	SR1	0.842	0.961	0.965	0.647
	SR10	0.826			
	SR11	0.771			
	SR12	0.840			
	SR13	0.821			
	SR14	0.821			
	SR15	0.722			
	SR2	0.820			
	SR3	0.853			
	SR4	0.840			
	SR5	0.842			
	SR6	0.823			
	SR7	0.721			
	SR8	0.764			
	SR9	0.742			

Table 2 Discriminant validity

	FCPA	FR	ITR	SR
FCPA				
FR	0.107			
ITR	0.411	0.155		
SR	0.473	0.076	0.536	

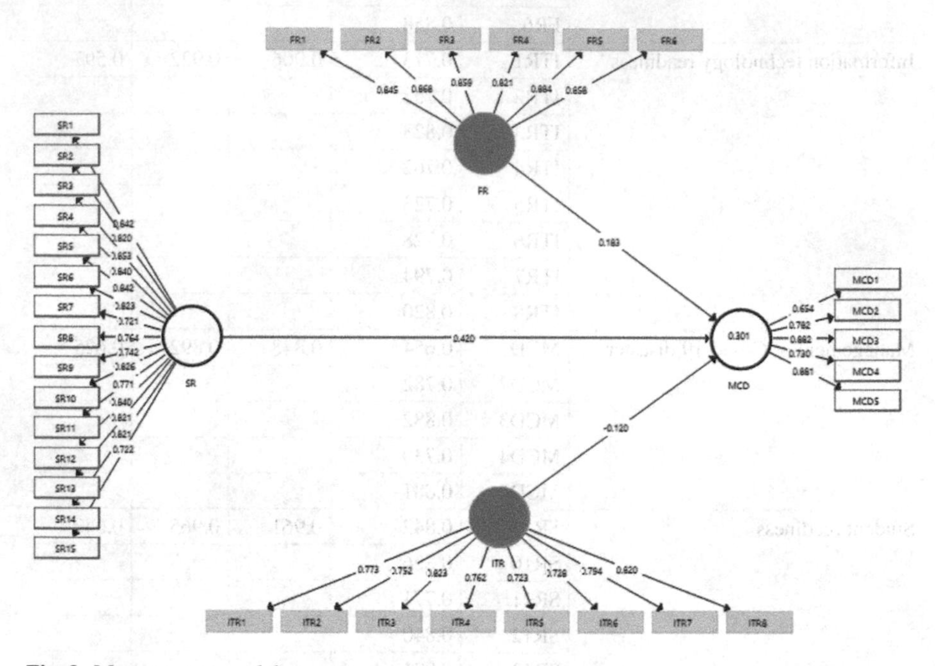

Fig. 3 Measurement model assessment

Table 3 Path analysis

Relationships	Beta	S.D	T statistics	P values	L.L	U.L
FR -> FCPA	0.183	0.059	3.092	0.001	0.069	0.256
ITR -> FCPA	−0.120	0.041	2.961	0.002	−0.216	−0.080
SR -> FCPA	0.420	0.061	6.928	0.000	0.337	0.534

the maintenance of information technology system. Such institutions have shown and are showing good education performance even in the period of covid-19 disaster, as they are able to continue their teaching classes online without even a short gap. Thus, the readiness of information technology has been helpful to educational institutions to cope with the adverse impacts of a covid-19 pandemic. These results are also supported by the recent literary work out of Yan et al. (2020), which investigates

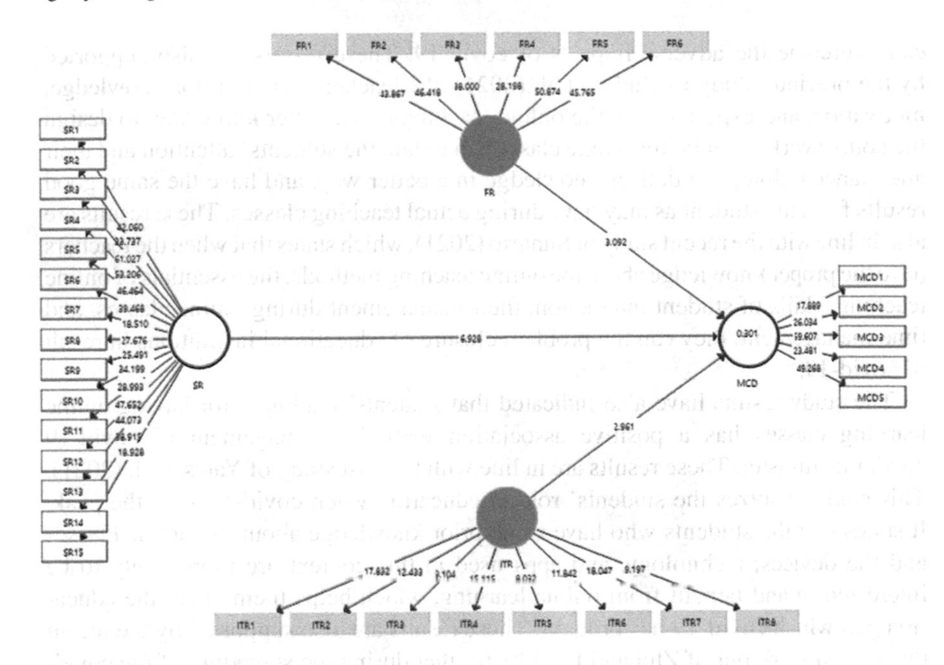

Fig. 4 Structural model assessment

the impacts of a pandemic disaster like covid-19 on the education of a country and the ways how to cope with those impacts. This study concludes that when an education institution has effective information technology (good quality hardware and software, effective network, and good quality infrastructure used in this regard) and takes great care for its maintenance, it can effectively use it for the exchange of information and knowledge. This readiness of information technology helps the institution to minimize the impacts of covid-19 on education because it makes it possible to provide education online. Thus, the readiness of information technology is of great importance to the management of the covid-19 pandemic.

The study results have indicated that the readiness of faculty at educational institutions has positive impacts on the management of the covid-19 pandemic. The study posits that the readiness of the academic staff at school, colleges, or universities for delivering online classes, enables the institution to arrange for delivering online classwork and thus continue their teaching classes. The spreading of the covid-19 pandemic in the country forces the government and the general public alike to close actual teaching or learning classes. In such a situation, the readiness of faculty for online classes helps to mitigate the adverse impacts of a covid-19 pandemic on education. These results are in line with the past study of Cutri et al. (2020). This elaborates that when the teaching staff at any educational institution have the motivation and willingness to prepare, effectively design, and assists courses within an online environment, and they have an interest in delivering online classes, sufficient relevant knowledge, and sufficient experience regarding the delivery of online classes, they can overcome the adverse impacts of covid-19. These results are also supported by the previous study of Budur et al. (2021), the teachers having prior knowledge, motivation, and experience in the online teaching class, better know how to design the coursework, arrange for online classes, maintain the students' attention and their attendance online, can deliver knowledge in a better way, and have the same good results from the student as may have during actual teaching classes. These results are also in line with the recent study of Sunarto (2021), which states that when the teachers have the proper knowledge about the online teaching methods, the essentials of online teaching, skills of student interaction, their management during online classes, and time management, they can the problem closure of educational institutions in result of covid-19.

The study results have also indicated that students' readiness for having online learning classes has a positive association with the management of covid-19 pandemic disaster. These results are in line with the past study of Yates et al. (2021). This study analyzes the students' role in education when covid-19 is at the peak. It states that the students who have some prior knowledge about the online classes and the devices, technology, and apps used in this context are more likely to be interested in and benefit from online learning, which helps them cover the education gap which covid-19 has brought. These results are also supported by the recent theoretical work out of Zhu and Liu (2020), that during the spreading of the covid-19 pandemic when most of the educational institutions have been closed for health safety, the mostly those students continue their study and thus, fight against the covid-19 pandemic disaster who have interest in the online education, knowledge, experience, and financial resources which are helpful in getting an online education. These results are also in line with the study of Dong et al. (2020), which elaborates that the student readiness to join and learn from online classes helps overcome the covid-19 disaster.

The current study has both theoretical and empirical implications. This study has great theoretical importance on account of its remarkable contribution to the literature on the Covid-19 pandemic. This study addresses the three essential factors of online education, such as information technology readiness, the readiness of faculty at educational institutions to deliver online classes, and the readiness of students for online learning and their impacts on the management of the covid-19 pandemic. The pandemic covid-19 not only effects the health of people but also affects their activities; it disturbs education and makes the regular actual classes impossible to be carried on at the educational institution. In such a situation, it becomes essential for educational institutions to carry online teaching classes. Since the start of the spreading of covid-19, many authors have dealt with the impacts of a covid-19 pandemic on the people's health, resulting in the close of activities at educational institutions along with other economic and social sectors, and the efforts of the management of covid-19. These studies have either checked the role of students' readiness for the online classes, or teachers' readiness to deliver online classes or the readiness of information technology in managing the covid-19 pandemic. This study proves to be an exception to the literature because it examines the role of information technology readiness, the readiness of faculty at educational institutions, and students' readiness for online classes to manage the covid-19 pandemic disasters simultaneously. This study has a great empirical significance in all the countries across the world in general and in developed countries like UAE in particular because it's a guideline for the education authority and any education institution management how to manage the covid-19 pandemic. This study suggests that the covid-19 pandemic disaster and its impacts on education can be overcome with the information technology readiness, readiness of faculty at education institution for online teaching, and the student's readiness for online learning.

6 Conclusion and Limitations

The current study analyzes the covid-19 pandemic disaster and its management in education institutions in UAE, a developed state of the world encompassing seven countries. In this regard, the study checks the role of three essential factors the information technology readiness, the readiness of faculty at education institutions for online teaching, and the student's readiness for online learning. The study posits that the readiness and effectiveness of information technology to be used in online classes is helpful for the educational institutions to cope with the disasters of covid-19. When all the components of the information technology such as computer hardware and software applications, digital devices, infrastructure, communication network, availability of fast internet, techniques, and processes to acquire, process, and exchange knowledge and information, makes it possible to overcome the negative impacts of covid-19 disaster on education by assisting online classes. The study suggests that the readiness of faculty at school, colleges, and universities for suddenly turning from actually teaching classes to online classes, is helpful to the organization to mitigate impacts of the covid-19 disaster. When the teachers have the motivation and willingness to prepare, effectively design, and assist course design within an online environment, have sufficient knowledge and experience to deliver online lectures, they can reduce the adverse impacts of covid-19. Similarly, the student's readiness for having online learning classes assists in managing the covid-19 pandemic. The organizations where the students already have initial knowledge of online classes, proper facility of online classes like computer system, relevant software apps, good infrastructure facilities, proper speaker, camera, room space, and time management skills, can effective manage the covid-19 disaster with the continuity of education.

Though the current study has great theoretical and empirical importance, it bears some limitations too, which are expected to be removed in the future by talented scholars. It examines only three factors, such as information technology readiness, the readiness of faculty at education institutions for online teaching, and the student's readiness for online learning while analyzing how to cope with the covid-19 disastrous effects on education. Besides these factors, there are several other factors, like government regulations, SOPs, financial resources of both institutions and parents, the culture of the students' household, teachers' commitment to an organization, and student learning commitment all affect the management of covid-19 pandemic. But all necessary factors have utterly been ignored by the author in this study. Future authors are recommended to pay attention to these factors along with information technology readiness, faculty readiness, and students' readiness for online classes. Though the covide-19 is a universal pandemic, affecting the educational sectors in most of the countries, this study has checked the management of covid-19 disaster with information technology readiness, the readiness of faculty at education institution for online teaching, and the student's readiness for online learning in the economy of UAE. This has made the validity and reliability of the study very limited. Thus, scholars in the future must increase the number of countries under discussion.

References

Ahmad, A., Alshurideh, M. T., Al Kurdi, B. H., & Salloum, S. A. (2021). Factors impacts organization digital transformation and organization decision making during Covid19 pandemic. *The Effect of Corona Virus Disease on Business Intelligence*, 95.

Aissani, R., & Awad ElRefae, G. (2021). Trends of university students towards the use of Youtube and the benefits achieved-field study on a sample of Al Ain University students. *AAU Journal Business Law* مجلة جامعة العين للأعمال والقانون, *5*(1), 26–53.

Akour, I., Baity, M. A. L., & Hamad, A. Q. A. L. (2020). A moderated mediation model of perceived e-learning within the UAE education sector: A practical study. *International Journal of Engineering and Research Technolnology, 9*(3), 541–550.

Akour, I., Alshurideh, M., Al Kurdi, B., Al Ali, A., & Salloum, S. (2021). Using machine learning algorithms to predict people's intention to use mobile learning platforms during the COVID-19 pandemic: Machine learning approach. *JMIR Medical Education, 7*(1), 1–17.

Al Khasawneh, M., Abuhashesh, M., Ahmad, A., Masa'deh, R., & Alshurideh, M. T. (2021a). Customers online engagement with social media influencers' content related to COVID 19. *The Effect of Corona Virus Disease on Business Intelligence, 334*, 404–385.

Al Khasawneh, M., Abuhashesh, M., Ahmad, A., Alshurideh, M. T., Masa'deh, R. (2021b). Determinants of e-word of mouth on social media during COVID-19 outbreaks: An empirical study. *Studies in Systems, Decision and Control, 334*, 347–366.

Al Kurdi, B., Alshurideh, M., & Salloum, S. (2020a). Investigating a theoretical framework for e-learning technology acceptance. *International Journal of Electrical and Computer Engineering, 10*(6), 6484–6496.

Al Kurdi, B., Alshurideh, M., & Al afaishata, T. (2020b). Employee retention and organizational performance: Evidence from banking industry. *Management Science Letters, 10*(16), 3981–3990.

Alameeri, K. A., Alshurideh, M. T., & Al Kurdi, B. (2021). The effect of Covid-19 pandemic on business systems' innovation and entrepreneurship and how to cope with it: A theatrical view. *The Effect of Corona Virus Disease on Business Intelligence, 334*, 275–288.

Al-Dmour, A., Al-Dmour, H., Al-Barghuthi, R., Al-Dmour, R., & Alshurideh, M. T. (2021a). Factors influencing the adoption of e-payment during pandemic outbreak (COVID-19): Empirical evidence. *The Effect of Corona Virus Disease on Business Intelligence, 334*, 154–133.

Al-Dmour, R., AlShaar, F., Al-Dmour, H., Masa'deh, R., & Alshurideh, M. T. (2021b). The effect of service recovery justices strategies on online customer engagement via the role of 'customer satisfaction' during the Covid-19 pandemic: An empirical study. *The Effect of Corona Virus Disease on Business Intelligence, 334*, 346–325.

AlHamad, M., Akour, I., Alshurideh, M., Al-Hamad, A., Kurdi, B., & Alzoubi, H. (2021). Predicting the intention to use google glass: A comparative approach using machine learning models and PLS-SEM. *International Journal of Data Network Science, 5*(3), 311–320.

Al-Hamad, M., Mbaidin, H., AlHamad, A., Alshurideh, M., Kurdi, B., & Al-Hamad, N. (2021). Investigating students' behavioral intention to use mobile learning in higher education in UAE during Coronavirus-19 pandemic. *International Journal of Data Network Science, 5*(3), 321–330.

AlHamad, A. Q., & Al Qawasmi, K. I. (2014). Building an ethical framework for e-learning management system at a university level. *Journal of Engineering in. Economic Development, 1*(1), 11.

AlHamad, A. Q. M., & AlHammadi, R. A. (2018). Students' perception of E-library system at Fujairah University. In *International Conference on Remote Engineering and Virtual Instrumentation* (pp. 659–670).

AlHamad, A. Q., Yaacob, N., & Al-Omari, F. (2012). Applying JESS rules to personalize learning management system (LMS) using online quizzes. In *2012 15th International Conference on Interactive Collaborative Learning (ICL)* (pp. 1–4).

Alhashmi, M., & Moussa-Inaty, J. (2021). Professional learning for Islamic education teachers in the UAE. *British Journal of Religious Education, 43*(3), 278–287.

Aljumah, A., Nuseir, M. T., & Alshurideh, M. T. (2021a). The impact of social media marketing communications on consumer response during the COVID-19: Does the brand equity of a university matter. *The Effect of Corona Virus Disease on Business Intelligence, 334*, 384–367.

Aljumah, A. I., Nuseir, M. T., & Alam, M. M. (2021b). Organizational performance and capabilities to analyze big data: Do the ambidexterity and business value of big data analytics matter? *Business Process Management Journal, 27*(4), 1088–1107.

Aljumah, A. I., Nuseir, M. T., & Alam, M. M. (2021c). Traditional marketing analytics, big data analytics and big data system quality and the success of new product development. *Business Process Management Journal, 27*(4), 1108–1125.

Al-Maroof, R. S., Alshurideh, M. T., Salloum, S. A., AlHamad, A. Q. M., & Gaber, T. (2021a). Acceptance of google meet during the spread of coronavirus by Arab university students. *Informatics, 8*(2), 24.

Al-Maroof, R., et al. (2021b). The acceptance of social media video for knowledge acquisition, sharing and application: A comparative study among YouYube users and TikTok users' for medical purposes. *International Journal of Data Network Science, 5*(3), 197–214.

Alomari, K. M., Alhamad, A. Q., Mbaidin, H. O., & Salloum, S. (2019). Prediction of the digital game rating systems based on the ESRB. *Opcion, 35*(Special Issue 19).

Alshamsi, A., Alshurideh, M., Kurdi, B. A., & Salloum, S. A. (2021). The influence of service quality on customer retention: A systematic review in the higher education. In *Advances in Intelligent Systems and Computing* (vol. 1261, pp. 404–416). AISC.

Alsharari, N. M., & Alshurideh, M. T. (2020). Student retention in higher education: The role of creativity, emotional intelligence and learner autonomy. *International Journal of Educational Management, 35*(1), 233–247.

Alshurideh, M. (2014). The factors predicting students' satisfaction with universities' healthcare clinics' services: A case-study from the Jordanian higher education sector. *Dirasat: Administrative Sciences, 161*(1524), 1–36.

Alshurideh, M. T., et al. (2021a). Factors affecting the use of smart mobile examination platforms by universities' postgraduate students during the COVID 19 pandemic: An empirical study. *Informatics, 8*(2), 32.

Alshurideh. (2017). A theoretical perspective of contract and contractual customer-supplier relationship in the mobile phone service sector. *International Journal of Business Management, 12*(7), 201–210.

Alshurideh, M. N., & Xiao, S. (2012). The effect of previous experience on mobile subscribers' repeat purchase behavior. *European Journal of Social Sciences, 30*(3), 366–376.

AlShurideh, M., Alsharari, N. M., & Al Kurdi, B. (2019). Supply chain integration and customer relationship management in the airline logistics. *Theoretical Economic Letters, 9*(02), 392–414.

Alshurideh, M., Salloum, S. A., Al Kurdi, B., & Al-Emran, M. (2019a). Factors affecting the social networks acceptance: An empirical study using PLS-SEM approach. In *8th International Conference on Software and Computer Applications* (pp. 1–5).

Alshurideh, M., Salloum, S. A., Al Kurdi, B., Monem, A. A., & Shaalan, K. (2019b). Understanding the quality determinants that influence the intention to use the mobile learning platforms: A practical study. *International Journal of Interactive Mobile Technologies, 13*(11).
Alshurideh, M., Al Kurdi, B., & Salloum, S. A. (2019c). Examining the main mobile learning system drivers' effects: A mix empirical examination of both the expectation-confirmation model (ECM) and the technology acceptance model (TAM). In *International Conference on Advanced Intelligent Systems and Informatics* (pp. 406–417).
Alshurideh, M., Al Kurdi, B., Salloum, S. A., Arpaci, I., & Al-Emran, M. (2020a). Predicting the actual use of m-learning systems: A comparative approach using PLS-SEM and machine learning algorithms. *Interactive Learning Environment*, 1–15.
Alshurideh, M., Gasaymeh, A., Ahmed, G., Alzoubi, H., & Al Kurd, B. (2020b). Loyalty program effectiveness: Theoretical reviews and practical proofs. *Uncertain Supply Chain Management, 8*(3), 599–612.
Alshurideh, M. T., Hassanien, A. E., & Masa'deh, R. (2021b). *The Effect of Coronavirus Disease (COVID-19) on Business Intelligence*. Springer.
Alshurideh, M. T., Al Kurdi, B., & Salloum, S. A. (2021c). The moderation effect of gender on accepting electronic payment technology: A study on United Arab Emirates consumers. *Review of International and Business Strategy*.
Amarneh, B. M., Alshurideh, M. T., Al Kurdi, B. H., & Obeidat, Z. (2021). The impact of COVID-19 on E-learning: Advantages and challenges. In *The International Conference on Artificial Intelligence and Computer Vision* (pp. 75–89).
Ashour, S. (2020). Quality higher education is the foundation of a knowledge society: Where does the UAE stand? *Quality in Higher Education, 26*(2), 209–223.
Baticulon, R. E., et al. (2021). Barriers to online learning in the time of COVID-19: A national survey of medical students in the Philippines. *Medical Science Educator, 31*(2), 615–626.
Batubara, B. M. (2021). The Problems of the World of Education in the Middle of the Covid-19 Pandemic. *Budapest International Research and Critics Institute: Humanities and Social Science, 4*(1), 450–457.
Bettayeb, H., Alshurideh, M. T., & Al Kurdi, B. (2020). The effectiveness of mobile learning in UAE universities: A systematic review of Motivation, Self-efficacy, usability and usefulness. *International Journal of Control and Automation, 13*(2), 1558–1579.
Blayone, T. J. B., Mykhailenko, O., Usca, S., Abuze, A., Romanets, I., & Oleksiiv, M. (2020). Exploring technology attitudes and personal–cultural orientations as student readiness factors for digitalised work. *Higher Education, Skills and Work-Based Learning, 11*(3), 649–671.
Budur, T., Demir, A., & Cura, F. (2021). University readiness to online education during Covid-19 pandemic. *International Journal of Social Science and Educational Studies, 8*(1), 180–200.
Chang, T.-Y., et al. (2021). Innovation of dental education during COVID-19 pandemic. *Journal of Dental Sciences, 16*(1), 15–20.
Chick, R. C., et al. (2020). Using technology to maintain the education of residents during the COVID-19 pandemic. *Journal of Surgical Education, 77*(4), 729–732.
Cutri, R. M., Mena, J., & Whiting, E. F. (2020). Faculty readiness for online crisis teaching: Transitioning to online teaching during the COVID-19 pandemic. *European Journal of Teacher Education, 43*(4), 523–541.
Daniel, J. (2020). Education and the COVID-19 pandemic. *Prospects, 49*(1), 91–96.
Dong, C., Cao, S., & Li, H. (2020). Young children's online learning during COVID-19 pandemic: Chinese parents' beliefs and attitudes. *Children and Youth Services Review, 118*, 105440.
Dray, B. J., Lowenthal, P. R., Miszkiewicz, M. J., Ruiz-Primo, M. A., & Marczynski, K. (2011). Developing an instrument to assess student readiness for online learning: A validation study. *Distance Education, 32*(1), 29–47.
Elsaadani, M. A. (2020). Investigating the effect of e-management on customer service performance. *AAU Journal of Business Law* مجلة جامعة العين للأعمال والقانون, *3*(2), 1.

Elshamy, A. M., Abdelghany, M. A., Alhamad, A. Q., Hamed, H. F. A., Kelash, H. M., & Hussein, A. I. (2017). Secure implementation for video streams based on fully and permutation encryption techniques. *International Conference on Computer and Applications (ICCA), 2017*, 50–55.

Hair, J. F., Jr., Babin, B. J., & Krey, N. (2017). Covariance-based structural equation modeling in the journal of advertising: Review and recommendations. *Journal of Advertising, 46*(1), 163–177.

He, W., Zhang, Z. J., & Li, W. (2021). Information technology solutions, challenges, and suggestions for tackling the COVID-19 pandemic. *International Journal of Information Management, 57*, 102287.

Her, M. (2020). How is COVID-19 affecting South Korea? What is our current strategy? *Disaster Medicine and Public Health Preparedness, 14*(5), 684–686.

Kurdi, B., Alshurideh, M., & Alnaser, A. (2020b). The impact of employee satisfaction on customer satisfaction: Theoretical and empirical underpinning. *Management Science Letters, 10*(15), 3561–3570.

Kurdi, B. A., Alshurideh, M., Salloum, S. A., Obeidat, Z. M., & Al-dweeri, R. M. (2020a). An empirical investigation into examination of factors influencing university students' behavior towards e-learning acceptance using SEM approach. *International Journal of Interactive Mobile Technologies, 14*(2).

Leo, S., Alsharari, N. M., Abbas, J., & Alshurideh, M. T. (2021). From offline to online learning: A qualitative study of challenges and opportunities as a response to the COVID-19 pandemic in the UAE higher education context. *The Effect of Corona Virus Disease on Business Intelligence, 334*, 203–217.

Liguori, E., & Winkler, C. (2020). *From offline to online: Challenges and opportunities for entrepreneurship education following the COVID-19 pandemic*. SAGE Publications Sage CA.

Martono, M., Satino, S., Nursalam, N., Efendi, F., & Bushy, A. (2019). Indonesian nurses' perception of disaster management preparedness. *Chinese Journal of Traumatology, 22*(1), 41–46.

McMinn, M., Dickson, M., & Areepattamannil, S. (2020) Reported pedagogical practices of faculty in higher education in the UAE. *Higher Educational*, 1–16.

Moralista, R., & Oducado, R. M. (2020). Faculty perception toward online education in higher education during the coronavirus disease 19 (COVID-19) pandemic. SSRN 3636438.

Nawaz, M. A., Seshadri, U., Kumar, P., Aqdas, R., Patwary, A. K., & Riaz, M. (2021). Nexus between green finance and climate change mitigation in N-11 and BRICS countries: Empirical estimation through difference in differences (DID) approach. *Environmental Science and Pollution Research, 28*(6), 6504–6519.

Nuseir, M. T., Ghaleb, A., & Aljumah, A. (2021b). The e-learning of students and university's brand image (Post COVID-19): How successfully Al-Ain university have embraced the paradigm shift in digital learning. *The Effect of Corona Virus Disease on Business Intelligence, 334*, 171–187.

Nuseir, M. T., Aljumah, A., & Alshurideh, M. T. (2021a). How the business intelligence in the new startup performance in UAE during COVID-19: The mediating role of innovativeness. In *Effective Coronavirus Disease Business Intelligence* (pp. 63–79).

Nuseir, M. T., Al Kurdi, B. H., Alshurideh, M. T., & Alzoubi, H. M. (2021c). Gender discrimination at workplace: Do artificial intelligence (AI) and machine learning (ML) have opinions about it. In *The International Conference on Artificial Intelligence and Computer Vision* (pp. 301–316).

Petko, D., Prasse, D., & Cantieni, A. (2018). The interplay of school readiness and teacher readiness for educational technology integration: A structural equation model. *Computers in the Schools, 35*(1), 1–18.

Pokhrel, S., & Chhetri, R. (2021). A literature review on impact of COVID-19 pandemic on teaching and learning. *Higher Education for the Future, 8*(1), 133–141.

Qasim AlHamad, A., Salameh, M., & Al Makhareez, L. (2014). Identifying students' trends toward personalizing learning management system (LMS) at Zarqa University (Extended). *Journal of Information Technology and Economic Development, 5*(1).

Rafique, G. M., Mahmood, K., Warraich, N. F., & Rehman, S. U. (2021). Readiness for online learning during COVID-19 pandemic: A survey of Pakistani LIS students. *The Journal of Academic Librarianship, 47*(3), 102346.

Rose, S. (2020). Medical student education in the time of COVID-19. *JAMA, 323*(21), 2131–2132.

Salloum, S. A., Alshurideh, M., Elnagar, A., & Shaalan, K. (2020). *Mining in educational data: Review and future directions* (vol. 1153). AISC.

Shah, S. F., Alshurideh, M. T., Al-Dmour, A., & Al-Dmour, R. (2021). Understanding the influences of cognitive biases on financial decision making during normal and COVID-19 pandemic situation in the United Arab Emirates. *The Effect of Corona Virus Disease on Business Intelligence, 334*, 274–257.

Sittig, D. F., & Singh, H. (2020). COVID-19 and the need for a national health information technology infrastructure. *JAMA, 323*(23), 2373–2374.

Sultan, R. A., Alqallaf, A. K., Alzarooni, S. A., Alrahma, N.H., AlAli, M. A., & Alshurideh, M. T. (2021). How students influence faculty satisfaction with online courses and do the age of faculty matter. In *The International Conference on Artificial Intelligence and Computer Vision* (pp. 823–837).

Sun, H., Awan, R. U., Nawaz, M. A., Mohsin, M., Rasheed, A. K., & Iqbal, N. (2021). Assessing the socio-economic viability of solar commercialization and electrification in south Asian countries. *Environment, Development and Sustainability, 23*(7), 9875–9897.

Sunarto, M. J. (2021). The readiness of lecturers in online learning during the Covid-19 pandemic at the faculty of information technology and the faculty of economics and business. *International Journal of Recent Educational Research, 2*(1), 54–63.

Taryam, M., Alawadhi, D., & Aburayya, A. (2020). Effectiveness of not quarantining passengers after having a negative COVID-19 PCR test at arrival to Dubai airports. *Systematic Reviews in Pharmacy, 11*(11), 1384–1395.

Theoret, C., & Ming, X. (2020). Our education, our concerns: The impact on medical student education of COVID-19. *Medical Education, 54*(7), 591–592.

Urbaczewski, A., & Lee, Y. J. (2020). Information technology and the pandemic: A preliminary multinational analysis of the impact of mobile tracking technology on the COVID-19 contagion control. *European Journal of Information Systems, 29*(4), 405–414.

Whitelaw, S., Mamas, M. A., Topol, E., & Van Spall, H. G. C. (2020). Applications of digital technology in COVID-19 pandemic planning and response. *The Lancet Digital Health, 2*(8), e435–e440.

Yan, A., Zou, Y., & Mirchandani, D. A. (2020). How hospitals in mainland China responded to the outbreak of COVID-19 using information technology–enabled services: An analysis of hospital news webpages. *Journal of the American Medical Informatics Association, 27*(7), 991–999.

Yates, A., Starkey, L., Egerton, B., & Flueggen, F. (2021). High school students' experience of online learning during Covid-19: The influence of technology and pedagogy. *Technology, Pedagogy and Education, 30*(1), 59–73.

Zhu, X., & Liu, J. (2020). Education in and after Covid-19: Immediate responses and long-term visions. *Postdigital Science Education, 2*(3), 695–699.

Evaluation of Blended E-learning from the Perspectives of the German Jordanian University Students

Safa Shweihat

Abstract To achieve e-learning sustainability, the current study aimed to investigate the efficiency of blended e-learning. The researcher developed a special questionnaire consisting of (50) items and two open questions, implemented on a random sample of 50% of the students studying the National Education Course at the German-Jordania University in 2021. The results showed that the total efficiency degree of blended E-learning was moderate: "Electronic Exams" ranked first, "Teaching through Videos" ranked second, "Technical use of the electronic platform" ranked third, "Appropriateness of blended e-learning" ranked four, and "Students' regular study through the electronic platform" ranked last. There were significant differences in favor of the students who dealt seriously with e-learning, with no significant differences in gender, experience, and academic year. The results also demonstrated several advantages and disadvantages of blended e-learning to achieve e-learning efficiency and sustainability, especially for the first-year university students of social studies and humanities. The recommendations focus on some skills that help students be integrated into e-learning, the teaching methods, technological infrastructure, and the model of the optimal education time distribution between e-learning and traditional learning.

Keywords Effectiveness · Blended E-learning · Advantages · Disadvantages · National education course · Students · German Jordanian university

1 Introduction

Like other countries, Jordan has suffered from the problematic learning of its various forms under the COVID-19 pandemic, which imposed using e-learning urgently, rapidly, and suddenly, regardless of educators' and learners' degree of experience skill, and knowledge of e-learning. It has become obligatory to employee-learning

---

S. Shweihat (✉)
School of Basic Sciences and Humanities, German Jordanian University, Amman, Jordan
e-mail: safa.shweihat@gju.edu.jo

M. Alshurideh et al. (eds.), *The Effect of Information Technology on Business and Marketing Intelligence Systems*, Studies in Computational Intelligence 1056,
https://doi.org/10.1007/978-3-031-12382-5_19

and relies on it entirely or partially in university education. So it is necessary to conduct evaluation studies to raise the efficiency of education reality in line with the challenges and obstacles of e-learning.

1.1 Theoretical Framework and Previous Studies

Technological progress has contributed to serving man in all aspects of life, especially in education, in terms of rapid performance, accuracy, and sum-up of time and effort. However, the interest rate of both faculty members and students in e-learning remains minimal, with e-mail, audio, and video broadcasting of lectures being of its most minor uses (Al-Abdallah, 2011), especially in the developing countries, who used to focus on face-to-face education, as some categorically reject the idea of online teaching based on many reasons. Moreover, the crisis of the COVID-19 pandemic obliged all these institutions to switch to virtual teaching (Dhawan, 2020).On the other hand, e-learning has widely spread as one of the modern forms of learning, especially in higher education institutions in America, Europe, and other countries that totally or partially rely on this type of education. Recently, e-learning has become a contributor to the income of some universities. For example, "The payment of Phoniex University in the USA from e-learning is (95.5%) of its total income of (1.28) billion dollars in 2004 (Qudhah, 2013). When looking for a competitive advantage for e-learning, the Polish Higher Education System created a legal basis and rules for the use of e-learning, based on the experience of John Paul II's Pontifical University in Krakow (Górska, 2016). Indeed, the difference in judging the spread of e-learning in its various forms is due to the different positioning of countries. Inevitably, the developing countries differ in capabilities and conditions from the developed countries, especially in technology production and use.

From the point of view of Carlines, e-learning is "Learning that occurs through the computer and any other computer-based resources that help in the teaching and learning process. The computer replaces the book or the teacher or both of them, where the educational material is displayed on the screen and the learner is presented the study material based on their response. The scientific material and accompanying tests can be simple, taking the form of an educational program, texts, static or animated graphics, audios or visuals, or all together. E-learning may consist of a course that includes lectures taking place through video conferencing at specific dates, precisely as in the case of traditional education. It can be a page on the internet accompanied by additional material that includes video activities, discussion outside the classroom via e-mail, and electronic tests, the results of which are automatically recorded in the student records (El-Madawy, 2021).

The American Federation for Distance Education has identified the advantages of e-learning that depends on the information network (the internet), as it provides the possibility of moving from a knowledge transfer model to a learner-directed education model, providing reciprocal communication, and encouraging the learner to participate in building knowledge through relying on higher-order thinking skills, as well as providing learners with participatory and interactive learning and teaching. On the other hand, e-learning may deprive students or learners of learning essential skills such as listening, handwriting, interaction with colleagues, speaking, dialogue, discussion, imitation, debate, simulation, and others (Abu-Aqail, 2014).

It is certain that activating e-learning, benefiting from its advantages, and compensating for the loss of face-to-face education is inevitably related to the experience and skill of those who lead the education process. It is not easy to attract students' attention and motivate them to learn while considering individual differences and their psychological state, and the requirements of people with special needs. Therefore, (Martin, 2020) states that "It is necessary to strengthen online student learning by providing clear instructions to the learner and selecting a well-organized and sequential study content divided into simple parts for the student to understand easily."

2 Literature Review

2.1 Field Studies That Aimed to Investigate the Obstacles and Challenges of E-Learning from Students' Perspectives

The results of a study carried out by Al-Issan (2007) about the perspective of 165 students who study e-courses at the Faculty of Education/Sultan Qaboos University indicated that one of the advantages of e-learning is activating cooperative learning among students, as well as narrowing the gap between student-instructor, and student-student. In addition to giving students the liberty to express their ideas and adequate opportunities to demonstrate their abilities and potentials through student participation, dialogue, and continuous scientific discussions. E-learning also helped in developing their computer skills.

The disadvantages of e-learning are the lack of and the insufficient number of computers for students in the College, and facing difficulties in connecting to the University website from remote places in Oman.

Abu-Aqail (2014) carried out a study on a random stratified sample of 404 students from Hebron University in Palestine. The results indicated that the number of technicians is not so enough for helping all students on how to deal with e-learning,students lack the knowledge of up-to-date technologies; the plentiful of courses students have at one semester, and the absence of e-learning workshops. There are statistical differences in estimating the size of obstacles in favor of females among the first and second-year students of the Faculties of Sharia and Administration.

2.2 Field Studies That Aimed to Investigate the Obstacles of E-Learning from the Instructors' Perspective

Mefleh al-Dahoun (2010) carried out a study on a random sample consisting of 47 male and 58 female teachers selected from the secondary stage teachers et al.-Kurah district in Jordan. The results indicated that the teacher-related obstacles ranked first, followed by administration-related obstacles, then infrastructure and equipment-related obstacles, and finally student-related obstacles. The results showed significant statistical differences in infrastructure and essential equipment in favor of male teachers and teachers with a master's degree and higher.

In a study carried out by Aldhafeeri (2017), a questionnaire was implemented on 1413 faculty members at Kuwait University in Kuwait State. The results show that academic staff at Kuwait University agreed on the necessity of implementing E-learning platforms into their teaching courses. In addition, the study found out that there were some cultural and technical challenges behind the reluctance to use e-learning platforms.

A study was carried out by Qudhah (2013) to identify e-learning challenges perceived by faculty members at the universities in Jordan. A random sample consisting of 113 faculty members from the Jordanian Universities participated in the study. The results indicated that the most critical challenges of e-learning are: scientific research, technical, financial, and administrative challenges, planning, design, and evaluation of e-learning, and no significant differences in challenges attributed to gender, experience, and College.

Al-Abdallah (2011) carried out a field study on a random sample of (113) faculty members and (774) male and female students at Tishreen University/Syria. The results indicated that the most important disadvantages of e-learning are: Sitting for a long time in front of a computer causes many diseases and the unavailability of rooms dedicated to e-learning. There were no significant differences between the mean grades of faculty members in terms of E-learning disadvantages according to the specialization variable. At the same time, there were differences in terms of the advantages and disadvantages of e-learning according to the student specialization variable.

Within the context of the studies that used the experimental approach, Oweis (2018) carried out a study to measure the effect of blended learning on achievement and motivation to learn the English language among the students of the German Jordanian University. An analysis of common variance (ANCOVA) showed statistical differences in student achievement as well as significant differences in the motivation to learn the English language in favor of the experimental group who studied the English language through a computer program mixed with the traditional method (Blended online learning) compared to the control group who studied the English language in the conventional way only. Moreover, a study carried out by Mefleh al-Dahoun (2010) showed a difference in the means of students' scores in the tests of mathematical power and reflective thinking skills in favor of the experimental group of 90 8th grade students in Jordan who used the e-learning platforms compared with those who used the traditional method.

The study of Al-Fraihat et al. (2020) aimed to evaluate e-learning systems as they are vital to ensuring effective use, successful delivery, and positive effects on learners. After reviewing some previous studies, the researchers developed a comprehensive model that provides a comprehensive picture and identifies different levels of success related to a wide range of determinants of success. By matching the model to data collected from 563 students participating in an e-learning system at a UK university, the model was experimentally validated by the Partial Least Squares Quantitative Method-Structured Equation Modeling (PLS-SEM). The results showed that the determinants of tangible e-learning satisfaction are information quality, technical system quality, support system quality, service quality, teacher quality, learner quality, and perceived usefulness, and these together explain 71.4% of the perceived satisfaction variance. In contrast, the drivers of perceived benefit are information quality, support system quality, technical system quality, teacher quality, and learner quality, and these explain 54.2% of the perceived benefit variance. Whereas four structures were found as determinants of e-learning use: the quality of the support system, the quality of the educational system, the quality of the learner, and the perceived benefit, together they represent 34.1%. Finally, 64.7% of the variance in e-learning benefits was explained by satisfaction, use, and benefit perception.

The study of Al Kurdi et al. (2020a) aimed to determine the factors that affect the acceptance of e-learning by students and to know the intention of students to use e-learning and how these factors are determined. The theoretical framework was developed based on the Technology Acceptance Model (TAM). The sample consisted of 270 university students who used the e-learning system, and a questionnaire was distributed to them. The results showed that "perceived pleasure, social impact, perceived usefulness, self-efficacy, and perceived ease of use" are the strongest and most significant predictors of students' intention towards e-learning systems. The research results provided valuable recommendations that serve as guidelines for the effective design of e-learning systems, and guidance for e-learning practices.

Al Kurdi et al, (2020b) conducted a study to validate the process of accepting and employing e-learning by university students in the United Arab Emirates. The study included a sample of 365 university students, and the structural equation modeling (SEM) method was used based on the Technology Acceptance Model (TAM). This standard structural model included computer self-efficacy, e-learning, system interaction, social impact, computer anxiety, enjoyment, technical support, perceived ease of use, perceived benefit, attitude, and behavioral intent to use e-learning. The results showed that understanding the acceptance of e-learning by users was done using (TAM), a theoretical tool that has proven to be effective. Computer self-efficacy, e-learning, system interaction, computer anxiety, social influence, enjoyment, technical support, perceived ease of use, attitude, and perceived usefulness, followed by behavioral intention to use e-learning, was the most critical construct to explain the causal process used in the model. The findings provide professionals, decision-makers, and developers with practical implications on how effective e-learning systems are.

The results of some recent field studies carried out during -COVID-19 pandemic showed a lack of financial support for employing e-learning, a lack of computers for some students at homes, with students and parents rejecting the e-learning culture (Aldhafeeri, 2017), and that the students suffered from the delay in sending the educational material, and delay in the arrival of educational videos (Chen et al., 2020). In addition, the poor internet connection, especially in remote areas, and the inappropriate study environment is not suitable due to not having separate rooms for them to study remotely from within their homes and across the educational platforms (Saleh, 2019). Students encountered significant challenges upon transition to e-learning during the COVID-19 pandemic, especially since the majority of students around the world do not have the capabilities that enable them to acquire the equipment that helps them attend their virtual classes, and that access to the internet in many regions of the world is difficult to almost impossible (Dhawan, 2020).

The limitations of writing mathematical symbols and the limited basic capabilities of the Learning Management System and multimedia software to support online learning represent other technical obstacles that learners and teachers encounter (Irfan, 2020).

The results of a study (Chen et al., 2020) showed that using e-learning platforms improves students' achievement and improves their level of satisfaction with online learning. The study of Kapadia (2020) found out that there is a large degree of positive impact of the use of distance education and that there are no differences according to the variable of gender. Irfan (2020d) concluded that the overall score for assessing the reality of distance education was moderate, with statistically significant differences in favor of females, with no differences depending on the variable of the educational stage. The results of a study conducted by Abuhassan (2020a) also showed that the degree of both the use of the e-learning management system (Moodle) and the obstacles to using it was moderate and that the trends towards the (Moodle) system were positive. The results of a study conducted by Meqdadi (2020) indicated that e-learning was rated high from the perspective of a random sample that included (120) students from the College of Education at Taibah University in Saudi Arabia, with statistically significant differences in favor of males and the effect of the interaction between gender and the study location.

In general, to the researcher's knowledge, it was noted that most of the previous studies adopted the quantitative research methodology, using the questionnaire, such as the studies carried out by Irfan (2020), Kapadia (2020), Meqdadi (2020). Meanwhile, there is a scarcity of using the qualitative research methodology except for Saleh (2019). There is also a scarcity of studies that combine quantitative and qualitative research. Previous studies dealt with evaluating e-learning in general, except for two studies that evaluated e-learning in specific study subjects, such as the studies carried out by Qudhah (2013) and Saleh (2019), in addition to the scarcity of studies that dealt with blended education except for the study carried out by Qudhah (2013).

The current study has benefited from the previous studies. However, it is distinguished by being the first to evaluate the effectiveness of blended education and its suitability for teaching the course of "National Education" through blended education for students of Jordanian universities on "Edraak" Platform, such as the University of

Jordan, the Hashemite University, Israa University, and LuminusTechnical University College. This means that it is possible to circulate the results of this study to the universities above to benefit from. This study is also considered the first study to evaluate the "Edraak" Platform, as far as the researcher's knowledge is concerned. The comprehensiveness of its five domains also characterizes the study. In addition to the variables used by previous studies, such as gender, experience, and academic level, the current study included the variable of seriousness in dealing with e-learning because the study is applied in a developing society to which E-learning is new. So there are many doubts about the seriousness of students' interaction with E-learning because of the current prevailing culture and the fact that a wide range of the Jordanian community rejects this type of education, including teachers, learners, and parents.

2.3 Study Problem

Despite having some positive trends towards employing e-learning in various educational institutions in Jordan for decades, its use was limited to conducting examina tions and monitoring test scores until the requirements of dealing with the COVID 19 pandemic have appeared. The blended e-learning raised problems and negative trends related to inputs, processes, and outputs of e-learning in the time of COVID-19. Academics, researchers, teachers, and students raised problems associated with the low level of required knowledge, experience, and skills. The researcher, being a faculty member in a Jordanian university, has been using blended learning for teaching the course of "National Education" since 2015, and she has an interest in modernizing and updating blended e-learning to keep up with the requirements of the modern age. The current study aims to evaluate the degree of effectiveness of blended E-learning and to investigate the advantages and disadvantages of using blended e-learning from the students' perspective.

2.4 Study Questions

1. What is the degree of effectiveness of blended e-learning for teaching the course of National Education on "Edraak" from the students' perspective?
2. Does the total degree of the effectiveness of blended e-learning for teaching the course of National Education on "Edraak" differ from the students' perspective, in terms of students' gender, academic level, previous experience, and seriousness in dealing with e-learning?
3. What are the advantages and disadvantages of using blended e-learning from the students' perspective?
4. What are the ways to develop the blended e-learning of the National Education course on "Edraak" from the students' perspectives?

2.5 Study Significance

The procedural significance of the current study stems from the importance of uncovering the degree of effectiveness of blended e-learning, in general, and its degree of relevance in teaching the course of "National Education" as a university requirement in Jordanian public and private universities. The current study results benefit both faculty members at Jordanian universities, the National Education curriculum developers, and the technicians responsible for developing Edraak's E-learning platform.

2.6 Procedural Concepts

Effectiveness Degree: The general mean of the study sample's estimates of the degree of effectiveness of blended e-learning, which is achieved by calculating the frequencies and percentages of the responses of the current study sample to the items of the questionnaire prepared for that purpose.

"National Education" Course: A university requirement approved by the Ministry of Higher Learning and Scientific Research in 2005 to teach three credit hours for all undergraduate students at all the Jordanian universities to form a common national and citizenship base for students according to the specifications of a good Jordanian citizen. The course consists of two main aspects: the national part, which focuses on the theoretical side of knowledge, values, and trend-building. And the citizenship part focuses on the behavioral and skills aspects.

Blended Education: A type of e-learning that mixes the existing traditional teaching methods and e-learning. The traditional part requires a physical presence that provides face-to-face interactive instructor-student meetings in the classrooms within the university campus. This type of education is approved for the practical side, which is concerned with developing scientific research skills in national issues, at a rate of 75 min per week for one time. As for e-learning, it takes place through the electronic platform through which the theoretical aspect of the study material is presented. This includes the study plan file and written files related to the study content and presented in a fragmented, sequential, and organized form, as well as video and audio recordings with key phrases through which instructors explain their courses. The videos and teaching files are displayed within a timeline of a weekly lecture that students follow on the platform, with the possibility of the student selecting the timing, place, and speed of viewing the lectures and following them up. The platform is also used to hold exams and monitor test scores. The platform also contains virtual rooms for dialogue and discussion among students and between students and instructors and other places to display pictures of students' activities related to the National Education course.

Edraak E-learning Platform: An Arabic, massive open online course (MOOC) platform, established as an initiative from Queen Rania Foundation for Education and Development (http://www.edraak.org).

The German Jordanian University (GJU): A public university, established in 2005, according to a Memorandum of Understanding between the Ministries of Higher Education and Scientific Research in Jordan and the Federal Republic of Germany, with an applied technological orientation. The university seeks to take serious steps towards internationalization. It aspires to meet all students' lifelong education and training needs in Jordan and the region. Among GJU policies: the enrollment of university students in the last year of study in German universities in Germany (to spend six months at the universities and six months in fieldwork at different worksites, each according to their academic specialization (http://www.gju.edu.jo).

3 Method

The current study has adopted a quantitative research method because it is appropriate for collecting students' data, identifying their evaluation of the effectiveness degree of blended education in various fields of study, and identifying statistical differences and significance. The qualitative research method was also adopted by asking open questions and analyzing students 'responses. We clarify that by detailed explaining upon talking about the study tool.

3.1 Study Variables

The current study includes four independent variables (gender, academic level, previous experience, and seriousness in dealing with blended e-learning) and one dependent variable: the effectiveness of blended e-learning for teaching the course of "National Education" on the" Edraak" Platform.

3.2 Study Limitations

The current study was carried out on undergraduate students at the German Jordanian University in the Governorate of Madaba, Jordan. They are enrolled to study the National Education course taught by blended e-learning in the first semester of the Academic Year 2020/2021. The generalization of the current study results is determined by the degree of effectiveness of the blended e-learning of the National Education course through the Edraak platform, according to the domains of the current study with the relevant items and open questions, as well as the results of the validity and reliability of the study tool. Furthermore, the study results are circulated to the population of students at the German Jordanian University and similar populations such as the students of other Jordanian universities who study the course of "National Education" through blended e-learning on the "Edraak" platform.

3.3 Study Population and Sample

The study population consisted of undergraduate students enrolled to study the National Education course taught by blended E-learning. They come from different academic specializations at the GJU in Jordan. The study questionnaire was posted on an electronic website and was e-mailed to all students distributed on 12 sections; Fifty percent of the study population responded to the questionnaire. Table 1 illustrates the distribution of the study subjects according to the study variables.

3.4 Study Tool

The literature of the topic and previous studies were reviewed and used to build the current study tool. In its initial form, the study questionnaire consisted of three parts: the first one is student's data, the second is (52) items to measure the degree of effectiveness of blended e-learning according to the Likert five-point scale, and the third is open-ended questions to identify the advantages and disadvantages of blended e-learning, and ways to develop it from the students' perspective.

Table 1 Distribution of the study subjects according to the study variables

Variable	Category	No.	Percentage
Gender	Male	123	49.8
	Female	124	50.2
Academic level	1st Year	146	59.1
	2nd Year	52	21.1
	3rd Year and Higher	49	19.8
Have you ever studied any blended online courses at school or the university?	Yes	148	59.9
	No	99	40.1
Do you write down notes or questions while following upon the course content through videos or reading the material?	Yes	149	60.3
	No	98	39.7

3.5 Subject Validity

The study tool was presented to a group of arbitrators specializing in measurement, evaluation, and social sciences. They were asked to express their opinion on the questionnaire items regarding linguistic integrity and the item's suitability to its domain. Some items were deleted, while others were amended based on the arbitrators' results. The items agreed upon by (80%) or more were adopted. In its final form, the questionnaire consisted of (40) items, divided into five domains: Appropriateness of blended e-learning for the National Education course: (1–9), student follow-up of the subject study through the electronic platform, (10–17), the competence of technical matters(18–26), the effectiveness of the electronic teaching method through the course videos (27–32), and effectiveness of electronic exams on Edraak platform, (33–40).

3.6 Reliability

The reliability of the study instrument was confirmed by using the internal consistency method using the Cronbach-Alpha equation. Table 2 shows the results.

Table 2 shows that the Cronbach Alpha reliability coefficients ranged between (0.916–0.690) and that the overall reliability coefficient of the questionnaire was (0.956). These values are statistically significant and acceptable for scientific research.

3.7 Construct Validity

The questionnaire construct validity was verified by calculating the Pearson Correlation Coefficient between the item's score with the domain to which it belongs on

Table 2 Reliability coefficients using Cronbach Alpha

No.	Domain	Cronbach Alpha value
1	Appropriateness of blended e-learning for the National Education course	0.916
2	Student follow-up the course through Edraak platform	0.835
3	Competence of technical use of Edraak platform	0.690
4	Effectiveness of teaching through the course videos/recordings on the Edraak platform	0.897
5	Efficacy of electronic exams on the Edraak platform	0.891
	The total degree of blended e-learning effectiveness	0.956

Table 3 The values of Pearson correlation coefficients for the questionnaire items with the domain to which they belong and with the total score of the questionnaire

No.	Item correlation with its domain	Item correlation with questionnaire total score	No.	Item correlation with its domain	Item correlation with questionnaire total score
1	0.754**	0.672**	21	0.753**	0.682**
2	0.827**	0.772**	22	0.764**	0.749**
3	0.797**	0.693**	23	0.703**	0.793**
4	0.776**	0.641**	24	0.747**	0.471**
5	0.808**	0.738**	25	0.634**	0.531**
6	0.776**	0.669**	26	0.499**	0.401**
7	0.755**	0.664**	27	0.875**	0.832**
8	0.831**	0.731**	28	0.852**	0.754**
9	0.663**	0.721**	29	0.845**	0.787**
10	0.793**	0.591**	30	0.780**	0.697**
11	0.761**	0.569**	31	0.736**	0.592**
12	0.578**	0.539**	32	0.804**	0.617**
13	0.409**	0.341**	33	0.728**	0.661**
14	0.683**	0.601**	34	0.710**	0.606**
15	0.776**	0.681**	35	0.771**	0.652**
16	0.706**	0.576**	36	0.761**	0.678**
17	0.725**	0.593**	37	0.761**	0.645**
18	0.740**	0.606**	38	0.735**	0.644**
19	0.766**	0.643**	39	0.773**	0.644**
20	0.819**	0.716**	40	0.802**	0.810**

** Significant correlation coefficient at (0.01) level

the one hand and the item score with the total score of the questionnaire on the other hand, as shown in Table 3.

Table 3 shows that the correlation coefficients of the questionnaire items with the domainto which they belong and with the total score of the questionnaire were positive and statistically significant. These values are acceptable for the purposes of the study. This also confirms the structural validity of the questionnaire.

4 Statistical Treatment

To answer the first question of the study questions, the means, standard deviations, rank, and score were used.

To answer the second question of the study questions, four-way ANOVA with the factorial design 2 * 3 * 2 * 2 was used, in addition to the Scheffe test for post hoc comparisons for the academic level variable because it showed statistically significant differences.

The frequencies and percentages were used to indicate the distribution of the study sample according to its variables and the responses to the two open questions. The frequencies of the responses were (203) and (170) to the first and second open questions, respectively. The responses dealing with the disadvantages or advantages were collected with each other, albeit the terms of the respondents are different.

Cronbach-Alpha was used to confirm the reliability of the study instrument through internal consistency.

Pearson correlation coefficient was used to confirm the construct validity of the questionnaire.

Each item of the questionnaire was given a graded weight according to Likert's five-point scale to measure the degree of severity of the obstacles. Very strongly agree (5) degrees, strongly agree (4) degrees, moderate agree (3) degrees, slightly agree (2) degrees, very slightly agree (1) degrees. To determine the cells of Likert five-point scale (the lower and upper limits) range was calculated $(5 - 4 - 1)$ and then divided by the number of cells to obtain the cell length, i.e. $(4 \div 3 = 1.33)$ and then this value was added to the lowest value in the scale $(1.33 + 1 = 2.33)$. Thus, the length of the cells was:—if the mean of the subject responses was (2.33 or less) to a small degree. If it is between (2.34–3.67), it is of a moderate degree, but if the mean is between (3.68–5.00), it is of a large degree.

5 Results

5.1 First Question: What is the Degree of Effectiveness of Blended E-Learning for Teaching the Course of "National Education" on the "Edraak" Platform from the Students' Perspective?

To answer the first question, the means, standard deviations, rank, and the degree of effectiveness of blended e-learning for teaching "National Education" on the " Edraak " Platform from the students' perspective were calculated, as shown in Table 4.

Table 4 indicates that the degree of effectiveness of blended e-learning for teaching "National Education" on the "Edraak" Platform from the students' perspective was moderate in general, where the mean was (3.38) with a standard deviation of (0.76). All the domains of the study instrument were moderate, as the means ranged between (3.21–3.64), with" Exam effectiveness through Edraak platform " ranking first with a mean of (3.64) and standard deviation of (0.93), and "Teaching effectiveness through study material videos/recordings on Edraak platform" ranking second with a mean of

Table 4 Means, standard deviations, rank, and the degree of effectiveness of blended e-learning for teaching "national education" on "Edraak" from the students' perspective arranged in a descending order

No.	Domain	M	SD	R	Effectiveness degree
5	Exam effectiveness through Edraak	3.64	0.93	1	Moderate
4	Teaching effectiveness through study material videos/recordings on the Edraak platform	3.50	0.95	2	Moderate
3	The efficiency of technical use of the Edraak platform	3.35	0.67	3	Moderate
1	Blended e-learning appropriacy to the National Education course	3.25	1.01	4	Moderate
2	Students follow up their courses through the Edraak platform	3.21	0.89	5	Moderate
The total degree of blended e-learning effectiveness		3.38	0.76		Moderate

3.50 and standard deviation of (0.95). Blended E-learning appropriacy to the National Education course" ranked before the last with a mean of (3.25) and standard deviation of (1.01). Finally, "Students follow up their courses through Edraak platform" ranked last with a mean of (3.21) and a standard deviation of (0.89).

The items of each domain were as follows.

5.1.1 First, Exam Effectiveness Through Edraak

Table 5 indicates that the degree of effectiveness of exams on the "Edraak" Platform from the students' perspective was moderate in general, where the mean was (3.64) with a standard deviation of (0.93). All the items of this domain were high and medium, as the means ranged between (3.30–4.15), with item 37 which states that "The multiple-choice questions are appropriate for the blended e-learning" ranking first with a mean of (4.15) and standard deviation of (1.14), and item38 which states that "The exam questions are varied" ranking second with a mean of(3.96) and standard deviation of (1.09). Item 34, which states that "The instructions for carrying out the exam are sufficient and accurate," ranked before the last with a mean of (3.42) and standard deviation of (1.35). Finally, item 40, which states that "I advise other students to study the course of National Education on Edraak platform in the same way I did." ranked last with a mean of (3.30) and a standard deviation of (1.37).

Table 5 Means, standard deviations, ranks, and degree of exam effectiveness through edraak platform from the students' perspective arranged in a descending order

No.	Item	M	SD	R	Effectiveness degree
37	The multiple-choice questions are appropriate for the blended E-learning	4.15	1.14	1	High
38	The exam questions are varied	3.96	1.09	2	High
36	The exam questions are from within the course material	3.71	1.17	3	High
39	The vocabulary of the exam questions is understandable	3.63	1.24	4	Moderate
33	The test guidelines are visible	3.49	1.31	5	Moderate
35	I trust the results of the National Education exam on the Edraak platform	3.46	1.23	6	Moderate
34	The instructions for carrying out the exam are sufficient and accurate	3.42	1.35	7	Moderate
40	I advise other students to study the course of National Education on the Edraak platform in the same way I did	3.30	1.37	8	Moderate
Degree of exam effectiveness through Edraak platform		3.64	0.93		Moderate

5.1.2 Second, Effectiveness of Teaching Through the Course Videos/Recordings on the Edraak Platform

Table 6 indicates that the degree of effectiveness of teaching through the course videos/recordings on the Edraak platform from the students' perspective was moderate in general, where the mean was (3.50) with a standard deviation of (0.95). All the items of this domain were moderate, as the means ranged between (3.30–3.65), with item29 which states that "Adequacy of accompanying phrases that appear on the screen synchronized with teachers' explanation in all video recordings" ranking first with a mean of (3.65) and standard deviation of (1.16), and item27 which states that "Adequacy of the content of explaining the material in the video recordings, including all the content of the course unit" ranking second with a mean of3.57 and standard deviation of (1.13). Item 32 states that "The method of teaching adopted by teachers is at a similar degree of competence." ranked before the last with a mean of (3.45) and standard deviation of (1.15). Finally, item31, which states that "I like

Table 6 Means, standard deviations, ranks, and degree of effectiveness of teaching through the course videos/recordings on edraak platform from the students' perspective arranged in a descending order

No.	Item	M	SD	R	Effectiveness degree
29	Adequacy of accompanying phrases that appear on the screen synchronized with teachers' explanation in all video recordings	3.65	1.16	1	Moderate
27	Adequacy of the content of explaining the material in the video recordings, including all the content of the course unit	3.57	1.13	2	Moderate
30	The teacher delivers the material at an appropriate speed	3.52	1.20	3	Moderate
28	Teaching the concepts of the National Education at a single level on the Edraak platform is within the diversity of teachers in each course unit	3.49	1.13	4	Moderate
32	The teaching method adopted by teachers is at a similar degree of competence	3.45	1.15	5	Moderate
31	I like the diversity of teachers in the recording videos of the National Education course	3.30	1.26	6	Moderate
Degree of the effectiveness of teaching through the course videos/recordings on the Edraak platform		3.50	0.95		Moderate

the diversity of teachers in the recording videos of the National Education course." ranked last with a mean of (3.30) and a standard deviation of (1.26).

5.1.3 Third, Efficiency of Technical Use of Edraak Platform

Table 7 indicates that the degree of efficiency of technical use of the Edraak platform from the students' perspective was moderate in general, where the mean was (3.35) with a standard deviation of (0.67). All the items of this domain were moderate, as the means ranged between (2.74–3.64), with item 23, which states that "The videos cover all the course content" ranking first with a mean of (3.64) and standard deviation of

Table 7 Means, standard deviations, ranks, and degree of efficiency of technical use of Edraak platform from the students' perspective arranged in a descending order

No.	Item	M	SD	R	Efficiency degree
23	The videos cover all the course content	3.64	1.18	1	Moderate
18	It is easy for students to open an account on the electronic platform	3.57	1.26	2	Moderate
21	I am familiar with the terminology used on the platform	3.56	1.14	3	Moderate
22	The course contents are adequately organized on the platform	3.54	1.19	4	Moderate
20	Dealing with the electronic platform is easy	3.47	1.25	5	Moderate
19	The instructions for using the platform are accurate and sufficient	3.42	1.32	6	Moderate
26	Matching the exam score that appears on the screen directly with the student's score report	3.29	1.01	7	Moderate
24	Sometimes I encounter problems with opening an Edraak website	2.95	1.37	8	Moderate
25	I encountered technical problems while taking the exam	2.74	1.48	9	Moderate
Degree of efficiency of technical use of Edraak platform		3.35	0.67		Moderate

(1.18), and item 18 which states that "It is easy for students to open an account on the electronic platform" ranking second with a mean of3.57 and standard deviation of (1.26). Item 24, "Sometimes I encounter problems with opening an Edraak website," ranked before the last with a mean of (2.95) and a standard deviation of (1.37). Finally, item 25, which states that "I encountered technical problems while taking the exam," ranked last with a mean of (2.74) and a standard deviation of (1.48).

5.1.4 Fourth, Appropriateness of Blended E-learning to the National Education Course

Table 8 indicates that the degree of appropriateness of blended e-learning to the National Education course on the Edraak platform from the students' perspective was moderate in general, where the mean was (3.25) with a standard deviation of (1.01). All the items of this domain were moderate, as the means ranged between (2.87–3.55), with item 9 which states that "The course material is upgraded on "Edraak" platform in terms of national developments and challenges" ranking first with a mean of (3.55) and standard deviation of (1.16). Item 1 states that "For me, the importance of the National Education course was not affected as it is taught by blended E-learning" and item 3 states that "In general, the National Education topics are commensurate with the blended e-learning" ranking second with a mean of3.38 and standard deviation of (1.30) and (1.24)respectively. Item 8, which states that "I enjoy following up the National Education course through e-learning," ranked before the last with a mean of (2.94) and standard deviation of (1.37). Finally, item 7, which states that "Studying national education through blended e-learning adds an element of diversity and keeps the feeling of boredom away," ranked last with a mean of (2.87) and a standard deviation of (1.43).

5.1.5 Fifth, Students' Following up the Course Material Through Edraak Platform

Table 9 indicates that the degree of students' following up the course material through the Edraak platform from the students' perspective was moderate in general, where the mean was (3.21) with a standard deviation of (0.89). All the items of this domain were moderate, except for one high item, as the means ranged between (2.73–3.70), with item 13, which states that "I prefer to follow the lessons through the attached written material" ranking first with a mean of (3.70) and standard deviation of (1.16), and item 14 which states that "When following the lessons, I sometimes, mix the reading material with the videos" ranking second with a mean of 3.55 and standard deviation of (1.22). Item 16, which states that "I shared conversations with my colleagues about the course material on Edraak platform," ranked before the last with a mean of (2.86) and a standard deviation of (1.40). Finally, item 17, which states that "I asked the instructor some questions through Edraak platform," ranked last with a mean of (2.73) and a standard deviation of (1.37).

5.2 Second Question: Does the Total Degree of the Effectiveness of Blended E-learning for Teaching National Education on "Edraak "Platform Differ from the Students' Perspective, in Terms of Students' Gender, Academic Level, Previous Experience in Using Blended E-learning, and Seriousness in Dealing with Blended E-learning?

Means and standard deviations of the total degree of the effectiveness of blended e-learning for teaching National Education on Edraak Platform from the students' perspective were calculated in terms of the variables of students' gender, academic level, previous experience in using blended e-learning, and student seriousness in dealing with blended e-learning, as shown in the following table.

Table 10 shows that there are apparent differences between the means of the total degree of the effectiveness of blended e-learning for teaching National Education on Edraak Platform from the students' perspective in terms of the variables of students'

Table 8 Means, standard deviations, ranks, and degree of appropriateness of blended. E-learning to the national education course on edraak platform from the student perspective arranged in a descending order

No.	Item	M	SD	R	Appropriacy Degree
9	The course material is upgraded on "Edraak" platform in terms of national developments and challenges	3.55	1.16	1	Moderate
1	For me, the importance of the National Education course was not affected as it is taught by blended e-learning	3.38	1.30	2	Moderate
3	In general, the National Education topics are commensurate with the blended e-learning	3.38	1.24	2	Moderate
2	The nature of the National Education objectives is commensurate with the blended e-learning	3.36	1.16	4	Moderate
4	National Education develops a sense of belonging to the country through blended e-learning	3.32	1.23	5	Moderate
5	National Education develops a sense of social responsibility through blended e-learning	3.30	1.27	6	Moderate
6	The video recordings of the course material make up for the teacher's explanation in class	3.13	1.52	7	Moderate
8	I enjoy following up the National Education course through blended e-learning	2.94	1.37	8	Moderate
7	Studying national education through blended e-learning adds an element of diversity and keeps the feeling of boredom away	2.87	1.43	9	Moderate

(continued)

Table 8 (continued)

No.	Item	M	SD	R	Appropriacy Degree
Appropriateness degree of blended e-learning to the National Education course		3.25	1.01		Moderate

Table 9 Means, standard deviations, ranks, and degree of students' following up the course material through edraak platform from the students' perspective arranged in a descending order

No.	Item	M	SD	R	Follow up degree
13	I prefer to follow the lessons through the attached written material	3.70	1.16	1	High
14	When following the lessons, I sometimes mix the reading material with the videos	3.55	1.22	2	Moderate
11	I study the course material following the schedule	3.42	1.27	3	Moderate
10	I follow up all the course videos on the Edraak platform according to the specific weekly schedule sequence	3.41	1.30	4	Moderate
12	I prefer to study through videos	2.99	1.35	5	Moderate
15	I viewed the supporting materials such as interviews with national figures and other materials on the Edraak platform	2.99	1.33	5	Moderate
16	I shared conversations with my colleagues about the course material on the Edraak platform	2.86	1.40	7	Moderate
17	I ask the instructor some questions through the Edraak platform	2.73	1.37	8	Moderate
Degree of students' following up the course material through Edraak platform		3.21	0.89		Moderate

gender, academic level, previous experience in using blended e-learning and student seriousness in dealing with blended E-learning. To determine whether the differences between the means are statistically significant at the significance level($\alpha = 0.05$), the four-way ANOVA analysis with the factorial design 2 * 3 * 2 * 2 was applied. Table 11 shows the variance analysis results.

Table 11 shows that there are no significant differences at the significance level ($\alpha = 0.05$) of the total degree of the effectiveness of blended e-learning for teaching National Education on the Edraak platform from the students' perspective in terms of the variable of students' gender, based on (F) value which is (1.490) with a significance level of (0.223). There are no significant differences in terms of the variable of using blended e-learning before, based on (F) value which is (0.012) with a significance level of (0.914). At the same time, there was a significant difference in terms of the variable of student seriousness of dealing with blended e-learning, based on (F) value which is (18.790) with a significance level of (0.000). The difference of this

Table 10 Means and standard deviations of the total degree of the effectiveness of blended E-learning for teaching national education on Edraak platform from the students' perspective in terms of the variables of students' gender, academic level, previous experience in using blended E-learning and student seriousness in dealing with blended e-learning

Variable	Category	No.	M	SD
Gender	Male	123	3.43	0.75
	Female	124	3.33	0.78
	Total	247	3.38	0.76
Academic level	1st year	146	3.25	0.79
	2nd year	52	3.55	0.73
	3rd year and higher	49	3.57	0.63
	Total	247	3.38	0.76
Previous experience in using blended E-learning	Yes	148	3.41	0.76
	No	99	3.34	0.77
	Total	247	3.38	0.76
Student seriousness in dealing with blended E-learning	Yes	149	3.55	0.64
	No	98	3.12	0.86
	Total	247	3.38	0.76

Table 11 The four-way ANOVA analysis to find out significance of differences between the means of the total degree of the effectiveness of blended E-learning for teaching national education on Edraak platform from the students' perspective in terms of the variables of students' gender, academic level, previous experience in using blended E-learning and student seriousness in dealing with blended E-learning

Source of variance	Total squares	Freedom degree	Square means	F value	Significance level
Gender	0.784	1	0.784	1.490	0.223
Academic level	4.700	2	2.350	4.464	0.012
Use of blended E-learning	0.006	1	0.006	0.012	0.914
The seriousness of dealing with blended E-learning	9.893	1	9.893	18.790	0.000
Error	126.890	241	0.527		
Total	143.299	246			

variable was in favor of the students who answered with (yes) since their mean was (3.55), which is higher than the mean of the students who answered with (No), which was (3.12). There were significant differences in terms of the variable of academic level, based on (F) value which was (4.464) with a significance level of (0.012).

Table 12 Results of Scheffe test for post hoc comparisons for differences in the total degree of the effectiveness of blended E-learning for Teaching National Education on Edraak Platform from the Students' Perspective in terms of the variable of students' academic level

Academic level	M	3rd year and higher	2nd year	1st year
		3.57	3.55	3.25
3rd year and higher	3.57	–	0.02	*0.32
2nd year	3.55		–	*0.30
1st year	3.25			–

Scheffe test for post hoc comparisons was used to determine whether any group's differences were in favor, as shown in Table (12).

Table 12 indicates that the difference was in favor of the students of the third year and higher and in favor of the second year students compared with the first-year students.

5.3 *What Are the Advantages of Using Blended E-learning Through the Edraak Platform?*

The frequencies and percentages of student responses were calculated to answer this question, as shown in Table 1).

Table 13 indicates that "Flexibility and the ability to open the platform at any time convenient for the student" gained the highest frequency of 75 and 37% of the total number of respondents, followed by" having no advantages" with a frequency of 30 and 14.8%. The advantages of "The material is presented in an organized manner," "having a second source of learning", and "Increasing the learning effectiveness and facilitates learning" were ranked last with a frequency of 1 and a percentage of 0.5% for each of them.

5.4 *What Are the Disadvantages of Using Blended E-learning Through the Edraak Platform?*

The frequencies and percentages of student responses were calculated to answer this question, as shown in Table 14.

Table 14 indicates that" lack of appropriate interaction between students and lecturer" gained the highest frequency of 53 and percentage of 31.2%of the total number of respondents, followed by" having no disadvantages" with a frequency of 31 and 18.2%. The disadvantages of "getting bored", change of instructors, and difficulty taking the exam through Edraak were ranked before the last item with a frequency of 5 and a percentage of 2.9% for each of them. Finally, the disadvantage

Table 13 Frequencies and percentages of student responses about the advantages of blended E-learning through Edraak

Responses	Frequency	Percentage
Flexibility and the ability to open the platform at any time convenient for the student	75	37
No advantages	30	14.8
Ease of access to material content	26	12.8
Save time and effort	19	9.4
Add some fun	17	8.4
Diversity in teaching methods	16	7.9
Comfort as you attend the lecture at home	11	5.4
A detailed explanation of material and whom it is presented with readings	2	1.0
Instructors understand the stress	2	1.0
I don't know	2	1.0
The material is presented in an organized manner	1	0.5
The second source of learning	1	0.5
Increase the learning effectiveness and facilitates learning	1	0.5
Total	203	100.0

Table 14 Frequencies and percentages of student responses about the disadvantages of blended E-learning through Edraak

Responses	Frequencies	Percentage
Lack of appropriate interaction between students and lecturer	53	31.2
No disadvantages	31	18.2
Instructions not clear	18	10.6
The website frequently breaks down	18	10.6
Exam time is not enough	16	9.4
One may get bored while attending	9	5.3
Plentiful	8	4.7
Getting bored	5	2.9
Change of instructors	5	2.9
Difficulty taking the exam through Edraak	5	2.9
Learning at home is distracting with a lack of concentration	2	1.2
Total	170	100.0

of "Learning at home is distracting with lack of concentration" came last with a frequency of 2 and a percentage of 1.2%.

6 Discussion

The current study aimed to investigate the total efficiency degree of blended e-learning in five domains and to identify the advantages and disadvantages of teaching the "National Education" course from the perspective of the German Jordanian University students. The results showed that the total efficiency degree of blended e-learning was moderate, which is somehow satisfactory, considering the instructors' and students' novel experience in el-learning and the efficiency level of e-learning infrastructure.

The results showed the following descending order of the efficiency degree of the study domains:

The domain of "Electronic Exam Efficiency" ranked first because the multiple-choice questions used on the electronic platform are preferred by students, as they are characterized by diversity, comprehensiveness, ease, and speed of performance, all of which conform with the university students' requirements. In addition, the software used in such kinds of exams does not consume high bundles of the internet and can be installed on smartphones, thus providing freedom to select where to take the exam without restrictions.

The domain of "Efficiency of Instruction through Videos" ranked second, as the videos provide students with the freedom to choose the time to attend the course electronically, and in a way that suits their circumstances, in addition to the ability to re-play certain clips, to review the material again and again, and the element of suspense in presenting the educational material in videos, audios, and motion. What gives weight to this analysis is the fact that the paragraph "Flexibility of e-learning…" got the highest rate of (37%) of the total number of respondents in terms of the advantages of blended e-learning (Table 13).

The domain of "Efficiency of Technical Use of the Electronic Platform" ranked third, where the students found out that the study material is presented in a way that keeps pace with modern time, as it is updated with the latest developments and national challenges, with the fact that the significance of the "National Education" course was not influenced by its being taught online.

The "Appropriateness of Blended E-learning to the National Education Course" domain ranked four, relatively low. This is because the students believe that the topics of the National Education "course require discussion and dialogue precisely like the direct teaching method in the classroom, which is an urgent necessity to achieve the goals of building values and attitudes and for students to form their opinions as it is in the social and human sciences. This indicates that the students need more direct traditional meetings, especially if the National Education credit hours are divided equally between digital online education and face-to-face education.

The domain of “Attending Courses on Platforms Regularly” ranked last. Here lies the defect of e-learning as students are free to select the time of attending their courses, which is a negative aspect when students don’t acquire the skills adequate for time management and self-management, with the result of irregular attendance, especially if this is accompanied with the lack of a technique that helps instructors print out attendance reports for students. Another factor is the number and duration of course videos since students don’t have enough time because of the huge study responsibilities students bear. Thus they tend to read the written material on Word documents available on the e-platform to save time. The results also showed no significant differences in estimating the total efficiency degree of blended e-learning in terms of students’ gender and academic level. This means that both male and female students from different academic levels and regardless of their previous experience in blended e-learning agree that the total efficiency degree of blended e-learning was intermediate. There were significant differences in student seriousness in dealing with blended e-learning, in favor of students who take down their notes of the online courses, making them more understanding and absorbed in the study materials than others.

Results shown in Table 14 indicate that the advantage that obtained the highest frequency of 75 with a rate of 37% of the total number of respondents is due to the flexibility of blended e-learning, followed by 14.8% of the respondents who consider that there are no advantages. This indicates having challenges accompanying e-learning related to the language, limited accessibility available to them such as a fast and robust network, or issues related to the devices they use (laptop or mobile), an electric power outage at some areas. Having no appropriate interactions between students and lecturers through blended e-learning was the highest disadvantage, with a rate of 31.2% of the total number of respondents.

Considering the features of the study objectives that distinguish it from previous studies, it can be said that there is an agreement between some of the details of the current study with previous studies and having points of difference. The results of this study are consistent with the results of the study carried out by Irfan (2020), regarding the result that the use of educational platforms leads to improving student outcomes and improving their level of satisfaction with the use of online learning. While the study carried out (Abuhassan, 2020a) concluded a positive impact of using online learning to a very large degree, the current study concluded that the total efficiency degree of teaching “National Education” course via blended e-learning is intermediate, with an agreement that there are no differences that can be attributed to the gender variable. Furthermore, the current study is consistent with the results of the study carried out by Al Kurdi et al. (2020a, b), Meqdadi (2020), which indicated that the total degree of estimating the reality of distance education came was intermediate, with no differences according to the variable of the study stage. Indeed, the time of carrying out the current study and the differences in terms of its population, sample, tool, and detailed objectives play an important role in determining the points of agreement or disagreement with the results of previous studies.

7 Conclusion

- Time management, self-management, and students dealing with blended e-learning seriously are significant factors in making blended e-learning succeed.
- The distribution of the "National Education" course into a model of one third for e-learning and two thirds for traditional interactive meetings between instructor and students is the most appropriate model for the instruction of "National Education" due to the students' need for direct dialogue and discussion, and the instructor's need for having a direct impact on their students, in terms of building attitudes and values.
- Flexibility in students' freedom of choice of watching online lectures is the most significant advantage of e-learning from students' point of view.

8 Recommendations

- Hold training courses for students to develop the skills of self-management and time management among students of e-learning because of their significance in raising the regularity of online attending of courses.
- Provide the faculty with a technique for monitoring students 'attendance of online lectures, as it provides instructors with the possibility of following up and urging absent students to adhere to attendance.
- Develop students' skills in preparing the schedules of study responsibilities and an appropriate time distribution that students can adhere to.
- Guide students to deal seriously with the blended e-learning system, as the student's taking down of notes makes them more absorbed into the educational material.
- Ensure the provision of suitable equipment for students and the establishment of adequate technological infrastructure in terms of quality, speed, and endurance.
- Adopt a model of one-third of the credit hours for blended e-learning, and two-thirds of
- direct instruction for its relevance in the teaching and learning social and national studies, instead of the half by half model, where the need for direct interaction is more.
- Provide sufficient opportunities for cooperative student work and the participation of students in dialogue and discussion in class meetings to compensate for the missing part in e-learning.
- Expand the study of the advantages and disadvantages of e-learning using the method of group interviews.
- Carry out future studies on the challenges of blended e-learning.

Funding This research obtained a Financial support from the German Jordanian University, Document No. 14/ 34/3/53/1 on 6/7/2020.

References

Abu-Aqail. I. (2014). Reality of E-learning and obstacles of implementing it in university education as perceived by students of Hebron university. *University of Palestine Journal for Research and Studies*.

Abuhassan, H. (2020a). Development of a new model on utilizing online learning platforms to improve student's academic achievements and satisfaction. *International Journal of Educational Technology in Higher Education, 17*(38), 1–23.

Al Kurdi, B., Alshurideh, M., & Salloum, S. A. (2020a). Investigating a theoretical framework for E-learning technology acceptance. *International Journal of Electrical and Computer Engineering, 10*(6), 6484–6496.

Al Kurdi, B., Alshurideh, M., Salloum, S., Obeidat, Z., & Al-dweeri, R. (2020b). An empirical investigation into examination of factors influencing university students' behavior towards E-learning acceptance using SEM approach. *International Journal of Interactive Mobile Technologies, 14*(2), 19–41.

Aldhafeeri, F. (2017). Investigating the opinions of academic staff members at Kuwait university about their preparedness for implementing web 2.0 educational platforms into their teaching courses. *International Journal for Research in Education, 41*(3).

Al-Fraihat, D., Joy, M., Masa'deh, R., & Sinclair, J. (2020). Evaluating E-learning systems success: An empirical study. *Computers in Human Behavior, 102*, 67–86.

Al-Issan, A.-A. W. (2007). The reality of E-learning from the perspective of the students of education college at Sultan Qaboos university. *Dirasat Educational Sciences, 34*(2).

Chen, T., Cong, G., Peng, L., Yin, X., Rong, J., & Yang, J. (2020). Analysis of user satisfaction with online education platforms in china during the covid-19 pandemic. *Healthcare (Switzerland), 8*(3). https://doi.org/10.3390/healthcare8030200.

Dhawan, S. (2020). Online learning: A Panacea in the time of COVID-19 crisis. *Journal of Educational Technology Systems, 49*(1), 5–22. https://doi.org/10.1177/0047239520934018

El-Madawy. Noor-book. Retrieved 20 Mar 2021.

Górska, D. (2016). E-learning in higher education. The person and the challenges. *The Journal of Theology, Education, Canon Law and Social Studies Inspired by Pope John Paul, 6*(2).

http://www.gju.edu.jo/content/about-gju-687.

https://www.edraak.org/about-us.

Hussamo & Al-Abdallah. (2011). The reality of E-learning in Tishreen University from the Perspective of Faculty Members and Students. *Damascus University Journal, 27*.

Irfan, M. (2020). Challenges during the pandemic: Use of e-learning in mathematics learning in higher education. *Infinity Journal, 9*(2), 147–158.

Kapadia, N. (2020). The impact of lockdown on learning status of undergraduate and postgraduate students during Covid-19 pandemic in West Bengal, India. *Children and Youth Services Review, 116*, 1–5.

Martin, A. (2020). How to optimize online learning in the age of coronavirus (COVID-19): A 5-point guide for educators. *UNSW Newsroom, 53*(9), 1–30.

Mefleh & al-Dahoun. (2010). Obstacles of the application of E-learning systems as viewed by secondary school teachers at Al-Kurah district. *Jordanian Journal of Educational Sciences, 6*(1).

Meqdadi, M. (2020). Perceptions of high school students in government schools in Jordan to use distance education in light of the corona crisis and its developments. *Arab Journal for Scientific Publishing, 19*(2), 96–114.

Oweis, T. (2018). Effects of using a blended learning method on students' achievement and motivation to learn English in Jordan: A pilot case study. *Educational Research International, 2018*.

Qudhah, M. B. (2013). E-learning challenges as perceived by faculty members at the private universities in Jordan. *Al-Manarah, 19*(3).

Saleh, J. (2019). The effect of using web quests and learning platforms for teaching mathematics to develop the mathematical power and reflective thinking among eighth grade students.

Shdeifat, M. (2020). The reality of employing distance education due to Covid-19 pandemic in Mafraq schools from the school principals' perspectives. *Arab Journal for Scientific Publishing, 19*(2), 185–207.

Evaluating Software Quality in E-Learning System by Using the Analytical Hierarchy Process (AHP) Approach

Case Study: Mu'tah University

Asmaa Jameel Al Nawaiseh

Abstract This study defines and analyzes essential quality characteristics that should be addressed in the creation and implementation of an E-learning system in institutes, as well as its System Content's Quality, Characterizes, and Attributes. The model was empirically validated by fitting it to data from 302 students who participated in an E-learning AHP. All of the evaluated end-users were compared using the pairwise comparison technique in one of Jordan's colleges. Student Oriented Domain, Online Services, and Content Quality are the three sub characteristics of System Content Quality; each sub characteristic has its own set of traits that are used to determine the evaluation's outcome. The Analytical Hierarchy Process (AHP) has been widely employed in multi-criteria decision-making for a variety of practical decision-making situations. The AHP model was used to collect pairwise comparison judgments from all evaluated end users.

Keywords Information system · E-learning system · COVID-19 epidemic · AHP · Priority pairwise comparison · Quality models · Academic · Information system · ISO 9126

1 Introduction

In our published paper (Al_Nawaiseh et al., 2020a), we proposed and built a new quality model dedicated to Academic Information System (AIS) to support a standard set of software quality characteristics.

Software Content must be evaluated as one of the software quality aspects to be considered in the E-Learning System to determine the level of System Content software. The same new model has been improved in. (Al_Nawaiseh et al., 2020a)

A. J. Al Nawaiseh (✉)
Department of Software Engineering, College Information Technology, Mu'tah University, Mu'tah, Jordan
e-mail: asma@mutah.edu.jo

M. Alshurideh et al. (eds.), *The Effect of Information Technology on Business and Marketing Intelligence Systems*, Studies in Computational Intelligence 1056,
https://doi.org/10.1007/978-3-031-12382-5_20

by identifying new attributes for the model's sub-characteristics and defining metrics rules to measure the quality of these new attributes.

The attributes are separated, as are the metrics and criteria that go with them. Furthermore, several existing methods for selecting E-Learning System components have been reviewed, and their shortcomings in comparison to the improved model have been identified.

E-Learning System Quality Evaluation is a paper that discusses an evaluation of the System Contents Quality characteristics on an E-learning system using our proposed software quality model (ESQE) (Al_Nawaiseh et al., 2020a; Almajali, 2016). The critical quality characterizes system content that should be taken into account to ensure the successful development and implementation of the E-learning system in institutes.

Student Oriented Domain, Online Services, and Content Quality are the three sub qualities of the Content Quality characteristic. Each sub characteristic has its own set of traits that are used to determine the evaluation's outcome. The enhanced model is tested with the support of AHP through an empirical study.

Mu'tah University is considered in this study to test and evaluate the quality of E-learning in its camps to aid system analysts, developers, and programmers in the implementation of their eLearning initiatives. Because of the evaluations, the recommendations will address both the construction and maintenance of the target system E-learning System, as well as the system's long-term consequences. The empirical study here follows the following strategy:

- For each of the quality attributes, an evaluation is made for an eLearning System in the sample study, and then the pairwise comparison judgment is determined.
- A questionnaire for the characteristics, sub-characteristics, and attributes is developed and deployed to obtain appropriate weight values that represent the relationships between each pair of attributes.

2 Literature Review

It would be beneficial to analyze the characteristics of actual QIS models and frameworks and then examine their applicability to the contextual production of software in universities. This paper summarizes and integrates the limited literature on models and frameworks of software quality and then constructs a new model to evaluate their relevance to the quality of the information system.

In Jordanian universities, (e-learning system). The quality of the software is the most important aspect of software development since the high quality will reduce the cost of maintenance and reuse of software. For developers, students, and staff, however, quality has very different definitions. The quality definitions and quality characteristics of several high-level institutes and organizations are (Al-Adwan et al., 2018a; Almajali, 2016).

Software Quality can be defined as: The researcher presents some definitions of international and standard organizations.

1. **ISO9126**: "Software quality characteristics is a set of attributes of the software product by which its quality is described and evaluated". (ISO/IEC TR, 9126-3, 2003).
2. **German Industry Standard Part II**: "Quality comprises all characteristics and significant features of a product or an activity which relate to the satisfying have given requirements".
3. **American National Standards Institute/American Society of Quality Control (ANSI/ASQC A3/1978)**: "Quality is the total of features and characteristics of a product or a service that bears on its ability to satisfy the given needs". (Alrawashdeh et al., 2013)
4. **Institute of Electrical and Electronics Engineers (IEEE)**: suggested some definition of quality.

> The composite characteristics of software that determine the degree to which the software in use will meet the expectations of the customer.

All these definitions give separate views on quality. Thus, the researcher needs to organize, clarify, and standardize a large number of quality-related definitions to obtain the best definitions for quality assurance.

2.1 *E-Learning as Comprehensive Information System*

E-Learning systems enable faculty members to manage their courses electronically and to use technology tools in teaching and communicating with their students. The E-Learning system is a relatively new set of software tools that have been used in educational settings for less than a decade.

Furthermore, E-Learning allows instructors to expand the classroom beyond its traditional time and space constraints (Al-Fraihat et al., 2020; Al-Shboul et al., 2013).

The example of E-Learning demonstrates that this system is a comprehensive system that requires the integration of all subsystems in universities as a platform that primarily addresses education, especially in this critical time of the COVID-19 Epidemic.

Now Administrative regulations, such as course enrolment, are included in E-Learning Systems, just like they are in traditional software. However, such a system would be unable to account for all of the restrictions and constraints. Online registration isn't error-free unless it's connected to administrative systems, which may demand manual enrollment list updates (Nagarajan & Jiji, 2010).

This necessitates a special need for interoperability, which includes the support of multiple formats as well as metadata access. In addition to describing the course content, metadata are data about data in other educational administration systems. As a result, there is a big demand for reusing teaching materials in digital learning systems (Al-Fraihat et al., 2020).

2.2 *Platform of E-Learning System at Mu'tah University*

Universities in various countries have responded to E-learning in their educational institutions in various ways, depending on the regulations. Before the COVID-19 Epidemic, Mu'tah University in Jordan's south adopted the blended learning approach, which is a hybrid of traditional learning and online learning, as did all Jordanian universities.

This means that students attend online lectures while also receiving all learning materials online from a variety of educational platforms, in addition to the Electronic Assessment for assignments and E-Exam. Mu'tah University has been using Moodle since 2005 to create a well-structured E-learning platform and Learning Management System (LMS).

LMS users at the Mu'tah University are classified into five categories according to their authorities: an editing-teacher, non-editing teacher, student administrator, creator, and guest where each one has different roles (Fig. 1).

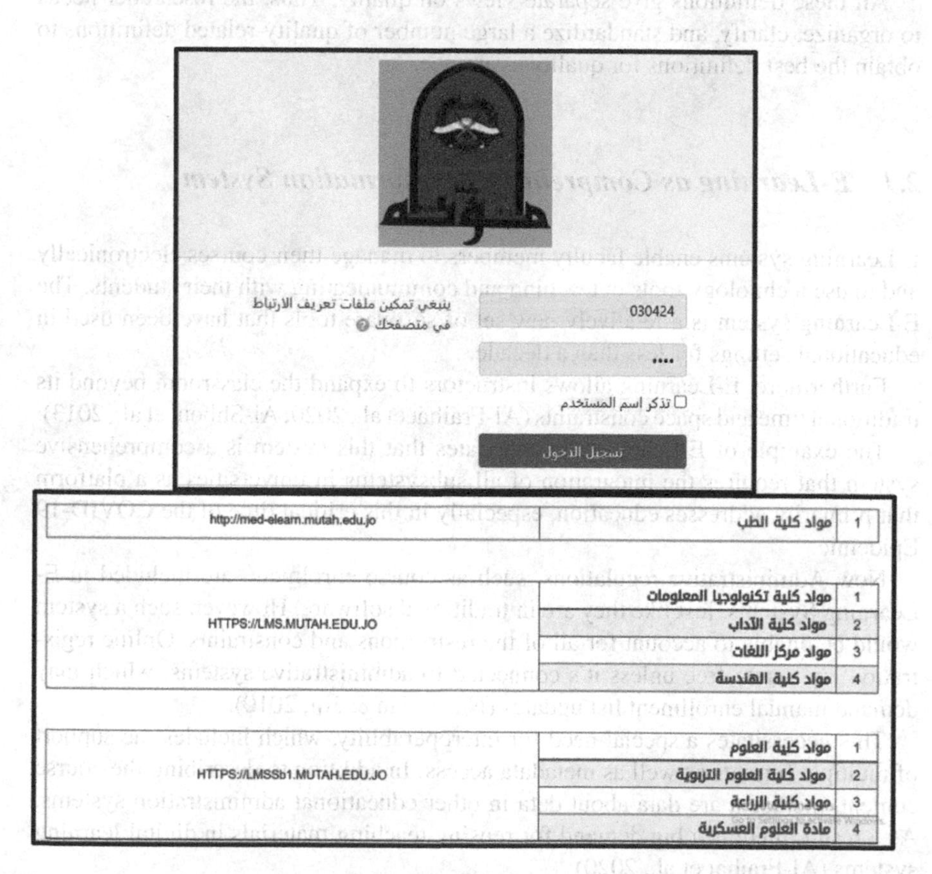

Fig. 1 Mu'tah University E-learning platform homepage

3 Research Objectives

The objectives that the research seeks to achieve are:

1. To clarify the concept of quality in information systems, as the information constitutes a unique resource No matter what companies must exploit it efficiently and successfully.
2. Identify basic attributes and their sub-attributes for measuring software quality in information systems, and identifying relationships between them in preparation for comparison, to detect potential weaknesses, and overcome them.
3. Study the integration of the basic and sub-factors of quality measurement in the applied information systems in the universities.
4. These suggest solutions and recommendations avoid most of the gaps and constraints that affect the effectiveness of measurement quality and success in the information systems applied in universities.
5. A common approach will be to establish measures and evaluate the quality of the AIS, consisting of various global quality levels of software, which will assist system analysts, system developers, and system programmers in their AIS projects
6. These create a standard approach that tests, assesses, and strengthens existing models of quality.
7. Save time and effort by choosing high-efficiency applications to support high-school information Systems.

4 Important of Research

All universities can benefit from this novel model by implementing an eLearning system that meets all of the qualities and conditions recommended in this research and serves as a measure of the efficiency of their systems; additionally, they will save time and effort in selecting high-efficiency software to support the Information Systems within their higher education sector.

Our novel model will represent the university's exterior image, so the university's systems will serve as a road map for visitors to the page, whether they are new students or their guardians. Another advantage is that it saves time and effort when it comes to picking high-efficiency software to support Information Systems in the higher education sector.

5 Previous Studies

Etaati et al., (2011) build their new E-Learning System model Based on ISO9126 attributes, their model consists of three key criteria, the (Wang model) and twenty-nine sub-criteria, "learning Community, System Contents and Interactivity." The sub-attributes of Device Material have (up-to-date, sufficient, and useful content). They evaluated their new quality model in three well-known universities in Iran by using standard AHP calculations for selecting the most appropriate E-Learning System features.

All of these Universities provided high-standard E-Learning Systems for students and staff, the researcher has captured information by conducting structured interviews with students and staff.

Baklizi and Alghyaline (2011) publish a study about how to evaluate the E-Learning website of Jordan universities based on ISO/IEC 9126 standard, which uses six main characteristics to evaluate software, and each characteristic includes its sub-characteristics. The results show that the average quality in E-Leaning websites is 65.45%. The results reflect the student's opinion about the website and might be used to improve the quality of E-Learning websites for those universities. This paper evaluates the quality of the E-Learning website for Jordan universities based on ISO/IEC 9126 standard by using the six characteristics: Functionality, Reliability, Usability, Efficiency, Maintainability, Portability, and its sub-characteristics. The respondents in this study are the students only; they do not have experience in educational software with academic information systems. If the academic staff with information developers participate also, the study will become more accurate.

Alrawashdeh et al., (2013) measuring the quality of the Enterprise Resource Planning System (ERP) to propose an acceptable software quality model for ERP systems, the author highlighted the most common software quality models in the literature; these are versions (McCall's, Dromey's Boehm, FURPS, and ISO\IEC 9126). The resulting quality model was used to check whether the ERP job in an educational institution evaluating the quality of the ERP from the user's side could succeed or failed. The same authors introduced a new ERP System Consistency Model (Alrawashdeh et al., 2013), called (ERPSQM). With its complexity and modularity, the Enterprise Resource Planning (ERP) framework has a different type of principles, real-time information and standardization is the essential function of the ERP system. Under the Flexibility, Usability, and Maintainability stage, four new sub-features were introduced, including (Modularity, Compatibility, Complexity, and Reusability). In their analysis, the Hierarchy Method (AHP) was used to assess the consistency of the proposed model, ranking its features and attributes by end-users.

Yuhana et al. (2014) A framework was proposed to assess the quality of web-based AIS using the perception approach of visitors, developers, and institutions. Their AIS quality instrument is based on a variation of the "WBA Quality Model" and "COBIT 4.1" (ISO/IEC 9126, ISO/IEC 25,010:2011) (Fig. 2).

The purpose of the evaluation is to determine the usability behaviors at the AIS administration module using ISO/IEC 9126 a standard software quality.

The results indicate that Learnability is not implemented. This framework was expected to produce a measurement of academic quality web-based information systems more accurately and provide detailed recommendations to gain a better system, especially to support the E-Business processes of AIS and web-based technology. This proposal works not implemented in a real-world case study.

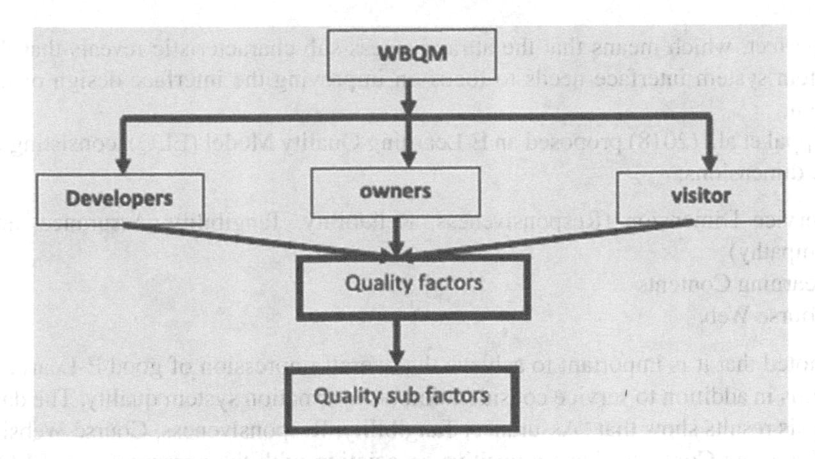

Fig. 2 Structure of WBA quality model. *Source* Yuhana et al. (2014)

Yuhana et al., (2014) publish a new AIS paper discussing the study of the usability function of the (IT'S ') system's management module as Academic Information to display academic data such as students, employees, classes, academic events, and information about the curriculum. Students, administration staff, faculty access this modularity feature, and department chairperson, some attributes generate a zero value like learnability.

This shows that the student administration module AIS is difficult to learn. Other sub-characteristics results have a non-zero value yet are not perfect.

The usability feature of the assessment is the lake of this study only as valuation instruments based on ISO/IEC 9126, whereas ISO/IEC 9126 has maximum measurement metrics. These overall results show that a technological acceleration and evolution in AIS is needed to boost the quality value and to be real.

Tabrizi et al., (2017) have conducted a study at their university to determine the usability of their systems derived from the Near East University (NEU) SIS's ISO/IEC 9126 (Understandability, Learnability, Operability, Attractiveness, and Compliance) because the ISO/IEC 9126 model has the most suitable metrics for assessing the internal usability of the systems. According to the developers, when considering development opportunities for potential iterations of the framework, the outcome of this research would be useful for the (NEU SIS) developer team.

One of the solutions that offer management facilities for educational and academic aspects is a Student Information System (SIS). All university students must log into the "Einstein System", as well as all levels of academic staff and university stakeholders, to access the system processes; as e-learning systems, it can be viewed as a robust academic system.

The study results revealed that the understandability of the system achieved the optimum value. The learnability sub-characteristic results indicate that the system is sufficiently learnable, Operability as an important sub-characteristic of the system yielded an acceptable result, Attractiveness and Usability compliance results were not perfect, which means that the attractiveness sub-characteristic reveals that the Einstein system interface needs to focus on improving the interface design of the system.

Uppal et al., (2018) proposed an E-Learning Quality Model (ELQ), consisting of three dimensions:

- Service Dimension (Responsiveness, Reliability, Tangibility, Assurance, and Empathy)
- Learning Contents
- Course Web.

We noted that it is important to achieve the overall impression of good E-Learning systems in addition to service consideration of information system quality. The data analysis results show that "Assurance, Tangibility, Responsiveness, Course Website and Learning Contents" have a positive association with the perception of (ELQ), students of E-Learning prefer a safe and simple E-Learning environment but do not perceive it. Understanding and reliability as the significant perception of ELQ.

Almarabeh et al., (2014) to determine the impact of the E-Learning Management System in the University of Jordan conducted another study. They surveyed students' opinions about this new educational system and addressed the challenges that students faced while using the E-Learning Management System.

The University of Jordan is attempting to employ Information and Communication Technology (ICT) in education and is progressing by implementing the latest version of E-Learning Management Systems (LMS) to stay up with the technological change in higher education, according to their report. The university uses the E-Learning framework, which was adapted from Khan's three-part structure, see Fig. 3.

Other issues that students have when utilizing the E-Learning System at the University of Jordan were uncovered in this study, including hardware resources and the university network. Other challenges students face include difficulty in seeking

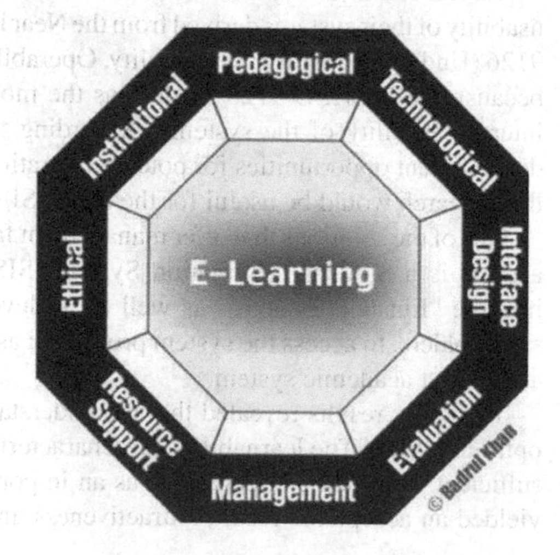

Fig. 3 Khan's E-learning framework. *Source* Almarabeh et al. (2014)

assistance with Moodle or resolving technical issues, such as learning a course from a computer screen.

Gürkut and Nat (2017) describe the Student Information System (SIS) as one of the primary systems for promoting the management and growth of higher education institutions (HEI).

The teacher to upload course material to students uses the SIS, students to register for online classes, review their schedule, test schedules, grades, and other activities use the SIS. The purpose of this study is to determine the impact of information system consistency on the satisfaction of SIS academic and administrative employees. They announced that System Consistency is one of the metrics for the software and hardware of the information processing system.

Al-Adwan et al., (2018b) publish his paper about the Jordanian universities on the advancement of E-learning in higher education systems. His research was aimed at evaluating the capacity of both Al-Zytonah University and Applied Science University students and staff to participate effectively in the E-Learning Systems Model.

They strongly believe that oral communication plays an important influence on Jordanian culture and education. Lecturers in this group strongly believed that oral communication is the most expressive way to deliver information to students.

The conclusion of the study has shown that full-time faculty members and students are not ready to get involved in E-Learning programs at (research time); many issues must be addressed according to their technological readiness and their attitudes before implementing E-Learning programs. This research varies from my research on the evolution of the E-learning System in the Jordanian universities' higher education systems. Our analysis on how to use the latest models of the software quality and our enhanced model of the framework in the E-Learning System using HEI, but the two studies complete each other in serving the end-users.

Cho et al., (2009) have conducted a study to look into the elements that influence how students use and behave with e-learning technologies. Their study explores the impact of performance expectancy, effort expectancy, hedonic motivation, habit, social influence, and trust on students' behavioral intentions, based on the TAM and UTAM's strong theoretical foundations and employing structural equation modeling (SEM) using AMOS 20.0.

However, behavioral intention and facilitating conditions accounted for 40% of the total, with significant favorable effects on students' e-learning system use behavior. On the other hand, both effort expectations and social influence had little effect on students' behavioral intentions.

A framework was proposed by Almajali (2016) to develop a theoretical framework toward e-learning adoption in Jordanian public and private universities through conducting a comprehensive review of literature, empirical studies and a survey questionnaire that will be distributed to a selected sample study on the students. Based on the technology acceptance model, the researchers developed a model designed to measure the impact of ease of use, perceived usefulness, and training on e-learning usage on the actual use of e-learning systems via the initial trend towards the use of e-learning systems and the intention towards the use of e-learning systems as mediating factors.

It focused on the way teachers interact with e-learning systems, where it discussed the issue within the obstacles and negatives that affected the interaction of students and teachers with e-learning systems.

In a study conducted by Tarhini et al. (2017), a conceptual framework was constructed by combining two more components, trust and self-efficacy, into the unified theory of technology adoption and usage (performance expectancy, effort expectancy, hedonic motivation, habit, social influence, price value, and facilitating conditions).

The findings revealed that, in order of influence, performance expectancy, social influence, habit, hedonic motivation, self-efficacy, effort expectancy, and trust strongly influenced behavioral intention and explained 70.6 percent of the variance in behavioral intention. Contrary to popular belief, enabling conditions and economic value had little effect on behavioral intention. The following characteristics are thought to be important in understanding technology adoption, but to the best of the authors' knowledge, no study has modeled all of these elements together.

As a result, by incorporating all of these variables and being the first to be tested in UK institutions, this study will contribute to the literature on social networking adoption.

Al Kurdi et al., (2020a) have conducted a study at higher educational institutes in the UAE.

The study involved a sample of 365 university students. To describe the acceptance process, the structural equation modeling (SEM) method was used. Based on the Technology Acceptance Model (TAM), the standard structural model that involves e learning, computer self-efficacy, social influence, and enjoyment, was developed. Computer Anxiety, Technical Support, Perceived Usefulness, Perceived Ease of Use, Attitude, and Behavioral Intention to Use e Learning were also developed.

The findings showed that TAM served as a suitable theoretical tool to comprehend the acceptance of e-learning by users. The most crucial construct to explain the causal process employed in the model was e-learning. Computer Self-Efficacy, Social Influence, Enjoyment, System Interactivity, Computer Anxiety, Technical Support, Perceived Usefulness, Perceived Ease of Attitude, followed by behavioral intention to use e-learning Practical implications.

Al Kurdi et al. (2020b) The goal of the study is to figure out what factors influence students' acceptance of e learning and how these aspects influence students' intention to use it. Based on the technological acceptance paradigm, a theoretical framework was created (TAM).

"Social influence, perceived enjoyment, self-efficacy, perceived usefulness, and perceived ease of use" were found to be the greatest and most important indicators of students' intention to utilize and attitudes toward e-learning systems.

The findings have practical significance for practitioners, legislators, and developers in the implementation of effective E-learning systems to improve university students' continued interests and activities in a virtual E-learning environment. The research findings provide useful recommendations for E-learning practices, which may turn out to be guides for the efficient design of E-learning systems.

The following steps are carried out to use the AHP approach in deciding the goals between various criteria:

Step 1: Decision-makers first use the AHP to break down their decision problem into multiple levels of sub-problems that are more easily evaluated and each of them can be independently analyzed.

Step 2: The target is at the top level of the root hierarchy; the hierarchy's elements can contribute to all aspects of the problem of decision.

Step 3: Identify alternatives to reach the goal.

The next level of the hierarchy contains the criteria. In Fig. 4, there are no criteria the decision-makers have determined to be important in reaching the goal. The last level of the hierarchy contains the alternatives, there are three alternatives considered in achieving the goal.

Step 4: Identify the evaluation criteria for alternatives.

In Fig. 4, the lines connecting each alternative to each criterion show how each alternative is evaluated according to each condition.

The evaluation is performed during the pairwise comparison process, described in step. The number of comparisons required to calculate priorities for a matrix of N elements is ((N2 − N)/2). The number of comparisons quadratically increases with a number of alternatives. At some point, AHP is no longer practical for large, if the decision-making problem consists of N criteria and alternatives; the decision matrix takes the form (Fig. 5).

Step 5: After the (AHP) method has been formed, a set of two pairwise comparisons are accomplished. The first set of pairwise comparisons is among the pairs of criteria

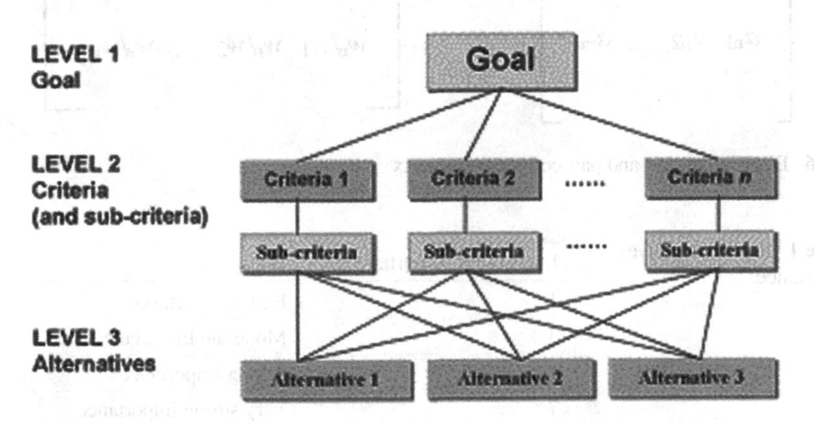

Fig. 4 AHP hierarchy

$$D = \begin{bmatrix} d_{11} & d_{12} & \dots & d_{1n} \\ d_{21} & d_{22} & \dots & d_{2n} \\ \vdots & & & \\ d_{m1} & d_{m2} & \dots & d_{mn} \end{bmatrix}$$

Fig. 5 Decision matrix

and the second set is between the pairs of alternatives. During the pairwise comparison process for the criteria, the researcher performing the comparisons assigns a relative importance weight to each criterion in the comparison (Fig. 6).

The importance of weight is evaluated in terms of the criterion's importance in reaching the goal. Criteria with an enormous impact on achieving the goal should receive higher importance weights than those with less of an impact. A scale frequently used is the nine-point scale, see Table 1.

After the pairwise comparisons have been conducted for the criteria, the weights provided in the comparisons have been structured as a matrix.

Step 6: Obtain Local and Global priorities for each alternative, then the local priority is calculated using the following equation:

$$\begin{bmatrix} a_{11} & a_{12} & \dots & a_{1n} \\ a_{21} & a_{22} & \dots & a_{2n} \\ \vdots & & & \\ a_{n1} & a_{n2} & \dots & a_{nn} \end{bmatrix} = \begin{bmatrix} w_1/w_1 & w_1/w_2 & \dots & w_1/w_n \\ w_2/w_1 & w_2/w_2 & \dots & w_2/w_n \\ \vdots & & & \\ w_n/w_1 & w_n/w_2 & \dots & w_n/w_n \end{bmatrix}$$

Fig. 6 Decision matrix and pair comparison matrix

Table 1 Scale of relative importance

Intensity of importance	Definition
1	Equal importance
3	Moderate Importance
5	Strong Importance
7	Very strong importance
9	Extreme of Importance
2, 4, 6, 8	Compromise between values

$$LP_i = \frac{\sum_{j=1}^{N} (RW_{ij})}{\sum_{i=1}^{N} \sum_{j=1}^{N} (RW_{ij})} \quad (1)$$

Alternatives, in which M is called alternative number and N is the criteria number. This equation then measures global priority or global weight (Aguarón et al., 2019), If (LPi) is a local priority criterion (I), (RWij) is criterion (I) relative to criterion (J), and (N) is criteria number (N) (Schwester, 2009).

To calculate the local priority for each alternative for each criterion, pair-wise comparisons are also performed. Local preference for alternatives uses the same equation as the criterion for local priority. The alternative, which complies more strongly with the comparison criterion,

Receipts have a higher weight. If an AHP hierarchy includes a criterion, the alternatives will have a comparison matrix. An (M × N) matrix builds criteria and criteria from local priorities.

$$GP_k = \sum_{j=1}^{N} (LPA_{kj}) \times (LP_j) \quad (2)$$

where GP_k is an alternative k's overall priority (n is criteria number) (LPA K_i) is an alternative K's local priority where ($1 \leq K \leq M$) and (j), and (LP_i) is a local criterion priority (j).

Using the global priority, the decision-maker determines which alternative should be selected; the alternative to the highest global weight is the best one. The eigenvectors of this matrix must be calculated, which would give the relative weights of the criteria. The relative weights obtained in the fourth step should satisfy the formula:

$$\lambda_{max} = A \times W \quad (3)$$

where (A) is the matrix of the pairwise, (W) is the weight and μmax is the higher values of its own (Rawashdeh & Matalkah, 2006). If the hierarchy level has attributes upwards, the weight vector has been determined by multiplying each factor by the parent at a higher level. If the hierarchy is up, this method continues until the height of the hierarchy is approaching, and the best choice is to take the alternative to the highest coefficient weight.

Step 7: Then determined by the AHP to represent the accuracy of the decisions of the decision-maker during the assessment period. The (C.R) consistency ratio, the estimate was rendered as follows both in the decision matrix and in the pairing matrix:

$$C \cdot I = \lambda_{\max} \frac{n}{n-1}$$
$$C \cdot R = \frac{C \cdot I}{R \cdot I} \tag{4}$$

where μmax represents the highest value on its own (explained in step 6 above). N represents several parameters, the greater the consistency if the consistency index is near zero to consider the AHP findings as consistent, the accuracy ratio should be lower than or equal to (0.10). If not, the decision-maker should return to steps three and four and make a further review and estimate.

Step 8: The number of the consistent column entries must therefore divide each factor. A uniform matrix in which all elements vector is summed up is obtained.

6 Structuring the Hierarchy for the Novel Model

This section formulates the appropriate hierarchy with the assistance of the enhanced model (Al_Nawaiseh, et al., 2020a) The hierarchy is expressed in a tabular format reflecting the four levels (Level 0 to Level 3) shown in Table 2. Such structuring makes the new model applies to the AHP process.

- The goal is to evaluate the Academic System (E-Learning System) according to our new model that meets end-user requirements, herein the goal is placed at the top level of the hierarchy (Level-0).
- The six main characteristics were identified to achieve our goal (Level 1).
- The sub-characteristics (Level 2) decompose the main characterizes of System Contents Quality in E-Learning (Level 1) into System Oriented Domain, On-line Services, and Content Quality.
- The Attributes (Level 4) consist of thirty-nine attributes, as listed in Table 2 and grouped in several subsets, each corresponding to one of the nineteen sub-characteristics.

6.1 *Evaluate System Content's Quality in E-learning System*

To strengthen our model, we added a new character, system content quality, which has four sub-characters: system domain, online services, and content quality with their attributes. The term "content quality" refers to the quality of a system used to display

Table 2 New proposed model (ESQE)

Goal (level-0)	Characteristics (level-1)	Sub-characteristics (level-2)	Attributes (level-3)
Choosing the best academic system	Functionality	Interoperability	Platform compatibility Data compatibility
		Sustainability	Application- domain Function specification changes
		Compliance	Standardization
		Security	Privacy Authentication
	Reliability	Recoverability	Time to recover Error reporting
		Availability	Planned down time
		Maturity	Fault tolerance evolvability
	Usability	Operability	Effort of operating Administrability
		Understandability	Documentation Training User support
		Learnability	Time to use Time to configure
		User interface	Consistency Simplicity User control
	Efficiency	Time behavior	Response time Scalability
		Resource behavior	Memory utilization Disk utilisation
	Maintainability	Changeability Testability Upgrading	Customizability Portability Self-test Test site Easy to upgrade
	System content's quality	System oriented domain	Content relevancy Student service information
		On-line services	Online assessment FTP service Connectively
		Content quality	Up to date content Content design Accessibility

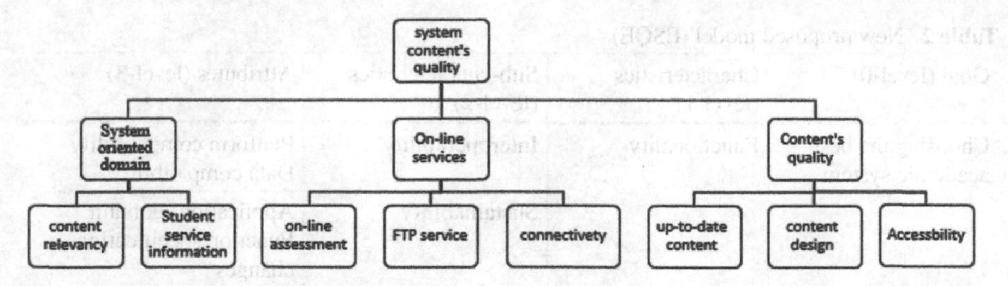

Fig. 7 System content's quality in ESQE

and layout information, as well as the presence of technological characteristics that influence students' perceptions of the system platform with online learning (Fig. 7).

The following are the sub characteristics of system content quality with their respective attributes that were used as a contribution to this study.

1. **Student Oriented Domain**: Academic system services tailored specifically to students, to assist them in their all-academic activities during their study years.
   - **Accessibility**: characterize is defined by Alkhattabi et al. (2010) as the quality features concerned with accessing distributed information. They believe relevancy to be a sub-attribute of accessibility, measured through sub-attributes such as response time and availability. The system in our approach provides relevant content that allows students or instructors to be informed. Correctly, to their right information is extracted from another system such as enrollment, academic degrees, and research.

     Student Services Information: This refers to how the system offers academic and service information to students, such as housing, healthcare, scholarships, and so on, and how it interfaces with other systems.
2. **On-Line Services**: Refers to any information and services provided over the Internet, for instance, enrollment, grades, fees, web services such as FTP, newsgroup, chatting, assessment…etc.

   Online Assessment: The system provides an online grading facility for E Exams and assignments, which sends the grade to students and instructors immediately after the attempt is completed.
3. **Content's Quality**: "Learning Contents" as implemented Combination of "Information" and "Course Website" to construct the entire framework. As up-to-date content in our enhancement model, "Learning Content" applies to all course materials, univ announcements, quizzes, and registered virtual classes in the current job. Factors of "learning material" were taken from previous research, mainly (Alla et al., 2013; Barnard et al., 2017).
   - **FTP Service**: end-users can simply transmit and receive files (download and upload) across the E-Learning system using FTP (Uppal et al., 2018).
   - **Connectivity**: the students connect with the instructors easily with communication tools provide by the system forums, chat, and e-mail.

- **Up-to-Date Content**: To keep all academic content up-to-date, the framework website must be periodically updated in terms of semester material, which includes recorded lectures as videos or live lectures, and updated assignments.
- **Content Design**: Significant factors influencing the quality of content, (Etaati et al., 2011) have been grouped and cited about interface design. Cho et al., (2009), navigation, attractiveness, and ease of use. Students can attend class lectures, log in without assistance to course websites, assuming the websites of the course are configured to be open and easy to reach, including adaptive screen reading and magnification control applications (Al_Nawaiseh et al., 2020b; Cidral et al., 2018)
- **Accessibility**: It defines access using distributions or measurements by sub-attributes to information or websites (relevancy, response time, and availability).

In this paper, accessibility refers to how end-users and developers can readily access their E-learning accounts using a variety of devices such as a computer, laptop, tablet, or mobile phone, and they can access the website most of the time.

Then, following the normal AHP approach, these additional characters will be evaluated by pairwise comparison of criteria and alternatives.

Using Saaty's scaling-table, the AHP six steps and the collected data from the questionnaire, the mean of responders' answers are described relative to the importance of characters to each other. As an Arithmetic mean value of all responses, a weight value was assigned to each of the characteristics: Functionality, Reliability, Usability, Efficiency, Maintainability, and system contents quality.

6.2 Weights Values Matrix for Mu'tah University as Case Study

Since its launch in 1993, the Computer Center has been considered the electronic portal of the University of Mutah, which would support all electronic services for students and workers at the university.

The Department of Publishing and E-Learning was established in 2008 to oversee the E-Learning website in terms of working on managing computerized exams, monitoring their progress and publishing electronic content related to materials for students, and providing protection and security for E-Learning data and recently. Later the University launched Moodle App services with these application features:

- Easily access course content even when offline, Connect with course participants.
- Keep up to date—receive instant notifications of messages and other events, Assignments, and Homeworks Submissions.
- Track student's progress -View the grades, check completion progress in courses and browse the learning plans.

- Complete activities anywhere, anytime—attempt quizzes, post in forums. Here are the pairwise comparison judgments for all E-Learning software attributes that have been evaluated by end-users.

6.3 Constructing the Weight Values Matrix for Sub-characteristics and Attributes for System Content Quality in Mu'tah University

This table describes responder's answers average relative to the importance of characters to each other. To conduct the comparison Table of size N × N for n criteria, also called the priority Table, we must find the ranking of priorities, namely the Eigen Vector, by normalizing the column entries by dividing each entry by the sum of the column entries, and then take the overall row average to get Priority Weight to see table below (Tables 3, 4, 5, 6, 7 and 8).

Table 3 Weights values for the sub-characteristics according to system content's quality

System content's quality	Student oriented domain	Online services	Content quality
System oriented domain	1.000	3.000	9.000
Online services	0.333	1.000	4.000
Content quality	0.111	0.200	1.000
Sum	1.444	4.200	14.000

Table 4 Pairwise comparisons judgment for the sub-characteristics according to system content's quality

System content's quality	Student oriented domain	Online services	Content quality	Priority weights
Student domain	0.692	0.714	0.643	0.683
Online services	0.231	0.238	0.286	0.252
Content quality	0.077	0.048	0.071	0.062
C.R = 0.01				∑Priority weights = 1

Table 5 Weights values for the attributes according to student oriented do main

Student oriented domain	Content relevancy	Student's services
Content relevancy	1.000	2.000
Student's services	0.500	1.000
Sum	1.500	3.000

Table 6 Weights values for the attributes according to online services

Online services	Online assessment	FTP services	Connectivity
Online assessment	1.000	3.000	3.000
FTP service	0.333	1.000	1.000
Connectivity	0.333	7.340	1.000
Sum	1.667	11.340	5.000

Table 7 Pairwise comparisons judgment for the attributes according to online services

Online services	Online assessment	FTP service	Connectivity	Priority weights
Online assessment	0.600	0.265	0.600	0.488
FTP service	0.200	0.088	0.200	0.163
Connectivity	0.200	0.647	0.200	0.349
C.R = 0				∑Priority weights = 1

Table 8 Weights values for the attributes according to content quality

Content quality	Up to date	Content design	Accessibility
Up to date content	1.000	2.000	1.000
Content design	0.500	1.000	0.333
Accessibility	1.000	3.000	1.000
Sum	2.500	6.000	2.333

6.4 Checking for Consistency

The next step is to calculate a Consistency Ratio (CR) to measure how consistent judgments have been relative to large samples of purely random judgments. To verify the consistency of the comparison table, Cho et al., (2009) proposed a Consistency Index (CI) and Consistency Ratio (CR) and should satisfy the condition of CR $< =$ 0.1 Where CR is the average consistency index.

Table 9 Random consistency index

N	1	2	3	4	5	6	7	8	9	10
R.I	0	0	0.52	0.88	1.11	1.25	1.35	1.40	1.45	1.49

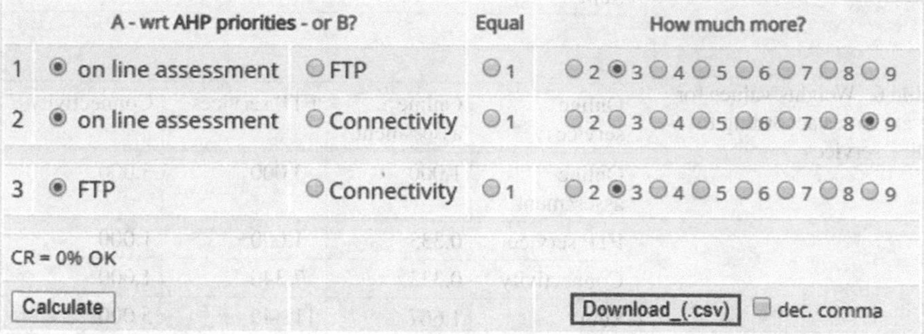

	A - wrt AHP priorities - or B?		Equal	How much more?
1	on line assessment	FTP	1	2 3 4 5 6 7 8 9
2	on line assessment	Connectivity	1	2 3 4 5 6 7 8 9
3	FTP	Connectivity	1	2 3 4 5 6 7 8 9

CR = 0% OK

Calculate   Download_(.csv)   dec. comma

Fig. 8 AHP priority calculator

C.R = C.I / R.I where R.I is the Random Index whose value depends on N (Table 9).

Consistency Index (C.I) = $((\lambda)_{max} - N)/N - 1$ where n is the compared elements. If the C.R is > 0.1 the judgments are untrustworthy because it is too close for comfort to randomness and the evaluation test is valueless must be repeated. Here is the consistency ratio is obtained by a free web-based AHP solution, the AHP priority calculator (Fig. 8).

The corresponding matrices for the sub-characteristics are constructed and shown below in Table 10, following the same procedures applied to the Characteristics, it's may notice that attributes are compared to itself because it is single sub-characteristic, therefore, the assigned priority value is one.

7 Results and Discussion

Table 10 presents all the calculations. It could be seen that the eigenvector of relative importance or in the other words the weights of the sub characteristics as the following:

- Student-oriented domain (0.753).
- On-Line Services (0.231).
- Content Quality (0.62).

Therefore, from students' and academicians' perspectives, the Student Oriented Domain of the E-learning system is the most important one in the other three Characteristics, followed by On-Line Services and Content Quality.

Table 10 Global weights for each characteristic in contents' quality in Mu'tah university

Characteristics (level-1)	Local weight	Sub characteristics (level-2)	Local weight	Attributes (level-3)	Local weight	Global weight
System content quality	0.316	Student oriented domain	0.753	Content relevancy	0.667	0.1587
				Student service information	0.333	0.0792
		On-line services	0.093	On-line assessment	0.488	0.0143
				FTP service	0.163	0.0047
				Connectivity	0.349	0.0102
		Content quality	0.622	Up to date content	0.321	0.0630
				Content design	0.225	0.0442
				Accessibility	0.454	0.0892
$\sum$ Global Weight = 1						

Honestly, the results were expected because the Academic system services are geared specifically to the students. To serve the students in their all-academic activities during their study years, so it must be oriented to the student and their instructors, this sub-character has the high eigenvectors or high priority weight; these services can be measured by their attributes and their priority weights.

- Content Relevance (0.667) and Student Service Information (0.333). It was clear from the evaluation of students and teachers of this feature that Contents Relevancy more important than sharing information with other systems such as registration information, grades information, payment, and university housing information and the consents relevancy gain a high score in user's view.
- From an educational perspective, the students consider the most vital aspects of an E-learning system is On-Line Assessment with priority weight (0.488), FTP Service (0.163), and Connectivity (0.349).
- The results revealed that On-Line Assessment achieved the optimum value corresponding to students. On-Line Assessment, what matters most to students and teachers in academic systems is the system of exams and scores especially now in Covid_19. Therefore, systems developers must improve and develop some services that are accurate and available to students especially now in COVID-19 Epidemic. That allows them to review their exam performance and to teachers improve their skills in assessing students fairly.
- secondly Connectivity (0.349) is a fetal service to students, to connect with their teachers and with each other's, to share materials and information, this feature requires a continuous uninterrupted network to enable students to continue communicating between each other and their teachers.

- According to respondents file transfer protocol FTP (0.163) is not an important service to students, they download course materials, upload Assignments, video and attach files to their teachers alas, it has a low score in this evaluation.
- Finally, the last three attributes Up-to-Date Contents (0.321), Content's Design (0.225), and Accessibility (0.454) are associated to achieve the sub-characterized Content's Quality.
- The most important one is Accessibility to E-learning, which means how the student can access correctly to their session? In addition, easy to access E-Learning through many devices PCs, laptops, Tablet, and Mobile phones, most academic systems and academic webs achieve this attribute.
- Secondly Up to Date Contents, in COVID-19 Epidemic the contents must be updated daily, while the students are not in camps.
- Then E-learning system design, the team needs to focus on improving the design of the content of the system, considering the design of contents, like elements, organized attractively and consistent, multimedia elements must be arranged in a consistently manner.
- AHP process has been applied to calculate eigenvectors and eigenvalue of sub-characteristics of quality characteristics of the target system.
- Therefore, this empirical study provides high institutes with the quality characteristics and their importance that should be taken into account in developing implementing their E-learning system. In which it can be assured that the E-learning Systems will be successfully implemented.

References

Aguarón, J., Escobar, M. T., Moreno-Jiménez, J. M., & Turón, A. (2019). AHP-group decision making based on consistency. *Mathematics, 7*(3), 242.

Al Kurdi, B., Alshurideh, M., Salloum, S., Obeidat, Z., & Al-dweeri, R. (2020a). An empirical investigation into examination of factors influencing university students' behavior towards e-learning acceptance using SEM approach.

Al Kurdi, B., Alshurideh, M., & Salloum, S. A. (2020b). Investigating a theoretical framework for e-learning technology acceptance. *International Journal of Electrical and Computer Engineering, 10*(6), 6484–6496.

Al_Nawaiseh, A. J., Helmy, Y., & Khalil, E. (2020a). A new software quality model for academic information systems, Case study e-learning system. *International Journal of Science Technology Research, 9*(01), 271–282.

Al_Nawaiseh, A. J., Helmy, Y., & Khalil, E. (2020b). Evaluating system content's quality in e-learning system. *International Journal of Psychosocial Rehabilitation, 24*(08).

Al-Adwan, A. S., Al-Madadha, A., & Zvirzdinaite, Z. (2018a). Modeling students' readiness to adopt mobile learning in higher education: An empirical study. *International Review of Research in Open and Distributed Learning, 19*(1).

Al-Adwan, A. S., Al-Madadha, A., & Zvirzdinaite, Z. (2018b). Modeling students' readiness to adopt mobile learning in higher education: An empirical study. *The International Review of Research in Open and Distance Learning, 19*(1), 221–241. https://doi.org/10.19173/irrodl.v19i1.3256

Al-Fraihat, D., Joy, M., & Sinclair, J. (2020). Evaluating E-learning systems success: An empirical study. *Computing Human Behaviour, 102*, 67–86.

Alkhattabi, M., Neagu, D., & Cullen, A. (2010). Information quality framework for e-learning systems. *Knowledge Management and E-Learning: an International Journal, 2*(4), 340–362.

Alla, M., Faryadi, Q., & Fabil, N. B. (2013). The impact of system quality in e-learning system. *Journal of Computer Science and Information Technology, 1*(2), 14–23.

Almajali, D. A. (2016). The role of information technology in motivating students to accept e-learning adoption in universities: A case study in Jordanian universities. *Journal of Business and Management, 4*(1), 36–46.

Almarabeh, T., Mohammad, H., Yousef, R., & Majdalawi, Y. K. (2014). The University of Jordan e-learning platform: State, students' acceptance and challenges. *Journal of Software Engineering and Applications, 7*(12), 999.

Alrawashdeh, T. A., Muhairat, M., & Althunibat, A. (2013). Evaluating the quality of software in erp systems using the iso 9126 model. *International Journal of Ambient System Application, 1*(1), 1–9.

Al-Shboul, M., Rababah, O., Al-Saideh, M., Betawi, I., & Jabbar, S. (2013). A vision to improve e-Learning at the University of Jordan. *World Applied Sciences Journal, 21*(6), 902–914.

Baklizi, M., & Alghyaline, S. (2011). Evaluation of E-Learning websites in Jordan universities based on ISO/IEC 9126 standard. In *2011 IEEE 3rd International conference on communication software and networks* (pp. 71–73).

Barnard, C., Bakkers, J. H., & Wünsche, S. (2017). The road to the digital future of SMEs. *Idc*, 14. https://www.virginmediabusiness.co.uk/pdf/InsightsGuides/TheSMEschangingtheworld.pdf.

Cho, V., Cheng, T. C. E., & Lai, W. M. J. (2009). The role of perceived user-interface design in continued usage intention of self-paced e-learning tools. *Computers & Education, 53*(2), 216–227.

Cidral, W. A., Oliveira, T., Di Felice, M., & Aparicio, M. (2018). E-learning success determinants: Brazilian empirical study. *Computers & Education, 122*, 273–290.

Etaati, L., Sadi-Nezha, S., & Makue, A. (2011). Using fuzzy group analytical network process and ISO 9126 quality model in software selection: A case study in e-learning systems. *Journal of Applied Sciences, 11*(1), 96–103.

Gürkut, C., & Nat, M. (2017). Important factors affecting student information system quality and satisfaction. *EURASIA Journal of Mathematics and Science Technology Education, 14*(3), 923–932.

Nagarajan, P., & Jiji, G. W. (2010). Online educational system (e-learning). *International Journal of u-and e-Service, Science Technology, 3*(4), 37–48.

Rawashdeh, A., & Matalkah, B. (2006). A new software quality model for evaluating COTS components. *Journal of Computer Science, 2*(4), 373–381.

Schwester, R. (2009). Examining the barriers to e-government adoption. *Electronic Journal of e-Government, 7*(1), 113–122.

Tabrizi, S. S., Tufekci, C., & Gumus, O. (2017). New trends and issues proceedings on humanities usability evaluation for near east university student information. 3(3), 235–243.

Tarhini, A., Al-Busaidi, K. A., Mohammed, A. B., & Maqableh, M. (2017). Factors influencing students' adoption of e-learning: A structural equation modeling approach. *Journal of International Education Business.*

Uppal, M. A., Ali, S., & Gulliver, S. R. (2018). Factors determining e-learning service quality. *British Journal of Educational Technology, 49*(3), 412–426.

Yuhana, U. L., Raharjo, A. B., & Rochimah, S. (2014). Academic information system quality measurement using quality instrument: A proposed model. In *Proceedings of 2014 international conference on data software engineering, ICODSE 2014*. https://doi.org/10.1109/ICODSE.2014.7062684.

Obstacles to Applying the E-Learning Management System (Blackboard) Among Saudi University Students (In the College of Applied Sciences and the College of Sciences and Human Studies)

Saddam Rateb Darawsheh, Muhammad Alshurideh, Anwar Saud Al-Shaar, Refka Makram Megli Barsom, Amira Mansour Elsayed, and Reham Abdullah Abd Alhameed Ghanem

Abstract The study aimed to identify the Obstacles to the implementation of applying the e-learning management system (Blackboard) among Saudi university students. To achieve the objectives of the study, the researcher adopted the descriptive approach. The study sample which was chosen randomly consisted of (250) female students from the College of Applied Sciences and the College of Sciences

S. R. Darawsheh (✉) · A. M. Elsayed
Department of Administrative Sciences, The Applied College, (Imam Abdulrahman Bin Faisal University), P.O. Box: 1982, Dammam, Saudi Arabia
e-mail: srdarawsehe@iau.edu.sa

A. M. Elsayed
e-mail: amabdelkhalik@iau.edu.sa

M. Alshurideh
Department of Management, College of Business Administration, University of Sharjah, Sharjah, United Arab Emirates
e-mail: malshurideh@sharjah.ac.ae; m.alshurideh@ju.edu.jo

Marketing Department, School of Business, The University of Jordan, Amman, Jordan

A. S. Al-Shaar
Department of Self Development, Deanship of Preparatory Year and Supporting Studies (Imam Abdulrahman Bin Faisal University), P.O. Box: 1982, Dammam 43212, Saudi Arabia

R. M. M. Barsom
Department Kindergarten, The Science and Huminites Collage, (Imam Abdulrahman Bin Fasil University), Jobil, Saudi Arabia
e-mail: rmbarsom@iau.edu.sa

R. A. A. A. Ghanem
Basic Science Department, Deanship of Preparatory Year and Supporting Studies, (Imam Abdulrahman Bin Faisal University), Dammam, Saudi Arabia
e-mail: raghanem@iau.edu.sa

M. Alshurideh et al. (eds.), *The Effect of Information Technology on Business and Marketing Intelligence Systems*, Studies in Computational Intelligence 1056,
https://doi.org/10.1007/978-3-031-12382-5_21

and Humanities at Imam Abdul Rahman bin Faisal University IAU. The researcher built the study tool "questionnaire" which covers four areas: Administrative barriers, technological barriers, financial (physical) barriers, and social barriers. The data was processed by the statistical method. The findings indicated that the areas of the barriers which face the implementation of the blackboard system among female students at the College of Applied Sciences and the College of Sciences and Humanities at IAU were as follows: Social barriers which ranked first and with a (high degree) followed by financial barriers (moderate degree), administrative barriers (moderate degree) and the technical barriers (low degree) and with the mean 2.60, 2.20, 2.13 and 1.37 respectively. The study recommended promoting the technical aspect in the university through using the blackboard system and providing specialists and technicians with suitable training regarding the use of this system and holding educational sessions for students and the College of Applied Sciences and the College of Sciences and Humanities members to enhance the concept of e-learning.

Keywords Barriers · E-Learning · Blackboard system

1 Introduction

Universities play a significant role in the education process. However, today they face many demands that are imposed on them by the rapid scientific and technological developments despite the limited potential and resources available to them. With the increasing demand for university education, they were keen to raise the level of the effectiveness and quality of education they provide to cope with the requirements of the current era and meet the needs of the labour market (Alshurideh et al., 2021; Kurdi et al., 2020). Therefore they sought to develop their development plans and human cadres to achieve the effectiveness of education in it in line with developments since the university education system must not be limited to the traditional method of teaching in the classroom (Alshurideh et al., 2019a, b; Iskandar, 2019), but rather modern developments in communication technology must be employed and used to provide education that its curricula reach university students in any time and any place, and they provide an opportunity to increase the absorptive capacity of universities for new students and to develop their competencies (; Alsharari & Alshurideh, 2020), skills that are necessary for achieving their success of individuals in social and professional life in the era of the knowledge revolution effectively (Alkmaishi, 2017; Al-Maroof et al., 2021a, b). Educators seek to benefit from the all-new available methods and find modern concepts consistent with scientific and knowledge development (Ghayad, 2016; Leo et al., 2021; Yasin, 2011). Global, Arab, and local interest in technology and technology implementation in education have increased in the light of developments in the fields of technology, communication, and invention accompanying the development of technological concepts and terminology in all disciplines.

The Ministry of Education in Saudi Arabia, especially universities, has started using the Learning Management System, known as the E-Learning Portal (Eduwave), which is one of the technological means in the educational communication process. Babu and Sridevi, (2018) and Al-Hamad et al. (2021) see that e-learning is important for education because it can improve the quality of the learning experience and extend the reach of every lecturer and tutor. It can help remove barriers to achievement, by providing new and creative ways of motivating and engaging pupils and learners of all abilities, enabling, and inspiring everyone to attain their educational potential (Al Kurdi et al., 2020; Amarneh et al., 2021). Where the use of these technological media as seen by (Zboun, 2015) contributes to raising the level of the teaching and learning process, and it helps to break out of the traditional education that is based on indoctrination, information preservation, and retrieval, to the interactive learning that leads to exploration, research, analysis, and explanation and enhances problem-solving skills (Alshurideh et al., 2019a, b; Bettayeb et al., 2020). This system has become one of the most prominent e-learning projects adopted by the Ministry and it aims to develop the teaching and learning processes as well as the educational outcomes (Akour et al., 2021; Tawfiq, 2010). Due to the importance of this method in the education process, many Arab and foreign countries have adopted it and started using the e-learning system in their universities (Saidi, 2016), taking advantage of Saudi expertise in the field of e-learning (Alhamad et al., 2021; Alshurideh et al., 2020). The use of the Saudi e-learning portal puts Saudi Arabia at the forefront of the Arab countries that employ technology in education, which contributes to improving the outputs of the educational process in Saudi universities, and makes Saudi Arabia one of the developed, modern, and exporting countries of distinguished human competencies, able to compete regionally and globally (Bedawi, 2017). The educational system has faced many academic, educational, social, and economic problems, and the most important of these problems is how to communicate with students, especially in the circumstances of the outbreak of the Corona pandemic which disrupted traditional education and call for shifting to e-learning (Alshurideh et al., 2021; Harahsheh et al., 2021). This kind of education is an integrated educational system consisting of the teacher, the student, the curriculum, the support, the Internet, and the material and social factors, all of which affect the virtual learning process (Ashal et al., 2021; Hamad, 2018). The interface of e-learning has many barriers, especially in universities, including administrative, technical, financial, and social barriers that hinder students' access to the required information, their academic courses, the Internet, a faculty member, and the university system (Alshurideh, 2014; Alshurideh et al., 2015). Therefore, the researcher has sought to examine and identify the barriers that hinder the implementation of the Blackboard System for female students at the Community College at Imam Abdul Rahman bin Faisal University.

1.1 Study Problems and Questions

The use of an e-learning portal such as BlackBerry in the learning process has exposed many issues, such as the gap in the ability of teachers to deal effectively with computerized curricula (Hayyani, 2019) and the weakness of technology infrastructure in universities, whether in the field of e-learning or the field of education administration, in addition to that faculty members do not make adequate use of the capabilities available in the e-learning portal. Bearing in mind the crucial role that faculty members play in the teaching and learning process and the importance of paying attention to the factors concerning them when preparing any educational development project in general, this study will be conducted to identify the barriers that hinder the use of the e-learning platform (Blackboard) from the female students' viewpoint in the community college. As a result of advances in the field of information and communication technology, the need to incorporate this technology in the field of education has arisen to create a generation capable of coping with this modern era, thus, educational institutions have faced more pressure to apply e-learning system especially because of the breakout of coronavirus and universities closure. During his work at IAU the College of Applied Sciences and the College of Sciences and Humanities, the researcher noted that students face some challenges in using the Blackboard e-learning management system, and in the light of the above, the study issue lies in trying to address the following key study question: What are the barriers faced by female College of Applied and the College of Sciences and Human students at Imam Abdul Rahman bin Faisal University when using the (blackboard) system. The following sub-questions stem from this question:

1. What are the administrative barriers that face female students applying to the e-learning management system (Blackboard) in Saudi universities students?
2. What are the technical, financial, and social barriers that female students at the College of Applied Sciences and the College of Sciences and Humanities at IAU face during using the Blackboard System?

1.2 The Importance of the Study

The importance of the study lies in the topic it covers at a time the world is focusing on employing modern technology in all its forms, integrating e-learning in the educational process, and providing systems for managing this important type of education to create a means of communication that is accessible and can be effectively used by the learner and the faculty member to achieve the desired learning. Therefore, it is necessary to identify the barriers facing students in using these systems to provide decision-makers at the university with information about these obstacles, in addition to the possibility to benefit from the results in other colleges and universities.

1.3 Study Objectives

The study aims to identify the barriers that face female students at the College of Applied Sciences and the College of Sciences and Humanities at IAU in using the blackboard e-learning management system and suggest some recommendations considering the results of the study.

1.4 Study Limits: The Study is Limited to the Following Limits

1. Time limits: in the second semester of the academic year 2019–2020
2. Spatial limits: the College of Applied Sciences and the College of Sciences and Humanities in Dammam and Jubail—(Imam Abdul Rahman bin Faisal University).
3. Objective Limits: Barriers to the implementation of the Blackboard System among female Students at the College of Applied Sciences and the College of Sciences and Humanities at IAU.
4. Human limits: Female students at the College of Applied Sciences and the College of Sciences and Humanities and those organized in the study.

1.5 Study Terms

Barriers: Procedurally in this study, is defined as a set of technical, material, administrative, and supervisory difficulties that prevent students from using e-learning in different educational situations.

E-learning: It is an innovative way to present an interactive, learner-centred, well-designed, pre-designed that is accessible environment by anyone, anywhere and anytime, using the Internet and digital technologies with educational design principles appropriate for open, flexible, and distributed learning environment (Badawi, 2015). Procedurally it is defined as an educational system in an interactive environment that the university provides to its students via electronic media, innovations, and virtual sessions.

The Blackboard System: It is an integrated system that manages the educational process synchronously and asynchronously, and provides a safe and easy-to-use environment, where faculty members upload their courses and lectures by adding multimedia (text, pictures, audio, video, graphics) in which learners meet to browse the content, each according to his needs, and communicate with each other through multiple communication tools (e-mail, forums) without being restricted by time and place factors, or through virtual classes that can be run from any type of smart device (Eryilmaz, 2015). Procedurally, it is an e-learning management system offered to students at Imam Abdul Rahman bin Faisal University (Community College), through which e-courses and interactive electronic applications are provided.

Female the College of Applied Sciences and the College of Sciences and Humanities: Female students enrolled in the College of Applied Sciences and the College of Sciences and Humanities at IAU in Saudi Arabia, for the academic year 2021/2022, and those who study in the College of Applied Sciences and the in the three departments: Administrative Sciences Department and, Financial Sciences Department, and Computer Science Department at Dammam Campus. And College of Sciences and Human Kindergarten undergraduate Students Department Jabil.

2 Literature Review

2.1 The Concept of E-Learning

E-learning is defined by Khalifa (2011) as the employment of electronic and computerized media in the process of transferring and communicating information (Abadi, 2014) and as a method of teaching or training that enables the learner to obtain education or application at anytime and anywhere in the world through Interactive communication and information technologies in a synchronized manner in the classroom and an asynchronous remote manner, depending on self-education and interaction between the teacher and the learner. Khalifa (2011) believes that e-learning is an interactive system that relies on an integrated digital environment and aims to build curricula in an ideal manner that is accessible by employing digital programs and applications that provide an ideal environment for integrating text with sound and image and offers the possibility of enriching information through links to information sources on various sites and e-libraries. Abadi (2014) believes that e-learning is a method that does not require the physical presence of the teacher and the student in the classroom, where they can interact remotely in a way that eliminates the place and time aspect. E-learning also was defined as using the Internet and digital technology to create an experience and education for learners (Mubark, 2018). Based on the foregoing, we conclude that all definitions considered e-learning as a method of learning that is achieved by employing communication networks and the Internet, where the information reaches recipients in all places and times. E-learning is a modern method that relies on digital technologies (the Internet, computers, electronic library, e-book, etc.) to deliver information without the physical presence of the teacher and the student in the classroom.

2.2 E-Learning Types

Different types of e-learning are represented in the following:

1. Synchronous Online Learning: It enables groups of students to participate in a learning activity together at the same time, from any place in the world (Ashi, 2017) and it involves teacher-students or students-students online chats and videoconferencing (Ibn Ishi and Ibn Ishi, 2018).
2. Asynchronous Online Learning: According to Soman (2011) it is online learning where students study independently at different times and locations from each other, without real-time communication taking place. And it is more student-centred learning where students determine the time and place of learning, and it is done through utilizing digital learning and applications or email. We believe that the World Wide Web combined Synchronous education with Asynchronous education, by storing lessons and courses so that the learners can use them at the appropriate time (Shehri, 2010).
3. Blended Learning: Blended learning includes many learning tools, such as online collaborative learning software, and online-based courses. A pragmatic definition of Blended Learning was proposed by (Blieck & Brussel, 2011): "a good blend is a mix of study materials, work forms and learning activities that contribute to the realization of the objectives, which motivate and challenge the students to show the best of themselves."

2.3 *Obstacles to Applying E-Learning*

Several obstacles hinder the employment of e-learning such as:

1. Technical obstacles: Scholtz (2015) revealed that the most prominently perceived barriers to e-learning include the personal sacrifice of time required, Internet speed, and the lack of on-demand assistance available when learning through the use of electronic media in isolation. The weak technical infrastructure necessary for the application of e-learning and the lack of a high-capacity network to ensure the speed down loading educational curricula and software that provides applications for this type of education leads to limiting the spread of e-learning and the difficulty of its application in educational fields (Kumpikaite & Duoba, 2012).
2. Financial obstacles: Financial obstacles also contribute to limiting the application of e-learning. learners need to have sufficient experience in using computers and the Internet platforms, and if they do not have this ability (Aggarwal, 2009) then, the educational institution has to bear the costs of involving students in training programs regarding access to e-learning platforms and e-learning applications. Also, the limited coverage of the Internet, its relative slowness, and its high price have a significant impact on not employing this type of education (Alberdi et al., 2012).
3. Human and administrative obstacles: Different human and administrative barriers limited the application of e-learning such as the failure to consider the staff

competencies when applying e-learning in the educational institution (Ahmadpour & Mirdamadi, 2010) and the lack of technical, educational, and administrative human cadres capable of advancing this type of education at the individual and community levels. And the lack of training courses offered by the educational institution for its staff and students, as there are few faculty members possessing e-learning skills in an integrated manner. Some students may feel frustrated because they believe that e-learning is not important and that it has no value. Another fact is that some students underestimate e-learning and see it as an unprofessional way of learning due to the lack of direct supervision in teaching. In addition to the negative attitude of some faculty members toward e-learning (Naveed et al., 2017) and the overloaded burden assigned to the faculty members by the administration also plays a crucial role in hindering the application of e-learning.

2.4 Blackboard System

E-learning management system (Blackboard) is Web-based server software that provides a virtual interactive learning environment that maintains the learning process synchronously and asynchronously without the restrictions of time and place, as it offers a secure and easy-to-use learning environment in which faculty members present their courses and lectures by the use of text and images, audio, and video multimedia. It helps learners to understand better and allows interaction between the faculty and students through the application of various communication tools. The blackboard system helps to achieve high efficiency in education as this system supports the publication and presentation of courses, management of student records, follow-up activities, the possibility of communication between students and teachers through special dialogue forums, and the publication of tests, evaluation, and presentation of lectures (Badawi, 2015). In 2013, Imam Abdul Rahman bin Faisal University began training its faculty members to use the Blackboard system. The training was based on cooperation between the University Education Development Center and the Deanship of E-Learning and Distance Education to urge faculty members to use the Blackboard electronic learning management system and to integrate it with the traditional education methods so that it can be benefited from its advantages in different circumstances as nowadays the university has benefited from this system in promoting the teaching-learning process considering the outbreak of the novel coronavirus.

2.5 *Advantages of the Blackboard Electronic Learning System*

E-learning systems have different features and characteristics, but they share a range of significant advantages (Dasimani, 2017; Bunyan, 2019).

1. A graphic interface to facilitate dealing with it through pictures, menus, and changing colours according to the desire of the teacher and the learner.
2. Registration allows students to register for courses and follow up in detail.
3. Customization as this feature allows controlling the setting of the learning system interface and this feature is important because it suits a very large number of users, both following their preferences and what was presented.
4. Helping create and deliver content. Where the system provides an easy-to-use interface with examples and explanations to help users perform any of the required steps in the system.
5. Scheduling and content management, which means distributing course content over the weeks of the year, determining exam dates, and handing over costs according to a tight timeline.
6. Communication supports communication between the faculty member and the student by sending electronic messages, chat rooms, and forums.
7. Virtual Classrooms: The system allows for easy use of classrooms via the Internet.
8. Reports: where the system contains reports related to students' performance, grades, and ratings in educational activities and costs, and can be displayed in the form of files or graphs.
9. Tests and Assessments: These systems enable different numbers of tests by grading them in levels of ease and difficulty while adapting to the individual differences of students.
10. Certificates: The system enables a faculty member to print certificates for students in any form.
11. Mobile learning, as the system adds features that support the use of smart devices to make it easier for students.
12. Using the system as a tool for blended learning, allows the course to be managed through the Internet with the ability to manage it or parts of it traditionally and regularly.

The Blackboard system provides five main systems and solutions interconnected with each other to make education more effective (Huthaili, 2012; Yasin, 2011) which are:

1. Blackboard Learn: It provides the faculty member and the student with the ability to maintain education easily and in any place and time and to follow up on the results of choices and costs in a sequential manner.
2. Blackboard Collaborate: It is a virtual classroom that provides a remote communication service between faculty members and students using a virtual classroom environment and applications that enhance the learning process. process.

3. Blackboard Mobile: It is the use of smartphones and iPads in the teaching and learning process in any place, which gives them greater opportunity to review the entire scientific material.
4. Blackboard Analytics: It helps in assessing the performance of students and groups and analyzing the results to evaluate the educational process to help and make many decisions that improve performance and achieve the required quality of learning through the results.
5. Blackboard Connect: It is a mass notification system that lets students and faculty members send updates and emergency alerts to everyone in the group through emails, phone calls, text messages, or social media channels.

Studies are a basic and valuable guide for the researcher, as they provide him with the valuable information required for the subject of his study and help him to build appropriate study tools. Several studies concerning the subject of the current study have been selected, and the following is a summary of the most important studies.

Shehri (2010) conducted a study that aimed to identify the use of e-learning and training programs and the most important obstacles that limit the application of these programs. These results indicated that e-learning and training programs have become widely used, and there were major obstacles, the most important of which are the technical, human, and educational aspects. Where in Jordan, Soman (2011) identify the barriers to the use of the e-learning portal (EDUWAVE) from the point of view of teachers at Jordanian public schools in Amman and their attitudes towards it. The findings showed that the teachers are satisfied with the e-learning portal site in terms of its design and components. It also showed negative attitudes towards the use of the e-learning portal in the education administration. Huthaili (2012) conducted a study aimed at identifying the most important obstacles facing university e-learning in Yemen, and the results indicated that obstacles to university education include obstacles related to academics and learners, financial and technical obstacles, and barriers specific to society. In a study aimed at identifying the obstacles facing the application of e-learning through an analytical study et al.-Hadba University College conducted by Abadi (2014) showed that there are various obstacles to e-learning, which are the technical, financial, and human obstacles (Ashi, 2017). The study of Al-Qahtani (2014) conducted a study aimed at evaluating the experience of King Khalid University in using the blackboard e-learning management system from the point of view of the faculty member and students at King Khalid University. The study tool covered three domains (patterns of using the blackboard system, the attitudes towards using the blackboard system, and the obstacles to using the blackboard system). The study sample, which was chosen randomly, consisted of (312) faculty members and (846) students. The researchers used the descriptive method through a questionnaire to collect the required information. The results indicated that the faculty members and students use of the different styles of the blackboard system was of a moderate degree and that there were some barriers that faculty and students face in using the blackboard system, and the attitude of the study sample members towards using the blackboard system was neutral. The results also indicate that there were no statistically significant differences attributed to the variable of specialization and training in using the blackboard system. The study recommended expanding the experience of King Khalid University by taking the viewpoints of students and faculty members, adding the standard of using the blackboard system to manage learning within the evaluation criteria of faculty members at King Khalid University. In addition to conducting a periodic survey of the training needs of faculty members from time to time to ensure that there are no needs that could impair the success of their teaching activities in various learning situations.

Ghayad (2016) conducted a study aimed at identifying e-learning as a strategic choice for Algerian universities, the results indicated the need to improve the efficiency and quality of universities, to benefit from the expertise and training of technicians and professionals in this field, and to benefit from the experiences of international countries in the field of e-learning and to overcome its obstacles. In 2017, Alkmaishi (2017) concluded the need to reform the educational system, especially e-learning because it faced various changes in its inputs and outputs and that the most obstacles it faces were the financial deficit. Also, in 2017, a study was conducted to identify the obstacles to using e-learning from the point of view of faculty members at Qassim University considering some variables. Bedawi (2017) built a questionnaire was prepared to collect data. The study found that the most prominent obstacles to using e-learning are the shortage of laboratories available to implement e-learning, the weak experience of the faculty member in using e-learning technology, the limited availability of technicians and specialists to solve technical problems. In addition to the fact that students lack motivation and direct support from the faculty member, and that the focus on the objectives of the online courses is not sufficient (Bedawi, 2017).

The study of Omar and Al-Masabi (2017) aimed to identify the effectiveness of using Blackboard Mobile Learn Application in developing the attitudes of female students at Najran University toward mobile e-learning and to check whether there were significant differences in learners' attitudes towards the use of the application due to the variables: (academic level, the skill of using electronic applications). It also sought to detect the most important obstacles facing students during the use of the application. The study was carried out on (26) female students in the fifth academic level who were enrolled in the "Computer in education" course. The experimental approach was used, the data collection tool was a questionnaire developed by researchers. The study findings indicated that students' attitudes were neutral in the pre-measurement, whereas they become positive in the post-measurement. The use of the Blackboard Mobile Learning Application was effective in developing students' attitudes towards Mobile Learning. Significant differences were noticed between the mean scores of participants in the pre-and post-measurement in favor of the post-measurement. The findings also revealed no significant differences among participants' attitudes towards Mobile Learning via the use of the Blackboard Mobile Learn application attributed to academic level and skill in using electronic applications. The effect of obstacles on the use of the application was moderate, and the most important obstacles were the difficulty of uploading files to the E-courses website and the difficulty of transferring and exchanging files among learners using smart devices.

Another study was conducted by Ashi (2017) to identify the importance of e-learning in achieving human development. The results indicated that developments in the fields of information and communication technology have made a qualitative leap or what is known as global transformations that have affected all educational processes, especially regarding teaching methods. It also indicated that e-learning becomes easy to employ and adapt to communications and information technology reduced social and cultural differences and overcomes the constraints of time and place and the scarcity of human resources. The methods of learning and e-learning also varied, and the need for future visions of the e-learning philosophy related to employing information and communication technology and its uses in all areas of the education system increased. Ruwaili (2018) conducted a study aimed at identifying the obstacles that face students at the College of Education at King Saud University in using the blackboard e-learning management system, and to achieve the objective of the study, the researcher used the quantitative descriptive approach and testing hypotheses and built a questionnaire to collect information. The study population consisted of all male students at the College of Education at King Saud University who were registered in the second semester of the year 1433–1432 AH. The findings showed that administrative obstacles were facing the students in using the blackboard e-learning management system, which came with a mean of (3.49 out of 5). In Dhali (2018) study which aimed to reveal the impediments to the use of e-learning from the point of view of faculty members of Najran University, the study sample consisted of (342) faculty members at 14 colleges of Najran University. The researcher adopted the descriptive method and to collect data, a questionnaire of 22 items was designed. The study results revealed impediments to a high degree regarding these problems: the difficulty of applying e-learning to some courses that require real observation, lack of encouraging incentives, lack of experience in using e-learning, lack of students' response to e-learning, ease of hacking into content and exams and students' lack of access to computer and internet. In addition to the poor internet service and less qualified technicians. Where the impediments to a moderate degree included insufficient counselling sessions and educational meetings, poor communication with students, lack of planning for simultaneous classes, and the difficulty of exam scoring and declaring results. Moreover, the paucity of specialists, poor technical support, and technical updates. Finally, the impediments of lower degree refer to extent of time and effort required for e-learning, lack of social perception towards it, and difficulty of dealing with it. On the other hand, no statistically significant differences were found regarding the variables of gender, educational qualification, and experience in the use of blackboards. The study Al Mubarak (2018) which aimed to identify the reality of applying e-learning in Sudanese universities in light of some recent trends of total quality, indicated that the application of comprehensive quality standards in e-learning available at the university came to a high degree and that the university's e-curriculum objectives consider comprehensive quality standards. It was of a high degree, and the evaluation in the university's electronic curriculum adheres to the comprehensive quality standards with a high mean.

Ibn Ishi and Ibn Ishi (2018) conducted a study aimed to detect the reality of applying e-learning in Algerian universities as a case study at the University of

Biskra, and the results showed that the obstacles to using e-learning lies in using and sitting in front of the computer for a large period and increasing students' burdens in terms of costs and tests. The high cost of maintaining computers, and the lack of technicians. The study of Hayyani (2019) aimed to identify the possibility of using e-learning in addressing students' learning problems through the viewpoints of faculty members and students, the research reached the following results: A positive development in the views of both teachers and students towards the use of e-learning in addressing Learning problems after watching the lecture, and the possibility of using e-learning to address students' electronic problems. Bunyan (2019) evaluated the experience of Umm Al Qura University in the use of the e-learning management system (Blackboard) from the point of view of the faculty members and to detect the obstacles encountered in the use of the system (Blackboard). The study sample was composed of (40) faculty members. The researcher used the descriptive method and a questionnaire as a tool to collect the data needed to answer the study questions. The results of the overall mean of the use patterns were: (average sample direction with a mean of 3.02 and standard deviation of 1.43 and 60.0%, while the results of the overall mean average of the obstacles: the mean of the sample was medium with a mean of 0.38 and a standard deviation of 1.38 and 67.6%. The study recommended encouraging faculty members to use the (Blackboard) in the educational and research process in Saudi universities by embedding them into the calendar entries and training courses. Iskandar (2019) conducted a study aimed at identifying the reality of the quality of the e-educational service provided to remote master's students while identifying the deficiencies in the system using the Ishikawa method to diagnose the problem to suggest methods for improving it. The findings showed that the students' perceptions about the quality standards provided in e-education were one of the deficiencies in addition to some financial obstacles.

The study of Dabab (2019) aimed at revealing the barriers to the attitudes towards digital education in Algerian schools. The results indicated that the main barriers were the financial aspect, lack of possession of an e-learning device, as well as the community's lack of confidence in e-learning. Meajel (2018) examined the faculty perceptions of barriers to using the Blackboard system in teaching and learning at King Saud University in Saudi Arabia. The researchers applied an online questionnaire that was circulated to (117) faculty members to determine how they perceive barriers to using Blackboard. The results showed that academic rank, experience, gender, and training had statistically significant effects on faculty perceptions regarding barriers to using Blackboard in teaching and learning. However, there were no statistically significant differences in faculty perceptions regarding barriers to the utilization of Blackboard triggered by the use or non-use of Blackboard. The current study also showed that 'technological barriers' and 'institutional barriers' were the most highly identified barriers among the four categories of barriers explored in this study. The student barriers category came as a third important factor, while the faculty barriers category ranked at the lower rank.

In Almaiah et al. (2020) study, the researchers explore the critical challenges that face the current e-learning systems such as the blackboard system and investigate the main factors that support the usage of e-learning systems during the COVID-19 pandemic. This study employed the interview method using thematic analysis through NVivo software. The interview was conducted with 30 students and 31 experts in e-learning systems at six universities in Jordan and Saudi Arabia. The findings showed that the factors that affect the usage of e-learning systems are (1) technological, (2) e-learning system quality factors, (3) trust factors, (4) self-efficacy factors, and (5) cultural aspects.

In 2020, a review conducted by Sharifov and Mustafa (2018) to provide an informed overview of e-learning platforms and review different features of Learning Management Systems (LMS), while exploring its implications, general issues, and challenges. The review adopted the Preferred Reporting Items for Systematic Reviews and Meta-Analyses (PRISMA) checklist for standard reporting. The results showed that one of the main barriers to institutional adoption and usage of specific e-learning platforms is that administrators are not permitted to constantly modify the system to better fit the user's requirements. And that the limitations of online learning systems are motivation, costs, feedback and assessment, authentication, and compatibility.

Shehri (2018) study aimed to reveal the reality of students' and English language teachers' use of the electronic learning management system (Blackboard) in the College of Education at King Khalid University, and the study used the descriptive approach. The sample of the study consisted of (51) students from the General Diploma Program in Education and specialists in the English language during the academic year (1436/1437 AH). The study tool was a questionnaire consisting of (26) items. The study concluded the tasks provided by the Blackboard system, which are classified according to use into three categories (high, medium, and rare use). Among the most frequently used tasks: See announcements, alerts, content, and course grades. In addition to handing over assignments. As for the most important tasks that enjoyed moderate use, they were reviewing the description, outline, and course terminology, reviewing the course professor's information, participating in the course's forum, performing electronic tests, and sending and receiving messages. The most notable tasks that were rarely used were creating and sharing blogs and diaries and using the calendar and address book.

Based on the foregoing and through a general review of previous studies we can notice that all the studies agreed on these barriers:

1. Obstacles related to the blackboard e-learning system and its evaluation (Bunyan, 2019) which evaluated the experience of King Khalid University in using the blackboard e-learning management system, and Eryilmaz (2015) study, which aimed to identify the impact of using the Blackboard Mobile learn application on the development Attitudes of female students of Najran University towards mobile e-learning, and a study conducted by Ruwaili (2018) aimed at identifying the obstacles facing students of the College of Education at King Saud University in using the blackboard e-learning management system. Dhali (2018) identifies the obstacles to using e-learning from the point of view of faculty members at

Najran University, and a study conducted by Bunyan (2019) evaluated the experience of Umm Al-Qura University in using the electronic learning management system "Blackboard" from the point of view of faculty members and the study of (Shehri, 2018; Almaiah et al., 2020).

2. Obstacles related to the qualifications of teachers and students to use the e-learning Blackboard (Al-Qahtani, 2014; Ibn Ishi and Ibn Ishi, 2018; Eryilmaz, 2015; Hayyani, 2019; Scholtz, 2015; Soman, 2011) aimed at knowing the reality of using Students are teachers of English for the E-Learning Management System (Blackboard) in the College of Education, King Khalid University.
3. The lack of adequate computer equipment and physical constraints (Alkmaishi, 2017; Ashi, 2017; Ghayad, 2016; Omar, & Al-Masabi, 2017; Shehri, 2018).
4. Some of them presented students' perceptions about e-learning and social obstacles (Almaiah et al., 2020; Blieck & Brussel, 2011; Shehri, 2018, 2010).

The current study has benefited from previous studies in crystallizing the problem of the study, defining its objectives, stating its importance, determining the appropriate statistical treatment, and linking its results with the results of previous studies. The current study also benefited from the rooting of theoretical literature and previous studies. And, this study is distinguished from previous studies, as none of the previous studies has studied the obstacles to implementing the Blackboard e-learning system in the Community College of Imam Abdul Rahman bin Faisal University, and this is the aspect that this study will cover.

3 Methods and Procedures

3.1 Study Methods and Procedure

The researcher adopted a descriptive and analytical approach by studying the research r problem and explaining it scientifically, to reach a reasonable interpretation, which allows the researcher to identify the framework for the research problem and assess the results of the research.

3.1.1 Study Population and Sample

The study population consisted of all students at the College of Applied Sciences and the College of Sciences and Humanities at IAU 2021/2022, and a simple random sample was chosen from the study population from College of Applied Sciences and College of Sciences and Human (Kindergarten undergraduate students), and the study sample consisted of (400) female students at the College of Applied Sciences and the College of Sciences and Humanities in the Dammam branch.

3.1.2 Study Tool

After reviewing the theoretical literature and previous studies related to the issue of obstacles to the use of e-learning in universities, schools, and educational institutions, the questionnaire was built to identify the barriers that hinder the implementation of the e-learning management system (Blackboard system) among female students at the Applied Colleges and the College of Sciences and Humanities at IAU, and the questionnaire consisted of (33) items distributed into four areas: the first area: administrative barriers (11) items, technical barriers and (9) items, financial barriers (7) items and social barriers (6) items. To examine the inferences of the construct validity of the measure, the correlation coefficients of the scale items with the area to which they belong were extracted, where the items of the measure were analyzed and the correlation coefficient of each item was calculated, as the correlation coefficient here represents the inferences of validity for each paragraph in the form of a correlation coefficient between each paragraph and the area that belongs to it as shown in Table 1.

It should be noted that all correlation coefficients were of acceptable scores and statistically significant, and therefore none of these items was omitted.

To check the content validity the tool was presented in its initial form to specialized and experienced arbitrators in education and science techniques in Jordanian and Saudi universities to verify the appropriateness of the items, and to recommend any omission, addition, or modification to the items of the questionnaire. Also, the internal consistency was verified. To verify the reliability of the tool, the internal consistency was calculated according to the Cronbach Alpha as shown in Table 2. These ratios were considered appropriate for this study.

Table 1 Correlation coefficients between paragraphs and the area to which they belong

No.	Correlation coefficients with the area	No.	Correlation coefficients with the area	No.	Correlation coefficients with the area	No.	Correlation coefficients with the area
1	0.740**	11	0.584**	21	0.713**	31	0.754**
2	0.493**	12	0.655**	22	0.758**	32	0.590**
3	0.724**	13	0.612**	23	0.548**	33	0.679**
4	0.645**	14	0.698**	24	0.706**		
5	0.698**	15	0.658**	25	0.761**		
6	0.767**	16	0.572**	26	0.489**		
7	0.609**	17	0.591**	27	0.631**		
8	0.730**	18	0.465**	28	0.730**		
9	0.563**	19	0.667**	29	0.752**		
10	0.705**	20	0.691**	30	0.755**		

*Statistical function at the significance level (0.05)
**Statistical significance at the level of significance (0.01)

Table 2 Cronbach Alpha internal consistency

Areas	Internal consistency	Items
Administrative barriers	0.760	11
Technical barriers	0.766	9
Financial barriers	0.847	7
Social barriers	0.863	6
Total	0.816	3

3.2 *Study Procedures*

The researcher followed these producers when conducting the study

- Reviewing the literature and previous studies concerning the e-learning portal website.
- Preparing the study tool, as previously explained.
- Presenting the questionnaire to specialized arbitrators and making any required modifications.
- Choosing the study sample female students at the Community College, using a simple random method from the colleges of the university.
- Applying the questionnaire to students at the selected community college for the second semester of the academic year 2019/2020.
- Using statistical analysis to analyze the responses.
- Extracting and interpreting the results and linking them to previous studies to answer the study question.

3.3 *Statistical Treatment*

The data obtained from the field study were analyzed statistically, using the Statistical Package for Social Sciences (SPSS—Statistical Package for Social Sciences) and to answer the study questions, the following statistical methods were used: The use of descriptive statistic measures, including the frequency and percentages.

4 Results

The researcher used the 3-point Likert scale to correct the study tool by giving each of its items one score out of three (high degree, moderate degree, low degree), which is represented numerically (3, 2, 1) respectively, and the following scale was adopted to analyze the results, and the interpretation of the results based on the percentages, and the mean according to the following estimates (Table 3).

Table 3 The 3-point Likert scale used to correct the study tool

Degree	Mean
Low	1.00–1.66
Moderate	1.67–2.33
High	2.34–3.00

4.1 Results of the First Question

What are the administrative obstacles that face female students the College of Applied Sciences and the College of Sciences and Humanities at IAU during the implementation of the Blackboard System? To answer this question, means, frequencies, and standard deviations of the estimates of the respondents were calculated on the items of the area of administrative barriers as indicated in Table 4.

Table 4 shows that the arithmetic means of the items ranged from (1.18–2.88), where Item (6) which states, "I think the college network connection is weak / the internet." ranked first with a mean of (2.88), while Item (9) which says, "I think that

Table 4 Mean and standard deviations of administrative barriers

Rank	No.	Items	Mean	SD	Degree
6	1	I think the college network connection is weak " the internet."	2.88	0.195	High
10	2	I think the Blackboard system is not always available in college	2.80	260	High
5	3	I think they lack readiness in the infrastructure for e-learning	2.70	0.192	High
1	4	I think there is a weakness in the capabilities of the technical support team to help female students on the Blackboard	2.66	0.290	High
8	5	I think the college does not have internet access for the student	2.40	0.240	High
11	6	I think there is no clear guideline for dealing with the Blackboard system	1.60	0.295	Moderate
4	7	I think there is insufficient support from the administration (department and college)	1.75	0.190	Moderate
2	8	I think the college does not offer intensive training courses on the degree of using the blackboard system	1.70	0.070	Moderate
3	9	I think there is a lack of computer labs to use the blackboard system	1.50	. 070	Low
7	10	I think there is centralization in the administrative system	1.22	0.088	Low
9	11	I think that there are no incentives for female students to the point of mastering the system	1.18	0.060	Low
		Total Administrative barriers	2.13	0.190	Moderate

Table 5 Mean and standard deviations of the degree of e-learning obstacles that College students face are arranged in descending

Rank	No.	Areas	Mean	SD	Degree
3	1	Technical barriers	1.37	0.210	Low
2	2	Financial barriers	2.20	0.320	Moderate
1	3	Social barriers	2.60	0.358	High
		Total	2.05	0.243	Moderate

there are no incentives for female students to the point of mastering the system." ranked last with a mean of (1.18). The overall means for the area " administrative obstacles" is (2.13.) and it came to a moderate degree. The results indicate that there are prominent administrative barriers in university education, and this is consistent with (Naveed et al., 2017), which showed that administrative barriers were facing the use of the blackboard e-learning management system. And affect the educational system (Scholtz, 2015), which showed the effect of administrative obstacles on students in acquiring the skills of blackboard (Dasimani, 2017) concluded that there were administrative obstacles facing students such as students' lack of computers and Internet, poor training of technicians, weakness and interruption The Internet disagreed with (Soman, 2011), which aimed to identify the importance of e-learning in achieving human development (Shehri, 2018).

4.2 Results of the Second Question

What are the technical, financial, and social barriers that female students at the Applied College and the College of Sciences and Humanities IAU face during using the Blackboard System? To answer this question, mean and standard deviations were extracted as shown in Table 5.

Table 5 shows that the means ranged between (2.30–1.38), where the area of social barriers came in first place with the highest mean of (2.60). Financial barriers ranked second with a mean of (2.20) followed by technical barriers which ranked last with a mean of (1.37). Where the overall mean records (2.05). This result is consistent with (Naveed et al., 2017), which showed that the technical obstacles have a major role in the student's failure to acquire blackboard skills and affect his performance in tests, assignments, etc., and with the study of (Hayyani, 2019) which indicated a positive development in the views of both teachers and students towards the use of e-learning in dealing with learning problems after watching the lecture, and the possibility of using e-learning in addressing students' electronic problems which solve students' problems in financial and social obstacles, and differed with (Blieck & Brussel, 2011) which pointed out that the obstacles facing faculty members in using the "blackboard" system, as well as financial, technical, and social obstacles. Also, the results of the study of (Almaiah et al., 2020; Sharifov & Mustafa, 2018; Shehri, 2018) support the result of the current study. The means and standard deviations of the estimates of the study sample individuals were calculated on the paragraphs of each field separately, as they were as follows.

4.2.1 Technical Barriers

Table 6 shows that the means ranged between (1.60 and 1.20). Item (9) which states "I find it difficult to deal with e-learning systems such as Blackboard and Zoom." came first place with a mean of (1.60), while Item (1) "I think the fast internet connection is weak." ranked last with a mean of (1.20). The overall mean for the technical obstacles is (1.37) as the technical obstacles play a major role in the student's receiving of some skills related to e-learning, especially the blackboard system. The overall mean came to a low degree because technical means need highly skilled technicians and high university support to enhance technical qualifications. This result is consistent with the study of (Sharifov & Mustafa, 2018; Shehri, 2018) which indicated the need for technical skills within the programs for preparing a faculty member and because technical skills are necessary for students in terms of preparation and qualification.

It is also consistent with the study Blackboard educational institution study (2010) which revealed that technical obstacles cause deterioration in the case of the academic student and the study of Eryilmaz (2015) which showed that one of the most important effects of technical obstacles is the difficulty of transferring and uploading large-sized files to smart devices, the difficulty of uploading files to the electronic courses website using smart devices, and the difficulty of transferring and exchanging files between learners using smart devices. The results are also consistent with (Sharifov & Mustafa, 2018; Bunyan, 2019; Shehri, 2018, 2010) which showed that the technical

Table 6 Means and standard deviations of the technical barriers, arranged in descending order according to the means

Rank	No.	Items	Mean	SD	Degree
1	12	I think the fast internet connection is weak	1.20	0.050	Low
2	13	I believe e-learning systems lack periodic maintenance	1.22	0.078	Low
3	14	I think the e-learning system faces many penetrations	1.30	0.070	Low
4	15	I frequently encounter interruptions on the internet	1.33	0.187	Low
9	16	I find it difficult to deal with e-learning systems such as Blackboard and Zoom	1.60	0.295	Moderate
5	17	I think e-learning infrastructure lack readiness	1.35	0.192	Low
6	18	I think curriculum development commensurate with e-learning is weak	1.40	0.194	Low
7	19	I think e-learning applications are not available in Arabic	1.45	low	Low
8	20	I think that the student is not prepared for e-learning	1.50	low	Low
		Total technical barriers	1.37	0.190	Low

obstacles have the most important impact on the e-learning process, especially the blackboard system.

4.2.2 Financial Barriers

Table 7 indicates that the means ranged between (1.94 and 2.20). Item (21) stating, "I think using e-learning systems outside the university is more costly." has the highest mean (2.20), and Item (27) "I think the incentives given to students for using e-learning systems are not sufficient" came last with a mean of (1.94). Table 7 illustrated that the overall mean of the items of the financial barriers is (2.20.) and it records a moderate degree. The researcher attributed this result to the fact that the financial obstacles have a significant impact on financial support for computer technologies and equipment and their provision, and rehabilitation through intensive courses given by skilled and specialized technicians. This cannot be achieved without the financial aspect as the lack of financial support leads to many problems for students in the classroom and outside, and these results are consistent with the study of (Ibn Ishi and Ibn Ishi, 2018).

One of the most important results indicated that there are financial barriers that lead to an increase in students' financial burdens in terms of costs of online learning, tests, the high cost of maintaining computers, and a lack of technicians. Ibn Ishi and Ibn Ishi (2018) and Shehri (2018) confirmed that the financial constraints hinder the application of comprehensive quality standards in e-learning, including students' possession of the required skills for using the blackboard system, and their possession of a computer to keep up with the e-learning education and its continuation. The results of this field are consistent with (Dasimani, 2017) which confirmed that the financial obstacles include technical support, and the weakness of technical updates to the e-learning system, which is reflected in the student's performance. This result

Table 7 Mean and the standard deviations of the financial barriers arranged in descending order

Rank	No.	Item	Mean	SD	Degree
1	21	I think using e-learning systems outside the university is more costly	2.20	0.350	Moderate
2	22	I think the student is overloaded with teaching	2.18	0.408	Moderate
3	23	I think there is a need for a human cadre of professionals and technicians in e-learning	2.17	0.453	Moderate
3	24	I think smart devices which support e-learning application are expensive	2.15	0.453	Moderate
5	25	Lack of laboratories and equipment to assist in e-learning	2.12	0.458	Moderate
6	26	I think not every university student	2.05	0.426	Moderate
7	27	I think the incentives given to students for using e-learning systems are not sufficient	1.94	0.463	Moderate
		Financial barriers	2.20	0.346	Moderate

is also consistent with the result of Alkmaishi (2017) and Ghayad (2016) which emphasized the role of financial obstacles in 'creating a wide gap in e-learning resulted from the weakness of the technicians, the lack of computer labs, and the financial support offered to the university students.

4.2.3 Social Barriers

Table 8 shows that the means of this area ranged between (2.40–2.90). Item (28) which came at the first rank with a mean of (2.90) states "I think that the educated members of the community are frustrated because of their lack of mastery of e-learning" with an average of (2.90), Where Item (33), which states "I think that body language and gestures are not present in e-learning and hinder the communication process" came at the last rank with a mean of (2.40). The overall mean of the social obstacles record (2.60) and a high degree. These results may be attributed to the fact that, this kind of learning is not highly accepted by communities, and that the society's culture only believes in education based on face-face interaction and physical presence in the classroom "traditional" and this is consistent with Dasimani (2017), which confirmed that social obstacles lead to underestimating this kind of learning and to the result which indicated the difficulty of dealing with the electronic system. It also agrees with Naveed et al. (2017), which confirmed that social obstacles have a great impact on the student's compatibility with e-learning. Badawi (2015) pointed out that students 'lack of motivation and direct support from the community affected their awareness to using e-learning in all its standards and types, and the study of Sharifov and Mustafa (2018) and Omar and Al-Masabi (2017), which emphasized the impact of social obstacles in the e-learning process.

Table 8 Means and standard deviations of the social barriers

Rank	No.	Items	Mean	SD	Degree
1	28	I think that the educated members of the community are frustrated because of their lack of mastery of e-learning	2.90	0.508	High
2	29	I think that society does not accept e-learning	2.85	0.515	High
3	30	I think the community awareness programs for e-learning are not sufficient	2.70	0.550	High
4	31	I think there is a difference between e-learning and the culture of society	2.60	0.530	High
5	32	I think there is a difficulty with the idea of e-learning by the community	2.50	0.533	High
6	33	I think that body language and gestures are not present in e-learning and hinder the communication process	2.40	0.535	High
		Total	2.60	0.530	High

5 Conclusion

The study concluded by answering the study's first question that, university students face administrative barriers, which came to a moderate degree and a mean of (2.13). The result also showed that the obstacles students face when using the blackboard system are represented in the following: technical barriers which came with a mean of (1.37) and a low degree, financial barriers which records a mean of (2.20) and a moderate degree. Where the area of the social barrier has the highest mean (2.60) and obtained a high degree. The study recommended promoting the technical side in universities by enhancing the use of e-learning, providing financial support for e-learning use through training of specialists and technicians, and holding educational sessions for students and community members to enhance the image of e-learning.

References

Abadi. W. (2014). Obstacles to the application of E-learning: An analytical study at Al-Hadba university college. An analytical study at Al-Hadba university college. *Journal, University of Mosul, College of Business and Economics, 36*(no.ue.116), 229–251.

Aggarwal, D. (2009). Role of e-learning in a developing country like India. In *Proceedings of the 3rd National Conference; INDIACom-2009 Computing For Nation Development, February 26–27* (pp. 3–6).

Ahmadpour, A., & Mirdamadi, M. (2010). Determining challenges in the application of E-learning in agricultural extension services in Iran. *Learning, 9*(3), 292–296.

Akour, I., Alshurideh, M., Al Kurdi, B., Al Ali, A., & Salloum, S. (2021). Using machine learning algorithms to predict people's intention to use mobile learning platforms during the COVID-19 pandemic: Machine learning approach. *JMIR Medical Education, 7*(1), 1–17. https://doi.org/10.2196/24032

Al Kurdi, B., Alshurideh, M., & Salloum, S. A. (2020). Investigating a theoretical framework for e-learning technology acceptance. *International Journal of Electrical and Computer Engineering (IJECE), 10*(6), 6484–6496.

Al Mubarak, H. (2018). The effect of the application of electronic education at the Sudanese universities according to some of the current trends of comprehensive quality. *Journal of Delta College of Science and Technology, 8*, 99–136.

Alberdi, A. Iribas., A. Martin., N. A. (2012). Collaborative web platform for rich media educational material creation. *International Journal of Educational and Pedagogical Sciences, 65778–782.*

Alhamad, A. Q. M., Akour, I., Alshurideh, M., Al-Hamad, A. Q., Kurdi, B. A., & Alzoubi, H. (2021). Predicting the intention to use google glass: A comparative approach using machine learning models and PLS-SEM. *International Journal of Data and Network Science, 5*(3). https://doi.org/10.5267/j.ijdns.2021.6.002.

Al-Hamad, M. Q., Mbaidin, H. O., Alhamad, A. Q. M., Alshurideh, M. T., Kurdi, B. H. A., & Al-Hamad, N. Q. (2021). Investigating students' behavioral intention to use mobile learning in higher education in UAE during Coronavirus-19 pandemic. *International Journal of Data and Network Science, 5*(3). https://doi.org/10.5267/j.ijdns.2021.6.001.

Alkmaishi, L. (2017). Employing information and communication technology in e-learning. *Al-Hikma Journal for Media and Communication Studies, 9*, 59–74.

Almaiah, A., Al-Khasawneh, A., & Althunibat, A. (2020). Exploring the critical challenges and factors influencing the E-learning system usage during COVID-19 pandemic. *Education and Information Technologies, 25*(6), 5261–5280.

Al-Maroof, R. S., Alshurideh, M. T., Salloum, S. A., AlHamad, A. Q. M., & Gaber, T. (2021b). Acceptance of Google Meet during the spread of Coronavirus by Arab university students. *Informatics, 8*(2), 24.

Al-Maroof, R., Ayoubi, K., Alhumaid, K., Aburayya, A., Alshurideh, M., Alfaisal, R., & Salloum, S. (2021a). The acceptance of social media video for knowledge acquisition, sharing and application: A comparative study among YouYube users and TikTok users' for medical purposes. *International Journal of Data and Network Science, 5*(3). https://doi.org/10.5267/j.ijdns.2021a.6.013.

Al-Qahtani, S. (2014). Evaluation of King Khalid University's Experience in Using Blackboard Education Management. *Arab Society for Educational Technology, 24*.

Alsharari, N. M., & Alshurideh, M. T. (2020). Student retention in higher education: the role of creativity, emotional intelligence and learner autonomy. *International Journal of Educational Management*.

Alshurideh, M. (2014). The factors predicting students' satisfaction with universities' healthcare clinics' services. *Dirasat Administrative Sciences, 41*(2), 451–464.

Alshurideh, et al. (2019a). Understanding the quality determinants that influence the intention to use the mobile learning platforms: A practical study. *International Journal of Interactive Mobile Technologies (IJIM), 13*(11), 157–183.

Alshurideh, M., Al Kurdi, B., Salloum, S. A., Arpaci, I., & Al-Emran, M. (2020). Predicting the actual use of m-learning systems: A comparative approach using PLS-SEM and machine learning algorithms. *Interactive Learning Environments*. https://doi.org/10.1080/10494820.2020.1826982

Alshurideh, M., Bataineh, A., Alkurdi, B., & Alasmr, N. (2015). Factors affect mobile phone brand choices—Studying the case of Jordan universities students. *International Business Research, 8*(3). https://doi.org/10.5539/ibr.v8n3p141.

Alshurideh, M., Al Kurdi, B., & Salloum, S. A. (2019b). Examining the main mobile learning system drivers' effects: A mix empirical examination of both the expectation-confirmation model (ECM) and the technology acceptance model (TAM). *International Conference on Advanced Intelligent Systems and Informatics*, 406–417.

Alshurideh, M.T., Al Kurdi, B., AlHamad, A. Q., Salloum, S. A., Alkurdi, S., Dehghan, A., Abuhashesh, M., & Masa'deh, R. (2021a). Factors affecting the use of smart mobile examination platforms by universities' postgraduate students during the COVID-19 pandemic: An empirical study. *Informatics, 8*(2). https://doi.org/10.3390/informatics8020032.

Alshurideh, M. T., Hassanien, A. E., & Masa'deh, R. (2021b). *The Effect of Coronavirus Disease (COVID-19) on Business Intelligence*. Springer.

Amarneh, B. M., Alshurideh, M. T., Al Kurdi, B. H., & Obeidat, Z. (2021). The impact of COVID-19 on E-learning: Advantages and challenges. In *The International Conference on Artificial Intelligence and Computer Vision* (pp. 75–89).

Ashal, N., Alshurideh, M., Obeidat, B., & Masa'deh, R. (2021). The impact of strategic orientation on organizational performance: Examining the mediating role of learning culture in Jordanian telecommunication companies. *Academy of Strategic Management Journal, Special Is* (Special Issue 6), 1–29.

Babu, G. S., & Sridevi, K. (2018). Importance of E-learning in higher education: A study. *International Journal of Research Culture Society, 2*(5), 1–8.

Badawi, M. (2015). Barriers in the use of faculty members at Menoufia University for management systems e-learning from their point of view. *Journal of Psychological and Educational Research, 30*(4), 96–146.

Bedawi, S. (2017). Obstacles in the use of e-learning from the faculty member's point of view at Qassim University in the light of some variables. *Journal of the Faculty of Education—Assiut University, College of Education, 33*(7379–445).

Bettayeb, H., Alshurideh, M. T., & Al Kurdi, B. (2020). The effectiveness of mobile learning in UAE universities: A systematic review of motivation, self-efficacy, usability and usefulness. *International Journal of Control and Automation, 13*(2), 1558–1579.

Blieck, Y., & Brussel, V. U. (2011). *Blended Learning for Lifelong Learners in a Multicampus Context Blieck , Y ., de Jong , M . and Vandeput , L . Leuven University College,. August*.

Bunyan, R. (2019). Evaluation of Umm Al-Qura university's experience in using the blackboard E-learning management system. *The Arab Journal of Educational and Psychological Sciences, Egypt, 8*, 75–98.
Buras, F. H., & Ashi, A. (2017). *The importance of electronic education in achieving development.*
Dabab, P. (2019). Obstacles to digital education in the Algerian school. *The Arab Journal of Literature and Human Studies, 7.*
Dasimani, S., & Amer, A. (2017). Evaluating King Saud university's experience in using the E-learning management system blackboard. *The Specialized International Educational Review. Dar Semat for Studies and Research: Jordan, 6*(3), 62–72.
Dhali, Z. (2018). Obstacles to using e-learning from the viewpoint of faculty members at Najran University. *The Arab Journal for Quality Assurance of University Education, 11*(36), 153–173.
Eryilmaz, M. (2015). The effectiveness of blended learning environment. Contemporary issus in education research. *Contemporary Issues in Education Research (CIER), 8*(4).
Ghayad, K. (2016). E-learning as a strategic choice for Algerian universities. *Derasat, 46*, 110–116.
Hamad, L. (2018). *The degree of using e-learning tools in teaching by faculty staff members in Jordanian universities and their attitudes toward it.*
Harahsheh, A. A., Houssien, A. M. A., & Alshurideh, M. T. (2021). The effect of transformational leadership on achieving effective decisions in the presence of psychological capital as an intermediate variable in private Jordanian. In *The Effect of Coronavirus Disease (COVID-19) on Business Intelligence* (pp. 243–221). Springer Nature.
Hayyani, B. (2019). The use of E-learning to address student learning problems. *The Arab Foundation for Education, Science and Literature, 8.*
Ishi, I., & Ibn Ishi, A. (2018). The reality of applying e-learning in Algerian universities: a case study of Biskra University. *The International Journal of Quality Assurance, Zarqa University, 1*(ue 1), 9–19.
Iskandar, S. (2019). Diagnosing the quality of the e-learning service provided to remote masters students: An exploratory study of a sample of Algerian University. *Journal of Economy and Human Development, 10*(2), 283–293.
Khalifa. (2011). Obstacles and problems in applying e-learning in university education from the viewpoint of faculty members at Jazan University in the light of some variables. *Journal of the College of Education, 20*(7129–154).
Kumpikaite, V., & Duoba, K. (2012). E-learning process: Students '. *Perspective, 27*, 130–134.
Kurdi, B. A., Alshurideh, M., Salloum, S. A., Obeidat, Z. M., & Al-dweeri, R. M. (2020). An empirical investigation into examination of factors influencing university students' behavior towards elearning acceptance using SEM approach. *International Journal of Interactive Mobile Technologies, 14*(2). https://doi.org/10.3991/ijim.v14i02.11115.
Leo, S., Alsharari, N. M., Abbas, J., & Alshurideh, M. T. (2021). From offline to online learning: A qualitative study of challenges and opportunities as a response to the COVID-19 pandemic in the UAE higher education context. *The Effect of Coronavirus Disease (COVID-19) on Business Intelligence, 334*, 203–217.
Meajel, M., & T. S. (2018). Barriers to using the blackboard system in teaching and learning: faculty perceptions. *Journal of Technology, Knowledge and Learning, 23*(2), 351–366.
Mubark, H., & F. (2018). The reality of electronic application in Sudanese universities considering some recent trends of total quality. *Journal of the Delta College of Science and Technology, Delta College, 18*, 136–199.
Naveed, Q. N., Muhammed, A., Sanober, S., Qureshi, M. R. N., & Shah, A. (2017). Barriers effecting successful implementation of E-learning in Saudi Arabian Universities. *International Journal of Emerging Technologies in Learning, 12*(6), 94–107. https://doi.org/10.3991/ijet.v12i06.7003
Omar, R., & Al-Masabi, Z. (2017). The effectiveness of using black board mobile learns application in developing the attitudes of Najran university students towards mobile-learning. *Specialized International Educational Journal 6*(7).
Qahwan, M., & Huthaili, S. (2012). *Obstacles to e-learning in university education in Yemen.* 289–329.

Ruwaili. (2018). Obstacles to using the E-learning management system, blackberry, among students at the College of Education at King Saud University. *Journal of the College of Education, 34.*

Saidi, O. (2016). Attitudes of faculty members at Majmaah University towards employing e-learning in the educational process. *Journal of Humanities and Administration Sciences, 2.*

Scholtz. (2015). *Barriers to e-learning in a developing country: An explorative study.*

Sharifov, M., & Mustafa, A. (2018). Review of prevailing trends, barriers, and future perspectives of learning management systems (LMS) in higher institutions. *The Online Journal of New Horizons in Education, 10*(3).

Shehri, A. (2010). *Launching e-learning and training programs: A proposed model. Institute of Public Administration. 90*(3).

Shehri. (2018). The reality of English language teachers' uses of the functions of the blackboard system at King Khalid University. *Journal of King Khalid University for Educational Sciences.*

Soman, I. (2011). Obstacles to using the EDUWAVE e-learning portal from the viewpoint of teachers at Jordanian public schools in Amman and their attitudes towards it. *Journal of Educational Sciences Studies—The University of Jordan, 20*(ue 33), 930–917.

Tawfiq, N. (2010). The roles of kindergarten teachers in e-learning and the obstacles to its application from their point of view. *Journal of Childhood and Education, 2*, 2.

Yasin, B. B. (2011). Barriers of using e-learning faced by teachers in Irbid first Directorate of Education. *Palestinian Journal of the College of Education, 35*, 115–136.

Zboun, M. (2015). The effect of teaching using the electronic course system (Moodle) on the achievement of students at the University of Jordan in computer skills and in developing their self-learning and social communication skills. *Unpublished Doctoral Thesis, University of Jordan, Amman: Jordan.*

Possession of Faculty Members and Students of Communication Skills and Their Reflection on Achievement in Saudi Universities

Saddam Rateb Darawsheh, Anwar Saud Al-Shaar, Kawther Abdelrahman Hassan, Lubna Abdullah Abass Almahdi, and Muhammad Turki Alshurideh

Abstract This study aimed to unveil the possession of faculty members and students of communication skills and their reflection on achievement in Saudi universities. The study population comprised (420) staff members representing all departments of the preparatory year. The sample comprised (136) male and female staff members which represent (43%) of the study population. To achieve the objectives of the study, a questionnaire was designed in compliance with the Likert five-point scale comprising of five-axis skills: speaking, listening, dialogue, persuasion and affecting, and negotiation. The most important findings were: Students' overall degree of practice verbal communication skills from the perspective of staff members was high. There were no differences with statistical significance at the statistical level (a $\geq$ 0.05) for staff member responses attributed to social type variable, but there were differences with statistical significance in staff members' responses regarding students'

S. R. Darawsheh (✉) · L. A. A. Almahdi
Department of Administrative Sciences, The Applied College, (Imam Abdulrahman Bin Faisal University), P.O. Box: 1982, Dammam, Saudi Arabia
e-mail: srdarawsehe@iau.edu.sa

L. A. A. Almahdi
e-mail: Laelmahdi@iau.edu.sa

A. S. Al-Shaar
Department of Self Development, Deanship of Preparatory Year and Supporting Studies (Imam Abdulrahman Bin Faisal University), P.O. Box: 1982, Dammam 43212, Saudi Arabia

K. A. Hassan
Department Library and Information Studies, College of Literature, (Imam Abdulrahman Bin Faisal University), Dammam, Saudi Arabia
e-mail: Kahassan@iau.edu.sa

M. T. Alshurideh
Department of Marketing, School of Business, The University of Jordan, Amman 11942, Jordan
e-mail: m.alshurideh@ju.edu.jo; malshurideh@sharjah.ac.ae

Department of Management, College of Business, University of Sharjah, 27272 Sharjah, United Arab Emirates

M. Alshurideh et al. (eds.), *The Effect of Information Technology on Business and Marketing Intelligence Systems*, Studies in Computational Intelligence 1056,
https://doi.org/10.1007/978-3-031-12382-5_22

academic departments in favour of self-development with a mean of (3.94) at the function level (a ≥ 0.05).

Keywords Oral communication skills · Teaching staff members · Saudi universities

1 Introduction

Both training and constant learning got an increasing interest during the second half of the twentieth century due to the unprecedented qualitative and quantitative revolution it witnessed. The success of any society has been measured by the extent of its adaptation to such developments. It has been proved that constant learning and training constitute one of the methods of achieving social development. Training programs via verbal and non-verbal communication skills are essential for educational plan development and for university students who need to practice such skills. Therefore, the researchers sought to make the ideal of sustainable professional development come true through developing self-esteem to upgrade university student levels by acquiring the necessary skills. As reported by Keyton (2017) humans conduct a great number of daily calls in which they exchange news, opinions, feelings, points of view, queries, and dialogues. Through such calls some give orders and others receive them. This kind of activity done by individuals is called communication skills (Aljumah et al., 2021; Alzoubi et al., 2020; Nazir et al., 2021). Thus, communication is an integrated process that depends on participation between two parties a sender and a receiver (Alshurideh & Shaltoni, 2014; Alyammahi et al., 2020; Kurdi et al., 2020). Where they exchange roles conveying their feelings, opinions, and trends (Al-Sugheir, 2003). It also found that communication to play its role, parties need to have active skills such as listening, speaking, persuasion, negotiation, and affecting (Artman, 2005). However, it was pointed in Al-Husami (2020) that relations among university students on campus play an important role in their academic performance and their social and cultural awareness. Most of these relations are academic in which the students exchange ideas and a variety of experiences (Alshamsi et al., 2020; Amarneh et al., 2021; Leo et al., 2021). On the other hand, the student's success in his relationships and contacts with others will have a positive effect on his psyche which accordingly affects his academic performance (Alhamad et al., 2021; Al-Hamad et al., 2021; Alsharari & Alshurideh, 2020). Despite the significance of communication process presented through verbal and non-verbal skills for exchanging information, ideas, trends, experiences, feelings, and opinions, we find many university students lack active communication skills, an issue that hampers achieving objectives of the communication skills resulting in many problems (Akour et al., 2021; Alshurideh et al., 2021; Sultan et al., 2021). Meetings and debates mostly end up with altercations and maybe into brawls. This is one of the factors that prompted writing this paper.

1.1 Significance of the Study

University staff members complain about disputes that happen among students themselves and between students and instructors due to a lack of communication skills. Thus, a need to secure a tool to monitor students' various communication skills became necessary to identify the degree of their practice of communication skills. Such a thing might end up with recommendations that urge holding training programs to develop students' skills by which they will be able to communicate and to have better awareness, socially and culturally (Al-Hamad et al., 2021; Al-Maroof et al., 2021). The significance of the study lies in providing libraries with the information needed for students, staff members, researchers, in addition to identifying communication skills and their positive influence on students at different stages of learning (Al Kurdi et al., 2020).

1.2 Problem and Questions of the Study

The problem of the study rises to from the apparent defect Possession of faculty members and students of communication skills and their reflection on achievement in Saudi universities. Besides, this defect has its negative impacts on students, on-campus or outside. The problem of the study is embedded in identifying the degree of practising oral skills from the perspective of staff members.

1.3 Questions of the Study

The questions might be outlined in the following:

1. What is the Possession of faculty members and students of communication skills and their reflection on achievement in Saudi universities?
2. From the Possession of faculty members and students of communication skills and their reflection on achievement in Saudi universities (a ≥ 0.05) which might be attributed to social category?

1.4 Objectives of the Study

The study attempts to find the answer to the following five questions it poses:

1. What is the degree of faculty members' and students' possession of communication skills and their reflection on achievement in Saudi universities?
2. Are there differences with statistical significance for the degree of practising verbal communication skills by preparatory year students Possession of faculty

members and students of communication skills and their reflection on achievement in Saudi universities (a ≥ 0.05) from the perspective of staff members which might be attributed to social category Variable?
3. Are there differences with statistical significance between the degree of students' practice at Imam's university at the statistical level (a ≥ 0.05), from the perspective of staff members, attributed to academic department variable?
4. Are there differences with statistical significance at the statistical level (a ≥ 0.05) between the practice of preparatory students, from the perspective of staff members, attributed to years of experience?
5. Are there differences with statistical significance between students' degree of practice regarding verbal communication skills at the statistical level (a ≥ 0.05), from the perspective of staff members regarding the interaction between social category, years of experience, and academic department?"

1.5 Limitations of the Study

Limitations of the study are:

1. Topic Limitation: What is the Possession of faculty members and students of communication skills and their reflection on achievement in Saudi universities
2. From the Possession of faculty members and students of communication skills and their reflection on achievement in Saudi universities (a ≥ 0.05) which might be attributed to social category?
3. Space limitation: students in Saudi universities—old building campus.
4. Huma limitation: The study included all teaching staff members (males and females) of all academic ranks.

1.6 Terminology of the Study

These are as follows:

Communication skill: The individual can send and successfully receive massages providing propitious feedback measured by the degree the individual gains via the tool.

Oral communication skill: It is the communication performed through words by which a voice message is given to a receiver and has several connotations. The language used, voice pitch, and place of articulation of words all play a significant role by adding additional meanings to the message.

Teaching staff members: they are the instructors who teach the preparatory year students at the University of Imam Abdul Rahman Bin Faisal.

Preparatory year at the university: the year composes three tracks: health track which includes faculties of medicine, dentistry, applied medical sciences, nursing and clinical pharmacy; engineering track includes faculties of architecture and planning, design engineering; scientific track includes faculties of computer sciences and information technology and business administration track which includes any other faculty which might be added up to the preparatory year following registration rules prescribed by university counsel.

The preparatory academic year comprises two semesters extendable to one more semester (Dammam university, preparatory year, 2016).

Male and female students of the preparatory year: they are male and female students who are admitted to the preparatory year after meeting registration conditions set by the deanship at the University of Imam Abdul Rahman Bin Faisal.

2 Theoretical Literature and Previous Studies

2.1 Theoretical Literature

This part of the study will focus on communication concepts, significance, elements, and types.

Communication can be defined as reported by Abu Isbaa and Saleh, (1996), a process or mechanism in which a certain person in a certain situation sends a message, with information, opinions, or feelings, to others to achieve a certain goal using a language or expressions irrespective of how the message was transmitted (Abu Arqoob & Salah, 1993; Al-Mahawi, 2004; Al-Tanoobi, 2001). While (Al-Olayyan & Al-Tobaji, 2005) believes that the communication process is just sending messages, receiving them, and getting feedback. From what preceded, one can see that communication is a process of exchanging, ideas, information, and feelings between parties to achieve a certain goal. This skill (communication) can be either verbal or non-verbal. Communication is very important for students to realize how significant it is. If listeners fail to understand your words or shun listening to you for feeling bored with your speech, then you lose the goal you aspire to achieve from communicating with your friends or acquaintances. Communication also plays an important role in the life of a university student as it helps him to positively learn from friends and teaching staff members, thus creating an environment of friendship and appreciation. Communication is the tool by which the human achieves his objectives. Its significance is manifested in the processes of speaking, listening, persuasion, negotiation, and leadership. In universities, communication facilitates and reinforces contacts between academic departments (Ismael, 2016; Sedeeq, 2015; Teska, 2003).

The communication process includes these elements: sender, message, means, and receiver. Two more elements were added byAl-Olayyan and Al-Tobaji (2005) namely, feedback and jamming.

The following illustrates such elements:

1. Sender: He is the source of information, initiator of the message who tries to successfully convey his message to the receiver. He might be a teacher who wants to convey his message to students, a student, or a physician. The good sender is the one who can transmit information depending on his good language.
2. Message: It is to the heart of the communication process and the basic axis on which the process rests. It takes numerous forms; it might be written words, oral, body language such as smiling, or griming; it might be clear or ambiguous (Abu Arqoob & Salah, 1993).
3. Receiver: He is the other party who receives and responds to the message if it fits, then he sends feedback. The message might include information or data.
4. Communication goes through different means or channels such as the message that is sent to the receiver. It might be personal talks, telephone calls, memoranda (Al-Mahawi, 2004).
5. Feedback: It is the reaction toward the message which reflects message acceptability or rejection (Abuhashesh et al., 2021; Al-Olayyan & Al-Tobaji, 2005).

Communication can be practised through these two main methods:

1. verbal communication which (Al-Oufi, 2012) is defined as how words, phrases, and sounds are sent by the sender orally through mouth and tongue and the receiver gets these sounds via his ears. Good speech enjoys the following: good quality, sound type, word choice, and adaptability.
2. Non-verbal communication: It is a method of conveying our feelings, opinions, and viewpoints through body movements (Abu Arqoob & Salah, 1993; Al-Khouri, 2000; Bodie, 2011; Stowe et al., 2012). It is a language that has its features, meanings, and implications. It includes movements of the (eye, face, hands, and distance).

Communication has various types including the following:

1. Speaking skill: it is an oral type of communication by which the speaker gets in direct contact with others. Weakness in this skill makes the speaker lose time and opportunities due to indifference and inaccurate talk (Alhamad et al., 2021; Hijab, 2003).
2. Listening skill: It is not only limited to what the ear hears in receiving voice vibrations, but it is an attempt to comprehend what the speaker says to be either sympathetic or adversary toward him and that requires putting ourselves in his shoes; thus, be imaginative.
3. Persuasion and affecting skill: It is the ability to influence others. It is a human action that is effective in proportion with its perfection.
4. Dialogue skill: it is a type of talk between two persons or two groups who debate in an equal manner.
5. Negotiation skill: It is a joint process to adapt conflicting interests to come up with a solution acceptable to both parties.

2.2 Previous Studies

This section reviews the relevant literature.

Al-Khasawneh (1998) conducted a study aiming to unveil hindrances in the process of communication. In that study, she attempted to identify communication problems between students of higher studies and teaching staff members at Yarmouk university from the perspective of the students. She tried to specify problems that lead to making the communication process defective. The study concluded that most of the problems in the process resulted from not providing students with chances to talk face to face to staff members, besides preferences given by staff members to certain students at the expense of others, in addition to hubbub outside classrooms. The study also concluded that young teaching staff members have more contact with students.

Al-Nathami (2002) conducted a study entitled "communication skills of teaching staff members in the faculty of education at Yarmouk university from the perspective of students" which aimed to determine whether teaching staff members have communication skills from the perspective of students. The study came up with the following results: staff members of the faculty of education possess that skill. It was also found that there were differences with statistical significance regarding communication skills among staff members in the faculty which might be attributed to the variable of the academic department, but there were no differences about the academic experience.

Rollins (2004) conducted a study entitled "Needed communication skills during initial employment as perceived by graduates of the West Virginia University Davis College of agriculture, forestry, and consumer sciences". The study aimed to determine the communication skills used during initial employment of graduates of Davis college who were registering in the communication course of agricultural resources. The study came up with the following conclusions, the foremost of which are: The graduates use communication skills, especially those of speaking and listening, but lack the skill of negotiation and persuasion. The study also revealed that there were differences with statistical significance between the means of cognition and students' use of communication skills from the perspective of managers which might be attributed to gender, but there were no differences with statistical significance between means of achievement by college students for verbal communication from the perspective of managers attributed to the variables of work type and job performance. Zhang (2011) conducted a study entitled "Developing communication skills of Chinese students in Canada" to verify the performance of oral communication of Chinese students in Canada. To achieve the objectives of the study, the researcher adopted the experimental descriptive method. The results revealed that there were statistical differences between the means of the degree of practice among Chinese students from the perspective of their instructors that might be attributed to qualification, academic department, and gender. But there were no differences with statistical significance between means of the degree of the practice of Chinese students that might be attributed to instructors' age and students having high communication skills that comply with skills of speech and negotiation. Ibn Jahlan (2009) conducted a study entitled "The effectiveness of a training program for mathematics teachers based on the mathematical communication standard-on achievement and development of oral and written communication skills among students at Saudi Intermediate Schools." The researcher designed a training program. The study concluded differences with statistical significance at the statistical level (a = 0.05) between means of sample members of achievement in favour of the experimental group. The researcher recommended that universities and faculties of education and those specialized in math prepare math teachers through verbal, and written communication skills. In 2015, Sharon (2015) conducted a study entitled "strategies for developing international communication skills for business students". The study aimed to reinforce students' communication skills through teaching staff and employees at Walden University and to ascertain the vision of teaching staff members and employees' communication skills of business graduates. The results indicated that students developed their communication skills after graduation. There were no differences with statistical significance between the means of the business of students' achievements from the perspective of instructors attributed to specialization and academic degree.

3 Methods and Procedures

This section discusses methods and procedures used in the study as presented in methodology, population, and sample.

3.1 *Study Approach*

The researcher used the descriptive analytical survey method that collects information from the population that comprises teaching staff members at the deanship of preparatory year of Imam Abdul Rahman Bin Faisal University. The information will be analyzed afterwards.

3.2 *Study Variables*

3.2.1 Independent Variables

1. Social type: It has two levels (male and Female)
2. Department: It has five levels (self-development, English language, computer, basic sciences, and Islamic culture).
3. Experience: It has four levels (From 1–5, 5–10, 10–15, 16 and above).

Table 1 Population distribution by social type, department, and years of experience

Variable	Levels	Number		
Social types	Social type	Male	Female	Total
Academic department	Self-development	38	31	69
	English language	52	66	118
	Islamic culture	12	12	24
	Computer	18	28	46
	Basic sciences	29	34	63
	Total	149	171	320
Experience	1–5	31	77	108
	6–10	65	70	135
	11–15	27	20	47
	16<	17	13	30
	Total	149	171	320
	Total	320		320

3.2.2 Dependent Variable

The communication skills possessed by the faculty members and students and their reflection on achievement in Saudi universities during the academic year 2022/2021 as perceived by teaching staff members. This was quantitively measured by the study tool.

3.3 Population and Sample

The population of the study comprised all teaching staff members of the preparatory year of Imam's university during the academic year 2022/2021. The number of staff members amounted (230). The sample of the study was collected using the purposive method. Tables 1 and 2 represent the population and sample of the study divided according to social type, academic department, and years of experience.

3.4 Sample of the Study

The sample was randomly selected from teaching staff members of the preparatory year at Imam's university the numbers. Table 2 shows the distribution of the sample (n = 136) by variables.

Table 2 Sample distribution by study variables

Variables	Levels	Number		Percentage	
Social type	Social type	Males	Females	Males	Females
		69	67	22	21
Total		**136**		**43**	
Self-development	Self-development	20	24	06	07
English language	English language	18	25	06	08
Islamic culture	Islamic culture	9	5	3	02
Computer	Computer	10	5	3	02
Total		69	67	22	2
		136		**43**	
	Basic sciences	12	8	4	2
	Total	69	67	22	21
	1–5 years	13	10	4	3
	6–10	14	17	5	5
	11- 15	29	30	9	9
	16<	13	10	4	4
	Total	69	67	22	21
Total		135		34	

3.5 *Tool of Study*

A questionnaire was prepared by the researchers to investigate the degree of practising verbal communication skills by preparatory year students from the perspective of teaching staff members at Imam's university during the academic year 2019/2020. Theoretical literature and relevant studies were reviewed. The questionnaire contains (5) fields regarding verbal communication skills (speaking, listening, dialogue, persuasion and affecting, and negation). All fields are covered by (39) items divided as shown in Table 3.

Table 3 Distribution of questionnaire items into five skills

Field	Items number
Speaking skill	7
Listening skill	10
Dialogue skill	5
Persuasion and affecting skill	9
Negotiation skill	8
Total	39

In compliance with the Likert five-point scale, each item included the following ranks: (very high, high, medium, little, and very little). They were given the following points: (5,4, 3,2,1) respectively.

3.6 Validity and Reliability of the Tool

The tool in its first version was sent to a group of arbitrators (n = 10) to examine clarity and drafting of the items, in addition to items deletion, amending and belonging to its file. Some items were amended and others deleted, but all judges agreed on the suitability of 80% of the items. Alpha Cronbach coefficient was used to calculate the scale coefficient. The reliability percentage was (0.94) which is considered acceptable for the study objectives. The study was applied to a pilot sample (n = 50) of teaching staff members who were randomly selected from the study population. (160) copies of the questionnaire were electronically administered to teaching staff members. (150) copies were retrieved, (14) were excluded due to incompletion. Data were tabulated in tables using SPSS to process the necessary statistics of the study.

3.7 Statistical Processing

Arithmetic means standard deviations, coefficients, and one-way ANOVA were used. (T-test) was also used to explain the differences in the degree level of students" use of communication skills following the variables: social type, academic department, and years of experience.

4 Study Results and Explication

Through limitations, study sample, and statistical processing used in the study, the researchers came up with the results presented in the following tables. Results were interpreted in ratio to percentages and arithmetic mean, according to the estimates in the following Table 4.

4.1 Results and Interpretations of the First Questions

Which is: "What is the degree faculty members' and students' possession of communication skills and their reflection on achievement in Saudi universities?", The results are presented in the following tables.

Table 4 Arithmetic mean

Mean	Level of response (practice)
1–1.80	Very low
1.80–2.60	Low
2.61–3.40	Medium
3.61–4.20	High
4.21–5	Very high

Table 5 reveals that the arithmetic means for all axes of the questionnaire ranged between (3.47–3.94) out of (5). The total grade for all axes was a mean of (3.66) high axes of speaking, dialogue, and negotiation also ranked high; persuasion and affecting ranked medium. This is due to students' practice of speaking, dialogue, and negotiation with each other and with staff members. This result agrees with those of Al-Oufi (2012) and Bodie (2012). As for persuasion skills, students didn't practice it much for they need skills of high rank while listening is a normal skill.

Table 6 reveals that the arithmetic means of all items ranged between (3.30–4.22) out of (5). All means are between very high and high except for item (quickly stops discussions that don't interest him) ranked medium; items: (busy himself with drawing…; evaluate some topics…; jots down notes….;) all ranked little. The table also shows the highest practised item was:

- Returns greeting ….
- Deals with others respectfully ….
- Defends his right in negotiation ….
- Respects feelings of the other party ….
- Achieves his goals through noble means….

The means of answering such items ranged between (30.2 and 4.22) out of (5) and that might be because students' level of communication skills for both males and females was high and associated with academic status. The students possess such skills and practice them on campus. The arithmetic means of the items: (returns greeting …; deals with others respectfully and defends his right while negotiating …) ranged between (4.6; 4.11, 4.22) indicating a very high or high degree of practice which might be attributed to the importance of shaking hands, a social behavioural

Table 5 Descriptive analysis of communication skills

	Axis	Arithmetic mean	Standard deviation	Degree
5	Negotiation	3.94	057,081	High
1	Speaking skill	3.68	70,584	High
3	Dialogue skill	3.66	064,934	High
4	Persuasion and affecting	3.56	082,785	Medium
2	Listening skill	3.47	079,305	Medium
	Total	3.66	063,253	High

Table 6 Arithmetic means and standard deviation for the degree of practising verbal communication skills by preparatory year students (N = 13)

	Item	Arithmetic means	Standard deviation	Degree practice	Rank
1	Returns greeting and shake hands with the other party	4.22	0.835	Very high	34
2	Deals with others respectfully while negotiating	4.11	1.023	High	37
3	Defends his right while negotiating politely and capably	4.06	0.727	High	38
4	Respects feelings of the other party while negotiating	4.04	0.682	High	35
5	Achieves his goals through using noble means and values	4.03	0.815	High	39
6	Easily accepts others' apologies when they err	4.01	0.71	High	19
7	Listens carefully and doesn't interrupt the other party	3.92	0.761	High	9
8	Negotiates others in a skillful manner	3.88	0.699	High	18
9	This shows that he enjoys listening to the other party	3.79	0.758	High	13
10	Repeats certain words and sentences which reflect interaction with the other party	3.78	1.267	High	11
11	Can understand other needs while talking to them	3.78	1.127	High	3
12	Allows others to present their opinions through dialogue	3.78	1.011	High	26
12	Enjoys persuading others on a certain issue	3.75	0.917	High	28

(continued)

Table 6 (continued)

	Item	Arithmetic means	Standard deviation	Degree practice	Rank
14	Enjoys discussing social issues with his instructors	3.74	1.059	High	25
15	The speaker notices listener's attention through answering yes or through humming	3.71	1.087	High	14
16	He cares about how his speech affects other	3.71	0.997	High	1
17	Tries to use simple words	3.68	0.851	High	4
18	Conforms with speaker's psychological case in pleasure or sorrow	3.66	1.08	High	17
19	Can easily understand the views of others	3.64	1.055	High	21
20	Cares about conformity between sound Pitch and the topic	3.63	1.047	High	6
21	Apologizes in case he makes a mistake	3.61	0.943	High	22
22	Friendly starts speaking to a foreigner	3.6	0.929	High	7
23	Asks the speaker to speak clearly to understand words	3.56	1.047	High	8
24	Provides orally the other party with convincing facts	3.55	1.002	High	27
25	Solves his problems with others without losing control over his emotions	3.55	1.002	High	29
26	Faces critiques of others with Pleasures	3.55	0.893	High	30

(continued)

Table 6 (continued)

	Item	Arithmetic means	Standard deviation	Degree practice	Rank
27	Chooses his sentences carefully to attract others	3.54	1.067	High	24
28	Calls his listeners with their names	3.49	1.018	High	5
29	Gives his opinion and comments even if not required	3.49	0.937	High	20
30	Enjoys sitting for a long to achieve his goals	3.43	0.677	High	33
31	Criticizes behavior, but not personality	3.4	0.954	High	31
32	Concludes his speech with sentences like enjoyed talking to you	3.31	1.034	High	10
33	Quickly stops discussions that don't interest him	2.66	0.9	Medium	23
34	Busy himself drawing on paper while talking to the other party	2.44	0.879	Low	12
35	Better evaluates topics after speech is over	2.40	0.847	Low	16
36	Jots down notes through talk to help in discussion later	2.30	0.837	Low	15
	Total	3.60	0.632	High	

principle, in addition to the importance of negotiation for the student as it helps achieve good relations with colleagues and staff members. Such a thing is reflected in students' achievements.

This finding agrees with that of Ibn Jahlan (2009). Where items (34, 35 and 36) obtained a low rank with the following means (2.44, 2.40, 2.30) respectively. This might be attributed to the difficulty of practising these items because they need high communication skills through evaluation and taking notes. The students care for subjects other than these in the verbal communication process, especially in listening which stands as an obstacle in the process of communication.

Table 7 Test results by the gender variable (N = 136)

Gender	Sample	Arithmetic mean	Standard deviation	Deviation error	Degree of practice	(T) value	Statistical function
Male	69	3.80	66.12	4.55	145	8.77	0.642
Female	67	3.75	76.12	12.22	52.70		

4.2 *Results of to the Second Question*

"Are there differences with the From Possession of faculty members and students of communication skills and their reflection on achievement in Saudi universities (a ≥ 0.05) which might be attributed to social category?

The test was used to explain the differences between the two viewpoints of staff members which might be attributed to the social variable of the aforementioned students.

Table 7 reveals that there were differences with statistical significance at the function level (a ≥ 0.05) between degrees of sample members attributed to social category. Male means were (3.80), while females were (3.75) (T) value was (8.77) and the function level (0.642), a value of statistical significance because (T) value at 134 degrees of freedom when (t) test is calculated for two unequal samples was (1.96) at the level (0.05) according to which the (t) value was considered functional.

5 Conclusion

The researchers concluded the following findings:

1. The degree of students' practices of verbal communication skills from the perspective of staff members in all items was similar except for four of them; means ranged between (2.30 and 4.22).
2. There were no statistical differences at the function level (a ≥ 0.05) for the variable of social category from the perspective of staff members.
3. There were differences with statistical differences in students' inclinations attributed to the academic department in favour of self-development with an arithmetic mean (3.94) and a statistical level (a ≥ 0.05)
4. There were no differences with statistical significance at the function level for staff members' answers attributed to the experience variable.

6 Recommendations

In light of the findings, the researcher recommends the following:

1. Giving more attention to communication skills, for students in all departments
2. Holding courses for students to acquire more skills that benefit them in their academic and practical life
3. Conducting future studies which focus on motivating students to attend conferences and training workshops in the field of self-development
4. Conducting more studies on communication skills and self-development
5. Training students on communication skills and on how to positively deal with their instructors.

References

Abu Arqoob, I., & Salah. (1993). *Human communication and it's role in social interaction*, 1st ed. Amman, Jordan: Dar Majdalawi.

Abu Isbaa, T., Saleh & Arja, A. (1996). *Communication and public relations*, 1st ed. Palestine: Al-Quds Open University.

Abuhashesh, M. Y., Alshurideh, M. T., & Sumadi, M. (2021). The effect of culture on customers' attitudes toward Facebook advertising: the moderating role of gender. *Review of International Business and Strategy*.

Akour, I., Alshurideh, M., Al Kurdi, B., Al Ali, A., & Salloum, S. (2021). Using machine learning algorithms to predict people's intention to use mobile learning platforms during the COVID-19 pandemic: Machine learning approach. *JMIR Medical Education, 7*(1), 1–17.

Al Kurdi, B., Alshurideh, M., Salloum, S., Obeidat, Z., & Al-dweeri, R. (2020). An empirical investigation into examination of factors influencing university students' behavior towards elearning acceptance using SEM approach.

Alhamad, A. Q. M., Akour, I., Alshurideh, M., Al-Hamad, A. Q., Kurdi, B. A., & Alzoubi, H. (2021). Predicting the intention to use google glass: A comparative approach using machine learning models and PLS-SEM. *International Journal of Data and Network Science, 5*(3). https://doi.org/10.5267/j.ijdns.2021.6.002.

Al-Hamad, M. Q., Mbaidin, H. O., Alhamad, A. Q. M., Alshurideh, M. T., Kurdi, B. H. A., & Al-Hamad, N. Q. (2021). Investigating students' behavioral intention to use mobile learning in higher education in UAE during Coronavirus-19 pandemic. *International Journal of Data and Network Science, 5*(3). https://doi.org/10.5267/j.ijdns.2021.6.001.

Al-Husami, W. (2020). The effect of using a computerized educational Program on the social communication skills, among the students of communication principles subject in the faculty of mass communication at the University of Petra. *Research Journal of Humanities and Social Sciences, 2*(12), 653–668.

Aljumah, A., Nuseir, M. T., & Alshurideh, M. T. (2021). The impact of social media marketing communications on consumer response during the Covid-19: Does the brand equity of a university matter. *Eff. Coronavirus Dis. Bus. Intell., 334*, 384–367.

Al-Khasawneh, B. (1998). Communication problems between higher studies students and teaching staff members at Yarmouk University. From students' perspective. Yarmouk University.

Al-Khouri, I. F. (2000). *Body language*, 1st ed. Beirut: Dar Al-Saqee.

Al-Mahawi, S. (2004). *Communication skills*, 1st ed. Amman, Jordan: Dar Yafa.

Al-Maroof, R. S., Alshurideh, M. T., Salloum, S. A., AlHamad, A. Q. M., & Gaber, T. (2021). Acceptance of Google meet during the spread of coronavirus by Arab university students. *Informatics, 8*(2), 24.

Al-Nathami, N. (2002). Communication skills of teaching staff members at college of education at Yarmouk University from the perspective of students. Yarmouk University.

Al-Olayyan, A. R., & Al-Tobaji. (2005). *Communication and foreign relations. Amman*, 1st ed. Amman, Jordan: Dar Safa for publishing and distribution.

Al-Oufi, A. L. (2012). *Basic skills in communication*, 1st ed. Riyadh: Al-Nashr Al-Ilmi wa Al-Matabi.

Alshamsi, A., Alshurideh, M., Al Kurdi, B., & Salloum, S. A. (2020). The influence of service quality on customer retention: A systematic review in the higher education. In *International conference on advanced intelligent systems and informatics* (pp. 404–416).

Alsharari, N. M., & Alshurideh, M. T. (2020). Student retention in higher education: the role of creativity, emotional intelligence and learner autonomy. *International Journal of Educational Management*.

Alshurideh, M. T., & Shaltoni, A. M. (2014). Marketing communications role in shaping consumer awareness of cause-related marketing campaigns. *International Journal of Marketing Studies, 6*(2), 163.

Alshurideh, M. T., et al. (2021). Factors affecting the use of smart mobile examination platforms by universities' postgraduate students during the COVID 19 pandemic: An empirical study. *Informatics, 8*(2), 32.

Alshurideh, M., Al Kurdi, B., & Salloum, S. A. (2019a). Examining the main mobile learning system drivers' effects: A mix empirical examination of both the expectation-confirmation model (ECM) and the technology acceptance model (TAM). In *International conference on advanced intelligent systems and informatics* (pp. 406–417).

Alshurideh, M., Salloum, S. A., Al Kurdi, B., Monem, A. A., & Shaalan, K. (2019b). Understanding the quality determinants that influence the intention to use the mobile learning platforms: A practical study. *International Journal of Interactive Mobile Technology, 13*(11).

Al-Sugheir Kuleib, A. (2003). *Constructing a training programme for the development of the communication skills of public school administrators in Irbid Governorate in light of their training needs*. Amman Arab University.

Al-Tanoobi, M. (2001). *Communication theories*, 1st ed. Alexandria, Egypt: Shuaa Library.

Alyammahi, A., Alshurideh, M., Al Kurdi, B., & Salloum, S. A. (2020). The impacts of communication ethics on workplace decision making and productivity. In *International conference on advanced intelligent systems and informatics* (pp. 488–500).

Alzoubi, H., Alshurideh, M., Al Kurdi, B., Inairat, M. (2020). Do perceived service value, quality, price fairness and service recovery shape customer satisfaction and delight? A practical study in the service telecommunication context. *Uncertain Supply Chain Management, 8*(3), 579–588. https://doi.org/10.5267/j.uscm.2020.2.005.

Amarneh, B. M., Alshurideh, M. T., Al Kurdi, B. H., & Obeidat, Z. (2021). The impact of COVID-19 on E-learning: Advantages and challenges. In *The international conference on artificial intelligence and computer vision* (pp. 75–89).

Artman, M. A. (2005). What we say and do: The nature and role of verbal and nonverbal communication in teacher-student writing conferences. The University of Wisconsin-Milwaukee.

Bodie, G. D. (2011). Development of oral communication skills by evidence of validity within the interpersonal domain. *Communication Quarterly, 39*(3), 277–295.

Hijab Mohammed, M. (2003). *Communication skills for media professionals, educators, and preachers, (8) 31*. Dar Al-Fajr for publishing and distribution.

Ibn Jahlan, A. (2009). The effectiveness of a training program for mathematics teachers-based on the mathematical communication standard-on achievement and development of oral and written communication skills among students at Saudi Intermediate Schools. Amman Arab University.

Ismael Mahmoud, A. (2016). *Communication skills* (1st ed.). Dar Al-Neel for disinhibition and publishing.

Keyton, J. (2017). Communication in organizations. *Annual Review of Organisational Psychology and Organisational Behaviour, 4*, 501–526.

Kurdi, B. A., Alshurideh, M., & Alnaser, A. (2020). The impact of employee satisfaction on customer satisfaction: Theoretical and empirical underpinning. *Management Science Letters, 10*(15). https://doi.org/10.5267/j.msl.2020.6.038.

Leo, S., Alsharari, N. M., Abbas, J., & Alshurideh, M. T. (2021). From offline to online learning: A qualitative study of challenges and opportunities as a response to the COVID-19 pandemic in the UAE higher education context. *Effective Coronavirus Disease Business Intelligence, 334*, 203–217.

Nazir, M. I. J., Rahaman, S., Chunawala, S., & AlHamad, A. Q. M. (2021). "Perceived factors affecting students academic performance. *Nazir, J., Rahaman, S., Chunawala, S., Ahmed, G., Alzoubi, H., Alshurideh, M., AlHamad, A. Perceived factors Affect. Students Acad. performance. Acad. Strateg. Manag. Journal, 21(Special Issue 4), 1–15.*, *21*(4), 1–15.

Rollins, J. R. (2004). *Needed communication skills during initial employment as perceived by graduates of the West Virginia University Davis College of Agriculture*. West Virginia University.

Sedeeq, M. (2015). *Communication skills: Theory and practice*, 1st ed. Egypt: Dar-Ilm wa Al-Iman for publishing and distribution.

Sharon, A. (2015). Strategies for developing interpersonal communication skills for business students. Amman Arab University.

Stowe, K., von Freymann, J., & Schwartz, L. (2012). Assessing active learning and skills training through alumni and current student views. *Journal of Case Studies in Accreditation and Assessment, 2*, 1.

Sultan, R. A., Alqallaf, A. K., Alzarooni, S. A., Alrahma, N. H., AlAli, M. A., & Alshurideh, M. T. (2021). How students influence faculty satisfaction with online courses and do the age of faculty matter. In *The International conference on artificial intelligence and computer vision* (pp. 823–837).

Teska, J. E. (2003). *The superintendency: Effective leadership through communication*. Eastern Michigan University.

Zhang, L. (2011). *Development of oral communication skills by Chinese students in Canada case studies* (1st ed.). LAP LAMBERT Academic Publishing.

Effectiveness of Supportive Services on Academic and Social Development of Students with Disabilities

Anas Mohammad Rababah, Jaber Ali Alzoubi, Saddam Rateb Darawsheh, Anwar Saud Al-Shaar, Muhammad Alshurideh, and Tareq Alkhasawneh

Abstract The aim of the current study is to identify supportive services' effectiveness on the academic and social development of students with disabilities at special education centers in Jordan from teachers' and parents' perspectives. The sample of the study consisted of (106) special education teachers, and (106) parents. To achieve the study objective, the supportive services and the academic and social development questioners were used. The results of the study revealed that the level of the supportive services provided to students with disabilities, as well as their academic and social development, was moderate from teachers' and parents' perspectives. Furthermore, a statistically significant positive correlation was found between the total score of

ملاحظة : بامكانك دكتور محمد إضافة ثلاث باحثين او اثنين بالتبادل معي .

A. Mohammad Rababah
Special Education – Ministry of Education, Sharjah, UAE
e-mail: anasrab93@hotmail.com

J. Ali Alzoubi
Speech Specialist – Jordan, Amman, Jordan
e-mail: Jaberjjj555@yahoo.com

S. Rateb Darawsheh (✉)
Department of Administrative Sciences, The Applied College, Imam Abdulrahman Bin Faisal University, P.O. Box: 1982, Dammam, Saudi Arabia
e-mail: srdarawsehe@iau.edu.sa

A. S. Al-Shaar
Deanship of Preparatory Year and Supporting Studies, Department of Self Development, (Imam Abdulrahman Bin Faisal University), P. O. Box: 1982, Dammam 43212, Saudi Arabia

M. Alshurideh
Department of Marketing, School of Business, The University of Jordan, Amman 11942, Jordan
e-mail: malshurideh@sharjah.ac.ae; m.alshurideh@ju.edu.jo

Department of Management, College of Business, University of Sharjah, 27272 Sharjah, United Arab Emirates

T. Alkhasawneh
Al Ain University, Sharjah, United Arab Emirates

M. Alshurideh et al. (eds.), *The Effect of Information Technology on Business and Marketing Intelligence Systems*, Studies in Computational Intelligence 1056,
https://doi.org/10.1007/978-3-031-12382-5_23

supportive services provided to students with disabilities and their academic and social development. In light of the results, the study recommends providing advanced levels of counseling and physiotherapy services for students with disabilities.

Keywords Supportive services · The academic and social development · Special education centers · Students with disabilities

1 Introduction

Gürgür et al. (2016) report that students with disabilities have cognitive, social, and emotional problems in addition to encountering many difficulties related to communication and language acquisition. This confirms the need to provide customized programs and services supporting their needs. According to the nature and severity of the disability, problems facing students with disabilities vary. Some cases may require the provision of short/long-term special and intensive services to develop their communication skills, speech, language, and their self-care (Hebbeler and Spiker, 2016). Due to the difficulties, they face in their academic, social and other various skills, students with disabilities need special services, especially in case of severe disability affecting their performance, even though they may have appropriate intellectual and mental skills. In view of the increasing number of children with disabilities, it is necessary to focus on providing them with supportive services address the problems they face in the academic, linguistic, mobility, developmental, emotional and social aspects, and to develop programs and services providing them with academic and social skills (Khasawneh, 2013).

Different bodies provide services to individuals with disabilities, including centers of special education, which support students with disabilities by providing various programs and services enabling them to acquire the necessary skills and knowledge to manage their disability and the problems they encounter (Din and Muhammad, 2020). Supportive services as contended by Kelly-Hall (2010) are among the services provided by special education centers for students with disabilities. These include the programs that provide students opportunities for achieving academic development and success and equip them with the necessary skills for continuing their education. These programs are defined as a set of services provided to students with disabilities on an ongoing basis targeting developing their skills and providing them with learning through special centers and within specified hours, or through specialists of special education centers providing home services, or through official affiliated-government bodies (Thombson et al., 2014). Furthermore, supportive services are considered among the means provided to students from specific groups in society to help them academically through directing them towards obtaining education, developing their skills and creating the suitable conditions according to their health and physical situation. Students are also supported through enabling them to acquire numeracy, language.

The Act of "Education of All Handicapped Children" indicated that the supportive services are the means used to provide children with disabilities with special educational services and the consequent services required for overcoming the difficulties and obstacles they face within the educational process (Zettle and Ballard, 2015). Special education centers have a key role in granting students with disabilities the right to learn through enhancing their cognitive and academic aspects; developing specialized educational programs based on technology; and equipping them with the necessary social and emotional skills enable them to interact with their families, teachers, peers, and society members. Thus, their present and future perceptions about life would be positively changed (Cheng and Lai, 2020).

Due to being one of the most important rights that should be secured for all children, many local and international bodies have been interested in securing the right of individuals with disabilities to help them obtaining their full rights, including the right to receive education. Academic development refers to an individual's excellence in a specific academic field to obtain expertise in a particular discipline through the physical efforts and interaction with curricular and extracurricular activities (Alsharari and Alshurideh, 2020; Alshurideh et al., 2021, 2021a). It also refers to an individual's acquisition of a number of qualities, such as self-confidence and punctuality that would help him in achieving the academic excellence (Al Kurdi et al., 2020a, 2020b; Alshurideh et al., 2019, 2019a; Merghany and Ibrahim, 2020).

Students with disabilities' enrollment in special education centers provides them the opportunity to learn, interact and learn new skills and techniques helping them to face the problems they encounter, contributing therefore to their academic and social development. Academic development is related to students' ability to develop their academic performance through providing the means and tools that help them develop their thinking and analysis skills, and their ability to organize time and concentrate on how to overcome the educational problems they encounter (AlHamad et al., 2021; Al-Hamad et al., 2021; Al-Maroof et al., 2021, 2021a; Grzegorz et al., 2017; Leo et al., 2021).

Students with disabilities' academic level is best developed by: ensuring the provision of various and appropriate teaching methods commensurate with the nature and severity of their disability; providing materially and morally appropriate resource rooms; providing electronic devices, programs and computers supporting their education process; providing the service of recording lectures on their devices; and ensuring specialists' and teachers' competency in helping students with disabilities develop their skills, obtain information and overcome the problems they encounter in the programs offered to them (Ahmad, 2017; Akour et al., 2021; Alshurideh et al., 2020).

Supportive services play an important role in promoting students with disabilities' academic development, during which programs and services, that help students to self-organize their knowledge, are provided through teaching them the skill of planning, controlling efforts and lifelong learning.

Social development of students with disabilities is important to enhance their self-value and self-concept. Integrating students with disabilities within the educational process, and the provision of appropriate educational programs contributes to enhancing their communication with their peers, teachers and specialists (Alshurideh et al., 2019b; Kurdi et al., 2021). In the resource rooms, teachers receive training and specialized courses on how to develop students with disabilities at different aspects, including the social one, through which students' behavioral patterns are modified, and an adequate counseling is provided to be able to control the different emotions and situations they encounter (Bettayeb et al., 2020; Leyser and Krick, 2007).

Social development has a main role in developing students with disabilities' skills related to building relationships and making friends through preparing them to deal with the real world; making them more accepting of diversity and difference between them and others; reinforcing their self-worth sense; influencing their emotions; changing their misbehaviors; and helping them to understand the problems they encounter (Amarneh et al., 2021; Dababneh, 2016). Students with disabilities are socially developed through the provision of recreational activities, through which they acquire communication skills, interact and cooperate with their peers and others as they complete a certain activity, develop their individual skills, and learn how to control their emotions (Ashal et al., 2021; Hilbert, 2014).

Furthermore, the supportive services also contribute to increasing the social interaction between students with disabilities and their peers with/without disabilities, in addition to enhancing their feeling that they have an effective role in society, and that they are able to attain achievements. Thus, differences between them and others would be reduced, in addition they will be more self-confident, more self-esteem, more acceptance to their disability and able to handle different problems and circumstances. Consequently, supportive services have a key role in achieving students with disabilities' social development (Hassan et al., 2017; Khasawneh et al., 2021a, 2021b).

In addition to the above, the supportive services also have a role in enhancing students with disabilities' verbal skills through the speech-language services and psychological counseling, increasing therefore their ability to understand others, and their verbal and behavioral communication as well. Thus, the supportive services affect the social development of students with disabilities and help them to control their emotions and achieve compatibility between their ideas and skills through structured and continuous training (Obeid and Al-Hadidi, 2018).

Given the above, supportive services are programs depend on a group of parties determining a student's status and his/her needs to have specific support and skills in order to be able to benefit from special educational programs, during which skills and experiences are developed to help them build social relationships and learn in a way suits the nature of their disability. Supportive services also include the parents' participation in activating these services, following them up at home and making notes to teachers and specialists in order to develop these services continuously, and keep pace with students with disabilities' needs of for academic and social development (Aljumah et al., 2021; Alwan and Alshurideh, 2022).

The Ministry of Education has sought to implement programs and plans related to students with disabilities' development; ensure the availability of qualified staffs; provide supportive and educational services; and ensure their development socially, intellectually and academically (Ministry of Education, 2015).

Students with disabilities' academic and social development depends on teachers' attitudes and their inclination to interact with them, and their ability to effectively implement the supportive services in a way ensuring the achievement of educational programs' objectives that are designed for this category of students. Providing the appropriate services to students with disabilities, preparing teachers and taking advantage of specialists in this field enhances teachers' ability to deal with students. It also makes them acquainted with the difficulties facing students and the required efforts to increase supportive services effectiveness in developing students' academic and social aspects (Alshamsi et al., 2020; Kang and Martin, 2017).

A study (Mulvey et al., 2016)of reported that teachers, with regard to students' existence within an educational environment, usually focus on the academic aspects, regardless the obstacles that hinder their progress and acquisition of knowledge. Students with disabilities situation differ according to the nature and severity of their disability, as it affects their response towards the educational programs and the provided curricula. It also reported that teachers do not focus on the social and emotional environment surrounding students with disabilities, which is considered a factor affecting students with disabilities' success and failure in life, compatibility with the surrounding community, and their acquisition of knowledge.

There are several studies examining the supportive services and the academic and social development, and the relationship between them. A study (Mohammad and Arafa, 2015), in Saudi Arabia, assessed the supportive services provided to students with special needs from faculties and supervisors' perspectives and their own perspectives as well among a sample of (13) male and female students with special needs from Majmaah University and (52) faculty members and supervisors from the Department of Special Education. The level of supportive services provided to students with special needs was "moderate", and that there were statistically significant differences in the level of supportive services provided to students with special needs from students' perspective, due to the gender, and in favor of males. (Gürgür et al., 2016) examined the effectiveness of the supportive services provided to students with hearing impairment at Turkey special education centers from the teachers' perspectives. The results also showed that there are, from the teachers' perspectives, some difficulties in providing supportive services, represented by the practical aspects, reports of special education centers, the time required to receive support, low levels of pre-service training and low levels of family participation.

In Saudi Arabia, (Hassan et al., 2017) examined effectiveness of supportive services on students with disabilities' social development from special education teachers' perspectives. Social development level was moderate from teachers' perspectives. There was a statistically positive correlation between the supportive services provided to students with disabilities and the development of the social and psychological skills of students with disabilities from teachers' perspectives. (Al-Blawi, 2018) investigated the effectiveness of the supportive services provided to students with hearing disabilities from their parents' and teachers' perspectives in Qurayyat city, in Jordan. The results of the study showed that the effectiveness of the supportive services was moderate from their parents' and teachers' perspectives. (Contreras et al., 2018) studied special education centers' role in the academic development of students with disabilities in the United States. The sample of the study consisted of (23) students with disabilities. To achieve the objective of the study, a quasi-experimental design was used. The results indicated that the academic development of students with disabilities was low.

2 Problem of the Study

The problem of the study stems from the recommendations indicated by the previous studies, such as (Al-Blawi, 2018) indicating that the effectiveness of the supportive services provided to students with disabilities was low. There is a need to ensure the appropriate provision of the programs and services that develop the intellectual and academic aspects of students with disabilities. As for the study (Din and Muhammad, 2020), it indicates that there are certain problems in the educational process of students with disabilities at special education centers, which came as a result to some problems in teachers' competence, in addition to the nature of the strategies used in teaching students with disabilities.

According to the researchers' field experience in Special Education, a lack in the supportive services provided to students with disabilities were observed, despite their importance in developing several social, behavioral and academic aspects. For example, the studies (Contreras et al., 2018; Hassan et al., 2017) confirmed that the provision of the supportive services can develop the academic and social aspects of students with disabilities regardless of the type and severity of their disability. This implies the importance of providing supportive services for this group of students, indicating a need to examine their impact on their various psychological and emotional aspects. According to the researchers' revision of previous literature related to special education, a clear knowledge gap was observed, as there is a scarcity of studies addressing the effectiveness of the supportive services on the academic and social development of students with disabilities. Thus, the researchers considered that there is a need to conduct a study identifying the nature of this effect. From the above, the problem of the current study stems from answering the following questions:

1. What is the level of the supportive services provided to students with disabilities in special education centers from their teachers' and parents' perspectives?
2. What is the level of academic and social development of students with disabilities at special education centers from their teachers' and parents' perspectives?
3. Is there a correlation between the supportive services and the academic and social development of students with disabilities at special education centers from their teachers' and parents' perspectives?

3 Significance of the Study

The importance of the current study stems from the fact it can shed light on the role of supportive services in developing students with disabilities academic and social skills. It is also hoped that this study will enable decision makers in the educational field in Jordan to identify the quality of the supportive services provided to students with disabilities, and the quality of other services that should be provided to them.

Definitions of Terms

Supportive services: A set of services provided to students with disabilities inside or outside the educational institutions to enable them getting benefit from the educational programs provided to them, including health, psychological, counseling, physical therapy, transportation, language and speech therapy, and other necessary services (Ahmad, 2017).

Academic development: Students' ability to develop their academic performance through the provision of means and tools help them develop their skills related to thinking, analysis, and ability to organize time and overcoming the educational problems they encounter within the educational process (Grzegorz et al., 2017).

Social development: Students' ability to develop their practices and behaviors in a way enabling them to achieve the adjustment with their peers and achieve balance and psychological stability through improving their ability to address problems, and control feelings of anxiety, depression and instability (Dessemontet and Bless, 2011).

4 Method and Procedures

To achieve the objectives of the study, descriptive analytical design was used, by collecting information related to the study and analyzing it to get the results that will help in understanding and developing the educational environment.

4.1 Study Sample

The study sample consisted of (106) special education teachers in addition to (106) parents. These were randomly selected from different geographic regions in Jordan.

4.2 Supportive Services Scale

The scale was developed by reviewing a set of previous studies such as (Alquraini, 2007; Alshamsan, 2008). The scale consisted in its primary format of (30) items distributed on (4) domains: Speech and language therapy services (7 items); occupational therapy services (8 items); physical therapy services (8 items); counseling services (7 items). To check the scale face validity, content validity was verified by asking (11) experts to give their remarks concerning the items suitability and adding any other appropriate item. (80%) of the proposed amendments by the juries were taken into consideration. The scale in its final format consisted (26) items distributed on (4) domains: Speech and language therapy services (6 items); occupational therapy services (6 items); physical therapy services (8 items); counseling services (6 items). Construct validity was verified by calculating correlation coefficients between the items and their domains were calculated through a pilot sample consisted of (20) parents and (20) special education teachers. Correlation coefficient of each item was calculated, as the correlation value indicates validity significance for each item since it indicates the correlation value between the item and the total score from one hand and between each domain and the total score on the other hand. The correlation coefficient ranged between (0.67–0.40), all values were larger than (0.20) which considered accepted for the purposes of the current study. It also can be noted that all correlation coefficients values were statistically significant and accepted for the purposes of this study. Scale reliability was checked using Test–Retest through a sample consisted of (20) parents and (20) special education teachers selected from the same population and out of the original sample. The scale was administrated and re-administered on the same sample after two weeks. Then, Pearson's Correlation Coefficient factor was calculated between their responses in both times. After that, Internal Consistency using Cronbach's Alpha was calculated for the sample scores on the first administration, Internal Consistency value was (0.815) which is a high coefficient indicating that the scale enjoys a high and acceptable stability coefficient, as seen in Table 1.

Table 1 Cronbach's alpha and person's correlation coefficient and the total score for the supportive services scale

Domain	Person correlation coefficient	Cronbach alpha
Speech and language services	0.903	0.831
Occupational therapy services	0.900	0.823
Physical therapy services	0.882	0.853
Counseling services	0.847	0.849
Total	0.888	0.815

4.3 Academic and Social Development Scale

The scale was developed by reviewing a set of previous studies such as Alquraini (2007), Dessemontet and Bless (2011), Grzegorz et al. (2017). The scale consisted in its primary format of (24) items distributed on (2) domains: Academic development (14 items) and social development (10 items). To check the scale validity, content validity was verified by asking (11) experts to give their remarks concerning the items suitability and adding any other appropriate item. (80%) of the proposed amendments by the juries were taken into consideration. The scale in its final format consisted (26) items distributed on (4) domains: Speech and language therapy services (6 items); occupational therapy services (6 items); physical therapy services (8 items); counseling services (6 items). Construct validity was obtained via calculating correlation coefficients between the items and their domains were calculated through a pilot sample consisted of (20) parents and (20) special education teachers. Correlation coefficient of each item was calculated, as the Correlation Coefficient value indicates validity significance for each item since it indicates the Correlation Coefficient value between the item and the total score from one hand and between each domain and the total score on the other hand. The Correlation Coefficient ranged between (0.71–0.43), all values were larger than (0.20) which considered accepted for the purposes of the current study. It also can be noted that all Correlation Coefficient values were statistically significant and accepted for the purposes of this study. Scale reliability was obtained using Test–Retest through a sample consisted of (20) parents and (20) special education teachers selected from the same population and out of the original sample. The scale was administrated and re-administered on the same sample after two weeks. Then, Pearson's Correlation Coefficient was calculated between their responses in both times. After that, Internal Consistency using Cronbach's Alpha was calculated for the sample scores on the first administration, Internal Consistency value was (0.823) which is a high coefficient indicating that the scale enjoys a high and acceptable stability coefficient, as seen in Table 2.

Table 2 Cronbach's alpha and person's correlation coefficient and the total score for academic and social services scale

Domain	Person correlation coefficient	Cronbach alpha
Academic development	0.892	0.841
Social development	0.880	0.829
Total	0.922	0.823

5 Results

5.1 Results of the First Question: "What is the Level of the Supportive Services Provided to Students with Disabilities in Special Education Centers from Their Teachers' and Parents' Perspectives?"

To answer this question, means and standard deviations were calculated for the level of supportive services provided to students with disabilities in special education centers from their teachers' and parents' perspective, as shown in Table 3.

Table 3 shows that the level of supportive services provided to students with disabilities in special education centers from their teachers' perspectives was moderate (M = 3.30, Std. Devi. = 0.490). Means of the scale domains ranged between (3.04–3.52) with a moderate level. Speech and language services domain ranked first (M = 3.52, Std. Devi. = 0.649) with a moderate level, while counseling services ranked last (M = 3.04, Std. Devi. = 0.774) with a moderate level.

Furthermore, the level of supportive services provided to students with disabilities in special education centers from their parents' perspectives was moderate (M = 3.25, Std. Devi. = 0.402). Means of the scale domains ranged between (2.97–3.77) with a moderate level. Speech and language services domain ranked first (M = 3.77, Std. Devi. = 0.703) with a moderate level, while physical therapy services ranked last (M = 2.97, Std. Devi. = 0.471) with a moderate level.

Table 3 Means and standard deviations for the level of supportive services provided to students with disabilities in special education centers from their Teachers' and Parents' perspective

No	Domain	Teachers' perspective		Parents' perspective		Level
		Mean	Std. Devi	Mean	Std. Devi	
1	Speech and language services	3.52	0.649	3.77	0.703	Moderate
2	Occupational therapy services	3.29	0.771	3.19	0.733	Moderate
3	Physical therapy services	3.34	0.681	2.97	0.471	Moderate
4	Counseling services	3.04	0.774	3.17	0.687	Moderate
Total		3.30	0.490	3.25	0.402	Moderate

Table 4 Means and standard deviations for the level of academic and social development of students with disabilities in special education centers from their Teachers' and Parents' perspective

No	Domain	Teachers' perspective		Parents' perspective		Level
		Mean	Std. Devi	Mean	Std. Devi	
1	Academic development	2.97	0.510	3.08	0.479	Moderate
2	Social development	3.31	0.588	3.17	0.470	Moderate
Total		3.15	0.420	3.12	0.364	Moderate

5.2 Results of the Second Question: "What is the Level of Academic and Social Development of Students with Disabilities' at Special Education Centers from Their Teachers' and Parents' Perspectives?"

To answer this question, means and standard deviations were calculated for the level of academic and social development of students with disabilities in the special education centers from their teachers' and parents' perspective, as shown in Table 4.

Table 4 shows that the level of academic and social development of students with disabilities in special education centers from their teachers' perspectives was moderate (M = 3.15, Std. Devi. = 0.420). Means of the scale domains ranged between (2.97–3.31) with a moderate level. Social development domain ranked first (M = 3.31, Std. Devi. = 0.588) with a moderate level, while counseling services ranked last (M = 2.97, Std. Devi. = 0.510) with a moderate level.

Furthermore, the level of academic and social development of students with disabilities in special education centers from their parents' perspectives was moderate (M = 3.12, Std. Devi. = 0.364). Means of the scale domains ranged between (3.08–3.17) with a moderate level. Social development domain ranked first (M = 3.17, Std. Devi. = 0.470) with a moderate level, while academic development domain ranked last (M = 3.08, Std. Devi. = 0.479) with a moderate level.

5.3 Results of the Third Question: "Is There a Correlation Between the Supportive Services and the Academic and Social Development of Students with Disabilities at Special Education Centers from Their Teachers' and Parents' Perspectives?"

To answer this question, Correlation Coefficients was calculated between supportive services and academic and social development of students with disabilities at special education centers, as follow.

Table 5 Pearson correlation coefficient between supportive services and academic and social development of students with disabilities in special education centers from Teachers' perspectives

Sample	Domains		Academic development domain	Social development domain	Total score of development
Teachers	Speech and language services	Correlation coefficient	0.378**	0.120	0.311**
		Sig.	0.000	0.219	0.001
		Sample	106	106	106
	Occupational therapy services	Correlation coefficient	0.202*	0.430**	0.419**
		Sig.	0.038	0.000	0.000
		Sample	106	106	106
	Physical therapy services	Correlation coefficient	0.082	0.559**	0.456**
		Sig.	0.404	0.000	0.000
		Sample	106	106	106
	Counseling services	Correlation coefficient	0.170	0.205*	0.254**
		Sig.	0.082	0.035	0.009
		Sample	106	106	106
	Total score	Correlation coefficient	0.286**	0.506**	0.535**
		Sig.	0.003	0.000	0.000
		Sample	106	106	106

Table 5 shows that there is a statistically positive correlation at ($\alpha = 0.05$) between the total score of supportive services and academic and social development of students with disabilities at social education centers from teachers' perspectives (Correlation Coefficient = 0.535**, Sig. = 0.000)

5.4 *Teachers' Perspective*

Correlation Coefficient was calculated between supportive services and academic and social development of students with disabilities at special education centers from teachers' perspectives, as shown in Table 5:

5.5 *Parents' Perspective*

Correlation Coefficient was calculated between supportive services and academic and social development of students with disabilities at special education centers from parents' perspectives, as shown in Table 6.

Table 6 Pearson correlation coefficient between supportive services and academic and social development of students with disabilities in special education centers from parents' perspectives

Sample	Domains		Academic development domain	Social development domain	Total score of development
Parents	Speech and language services	Correlation coefficient	0.355**	0.201*	0.364**
		Sig.	0.000	0.039	0.000
		Sample	106	106	106
	Occupational therapy services	Correlation coefficient	0.243*	0.248*	0.320**
		Sig.	0.012	0.010	0.001
		Sample	106	106	106
	Physical therapy services	Correlation coefficient	0.079	0.510**	0.382**
		Sig.	0.421	0.000	0.000
		Sample	106	106	106
	Counseling services	Correlation coefficient	0.103	0.122	0.147
		Sig.	0.001	0.000	0.000
		Sample	106	106	106
	Total score	Correlation coefficient	0.314**	0.417**	0.477**
		Sig.	0.001	0.000	0.000
		Sample	106	106	106

Table 6 shows that there is a statistically positive correlation at ($\alpha = 0.05$) between the total score of supportive services and academic and social development of students with disabilities at social education centers from parents' perspectives (Correlation Coefficient = 0.477**, Sig. = 0.000).

6 Discussion

The study results showed that the level of the supportive services provided to students with disabilities in special education centers from their teachers' perspectives was moderate. This finding can be explained by the fact that special education centers do not have the adequate resources to provide supportive services. Additionally, the nature of the supportive services needs human resources of workers, specialists and counselors, which means that there is a need to provide many financial resources to attract and employ them and this is what special education centers lacks. This result can also be explained by the fact that the supportive services are diverse, numerous, and differ according to the nature of the disability that the child suffer from; this dictates to train the specialists in various disciplines in order to offer high-quality supportive services. Furthermore, the results showed that the level of the supportive services provided to students with disabilities in special education centers from their parents' perspectives was moderate. It can be explained by the fact that the parents of students with disabilities are aware of the importance of supportive services, the need for them in order to ensure better education for the students, and the ability to develop a set of motor and performance skills, which can help in evolving their performance even though their level was moderate. Besides, parents are well informed about the difficulty of providing supportive services by special education centers which lack the adequate material and human resources capable of providing this type of services. This result can be also explained by the fact that PWD Act of the Government of Jordan emphasized the importance of providing the highest level of supportive services for the parents of individuals with disability to enable them to provide a closer to normal environment for their children. Furthermore, private enterprise become aware about the importance of serving individuals with disabilities and their parents and that encouraged many investors to establish private enterprises concerned with providing residential and supportive services for individuals with disabilities. Additionally, the workers in public and private enterprises which focus on individuals with disabilities have sufficiency, knowledge and experience that enable them to understand the importance of supportive services for the parents of individuals with disabilities to help them to provide a normal environment for their children. This result is consistent with the results of Mohammad and Arafa (2015) study, which showed that the level of supportive services for students with disabilities was moderate. It is also consistent with the results of Al-Blawi (2018) study which showed that the level of supportive services for students with hearing disabilities was moderate from parents and teachers' perspectives.

Additionally, the results of the study indicated that the level of academic and social development of students with disabilities at special education centers from teachers' perspective was moderate. This result can be explained by the fact that teachers are aware of the importance of improving students academically and socially, this makes them look positively at any academic or social improvement achievement by the child with disability. Also, the contribution of supportive services in academic development is conditional on the capacity of the special education center to provide the best level of supportive services, which dictates that the center should provide great human potentials from specialists in different disciplines, in addition to providing the necessary infrastructure. Therefore, if the supportive services properly supplied, they would contribute significantly in achieving academic and social improvement for the child with disability; as they help him to adapt with the nature of his disability through what they provide of environmental adaptations capable of meeting this child needs. Merghany and Ibrahim (2020) confirmed that the effectiveness of the programs provided to students with disabilities in special education centers is influenced by the level of quality commitment in the delivery of such programs, and their compatibility level with global criteria and indicators designed to implement these programs. Therefore, this result can be explained by the fact that teachers are aware of the importance of achieving academic and social development; however their awareness level about the criteria and indicators could be one of the factors affecting planning and management processes, providing an appropriate educational environment for students, and implementing programs adequately. This means that teachers seek to improve the level of supportive services as they are aware of their importance in helping child with disability in achieving higher levels of academic and social improvement. As if working on providing supportive services contributes significantly to help child with disability in understanding the nature of his disability and employing his skills and potentials to achieve higher levels of life works. This result can be due to those teachers in special education centers use the different teaching methods however they are limited. Din and Muhammad (2020) emphasizes that teachers ability on defining the adequate teaching methods for the student with disability increases the ability of achieving significant academic improvement. Consequently, teachers may not tend to diversify teaching methods, and not tend to innovate extensively, which requires ensuring that teachers are trained before and during service and ensuring that the programs and electronic services are which enhance the students' academic level are available. The results also found that the level of academic and social development of students with disabilities at special education centers from parents' perspective was moderate. This result can be explained by the fact that parents watching their children with disabilities daily and notice the level of academic and social improvement they can achieve by using the provided supportive services in special education centers. Also, their experience with the child with disability made them seek for better levels of services, which made them able to provide substantive provisions concerning supportive services ability on achieving academic and social development for the child with disability. It also can be contended that parents are more aware about the nature of the provided supportive services for their children with disabilities, and this makes them make relatively good judgments concerning the level of these services and their ability to meet some basic needs of the child with disability. Moreover, parents want to get more information about the different services they can offer for their children that can help them to overcome some functional problems that they face due to their disability. The result is consistent with the results provided by Hassan et al. (2017) study which found that the level of social development was moderate among students with disabilities from teachers' perspectives. While it differs from Contreras et al. (2018) study which found that the level of academic development was low among students with disabilities.

In addition, a positive significant correlation was found between supportive services and academic and social services for the students with disabilities in special education services from teachers' and parents' perspectives. This result can be explained to the fact that working on providing supportive services is an indicator that public and private educational enterprises take into consideration the importance of supportive services and their role in achieving academic and social development for the children with disabilities. Furthermore, providing services such as speech and language therapy services, occupational therapy services, physical therapy services, and counseling services contribute significantly in helping children with disabilities to adapt with the nature of their disability and employing their skills, and their mental and physical abilities in achieving higher levels of academic and social improvement. Also, the academic needs of students with disabilities are vary from normal students; as the fact of just providing books and school friendly facilities for students with disabilities does not mean their positive effect of the academic aspects; there is a need for providing supportive services able to help students with disabilities to live relatively normal through these services contribution in making them more willing to live a normal life, which means that they are able to reach relatively acceptable levels of academic achievement and to live a relatively normal social life; and this means that providing supportive services is related significantly to helping children with disabilities in achieving academic and social development.

7 Recommendations

In light of the results, the study recommends providing advanced levels of counseling and physiotherapy services for students with disabilities. Also, there is a need to create the optimal conditions to achieve the academic and social development of students with disabilities, such as preparing school and special education institutions to receive them. Moreover, additional light should be shed on conducting more studies addressing the effectiveness of the supportive services on other variables, such as parents' mental health and students with disabilities' social skills.

References

Ahmad, A. (2017). The status co of support services and their relationship to the level of satisfaction of students with visual disabilities about university life at the College of Education - Qassim University. *Journal of Educational Sciences, 1*, 170–207.

Akour, I., Alshurideh, M., Al Kurdi, B., Al Ali, A., & Salloum, S. (2021). Using machine learning algorithms to predict people's intention to use mobile learning platforms during the COVID-19 pandemic: Machine learning approach. *JMIR Medical Education, 7*(1), 1–17.

Al Kurdi, B., Alshurideh, M., & Salloum, S. A. (2020a). Investigating a theoretical framework for e-learning technology acceptance. *International Journal of Electrical and Computer Engineering (IJECE), 10*(6), 6484–6496.

Al Kurdi, B., Alshurideh, M., Salloum, S., Obeidat, Z., & Al-dweeri, R. (2020b). An empirical investigation into examination of factors influencing university students' behavior towards elearning acceptance using SEM approach. *International Journal of Interactive Mobile Technologies, 14*(2), 19–24.

Al-Blawi, M. (2018). Degree of parents and teachers' satisfaction about supportive services provided to hearing impaired students in Al –Qurayyat. Unpublished M.A Thesis, Yarmouk University, Jordan.

AlHamad, M., Akour, I., Alshurideh, M., Al-Hamad, A., Kurdi, B., & Alzoubi, H. (2021). Predicting the intention to use google glass: A comparative approach using machine learning models and PLS-SEM. *International Journal of Data and Network Science, 5*(3), 311–320.

Al-Hamad, M., Mbaidin, H., AlHamad, A., Alshurideh, M., Kurdi, B., & Al-Hamad, N. (2021). Investigating students' behavioral intention to use mobile learning in higher education in UAE during Coronavirus-19 pandemic. *International Journal of Data and Network Science, 5*(3), 321–330.

Aljumah, A., Nuseir, M. T., & Alshurideh, M. T. (2021). The impact of social media marketing communications on consumer response during the COVID-19: Does the brand equity of a university matter. In *The effect of coronavirus disease (COVID-19) on business intelligence* (pp. 367–384).

Al-Maroof, R., Ayoubi, K., Alhumaid, K., Aburayya, A., Alshurideh, M., Alfaisal, R., & Salloum, S. (2021b). The acceptance of social media video for knowledge acquisition, sharing and application: A comparative study among YouYube users and TikTok users' for medical purposes. *International Journal of Data and Network Science, 5*(3), 197–214.

Al-Maroof, R. S., Alshurideh, M. T., Salloum, S. A., AlHamad, A. Q. M., & Gaber, T. (2021a). Acceptance of Google Meet during the spread of Coronavirus by Arab university students. *Informatic, 8*(2), 1–17; Multidisciplinary Digital Publishing Institute.

Alquraini, T. (2007). The availability and effectiveness of the supportive in supporting the educational process for students of intellectual education. Unpublished M.A Thesis, King Saud University, KSA.

Alshamsan, A. (2008). Evaluation related services for mentally disabled children at the institutes of intellectual education in Saudi Arabia. Unpublished M.A Thesis, Jordan University, Jordan.

Alshamsi, A., Alshurideh, M., Al Kurdi, B., & Salloum, S. A. (2020, October). The influence of service quality on customer retention: a systematic review in the higher education. In *International Conference on Advanced Intelligent Systems and Informatics* (pp. 404–416). Springer, Cham.

Alsharari, N. M., & Alshurideh, M. T. (2020). Student retention in higher education: the role of creativity, emotional intelligence and learner autonomy. *International Journal of Educational Management. International Journal of Educational Management, 35*(1), 233–247.

Alshurideh, M., Salloum, S. A., Al Kurdi, B., Monem, A. A., & Shaalan, K. (2019). Understanding the quality determinants that influence the intention to use the mobile learning platforms: A practical study. *International Journal of Interactive Mobile Technologies, 13*(11), 183–157.

Alshurideh, M. T., Al Kurdi, B., & Salloum, S. A. (2021). The moderation effect of gender on accepting electronic payment technology: A study on United Arab Emirates consumers. *Review of International Business and Strategy, 31*(3), 375–396.

Alshurideh, M., Al Kurdi, B., & Salloum, S. A. (2019a). Examining the main mobile learning system drivers' effects: A mix empirical examination of both the Expectation-Confirmation Model (ECM) and the Technology Acceptance Model (TAM). In *International Conference on Advanced Intelligent Systems and Informatics* (pp. 406–417). Springer, Cham.

Alshurideh, M., Salloum, S. A., Al Kurdi, B., & Al-Emran, M. (2019b). Factors affecting the social networks acceptance: an empirical study using PLS-SEM approach. In *Proceedings of the 2019b 8th International Conference on Software and Computer Applications* (pp. 414–418).

Alshurideh, M., Al Kurdi, B., Salloum, S. A., Arpaci, I., & Al-Emran, M. (2020). Predicting the actual use of m-learning systems: a comparative approach using PLS-SEM and machine learning algorithms. *Interactive Learning Environments*, 1–15.

Alshurideh, M. T., Kurdi, B. A., AlHamad, A. Q., Salloum, S. A., Alkurdi, S., Dehghan, A., & Masa'deh, R. E. (2021a). Factors affecting the use of smart mobile examination platforms by universities' postgraduate students during the COVID 19 pandemic: an empirical study. In Informatics, 8 2), 1–21. Multidisciplinary Digital Publishing Institute.

Alwan, M., & Alshurideh, M. (2022). The effect of digital marketing on purchase intention: Moderating effect of brand equity. *International Journal of Data and Network Science, 10*(3), 1–12.

Amarneh, B. M., Alshurideh, M. T., Al Kurdi, B. H., & Obeidat, Z. (2021, June). The impact of COVID-19 on e-learning: advantages and challenges. In *The International Conference on Artificial Intelligence and Computer Vision* (pp. 75–89). Springer, Cham.

Ashal, N., Alshurideh, M., Obeidat, B., & Masa'deh, R. (2021) The impact of strategic orientation on organizational performance: Examining the mediating role of learning culture in Jordanian telecommunication companies. *Academy of Strategic Management Journal, 21*(Special Issue 6), 1–29.

Badr Al- Din, A., & Muhammad, A. (2020). Status of using teaching strategies by teachers for the mentally disabled students at special education centers of Erbil Governorate. In *Researches of the 2nd International Scientific Conference of the Academic-Strategic Development Center and Salahaddin University, Erbil – Iraq, on 10/11/2020* (pp. 957–978).

Bettayeb, H., Alshurideh, M. T., & Al Kurdi, B. (2020). The effectiveness of mobile learning in UAE universities: A systematic review of motivation, self-efficacy, usability and usefulness. *International Journal of Control and Automation, 13*(2), 1558–1579.

Cheng, S., & Lai, C. (2020). Facilitating learning for students with special needs: A review of technology-supported special education studies. *Journal of Computers in Education, 7*, 131–153.

Contreras, S., Cedillo, I., Rodríguez, S., Ramírez, A., & Barrera, V. (2018). Influence of type of school (special or general) on the achievement of students with special educational needs. *Universitas Psychologica, 17*(1), 1–11.

Dababneh, k. (2016). The extent of satisfaction of parents of children with learning difficulties with the level of educational services provided to their children in the resource rooms within the integration program in Jordan and the factors affecting the extent of satisfaction. *The Jordanian Journal of Educational Sciences, 12*(2), 269–286

Dessemontet, R., & Bless, G. (2011). Effects of inclusion on the academic achievement and adaptive behaviour of children with intellectual disabilities. *Journal of Intellectual Disability Research, 56*(6), 579–587.

Grzegorz, S., Smogorzewska, J., & Karwowski, M. (2017). Academic achievement of students without special educational needs in inclusive classrooms: A meta-analysis. *Educational Research Review, 21*, 22–54.

Gürgür, H., Büyükköse, D., & Kol, Ç. (2016). Özel eğitim ve rehabilitasyon merkezlerinde işitme kayipli öğrencilere sunulan destek hizmetler: Öğretmen görüşleri. *Ilkogretim Online, 15*(4), 1234–1253.

Hassan, A., Adhabi, E., & Jones, L. (2017). The impact of inclusion setting on social interaction and psychological adjustment of students with disabilities. *IJRST, 3*(4), 121–128.

Hebbeler, K., & Spiker, D. (2016). Supporting young children with disabilities. *The Future of Children, 26*(2), 185–205.

Hilbert, D. (2014). Perceptions of parents in young children with and without disabilities attending inclusive preschool programs. *Journal of Education & Learning, 3*(4), 49–45.

Kang, D., & Martin, S. (2017). Improving learning opportunities for special education needs (SEN) students by engaging pre-service science teachers in an informal experiential learning course. *Asian Pacific Journal of Education, 38*(3), 319–347.

Kelly-Hall, C. (2010). The role of student support services in encouraging student involvement and its impact on students perceptions and academic experiences. Unpublished Ph.D. Thesis, Clemson University, USA.

Khasawneh, M. (2013). The effectiveness of educational services in the resource room provided to students with learning disabilities in the primary stage in Irbid district from the parents' point of view. *Journal of Al-Quds Open University, 3*(1), 51–76.

Khasawneh, M. A., Abuhashesh, M., Ahmad, A., Alshurideh, M. T., & Masa'deh, R. (2021a). Determinants of e-word of mouth on social media during COVID-19 outbreaks: An empirical study. In *The effect of coronavirus disease (COVID-19) on business intelligence* (pp. 347–366). Springer, Cham.

Khasawneh, M. A., Abuhashesh, M., Ahmad, A., Masa'deh, R., & Alshurideh, M. T. (2021b). Customers online engagement with social media influencers' content related to COVID 19. In *The effect of coronavirus disease (COVID-19) on business intelligence* (pp. 385–404). Springer, Cham.

Kurdi, B. A., Alshurideh, M., Nuseir, M., Aburayya, A., & Salloum, S. A. (2021, March). The effects of subjective norm on the intention to use social media networks: an exploratory study using PLS-SEM and machine learning approach. In *International Conference on Advanced Machine Learning Technologies and Applications* (pp. 581–592). Springer, Cham.

Leo, S., Alsharari, N. M., Abbas, J., & Alshurideh, M. T. (2021). From offline to online learning: A qualitative study of challenges and opportunities as a response to the COVID-19 pandemic in the UAE higher education context. In *The effect of coronavirus disease (COVID-19) on business intelligence* (pp. 203–217). Springer, Cham.

Leyser, Y., & Krick, R. (2007). Evaluating inclusion: An examination of parent views and factors influencing their perspectives. *International Journal of Disability Development & Education, 51*(3), 271–285.

Merghany, S., & Ibrahim, H. (2020). Evaluation of Educational programs provided to students with Disabilities in Wadi Al- Dawasir Province in light of the international standards of special education. *Journal of Educational & Psychological Sciences, 4*(28), 117–137.

Ministry of Education. (2015). *Ministry of education experiment in addressing student with special needs*. Ministry of Education.

Mohammad, A., & Arafa, A. (2015). Assessment of the supportive services provided to students with special needs at Majmaah University (A Fiels Study). *Journal of Educational Sciences, 4*, 163–199.

Mulvey, B., Chiu, J., Ghosh, R., & Bell, R. (2016). Special education teachers' nature of science instructional experiences. *Journal of Research in Science Teaching, 53*(4), 554–578.

Obeid, M., & Al-Hadidi, M. (2018). The effect of a training program based on palliative education in developing the cognitive and behavioral skills of people with mental disabilities. *Studies-Educational Sciences, 45*(4), 354–376.

Thombson, J., Schalock, R., Agosta, J., Teninty, L., & Fortune, J. (2014). How the support paradigm is transforming the developmental disabilities service system. *Inclusion, 2*(2), 86–99.

Zettle, J., & Ballard, J. (2015). The education for all handicapped children act of 1975 PL 49–142: Its history, origins, and concepts. *The Journal of Education, 161*(3), 5–22.

The Effectiveness of the Performance of Principals of Basic Education Schools in the Sultanate of Oman in Managing Change in Light of the Corona Pandemic

Ahmed Said Alhadrami, Saddam Rateb Darawsheh, Anwar Sadu Al-Shaar, and Muhammad Alshurideh

Abstract The current study aimed to identify the effectiveness of the performance of principals of basic schools, the first cycle, in change management in light of the Corona pandemic. The researcher used the descriptive-analytical method. A questionnaire was applied to (400) male and female teachers. The effectiveness of the principals' performance for change management fields was as follows: Follow-up and implementation ranked first with a mean (3.83), followed by planning (3.73), and evaluation (3.60). The study showed that all the statements of the three axes regarding the effectiveness of the performance of principals of basic education in the axes of change management came to a high degree, except one item in evaluation axe "Item 3", which reads (the principal focuses on evaluating qualitative performance and neglects quantitative performance). The study showed statistically significant differences at the significance level (0.05) in all fields in favour of females. The study also indicated that there were no statistically significant differences between the means due to the academic qualification variable. The researchers recommended that the Ministry of Education should adopt the change project for all school principals in its plans as a basic project in its organizational structure.

A. Said Alhadrami (✉)
Assistant Professor of Education Administration, A'Sharqiyah University- Sultanate of Oman, Ibra, Oman
e-mail: Ahmed.alhadrami@asu.edu.om; Asnh7887@gmail.com.2022

S. Rateb Darawsheh
Department of Administrative Sciences, The Applied College, Imam Abdulrahman Bin Faisal University, P.O. Box: 1982, Dammam, Saudi Arabia
e-mail: srdarawsehe@iau.edu.sa

A. Sadu Al-Shaar
Deanship of Preparatory Year and Supporting Studies, Department of Self Development, (Imam Abdulrahman Bin Faisal University), P.O. Box: 1982, Dammam 43212, Saudi Arabia

M. Alshurideh
Department of Management, College of Business Administration, University of Sharjah, Sharjah, United Arab Emirates
e-mail: malshurideh@sharjah.ac.ae

M. Alshurideh et al. (eds.), *The Effect of Information Technology on Business and Marketing Intelligence Systems*, Studies in Computational Intelligence 1056,
https://doi.org/10.1007/978-3-031-12382-5_24

Keyword Change management · Al Dakhiliyah and Eastern Governorates · Corona Covid 19 pandemic

1 Introduction

The world is witnessing rapid changes and transformations in all aspects of scientific and practical life, whether at the economic, social, cultural, or political level. Therefore, various international institutions have faced several challenges represented in the requirement to bring about innovative change (Nuseir et al., 2021a; Alzoubi et al., 2022;Alameeri et al., 2021). The nature of life in this era requires these institutions to be strong and able to show unique competition. Indeed, this only comes through the perfect competencies of quality and excellence. Despite the rapid and significant development performed by the various institutions on dealing with the change in the work environment, these efforts are still confined to a very narrow field (Alshurideh et al., 2019a; Alameeri et al., 2021; Nuseir et al., 2021b). In a time of high competition and institutional and technological change, the emergence of methodologies for managing change and its multiple theories has contributed to the provision of many advanced modern administrative applications with scientific and practical foundations that institutions adopt as a kind of guidance for solving various organizational problems within the institution (Almaazmi et al., 1261; Al Suwaidi et al., 2021; Tariq et al., 2022a). Integration of change methods plays a vital role in creating and coordinating modern organizational initiatives (Hayajneh et al., 2021). Working to develop a culture and approach to change within institutions necessitates several activities, the most important of which is focusing on making radical and fundamental changes in many effective administrative systems that have a significant impact on human forces in terms of their scientific and technical skills, as well as proposing policies concerning the implementation of work within the institution and the creation of organizational policies (Shehata, 2021; Abuhashesh et al. 2021; Odeh et al., 2021).

The process of change is not limited to private organizations and businesses, but also to scientific and educational institutions, which are under a lot of pressure to change because they are in charge of the education process and its development, as well as working to modernize its educational processes (Kurdi et al., 2020a; Kurdi et al. 2020; Alshurideh, et al., 2021). Whereas the educational process is the first basic pillar of the educational process to which the teaching and learning processes are entrusted, it is what is relied on to perform a lot of work and organizations for this process. As a result, it is difficult for these educational institutions to adapt to change. They and their leaders must experience numerous significant challenges during the change process. However, successful educational leaders and educational institutions are the ones who can overcome the challenges of managing change (Al-Wasmi, 2006; Alshurideh et al., 2020; Akour et al., 2021).

Al-Azzam (2016) claims that to ensure that these institutions play an active and entrusted role in the development of society and the revitalization of the economy, education, and culture in the current era, all of the various educational institutions throughout the Arab world must apply and use the management of change and excellence that is appropriate for this age. However, if these institutions adopt relevant regulations and trends in response to the world's development in its different disciplines, they will be capable to boost their levels of innovation and institutional creativity in the fields of education. Change has become a feature of the era that requires a lot of human, material and intellectual costs. Therefore, when change occurs, it will seek to achieve many lofty goals, including the following as mentioned by Jilali and Ali (2019): working on establishing models for openness in the organization from employees to handle all of the institution's problems directly, and to offer an acceptable climate for them based on trust and good treatment among them. In addition to working on developing complementary relationships among employees within the organization as individuals and groups to encourage competition, innovation, and team spirit, which will lead to motivating them and increasing their confidence and interaction in the work they perform (Al Kurdi et al., 2020; Al Kurdi et al., 2020b; Alshurideh et al., 2019b).

Change necessitates deliberate and empowered leadership (AlShehhi et al., 1261). The inspired and willing leaders to manage change in the new twenty-first century, whose primary goal is to contemplate the change process, continually look for poten tial chances for change, and try to generate change in a novel and successful method within educational institutions (Al-Dhuhouri et al., 1261; Harahsheh et al., 2021; Alshurideh et al., 2019c). This necessitates some serious steps, the most important of which are: working on developing future-oriented policies, searching for modern methodologies for the processes of searching for and anticipating the type of change, searching for the best appropriate methods for introducing the change process inside and outside the institution, and developing modern scenarios to find various policies to ensure a balanced process of change.

Just as change is closely related to the administrative and technical structure within educational institutions, leaders must have viable plans within their institutions, not just a written future vision. It is necessary to work well and participate with other employees in delegating powers and empowering you, as this will give a strong impetus towards the process of change within educational institutions, especially on the part of employees, making them think about issues that affect the educational aspect. Thus, the principle of participation with leadership in the process of change will be activated in many aspects. Also, the growth of the technological aspect within educational institutions forced them to take many practical steps towards this aspect, including transparency towards accountability in the processes of attendance, absence, dereliction of work, and…etc. Change management is one of the most successful administrative trends, where most of the employees or the organization are transferred to hoped-for drawings and strategies in the future if the goal of change is to ensure the success of the institution and achieve the best results at the regional and global levels and work to provide an appropriate climate for all workers in those institutions without any pressure. Therefore, many studies have dealt with the accountability of change management as it is a followed and basic approach to the process of developing institutions of all kinds (Al Batayneh et al., 2021; Alshurideh,

2022; Tariq et al., 2022b). Among the studies that dealt with change management are the following:

The study of Al-Muslat (2021), aimed to explore the role of the educational supervisors' practice in change management and its relationship to teachers' job commitment in secondary schools in the Asir region. The researcher adopted the descriptive-analytic approach. A questionnaire was applied to a sample (n = 341) of male and female teachers. It was found that the educational supervisors practice their role in change management to a high degree with a mean of (3.38). Teachers' job commitment came to a high degree with a mean of (4.22). No statistically significant differences were found regarding the mean degree of the educational supervisor's practice of their role in the change management process and the mean job commitment of teachers attributed to the variables: specialization - practical experience - location. The study also indicated that there was a kind of positive correlation between the role that plays. The educational supervisor in change management showed the average degree of job commitment (Allozi et al., 2022; Alketbi et al., 2020; Alshurideh et al., 2017).

The study of Shehata (2021) aimed to identify the availability of organizational change processes, knowledge and reality of institutional excellence within institutions, and to identify the strategic dimensions of the training and sustainable development process in various commercial banks in the Republic of Egypt. (107) employees participated in the study. The findings showed a positive relationship effect in the process of organizational change on the applied strategy for the training and development processes, as well as the existence of a relationship with a positive impact on the quality of training and development in institutional excellence. The result also revealed that the applied strategy for the process of training and development has an active role in the partial mediation process in the relationship of the process of organizational change and institutional excellence and that there were significant differences in the impact of the organizational change process on the existing strategy and practice in training and development.

Al-Otaibi and Al-Anzi (2021) sought to identify the terms of change management and its multiple concepts, as well as the practical and theoretical importance within educational institutions.. The researcher used the analytical method. After reviewing and analysing the studies and literature related to the study, the following was concluded: Change management is a philosophy that exists within the organization among the strategic plans scheduled to be achieved in one of the aspects within the organization to increase the effectiveness of the organization and work to achieve the required balance. Also, organizational loyalty is one of the biggest factors affecting the effectiveness and production of the institution, and it is one of the aspects that preoccupied many institutions and organizations. The study also clarified that among the reasons that prompted leaders in the State of Kuwait to adopt change management is to increase trust and positive interaction among members of the administrative leaders. Where the study of Njuguna and Muathe (2016) sought to identify the challenges facing organizations in the process of change. The researchers used the analytical method. The questionnaire was applied to a random sample of various operating institutions. The study concluded that there are great challenges that hinder the process of managing change in institutions and that institutions must stand tall to survive and maintain their level by taking a set of basic steps that help them to do so. The study also made it clear that institutions should consider the needs of their employees, help them achieve their goals, and encourage them in various fields of work, to be able to compete with their peers in global markets. The results showed that the change management process is closely related to the human elements within the single organization to be able and successful to bring about change within the organization. The study of Hallinger (2010) aimed to explore the most important factors that enable change management and improve its ability in educational institutions by linking some leadership styles: transformational leader style, directed leader style, with their strategic role to bring about the change process within educational schools. The analytical method was employed the researcher. The study yielded the following findings: The results prove that the effectiveness of both transformational and directed leadership styles affect change inside educational institutions. It also proved that the real leader is that person who can work in the spirit of the group within his organization with all the employees working with him and is also able to ignite the spirit of enthusiasm and distribute institutional trust among the employees; To enhance organizational loyalty within the institution and thus help to bring about radical change within the educational schools.

1.1 Problem Statement

The Sultanate of Oman, like other Arab countries, seeks a continual development process in all sectors of life inside the Sultanate. Perhaps this is what was indicated in the national strategy 20/40 and the future vision for Oman 20/40, both of which have been meticulously put out. The Sultanate began deploying it in early 2021. This vision addressed major axes that affect the Omani person's life (economic, social, political, cultural). These axes include Omani society and individuals in their legal capacity, growth in all its forms and facets, performance and governance, and a clean and healthy environment. The ultimate goal of this vision is to develop all the main activities and areas so that the Sultanate enters into the various classifications in providing services and welfare to the people. And also, to compete in obtaining global indicators that prove the interest in the Omani individual and his well-being and to provide all things to lead a happy life following what is stated in international standards. And all this helped to work in bringing about some radical changes in the life matters that are practised on the ground, which showed some obstacles to achieving many things. Thus, we find that the process of change is indispensable, as it is a necessity to achieve justice, prosperity and growth. This is what was stated in the text of Sultan Haitham's speech (2020) that the next stage will be a stage of growth, progress and change in many of the issues imposed by the current challenges.

The Ministry of Education in the Sultanate of Oman make different administrative and educational changes. Among these changes were the changes of the basic education system in 1996 when it was applied to some students, then in 2006, it includes various schools. Another change was the establishment of an online portal concerned with the educational process, which is known as the educational portal, which is one of the best innovations that keep pace with technical and technological progress. Another system was introduced, known as the educational indicators system, which is also a qualitative leap in the field of education, which diagnoses the educational process in all its aspects. In addition to the introduction of another program known as the School Management Program, which is also a qualitative leap, this was proven by many Omani studies, which emphasized the importance of change management in the Ministry of Education in the Sultanate of Oman. Among those studies are the following: The study of Al-Jaraydah (2021) which linked change management to the training needs of the Ministry of Education, and the study of Al-Ghailanih (2015) which tried to identify the perspective of the employees of the Ministry's general office in the Sultanate of Oman on managing change. And the study of Al-Ghailanih (2015) aimed to identify the degree of the practice of administrative supervisors to change management and its relationship to job satisfaction. Therefore, this study came to answer the following questions:

1. What is the effectiveness of basic school principals' role in managing change in light of the Corona pandemic from the teachers' perspective?
2. What is the order of the areas of management change according to the performance of principals of basic schools in light of the Corona pandemic, from the teachers' perspective?
3. Are there statistically significant differences at the significance level (0.005) in the effectiveness of the performance of principals of basic schools in the Sultanate of Oman in the areas of management change during the Corona pandemic due to the variable (gender, qualification)?

1.2 Study Significance

Theoretical significance: The theoretical importance of the study lies in its subject, which is the management of change, its terminology, its objectives, its importance and the necessity of its presence inside and outside educational institutions. Being an indispensable variable. The study also seeks through this work to enrich the Arabic and Omani literature with a kind of modern studies on the effectiveness of the performance of principals of basic education in the Sultanate of Oman in the areas of change management.

Practical Significance: It is represented in the results obtained through the actual application of the questionnaire, as well as in the study's recommendations to implement the change management process in educational contexts and by standing on some crucial issues. The study also aims to determine the areas and types of areas involved in the practice of change management. This is expected to assist decision-makers in the Ministry of Education in determining the most significant and appropriate main activities for their implementation.

1.3 Goals

1. Measuring the effectiveness of basic education principals' performance in the Sultanate of Oman in the areas of change management during the Corona epidemic from the perspective of teachers.
2. Identifying the order of change management areas based on the performance of basic education managers in light of the Corona epidemic from the perspective of teachers
3. Revealing if there are statistically significant differences in the efficacy of basic education school principals' performance in the Sultanate of Oman in the areas of change management during the Corona epidemic, according to the variable (gender, qualification).

1.4 Study Limitations

1. Objective limits: including change management
2. Spatial boundaries: all primary education schools in the Sultanate of Oman.
3. Time limits: 2022/2023
4. Human limits: basic schoolteachers.

1.5 Study Terminology

Basic education: It is the educational system that started in 1996 for the first time. It is the unified education provided by the government to all the children of the Sultanate of Oman who has reached school age. The duration of this type of education is ten years, during which the most important basic educational needs such as information, knowledge and skills. It works on the development of various behavioural trends and human values through which learners are enabled to continue education and training according to their readiness, inclinations and ability. This type of education aims to work in the development of individuals to face the various future challenges that may confront them (42.The Ministry of Education, 2022).

Change management: It is a continuous and comprehensive process carried out by the leader through employing all of the skills and technical expertise on leadership to make a better influence. And to bring about an effective change process in all parts of the educational institution by adapting to the course of the future vision of education, which affects the nature of employee performance (Alhamad et al., 2021; Alsuwaidi et al., 2021; Muia, 2015; Okemba, 2018) defined it as the system that facilitates the process of shifting individuals within organizations from one position to a better future position to meet the corporate strategy objectives. Alternatively, it is an organized process in which change processes are carried out to attain the best results possible, using organized administrative procedures to accomplish the change management process.

Corona pandemic: The World Health Organization defines it as a disease that affects the human body and causes acute pneumonia, leaving traces of exhaustion, fatigue, and sometimes death, and was also known as (covid-19), which was declared by the World Health Organization as a global disease that has outbroken globally, causing a wide disaster for the entire world (AL-Hadhrami and Al Shaqsia, 2021; Al-Dmour et al., 2021a; Al-Dmour et al., 2021b; Al Khasawneh et al., 2021a; Al Khasawneh et al., 2021b).

2 Method and Procedures

The methodological procedures of the research consist of the following steps:

2.1 Study Population

The study population was selected from all categories of male and female teachers in basic education schools representing the first cycle of (1–4) in the Sultanate of Oman, which counted (15, 032) teachers according to the statistics of the Ministry of Education's statistical book for the year 2021.

2.2 Study Sample

The study sample (n = 500) teachers who were selected using the simple random method, which amounted to approximately (500) teachers and teachers, and due to the current conditions and lack of awareness, the researchers were unable to cover the representative sample of the study population.

2.3 Study Instrument:

The researcher used a five-point Likert scale, which is represented by the following five rates: (very few - few - medium - high - very high). The results were to the scale as shown in Table 1.

Table 1 The study instrument

Mean	1–1.8	1.81–2.60	2.61–3.40	3.41–4.20	4.21–5.00
Rate	Very low	Low	Medium	High	Very high

Table 2 The instrument reliability coefficient

	Field	Items (N)	Reliability coefficient
1	Planning	9	0.80
2	Follow up and implementation	9	0.79
3	Evaluation	7	0.77
Total			0.79

2.4 *Instrument Validity*

The researcher developed the study instrument by referring to previous studies related to change management. The study tool was presented to 7 experienced arbitrators in the field of management. Where the initial form of the questionnaire consisted of (3) basic fields and (35) statements distributed as follows: (1) Planning including (13) statements and (2) Follow-up and implementation (15) statements and field (3) The evaluation (10) statements. All the required modifications were considered, and the tool reached in its final form (25) statements divided into the three fields of planning (9) statements, follow-up and implementation (9) statements and evaluation (7) statements.

2.5 *Instrument Reliability*

The instrument reliability was tested using the Cronbach alpha method. The tool was administered to a sample (n = 40) of male and female teachers from different basic schools in the Sultanate of Oman, as indicated in Table2.

Table 2 shows that the instrument reliability coefficient was (0.76), which is a high degree of reliability, and this indicates the reliability of the overall scale.

3 Results

3.1 *Results of the First Question*

"What is the Effectiveness of Basic School Principals' Role in Managing Change in Light of the Corona Pandemic from the Teachers' Perspective ?"

Table 3 Descriptive analysis for the field "Planning"

	Statements	Mean	SD	Rank	Degree
1	The principal implements school plans flexibly according to changes	3.82	0.932	4	High
2	The principal follows up on the school's issues according to a set schedule	3.87	0.993	2	High
3	The principal continues to study the factors of change on the educational plan applied within the school	3.76	0.992	5	High
4	The principal considers changes in the field while setting the school schedule	3.88	1.089	1	High
5	The principal sets his school plans in light of the expected future changes	3.71	0.965	6	High
6	The principal has initiatives in developing plans to keep pace with future changes	3.65	1.016	7	High
7	The principal has coordinated plans to upgrade teachers to keep pace with future changes	3.59	0.942	8	High
8	The principal is keen in his planning to incorporate modern technological changes into schoolwork	3.84	1.084	3	High
9	The principal has modern communication systems that keep pace with modern changes	3.45	1.137	9	High

To answer this question, the descriptive analysis was used (arithmetic means & standard deviations) to reveal the effectiveness of the performance of principals of basic schools for all aspects of change management. The descriptive analysis for each field is as follows:

First: Planning

The arithmetic means and standard deviations were obtained to reveal the effectiveness of the performance of the principals of basic schools in the change management in the field "planning" (see Table 3).

Table 3 demonstrates that the arithmetic means of the field's statements ranged between (3.45–3.87). Statement (4) reads "The principal considers changes in the field while setting the school schedule." The researcher attributes this to the principals' awareness of what is happening in the educational field and the wide awareness campaigns. While Statement (9), "The principal has modern communication systems that keep pace with modern changes" ranked last in a high degree and a mean of (3.45). The researcher attributed this to the lack of provision of electronic devices and communication systems within the school, and this may also be due to the weak infrastructure of communication systems.

Second: Follow up and implementation

The arithmetic means and standard deviations were extracted to reveal the effectiveness of school principals' performance of the basic fields of change management in the field of follow up "implementation", as indicated in Table 4.

Table 4 Descriptive analysis of the "Follow up and implementation" field

	Statements	Mean	SD	Rank	Degree
1	The principal explains the school's strategic plan to teachers	3.71	1.101	7	High
2	The principal seeks to constantly develop the skills of teachers and train them to keep pace with change	3.49	1.065	9	High
3	The principal encourages teachers to come up with creative ideas to advance the teaching process	4.14	1.000	1	High
4	The principal overlooks minor errors as they are remediable within the school	3.92	1.055	4	High
5	The principal rewards teachers with creative solutions associated with change	3.65	1.214	8	High
6	The principal encourages the use of modern technologies in the teaching process within the school	3.98	1.049	3	High
7	The principal monitors change processes within the school to provide feedback to teachers	3.76	1.176	6	High
8	The principal strives to provide the best buildings and technical equipment for the school to keep pace with the changing processes	4.02	1.068	2	High
9	The manager activates the means of communication in all its branches to clarify the change process	3.84	1.046	5	High

Table 4 shows that the means of this field ranged between (3.49–4.14). Statement (3) which stated, "The principal encourages teachers to come up with creative ideas to advance the teaching process." came in the first place with a high degree (4.14). The researcher attributes this result to the good awareness that has taken place in the educational field and a large number of training programs in the same field, while Statement (2) "The principal seeks to constantly develop the skills of teachers and train them to keep pace with change." ranked last and to a high degree, with a mean (3.49). The researcher attribute this to what the educational system was exposed to in schools during the Corona pandemic.

Third: Evaluation

Descriptive analysis was extracted to reveal the effectiveness of basic school principals' performance of change management in the field of evaluation, as seen in Table 5.

Table 5 indicates that the arithmetic means of this field range between (3.33–3.75). Statement (5) which reads "The principal is keen to highlight the strengths and weaknesses of each teacher" came in the first rank with a high degree and a mean of (3.75). The researcher attributes this to the career culture followed by the principals, while statement (4) "The principal focuses on evaluating qualitative performance and neglects quantitative performance." ranked last with a medium degree and a mean (3.33). The researcher attributes this to the behaviour of some principals and

Table 5 Descriptive analysis of the field "Evaluation"

	Statements	Means	SD	Rank	Degree
1	The principal follows real standards for evaluating the educational process	3.69	0.927	3	High
2	The principal depends on unconventional standards compatible with technical development	3.71	0.879	2	High
3	The principal uses a modern system for the process of motivation and creativity within the school	3.59	0.983	5	High
4	The principal focuses on evaluating qualitative performance and neglects quantitative performance	3.33	0.993	7	Medium
5	The principal is keen to highlight the strengths and weaknesses of each teacher	3.75	0.935	1	High
6	The principal displays the results of the teacher's evaluation after each evaluation process	3.67	0.931	4	High
7	The principal has an accounting system by which the negligent in his work is held accountable	3.45	0.966	6	High

their lack of full knowledge of quantitative performance and its importance in the educational process.

3.2 *Results of the Second Question*

What is the order of the areas of management change according to the performance of principals of basic schools in light of the Corona pandemic, from the teachers' perspective? Descriptive analysis was conducted to answer this question.

Table 6 indicates that the total arithmetic mean of the effectiveness of school principals' performance of change management was (3.73) and it came to a high degree. The follow-up and implementation field ranked in the first place with a high degree and a mean of (3.83), followed by the Planning field, which obtained a high degree and a mean of (3.73). Where the field Evaluation came last with a high degree and a mean (3.60).

Table 6 Descriptive analysis for the effectiveness of the performance of principals of basic schools for managing change, according to the order of the fields

	Field	Mean	SD	Rank	Degree
1	Planning	3.73	0.843	2	High
2	Follow up and implementation	3.83	0.907	1	High
3	Evaluation	3.60	0.740	3	High
Total			3.73	0.803	High

3.3 Results of the Third Question

Are there statistically significant differences at the significance level (0.005) in the effectiveness of the performance of principals of basic schools in the Sultanate of Oman in the areas of management change during the Corona pandemic due to the variable (gender, qualification)? To answer this question, the arithmetic means and standard deviations of the effectiveness of the performance of principals of basic education schools for managing change were calculated according to the variable (type). the "t" value was calculated for the arithmetic means as shown in Table 7.

Table 7 shows that there are no statistically significant differences at the significance level (0.05α ≤) to the degree to which school principals practice change management attributable to the governorate's variable in the three fields and the overall tool.

Gender and Qualification variables

The arithmetic means and standard deviations were calculated to reveal the degree of school principals' practice of managing change by gender variable, and the "t" value was calculated for the arithmetic means as shown in Table 7.

Table 8 shows statistically significant differences at the significance level (0.05α ≤) in all fields and the overall tool favour of females. The researcher attributes this

Table 7 Descriptive analysis, and "t" value for the degree of school principals' practice of change management according to the governorate's variable (the middle and eastern)

Field	Gender	N	Mean	SD	T-value	Sig
Planning	Male	190	3.77	0.907	0.290	0.786
	Female	210	3.70	0.997		
Follow-up and implementation	Male	190	3.88	0.980	0.288	0.766
	Female	210	3.80	0.840		
Evaluation	Male	190	3.59	0.920	0.145	0.881
	Female	210	3.55	0.540		
Total	Male	190	3.76	0.907	0.267	0.785
	Female	210	3.69	0.712		

Table 8 Descriptive analysis and the value of "t" for the effectiveness of the performance of principals of basic education schools for managing change by gender variable

Field	Gender	N	Mean	SD	T value	Sig
Planning	Male	90	3.12	1.074	3.774	0.001
	Female	110	3.99	0.570		
Follow up and implementation	Male	90	3.15	1.155	3.972	0.001
	Female	110	4.12	0.595		
Evaluation	Male	90	3.11	0.974	3.283	0.002
	Female	110	3.80	0.513		
Total	Male	90	3.13	1.041	3.930	0.001
	Female	110	3.98	0.516		

result to the excessive care in the teaching and learning process, and the clear interest of female teachers in the education processes.

Table 9 indicates that there are differences between the effectiveness of the performance of basic school principals for change management fields in the Sultanate of Oman, attributed to the academic qualification variable, and to determine if there are statistically significant differences between the means of the fields at the level of significance (0.05 The ANOVA analysis was conducted as indicated in Table 9.

Table 10 shows that there are no statistically significant differences at the level (0.05) between the degrees of effectiveness of the performance of principals of basic schools for the fields of change management attributable to the academic qualification variable. The "f" value reached (0.657) with a level of significance (0.602). As shown the value of the significance is greater than (0.05), which indicates that there are no statistically significant differences for the academic qualification variable.

Table 9 Descriptive analysis and the "t" value for the effectiveness of the performance of principals of basic school by qualification variable

Variables	Mean	SD
Bachelor's	3.67	0.89
Bachelor's + Higher Diploma	3.56	0.95
Master's degree and above	3.45	0.93
Total	3.52	0.91

Table 10 ANOVA analysis results

Source of variance	SS	DF	MS	F value	Sig
Between groups	0.872	2	0.499	0.657	0.602
Within groups	125.54	209	0.678		
Total	126.412	211			

4 Conclusion

The study concluded that the order of the study fields regarding the effectiveness of the performance of the principals of basic school for change management according to the means of the responses was as follows: Follow-up and implementation (3.83), planning (3.73), and evaluation (3.60). The three fields obtained a high degree. It was found that all the statements of the change management came to a high degree except the statement which reads "The principal focuses on evaluating qualitative performance and neglects quantitative performance" under the field "Evaluation," which came to a medium degree. The researcher attributed this to the fact that principals an evaluation method in only one aspect, which is under the administrative regulations and laws and the ones stated in job descriptions. The study proved that there are statistically significant differences at the significance level (0.05) in all areas and the overall scale in favour of females and that there are statistically significant differences between the means attributed to the academic qualification variable.

5 Recommendations

1. Urging the Ministry of Education to adopt the process of change as a future project for all school leaders.
2. Develop an executable plan for selecting school leaders with appropriate specifications and conditions so that they are able to practice the process of change.
3. Continuing to implement qualitative programs for all employees of the Ministry of Education, especially teachers, to accept the change process smoothly.
4. Involve all school principals working in the Ministry of Education in special qualitative programs to measure the extent of the competencies they possess during the change process.

References

Abuhashesh, M. Y., Alshurideh, M. T., & Sumadi, M. (2021). The effect of culture on customers' attitudes toward Facebook advertising: The moderating role of gender. *Review of International Business and Strategy*.

Akour, I., Alshurideh, M., Al Kurdi, B., Al Ali, A., & Salloum, S. (2021). Using machine learning algorithms to predict people's intention to use mobile learning platforms during the COVID-19 pandemic: Machine learning approach. *JMIR Medical Education, 7*(1), 1–17. https://doi.org/10.2196/24032.

Al Batayneh, R. M., Taleb, N., Said, R. A., Alshurideh, M. T., Ghazal, T. M., & Alzoubi, H. M. (2021). IT governance framework and smart services integration for future development of Dubai infrastructure utilizing AI and big data, its reflection on the citizens standard of living. In *The International Conference on Artificial Intelligence and Computer Vision* (pp. 235–247).

Al Khasawneh, M., Abuhashesh, M., Ahmad, A., Alshurideh, M. T., & Masa'deh, R. (2021a). *Determinants of E-word of mouth on social media during COVID-19 outbreaks: An empirical study* (Vol. 334).

Al Khasawneh, M., Abuhashesh, M., Ahmad, A., Masa'deh, R., & Alshurideh, M. T. (2021b). *Customers online engagement with social media influencers' content related to COVID 19* (Vol. 334).

Al Kurdi, B., Alshurideh, M., & Al afaishata, T. (2020). Employee retention and organizational performance: Evidence from banking industry. *Management Science Letters, 10*(16), 3981–3990.

Al Suwaidi, F., Alshurideh, M., Al Kurdi, B., & Salloum, S. A. (2021). *The impact of innovation management in SMEs performance: A systematic review* (Vol. 1261). AISC.

Alameeri, K., Alshurideh, M., Al Kurdi, B., & Salloum, S. A. (2021). *The effect of work environment happiness on employee leadership* (Vol. 1261) AISC.

Alameeri, K. A., Alshurideh, M. T, & Al Kurdi, B. (2021). *The effect of Covid-19 pandemic on business systems' innovation and entrepreneurship and how to cope with it: A theatrical view* (Vol. 334).

Al-Azzam, M. (2016). The degree of public secondary school principals' practice in Irbid governorate to lead change from the teachers' point of view. *UJ Dirasat Journals, 43*(3), 1283–1297.

Al-Dhuhouri, F. S., Alshurideh, M., Al Kurdi, B., & Salloum, S. A. (2021). *Enhancing our understanding of the relationship between leadership, team characteristics, emotional intelligence and their effect on team performance: A critical review* (Vol. 1261). AISC.

Al-Dmour, A., Al-Dmour, H., Al-Barghuthi, R., Al-Dmour, R., & Alshurideh, M. T. (2021a). *Factors influencing the adoption of E-payment during pandemic outbreak (COVID-19): Empirical evidence* (Vol. 334).

Al-Dmour, R., AlShaar, F., Al-Dmour, H., Masa'deh, R., & Alshurideh, M. T. (2021b). *The effect of service recovery justices strategies on online customer engagement via the role of "Customer Satisfaction" during the Covid-19 pandemic: An empirical study* (Vol. 334).

Al-Ghailanih, S. (2015). *Degree of practice administrative supervisors in change management dimensions and its relationship to job satisfaction of headmasters and their assistants in South Sharqia Governorate in Sultanate of Oman*. Uninersity of Nizwa.

Alhamad, A. Q. M., Akour, I., Alshurideh, M.,. Al-Hamad, A. Q, Al Kurdi, B., & Alzoubi, H. (2021). Predicting the intention to use google glass: A comparative approach using machine learning models and PLS-SEM. *International Journal of Data and Network Science, 5*(3), 311–320. https://doi.org/10.5267/j.ijdns.2021.6.002.

AL-Hadhrami, T., Al Shaqsia, A. (2021). The impact of the use of technology and communication means for children during the corona pandemic from the point of view of families. *Journal of Human Development Education and Special Education, 7*(4), 123–156.

Al-Jaraydah, M. (2021). Training needs for school principals and their assistants in Al Buraimi governorate, Sultanate of Oman in the light of change management. *Educational Journal, 138*(3), 105–108. https://doi.org/10.34120/0085-035-138-008.

Alketbi, S., Alshurideh, M., & Al Kurdi, B. (2020). The influence of service quality on customers' retention and loyalty in the UAE hotel sector with respect to the impact of customer' satisfaction, trust, and commitment: A qualitative study. PalArch's Journal of Archaeology of Egypt/Egyptology, *17*(4), 541–561.

Allozi, A., Alshurideh, M., AlHamad, A., & Al Kurdi, B. (2022). Impact of transformational leadership on the job satisfaction with the moderating role of organizational commitment: case of UAE and Jordan manufacturing companies. *Academy of Strategic Management Journal, 21*, 1–13.

Almaazmi, J., Alshurideh, M., Al Kurdi, B., & Salloum, S. A. (2021). *The effect of digital transformation on product innovation: A critical review* (Vol. 1261). AISC.

Al-Muslat, S., & Al-Mklafe, M. (2021). The role of the educational supervisor in change management and its relationship to the level of teachers' job commitment in secondary schools in Asir Region. *Al-Adab Journal of Psychology Educational Studies, 9*(2), 171–112.

Al-Otaibi, A., & Al-Anzi, M. (2021). Change management and its relationship to organizational loyalty in educational institutions in the State of Kuwait. *Jaya College of Education, 31*(1), 63–81. https://doi.org/10.21608/jealex.2021.155213.

AlShehhi, H., Alshurideh, M., Kurdi, B. A., & Salloum, S. A. (2021). *The impact of ethical leadership on employees performance: A systematic review* (Vol. 1261). AISC.

Alshurideh, M. (2022). Does electronic customer relationship management (E-CRM) affect service quality at private hospitals in Jordan? *Uncertain Supply Chain Management, 10*(2), 1–8.

Alshurideh, M., Al Kurdi, B., Abu Hussien, A., & Alshaar, H. (2017). Determining the main factors affecting consumers' acceptance of ethical advertising: A review of the Jordanian market. *Journal of Marketing Communications, 23*(5). https://doi.org/10.1080/13527266.2017.1322126.

Alshurideh, M., Al Kurdi, B., Shaltoni, A. M., Ghuff, S. S. (2019a). Determinants of pro-environmental behaviour in the context of emerging economies. *International Journal of Sustainable Society, 11*(4). https://doi.org/10.1504/IJSSOC.2019.104563.

Alshurideh, M., Salloum, S. A., Al Kurdi, B., & Al-Emran, M. (2019b). Factors affecting the social networks acceptance: An empirical study using PLS-SEM approach. In *Proceedings of the 2019b 8th International Conference on Software and Computer Applications* (pp. 414–418).

Alshurideh, M., Salloum, S. A., Al Kurdi, B., Monem, A. A., & Shaalan, K. (2019c). Understanding the quality determinants that influence the intention to use the mobile learning platforms: A practical study. *International Journal of Interactive Mobile Technologies, 13*(11). https://doi.org/10.3991/ijim.v13i11.10300.

Alshurideh, M., Al Kurdi, B., Salloum, S. A., Arpaci, I., & Al-Emran, M. (2020). Predicting the actual use of m-learning systems: A comparative approach using PLS-SEM and machine learning algorithms. *Interactive Learning Environments*, 1–15.

Alshurideh, M. T., Al Kurdi, B., Masa'deh, R., & Salloum, S. A. (2021). The moderation effect of gender on accepting electronic payment technology: A study on United Arab Emirates consumers. *Review of International Business and Strategy, 31*(3). https://doi.org/10.1108/RIBS-08-2020-0102.

Alshurideh, M. T. et al. (2021). Factors affecting the use of smart mobile examination platforms by universities' postgraduate students during the COVID-19 pandemic: An empirical study. *Informatics, 8*(2). https://doi.org/10.3390/informatics8020032.

Alsuwaidi, M., Alshurideh, M., Al Kurdi, B., & Salloum, S. A. (2021). *Performance appraisal on employees' motivation: A comprehensive analysis* (Vol. 1261). AISC.

Al-Wasmi, S. (2006). The degree of educational leaders' contribution to managing change in educational institutions in the State of Kuwait. Amman Arab University for Graduate Studies.

Alzoubi, H., Alshurideh, M., Kurdi, B., Akour, I., & Aziz, R. (2022). Does BLE technology contribute towards improving marketing strategies, customers' satisfaction and loyalty? The role of open innovation. *International Journal of Data and Network Science, 6*(2), 449–460.

Hallinger, P. (2010). Leading Educational Change: Reflections on the practice of instructional and transformational leadership. *Cambridge Journal of Education, 33*(3), 329–352. https://doi.org/10.1080/0305764032000122005

Harahsheh, A. A., Houssien, A. M. A., Alshurideh, M. T., & Mohammad, A. M. (2021). *The effect of transformational leadership on achieving effective decisions in the presence of psychological capital as an intermediate variable in private Jordanian Universities in light of the corona pandemic* (Vol. 334).

Hayajneh, N., Suifan, T., Obeidat, B., Abuhashesh, M., Alshurideh, M., & Masa'deh, R. (2021). The relationship between organizational changes and job satisfaction through the mediating role of job stress in the Jordanian telecommunication sector. *Management Science Letters, 11*(1), 315–326.

Jilali, S., & Ali, F. (2019). *The impact of change management on improving institutional performance case study of the (Batimetal) foundation.* University of Djilali Bounama.

Kurdi, B. A., Alshurideh, M., & Salloum, S. A. (2020a). Investigating a theoretical framework for e-learning technology acceptance. *International Journal of Electrical and Computer Engineering, 10*(6). https://doi.org/10.11591/IJECE.V10I6.PP6484-6496.

Kurdi, B. A., Alshurideh, M., Salloum, S. A., Obeidat, Z. M., & Al-dweeri, R. M. (2020b). An empirical investigation into examination of factors influencing university students' behavior towards e-learning acceptance using SEM approach. *International Journal of Interactive Mobile Technologies, 14*(2). https://doi.org/10.3991/ijim.v14i02.11115.

Kurdi, B. A., Alshurideh, M., & Alnaser, A. (2020c). The impact of employee satisfaction on customer satisfaction: Theoretical and empirical underpinning. *Management Science Letters, 10*(15). https://doi.org/10.5267/j.msl.2020.6.038.

Muia, C. M. (2015). Change management challenges affecting the performance of employees: A case study of Kenya Airports Authority, NAIROBI. The Management University of Africa.

Njuguna, E. N., & Muathe, S. M. A. (2016). Critical review of literature on change management on employees performance. *International Journal of Research In Social Sciences, 6*(3), 9–22.

Nuseir, M. T., Aljumah, A., & Alshurideh, M. T. (2021a). *How the business intelligence in the new startup performance in UAE during COVID-19: The mediating role of innovativeness* (Vol. 334).

Nuseir, M. T., Al Kurdi, B. H., Alshurideh, M. T., & Alzoubi, H. M. (2021b). Gender discrimination at workplace: Do Artificial Intelligence (AI) and Machine Learning (ML) have opinions about it. In *The International Conference on Artificial Intelligence and Computer Vision*, 2021b(pp. 301–316).

Odeh, R., Obeidat, B. Y., Jaradat, M. O., Masa'deh, R., & Alshurideh, M. T. (2021). The transformational leadership role in achieving organizational resilience through adaptive cultures: the case of Dubai service sector. *International Journal of Productivity and Performance Management*. https://doi.org/10.1108/IJPPM-02-2021-0093.

Okemba, S. (2018). The impact of change management on organizational success. Vaasan Ammattikorkeakoulu/University of Applied Sciences.

Shehata, Y. (2021). The impact of organizational change on institutional excellence: Analysis of the mediating role of a training and development strategy, an applied comparative study on commercial banks. *Science Journal of Financial Commercial Studies Research, 2*(2), 101–147. https://doi.org/10.21608/CFDJ.2021.171142

Tariq, E., Alshurideh, M., Akour, I., & Al-Hawary, S. (2022a). The effect of digital marketing capabilities on organizational ambidexterity of the information technology sector. *International Journal of Data and Network Science, 6*(2), 401–408.

Tariq, E., Alshurideh, M., Akour, I., Al-Hawary, S., & Al, B. (2022b). The role of digital marketing, CSR policy and green marketing in brand development. *International Journal of Data and Network Science, 6*(3), 1–10.

The Ministry of Education. (2022). Education system: The educational portal is the beating heart of education.

The Effectiveness of Mobile Phones Applications in Learning English Vocabularies

Ibrahim Fathi Huwari, Saddam Rateb Darawsheh, Anwar Saud Al-Shaar, and Hevron Alshurideh

Abstract There is an increasing trend of moving away from traditional technology in the learning process and toward mobile technologies. Today students are well-equipped with this technology, which may facilitate this transition. The purpose of this study is to determine how frequently Jordanian students use their smartphones to learn vocabulary, as well as which application is the most and least preferred for learning vocabulary, and whether they believe that utilizing smartphones will help them enhance their vocabulary. The quantitative research approach was used in this research. The researchers adopted the questionnaire to collect the research data. (135) fourth-year students from Zarqa University (ZU), who studies English language and Translation studies, participated in this research. The findings showed that the majority of Jordanian bachelor students at ZU use their Smartphones for learning vocabulary. IOS and Android were the most used applications among students at ZU. While the least preferred applications were Skype, Viber, Imo, and Hi. The majority of the students believe that using Smartphone help them to improve their vocabularies while searching for the meaning, spelling, pronunciation of the words, listening to English videos through YouTube, English channels, and English music, and using social media. Based on the findings, the researchers provide a number

I. F. Huwari (✉)
Department of English Language and Literature and Translation, Faculty of Arts, Zarqa University, Zarqa, Jordan
e-mail: ihuwari@zu.edu.jo

S. R. Darawsheh
Department of Administrative Sciences, The Applied College, Imam Abdulrahman Bin Faisal University, P.O. Box: 1982, Dammam, Saudi Arabia
e-mail: srdarawsehe@iau.edu.sa

A. S. Al-Shaar
Deanship of Preparatory Year and Supporting Studies, Department of Self Development, (Imam Abdulrahman Bin Faisal University), P.O. Box: 1982, Dammam 43212, Saudi Arabia

H. Alshurideh
English Literature Department, School of Foreign Languages, University of Jordan, Amman, Jordan

M. Alshurideh et al. (eds.), *The Effect of Information Technology on Business and Marketing Intelligence Systems*, Studies in Computational Intelligence 1056,
https://doi.org/10.1007/978-3-031-12382-5_25

of recommendations and more studies that may aid academics in understanding the innovations needs in the area.

Keywords Mobile phones · Learning vocabularies · Jordanian students

1 Introduction

Technology has been a particularly important source of information in modern civilization during the last fifty years, offering up a range of new options in a variety of activities (Almaagbh and Huwari, 2021). Technology is changing the way people interact, play, shop, study, do business, and communicate (Ababneh, 2017). In the past, explaining information and giving lectures were the method that teachers used in teaching, but nowadays, students try to use other ways such as mobile phones in learning (Hameed, 2019). According to studies, pupils who utilise mobile devices are more motivated to learn (Tayan, 2017; Zheng et al., 2017). For example, Tayan (2017) reported that 85% of university students and 100 percent of professors said mobile learning may be used to motivate students to study a foreign language. This is also connected to the fact that students generally prefer the use of mobile devices in foreign language lessons (Kwangsawad, 2019). Traditional techniques of language learning, such as traditional classroom instruction, are put under duress as a result of these developments. They also provide us excellent opportunities to rethink how we teach and learn English (Murray and Kidd, 2017). Mobile phones have become an almost necessary part of daily life since the 1990s. Mobile technology's mainstreaming has resulted in mobile devices that are increasingly functioning as superb learning platform (Al-Sofi, 2021; Muhammad Alshurideh et al., 2012, 2019). According to a national study conducted in 2010, mobile phones are the most significant mode of communication for young people (Alshurideh, 2012, 2016; Alshurideh et al., 2020a, 2020b; Al-Sofi, 2021). Mobile devices, on the other hand, allow students to interact with one another in both informal and formal English learning settings. Also, using technology in teaching for the purpose to understanding is a positive step and it is agreed by most of the schools, colleges, and universities all around the world (Alshurideh, 2012; Alshurideh et al., 2021). Technology can play dynamic role in improving education system (Alshurideh et al., 2020a, 2020b; Reiser and Dempsey, 2012). Through our history, the global amounts of books were our controlling source of knowledge. There is no doubt that technology has advanced highly in the last half-century. Technology has provided unusual improvements to our world, being an eternal source of entertainment and continuously communication anywhere at any time (Douglas, 2001). For that reason, using mobile devices in teaching becomes more popular (Seraj et al., 2021). Nowadays, students have shifted using traditional technologies towards mobile/smart phones (Akour et al., 2021; Alhamad et al., 2021; Hongmei, 2021).

The bulk of mobile applications, according to Heil et al. (2016) research, are focused on vocabulary acquisition. They observed that 84 percent of the top commercial language learning applications focused on practising vocabulary items as isolated units, whereas 53 percent assessed vocabulary in context, which is critical for building students' vocabulary knowledge through inadvertent repeated exposure to new terms. It is critical for pupils to study vocabulary to improve their understanding of a second language. One of the most crucial parts of language acquisition is vocabulary. It is important in the development of language. Furthermore, vocabulary is one of the most important aspects of language acquisition. It is essential in the development of language (Al-Sofi, 2021). Furthermore, many English as a Foreign Language (EFL) students may not be able to practise the language with native speakers. As a result, smartphone applications are expected to bridge this gap by introducing new vocabulary to users and to play a significant role in facilitating communication in virtual circumstances. Furthermore, despite the widespread use of smartphone apps and their importance in vocabulary acquisition, the use of these pre-existing apps for learning purposes is still being researched (Al-Hamad et al., 2021; Leo et al., 2021; Rahimi and Miri, 2014). In addition, using mobile phones in learning is more effective than using tradition technologies such as laptop or personal computer (Asyifa et al., 2019, Al Kurdi et al., 2020a, 2020b; Amarneh et al., 2021). This statement was supported by a research done by (Lu, 2008) on Taiwanese students. The result was students are preferred to use mobile phones rather than other traditional technologies because mobile phones are easily to use and find always with them. The advantages of using a mobile phone in learning a second language include:

1. A more rewarding and engaging learning experience. The usage of mobile devices improves traditional learning by making it more dynamic, exciting, and entertaining, making the learning experience more engaging and rewarding overall (Al Kurdi et al., 2020; Ciampa, 2014).
2. The capacity to learn in a flexible, cost-effective, and timely manner.
3. It provides a one-of-a-kind blend of convenience and social engagement, allowing for face-to-face connection with other students as well as instructors/educators (Attewell and Savill-Smith, 2003; Bettayeb et al., 2020).
4. Improved interaction between students and instructors. Amazingly, communication between learners and instructors is considerably superior to that observed in conventional learning situations when technology is not used in classrooms (Attewell and Savill-Smith, 2003; Sultan et al., 2021).
5. Kids prefer to use mobile phones because it's advantages such as applications, and games (El Hariry, 2015; Nazir et al., 2021).
6. Improving students' ability to analyze, explore, discover, and choose activities that are real and meaningful (El Hariry, 2015; Al-Maroof et al., 2021).
7. Using a fun way in learning by using mobile phones (Alsharari and Alshurideh, 2020; El Hariry, 2015).

Many technological breakthroughs are now permeating our daily lives. The smartphone is one of the most contemporary technological devices. The smartphone, which is becoming increasingly popular, has become an essential component of many people's life, particularly among the younger age. In addition to mobile communication, smartphones offer a number of services such as Viber, WhatsApp, Instagram, tango, and so on. These qualities boost smartphone popularity and make them appealing and desired to a diverse group of people (Güdücü, 2016). The use of mobile/smart phones in learning is critical for both students and teachers. As a consequence, smartphones may assist in English language development and provide our students with a rich learning environment (Hsu and Ching, 2012). They advocate for mobile-based learning, which allows students to benefit anytime and wherever they wish (Sandberg et al., 2011). Mobile phones, according to Hsu and Ching (2012) can assist pupils improve their vocabulary.

Several studies explored the importance of mobile phones in learning and teaching. The most recent studies were done by Ta'amneh (2021). He investigated the difficulties of using smart phones among Saudi students at Taibah University (Badr Branch) studied English language through students' perspectives. The sample of this study is 151 students for the academic year 2019/2020. The researcher used a questionnaire adopted from other studies to achieve his goals. The results of his study were students showed positive with moderate usage of their Smartphone in their attempt to learn English. At the same time, there were no difficulties faced by students in learning through smartphones. In Jordanian context, a study done by Ababneh (2017), the purpose of this study is to find out how Jordanian EFL students feel about using their phones to learn English as a foreign language. The study's goal is to learn more about Jordanian EFL students' knowledge of the need of using technology (mobile phones) correctly in English learning. Furthermore, it intends to investigate the impact of their gender and academic major (English—Translation) on their attitudes regarding utilizing mobile phones to learn English as a second language. The study included 101 students enrolled in English (417), a semantics course offered by the English Department at Yarmouk University in Jordan. The subjects' attitudes toward and use of mobile phones were investigated using a questionnaire. The results showed that the subjects' average score for using their phones while learning English was 3.85 out of 4, indicating that mobile phones are widely used in English learning. The students are also enthusiastic about using their mobile phones to improve their English skills. The results also reveal that both the gender of the individuals and their academic major had a significant impact on their use. Another study was done by Al Aroud and Yunus (2020) examined English as foreign language students at Yarmouk University engagement in utilizing e-learning apparatuses in learning the English. The researchers used a mixed-method approach included 1st first-year English Language and Literature students who were chosen randomly. The findings of the study showed that past experiences affected EFL Jordanian undergraduate students at Yarmouk University's usage of e-learning technologies / MOOCs when studying English.

An Indonesian study done by Asyifa et al. (2019), they looked into the characteristics of mobile phones that EFL students could utilize, as well as their perceptions of such aspects. This study, which took place at one private institution in Jakarta aimed to address two primary questions: (1) what are the characteristics of mobile phones that EFL students most commonly utilize to aid their English learning? (2)

What are EFL students' attitudes about the usage of mobile phone features in the context of English learning? The researchers used a questionnaire in this study. The results showed that students use mobile phone features for learning English, with MP3 players being the most commonly used function for listening to English music, while the least was video recorder. Meanwhile, the students clearly said that they loved learning English with the use of mobile phone characteristics. They improved mobile phone capabilities to aid self-learning rather than conventional learning.

2 Vocabulary Learning

Vocabulary development is critical for foreign or second language learners to achieve proficient communication skills (Al-Khasawneh, 1998; Al-Khasawneh and Huwari, 2014; Huwari et al., 2021) As the attention in vocabulary acquisition expands, many types of tools are now available to learners and students, and one of those is mobile applications, which is one the most efficient, and famous. Furthermore, there are two types of vocabulary learning: intentional and accidental. Any action aimed at committing lexical information to memory is defined as intentional vocabulary acquisition. It involves investing the necessary mental effort and memorizing the words until learners know their meanings (Basal et al., 2016). It is claimed that adopting mobile technology to study English as a second language will improve teaching and learning (Basal et al., 2016). The device's mobility allows students to learn whenever and wherever they choose. Mobile devices, which are frequently utilised among students, may be employed in education and can serve as a motivating tool for pupils (Govindasamy et al., 2019). Learners who acquire language on their phones are subjected to spaced repetition, which is thought to be more addictive than massed repetition. This kind of learning is advantageous to vocabulary development because it allows learners to learn further than linguistic form (Zhang et al., 2011). Hwang and Fu (2019) analysed 93 publications and found that the majority of research agreed that the use of mobile applications for increasing language abilities, particularly vocabulary learning, had potential efficacy. Similarly, Al-Sofi (2021) performed a study to evaluate Saudi learners' perceptions of the applicability and usefulness of smartphone apps (apps) in enhancing vocabulary learning. An online questionnaire and the researcher's observation were utilised to collect data from 270 English majors at the University of Bisha in Saudi Arabia. The SPSS and NVivo software programmes were used to analyse the data. Overall, respondents believe that smartphone applications are effective in improving vocabulary learning because they play a transforming role in providing users with suitable vocabulary input. Fageeh (2013) carried out research among undergraduate Saudi EFL learners to investigate the effects of mobile phone applications in terms of boosting vocabulary learning and motivation. These applications improved vocabulary acquisition, according to the findings.

A study done by Rezaei et al. (2014) investigated the usage and efficacy of mobile applications in the learning of English vocabulary. This study compares the performance of intermediate-level English learners before and after they were exposed to mobile applications as an intervention. It looks at the effects of multimedia courseware on vocabulary development in second language learning. The quantitative results showed that the apps improved learners' performance, and the questionnaire analysis revealed that utilising the applications improved vocabulary learning, confidence, and class involvement, and that students had a good attitude about multimedia use.

The impacts of YouTube videos on student vocabulary development were investigated by Kabooha and Elyas (2018). According to the findings, students thought these tools were effective for learning vocabulary and that the website was simple to use. In another study, Zhang et al. (2011) studied the effectiveness of vocabulary acquisition using mobile phones. From two entire classes of sophomores at a Chinese university, two groups of students (N = 78) were formed: the SMS group (the experimental group) and the paper group (the control group). Following that, they were given a pre-test to determine their prior vocabulary knowledge level. According to the statistics, there was no significant difference ($p > 0.05$) between the SMS group (Mean = 33.34, SD = 14.30) and the paper group (Mean = 37.13, SD = 15.21). They were then assigned to one of two intervention scenarios. The SMS group studied a pre-selected list of vocabulary using text messages delivered to their mobile phones, whereas the paper group worked independently on the same list of vocabulary using paper materials. In the post-tests, there was a significant difference (p.05) between the two groups, but not in the delayed tests ($p > 0.05$). The study concludes that each of these vocabulary acquisition approaches is good, but that combining them may be more beneficial in terms of long-term retention rates. These findings were reinforced by Bloom and Shuell (1981) empirical research of two groups of students (N = 56) learning French vocabulary items. The pupils were divided into two groups: those who got scattered (spaced) practise and those who received massed practise. The post-test findings showed no significant difference between the two groups, but a four-day delayed test indicated that the group with spaced practise fared much better than the group with massed practise. The experimental group had the chance to "practise [both] the vocabulary terms themselves… [and] their recall from long-term memory", but the control group could only "recall information from short-term memory during learning" (Hwang and Fu, 2019, p.247).

A difficulty that has hindered good communication in an L2 is a limited vocabulary. Learners, like smartphone users, may come across unfamiliar language while carrying out the planned operations. Students were unable to adequately communicate their ideas and opinions using an L2 due to a lack of vocabulary. They resorted to utilising their L1 (Arabic) as the type of code-mixing approach to express the intended message and relied on nonverbal communication abilities. Such requirements may serve as an internal motivator for students to use smartphone applications favourably for L2 vocabulary development. To put it another way, pupils are no longer limited to traditional vocabulary memorising. The current study sought to answer the following questions:

1. How often do Jordanian students use their smartphones for learning vocabulary?
2. From your point of view, what is the most and the least preferable application in learning vocabulary?
3. Applications for IOS, and Android, Online dictionaries, WhatsApp, others.
4. Do you think that using smartphones help learners to improve their vocabulary? If yes, how?

3 Research Methodology

3.1 Research Design

This study employed a quantitative research strategy, with data gathered via questionnaires. The more the researcher focus on guaranteeing the principles of a sound research design, the greater the researcher's and, more importantly, the study's user's trust in the research's recommendations and eventual usefulness (Creswell, 2012). This emphasis on data quality may be applied to any study. And, while it is most commonly discussed by survey researchers, quantitative researchers are increasingly discussing data quality issues as well.

3.2 Participants of the Research

Thirty-four individuals participated in this study. The whole participants were undergraduate fourth-year students at Zarqa University in Jordan. However, the major of the participants included both majors' English literature and Translation. The average is between 19 to 22 years old. In condition, each participant in the study owned a mobile phone.

3.3 Data Collection

In this research paper, the researchers employed a questionnaire to gather data. Ababneh (2017) used a five-point Likert Scale to develop the questionnaire (Strongly Agree, Agree, Neutral, Disagree and Strongly Disagree). Students were asked to rate how much they agreed with each of the sixteen items on the questionnaire. The participants were invited to express their true feelings. Furthermore, the researchers answered all questionnaire-related queries from participants to ensure that their responses were correct.

3.4 Data Analysis

The researchers utilised a manual analysis technique in this investigation since they uncovered a tiny number of databases, which were represented in less than 500 pages of transcripts. This merely aids in the location of text sections and the organisation of files. Furthermore, the researcher intended to avoid machine intrusion and have a hands-on sense of data. To become acquainted with the material obtained, the researchers transcribed all of the interviews. According to Lincoln and Guba (1985), the transcribing process aids the researchers in becoming acquainted with the data. The following steps were taken by the researchers of this study when it came to the verbatim transcription of the interview. First, the researchers listened intently to the audio recording of the interview guide questions. Second, on a blank sheet of paper, they typed down the exact words used by the participants when they were obliged to express their ideas. When required, the researchers replay the recordings to find the exact data. Finally, the recordings are replayed to check that the results are correct.

4 Findings/Results

The objectives of the paper were investigating about the frequency of Jordanian students using their smart phones for learning vocabulary, and then what is the most and the least preferable application in learning vocabulary and lastly whether you think using smart phones will help learners to improve their vocabularies.

Q1: Do Jordanian EFL learners use smartphones for learning English?

As is illustrated in Fig. 1, the majority of Jordanian bachelor students at ZU use their Smartphones for learning English. 32 students out of 34 students (94.1%) use their Smartphones for learning English. On the other hand, only 2 students (5.9%) showed that they do not use their Smartphones for learning English.

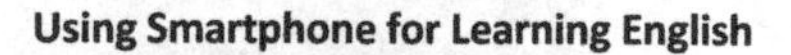

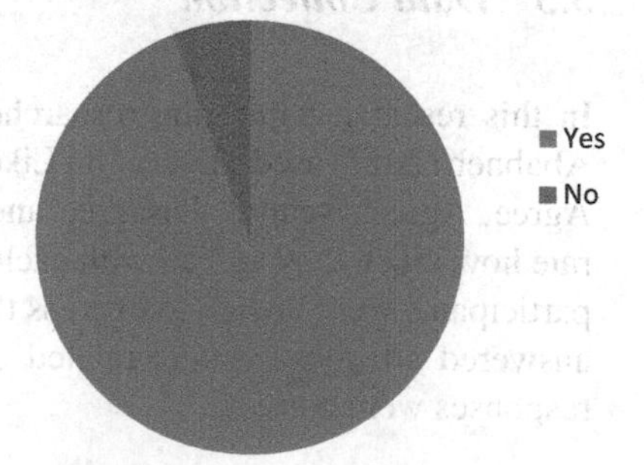

Fig. 1 The percentage of learners using Smartphones for learning vocabulary

Q2: How often do you use your Smartphone for learning vocabulary?

As it is illustrated in the Fig. 2, half of the participants (50%) of Jordanian bachelor students at ZU (17 students out of 34) declare that they sometimes use their Smartphone's for learning vocabularies. Besides, 14 out of 34 students (41.2%) declare that they usually use their Smartphones for learning vocabulary. While the rest of the students 2 out of 34 students declare that they are rarely used their Smartphone's for learning vocabularies. The last colon, only 1 student (2.9%) showed that he is very often used his Smartphone's to learn vocabulary.

Q3: From your point of view, what are the most and the least preferable application in learning vocabulary? Applications for IOS, and Android, Online dictionaries, WhatApp, others.

As illustrated in Fig. 3 about the most and the least preferable application in learning vocabulary, students did not show a huge gap between the applications that they use in learning vocabularies. Applications for IOS and Android was mentioned by 11 students out of 34, using online dictionaries was used by 10 students out of 34, WhatsApp was also preferred to be used by 8 students out of 34, while the least preferred application was others such as Skype, Viber, Imo, Hi, were mentioned by 5 students only.

Q4: Do you think using Smartphones help you to improve your vocabulary? If yes, how?

In regarding to the fourth question, almost all students (32 out of 34) believe that using Smartphone's help them to improve their vocabularies. The way that students use their Smartphone's helps them to improve their vocabularies such as searching for the meaning, spelling, pronunciation of the words, listening to English videos through YouTube, English channels, and English music, and using social media.

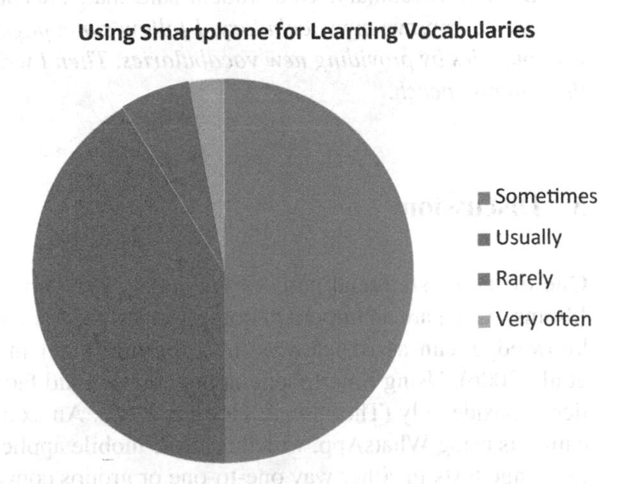

Fig. 2 The percentage of learners using Smartphone's for learning vocabularies

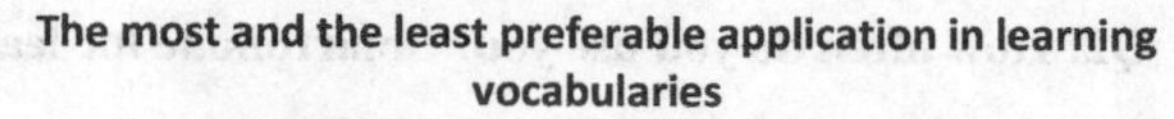

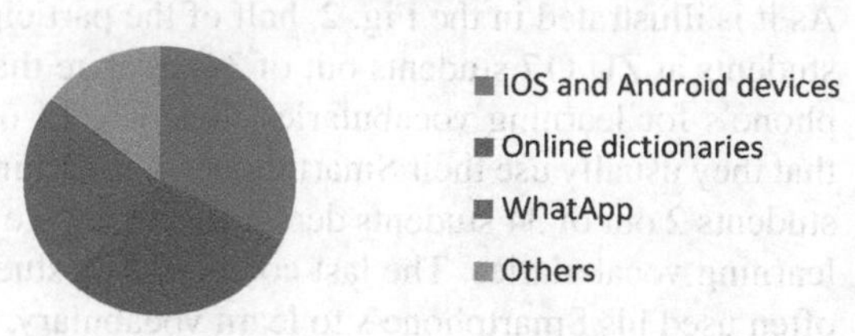

Fig. 3 The most and the least preferable application in learning vocabulary?

The first theme was searching for the meaning, spelling, and pronunciation of the new words. At this point, the majority of the students showed that looking for a new word using Smartphone is much easier than the traditional way. One of the students said that: *Since I have started using my phone to gain more vocabulary, the number of vocabulary that I have is already increased, I think. As I can get the pronunciation as well.*

Spelling is also important in Smartphone. One student said: *I am sure my phone helps me to improve my spelling, as I was writing the new word on a piece of paper so that I avoid spelling mistakes once I come to English composition.*

The second theme was listening to English videos through YouTube, English channels, and English music which help students to improve their vocabulary. In this theme, the majority of the students showed that listening to English audios was a way to improve their vocabulary. One student said: *The best way for me to increase the number of vocabulary was listening. I keep watching English movies then I write down the new vocabularies that I gain on each video or film.*

The third theme was using social media platforms which helps students to improve their vocabulary. This theme was mentioned by half of the whole population in this study. Social media like Facebook or Twitter was a way to improve students' ability to gain new vocabulary. One student said that: *Facebook is one of the ways that help me to improve my vocabulary. I follow some pages which focused on English vocabularies by providing new vocabularies. Then I write down them and try to use them in my speech.*

5 Discussion

One of the most crucial parts of language instruction is vocabulary development. Idioms, which are an important component of vocabulary knowledge, and idiomatic knowledge, can assist learners in becoming fluent in the target language (Boers et al., 2006). Using Smartphone applications could facilitate mobile learning practices considerably (Thornton & Houser, 2005). An example of a smartphone application is using WhatsApp, which is a free mobile application that allows its users to exchange texts in either way one-to-one or groups conversation and make calls.

In terms of the usage of smartphones for learning English, this study agrees with Güdücü (2016) about the use of smartphones in learning English among different nationalities studying for an undergraduate degree in Turkey. According to his research, 63.3% of students use their smartphones for instructional purposes. Lincoln and Guba (1985) proved that learners utilize their smartphones for language acquisition with the support of mobile web apps. According to Kim et al. (2013), ICT may be an effective instrument for providing diversity to learners and assisting them in learning English. Every learner is unique; they all have distinct learning methods (such as visual/auditory/verbal/musical). The findings of the interviews demonstrate that learners have learning differences and have good attitudes regarding the usage of smartphones since it includes visuals that help them comprehend better.

Previous research indicated that students had favorable attitudes toward the use of mobile applications for vocabulary acquisition. Hayati et al. (2013) investigated three approaches of teaching vocabulary (self-study, SMS, in class). They observed that the SMS group advanced significantly quicker than the other groups that received positive feedback on using mobile phones to learn vocabulary. Lu (2008) made a similar discovery, demonstrating that the SMS group identified more vocabulary than the printed materials. According to [15], using a mobile phone to communicate vocab ulary is more effective than using other approaches such as flashcards. A study by Güdücü (2016) investigated the usage of smartphones in acquiring English vocabulary among students of various ethnicities in Turkey pursuing an undergraduate degree. It was revealed that preparatory students frequently use their cell phones to acquire language. Online dictionaries, such as Merriam-Webster, are preferred by students over paper dictionaries. One of the participants, for example, states that he consults the Merriam-Webster dictionary to determine the definition of a word. To remember the terms for the future, he prefers English to English dictionaries. According to an examination of the questionnaire data on smartphone use, 13% of learners always use their cellphones as an electronic dictionary, as shown below. 43,3% of them do so often, and 40% do so sometimes. The data reveals that smartphones were utilized by preparatory learners to boost vocabulary acquisition.

The current study's findings are consistent with those of Lu (2008) and Zhang et al. (2011), who discovered that short-term spaced vocabulary acquisition using mobile phones can be more effective than massed vocabulary learning via paper. This might be due to the students' convenient access to the mobile device, resulting in their recurrent exposure to and frequent practice of the vocabulary items daily in a spaced manner. This type of learning technique encourages enhanced vocabulary learning (Waring and Nation, 2004). The findings of the questionnaire and interviews show that using smartphones has a significant impact on learners' vocabulary development.

6 Conclusion

Results of this study indicate that majority of Jordanian bachelor students at ZU use their Smartphones for learning vocabulary. 32 students out of 34 students (94.1%) use their Smartphones for learning vocabulary. In addition, the second research question was declared that half of the participants (50%) of Jordanian bachelor students at ZU (17 students out of 34) declare that they sometimes use their Smartphones for learning vocabulary. Besides, 14 out of 34 students (41.2%) declare that they usually use their Smartphones for learning vocabulary. While the rest of the students 2 out of 34 students declare that they are rarely used their Smartphones for learning vocabulary. In the last colon, only 1 student (2.9%) showed that he is very often used his Smartphones to learn vocabulary. In the last research question, almost all students (32 out of 34) believe that using Smartphones help them to improve their vocabulary. The way that students use their Smartphones helps them to improve their vocabularies such as searching for the meaning, spelling, pronunciation of the words, listening to English videos through YouTube, English channels, and English music, and using social media platforms.

7 Recommendations

Learners' understanding of the importance of vocabulary acquisition in language learning, as well as the usefulness of smartphone applications in vocabulary development, is critical. Being acclimated to the good use of smartphone applications, in particular, prepares students sufficiently for the work market following graduation. Instructors and policymakers should motivate learners to use such self-led and self-directed apps to effectively serve their learning requirements and personal choices that complement in-class assignments and activities, while real use of smartphone applications for vocabulary acquisition may be inadequate. They should be encouraged to keep up with the latest apps and to perceive smartphone devices as more than simply social and/or entertaining tools, but also as complementary learning tools that help them enhance their fluency. While developing free educational applications, designers should also answer the communicatively authentic wants of users. Vocabulary knowledge is a difficult task that must be built gradually. Once the motivating and promising status of smartphone applications is established, it is envisaged that using them for learning purposes will become commonplace and a deliberate decision in the learning process.

8 Limitations

This study has limitations concerning the number of participants and the method used to collect data. A large-scale study is required to assess the benefits and drawbacks of vocabulary learning using cell phones. Furthermore, additional research is required to evaluate the strength of the relationship between mobile applications and their effectiveness in vocabulary learning.

References

Ababneh, S. (2017). Using mobile phones in learning English: The case of Jordan. *Journal of Education and Human Development, 6*(4), 120–128.

Akour, I., Alshurideh, M., Al Kurdi, B., Al Ali, A., & Salloum, S. (2021). Using machine learning algorithms to predict people's intention to use mobile learning platforms during the COVID-19 pandemic: Machine learning approach. *JMIR Medical Education, 7*(1), 1–17. https://doi.org/10.2196/24032

Al-Hamad, M. Q., Mbaidin, H. O., Alhamad, A. Q. M., Alshurideh, M. T., Kurdi, B. H. A., & Al-Hamad, N. Q. (2021). Investigating students' behavioral intention to use mobile learning in higher education in UAE during Coronavirus-19 pandemic. *International Journal of Data and Network Science, 5*(3). https://doi.org/10.5267/j.ijdns.2021.6.001.

Al-Khasawneh, B. (1998). *Communication problems between higher studies students and teaching staff members at Yarmouk University. From students' perspective*. Yarmouk University.

Al-Khasawneh, F. M., & Huwari, I. F. (2014). The effect of metacognitive strategies instruction on vocabulary learning among Jordanian University students. *International Journal of English and Education, 3*(3), 102–113.

Al-Maroof, R. S., Alshurideh, M. T., Salloum, S. A., AlHamad, A. Q. M., & Gaber, T. (2021). Acceptance of google meet during the spread of coronavirus by Arab university students. *Informatics, 8*(2). https://doi.org/10.3390/informatics8020024.

Al-Sofi, B. B. M. A. (2021). Student satisfaction with e-learning using blackboard LMS during the Covid-19 circumstances: realities, expectations, and future prospects. *Pegem Journal of Education and Instruction, 11*(4), 265–281.

Al Aroud, B., & Yunus, K. (2020). The use of e-tools in learning English by EFL students at Yarmouk University. *International Journal of Education, Psychology and Counselling, 5*(37), 72–81.

Al Kurdi, B, Alshurideh, M., & Salloum, S. (2020a). Investigating a theoretical framework for e-learning technology acceptance. *International Journal of Electrical and Computer Engineering (IJECE), 10*(6), 6484–6496.

Al Kurdi, B., Alshurideh, M., Salloum, S., Obeidat, Z., & Al-dweeri, R. (2020b). *An empirical investigation into examination of factors influencing university students' behavior towards elearning acceptance using SEM approach*.

Alhamad, A. Q. M., Akour, I., Alshurideh, M., Al-Hamad, A. Q., Kurdi, B. A., & Alzoubi, H. (2021). Predicting the intention to use google glass: A comparative approach using machine learning models and PLS-SEM. *International Journal of Data and Network Science, 5*(3). https://doi.org/10.5267/j.ijdns.2021.6.002.

Almaagbh, I. F., & Huwari, I. F. (2021). E-English learning after Covid-19 pandemic: Problems and challenges. *Review of International Geographical Education, 11*(8), 2854–2858.

Alsharari, N. M., & Alshurideh, M. T. (2020). Student retention in higher education: the role of creativity, emotional intelligence and learner autonomy. *International Journal of Educational Management*.

Alshurideh, M. (2012). The effect of previous experience on mobile subscribers' repeat purchase behaviour. *European Journal of Social Sciences, 30*(3).

Alshurideh, M., Al Kurdi, B., & Salloum, S. A. (2020a). Examining the main mobile learning system drivers' effects: A mix empirical examination of both the Expectation-Confirmation Model (ECM) and the Technology Acceptance Model (TAM). In *Advances in Intelligent Systems and Computing* (Vol. 1058). https://doi.org/10.1007/978-3-030-31129-2_37.

Alshurideh, M., Al Kurdi, B., Salloum, S. A., Arpaci, I., & Al-Emran, M. (2020b). Predicting the actual use of m-learning systems: A comparative approach using PLS-SEM and machine learning algorithms. *Interactive Learning Environments*. https://doi.org/10.1080/10494820.2020.1826982

Alshurideh, M. T., Al Kurdi, B., AlHamad, A. Q., Salloum, S. A., Alkurdi, S., Dehghan, A., et al. (2021). Factors affecting the use of smart mobile examination platforms by universities' postgraduate students during the COVID-19 pandemic: An empirical study. *Informatics, 8*(2). https://doi.org/10.3390/informatics8020032.

Alshurideh, M. (2016). Scope of customer retention problem in the mobile phone sector: A theoretical perspective. *Journal of Marketing and Consumer Research, 20*, 64–69.

Alshurideh, M., Masa'deh, R. M. d. T., & Alkurdi, B. (2012). The effect of customer satisfaction upon customer retention in the Jordanian mobile market: An empirical investigation. *European Journal of Economics, Finance and Administrative Sciences, 47*(47), 69–78.

Alshurideh, M., Salloum, S. A., Al Kurdi, B., Monem, A. A., & Shaalan, K. (2019). Understanding the quality determinants that influence the intention to use the mobile learning platforms: A practical study. *International Journal of Interactive Mobile Technologies, 13*(11).

Amarneh, B. M., Alshurideh, M. T., Al Kurdi, B. H., & Obeidat, Z. (2021). The Impact of COVID-19 on E-learning: Advantages and challenges. In *The International Conference on Artificial Intelligence and Computer Vision* (pp. 75–89).

Asyifa, D. I., Warni, S., & Komara, C. (2019). The use of mobile phone features by efl students in a private university in Jakarta. In *UICELL Conference Proceeding* (pp. 46–56).

Attewell, J., & Savill-Smith, S. (2003). *Young people, mobile phones and learning*. Learning and Skills Development Agency.

Basal, A., Yilmaz, S., Tanriverdi, A., & Lutfiye, S. (2016). Effectiveness of mobile applications in vocabulary teaching. *Contemporary Educational Technology, 7*(1), 47–59.

Bettayeb, H., Alshurideh, M. T., & Al Kurdi, B. (2020). The effectiveness of mobile learning in UAE universities: A systematic review of motivation, self-efficacy, usability and usefulness. *International Journal of Control and Automation, 13*(2), 1558–1579.

Bloom, K. C., & Shuell, T. J. (1981). Effects of massed and distributed practice on the learning and retention of second-language vocabulary. *The Journal of Educational Research, 74*(4), 245–248.

Boers, F., Eyckmans, J., Kappel, J., Stengers, H., & Demecheleer, M. (2006). Formulaic sequences and perceived oral proficiency: Putting a lexical approach to the test. *Language Teaching Research, 10*(3), 245–261.

Ciampa, K. (2014). Learning in a mobile age: An investigation of student motivation. *Journal of Computer Assisted Learning, 30*(1), 82–96.

Creswell, J. W. (2012). *Educational research: Planning, conducting, and evaluating quantitative and qualitative research* (4th ed.). Pearson Boston.

El Hariry, N. A. (2015). Mobile phones as useful language learning tools. *European Scientific Journal, 11*(16), 298–317.

Fageeh, A. A. I. (2013). Effects of MALL applications on vocabulary acquisition and motivation. *Arab World English Journal, 4*(4).

Govindasamy, P., Yunus, M. M., & Hashim, H. (2019). Mobile assisted vocabulary learning: Examining the effects on students' vocabulary enhancement. *Universal Journal of Educational Research, 7*(12), 85–92.

Güdücü, M. (2016). *The uses of smartphones among EFL learners and their effects on vocabulary learning in foreign language English preparatory school*. Eastern Mediterranean University (EMU)-Doğu Akdeniz Üniversitesi (DAÜ).

Hameed, D. T. (2019). The impact of using mobile phone technology in learning English. *Journal of Al-Frahedis Arts, 11*(38), 569–586.
Hayati, A., Jalilifar, A., & Mashhadi, A. (2013). Using Short Message Service (SMS) to teach English idioms to EFL students. *British Journal of Educational Technology, 44*(1), 66–81.
Heil, C. R., Wu, J. S., Lee, J. J., & Schmidt, T. (2016). A review of mobile language learning applications: Trends, challenges, and opportunities. *The EuroCALL Review, 24*(2), 32–50.
Hongmei, W. (2021). Application of smart-phone apps by Chinese college students in learning English. *Sino-US English Teaching, 18*(5), 116–120.
Hsu, Y.-C., & Ching, Y.-H. (2012). Mobile microblogging: Using Twitter and mobile devices in an online course to promote learning in authentic contexts. *International Review of Research in Open and Distributed Learning, 13*(4), 211–227.
Hwang, G.-J., & Fu, Q.-K. (2019). Trends in the research design and application of mobile language learning: A review of 2007–2016 publications in selected SSCI journals. *Interactive Learning Environments, 27*(4), 567–581.
Kabooha, R., & Elyas, T. (2018). The effects of YouTube in multimedia instruction for vocabulary learning: Perceptions of EFL students and teachers. *English Language Teaching, 11*(2), 72–81.
Kim, D., Rueckert, D., Kim, D.-J., & Seo, D. (2013). Students' perceptions and experiences of mobile learning. *Language Learning and Technology, 17*(3), 52–73.
Kwangsawad, T. (2019). University students' perceptions of MALL in EFL classes. *Studies in English Language Teaching, 7*(1), 75–82.
Leo, S., Alsharari, N. M., Abbas, J., & Alshurideh, M. T. (2021). From offline to online learning: A qualitative study of challenges and opportunities as a response to the COVID-19 pandemic in the UAE higher education context. *The Effect of Coronavirus Disease (COVID-19) on Business Intelligence, 334*, 203–217.
Lincoln, Y. S., & Guba, E. G. (1985). *Naturalistic inquiry*. SAGE Publications Sage UK.
Lu, M. (2008). Effectiveness of vocabulary learning via mobile phone. *Journal of Computer Assisted Learning, 24*(6), 515–525.
Hawari, A., Moh'd, O., & Huwari, I. F. (2021). Patterns of vocabulary learners strategies employed by Saudi EFL learners. *Review of International Geographical Education Online, 11*(3), 797–807.
Murray, J., & Kidd, W. (2017). *Using emerging technologies to develop professional learning*. Routledge.
Nazir, M. I. J., Rahaman, S., Chunawala, S., & AlHamad, A. Q. M. (2021). Perceived factors affecting students academic performance; Nazir, J., Rahaman, S., Chunawala, S., Ahmed, G., Alzoubi, H., Alshurideh, M., et al. (2022). Perceived factors affecting students academic performance. *Academy of Strategic Management Journal, 21*(4), 1–15; 21(4), 1–15.
Rahimi, M., & Miri, S. S. (2014). The impact of mobile dictionary use on language learning. *Procedia-Social and Behavioral Sciences, 98*, 1469–1474.
Reiser, R. A., & Dempsey, J. V. (2012). *Trends and issues in instructional design and technology*. Pearson Boston.
Rezaei, A., Mai, N., & Pesaranghader, A. (2014). The effect of mobile applications on English vocabulary acquisition. *Jurnal Teknologi, 68*(2).
Sandberg, J., Maris, M., & De Geus, K. (2011). Mobile English learning: An evidence-based study with fifth graders. *Computers & Education, 57*(1), 1334–1347.
Seraj, P. M. I., Klimova, B., & Habil, H. (2021). Use of mobile phones in teaching English in Bangladesh: A systematic review (2010–2020). *Sustainability, 13*(10), 5674.
Sultan, R. A., Alqallaf, A. K., Alzarooni, S. A., Alrahma, N. H., AlAli, M. A., & Alshurideh, M. T. (2021). How students influence faculty satisfaction with online courses and do the age of faculty matter. In *The International Conference on Artificial Intelligence and Computer Vision* (pp. 823–837).
Ta'amneh, M. A. A. A. (2021). The use of smartphones in learning english language skills: A study of university students' perspectives. *International Journal of Applied Linguistics and English Literature, 10*(1), 1–8.

Tayan, B. M. (2017). Students and teachers' perceptions into the viability of mobile technology implementation to support language learning for first year business students in a Middle Eastern university. *International Journal of Education and Literacy Studies, 5*(2), 74–83.
Thornton, P., & Houser, C. (2005). Using mobile phones in English education in Japan. *Journal of Computer Assisted Learning, 21*(3), 217–228.
Waring, R., & Nation, I. S. P. (2004). Second language reading and incidental vocabulary learning. *Angles on the English Speaking World, 4*, 97–110.
Zhang, H., Song, W., & Burston, J. (2011). Reexamining the effectiveness of vocabulary learning via mobile phones. *Turkish Online Journal of Educational Technology-TOJET, 10*(3), 203–214.
Zheng, Q., Chen, T., & Kong, D. (2017). An empirical study on context awareness integrated mobile assisted instruction and the factors. *Eurasia Journal of Mathematics, Science and Technology Education, 13*(6), 1737–1747.

The Effect of Using SCAMPER Strategy on Developing Students' Achievement

Nofah Sameh Almawadeh, Saddam Rateb Darawsheh, Anwar Saud Al-Shaar, and Hevron Alshurideh

Abstract The study aimed to find out the effect of using SCAMPER strategy on developing students' some habits of mind.. The participants were chosen from the Fifth-grade students prep stage and they were divided into two groups, experimental and control group. The study tools prepared by the researcher (achievement test and habits of mind questionnaire) were pre- administrated. The experimental group learnt the presented material using SCAMPER strategy and the control group studied in the traditional method. The study tools were post applied. The results showed that there were statistically significant differences at the level (0.01) between the mean scores of experimental and control groups in the post application of the tools in favor of the experimental group.

Keywords SCAMPER strategy · Students' achievement

N. S. Almawadeh (✉)
Faculty Educational Sciences, Jerash University, Jerash, Jordan
e-mail: n.mawadeh@jpu.edu.jo

S. R. Darawsheh
Department of Administrative Sciences, The Applied College, Imam Abdulrahman Bin Faisal University, P.O. Box: 1982, Dammam, Saudi Arabia
e-mail: srdarawsehe@iau.edu.sa

A. S. Al-Shaar
Deanship of Preparatory Year and Supporting Studies, Department of Self Development, (Imam abdulrahman Bin Faisal University), P. O. Box: 1982, Dammam 43212, Saudi Arabia

H. Alshurideh
English Literature Department, School of Foreign Languages, University of Jordan, Amman, Jordan
e-mail: HevronAlshurideh@gmail.com

M. Alshurideh et al. (eds.), *The Effect of Information Technology on Business and Marketing Intelligence Systems*, Studies in Computational Intelligence 1056,
https://doi.org/10.1007/978-3-031-12382-5_26

1 Introduction

The philosophy of the SCAMPER strategy is based on the idea of helping students generate new or alternative ideas. It enables the learner to access, analyze and evaluate the text as well as to think creatively and critically to find solutions. The name came from the beginning of the first letter of the word SCAMPER (Switch, add, adapt, modify, dig, put other uses, delete, reverse, rearrange) (Antonacci et al., 2014). There are a number of pillars upon which the Scamper strategy is based, namely that training in imagination in the manner of fun and play, and conducting mental treatments through the Spurring Checklist contribute to the development of creative thinking and imagination, and this is done By either providing programs and activities that aim to teach thinking independently of the regular school curricula, and as a single curriculum as an independent enrichment program for the development of creative thinking, or by indirectly providing activities and developing creative thinking and presenting the strategy within the content of the regular curriculum, and the owners of this view. The tendency is that mental operations are learned in this direction through teaching with the scamper strategy (Horn, 2013). When using the scamper strategy, it is not necessary to use the seven mentioned steps, but some of them can be chosen according to the nature of the lesson presented, and the required skill: this means that the use of the strategy will depend on what is appropriate for the topic of the lesson (Animasahun, 2014).

Gladding (2007) defines SCAMPER strategy as a technique to solve problems which can be used to stimulate individuals when facing a problem by helping them to use or recycle their present ideas to reach a solution, this solution can be used in similar situations. everything new is just a modification to what already exists (Georgiakakis et al., 2010; Alshurideh et al., 2020; Alshurideh et al., 2021). SCAMPER strategy guides the learners to find creative ideas. Several studies have showed that creativity can be developed using SCAMPER strategy like (Zomerdijk, 2010; Folashade, 2012; Moreno and Yang, 2014). In today's classrooms, it is important to engage thinking and creative activities that enable students to go beyond the simple memorization of facts. Developing these thinking dispositions help students to understand, to reason and apply information to solve problems both in and out of the classroom. Particularly in English as Second Language (ESL) classrooms, verbal interaction or communication is a major key to learning a second language. Thus, there is a need to move from the traditional focus on how many answers a student knows to a focus on engaging habits of mind as the learning outcome in a teaching and learning classroom (Costa, 1985).

2 Problem of the Study

The problem of the study is represented by the weakness of Fifth grade students in achievement in the study of artificial intelligence methods. In an attempt to solve this problem, the researchers used the Scamper strategy to find out its impact on developing the achievement of Fifth graders in the chapter of artificial intelligence.

3 Purpose of the Study

The purpose of the study was to develop Fifth grade students' achievement.

4 Questions of the Study

The present study attempted to find answer to the following main question:

What is the effect of using SCAMPER strategy on developing Fifth grade students' achievement?

5 Hypotheses of the Study

This study sought to verify the following hypotheses:

1. There will be a statistically significant difference between the mean scores of the achievement test of the experimental and control groups in the post administration of the test in favor of the experimental group.
2. There will be statistically significant differences between the mean scores of the experimental group in the pre- and post-administration of the achievement test in favor of the post test.

6 Aims of the study

The present study aimed at identifying the effect of using SCAMPER strategy on developing Fifth grade students' achievement.

6.1 Variables of the Study

1-Independent variable

Using SCAMPER strategy to develop Fifth grade students' achievement.

2-Dependent variable

- Fifth grade students' achievement.

6.2 Delimitations of the Study

The study is delimited to:

- Two Fifth grade students' classes from madaba preparatory school for boys in Al madaba district in madaba during the first term of the academic year 2021–2022.
- Chapter 6 artificial intelligence

6.3 The Tools of the Study

Treatment materials:

1. A teacher's guide to present the Chap. 6 (artificial intelligence) using SCAMPER strategy.
2. Activity papers for students to improve their achievement.

   Measurement tools:
   A pre-post achievement test.

7 Significance of the Study

The present study is hoped to achieve the following:

- Provide English language specialists and curricula designers with SCAMPER strategy to be taken into consideration in planning and constructing activities for English language students.
- Provide valid tools for testing achievement
- Support using modern teaching methods and strategies to develop language

8 Definition of Terms

The following terms refer to the operational definition in the current study:

Strategy: A group of techniques and procedures followed by the teacher to carry out the learning process inside or outside the classroom (Alshurideh et al., 2019a, 2019b; Al Kurdi et al., 2020; Kurdi et al., 2020) to chapters (6) of the (artificial intelligence) for Fifth grade Fifth grade. It aims at developing achievement. SCAMPER strategy is planned to be used in sequence steps for the teacher and students to achieve the education objectives with less effort and time.

SCAMPER strategy: A group of procedures and techniques to help the learners to work their imagination and reach creative ideas using seven skills: S(substitute), C (combine), A (adapt), M (magnify), P (modify), P (put to other uses), E (eliminate) and reverse (Eberle, 2008).

9 Theoretical Background

9.1 SCAMPER Strategy Historical Background

SCAMPER is a mnemonic acronym that provides a planned way to help students and teachers with understanding creative problem solving and improving extension-building activities based on prior ideas and processes (Toraman and Altun, 2013). As stated by Fahmy et al. (2017), SCAMPER strategy was first proposed by Alex Osborne in 1953 who designed a spurring checklist, the first letters in the words of the checklist formed the word SCAMPER. This thinking strategy was further developed by Frank Williams who suggested some techniques to provoke imagination and creative expression for children. His techniques relied on two dimensions:

Cognitive processes such as: fluency, flexibility, originality, and tendency to having details.

Affective processes such as: curiosity, readiness to take risks, preferring complexity and gut instinct.

Finally, Bob Eberle mixed the previous experiences especially those related to generating idea and defined operationally each of the strategies forming the word SCAMPER. In his 1971 book, SCAMPER: Games for Imagination Development. Eberle states that in the same way as the word SCAMPER suggests "running playfully about as a child", the strategy SCAMPER may also require the need "to run playfully about in one's mind in search of ideas" (Oakshott et al., 2011). He formed different games and activities and pended it together to construct SCAMPER strategy to develop creative imagination. Eberle added three principles to the strategy: Magnify, minify, and re-arrange, it has become ten creative principles in the strategy instead of seven principles.

The importance of teaching using SCAMPER is due to the following (Vidal, 2005).

- Develop imagination, especially the creative imagination of students,

Providing students and teaching them to practice the methods of generating ideas included in the Scamper strategy.

Developing students' thinking skills in general and productive thinking in particular.

Enable students to generate creative ideas about the issues presented to them.

Enhancing self-concept and creating high levels of ambition among students.

Arousing curiosity, risk tolerance, and a preference for complexity among students.

Building group spirit and increasing students' attention spans.

Opening the horizons of divergent thinking among students.

Helping students to generalize the experiences gained in different life situations, after presenting them to them in a variety of contexts.

This strategy consists of the following steps and questions as mentioned by (Toraman and Altun, 2013):

First letter: S: Substitute.

The second letter: (to combine or combine): C: Combine, which is to combine or combine things.

Third letter: (adaptive): A: Adapt.

Fourth letter: (modification and development): M: Madifty.

Fifth character: (other uses): P: Put to other uses.

Sixth character: (eliminate): E: Eliminate.

Seventh letter: (rearrange or reverse): R: Reverse-Rearrange.

9.2 The Steps of Applying SCAMPER Strategy

Reviewing previous studies and literature about SCAMPER strategy led the researcher to identify two directions to apply it. The following steps are mentioned by Feather (2008) as:

- The instructor explains to the students SCAMPER strategy to encourage them to generate new and creative ideas. He also explains the guiding questions used in every strategy.
- The instructor gives the learners the text to read and understand carefully and then to generate new ideas.
- The learners choose any of the SCAMPER strategy and the instructor encourages them to write all the creative ideas they reach. Another strategy is chosen and the same steps are repeated.
- The teacher revises the list of new ideas created by the learners to identify the learners' response to the strategy.

The researcher benefited from this part in designing the chapters of artificial intelligence according the steps of the second direction of applying SCAMPER strategy and also in preparing the teacher's guide.

9.3 The Learner's Role in SCAMPER Strategy

The role of the learner in SCAMPER strategy is not a traditional one, he does not only listen to the teacher or answer questions in the classroom. The learner becomes a knowledge seeker who substitutes, combines, adapts, modifies, puts to other uses, eliminates and rearranges the information to generate creative and new ideas. The role of the learner can be summed in:

- He searches for the information and may be a source for it.
- He participates effectively in presenting ideas.
- He transfers his experience to similar situations.
- He chooses the suitable SCAMPER strategy to solve the problem in a creative way.
- He discusses the new ideas with the teacher and learns to pose questions to his teacher.
- He learns to accept the others' points of view and solutions.
- He creates the largest number of creative solutions he can reach.

9.4 The Teacher' Role in SCAMPER Strategy

The role of the learner will not be achieved without having a teacher who works as a facilitator and guide for the learning process. The teacher' role can be identified in the following points:

- He identifies the objectives that learners should achieve.
- He presents the problem or situation which provokes the learners' brains to search.
- He guides the students by explaining SCAMPER strategy and the guiding questions related to it.
- He presents suitable atmosphere in the class for discussion and cooperative work.
- He responds and answers the questions that students ask.
- He organizes the time to achieve the lesson objectives.
- He supplies feedback for the learners at every stage of the lesson.
- He evaluates the answers and solutions presented by the learners according to criteria known for the learners.
- He gives confidence to the learners and encourages them to present creative ideas.

10 Method

The design of the current study is the two groups' pre-post testing experimental design. To answer the study questions, the following procedures were followed:

1-Identifying the organization and reconstructing the content of Fifth grade (Chap. 6 artificial intelligence) **using SCAMPER strategy**

**Choosing the content*

The researcher has chosen the Chap. 6 (the course of the first term) (artificial intelligence) to be presented using SCAMPER strategy for the following reasons:

- There are also a lot of description and adventure that encounters the characters of the story.
- The Chap. 6 are to be presented through the whole second term which would give enough time to improve achievement

**Reconstructing the content*

The content was reconstructed using SCAMPER strategy as follows:

<u>Preparing the students' activity papers</u>

In the light of the goals and the content of the story, the researcher prepared the students' activity papers according to SCAMPER strategy. It included the content of the chapters divided into parts and followed by different activities. Different activities are presented followed by guiding questions to help the learners choose and apply different SCAMPER strategy skills. Every activity is followed by the steps that should be followed by the learners. Spaces for writing down the notes, explanations or new ideas are supplied.

<u>Preparing the teacher's guide</u>

The researcher has prepared the teacher's guide to teach the Chap. 6 of (artificial intelligence) using SCAMPER strategy. The guide includes introduction about SCAMPER strategy, its philosophy, skills and the steps to use it. It also gives general directions for the teacher to present the lessons and activities with the required teaching aids. Distribution of the lessons and time plan are included.

<u>Time plan for presenting the content</u>

The presentation of the content lasted for 10 weeks (two months and half). The teacher met the learners once every week for 90 min (one school period). The first two periods were devoted to explaining SCAMPER strategy, its skills, guiding questions and importance for the learners. The learners also were told how the content will be presented and their expected role.

2- Constructing the tools of the study

The following steps were followed to design and construct the two measurement tools of the study. The jury who kindly validate them were three staff members from the Department of Curricula and Methods of Teaching, Girls' College, Ain Shams University and two staff members from other faculties.

a. **A pre-post achievement test**

The test was designed as follows:

The test aim: It aimed at pre-post measuring achievement in the Chap. 6 –the first term (artificial intelligence) of the members of both the experimental and control groups of the current study.

Constructing the test: the test consisted of 30 questions, the first ten questions were "wh" questions which required complete answers from the students. The second ten questions were completion questions and the last ten questions were multiple choice questions.

The test validity: the first version of the test was submitted to jury members to evaluate the test in terms of (a) number of questions and appropriateness to the content and (b) suitability of the test to the Fifth grade. The jury members made some modifications which were taken seriously into consideration and were done accordingly. After the modifications were made, the test was mostly a valid one, as it showed that it measured what it was intended to measure as stated by the jury.

Reliability of the test: The test–retest method was used to determine the reliability of the test. The test was administered to 35 Fifth grade in madaba school and was re administered by an interval of fifteen days to the same group. Then, the Pearson Correlation Coefficient between the test/re-test was calculated. The reliability coefficient was 0.85 which is relatively high. So, the test was reliable.

Timing the test: During the piloting administration of the test on the same 35 students, the researcher could decide the timing of the test. The following formula was used: the time taken by all the students divided by the number of the students, (summation of the time taken by all the students)/(the number of the students) = 2100/35 = 60 min.

Scoring the test: The total mark of the test was 40 marks divided as follows: 2 marks for the complete answer of each of the ten "wh" questions (20 marks). One mark was given for the correct completion for the ten completion items (10 marks) and one mark for the correct choice in the following ten MCQ (10 marks).

Participants

The participants in the current study were two groups: experimental group and control group. Each group had 28 Fifth grade students during the first term of the academic year 2021/2022. Two classes were chosen randomly from Madaba school for boys, students of the first class were chosen as experimental group while students of the other class were the members of the control group after being sure that they had the same achievement level from their teachers.

Table 1 T-test results of the post administration of the achievement test on the experimental and control groups

Group	N	M	S.D	Difference between means	Degree of freedom	t- value	Significance level
Experimental	28	35.46	3.45	14.61	56	23.80	0.01 Significance
Control	28	20.85	2.88				

11 Data Analysis

The statistical analysis of data and the results are interpreted in terms of the research hypotheses. For this purpose, each hypothesis is presented together with the findings related to it.

11.1 The First Hypothesis

There will be a statistically significant difference between the mean scores of the achievement test of the experimental and control groups in the post administration of the test in favor of the experimental group. A t-test was thus applied to the data; (see Table 1).

Table 1 shows that the mean score of the experimental group in the post administration of the achievement test (35.46) is more than the mean scores of the control group (20.85). The difference is (14.61) marks in favor of the experimental group. The above table shows that the estimated t-value (23.80) is statistically significance at 0.01 level. The effect size was large (60.40). This shows that the members of the experimental group's achievement developed due to the positive impact of using SCAMPER strategy.

11.2 The Second Hypothesis

There will be statistically significant differences between the mean scores of the experimental group in the pre and post administration of the achievement test in favor of the post test.

Table 2 shows that the mean score of the experimental group in the post administration of the achievement test (35.46) is more than the mean scores of the pre administration (18.65). The difference is (16.81) marks in favor of the post administration. The above table shows that the estimated t-value (19.29) is statistically significance at 0.01 level. The effect size was large (7.34). This shows that the members of the experimental group's achievement developed due to the positive impact of using

Table 2 T-test results of the pre-post administration of the achievement test on the experimental

Application	N	M	S.D	Difference between means	Degree of freedom	t- value	Significance level
Post-test	28	35.46	3.45	16.81	29	19.29	0.01 Significance
Pre-test	28	18.65	5.83				

SCAMPER strategy. Thus, the second hypothesis of the study was verified and the second sub question of the study was answered.

12 Discussion of the Results

The results of the achievement test

The results showed that the use of the scamper strategy makes the learner more active than the traditional method and makes him more willing to express new ideas and thus increases his willingness and motivation to learn because he is more enjoying it. In addition, the use of different forms of learning motivates the learner to learn.

13 Recommendations

In light of the results of the study, it is necessary to continue using the scamper strategy after the teacher has been qualified and a guide to use this strategy has been prepared. And start applying this strategy at an early stage of the study. In addition to the need to measure the use of this strategy and know its impact on all English language skills.

References

Al Kurdi, B., Alshurideh, M., & Salloum, S. (2020). Investigating a theoretical framework for e-learning technology acceptance. *International Journal of Electrical and Computer Engineering (IJECE), 10*(6), 6484–6496.

Alshurideh, M., Al Kurdi, B., & Salloum, S. (2019a). Examining the main mobile learning system drivers' effects: a mix empirical examination of both the Expectation-Confirmation Model (ECM) and the technology acceptance model (TAM). In *International Conference on Advanced Intelligent Systems and Informatics* (pp. 406–417).

Alshurideh, M., Salloum, S. A., Al Kurdi, B., Monem, A. A., & Shaalan, K. (2019b). Understanding the quality determinants that influence the intention to use the mobile learning platforms: A practical study. *International Journal of Interactive Mobile Technologies, 13*(11).

Alshurideh, M., Al Kurdi, B., Salloum, S. A., Arpaci, I., & Al-Emran, M. (2020). Predicting the actual use of m-learning systems: a comparative approach using PLS-SEM and machine learning algorithms. *Interactive Learning Environments*, 1–15.

Alshurideh, M. T., Kurdi, B. A., AlHamad, A. Q., Salloum, S. A., Alkurdi, S., et al. (2021). Factors affecting the use of smart mobile examination platforms by universities' postgraduate students during the COVID 19 pandemic: an empirical study. *Informatics*, *8*(2), 32.

Animasahun, R. A. (2014). Effects of scamper creativity training in the prevention of social problems among selected inmates in Nigeria prisons. *Journal of Emerging Trends in Educational Research and Policy Studies*, *5*(3), 301–305.

Antonacci, P. A., O'callaghan, C. M., & Berkowitz, E. (2014). *Developing content area literacy: 40 strategies for middle and secondary classrooms*. Sage Publications.

Costa, A. L. (1985). *Developing minds: A resource book for teaching thinking*. ERIC.

Eberle, B. (2008). *Creative games and activities for imagination development* (1st edn.). Prufrock Press.

Fahmy, G. A., Qoura, A.A.-S., & Hassan, S. R. (2017). YUsing SCAMPER-based activities in teaching story to enhance EFL primary stage pupils' speaking skills (an exploratory study). *Journal of Research in Curriculum Instruction and Educational Technology, 3*(4), 11–33.

Feather, W. (2008). *Infusing habits of mind into lessons*.

Folashade, A. (2012). Including creativity in basic science and technology classroom for effective learning of process skills. *IRCAB Journal of Arts and Education*, *25*(1), 297–301.

Georgiakakis, P., Retalis, S., & Psaromiligkos, Y. (2010). Design patterns for inspection-based usability evaluation of e-learning systems. In *Technology-enhanced learning* (pp. 167–182). Brill.

Gladding, S. (2007). Becoming creative as a counselor: The SCAMPER model (DVD). Microtraining Associates.

Horn, C. (2013). *Inspire creative brain storming with scamper strategy* (1st ed.). Horns book.

Kurdi, B. A., Alshurideh, M., Salloum, S. A., Obeidat, Z. M., & Al-dweeri, R. M. (2020). An empirical investigation into examination of factors influencing university students' behavior towards elearning acceptance using SEM approach. *International Journal of Interactive Mobile Technologies*, *14*(2). https://doi.org/10.3991/ijim.v14i02.11115.

Moreno, D. P., & Yang, M. C. (2014). Creativity in transactional design problems: non-intuitive findings of an expert study using SCAMPER. In *DS 77: Proceedings of the DESIGN 2014 13th International Design Conference* (p. 569–578).

Oakshott, S., McNeely, T., & Klattenhoff, B. (2011). 21st Century skills in career and technical education resource manual. *Mevlana International Journal of Education (MIJE)*, *3*(4), 166–185.

Toraman, S., & Altun, S. (2013). Application of the six thinking hats and scamper techniques on the 7th grade course unit Human and environment: An exemplary case study. *Mevlana International Journal of Education*, *3*(4), 166–185.

Vidal, R. V. V. (2005). Creativity for operational researchers. *Investigacao Operacional*, *25*(1), 1–24.

Zomerdijk, S. (2010). *How to create and transfer knowledge for the development of standard and custom products?* University of Twente.

The Influence of Metacognitive Strategies Training on the Writing Performance in United Arab Emirate

Hani Yousef jarrah, Saud alwaely, Tareq Alkhasawneh, Saddam Rateb Darawsheh, Anwar Al-Shaar, and Muhammad Turki Alshurideh

Abstract Writing is one of the most difficult components of learning a foreign language. Writing is the most difficult and time-consuming component of the language system. Writing necessitates a combination of orthographic, graphomotor, and other language abilities, as well as semantics, grammar, spelling, and writing standards, which are not limited to the aforementioned talents. The advancement of cognitive psychology, particularly metacognition, has attracted the attention of a growing number of scholars and prepared the way for new dimensions in EFL writing, notably in the area of writing achievement. This study seeks to evaluate the impact of applying metacognitive methods on United Arab Emirate EFL learners' writing performance due to the method's high executive aptness, which includes formulation, supervision, and assessment. Sixty-eight students were chosen at random from secondary school to participate in the study's experimental control. The experimental

H. Yousef jarrah (✉) · S. alwaely · T. Alkhasawneh
Al Ain University, Abu Dhabi, United Arab Emirates
e-mail: Hani.jarrah@aau.ac.ae

S. alwaely
e-mail: Suad.alwaely@aau.ac.ae

T. Alkhasawneh
e-mail: Tareq.alkhasawneh@aau.ac.ae

S. R. Darawsheh
Department of Administrative Sciences, The Applied College, Imam Abdulrahman Bin Faisal University, P.O. Box: 1982, Dammam 43212, Saudi Arabia
e-mail: srdarawsehe@iau.edu.sa

A. Al-Shaar
Deanship of Preparatory Year and Supporting Studies, Department of Self Development, (Imam Abdulrahman Bin Faisal University), P.O. Box: 1982, Dammam 43212, Saudi Arabia

M. Turki Alshurideh
Department of Marketing, School of Business, The University of Jordan, Amman 11942, Jordan
e-mail: m.alshurideh@ju.edu.jo; malshurideh@sharjah.ac.ae

Department of Management, College of Business, University of Sharjah, 27272 Sharjah, United Arab Emirates

M. Alshurideh et al. (eds.), *The Effect of Information Technology on Business and Marketing Intelligence Systems*, Studies in Computational Intelligence 1056,
https://doi.org/10.1007/978-3-031-12382-5_27

group was given metacognitive strategies-based writing training, whereas the control group was given merely regular writing instruction. Both groups were post-tested after five weeks of teaching, and the students were given another post-test at the end of the twelve-week programme. The Wilcoxon Signed-Rank test was used to analyse the data after the independent Mann–Whitney U test. The experimental group's writing performance improved as a result of the study's findings. This study's findings have consequences for both education and future research.

Keywords Control group · Experimental group · Mann–Whitney U test · Metacognitive strategies · Writing performance

1 Introduction

In the United Arab Emirate (UAE), English plays an essential part in education, and students are expected to communicate successfully in institutions where English is the language of instruction. According to Agbaria and Bdier (2021), English is taught as a foreign language in the UAE from kindergarten through the second year of secondary school, and it is a required subject. English is crucial for worldwide communication, social growth, and the adoption of new technologies, in addition to schooling (Li et al., 2021). The Ministry of Education of the United Arab Emirate pays close attention to English education, particularly curriculum and teacher training (Alsharari & Alshurideh, 2020; Alshamsi et al., 2020; Leo et al. 2021; Alshurideh et al., 2019). UAE students, on the other hand, continue to lack good English skills, notably in writing, which is regarded as the most challenging of the four language skills. Despite attempts by educators to remedy this shortcoming, English as a foreign language (EFL) students in the UAE have poor writing abilities (Ibrahim, 2021, Alhamad et al., 2021a, 2021b; Al Kurdi et al., 2020a). Traditional learning methods, such as emphasising grammatical rules, rote learning, and textbook-bound lectures, might be claimed to have demotivated students into passive learners who wait for their lecturers to deliver explicit teaching in the classroom (Fung, 2017; Alshurideh et al., 2020a, 2021).

English teachers assign numerous projects to their pupils to develop their writing skills without teaching them how to apply writing tactics. Before enrolling in any university in the UAE, students must pass English tests in secondary school. As a result, the UAE's principal educational concept is to use curriculum and instructional methods that give pupils fundamental conversational abilities (Frache, 2020; Sultan et al., 2021; Nazir et al., 2021). Because of its importance as an international language, the Ministry of Education approved in the curriculum in 2000 that English would be taught beginning at the age of six (Nguyen et al., 2018). Despite these and other initiatives, however, teaching excellent writing to UAE secondary school students remains a difficulty (Thomure, 2019). Writing is now seen as a means of expressing thoughts and feelings (Monagban and Saul, 2018; Al Kurdi et al., 2020; Alshurideh et al., 2020a, 2020b). It is a difficult and complex activity, but it is a crucial skill for EFL students to learn (Nasser, 2018). Its complexity stems from the need to communicate a concept, gain support for it, and then organise and revise information to improve the work's quality and eliminate faults (Geta and Olango, 2016; Amarneh et al., 2021; Akour et al., 2021).

Writing is a methodical, yet recursive, process that includes various sub-skills such as brainstorming, writing, summarising, and editing, whether in the native or second/foreign language (Alazemi et al., 2019; Al-Hamad et al., 2021; Bettayeb et al., 2020). Writing is particularly important in a person's everyday life as well as for the development of other language sub-skills (spelling, vocabulary acquisition, punctuation, expressing thoughts, and accurate grammar usage; Geta and Olango, 2016). Furthermore, writing is typically the major means of evaluating student success. As a result, poor writing abilities have been highlighted as a primary cause of English test failure, particularly when English is studied as a second or foreign language. Poor writing abilities have an impact on performance in practically all disciplines, not just English (Grabe and Kaplan, 2014). From an educational standpoint, research has demonstrated that writing has a major influence on students' academic success (Carmichael et al., 2018). Even when writing is acknowledged as necessary, teachers frequently overlook it, especially in secondary schools in the UAE (Bahari et al., 2021). Writing in English is essential not just for academic objectives, but also for international communication in which ideas and sentiments are expressed sensibly (Namaziandost et al., 2019). Writing improves language acquisition while also improving cognitive abilities due to its multi-process and creative character. Learners engage in both higher and lower-level abilities when writing. (Hoeve, 2022) described the personalities of 'excellent' writers as they plot (brainstorming), reread (rereading), rewrite (editing), and present their work (drafting). Lock et al. (2021) provided similar characteristics, such as planning (brainstorming), translation (putting thoughts into words), and review (evaluation). Writing processes is referred to in Healey and Gardner (2021), as norms, which emphasises that meaning may be formed by developing concepts, choosing the right phrase, and using the right language. According to Vega and Pinzón (2019), the writing process includes content, punctuation, the writer's method, students' ideas, audience, purpose, word choice, spelling, rhetorical structure, mechanics, grammar, and syntactic characteristics. In a similar spirit, Alazemi et al. (2019) described writing processes as involving many exchanges between students aimed at the meaning, structure, method, and arrangement of the message. Numerous studies have shown that EFL students in the UAE have poor writing abilities. Therefore, this study examined the influence of metacognitive strategies training on the writing performance among secondary school learners in United Arab Emirate.

2 Literature Review

2.1 Writing Skills

Writing is a multi-level cognitive activity that requires synchronised attention on various levels: theme, paragraph, phrase, grammatical, and lexical. It is regarded as a mark of competence on a global scale (Kim et al., 2021). Speech is more than the creation of sounds; therefore, it is a talent developed beyond the production of visual symbols. Writing, according to Indrilla and Ciptaningrum (2018), is both a process and a result that involves imagination, focus, and perseverance. It's considered one of the productive language skills since it demands students to use the language, they've learned to create something. Writing is a key language skill, and it is one of the four major language skills: writing, reading, speaking, and listening (Balta, 2018). Writing competence, according to Kim et al. (2021), is the hardest output ability for an EFL learner to achieve. It is the most comprehensive way of evaluating pupils' performance. Poor writing abilities have been highlighted as one of the key issues for students' failures in English language exams, particularly in circumstances where English is taught as an ESL or EFL. It's one of the most difficult aspects of language learning (Khair & Misnawati, 2022). Poor writing abilities have an impact on students' success in practically all disciplines, not just English (1). Additionally, writing is one of the four talents of the English language that must be considered and not overlooked (Baisov, 2021). Most significantly, writing is seen as a tool for language growth, creative development, and learning extension in all disciplines (Abdul Rahman, 2013). Writing, according to Hyland (2015), is a process in which writers explore their thoughts and ideas and make them visible for others to read. Spelling, punctuation, mechanics, vocabulary, substance, structure, purpose, and, most importantly, the audience should all be addressed in good writing (Hyland, 2015). The inability of non-native students to write decent English can be linked to a lack of language knowledge (Jahin & Idrees, 2012). In other words, writing is the final step in evaluating a student's achievement at practically all stages of education, particularly for a transition, grading, and admission tests. It is a talent that allows pupils to best display their language proficiency and control.

Students' weak writing abilities have been recognised as one of the key causes influencing students' failures in English Language tests in the previous study in SL/FL situations (Bennui, 2016). It may be claimed that poor writing ability has an impact on students' success in practically all of the areas they study as an applied phenomenon (Daud et al., 2016). The issue of students failing to study or performing poorly in exams in connection to writing piques the interest of second language researchers. Researchers have long held the belief that learning to write is a tough, sophisticated, and complex process that necessitates rigorous practises that may be learnt directly via experiences (Aliyu et al., 2016; Buyukyavuz & Cakir, 2014).

2.2 Learning Strategies

O'malley et al., 1990) describe learning strategies as "unique ideas or behaviours that individuals utilise to perceive, remember, or retain new knowledge". "Actions, behaviours, procedures, or strategies pupils adopt, sometimes subconsciously, to assist their development in apprehending, internalising, and applying the L2," according to Oxford (1994). Different learning techniques were presented and tried during the learning process. In the 1980s and early 1990s, research mostly concentrated on classifying the tactics discovered in previous decade's studies. As a result, numerous taxonomies for classifying them have been proposed, including classifications of language learning techniques in general and language sub-skills strategies specifically. For example, O'malley et al. (1990) categorised the strategies into three categories: cognitive, metacognitive, and socio-affective, each of which comprises a variety of sub-strategies including rehearsing, organising, summarising, deducing, and visualisation. (Oxford, 1994), on the other hand, presented a more complete model with six components divided into two groups: direct and indirect. Memory, cognitive, and compensatory techniques are direct strategies, whereas metacognitive, emotional, and social strategies are indirect strategies. The social and affective tactics are less common in L2 research, according to Oxford and Shearin (1994). This might be because L2 researchers don't often study these behaviours, and learners aren't used to paying attention to their own sentiments and social interactions as part of the L2 learning process. Finally, according to O'malley et al. (1990), cognitive (e.g., translating, analysing) and metacognitive (e.g., planning, organising) methods are frequently utilised in tandem, supporting one another. The notion is that combining methods has a greater impact than adopting single techniques. The contrasts between cognitive and metacognitive strategies are significant, as (Schraw and Graham, 1997) pointed out, partly because they assist us to determine which strategies are most relevant in determining the success of learning. (Schraw and Graham, 1997) argued that metacognitive methods, which enable students to plan, manage, and assess their learning, play the most important role in learning progress. "Developing metacognitive awareness may potentially lead to the development of greater cognitive skills," (Anderson, 2002) contended.

2.3 Metacognitive Strategies

2.3.1 Planning

One of the distinctive metacognitive tools for boosting learning is planning. The use of planning as a method improves students' goals and the effectiveness with which they may be achieved (Flavell, 1987; Okoro, 2011). For this purpose, proper learning may be achieved through the wise use of resources that have an impact on performance. Setting objectives, reviewing appropriate information, questioning, and task analysis are all part of the planning process. As a result, students arrange their studies to create the ideal learning environment. This allows kids to concentrate on their studies and build tactics that are relevant and suited for them (Suskie, 2018). In other words, planning comprises the use of strategic approaches in the allocation of resources to influence performance, as well as effectively selecting strategies and distributing resources to have a favourable impact on task performance. As (Steiner, 2010) correctly pointed out, predictions may be formed before writing, planning sequencing, or allocating time to properly complete any activity. Kellogg (Kellogg, 2008) focused on a detailed examination of the differences between good and bad writing. According to the findings, more experienced writers can plan successfully regardless of text "content," but weak writers can't. These findings have implications for the regulating metacognition developmental process.

2.3.2 Monitoring

Monitoring denotes a process of thorough awareness and follow-up in the completion of work. Monitoring is associated with the capacity to gradually develop pupils in research (Kazemi, Franke and Lampert, 2009). Several recent research, however, has discovered a robust link between metacognitive knowledge and monitoring accuracy. (Pearson, 2014) found that people's ability to predict how well they will grasp a text before reading was associated with monitoring accuracy during a post-reading understanding assessment. Similarly, Hoffman and Spatariu (2008) looked at the abilities of fifth and sixth graders to solve computer tasks. This study included three groups, with the findings revealing that the monitored problem-solving group answered more complicated issues faster than the other two. Monitoring, according to Andrade and Cizek (2010), is keeping track of progress toward a goal of improving performance. It's also important to maintain track of the learning process. Understanding pupils' comprehension issues and how to successfully address them may be determined with the use of monitoring (Sun, 2013).

2.3.3 Evaluating

The process of assessing the success of a particular strategy in accomplishing organisational objectives and taking corrective action when necessary is referred to as evaluation. The task evaluation is the final phase in the strategic management process. Examining the basic foundations of a task approach, comparing actual outcomes to predicted results, and taking remedial actions are the main assessment tasks. The purpose of the evaluation is to ensure that the organisational strategy, as well as its execution, achieves the goals of the organisation. Strategy evaluation is just as important as strategy design since it reveals the comprehensive plans' efficiency and effectiveness in reaching the targeted outcomes. In the assessing stage, students reflect on their work and the techniques, tools, resources, and/or procedures they used. According to Schmitz et al. (2004), reflection is transmitted when it is carried out, or "reflection-on-action."

2.3.4 Metacognitive Strategies and Writing

Expert reaction to problem-solving in general (Sternberg, 1984), as well as expert writing, has been identified as a component of metacognitive methods (Flavell, 1979; Scardamalia and Bereiter, 1984; Brown, 1982). In comparison to inefficient writers, proficient writers are more mindful of what they write, make more decisions about planning and regulating as they write, and are more likely to self-evaluate their writing as they write. It's crucial to think about the link between cognitive and metacognitive techniques that a strategic writer uses to understand the importance of metacognitive strategies in writing. When it comes to performing writing assignments, an effective writer employs a variety of cognitive methods. Learners can use cognitive techniques to assist them reach their cognitive goals (Flavell, 1979). Brainstorming ideas, constructing an outline, prewriting, writing the first draft, composing effective sentences, and correcting for grammatical problems are all examples of cognitive methods for crafting. Metacognitive techniques, in contrast to cognitive strategies, are intended to track cognitive growth. To coordinate their participation in the writing process, student writers use metacognitive processes or methods (Butler, 1998). Analyzing policy, the writing task to determine what is required, making plans underwriting strategies to use in a given writing task (e.g. determining whether brainstorming is required), monitoring success (e.g. judging whether sufficient ideas were generated during brainstorming), and selecting remedial strategies are examples of metacognitive strategies for writing (e.g. deciding that more research is needed to gather ideas). In a nutshell, strategic writers (and learners) employ cognitive methods to accomplish a specific writing objective and metacognitive strategies to guarantee that the cognitive writing goal is reached (Butler, 1998; Livingston, 1996). While performing writing tasks, efficient writers may switch between cognitive and metacognitive activity. As a result, teachers should assist students in developing metacognitive techniques in order for them to become effective writers. Instead of offering a list of teaching tactics and exercises, the researcher in this study would want to offer metacognitive writing module recommendations to EFL writing instructors who want to include metacognition into their writing classes.

2.3.5 Writing Performance in EFL Context

In an EFL framework, writing is the most important ability (Cai, 2017). Writing is a talent that is used to evaluate students' performance at all stages of education: primary, secondary, and tertiary, as well as admission tests, application letters, and research thesis, and for a variety of other purposes (Ei et al., 2016; Tabatabaei and Assefi, 2012). However, grammatical accuracy determines the efficiency of any work, as syntactic ambiguity leads to semantic uncertainty. Mechanical issues in the script;

correctness of grammar and lexis; style of writing to the needs of the scenario; and developing the ease and comfort in expressing writing form are some of the causes for incapacity to write (Bilal et al., 2016). For foreign languages, the level of difficulty in writing is substantially higher (FL). When writing, differences in linguistic patterns, ways of expressing concepts, writing styles, and culture are all important considerations. Without a doubt, writing is a cognitive ability that takes into consideration mental functioning and requires the administration of a diverse set of abilities, writing processes, and writing methods (Willinsky, 2017). However, current procedures in UAE secondary schools seldom take these complicated sets of abilities into account (Al-Abed Al-Haq and Al-Sobh, 2012). Furthermore, some teachers provide written assignments to students without providing them with clear directions on how to write or writing skills that would aid their learning (Alhabahba et al., 2016). Students skipped the writing part in the Tawjihi English Exam, according to Tahaineh and Daana (2013) and (Nazzal, 2020), since the exam does not aid them in the future. The scores of the English Tawjihi English Exam were poor, and student's performance in the writing work improved slowly.

Many elements can contribute to students' difficulty in writing TEE, including their competency in English language skills, particularly writing as a primary strategy, measuring students' performance, motivation, anxiety, and L1 transfer. Students who take the Tawjihi English Exam usually grumble about how difficult the writing component is (Al-Sawalha and Chow, 2012). Writing tasks, according to Magogwe (2013), is a significant barrier for secondary school students during their studies, since pupils struggle to cope with learning challenges, notably during the TEE. As a result, because writing is a component of TEE, it cannot be overlooked. The TEE comprises two papers, according to Dweik and Awajan (2013), the first test includes two primary sections, reading comprehension and vocabulary. The structure and content are based on the second language. A lot of academics have argued that pupils need additional assistance with their writing abilities (Hanjani and Li, 2014; Montgomery and Baker, 2007). This will also assist them further their studies and professional growth in the future. As Price-Dennis et al. (2014) pointed out, many students avoid taking the English test, particularly the written composition element, which is thought to be the most difficult. This issue, however, is not restricted to public schools; it also affects private educational institutions (Silva and Graham, 2015). Students generally overlook the writing segment, according to Khoshsima et al. (2014), due to their incapacity or lack of confidence in producing strong and proper sentences, cohesive paragraphs, and a well-structured written expression. Similarly, (Salem, 2018) concluded that the majority of UAE students lament their low English writing skills. According to Salem (2018), it is extremely difficult for UAE secondary school students to produce a decent sentence since they lack the necessary writing skills. For both native and non-native pupils, writing properly is a significant task (Muslim, 2014). Writing in English has become a valuable talent in Arab nations, even though it is more difficult for Arab pupils (Al-Bahrani, 2014). For Arab pupils, it is a foreign language that is very different from their tongue. Similarly, because writing is such a tough and complex endeavour, it is the most difficult of all skills to master. However, writing in English among Arab students is difficult. They are more likely to make mistakes (Muthanna, 2016). As a result, Arab English learners, including UAE, have significant difficulties with writing abilities. Researchers such as (Salem, 2018) and (Al-Bahrani, 2014) have explored this issue (2014). The main obstacle to successful writing among Arab students, particularly in the UAE, is a lack of meaningful and cohesive ideas (Abdo and Breen, 2010; Bani-Khaled, 2013).

3 Methodology

The participants in this study were 68 EFL students in their last year of secondary school at Rashid School for Boys in Dubai. They were in secondary school at the time. Two researchers used a holistic scoring system to measure their language skills. For 12 weeks, the class met for around two hours each week. A variety of instruments were employed in this study, including a pre-test, an immediate post-test, and a delay test. The test question was taken from the course materials, which they understood to be their current study's sourcebook. Because their educational programme had pre-placed the students in two separate interactive courses, one was designated as the control group and the other as the experimental group. The experimental group got writing instruction based on metacognitive methods, whereas the control group received just ordinary writing instruction. Both groups were promptly post-tested after five weeks of teaching. At the end of week twelve of the teaching programme, both groups are given a delayed post-test. To determine the impact of the metacognitive technique on writing performance among EFL learners, the data were subjected to a Mann-whiny u test followed by a Wilcoxon signed-rank test. All participants in this study were given a writing performance pre-test. The students were given a topic to write on in an essay format of roughly 150 words. After that, the experimental group received 35 min of instruction on metacognitive learning processes.

4 Results

Table 1 showed descriptive statistics based on mean, standard deviation, and standard error for total scores for general writing skills in both groups. According to their writing performance at the start of the study, the Pre-Test mean scores of the experimental group (M = 4.08, SD = 1.651, SDE = 0.321) and the control group (M = 3.11, SD = 0.631, SDE = 0.128) were not significantly different. After five weeks, an intermediate post-test was conducted with the experimental group mean scores of the experimental group (M = 4.84, SD = 1.660, SDE = 0.354) and the control group (M = 3.66, SD = 1.602, SDE = 0.341). Then, at the end of the intervention, a delayed post-test was conducted with the experimental group (M = 4.96, SD = 2.373, SDE = 0.273) and the control group (M = 3.90, SD = 1.565, SDE = 0.334). These results showed a difference in the mean scores between groups.

Table 1 Descriptive statistics for writing proficiency

	Group	N	M	SD	Std. Error
Pre-test	Experimental group	34	4.08	1.651	0.321
	Control group	34	3.11	0.631	0.128
	Total	68			
Immediate post-test	Experimental group	34	4.84	1.660	0.354
	Control group	34	3.66	1.602	0.341
	Total	68			
Delayed post-test	Experimental group	34	4.96	2.373	0.273
	Control group	34	3.90	1.565	0.334
	Total	68			

Further analysis was conducted to determine the effect size of metacognitive strategies on the overall performance of participants. Since the scores were not normally distributed, the Mann–Whitney U test, the non-parametric counterpart of a one-way independent ANOVA, was used to find any possible difference between the groups at the beginning and end of the program. The Wilcoxon Signed-Rank test was used to examine within-group differences. To establish the homogeneity of the groups in terms of their overall writing performance, a Mann–Whitney U test was carried out.

As illustrated in Table 2, this test found no significant difference in the overall pre-test scores of the two groups ($U = 148.54$, $z = -2.051$, $p = 0.045$, $r = -0.351$). The mean score margin between the two groups was not reasonable. Meanwhile, the effect size ($r = -0.351$) indicated a small effect between the groups considering the threshold of Cohen's benchmark for large effect size.

Table 3 presented the results obtained from the immediate post-test. This test was conducted to determine if there was a significant difference between the experimental group, which received metacognitive instruction, and the control group, which did not, at the intermediate stage of the training program.

The Mann–Whitney U test (independent sample U test) was employed to measure the difference in mean scores between the two independent groups to illustrate the effect size of the differences between groups. The results revealed a significant difference in the overall scores between groups ($U = 25.876$, $z = -4.889$, $p = 0.000$, $r = -0.838$). The difference observed in the mean rank value though comparing the two groups' performance revealed the difference was significant. Meanwhile, the

Table 2 Between-group overall scores for pre-test

Group	N	Mean Rank	U value	z value	p-value	r value
Experimental Group	34	36.87	148.54	−2.051	0.045	0.351*
Control group	34	21.65				
Total	68					

Note. * $p < 0.05$

Table 3 Between-Group Overall Scores for Immediate Post-Test

Group	N	Mean Rank	U value	z value	p-value	r value
Experimental Group	34	44.62	25.87	-4.889	0.000	−0.838*
Control group	34	30.22				
Total	68					

Note. * $p < 0.05$

r-value (−0.838) showed a medium-to-large effect size. The delayed post-test was the final test administered to both groups in Table 4. It aimed to examine the impact of metacognitive strategies on the experimental group and their effects on the writing performance of EFL students in general. The Mann–Whitney U test results revealed a significant difference in the performance between groups ($U = 11.65$, $z = -5.478$, $p = 0.000$, $r = -0.939$). The difference in mean rank value between the experimental group ($M = 56.77$) and control group ($M = 48.91$) was significant, and the wide margin between the two means indicated the impact of the metacognitive instruction received by the experimental group. Meanwhile, the r-value (−0.939) showed a medium-to-large effect size.

The Wilcoxon Signed-Rank test in Table 5 showed a significant difference in the gain score of the experimental group's performance on the pre-test ($t = 1.988$, $z = -2.312$, $p = 0.038$, $r = 0.393$) before and after metacognitive instruction. The results showed that instruction with metacognitive strategies had a significant effect on the writing performance of this group. The p-value (0.000) indicated the relationship was statistically significant. According to these results, the experimental group made significant progress between the pre-test and immediate post-test. Furthermore, there was a large effect size ($r = 0.393$) from metacognitive strategies.

Table 4 Between-group overall scores for delayed post-test

Group	N	Mean Rank	U value	z value	p-value	r value
Experimental group	34	56.77	11.65	−5.478	0.000	−0.939*
Control group	34	48.91				
Total	68					

Note. * $p < 0.05$

Table 5 Within-group overall scores for experimental group (pre-test with immediate post-test)

	N	t value	z value	p value	r value
Before	34	1.988	−2.312	0.038	0.393*
After	34				
Total	68				

Note. * $p < 0.01$

Table 6 showed the results of comparing the overall scores of the experimental group between the pre-test and delayed post-test. The Wilcoxon Signed-Rank test was used to examine the difference in the gain score and to find the effect size of metacognitive strategies on writing performance. The results ($t = 2.150, z = -2.769, p = 0.000, r = 0.475$) revealed that metacognitive instruction had a significant effect on the group's overall writing scores. In addition, there was a large effect size ($r = 0.475$) from metacognitive strategies.

These results revealed statistically significant differences in the experimental group's performance during and after the intervention, $p < 0.05$, with a large effect size ($r = -0.628, -0.630$). This indicated tremendous improvement by the end of the course and that metacognitive instruction had a major impact on the writing performance of EFL learners. A Wilcoxon Signed-Rank test was likewise run to compare the control group's gain score before and after treatment. As shown in Table 4.14, there was a significant difference in the group's gain score in the immediate post-test ($t = 2.871, z = -3.632, p = 0.000, r = -0.623$). These results indicated that non-metacognitive methods could also improve the writing of EFL learners in UAE high schools (Tables 7 and 8).

Another Wilcoxon Signed-Rank test examined if there were statistically significant differences in the control group's performance in the delayed post-test (see

Table 6 Within-group overall scores for experimental group (pre-test)

	N	*t* value	*z* value	*p* value	*r* value
Before	34	2.150	−2.769	0.000	0.475*
After	34				
Total	68				

Note. * $p < 0.05$

Table 7 Within-group overall scores for control group (pre-test with immediate post-test)

	N	*t* value	*z* value	*p* value	*r* value
Before	34	2.871	−3.632	0.000	0.623
After	34				
Total	68				

Note. * $p < 0.05$

Table 8 Within-group overall scores for control group (pre-test with delayed post-test)

	N	*t* value	*z* value	*p* value	*r* value
Before	34	3.880	−4.621	0.000*	0.792*
After	34				
Total	68				

Note. * $p < 0.05$

Table 4.15). The results ($t = 3.880$, $z = -4.621$, $p = 0.000$, $r = -0.561$) showed a significant difference between the gain score before and after the teaching program and a small effect size ($r = -0.792$).

5 Conclusion

The rationale for this research was that teaching meta-cognitive learning methods may help students enhance their writing skills. The course lasted for a total of twelve weeks. During this period, the researcher used metacognitive learning techniques and taught the experimental group members how to use metacognitive tactics in their writing. During their writing skill practise, the participants in the control group, on the other hand, received no training on how to employ these tactics. The findings of the post-test revealed that teaching meta-cognitive learning techniques increased the learners' writing ability. The experimental group that employed meta-cognitive learning methodologies outperformed the control group in terms of writing ability. Metacognition was found to be a key driver of English writing proficiency in recent research. implementing metacognitive activities entails delegating some obligations to students, which may increase their stress levels, particularly among the less capable. As a result, it is advised that clear and simple instructions and modelling, as well as monitored practise, be easily available. When teaching metacognitive EFL writing, the teacher should encourage and motivate students, as well as listen to their voices from various locations to manage and access the teaching tactics used. Furthermore, cultural influences may complicate the scenario of putting metacognition into practice, necessitating greater research in this area. Promoting metacognitive awareness among students in an integrative EFL writing class entails that the process and consequence of methods are not seen as adversarial, but rather as complementary (Mbato, 2013). It also aids teachers and students in accessing and regulating their efforts, resulting in a more productive EFL writing session. Nevertheless, as students' understanding of and experience with techniques grows, genre approaches may be linked as "a continuation of product approaches" to link writing with distinct social settings, according to Byrnes (2012). Writing ability is critical, yet it is a difficult academic talent to master for secondary and college students. An in-depth evaluation of students' writing abilities, with a focus on persuasive writing assignments, provides additional insight into the cognitive processes of writing performance as they improve.

References

Abdo, I. B., & Breen, G.-M. (2010). Teaching EFL to Jordanian students: New strategies for enhancing English acquisition in a distinct Middle Eastern student population. *Creative Education, 1*(1), 39–50.

Abdul Rahman, Z. A. A. (2013). The use of cohesive devices in descriptive writing by Omani student-teachers. In *Sage Open* (Vol. 3, no. 4), p. 2158244013506715.

Agbaria, Q., & Bdier, D. (2021). The role of parental style and self-efficacy as predictors of internet addiction among Israeli-Palestinian college students in Israel. *Journal of Family Issues* 0192513X21995869.

Akour, I., Alshurideh, M., Al Kurdi, B., Al Ali, A., & Salloum, S. (2021). Using machine learning algorithms to predict people's intention to use mobile learning platforms during the COVID-19 pandemic: Machine learning approach. *JMIR Medical Education, 7*(1), 1–17. https://doi.org/10.2196/24032.

Al Kurdi, B., Alshurideh, M., Salloum, S., Obeidat, Z., & Al-dweeri, R. (2020a). An empirical investigation into examination of factors influencing university students' behavior towards e-learning acceptance using SEM approach.

Al Kurdi, B., Alshurideh, M., & Salloum, S. (2020b). Investigating a theoretical framework for e-learning technology acceptance. *International Journal of Electrical and Computer Engineering, 10*(6), 6484–6496.

Al-Abed Al-Haq, F., & Al-Sobh, M. (2012). Online linguistic messages of the Jordanian secondary students and their opinions toward a web-based writing instructional EFL program. *International Journal of Humanities and Social Science, 2*(6), 228–299.

Alazemi, A. F., Sa'di, I. T., & Al-Jamal, D. A. H. (2019). Effects of digital citizenship on EFL students' success in writing. *International Journal of Learning, Teaching and Educational Research, 18*(4), 120–140.

Al-Bahrani, M. A. (2014). A qualitative exploration of help-seeking process. *Advances in Applied Sociology*.

Alhamad, A. Q. M., Akour, I., Alshurideh, M., Al-Hamad, A. Q., Kurdi, B. A., & Alzoubi, H. (2021a). Predicting the intention to use google glass: A comparative approach using machine learning models and PLS-SEM. *International Journal of Data and Network Science, 5*(3). https://doi.org/10.5267/j.ijdns.2021.6.002.

AlHamad, M. Q., Mbaidin, H. O., Alhamad, A. Q. M., Alshurideh, M. T., Kurdi, B. H. A., & Al-Hamad, N. Q. (2021b) Investigating students' behavioral intention to use mobile learning in higher education in UAE during Coronavirus-19 pandemic. *International Journal of Data and Network Science, 5*(3). https://doi.org/10.5267/j.ijdns.2021.6.001.

Aliyu, M. M., Fung, Y. M., Abdullah, M. H., & Hoon, T. B. (2016). Developing undergraduates' awareness of metacognitive knowledge in writing through problem-based learning. *International Journal Applied Linguistics and English Literature, 5*(7), 233–240.

Al-Sawalha, A. M. S., & Chow, T. V. V. (2012). The effects of writing apprehension in English on the writing process of Jordanian EFL students at Yarmouk University. *International Interdisciplinary Journal of Education, 1*(1), 6–14.

Alshamsi, A., Alshurideh, M., Al Kurdi, B., Salloum, S. A. (2020). The influence of service quality on customer retention: A systematic review in the higher education. In *International Conference on Advanced Intelligent Systems and Informatics* (pp. 404–416).

Alsharari, N. M., & Alshurideh, M. T. (2020). Student retention in higher education: The role of creativity, emotional intelligence and learner autonomy. *International Journal of Educational Management*.

Alshurideh, M., Salloum, S. A., Al Kurdi, B., Monem, A. A., & Shaalan, K. (2019). Understanding the quality determinants that influence the intention to use the mobile learning platforms: A practical study. *International Journal of Interactive Mobile Technologies, 13*(11).

Alshurideh, M., Al Kurdi, B., & Salloum, S. A. (2020a). *Examining the main mobile learning system drivers' effects: a mix empirical examination of both the Expectation-Confirmation Model (ECM) and the Technology Acceptance Model (TAM)* (Vol. 1058).

Alshurideh, M., Al Kurdi, B., Salloum, S. A., Arpaci, I., & Al-Emran, M. (2020b). Predicting the actual use of m-learning systems: A comparative approach using PLS-SEM and machine learning algorithms. *Interactive Learning Environment*. https://doi.org/10.1080/10494820.2020.1826982.

Alshurideh, M. T. et al. (2021). Factors affecting the use of smart mobile examination platforms by universities' postgraduate students during the COVID-19 pandemic: An empirical study. *Informatics, 8*(2). https://doi.org/10.3390/informatics8020032.

Amarneh, B. M., Alshurideh, M. T., Al Kurdi, B. H., & Obeidat, Z. (2021). The impact of COVID-19 on e-learning: advantages and challenges. In *The International Conference on Artificial Intelligence and Computer Vision* (pp. 75–89).

Anderson, N. J. (2002). The role of metacognition in second language teaching and learning. ERIC Digest.

Andrade, H., & Cizek, G. J. (2010). *Handbook of formative assessment*. Routledge.

Bahari, A. A., Kussin, H. J., Harun, R. N. S. R., Mohamed, M., & Jobar, N. A. (2021). The limitations of conducting collaborative argumentation when teaching argumentative essays in Malaysian secondary schools. *Study of the English Language Education, 8*(3), 1111–1122.

Baisov, A. S. (2021). The effectiveness of the method talk for writing in developing writing skills of efl students. *Academic Research in Educational Sciences, 2*(2).

Balta, E. E. (2018). The relationships among writing skills, writing anxiety and metacognitive awareness. *Journal Education Learning, 7*(3), 233–241.

Bani-Khaled, T. (2013). Learning English in difficult circumstances: The case of north Badiah disadvantaged schools in Jordan. *Australian Journal of Basic and Applied Sciences, 7*(8), 269–284.

Bennui, P. (2016). A study of L1 intereference in the writing of Thai EFL students. *Malaysian Journal of ELT Research, 4*(1), 31.

Bettayeb, H., Alshurideh, M. T., & Al Kurdi, B. (2020). The effectiveness of mobile learning in UAE universities: a systematic review of motivation, self-efficacy, usability and usefulness. *International Journal of Control, Automation, 13*(2), 1558–1579.

Bilal, M., Israr, H., Shahid, M., & Khan, A. (2016). Sentiment classification of Roman-Urdu opinions using Naïve Bayesian, decision tree and KNN classification techniques. *Journal King Saud University and Information Science, 28*(3), 330–344.

Brown, A. L. (1982). Learning, remembering, and understanding. Technical Report No. 244.

Butler, D. L. (1998). The strategic content learning approach to promoting self-regulated learning: A report of three studies. *Journal of Educational Psychology, 90*(4), 682.

Buyukyavuz, O., & Cakir, I. (2014). Uncovering the motivating factors behind writing in English in an EFL context. *Anthropol., 18*(1), 153–163.

Byrnes, H. (2012). Conceptualizing FL writing development in collegiate settings: A genre-based systemic functional linguistic approach. In *L2 writing development multiple perspectives* (pp. 190–218).

Cai, L. J. (2017). Students' perceptions of academic writing: A needs analysis of EAP in China. In *Asian-focused ELT research and practice voices from far edge* (pp. 127–151).

Carmichael, M., Reid, A., & Karpicke, J. D. (2018). Assessing the impact of educational video on student engagement, critical thinking and learning. *Sage Publ.*

Daud, N. S. M., Daud, N. M., & Kassim, N. L. A. (2016). Second language writing anxiety: Cause or effect? *Malaysian Journal of ELT Research, 1*(1), 19.

De Silva, R., & Graham, S. (2015). The effects of strategy instruction on writing strategy use for students of different proficiency levels. *System, 53*, 47–59.

Dweik, B. S., & Awajan, N. W. (2013). Factors that enhance English language teachers' motivation in Jordanian secondary schools. *English Linguistics Research, 2*(1), 33–42.

Flavell, J. H. (1979). Metacognition and cognitive monitoring: A new area of cognitive–developmental inquiry. *American Psychologist, 34*(10), 906.

Flavell, J. H. (1987). Speculations about the nature and development of metacognition. *Metacognition, Motivation and Understanding.*

Frache, G. (202). Developing constructively aligned learning-by-doing model incorporating 21st century skills for enhancing the teaching of engineering and science curriculum in higher education. Εθνικό και Καποδιστριακό Πανεπιστήμιο Αθηνών (ΕΚΠΑ). Σχολή Θετικών Επιστημών.

Fung, D. (2017). *A connected curriculum for higher education*. UCL Press.
Geta, M., & Olango, M. (2016). The impact of blended learning in developing students' writing skills: Hawassa university in focus. *African Educational Research Journal, 4*(2), 49–68.
Grabe, W., & Kaplan, R. B. (2014). *Theory and practice of writing: An applied linguistic perspective*. Routledge.
Hanjani, A. M., & Li, L. (2014). Exploring L2 writers' collaborative revision interactions and their writing performance. *System, 44*, 101–114.
Healey, B., & Gardner, P. (2021). Writing, grammar and the body: A cognitive stylistics framework for teaching upper primary narrative writing. *Literacy, 55*(2), 125–135.
Hoeve, S. J. (2022). *Teach writing with growth mindset: Classroom-ready resources to support creative thinking, improve self-talk, and empower skilled*. Simon and Schuster.
Hoffman, B., & Spatariu, A. (2008). The influence of self-efficacy and metacognitive prompting on math problem-solving efficiency. *Contemporary Educational Psychology, 33*(4), 875–893.
Hyland, K. (2015). *Teaching and researching writing*. Routledge.
Ibrahim, O. (2021). Academic writing teaching method for Arab students: A case study in a health sciences university in the UAE. *Journal of Critical Studies Language and Literature, 2*(3), 8–19.
Ibrahim, M. E. E., Eljack, N. S. A., & Mohammed Elhassan, I. B. (2016) The effect of argumentative essay writing strategies on enhancing English as a foreign language learners critical thinking skills.
Indrilla, N., & Ciptaningrum, D. S. (2018). An approach in teaching writing skills: Does it offer a new insight in enhancing students' writing ability. *LLT Journal: A Journal Language and Language Teaching, 21*(2), 124–133.
Jahin, J. H., & Idrees, M. W. (2012). EFL major student teachers' writing proficiency and attitudes towards learning English. *J Umm Al-Qura University Journal of Educational Psychologic Sciences, 4*(1), 10–72.
Kellogg, R. T. (2008). Training writing skills: A cognitive developmental perspective. *Journal Writing Research, 1*(1).
Khair, U., & Misnawati, M. (2022). Indonesian language teaching in elementary school: Cooperative learning model explicit type instructions chronological technique of events on narrative writing skills from interview texts. *Linguistics and Culture Review, 6*, 172–184.
Khoshsima, H., Saed, A., & Ghassemi, P. (2014). The application of ESP principles on course design: The case of English for students of Management and Fisheries. *International Journal of Language Learning Applied Linguistics World, 5*(2), 163–175.
Kim, Y.-S.G., Yang, D., Reyes, M., & Connor, C. (2021). Writing instruction improves students' writing skills differentially depending on focal instruction and children: A meta-analysis for primary grade students. *Educational Research Review, 34*, 100408.
Leo, S., Alsharari, N. M., Abbas, J., & Alshurideh, M. T. (2021). From offline to online learning: A qualitative study of challenges and opportunities as a response to the COVID-19 pandemic in the UAE higher education context. *The Effect of Coronavirus Disease (COVID-19) on Business Intelligence, 334*, 203–217.
Li, C., He, L., & Wong, I. A. (2021). Determinants predicting undergraduates' intention to adopt e-learning for studying English in Chinese higher education context: A structural equation modelling approach. *Education and Information Technologies, 26*(4), 4221–4239.
Livingston, J. A. (1996). Effects of metacognitive instruction on strategy use of college students. *Unpubl. manuscript, State University New York Buffalo.*
Lock, J., Becker, S., & Redmond, P. (2021). Teachers conceptualizing and developing assessment for skill development: Trialing a maker assessment framework. *Research Evaluation, 30*(4), 540–551.
Magogwe, J. M. (2013). Metacognitive awareness of reading strategies of University of Botswana English as Second Language students of different academic reading proficiencies. *Reading and Writing-Journal of the Reading Association South Africa, 4*(1), 1–8.
Mbato, C. L. (2013). *Facilitating EFL learners' self-regulation in reading: Implementing a metacognitive approach in an Indonesian higher education context*. Southern Cross University.

Monagban, E. J., & Saul, E. W. (2018). The reader, the scribe, the thinker: A critical look at the history of American reading and writing instruction. In *The formation of school subjects*, Routledge (pp. 85–122).

Montgomery, J. L., & Baker, W. (2007). Teacher-written feedback: Student perceptions, teacher self-assessment, and actual teacher performance. *Journal of Second Language Writing, 16*(2), 82–99.

Muthanna, A. (2016). Teaching and learning EFL writing at Yemeni universities: A review of current practices. In *Teaching EFL Writing 21st Century Arab World* (pp. 221–243).

Namaziandost, E., Neisi Kheryadi, Nasri, M. (2019). Enhancing oral proficiency through cooperative learning among intermediate EFL learners: English learning motivation in focus. *Cogent Education, 6*(1), 1683933.

Nasser, S. M. (2018). Iraqi EFL students' difficulties in writing composition: An experimental study (University of Baghdad). *International Journal of English Linguistics, 9*(1), 178–184.

Nazir, M. I. J., Rahaman, S., Chunawala, S., & AlHamad, A. Q. M. (2021). Perceived factors affecting students academic performance; Nazir, J., Rahaman, S., Chunawala, S., Ahmed, G., Alzoubi, H., Alshurideh, M., AlHamad, A. (2021). Perceived factors affect. students Acad. performance. *Academy of Strategic Management Journal, 21*(Special Issue 4), 1–15, *21*(Special Issue 4), 1–15.

Nazzal, J. S. (2020). *Writing proficiency and student placement in community college composition courses*. University of California.

Nguyen, H. T. M., Nguyen, H. T., Van Nguyen, H., & Nguyen, T. T. T. (2018). Local challenges to global needs in English language education in Vietnam: The perspective of language policy and planning. *Un Lang. Plan. a Glob. world Mult. levels Play. Work* (pp. 214–233).

Okoro, C. O. (2011). Metacognitive strategies: A viable tool for self–directed learning. *Journal Educational and Social Research, 1*(4), 71.

O'malley, J. M., O'Malley, M. J., Chamot, A. U., O'Malley, J. M. (1990). *Learning strategies in second language acquisition*. Cambridge University Press.

Oxford, R. (1994). Language learning strategies: An update. ERIC Digest.

Oxford, R., & Shearin, J. (1994). Language learning motivation: Expanding the theoretical framework. *The Modern Language Journal, 78*(1), 12–28.

Pearson, P. D. (2014). The roots of reading comprehension instruction. In *Handbook of research on reading comprehension* (pp. 27–55). Routledge

Price-Dennis, D., Wiebe, M. T., & Fowler-Amato, M. (2014). Learning to develop a culturally relevant approach to 21st century writing instruction. *Teaching/Writing the Journal of Writing Teacher Education, 3*(2), 5.

Salem, A. A. (2018). Engaging ESP university students in flipped classrooms for developing functional writing skills, HOTs, and eliminating writer's block. *English Language Teaching, 11*(12), 177–198.

Scardamalia, M., & Bereiter, C. (1984). Development of strategies in text processing. Centre for Applied Cognitive Science, Ontario Institute for Studies in Education.

Schmitz, T. W., Kawahara-Baccus, T. N., & Johnson, S. C. (2004) Metacognitive evaluation, self-relevance, and the right prefrontal cortex. *Neuroimage, 22*(2), 941–947.

Schraw, G., & Graham, T. (1997). Helping gifted students develop metacognitive awareness. *Roeper Review, 20*(1), 4–8.

Steiner, G. A. (2010). *Strategic planning*. Simon and Schuster.

Sternberg, R. J. (1984). What should intelligence tests test? Implications of a triarchic theory of intelligence for intelligence testing. *Educational Researcher, 13*(1), 5–15.

Sultan, R. A., Alqallaf, A. K., Alzarooni, S. A., Alrahma, N. H., AlAli, M. A., & Alshurideh, M. T. (2021). How students influence faculty satisfaction with online courses and do the age of faculty matter. In *The International Conference on Artificial Intelligence and Computer Vision* (pp. 823–837).

Sun, L. (2013). The effect of meta-cognitive learning strategies on English learning. *Theory Practice in Language Studies, 3*(11), 2004.

Suskie, L. (2018). *Assessing student learning: A common sense guide*. Wiley.
Tabatabaei, O., & Assefi, F. (2012). The Effect of Portfolio Assessment Technique on Writing Performance of EFL Learners. *English Language Teaching, 5*(5), 138–147.
Tahaineh, Y., & Daana, H. (2013). Jordanian undergraduates' motivations and attitudes towards learning English in EFL context. *International Review of Social and Sciences Humanities, 4*(2), 159–180.
Thomure, H. (2019). Arabic language education in the UAE: Choosing the right drivers. In *Education in the United Arab Emirates* (pp. 75–93). Springer.
Vega, L. F. S., & Pinzón, M. M. L. (2019). The effect of the process-based approach on the writing skills of bilingual elementary students. *Latin American Journal of Content and Language Integrated Learning, 12*(1), 72–98.
Willinsky, J. (2017). *The new literacy: Redefining reading and writing in the schools*. Routledge.

Perspectives of Online Education in Pakistan: Post-covid Scenario

Moattar Farrukh, Tariq Rahim Soomro, Taher M. Ghazal, Haitham M. Alzoubi, and Muhammad Alshurideh

Abstract The outbreak of the famous Corona Virus in the world at the end of the year 2019 has triggered the world to bring major changes to the everyday walks of life. This virus entered Pakistan in February 2020 and resulted in the lockdown to be enforced throughout the country. Just as many other fields of life, the education sector also had to tolerate a massive hit of this closing down. The traditional approach of education based on the face-to-face interactions between the students and instructors went dependent on online platforms. This was a major and sudden switch, which was found extremely difficult throughout the world as well as in Pakistan. The concept of online education was not anything new. It prevailed earlier also, but that was the one conducted after proper planning, which was not at all in the case Covid-19. This research paper first discusses the perspectives of parents and teachers on the effectiveness of online education by qualitative approach. Secondly, the results of

M. Farrukh (✉) · T. R. Soomro
Institute of Business Management, CCSIS, Karachi, Sind, Pakistan
e-mail: std_27018@iobm.edu.pk

T. R. Soomro
e-mail: tariq.soomro@iobm.edu.pk

T. M. Ghazal
Center for Cyber Security, Faculty of Information Science and Technology, Universiti Kebansaan Malaysia (UKM), Bangi, Selangor, Malaysia
e-mail: taher.ghazal@skylineuniversity.ac.ae

Skyline University College, Sharjah, UAE

H. M. Alzoubi
School of Business, Skyline University College, Sharjah, UAE
e-mail: haitham.alzubi@skylineuniversity.ac.ae

M. Alshurideh
Department of Marketing, School of Business, University of Jordan, Amman, Jordan
e-mail: m.alshurideh@ju.edu.jo; malshurideh@sharjah.ac.ae

Department of Management, College of Business Administration, University of Sharjah, Sharjah, UAE

M. Alshurideh et al. (eds.), *The Effect of Information Technology on Business and Marketing Intelligence Systems*, Studies in Computational Intelligence 1056,
https://doi.org/10.1007/978-3-031-12382-5_28

the conducted surveys are evaluated in the quantitative part of the paper. Further, the results of both the research types are compared to discuss and reason the overall scenario of online education in Pakistan.

Keywords Online education · Education · Post-Covid · Stakeholder's engagement · Scenario planning · Survey

1 Introduction

The global spread of the notorious Coronavirus has transformed the way things used to work out (Alshurideh et al., 2021a; 2021b). All the major fields linked with the economic growth of countries have been highly affected by this life-threatening virus. One of these affected sectors, is educational sector (Alshamsi et al., 2020; Alsharari and Alshurideh, 2020). At the start of Covid-19, the provision of education seemed to be a huge challenge (Al-Hamad et al., 2021; Leo et al., 2021). Schools, colleges, and Universities were shut down immediately after the outbreak of this epidemic (Akour et al., 2021; Alshurideh, et al., 2021c). Then, just as everything else, education went online, but it did never resolve all the related issues anywhere around the world (Al-Maroof et al., 2021; Alshurideh et al., 2021c). Never before have we witnessed educational disruption on such a large scale (Demuyakor, 2020; Sultan et al., 2021). This statement coming from a man on such an effective post indicates that the sudden change from traditional to distance learning had been a huge dare for all the countries around the globe (Al Kurdi et al., 2020a; Nazir et al., 2021). Different countries had to face distinguished challenges for assuring the continuation of providing education to its students (Alshurideh et al., 2019; Kurdi et al., 2020b).

The Covid-19 confirmed entering Pakistan when its first case was filed on 26th February 2020 (Abid et al., 2020). Although the growth of these cases was at a slow pace, as a precautionary measure, the Government of Pakistan soon declared the closure of all the educational institutions. Initially, everything appeared to be a blackout. Then, few private institutes started to shift from conventional to distance learning within their limited resources. But most institutes in Pakistan were not able to manage and tolerate this major switch. For a developing country like Pakistan, where education is already under some extensive considerations, Covid-19 proved to be a nightmare. A great many issues had to be endured in making online education possible (Al Kurdi et al., 2021, 2021b; Alshurideh et al., 2020a, 2020b, 2020c). Most of the Government institutes, were not able to go online. They simply did not have the required resources. Even some of the well-known universities of the country completely called off education for a few months until they assured the availability of adequate resources. Many institutes did their best for the continuance of providing education to their students, but they also had to cross some major barriers. There exist some marked commons amongst the institutes, who completely got closed for education and those who went online (Bettayeb et al., 2020; Salloum et al., 2020). The availability of resources is the most highlighted amongst these common issues. All the institutions had to manage their resources smartly. Apart from the accessibility of resources, there lived some other crucial factors, which markedly affected the education system and the quality of education. (Alnazer et al., 2017; Alshurideh et al., 2020a, 2020b, 2020c; Alzoubi and Ahmed, 2019; Mehmood et al., 2019). This research paper will uncover all these occurred issues and their impact levels in various situations.

This research study intends to examine the overall scenario of online education in Pakistan after the occurrence of Covid-19. This paper focuses on the impacts put up by the unexpected shift from traditional to online education on the stakeholders in many diversified ways. For this purpose, surveys were conducted to learn what do most Pakistani parents and teachers have to say about the overall scenario of distance learning. The overall scenario means the combination of all the related factors that have been associated with online education in Pakistan. These factors include finance, accessibility to the resources and the seriousness of students towards education, while sitting at their homes, connectivity issues, communication challenges, and most importantly the effectiveness of education.

In Pakistan, not all the parents are well-educated, so, when it comes to online education of children especially the younger ones, it is extremely important to figure out that how much computer literate the parents are because they can guide their kids in accessing the devices and facing various associated difficulties, only when they themselves know how to tackle with the possible situations (Alzoubi and Yanamandra, 2020; Alzoubi et al., 2020a, 2020b). Also, in the case of online education, where teacher-student communication has been curtailed in comparison to live-learning, the role of parents in helping their children in their studies has grown high. This issue is not only on the parent-end, but also many teachers are not much habitual of using the digital platforms (Akhtar et al., 2021; Al Ali, 2021; Alhamad et al., 2021; Joghee et al., 2020b). So, finding out the education level and computer literacy rate of parents and teachers of different levels and relating them with the flow of online classes is important. Moreover, the availability of resources like digital devices, and the Internet, which are also the precautionary measures to avail as well as conduct online education also bring various issues related to finance, access, connectivity, and experience. So, we have viewed online education in consideration of such crucial factors (Ahmad et al., 2021; Alameeri et al., 2021; Nuseir et al., 2021a, 2021b). Furthermore, getting to learn and understand the psychological approach and interest level of students regarding education while sitting at home is also one major consideration covered in this research work. So, based on the literature review and the received survey responses, this research paper argues upon many diversified factors that have their impact on the effectiveness and fruitfulness of online education in Pakistan. Different education levels have their distinguished as well as common problems and the impact levels of these problems on the quality and potency of education are also quite different. This paper does not discuss the point of view of the organizations that how do the management of educational institutes view online education in the post Covid-19 scenario.

2 Literature Review

Education is an essential human virtue, a necessity of society, the basis of good life, and the sign of freedom. Education is important for the integration of separate entities. It plays a significant role in designing a successful life for an individual. Not only does it open the doors of financial stability in an individual's life but, it also provides him/her with the platform to grow socially, and ethically. It also develops a sense of responsibility, self-respect, and strength in an individual (Bhardwaj, 2016). Also, education is strongly associated with a country's economic development. Pakistan is a developing country, where the educational sector is already struggling to prosper. There exist many considerable issues in the educational system of the country. The year 2020 added one more to the list of these concerns in the face of Covid-19 (Alshurideh et al., 2020a, 2020b, 2020c; Amarneh et al., 2021; Ashal et al., 2021).

Coronavirus is a powerful pandemic that forced even the strongest of economies to get themselves shut (Aljumah et al., 2021a; Nuseir et al., 2021a, 2021b). Locking down the cities was one ultimate possible solution to hold the spread of Covid-19 (Al-Dmour et al., 2021; Al Khasawneh et al., 2021a, 2021b). Governments of almost all the countries throughout the world decided to declare shutting down every-day life activities so that people could stay at their homes, which was necessary to reduce the spread of the concerned epidemic (Al Khasawneh et al., 2021a, 2021b; Murphy, 2020). This very situation rose the switch of traditional approaches to move online. Just as many other aspects of every-day life, education also shifted from live to online (Ali et al., 2021; Alnuaimi et al., 2021). Schools, colleges, and universities around the world were closed and students were directed to avail distance learning. This sudden switch between conventional to virtual learning was not easy at all. Different educational institutes took diversified amounts of time to accomplish this task in accordance with their capacities (Bestiantono et al., 2020). But, since this epidemic is completely unpredictable, no one knew when it would end, therefore despite all the obstacles, it was extremely necessary to take all the possible measures to provide students with online education (Crawford et al., 2020).

Online education is education that is carried out over the internet and students receive education at their homes. It is often known as e-learning and is the newest form of distance learning (Stern, 2020). It is focused on the concept of facilitating and earning education remotely. For this very purpose, different on-net platforms were used including Zoom, Skype, Microsoft Teams, Google Meet, etc. This research study is most concerned about the level of effectiveness of education, which has more to do with deep thinking and successful stakeholders rather than the involvement or delusion of technology (Serdyukov, 2020). The two most concerned groups involved in the process of education are of teachers and students. These two groups are highly emphasized when any discussion regarding education is carried out (Alsharari, 2021; Alzoubi, 2021a, 2021b).

A famous educator Andrew suggested that technology is human dependent. It is up to the instructor that how does he/she uses it to deliver the learning concepts (Kemp et al., 2014). Tools are there but every teacher uses it in diversified ways in order to convey knowledge to the students. Also, it depends on the students that how do they perceive learning online. (Semler, 2010) suggested that online education is effective for the students who are matured, prompted, and self-mastered. Even in the traditional learning environment, the promptness of students is highly notifiable (Serdyukov, 2020), but when it comes to distance learning, the students must make sure to enhance their time-management and self-management skills (Semler, 2010). When talking about online education, one must consider this very fact that it falls under two categories (Alzoubi and Aziz, 2021). The first one is normal online learning, which is properly pre-planned and the second one is emergency online learning (Hodges et al., 2020). The scenario faced after the global pandemic of Coronavirus is of "*emergency online learning*" which happened suddenly without any prior notice period (Alnuaimi et al., 2021). One important thing to notice is that even the seriousness, motivation, and interest levels of the teachers as well as the students are controlled by various psychological, environmental, and material drivers (Alzoubi et al., 2021; Aziz and Aftab, 2021; Cruz, 2021). Some of these driving factors are discussed below and their impacts on the effectiveness of education are grouped by primary, secondary, and higher education levels.

2.1 Students' and Parents' Perspectives

This part of the paper focuses on the perspectives of students and their parents, when it comes to the switch from traditional to online education. Students enrolled in different education levels as well as their parents had to face contrastive sorts of problems. The most common amongst these are discussed here.

2.1.1 Learning Environment

The learning environment comprises all the things that are around a learner, while he/she is involved in the process of gaining education. In the traditional approach, mostly the classroom environment is observed. The classroom environment is formed by combining various classroom standards including the presence of table-chair discipline, the proper seating arrangement of the students, such that, each one of them could view and acknowledge the instructions easily and clearly, and the presence of teacher as an instructor as well as an authorized controller (Sieberer-Nagler, 2015). These pennants are followed to minimize disruptions, which ultimately leads to maximized learning as such factors are closely associated with the involvement and intellectual activeness of the students as well as of the teachers. Research shows that the students are emotionally attached to their classrooms. This emotional devotion has a strong impact on the performance and understanding levels of the students. The level of this sentimental bond of students with the classroom varies amongst students of different education levels (Khajavy et al., 2018; Sandilos et al., 2017). It is more likely believed that the students of primary classes have a much warmer relationship with their classrooms in comparison to the students of higher education levels. Research shows that the classroom environment formed by the kids and the kindergarten teachers helps in establishing a positive cognitive climate (Gkloumpou and Germanos, 2020). Younger kids are more into asking questions, responding back to the questions of their teachers, and in expressing their thoughts and opinions when they are in a productive and welcoming classroom environment (Gkloumpou and Germanos, 2020; Johnson and Johnson, 2002). The active participation really adds up a lot in the learning process. The students of secondary and higher education levels also have a close relationship with their institutes and classrooms. The relationship between classroom and the learning environment is marked significant because the fashioned environment holds a huge impact on the behavior and performance of the individuals (Castilla et al., 2017; Scott-Webber et al., 2000)The classroom environment comes under traditional teaching methodology but, when it comes to online education, the classroom setup disappears. Now, students and their house members are held responsible for creating a suitable environment for learning at homes. For the kids of primary education level, especially till grade II, establishing a learning environment for themselves on their own is quite impossible. The dependency of younger students on their elders really puts many more points of consideration. For a healthy learning environment, it is extremely significant that everything goes smoothly (Chen et al., 2020) and that in the case of online education is possible only when people are good at dealing with computer and connectivity-related issues (Eli, 2021). In Pakistan, most parents do not use digital devices too often. Hence, they are not much familiar and prompt in resolving related problems. During online classes, if students face disruptions, they are more likely to get distorted and lose interest in even attending their classes. Furthermore, following a proper table-chair discipline is also considered important, which has been found lacked during the online classes of school students. Secondary and higher education students are grown-up and are expected to solve the online classes' related issues themselves with little or no help from the elders (Farouk, 2021; Ghazal et al., 2021). This reduces the influence of parents being less aware. Moreover, while the students are gaining education by staying at their homes, it is a must that proper disciplined study environment is provided to the adult learners because higher education demands students to be focused (Kara et al., 2019). Here comes the responsibility of the parents and other house members that they aid in creating such a healthy environment in homes that students can easily focus on their studies. Maintaining a proper table-chair discipline is important. Discipline is part of the values of the learning environment and discipline is extremely significant for staying focused, and productive "ecole_admin".

2.1.2 Student–Teacher Relationship

Student teacher relationship is extremely significant in enhancing the abilities and strengths of students especially during school years. Teachers have a huge impact on student's social, emotional, and intellectual experiences. For being able to promote the cognitive processes effectively, it is important that the teachers understand the emotional behavior, strengths, and weaknesses clearly. School students especially those enrolled in primary classes are found to have a very friendly and strong relationship with their teachers. During the conduction of live classes, this relationship is strengthened day-by-day and such a learning environment is created which encourages the learning process. School students spend a good amount of time with their teachers at schools where they are socially, emotionally, and intellectually understood and trained. In the presence of their teachers, school students tend to pay more attention to their studies than they do in their absence. The presence of teachers puts a positive impact on young minds and holds a huge influence on the way they behave and think (Koca, 2016) (Vandenbroucke et al., 2018). Secondary and higher education level students need more understanding of concepts. This is found to be more convenient when the teachers and students are having face-to-face interaction because students feel motivated and encouraged when their teacher/instructor is in front of them (Vandenbroucke et al., 2018). Higher education is found to be more impactful when there is active participation of the students during their classes. After the shift of traditional classes to online classes, students and teachers are now left only digitally connected. Various problems like dis connectivity issues and communication problems have been found to weaken the bond between the students and instructors (Joghee et al., 2020a). School student's especially primary level kids have missed this all-important teacher–child relationship during online classes. The gaps between students and teachers have forced parents to get more focused on the education of their kids. Effective online education has put up the demand for parents to be more tolerant, responsible, helpful, and observant towards the education of their children.

2.1.3 Access to Online Education

Attaining online education requires the availability of some resources including digital devices, fast and stable internet connection, and a sustained supply of electricity. Without these resources, it gets impossible to gain access to online classes. This platform-associated apprehension holds a significant effect on students' as well as parents' satisfaction and interest in attending the online classes which ultimately affects the effectiveness of education (Chen et al., 2020). In Pakistan, issues regarding the non-availability of electronic gadgets and non-attainability of internet connection are common. Pakistani parents had to face multiple difficulties in assuring the availability of devices and accessibility of the internet to facilitate the education of their children. This non-synchronous availability of resources amongst the students creates an atmosphere of socio-economic inequality (Hasan et al., 2021; Qureshi et al., 2012). The financial burden is one major concern faced by most Pakistani people. Arranging suitable digital devices and good internet connection with appropriate bandwidth is one big concern of online education (Qureshi et al., 2012). In case of non-availability of appropriate gadgets or bad internet connectivity where dis-connectivity and distortion kind of issues are faced too often, makes the students of all ages unfasten their interest in attending the online classes. This irritation, dissatisfaction, and perplexity amongst the students conclusively result in the lack of effectiveness of education. One big issue in Pakistan is load-shedding which cannot be neglected when it comes to online education. Uninformed load-shedding results in many related problems like devices remaining uncharged, internet not working because of the unavailability of devices, etc. The only solution to this which can be adopted by the parents is to arrange generators or UPS at their homes so that the classes if their children do not get missed in case of load-shedding. This again demands a strong financial background which is not common in developing countries like Pakistan (Qureshi et al., 2012).

2.1.4 Level of Awareness and Computer Literacy

The level of expertise in computer technology is an important factor in successfully adopting technology. Research suggests that the more proficient the user is in using the computer and technology, the more welcoming would he/she be towards e-learning (Qureshi et al., 2012). Hence in the case of online education especially when it is the case of an emergency switch, it is extremely important to examine the level of awareness and computer knowledge of the parents and the students. When it comes to younger students who are not very familiar with the requirements, the parents are expected to be very good at using the computers and the internet because only then would they be able to provide assistance to their kids. In Pakistan, most of the parents are not efficient at using computers. Especially the mothers who are at homes during the conductance of online classes do not know much about dealing with the technology. This causes puzzlement and the process of attainment of education is hindered which results in the irritation of young minds. This ultimately results in the lack of effectiveness of learning. The students of secondary and higher education levels are much familiar with the technology and are expected to handle the issues on their own.

2.2 Teachers' Perspectives

This part of the paper focuses on the perspectives of Pakistani teachers related to online education post-Covid-19. Teachers own a prominent position, when it comes to education. Hence, examining the satisfaction-level of the teachers teaching various education levels is extremely crucial in determining the effectiveness of online education. The measure of this effect depends on various factors. Some of them are analyzed here.

2.2.1 Face to Face Interaction with Students

Face-to-face interaction with students is one of the most elementary aspects of the learning environment. Research shows that it is important for teachers to examine the facial expressions of the students. Different students take things differently and teachers get an idea about that via students' body language. Some students listen, think, and learn instantly whereas others need more clarification, and they appear to be perplexed. Hence, there exist many behavioral possibilities of the students, which can only be understood by the teachers precisely when they have face-to-face interaction with their students. Teachers can help the students understand better only when they have clear know-how about their feelings and level of understanding (Guergov and Radwan, 2021; Hamadneh et al., 2021; Hanaysha et al., 2021a, 2021b). Although, through camera faces of students can be seen by the teachers, but due to many possible issues it cannot replace the significance of face-to-face interaction in classrooms. Pre-primary and primary teachers are more likely to face interaction issues during online classes where they are far distant from their little students. Young students demand their teachers to understand them in all the possible ways and that for teachers is possible only when there is barrier-free contact with their students. The college and university teachers also face this issue because in higher studies discussion is extremely important for extended learning. This, in online education, is feasible only in the case of proper planning of courses which is not possible in case of emergency online teaching (Hodges et al., 2020).

2.2.2 Emergency Case: Sudden Shift

The switch from conventional to online teaching was a major one. The lectures and code of conduct are always pre-prepared and that also in consideration with the medium of lecture delivery and learning environment. This all demands a suitable amount of time which was not the scenario post-Covid-19 (Kashif et al., 2021; Khan, 2021). After the shutdown of educational institutions due to the global pandemic, teachers were abruptly asked to play the role of both designers and tutors and that also by using such tools which very few of them had used earlier (Rapanta et al., 2020). Online education is very different from traditional teaching and requires proper training, which did not happen in the concerned case.

2.2.3 Assessing the Performance of Students

In academia, assessment is considered the most vital ingredient to keep track of the performance of the students. Assessing the performance of students is important for determining, examining, and fixing the problems students are facing. Research shows that teachers consider taking assessments of the students necessary due to three main reasons, which include enhancing and facilitating the learning process, preparing learners for the upcoming level, and examining the effectiveness of the teaching methodology (Schut et al., 2020). It is the prompt way of getting access to the problem areas of individual students so that they could be helped by the teachers in a better way (Rapanta et al., 2020). Assessing the performance of students during online classes is one frustrating task. One thing to be highlighted here is that here we are basically talking about *emergency* online teaching, not about the general one. At the time, the lockdown of Covid-19 happened, the annual year at schools and colleges and the semester at universities were all going-on and shifting instantly from traditional to online was one tough task, which comes under emergency situation. In such scenarios, proper planning and instant re-designing of the courses is impossible Online assessments can be made fruitful but with proper evaluation from the scratch which was not possible in the argued scenario (Hodges et al., 2020; Rapanta et al., 2020).

3 Materials and Methods

This research was carried out qualitatively as well as quantitatively. The idea behind doing so was to reach out to strong justification points regarding the effectiveness of online education in Pakistan after the Covid-19 outbreak caused lockdown to happen across the world.

3.1 Qualitative Method

The related research papers were gathered to argue upon the factors affecting online education which were then related to education in Pakistan in the Covid-19 scenario. This section of qualitative research was divided into two subdomains covering the perspectives of students and parents and the other one dealing with the viewpoints of the teachers. This research paper focuses on the students and teachers of pre-primary to doctorate level. A comparison regarding the difficulties faced and effectiveness amongst various education levels has also been carried out.

3.1.1 Students and Parents Perspectives

When it comes to the viewpoint of students and parents, the following are the four major factors that this paper discusses having a heavy impact on the effectiveness of online education in the post-Covid-19 scenario:

- Learning Environment
- Students' teacher relationship
- Access to online education
- Level of Awareness and Computer Literacy

These four factors are reviewed in a very detailed manner. The social, psychological, financial, as well as cultural; all impacts were covered in this research. The learning environment owns immense importance, when talking about educational potency (Lee and Ahmed, 2021; Mehmood, 2021). Considering this very fact, educational institutions and classrooms are designed in such a manner to facilitate learning for students of all levels. When everything went online all of a sudden after the unprecedented situation of Covid-19, it appeared to be a major challenge to provide students with a healthy learning environment at home. The learning habitat holds a strong impact on the interest levels of the students (AlHamad et al., 2022; Ali et al., 2022; Radwan and Farouk, 2021). Also, Pakistani culture strongly supports a joint family system, which results in the presence of many people at home. All these related points have their psychological, cultural, and environmental considerations which made us review the significance of the learning environment and the problems which were faced by the parents in providing a healthy ambiance for benefiting from online education. The sudden lockdown due to Covid-19 created a barrier between students and their teachers (Lee et al., 2022a, 2022b; K. Lee, Romzi, et al., 2022a, 2022b). This relationship was then entertained by digital devices and the internet. Although it is considered to be one major benefit of technology that students get to attain education even in these difficulties and they are somehow connected to their teachers and trainers but still, the teacher-student relationship has surely been affected (Miller, 2021; Mondol, 2021; Obaid, 2021). Also, this has created an additional responsibility on the parents to cope-up with the semi-absence of the teachers. For acquiring maximum benefits of education, while being online, it is extremely important to have proper access to digital gadgets and the internet. Also, we know that load-shedding is one of the major issues in Pakistan. Also, the difficulties faced by the Pakistani parents in making the availability of these online necessities possible for the education of their children are discussed. The impacts of having a good or bad Internet connection and having or not having digital appliances over the education of the students of various education levels have been discussed. Keeping in mind the Pakistani society and culture, it was crucial to review the impact of the awareness level and computer literacy of parents, when it comes to online education.

3.1.2 Teachers' Perspectives

Section 2.2. of this paper touches three major concerns of the teachers of all education levels.

- Face-to-face interaction with students
- Emergency case: Sudden shift
- Assessing the performance of students

These factors were thought of as being important for the teachers of all education levels. Impact levels might differ as we move on from pre-primary to doctorate level. It is considered necessary for the satisfaction of teachers to know whether the students have understood the delivered concepts or not. One of the methods to het this idea is by reading the body language of the students (Alshurideh et al., 2012; Alzoubi et al., 2022; Al Kurdi et al., 2020a, 2020b). The significance of the face-to-face interaction of teachers with their students was discussed. Online education somehow existed before the Covid-19 lockdown also, but it was less common and carried out with proper planning. In case of Covid-19, education system had to switch immediately from traditional to online system which surely was not easy for the stakeholders. The difficulties faced by the teachers in the concerned emergency situation and the impact of this sudden major switch on the quality of education had also been reviewed. Conducting various exams at different levels is one important element of education. It was much needed to examine the significance of assessing the performance of students. One major concern of teachers related to emergency online education was to come-up with a way for trying to evaluate the performance of students in the most effectual manner. The qualitative part of the paper basically relates some serious considerations of education system with emergency online education which here is Covid-19.

3.2 Quantitative Method

It was extremely significant to gather the opinions of Pakistani parents and teachers while carrying out the research regarding the effectiveness of online education. For this very purpose, two separate survey forms were prepared to be filled by the parents and teachers, respectively. These survey forms were designed using Google Forms and were digitally sent to the parents and teachers via different social forums.

3.2.1 Parents' Perspectives

The survey form for the parents consisted of sixteen questions, whereas the survey form for the teachers comprised of twenty questions. All the questions of both the questionnaires were kept close-ended to get to-the-point results. While designing the questionnaire to be filled by the parents, the idea was to cover all the points that appeared to have the maximum relevancy at the end of parents and students. The education level and extent of computer literacy of the parents was acquired for the sake of assessing the participation of parents in making the online education of their kids easy and fruitful. The education level of the children was also interrogated so as to examine the differences as well as the reasons of those differences in the effectiveness of online education at different levels of education. Some questions were targeted at the availability of devices and accessibility of internet as these two are considered the basic instruments of online education. Few questions were kept in order to cover the financial aspect of online education as well. Also, the interest level of students and their progress scale after the shift of traditional to online education, as well as their willingness to continue with the online education in future was tried to be examined through the survey.

3.2.2 Teachers' Perspectives

The questions included in the survey form outlined to be filled up by the teachers of different education levels aimed at gathering the practical responses to all the aspects covered in theory. The computer literacy of the teachers was interrogated to examine the level of struggle they had to put for entertaining the education online. Covering the financial and social elements was also important so few questions were kept for that also. Questions regarding the internet connection and digital devices were also incorporated in the questionnaire for teachers also. Moreover, few questions were targeted at the involvement of students during online classes and regarding assessing their performance. We know that each curriculum course is different from the other and hence is subjected to be taught differently, keeping that in mind, a question regarding the subjects taught by the teachers was also asked so that it could be assessed that which subjects were easier to be taught comparative to the others.

4 Results

This section of the paper first discusses the results and findings accumulated after carrying out the qualitative and quantitative research of the perspectives of parents regarding the effectiveness of online education of their children in the post-Covid-19 scenario. Results from both types of research are discussed separately. Further, the comparison of their results is carried out. Same is then done for the results received regarding the effectiveness of online education in the post-covid-19 scenario from the all-important perspectives of the teachers.

4.1 Perspectives of Parents

Qualitative research suggests that in a country like Pakistan, where the use of technology is still not very common the sudden shift of the education mode was not supposed to be easily accepted by the students and parents. Having frequent access of children to technology is still not much found in majority areas of the country. Even the parents do not use it very often in many areas. In such cases, it was difficult to manage acquiring the online education in the emergency situation after the lockdown of Covid-19. Only those parents could help their children in a better way who themselves are well educated and technology users. Also, providing disciplined and fruitful study environment at homes appears to be quite difficult for the Pakistani parents as in majority of the areas of the country, joint family system is very common. The survey conducted to gather the opinions of parents on the effectiveness of online education received 73 responses. The following Fig. 1 shows the education level of the respondents' children, while Fig. 2 shows education level of parents. Figure 3 depicts the uses of digital devices by parents.

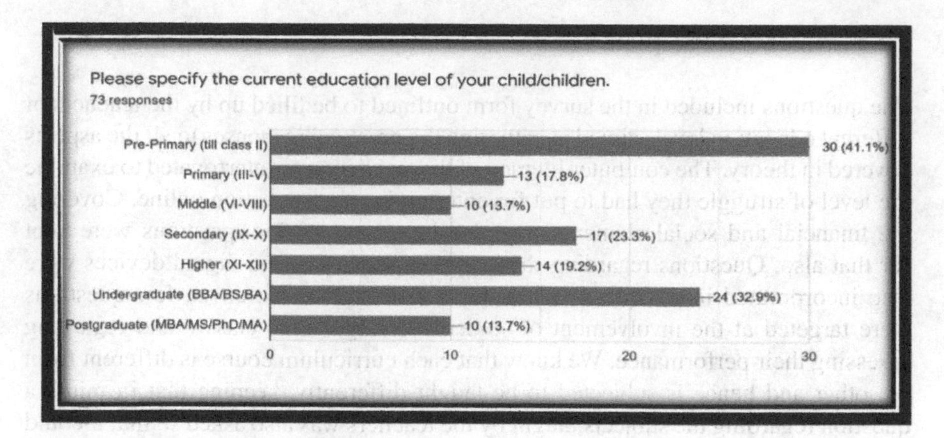

Fig. 1 Education level of students

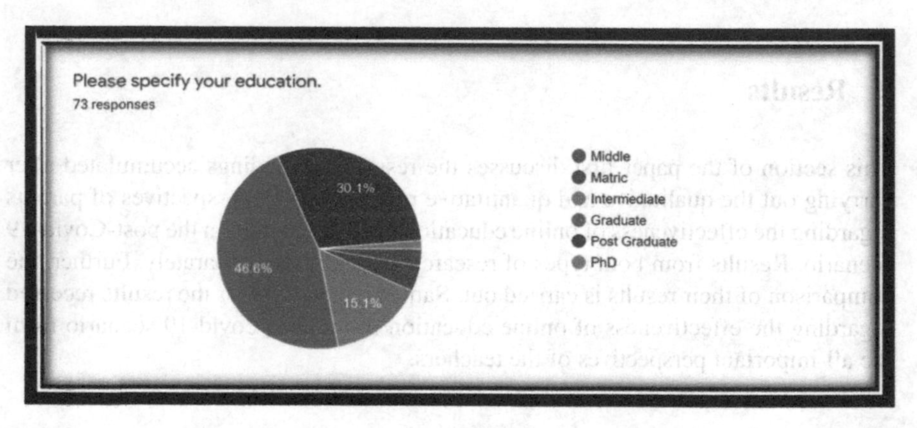

Fig. 2 Education level of parents

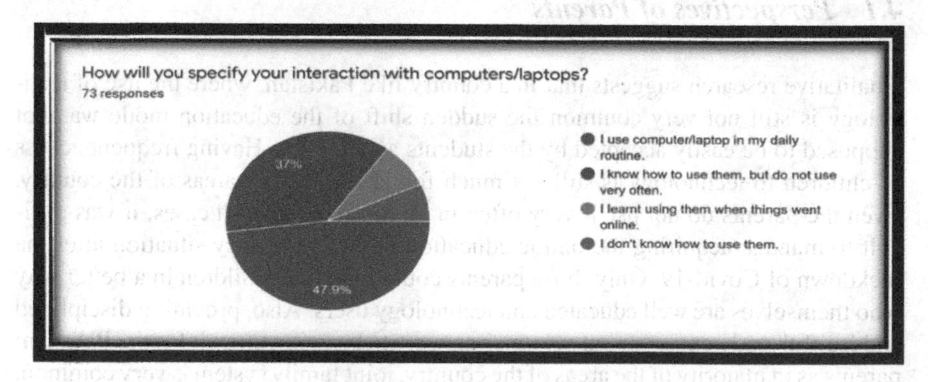

Fig. 3 Usage of digital devices by the parents

The conducted surveys showed that the parents of the students acquiring the online education are educated enough to have helped their children in getting access to the online education in the best possible way. The Figs. 2 and 3 as shown above depict that 46.6% of the parents are graduated and 47.9% use digital devices in their daily routine. This shows that parents who are well educated and prompt at using the technology can help their students in getting smooth access to online education. Figure 4 shown above suggests that only 39.7% parents had to buy the digital devices for entertaining the online education of their kids. Rest of them already had the resources. Further, Fig. 5 clarifies that even out of those 39.7% parents, 60.3% had to buy only a single device. Figure 6 shows that more than 60% of the parents were sure about not having any sort of increased financial burden caused due to online education. These results suggest that there was not much of the increased financial burden on the majority of the parents after the switch of traditional to emergency online learning.

The Fig. 7 suggests that the dis-connectivity issues were faced by majority of students, but not most often. So, these results suggest the situation which can be negotiated well. Figure 8 shows that majority of the children followed proper table-chair discipline at homes during their online classes, which suggests the increase in the probability of the effectiveness of online education. The Figs. 9 and 10 propose that the interest levels of the children, while taking their online classes were not good

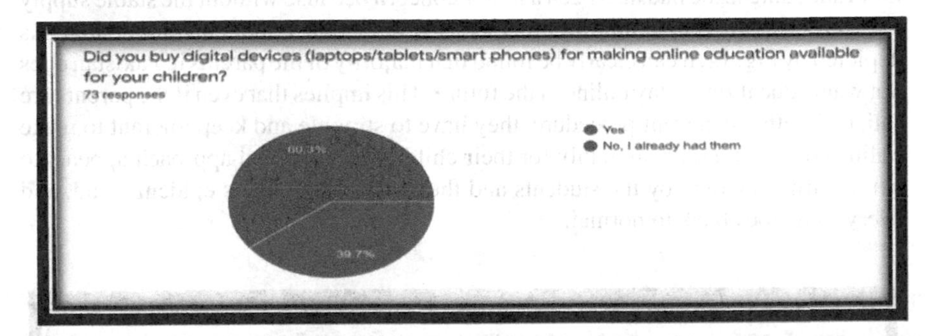

Fig. 4 Parents who had to buy the digital devices for the purpose of online education of their kids

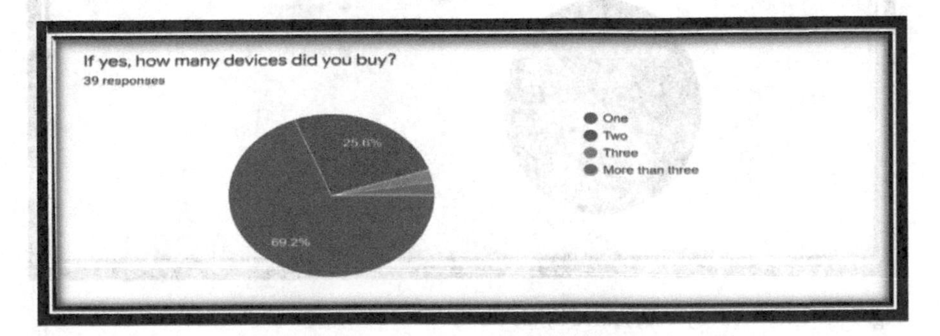

Fig. 5 Number of digital devices purchased by parents to facilitate online education of their kids

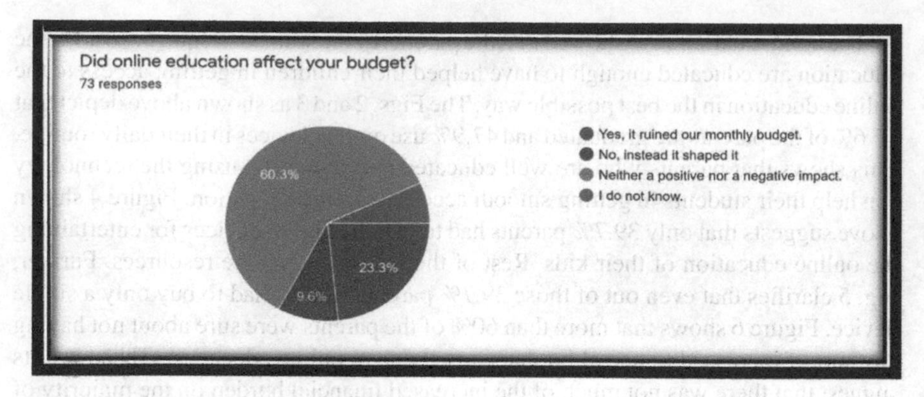

Fig. 6 Effect of online education on the budget

enough. For the fruitfulness of education and better performance of the students, it is extremely significant in the process of effective education.

Figure 11 shows that some serious issues had their impact on the effectiveness of online education. Communication of students with their teachers and motivating the students to take online classes seriously, have been found difficult to be managed. Load shedding issue had also been a major concern because without the stable supply of electricity, anything online won't be considered much smooth and influential. As depicted by Fig. 12, it can clearly be found that majority of the parents in Pakistan does not want education to stay online in the future. This implies that even if the parents are satisfied with the current procedure, they have to struggle and keep tolerant to make online education go on smoothly for their children. Traditional approach appears to get a warm welcome by the students and their parents once this epidemic end, and everything goes back to normal.

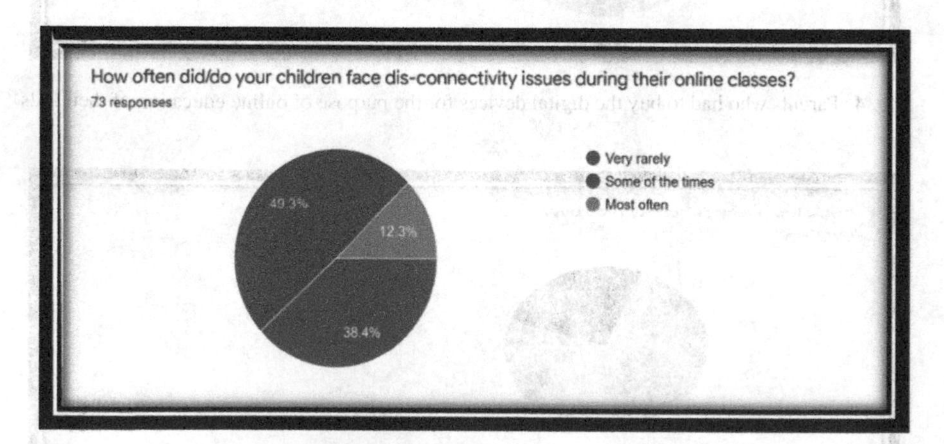

Fig. 7 Dis-connectivity issues faced by the students

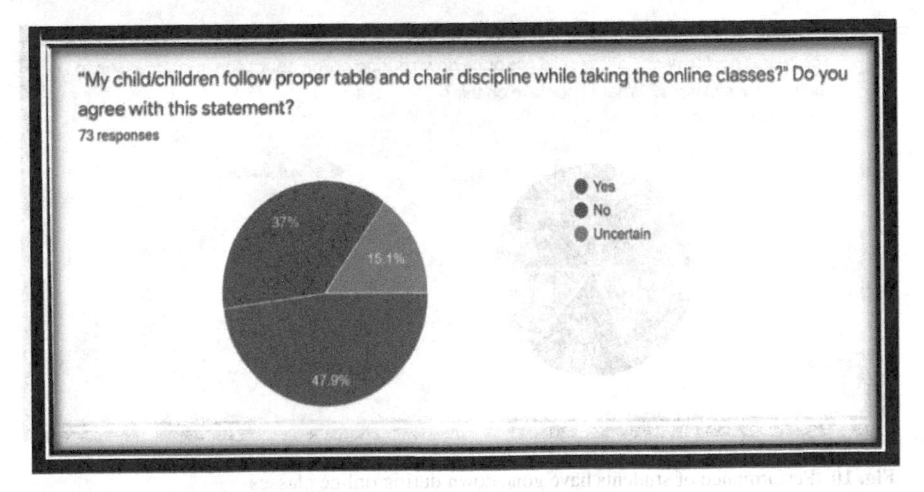

Fig. 8 Children were disciplined while taking online classes

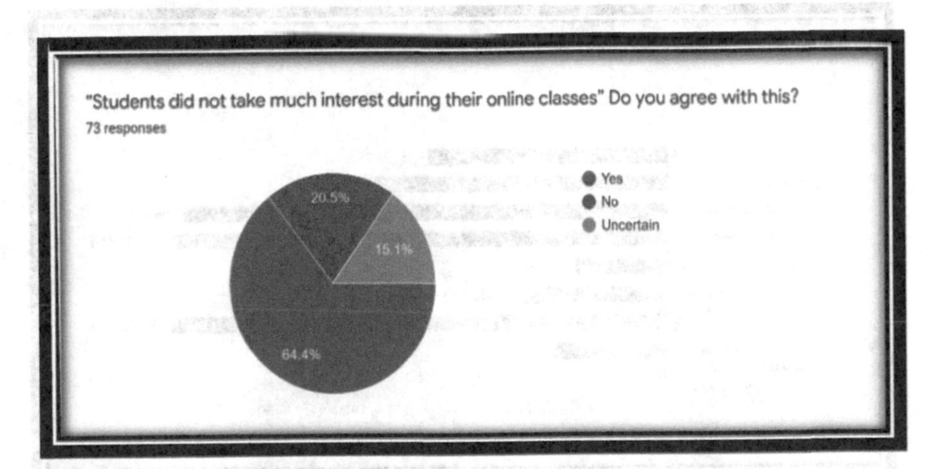

Fig. 9 Students were not interested during the online classes

Both, qualitative as well as the quantitative research suggests that online education is fruitful, where there the resources are properly managed, a disciplined learning environment is maintained, and parents are well educated and IT literate. In comparison to qualitative research, which suggests many difficulties with respect to the financial and educational backgrounds of the parents and the cultural values being followed in the country, to be faced by the Pakistani parents, quantitative results are very much normalized. Also, it has been found that the parents of pre-primary and primary classes have faced more difficulties in comparison to those faced by the parents of children enrolled in higher education levels.

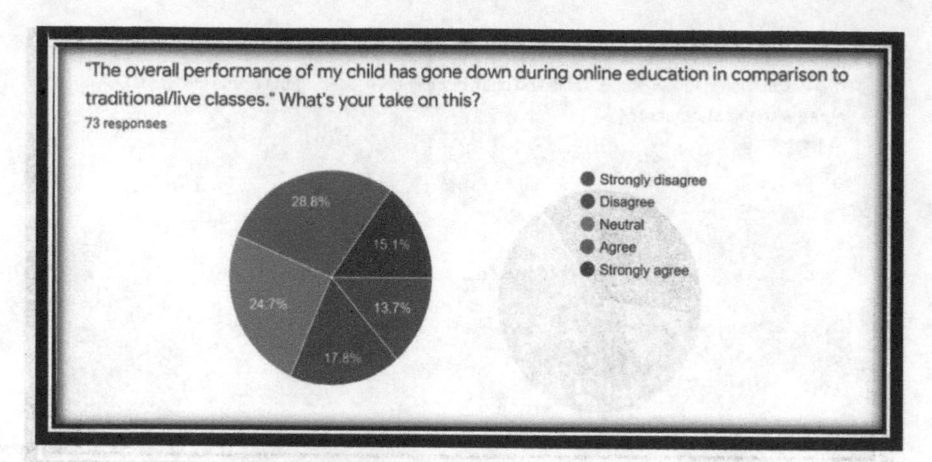

Fig. 10 Performance of students have gone down during online classes

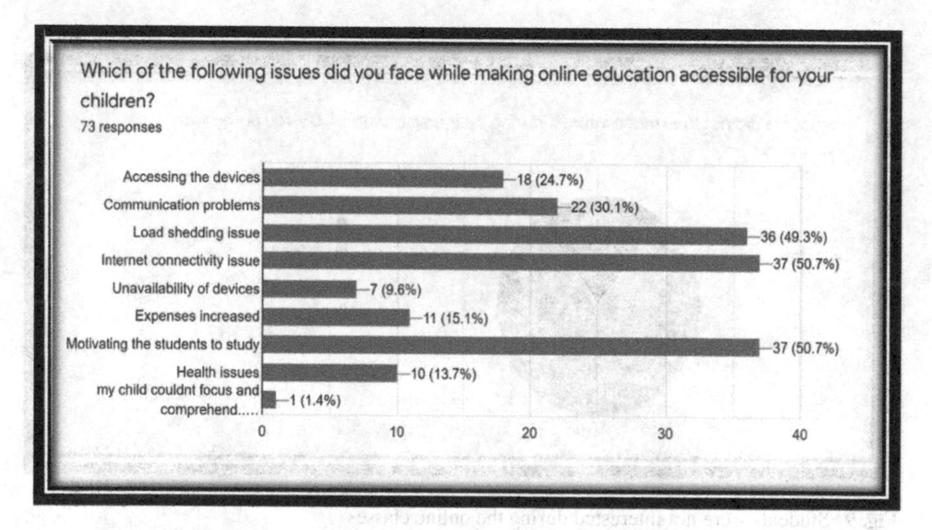

Fig. 11 Issues faced during online classes

4.2 Perspectives of Teachers

The qualitative part of the perspectives of the teachers suggests that since the switch of traditional education to online education after Covid-19 was an emergency case, teachers did not get the required amount of time to do proper planning regarding the conduction of online classes. Due to this, the courses could not get delivered in a much effective manner. Teacher satisfaction levels were found to be much affected as they could not connect to the students in a way they normally used to do during the online classes (Aburayya et al., 2020; Alshurideh et al., 2022). The issues regarding

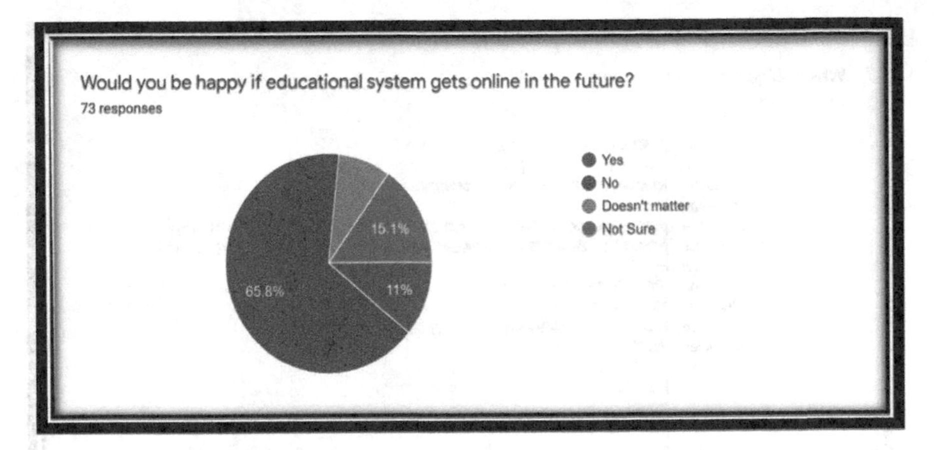

Fig. 12 Willingness of parents for the continuation of online classes

the accessibility of the devices have also been there which had found to put a good amount of impact on the motivation level of the teachers also. The survey conducted to gather the opinions of teachers on the effectiveness of online education received 60 responses. The following Fig. 13 shows the education levels to which the respondent teachers teach, and Fig. 14 shows the subjects taught by those teachers.

The results as shown in Fig. 15 depicts that more than 75% of the Pakistani teachers were familiar with the use of computers before the global epidemic caused the shutdown to take place. This is quite a good percentage which shows that accessing the devices was not a big issue for majority of the teachers. Figure 16 shows that

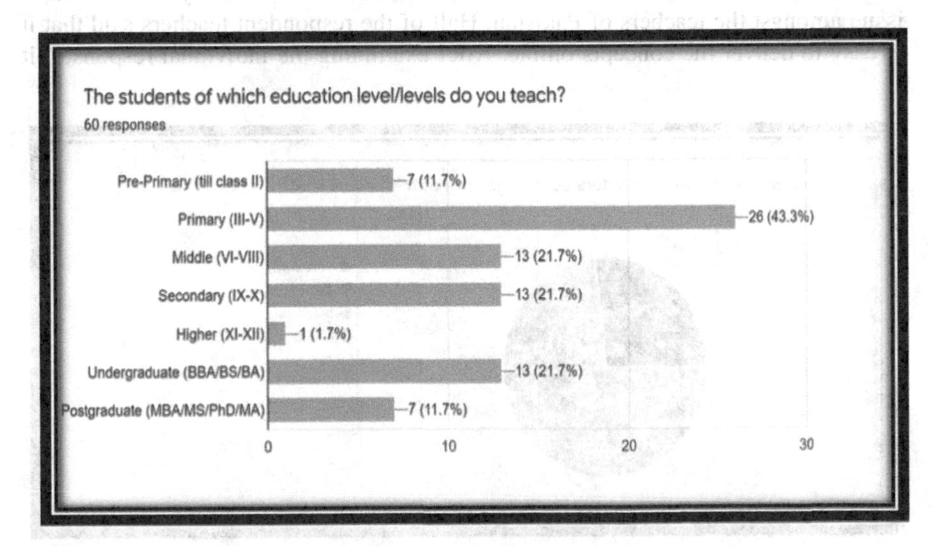

Fig. 13 Teachers teaching different education levels

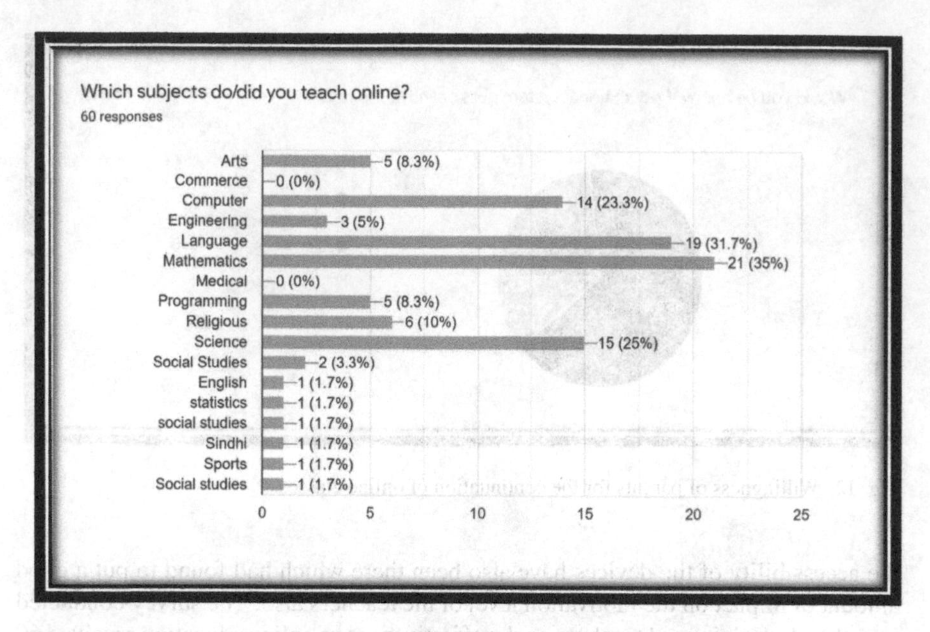

Fig. 14 Subjects taught by the respondent teachers

only 31.7 teachers had to buy digital equipment for conducting online classes. As presented in Fig. 17, out of these 31.7 teachers 83.3% declared that they did not receive any sort of financial support in buying the devices from their employers. This might have caused some disturbance in the budget management for such teachers.

The Figs. 18 and 19 shows that tolerating Internet dis-connectivity was one major issue amongst the teachers of Pakistan. Half of the respondent teachers said that it is easy to deliver the concepts online. After examining the individual responses, it

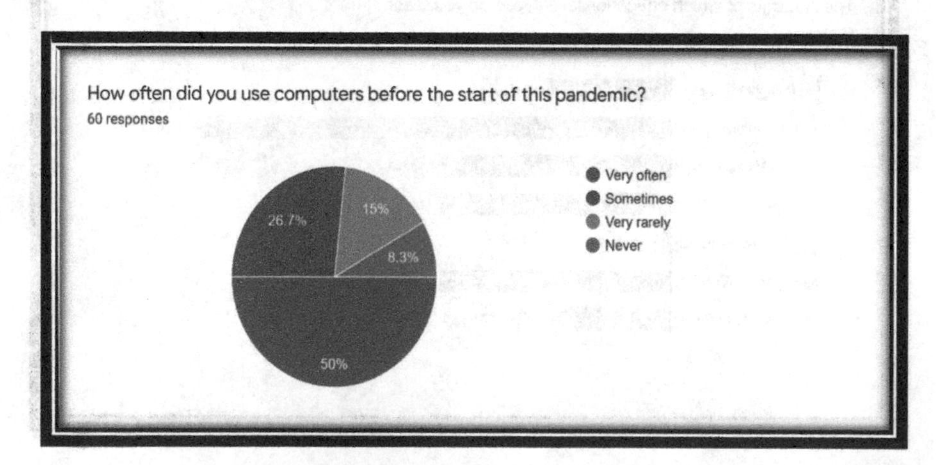

Fig. 15 Fluency of teachers' usage of technology

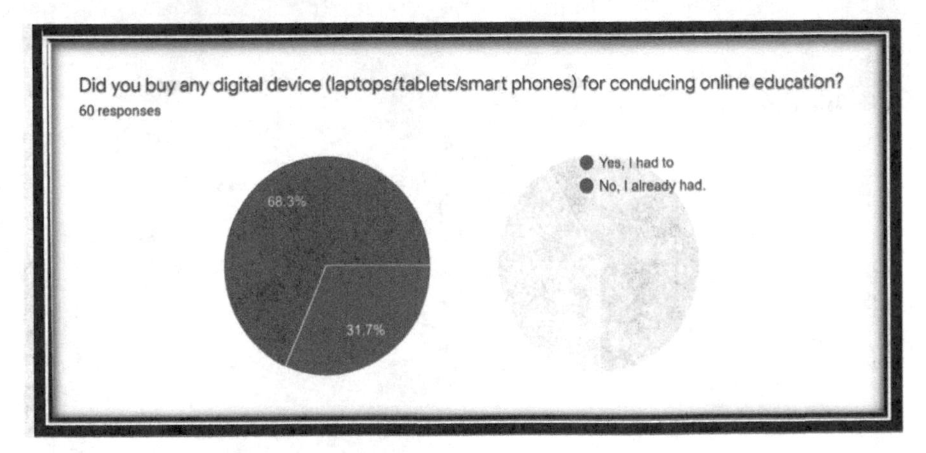

Fig. 16 Non-availability of devices for teachers

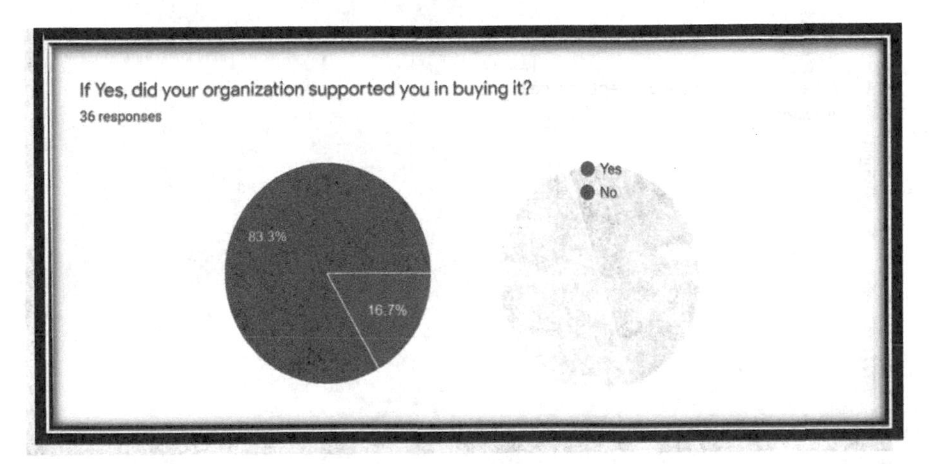

Fig. 17 Organizations' support in making the digital devices available for the teachers

was concluded that this varies from subject to subject. Language, theoretical, and computer related subjects were declared easy to be carried out whereas most of the mathematical teachers suggested that delivering the concepts is not easy. As shown in Fig. 20, 41.7% of the respondent teachers does not feel satisfied after conducting their online classes. Teaching is that profession in which inner satisfaction of a teacher says a lot about the quality of educational process that is going on.

The results shown in the Figs. 21, 22, and 23 suggest that majority of the teachers does not feel much calm at conducting the online classes as they miss the physical appearance and active participation of students.

The results of Fig. 24 clearly show that the Pakistani teachers have found it extremely difficult to scrutinize the progress and performance of students. Figure 25 declares many issues, which were found to be faced by most of the Pakistani teachers.

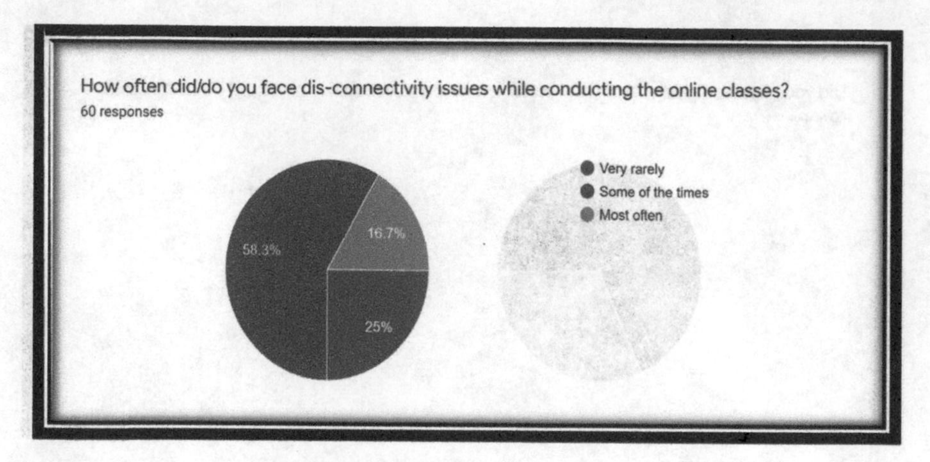

Fig. 18 Dis-connectivity issues faced by the teachers

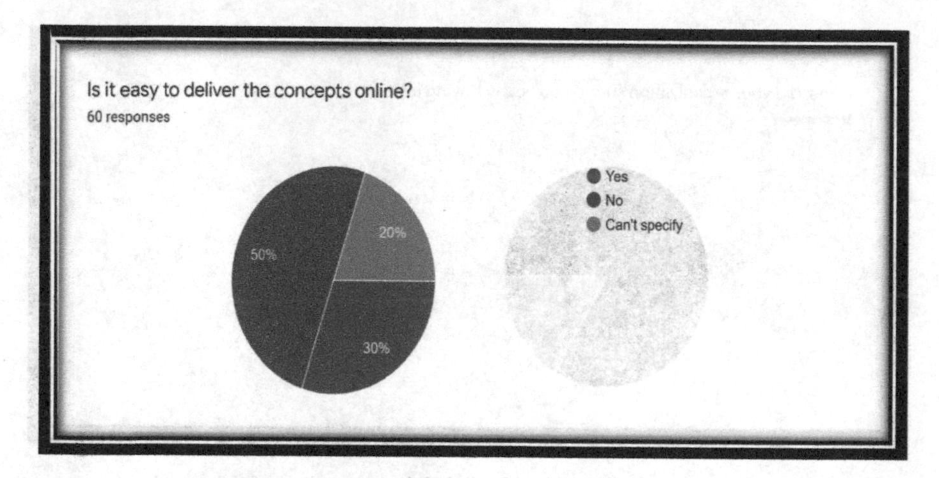

Fig. 19 Delivering the concepts online

Accessing the technology by teachers as well as the students, communicating effectively with the students, increased workload, and time management have been the common issues addressed by the majority of the respondent teachers.

48.3% of the Pakistani teachers are against the continuation of online education in future, depicted in Fig. 26. This shows that they do not find online education much effective to be adopted as a tradition in the future.

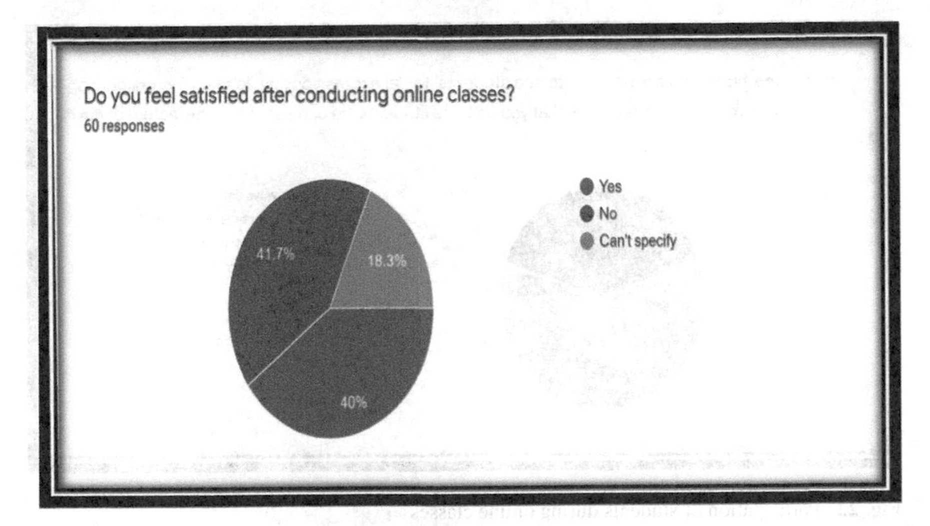

Fig. 20 Satisfaction of teachers after conducting online classes

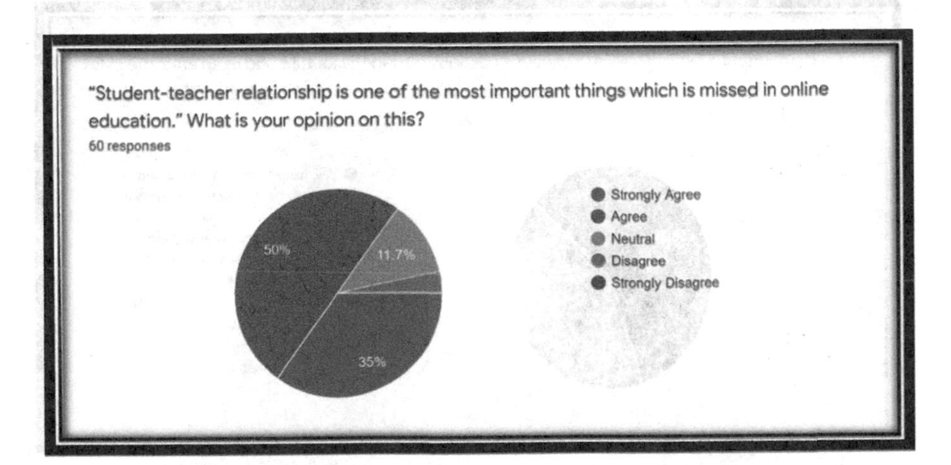

Fig. 21 Student–teacher relationship in online education

5 Discussion and Future Work

The results of the qualitative and quantitative research carried out on the effectiveness of online education suggests that online education could be extremely effective and beneficial if planned properly which was not the case post Covid-19. Although the issues discussed in the literature review were not found much problematic in the quantitative results, but still there are many people who could not get proper access to education after it went online. The level of effectiveness of online education appears to be varying in many ways. Students of higher education levels found accessing

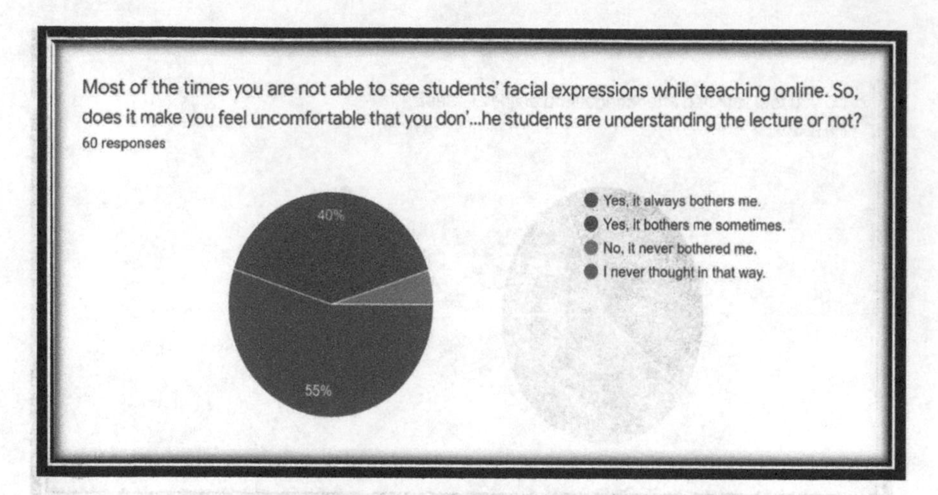

Fig. 22 Participation of students during online classes

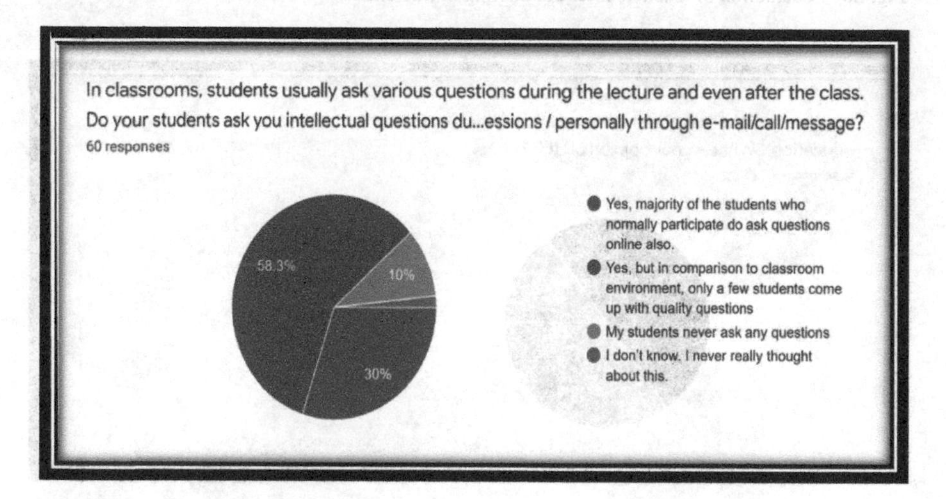

Fig. 23 Engagement of students during online classes

the technology and managing the related issues in a better way in comparison to that of the school level students. Parents owning different educational and financial backgrounds had to face distinguished issues and at diversified extents. Similarly, some teachers were able to connect in a way better than the others. This research paper discussed online education only from the viewpoint of parents and teachers. In future, researchers must cater the opinions of the owners and management of educational institutions.

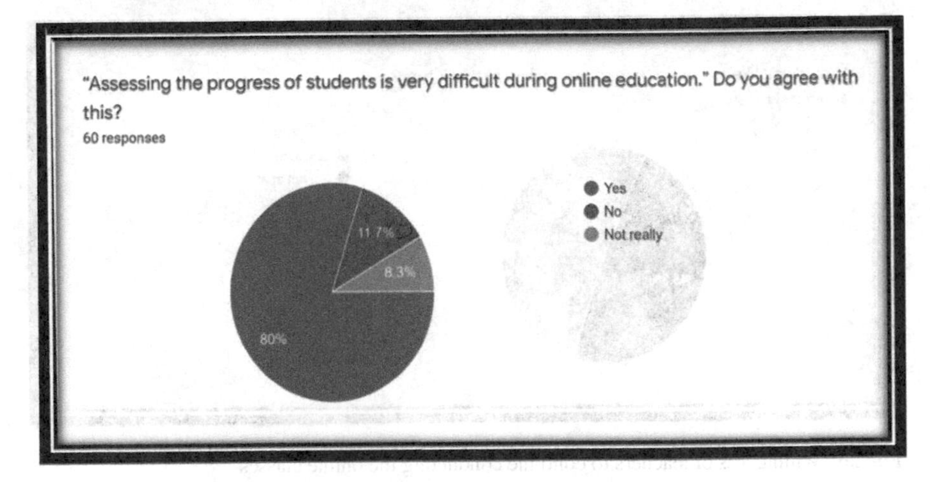

Fig. 24 Assessing the progress of students

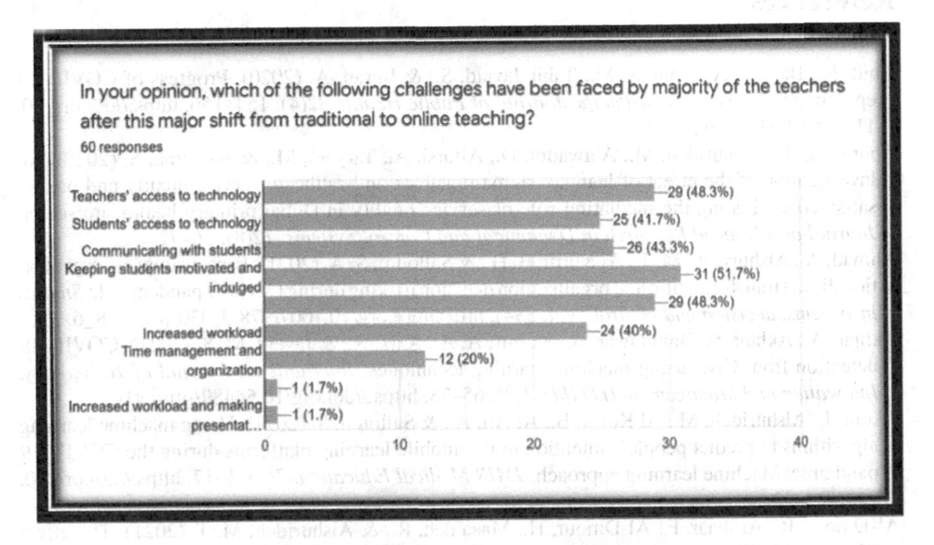

Fig. 25 Major problems encountered during online education in the opinion of Pakistani teachers

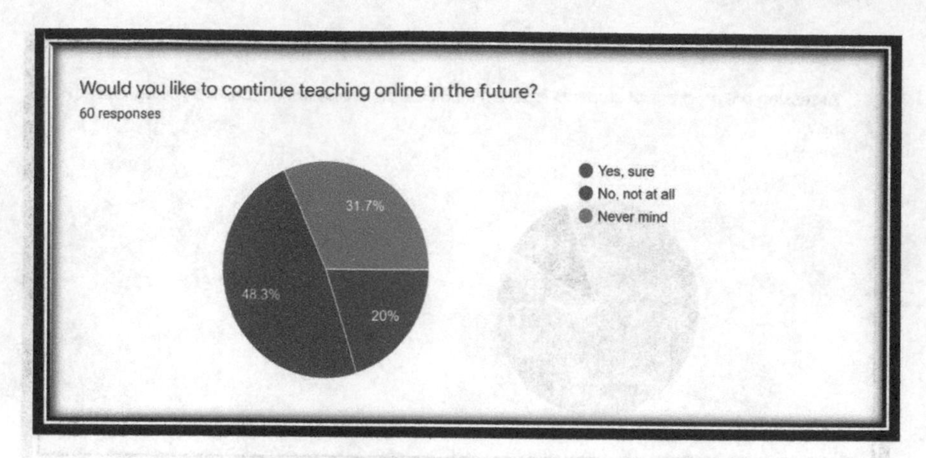

Fig. 26 Willingness of teachers to continue conducting the online classes

References

Abid, K., Bari, Y. A., Younas, M., Tahir Javaid, S., & Imran, A. (2020). Progress of COVID-19 epidemic in Pakistan. *Asia-Pacific Journal of Public Health, 32*(4), 154–156. https://doi.org/10.1177/1010539520927259

Aburayya, A., Alshurideh, M., Alawadhi, D., Alfarsi, A., Taryam, M., & Mubarak, S. (2020). An investigation of the effect of lean six sigma practices on healthcare service quality and patient satisfaction: Testing the mediating role of service quality in Dubai primary healthcare sector. *Journal of Advanced Research in Dynamical and Control Systems, 12*(8), 56–72.

Ahmad, A., Alshurideh, M. T., Al Kurdi, B. H., & Salloum, S. A. (2021). Factors impacts organization digital transformation and organization decision making during Covid19 pandemic. In *Studies in systems, decision and control* (Vol. 334). https://doi.org/10.1007/978-3-030-67151-8_6.

Akhtar, A., Akhtar, S., Bakhtawar, B., Kashif, A. A., Aziz, N., & Javeid, M. S. (2021). COVID-19 detection from CBC using machine learning techniques. *International Journal of Technology, Innovation and Management (IJTIM), 1*(2), 65–78. https://doi.org/10.54489/ijtim.v1i2.22.

Akour, I., Alshurideh, M., Al Kurdi, B., Al Ali, A., & Salloum, S. (2021). Using machine learning algorithms to predict people's intention to use mobile learning platforms during the COVID-19 pandemic: Machine learning approach. *JMIR Medical Education, 7*(1), 1–17. https://doi.org/10.2196/24032

Al-Dmour, R., AlShaar, F., Al-Dmour, H., Masa'deh, R., & Alshurideh, M. T. (2021). The effect of service recovery justices strategies on online customer engagement via the role of "Customer Satisfaction" during the Covid-19 pandemic: An empirical study. In *Studies in systems, decision and control* (Vol. 334). https://doi.org/10.1007/978-3-030-67151-8_19.

Al-Hamad, M. Q., Mbaidin, H. O., Alhamad, A. Q. M., Alshurideh, M. T., Kurdi, B. H. A., & Al-Hamad, N. Q. (2021). Investigating students' behavioral intention to use mobile learning in higher education in UAE during Coronavirus-19 pandemic. *International Journal of Data and Network Science, 5*(3). https://doi.org/10.5267/j.ijdns.2021.6.001.

Al-Maroof, R. S., Alshurideh, M. T., Salloum, S. A., AlHamad, A. Q. M., & Gaber, T. (2021). Acceptance of google meet during the spread of coronavirus by Arab university students. *Informatics, 8*(2). https://doi.org/10.3390/informatics8020024.

Al Ali, A. (2021). The impact of information sharing and quality assurance on customer service at UAE banking sector. *International Journal of Technology, Innovation and Management (IJTIM), 1*(1), 01–17. https://doi.org/10.54489/ijtim.v1i1.10.

Al Khasawneh, M., Abuhashesh, M., Ahmad, A., Alshurideh, M. T., & Masa'deh, R. (2021a). Determinants of E-word of mouth on social media during COVID-19 outbreaks: An empirical study. In *Studies in systems, decision and control* (Vol. 334). https://doi.org/10.1007/978-3-030-67151-8_20.

Al Khasawneh, M., Abuhashesh, M., Ahmad, A., Masa'deh, R., & Alshurideh, M. T. (2021b). Customers online engagement with social media influencers' content related to COVID 19. In *Studies in systems, decision and control* (Vol. 334). https://doi.org/10.1007/978-3-030-67151-8_22.

Al Kurdi, B., Alshurideh, M., Nuseir, M., Aburayya, A., & Salloum, S. A. (2021a). The effects of subjective norm on the intention to use social media networks: An exploratory study using PLS-SEM and machine learning approach. *Advanced Machine Learning Technologies and Applications: Proceedings of AMLTA, 2021*, 581–592.

Al Kurdi, B., Alshurideh, M., Salloum, S., Obeidat, Z., & Al-dweeri, R. (2020b). *An empirical investigation into examination of factors influencing university students' behavior towards elearning acceptance using SEM approach.*

Alameeri, K. A., Alshurideh, M. T., & Al Kurdi, B. (2021). The effect of Covid-19 pandemic on business systems' innovation and entrepreneurship and how to cope with it: A theatrical view. In *Studies in Systems, Decision and Control* (Vol. 334). https://doi.org/10.1007/978-3-030-67151-8_16.

AlHamad, A., Alshurideh, M., Alomari, K., Kurdi, B., Alzoubi, H., Hamouche, S., et al. (2022). The effect of electronic human resources management on organizational health of telecommunications companies in Jordan. *International Journal of Data and Network Science, 6*(2), 429–438.

Alhamad, A. Q. M., Akour, I., Alshurideh, M., Al-Hamad, A. Q., Kurdi, B. A., & Alzoubi, H. (2021). Predicting the intention to use google glass: A comparative approach using machine learning models and PLS-SEM. *International Journal of Data and Network Science, 5*(3). https://doi.org/10.5267/j.ijdns.2021.6.002.

Ali, N., Ahmed, A., Anum, L., Ghazal, T. M., Abbas, S., Khan, M. A., et al. (2021). Modelling supply chain information collaboration empowered with machine learning technique. *Intelligent Automation and Soft Computing, 30*(1), 243–257. https://doi.org/10.32604/iasc.2021.018983.

Ali, N. M., Ghazal, T., Ahmed, A., Abbas, S., Khan, M. A., Alzoubi, H., et al. (2022). Fusion-based supply chain collaboration using machine learning techniques. *Intelligent Automation & Soft Computing, 31*(3), 1671–1687. https://doi.org/10.32604/iasc.2022.019892.

Aljumah, A., Nuseir, M. T., & Alshurideh, M. T. (2021). The impact of social media marketing communications on consumer response during the COVID-19: Does the brand equity of a university matter. *The Effect of Coronavirus Disease (COVID-19) on Business Intelligence, 334*, 384–367.

Alnazer, N. N., Alnuaimi, M. A., & Alzoubi, H. M. (2017). Analysing the appropriate cognitive styles and its effect on strategic innovation in Jordanian universities. *International Journal of Business Excellence, 13*(1), 127–140. https://doi.org/10.1504/IJBEX.2017.085799

Alnuaimi, M., Alzoubi, H. M., Ajelat, D., & Alzoubi, A. A. (2021). Towards intelligent organisations: An empirical investigation of learning orientation's role in technical innovation. *International Journal of Innovation and Learning, 29*(2), 207–221. https://doi.org/10.1504/IJIL.2021.112996

Alshamsi, A., Alshurideh, M., Al Kurdi, B., & Salloum, S. A. (2020). The influence of service quality on customer retention: A systematic review in the higher education. In: *International Conference on Advanced Intelligent Systems and Informatics* (pp. 404–416).

Alsharari, N. (2021). Integrating blockchain technology with internet of things to efficiency. *International Journal of Technology, Innovation and Management (IJTIM), 1*(2), 1–13.

Alsharari, N. M., & Alshurideh, M. T. (2020). Student retention in higher education: the role of creativity, emotional intelligence and learner autonomy. *International Journal of Educational Management*.

Alshurideh, M., Al Kurdi, B., & Salloum, S. A. (2020a). Examining the main mobile learning system drivers' effects: A mix empirical examination of both the expectation-confirmation model (ECM)

and the technology acceptance model (TAM). In *Advances in intelligent systems and computing* (Vol. 1058). https://doi.org/10.1007/978-3-030-31129-2_37.
Alshurideh, M., Al Kurdi, B., Salloum, S. A., Arpaci, I., & Al-Emran, M. (2020b). Predicting the actual use of m-learning systems: A comparative approach using PLS-SEM and machine learning algorithms. *Interactive Learning Environments*. https://doi.org/10.1080/10494820.2020.1826982
Alshurideh, M., Gasaymeh, A., Ahmed, G., Alzoubi, H., & Kurd, B. A. (2020c). Loyalty program effectiveness: Theoretical reviews and practical proofs. *Uncertain Supply Chain Management, 8*(3). https://doi.org/10.5267/j.uscm.2020.2.003.
Alshurideh, M., Masa'deh, R. M. T., & Alkurdi, B. (2012). The effect of customer satisfaction upon customer retention in the Jordanian mobile market: An empirical investigation. *European Journal of Economics, Finance and Administrative Sciences, 47*.
Alshurideh, M. T., Al Kurdi, B., Masa'deh, R., & Salloum, S. A. (2021a). The moderation effect of gender on accepting electronic payment technology: A study on United Arab Emirates consumers. *Review of International Business and Strategy, 31*(3). https://doi.org/10.1108/RIBS-08-2020-0102.
Alshurideh, M. T., Hassanien, A. E., & Masa'deh, R. (2021b). *The effect of coronavirus disease (COVID-19) on business intelligence*. Springer.
Alshurideh, M. T., Kurdi, B. A., AlHamad, A. Q., Salloum, S. A., Alkurdi, S., Dehghan, A., et al. (2021c). Factors affecting the use of smart mobile examination platforms by universities' postgraduate students during the COVID 19 pandemic: An empirical study. *Informatics, 8*(2), 32.
Alshurideh, M., Salloum, S. A., Al Kurdi, B., Monem, A. A., & Shaalan, K. (2019). Understanding the quality determinants that influence the intention to use the mobile learning platforms: A practical study. *International Journal of Interactive Mobile Technologies, 13*(11).
Alshurideh, M. T., Al Kurdi, B., Alzoubi, H. M., Ghazal, T. M., Said, R. A., AlHamad, A. Q., et al. (2022). Fuzzy assisted human resource management for supply chain management issues. *Annals of Operations Research*, 1–19.
Alzoubi, A. (2021a). The impact of process quality and quality control on organizational competitiveness at 5-star hotels in Dubai. *International Journal of Technology, Innovation and Management (IJTIM), 1*(1), 54–68. https://doi.org/10.54489/ijtim.v1i1.14.
Alzoubi, A. (2021b). Renewable Green hydrogen energy impact on sustainability performance. *International Journal of Computations, Information and Manufacturing (IJCIM), 1*(1), 94–110. https://doi.org/10.54489/ijcim.v1i1.46.
Alzoubi, H. M., & Aziz, R. (2021). Does emotional intelligence contribute to quality of strategic decisions? The mediating role of open innovation. *Journal of Open Innovation: Technology, Market, and Complexity, 7*(2), 130. https://doi.org/10.3390/joitmc7020130
Alzoubi, H. M., Vij, M., Vij, A., & Hanaysha, J. R. (2021). What leads guests to satisfaction and loyalty in UAE five-star hotels? AHP analysis to service quality dimensions. *Enlightening Tourism, 11*(1), 102–135. https://doi.org/10.33776/et.v11i1.5056.
Alzoubi, H. M., & Yanamandra, R. (2020). Investigating the mediating role of information sharing strategy on agile supply chain. *Uncertain Supply Chain Management, 8*(2), 273–284. https://doi.org/10.5267/j.uscm.2019.12.004
Alzoubi, H., Ahmed, G., Al-Gasaymeh, A., & Kurdi, B. (2020a). Empirical study on sustainable supply chain strategies and its impact on competitive priorities: The mediating role of supply chain collaboration. *Management Science Letters, 10*(3), 703–708.
Alzoubi, H., Alshurideh, M., Kurdi, B., Akour, I., & Aziz, R. (2022). Does BLE technology contribute towards improving marketing strategies, customers' satisfaction and loyalty? The role of open innovation. *International Journal of Data and Network Science, 6*(2), 449–460.
Alzoubi, H., & Ahmed, G. (2019). Do TQM practices improve organisational success? A case study of electronics industry in the UAE. *International Journal of Economics and Business Research, 17*(4), 459–472. https://doi.org/10.1504/IJEBR.2019.099975
Alzoubi, H., Alshurideh, M., Kurdi, B. A., & Inairat, M. (2020b). Do perceived service value, quality, price fairness and service recovery shape customer satisfaction and delight? A practical

study in the service telecommunication context. *Uncertain Supply Chain Management, 8*(3), 579–588. https://doi.org/10.5267/j.uscm.2020.2.005
Amarneh, B. M., Alshurideh, M. T., Al Kurdi, B. H., & Obeidat, Z. (2021). The Impact of COVID-19 on e-learning: Advantages and Challenges. *The International Conference on Artificial Intelligence and Computer Vision* (pp. 75–89).
Ashal, N., Alshurideh, M., Obeidat, B., & Masa'deh, R. (2021). The impact of strategic orientation on organizational performance: Examining the mediating role of learning culture in Jordanian telecommunication companies. *Academy of Strategic Management Journal, Special Is*(Special Issue 6), 1–29.
Aziz, N., & Aftab, S. (2021). Data mining framework for nutrition ranking: Methodology: SPSS modeller. *International Journal of Technology, Innovation and Management (IJTIM), 1*(1), 85–95.
Bestiantono, D. S., Agustina, P. Z. R., & Cheng, T.-H. (2020). How students' perspectives about online learning amid the COVID-19 pandemic? *Studies in Learning and Teaching, 1*(3), 133–139. https://doi.org/10.46627/silet.v1i3.46.
Bettayeb, H., Alshurideh, M. T., & Al Kurdi, B. (2020). The effectiveness of mobile learning in UAE universities: A systematic review of motivation, self-efficacy, usability and usefulness. *International Journal of Control and Automation, 13*(2), 1558–1579.
Bhardwaj, A. (2016). Importance of education in human life. *International Journal of Science and Consciousness, 2*(2), 23–28.
Castilla, N., Llinares, C., Bravo, J. M., & Blanca, V. (2017). Subjective assessment of university classroom environment. *Building and Environment, 122*, 72–81. https://doi.org/10.1016/j.buildenv.2017.06.004
Chen, T., Peng, L., Yin, X., Rong, J., Yang, J., & Cong, G. (2020). Analysis of user satisfaction with online education platforms in China during the COVID-19 pandemic. *Healthcare, 8*(3), 200. https://doi.org/10.3390/healthcare8030200
Crawford, J., Henderson, K. B., Rudolph, J., Malkawi, B., Glowatz, M., Burton, R., Magni, P. A., & Lam, S. (2020). Journal of Applied Learning & Teaching COVID-19: 20 countries ' higher education intra-period digital pedagogy responses. *Journal of Applied Learning & Teaching, 3*(1), 1–20.
Cruz, A. (2021). Convergence between blockchain and the internet of things. *International Journal of Technology, Innovation and Management (IJTIM), 1*(1), 35–56.
Demuyakor, J. (2020). Coronavirus (COVID-19) and online learning in higher institutions of education: A survey of the perceptions of ghanaian international students in China. *Online Journal of Communication and Media Technologies, 10*(3), e202018. https://doi.org/10.29333/ojcmt/8286.
Eli, T. (2021). StudentsPerspectives on the use of innovative and interactive teaching methods at the university of Nouakchott Al Aasriya, Mauritania: English department as a case study. *International Journal of Technology, Innovation and Management (IJTIM), 1*(2), 90–104.
Farouk, M. (2021). The universal artificial intelligence efforts to face coronavirus COVID-19. *International Journal of Computations, Information and Manufacturing (IJCIM), 1*(1), 77–93. https://doi.org/10.54489/ijcim.v1i1.47.
Ghazal, T. M., Hasan, M. K., Alshurideh, M. T., Alzoubi, H. M., Ahmad, M., Akbar, S. S., Al Kurdi, B., & Akour, I. A. (2021). IoT for smart cities: Machine learning approaches in smart healthcare—a review. *Future Internet, 13*(8), 218. https://doi.org/10.3390/fi13080218
Gkloumpou, A., & Germanos, D. (2020). The importance of classroom cooperative learning space as an immediate environment for educational success. An action research study in Greek Kindergartens. *Educational Action Research, 00*(00), 1–15. https://doi.org/10.1080/09650792.2020.1771744.
Guergov, S., & Radwan, N. (2021). Blockchain convergence: Analysis of issues affecting IoT, AI and blockchain. *International Journal of Computations, Information and Manufacturing (IJCIM), 1*(1), 1–17. https://doi.org/10.54489/ijcim.v1i1.48.
Hamadneh, S., Pedersen, O., & Al Kurdi, B. (2021). An investigation of the role of supply chain visibility into the scottish bood supply chain. *Journal of Legal, Ethical and Regulatory Issues, 24*(Special Issue 1), 1–12.

Hanaysha, J. R., Al-Shaikh, M. E., Joghee, S., & Alzoubi, H. (2021a). Impact of innovation capabilities on business sustainability in small and medium enterprises. *FIIB Business Review*, 1–12. https://doi.org/10.1177/23197145211042232.

Hanaysha, J. R., Al Shaikh, M. E., & Alzoubi, H. M. (2021b). Importance of marketing mix elements in determining consumer purchase decision in the retail market. *International Journal of Service Science, Management, Engineering, and Technology (IJSSMET), 12*(6), 56–72.

Hasan, S. M., Rehman, A., & Zhang, W. (2021). Who can work and study from home in Pakistan: Evidence from a 2018–19 nationwide household survey. *World Development, 138*, 105197. https://doi.org/10.1016/j.worlddev.2020.105197

Hodges, C., Moore, S., Lockee, B., Trust, T., & Bond, A. (2020). The difference between emergency remote teaching and online learning. *Educause*, 1–12.

Joghee, S., Alzoubi, H., & Dubey, A. (2020a). Decisions effectiveness of FDI investment biases at real estate industry: Empirical evidence from dubai smart city projects. *International Journal of Scientific & Technology Research, 9*(3), 1245–1258.

Joghee, S., Alzoubi, H. M., & Dubey, A. R. (2020b). Decisions effectiveness of FDI investment biases at real estate industry: Empirical evidence from Dubai smart city projects. *International Journal of Scientific and Technology Research, 9*(3), 3499–3503.

Johnson, D. W., & Johnson, R. T. (2002). Learning together and alone: Overview and meta-analysis. *Asia Pacific Journal of Education, 22*(1), 95–105. https://doi.org/10.1080/0218879020220110

Kara, M., Erdoğdu, F., Kokoç, M., & Cagiltay, K. (2019). Challenges faced by adult learners in online distance education: A literature review. *Open Praxis, 11*(1), 5. https://doi.org/10.5944/openpraxis.11.1.929

Kashif, A. A., Bakhtawar, B., Akhtar, A., Akhtar, S., Aziz, N., & Javeid, M. S. (2021). Treatment response prediction in hepatitis C patients using machine learning techniques. *International Journal of Technology, Innovation and Management (IJTIM), 1*(2), 79–89. https://doi.org/10.54489/ijtim.v1i2.24.

Kemp, A. T., Preston, J., Page, C. S., Harper, R., Dillard, B., Flynn, J., et al. (2014). Technology and teaching: A conversation among faculty regarding the pros and cons of technology. *Qualitative Report, 19*(3). https://doi.org/10.46743/2160-3715/2014.1284.

Khajavy, G. H., MacIntyre, P. D., & Barabadi, E. (2018). Role of the emotions and classroom environment in willingness to communicate. *Studies in Second Language Acquisition, 40*(3), 605–624. https://doi.org/10.1017/S0272263117000304

Khan, M. A. (2021). Challenges facing the application of IoT in medicine and healthcare. *International Journal of Computations, Information and Manufacturing (IJCIM), 1*(1), 39–55. https://doi.org/10.54489/ijcim.v1i1.32.

Koca, F. (2016). Motivation to learn and teacher-student relationship. *Journal of International Education and Leadership, 6*(2).

Kurdi, B. A., Alshurideh, M., & Alnaser, A. (2020a). The impact of employee satisfaction on customer satisfaction: Theoretical and empirical underpinning. *Management Science Letters, 10*(15). https://doi.org/10.5267/j.msl.2020.6.038.

Kurdi, B. A., Alshurideh, M., & Salloum, S. A. (2020b). Investigating a theoretical framework for e-learning technology acceptance. *International Journal of Electrical and Computer Engineering, 10*(6). https://doi.org/10.11591/IJECE.V10I6.PP6484-6496.

Lee, C., & Ahmed, G. (2021). Improving IoT privacy, data protection and security concerns. *International Journal of Technology, Innovation and Management (IJTIM), 1*(1), 18–33. https://doi.org/10.54489/ijtim.v1i1.12.

Lee, K., Azmi, N., Hanaysha, J., Alzoubi, H., & Alshurideh, M. (2022a). The effect of digital supply chain on organizational performance: An empirical study in Malaysia manufacturing industry. *Uncertain Supply Chain Management, 10*(2), 495–510.

Lee, K., Romzi, P., Hanaysha, J., Alzoubi, H., & Alshurideh, M. (2022b). Investigating the impact of benefits and challenges of IOT adoption on supply chain performance and organizational performance: An empirical study in Malaysia. *Uncertain Supply Chain Management, 10*(2), 537–550.

Leo, S., Alsharari, N. M., Abbas, J., & Alshurideh, M. T. (2021). From offline to online learning: A qualitative study of challenges and opportunities as a response to the COVID-19 pandemic in the UAE higher education context. In *Studies in systems, decision and control* (Vol. 334). https://doi.org/10.1007/978-3-030-67151-8_12.

Mehmood, T. (2021). Does information technology competencies and fleet management practices lead to effective service delivery? Empirical evidence from e-commerce industry. *International Journal of Technology, Innovation and Management (IJTIM), 1*(2), 14–41.

Mehmood, T., Alzoubi, H. M., & Ahmed, G. (2019). Schumpeterian entrepreneurship theory: evolution and relevance. *Academy of Entrepreneurship Journal, 25*(4).

Miller, D. (2021). The best practice of teach computer science students to use paper prototyping. *International Journal of Technology, Innovation and Management (IJTIM), 1*(2), 42–63. https://doi.org/10.54489/ijtim.v1i2.17.

Mondol, E. P. (2021). The impact of block chain and smart inventory system on supply chain performance at retail industry. *International Journal of Computations, Information and Manufacturing (IJCIM), 1*(1), 56–76. https://doi.org/10.54489/ijcim.v1i1.30.

Murphy, M. P. A. (2020). COVID-19 and emergency eLearning: Consequences of the securitization of higher education for post-pandemic pedagogy. *Contemporary Security Policy, 41*(3), 492–505. https://doi.org/10.1080/13523260.2020.1761749

Nazir, M. I. J., Rahaman, S., Chunawala, S., & AlHamad, A. Q. M. (2021). Perceived factors affecting students academic performance; Nazir, J., Rahaman, S., Chunawala, S., Ahmed, G., Alzoubi, H., Alshurideh, M., AlHamad, A. (2022) Perceived factors affecting students academic performance. *Academy of Strategic Management Journal, 21(Special Issue 4),* 1–15, *21*(Special Issue 4), 1–15.

Nuseir, M. T., Aljumah, A., & Alshurideh, M. T. (2021a). How the business intelligence in the new startup performance in UAE during COVID-19: The mediating role of innovativeness. In *Studies in Systems, Decision and Control* (Vol. 334). https://doi.org/10.1007/978-3-030-67151-8_4.

Nuseir, M. T., Al Kurdi, B. H., Alshurideh, M. T., & Alzoubi, H. M. (2021b). Gender discrimination at workplace: Do Artificial Intelligence (AI) and Machine Learning (ML) have opinions about it. In *The International Conference on Artificial Intelligence and Computer Vision* (pp. 301–316).

Obaid, A. J. (2021). Assessment of smart home assistants as an IoT. *International Journal of Computations, Information and Manufacturing (IJCIM), 1*(1), 18–36. https://doi.org/10.54489/ijcim.v1i1.34.

Qureshi, I. A., Ilyas, K., Yasmin, R., & Whitty, M. (2012). Challenges of implementing e-learning in a Pakistani university. *Knowledge Management and E-Learning, 4*(3), 310–324. https://doi.org/10.34105/j.kmel.2012.04.025.

Radwan, N., & Farouk, M. (2021). The growth of internet of things (IoT) in the management of healthcare issues and healthcare policy development. *International Journal of Technology, Innovation and Management (IJTIM), 1*(1), 69–84. https://doi.org/10.54489/ijtim.v1i1.8.

Rapanta, C., Botturi, L., Goodyear, P., Guàrdia, L., & Koole, M. (2020). Online university teaching during and after the Covid-19 crisis: Refocusing teacher presence and learning activity. *Postdigital Science and Education, 2*(3), 923–945. https://doi.org/10.1007/s42438-020-00155-y

Salloum, S. A., Alshurideh, M., Elnagar, A., & Shaalan, K. (2020). Mining in educational data: Review and future directions. *AICV*, 92–102.

Sandilos, L. E., Rimm-Kaufman, S. E., & Cohen, J. J. (2017). Warmth and demand: The relation between students' perceptions of the classroom environment and achievement growth. *Child Development, 88*(4), 1321–1337. https://doi.org/10.1111/cdev.12685.

Schut, S., Heeneman, S., Bierer, B., Driessen, E., van Tartwijk, J., & van der Vleuten, C. (2020). Between trust and control: Teachers' assessment conceptualisations within programmatic assessment. *Medical Education, 54*(6), 528–537. https://doi.org/10.1111/medu.14075

Scott-Webber, L., Abraham, J., & Marini, M. (2000). Higher education classroom fail to meet needs of faculty and students. *Journal of Interior Design, 26*(2), 16–34. https://doi.org/10.1111/j.1939-1668.2000.tb00356.x

Semler, J. (2010). Pros and cons of online research. *AgriMarketing, 48*(4), 55.

Serdyukov, P. (2020). Online education. *Challenges, 2012*, 1146–1151.
Sieberer-Nagler, K. (2015). Effective classroom-management & positive teaching. *English Language Teaching, 9*(1), 163. https://doi.org/10.5539/elt.v9n1p163
Stern, J. (2020). Introduction to online teaching and learning. *The TESOL Encyclopedia of English Language Teaching*. https://doi.org/10.1002/9781118784235.eeltv06b
Sultan, R. A., Alqallaf, A. K., Alzarooni, S. A., Alrahma, N. H., AlAli, M. A., & Alshurideh, M. T. (2021). How students influence faculty satisfaction with online courses and do the age of faculty matter. In *The International Conference on Artificial Intelligence and Computer Vision* (pp. 823–837).
Vandenbroucke, L., Spilt, J., Verschueren, K., Piccinin, C., & Baeyens, D. (2018). The classroom as a developmental context for cognitive development: A meta-analysis on the importance of teacher–student interactions for children's executive functions. *Review of Educational Research, 88*(1), 125–164. https://doi.org/10.3102/0034654317743200

The Role of Distance Learning Technology in Mitigating Unknown-Unknown Risks: Case of Covid-19

Mounir El khatib, Khalil Al Abdooli, Rashid Alhammadi, Fatma Alshamsi, Najma Abdulla, Amena Al Hammadi, Haitham M. Alzoubi, and Muhammad Alshurideh

Abstract With the changing circumstances across the world due to the pandemic COVID-19, the concept of distance learning and the use of technology has increased at a very fast pace. This has resulted different online classes and different associated risks related to security with these online classes for the organizations. However, with the passage of time, it has been realized that COVID-19 has now become a part of our lives, and what we considered a matter of days or weeks might last for a very long period, which is still unknown to us. Most importantly, this new phase of life has become a new normal for the people today. It has now become important for us to change the course of our lives and to hold different activities according to this new normal. Most of the business activities are also now being adapted through this social distancing and many other activities involving the education system to be on top of everything else. The methodology adopted for the study was qualitative study and the method adopted for the study was interviewed and generating the study through literature reviews.

Keywords Distance learning technology · Unknown risks · Risk mitigation · Project risk management · Covid-19

M. El khatib
Program Chair, Hamdan Bin Mohamad Smart University, Dubai, UAE

K. Al Abdooli · R. Alhammadi · F. Alshamsi · N. Abdulla · A. Al Hammadi
Hamdan Bin Mohamad Smart University, Dubai, UAE

H. M. Alzoubi (✉)
School of Business, Skyline University College, Sharjah, UAE
e-mail: haitham.alzubi@skylineuniversity.ac.ae

M. Alshurideh
Department of Marketing, School of Business, University of Jordan, Amman, Jordan
e-mail: m.alshurideh@ju.edu.jo; malshurideh@sharjah.ac.ae

Department of Management, College of Business Administration, University of Sharjah, Sharjah, UAE

M. Alshurideh et al. (eds.), *The Effect of Information Technology on Business and Marketing Intelligence Systems*, Studies in Computational Intelligence 1056,
https://doi.org/10.1007/978-3-031-12382-5_29

1 Introduction

Today with the development of the internet and especially with the emergence of COVID-19, people have now started taking online courses and have started the distance learning (Akour et al., 2021; Alshurideh et al., 2021a, 2021b). People are now more divulging in the distance learning (Amarneh et al., 2021; Leo et al., 2021). Distance learning defined as a type of learning which allows the delivery of education through the exchange of resources over the communication network (Jiang and Ryan, 2020). Nowadays, people are focused on using the web-based technologies to help the learners and the instructors to interact with each other more effectively (AlHamad et al., 2022; Alzoubi et al., 2022; Lee et al., 2022a, 2022b). Different web-based technologies and social networking sites have inspired the learners to continue with their education, even with this ongoing pandemic (Barween Al Alshurideh et al., 2019a, 2019b; Al Kurdi et al., 2020, 2020b). Most recently with the recent pandemic, several online courses are offered to students (Alshurideh et al., 2020a, 2020b, 2020c; Al Kurdi et al., 2020, 2020b). Therefore, the number of institutes has gained the attention of the students offering higher education. Distance learning have adopted online platforms to provide effective online learning (Alshurideh et al., 2020a, 2020b, 2020c). Though, the concept of security holds a great importance. The process of online learning involves some threats, hacking or attacks which need to be tackled (Al-Maroof et al., 2021a, 2021b, 2021c; Alshurideh et al., 2021a, 2021b). The institutes provide services with awareness of the unauthorized modification, destructions or harm to other educational assets available online (Al Kurdi et al., 2021; Al-Maroof et al., 2021a, 2021b, 2021c). It's essential to the concern of the online learning activities and to consider the inherent security risks on the internet (Bettayeb et al., 2020; Nuseir et al., 2021a, 2021b). Some of the important risks involve identity theft, impersonating or inadequate authentication (Al-Hamad et al., 2021; Alsharari and Alshurideh, 2021). Online learning systems are under the attention of the cyber criminals who often hack into these systems or harm the purpose of the learning process for different means (Alhashmi et al., 2020; Salloum et al., 2020). The businesses and the education sectors are mostly concerned and aware of the potential risks involved and often design dynamic and highly regulated means to fight against these risks. Nevertheless, with the advent of the technology, level of threats and the potential of the risks is increasing. A concept of knowing known and the unknown unknowns have also been introduced. The unknown unknown risks are an organization unawareness of certain kinds of risks existence. Unawareness of the risks may occur new challenging and management difficulties for an organization.

As the COVID-19 brought advanced ways to overcome the risks and continue with social distancing and the online forums or platforms that are being used with new and creative ideas in distance learning (Ahmad et al., 2021; Nuseir et al., 2021; Taryam et al., 2020). Thus, the use of the new modes of learning, increased the risk of unknown unknowns as this is an unexplored territory for everyone.

This report would identify the importance of such risks and their importance in the education system and would also identify whether the education sector is prepared to fight against these risks or mitigate these risks efficiently.

2 Literature Review

According to Liu, distance learning can be described as a learning process, where both the tutor and students are separated by geographical location and time (Liu & (Lu), 2008). In a nutshell, distance learning is a type of study program, where learning is delivered entirely off-campus. Before the advent of the Internet, distance learning is delivered through two approaches to distance learning, which can either be combined or used separately (Ali et al., 2022; Alshurideh et al., 2022; Lee et al., 2022a, 2022b). The first approach is the paper-reliant approach. In this approach, the student receives a package of learning materials via mail or postal service, completes everything necessary in the package, returns whatever needs to be submitted through mail or postal service, and then receives feedback. The second approach is the audio-visual approach, where television programmes and instructional videos are used. In this approach, distance learning students view a pre-recorded lecture, complete the necessary tasks, returns the task through mail or postal service, and then receive feedback (Nazir et al., 2021; Totaro et al., 2005). Most times, the audio-visual approach is enhanced with visual aids and documentary style materials. Nowadays, distance learning is being delivered via online course materials, using social network platforms, emails, and websites as support (Alzoubi, 2021a).

Lake noted that the advent of distance learning technology has enabled distance learning to be delivered in new ways without much constraints (Lake, 1999; Sultan et al., 2021). He said, "technological change has enabled the materials of education to be dispersed economically in new ways with fewer temporal and spatial constraints." With the distance learning technology, there is no longer need for students to attend regular lectures delivered in brick-and-mortar lecture rooms or theatres at a predetermined location and time. Students receive lectures in a different learning environment – online class (Alhamad et al., 2021; Ashal et al., 2021; Kelly, 2020).

Almost all the Universities and institutions of learning have embraced the relatively new distance learning technology not only for its potential for increasing the number of students without the need to accommodate them in a physical lecture hall/theatre but also for the safety and convenience it holds, especially in this period of Covid-19 pandemic (Akhtar et al., 2021; Kashif et al., 2021). Currently, distance learning has become an increasingly popular way of delivering higher education in higher education establishments (Baber, 2020).

According to Liu, the term "distance learning" is simply education delivered at a distance (Liu and (Lu), 2008). It is also referred to as "distance education," "on-line learning," and "e-learning." Some literatures have taken distance learning to incorporate online-based education programmes with face-to-face contact sessions (Al Ali,

2021; Muhammad Alshurideh et al., 2019a, 2019b). However, such incorporated programmes are mostly referred to as "blended learning" programmes and is beyond the scope of this paper (AlHamad and AlHammadi, 2018). This study focuses mainly on distance learning as a learning process where the learning courses and materials are remotely delivered to students from the university or hired e-learning facility, and lecturers and students do not need face-to face encounter (AlHamad and Al Qawasmi, 2014; Al-Mashaqbeh and Al Hamad, 2010).

2.1 *Distance Learning in Higher Education*

The COVID-19 pandemic caused several institutions of learning to shut down temporarily, though some have resumed physical classes, some are still temporarily closed. Hence, face-to-face learning has halt in several schools, colleges, and universities. This will subsequently have negative impacts on learning activities if nothing is done. As a result, most institutions of learning are trying alternatives ways of managing this difficult circumstance to prevent interruption to education (Dhawan, 2020; Otair and Hamad, 2005). This stimulated the growth of distance learning and many schools, colleges, and universities have been trying out how best to offer distance learning (Al-Maroof et al., 2021a, 2021b, 2021c; Mukhtar et al., 2020).

It is also expected that the Covid-19 pandemic would make the new distance learning technology adopted by institution that were previously resistant to accept the method. Before the Covid-19 pandemic, distance learning was underutilized, especially in developing countries. However, the on-going crisis of the COVID-19 pandemic has enforced almost all the nations of the entire world to rely on distance learning for education (Eli, 2021).

Distance learning, also referred to as e-learning or online e-learning, is described as a learning process using several different electronic devices with internet availability or data connectivity in synchronous or asynchronous environmental conditions. These devices include computers, laptops, smart devices, and so on. Distance learning could be a platform that will make the education process more student-centred, flexible, and creative (Singh and Thurman, 2019). It is easily accessible and cost-effective, especially when the receiving students are far away, in rural, or in remote areas (AlHamad et al., 2014; Dhawan, 2020). Distance learning technology brought tools to implement several creative strategies to combat unknown risks and deliver online courses. These tools include various software applications like Zoom, TeamViewer, Google Classroom, Microsoft Teams, and so on.

It is expected that the implementation of distance learning technology should not only transform education process from the traditional teacher-centric method to student-centric model but also help in mitigating unknown risks (Alsharari, 2021; Mehmood, 2021; Miller, 2021). If this true, then distance learning technology can be accepted as an education tool within institution of learning that if implemented would help in future endeavours during pandemic, and even in non-pandemic situations, to smoothen learning process.

2.2 Implementation of Distance Learning Technology in Mitigating Risk

Today the technology plays a very important part in introducing the world to new and more advanced ways of performing different types of activities especially in the new crisis of COVID-19 virus (Alameeri et al., 2021; Obaid, 2021). First, this disease was a newfound threat, which was unheard of before, and all of the people were not aware about the dangers of it. It was not expected that it will last for this long. Through this crisis nearly everyone from all around the world were affected. Therefore, the most successful organizations were the one who was having a better methodology to deal with the known and unknown risk (Wetsman, 2020).

Distance learning is one of the most important fields in the education which focuses on the teaching methods and technology in order to deliver information and teaching. This type of teaching is when the students are not physically present in a classroom (Bušelić, 2017). In other words, distance learning defined as instructional learning which force the time/geographic situation to have no contact between the student and instructor or teacher (AlHamad, 2020; King et al., 2001). Distance learning offers a huge number of advantages which might be evaluated according to technical, social and economic standards (Aziz and Aftab, 2021; Radwan and Farouk, 2021). Therefore, it supports the quality and diversity of the educational structures, to improve and support capacity and ability (Bušelić, 2017). Moreover, this learning method helps the organization to save money and time and both learner and educator can use it wherever they want (Bijeesh, 2017). On the other hand, there are lots of issues or disadvantage for this type of learning just like it is unlike the traditional learning because there isn't immediate feedback for the learner through it (Alzoubi, 2021b; Farouk, 2021). Moreover, oral communication skills don't have the opportunity to appear in the distance learning and the learner will not have the chance to use verbal interaction between each other (Bušelić, 2017).

Based on the information that Kim (2017) mentioned in his research that unidentified risk or event are the one which is not identified because there wasn't any access to the knowledge or information about it. Some of them are unexpected or not imagined because there not enough information about it. In other words, the unknown risk shown as a cause of some events than cannot be determined or to identify their factors. As an example of that the political events and some types of liquidity risks that hard to be measured properly (Anon, 2020). One of the main unknown risks that all the world is facing in these days is COVID-19 risk and to manage the known and unknown risks, the risk professionals have to change their priorities and work habits to have the best way to deal with this "new normal. From the beginning of March this year, all educational institutions from all around the world and UAE had closed, sent all the students' home, and moved for online learning. Therefore, all the educational institution plan to manage all the risk issues that COVID-19 cause of the learning through few basic steps such as, risk assessment triage by ordering which risk categories will need reassessment first, and which risk can wait (Cruz, 2021). Then to monitor and repeat which means to monitor the risks case, analyse the available data, and re-evaluate everything again, as required. COVID-19 keeps people apart, so the technology and the distance learning give the chance for the people now to work remotely by provides easy-to-use communication tools.

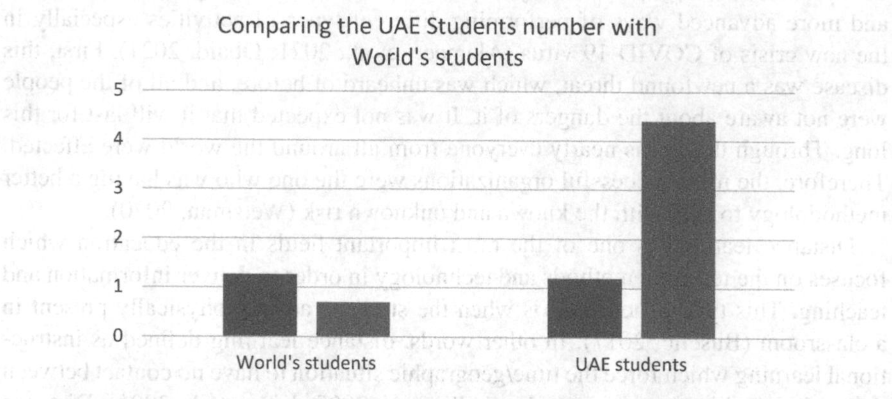

As an example of the importance that distance learning mitigating this crisis that some countries like French have reopened the schools after the lockdowns very fast and that cause to have flare-up of about 70 COVID-19 cases between students or teachers after one week only (Cole, 2020). This case explains about how importance to deal with this crisis by using learning distance to continue the academic year. According to the H.E. Al Hammadi the UAE had adopted the necessary actions to complete the academic year through using a smart learning system for almost 1.2 million students from the UAE's schools and universities. While there are 70% of the world's students around 1.26 billion who are facing a lot of difficulties of completing their academic year in 2020 because of Covid-19 (Masoud and Bohra, 2020).

The educator and educational institutions need to consider some principles during using the distance learning through this crisis COVID-19 just like to set new and different learning goals and process which fit the new situation, design the distance learning activities based on the online environment (Alameeri et al., 2021; Guergov and Radwan, 2021). Also, to be prepared very well to deal with all the challenges that will face the learner (Folsom, 2020). One more important thing is to keep these main three requirements in the online learning which are confident of both the learners and educators, integrity for both and to ensure the easy availability of both ends.

Confidentiality would ensure to include and protect the sensitive private information for both educators and learners of being accessed by any unauthorized person and to make sure that no such information is disclosed to any unauthorized person (Khan, 2021; Lee and Ahmed, 2021; Mondol, 2021). It is required to design a login system with a strong delimitation to safeguard the registered users and to protect them. The organizations develop different resources to ensure the confidentiality of the data and to protect and provide the data completely. Availability would ensure that access of the information should be easily granted to the authorized users whenever required.

3 Research Methods

Selecting the right research method is extremely vital for any research study. The research method selected for this study is qualitative method. This chapter is focused mainly on implementing the learning objectives which would help in evaluating the role of information technology and how it can help in influencing the distance learning and eventually mitigate the possible risks associated with the distance learning.

This method is more based upon the factual data and collection of information rather than collecting the numerical data. The data collection methods adopted in this study would be to gather information from different authentic literature available and also to collect the information through collecting interviews from five different organization as follows: a Marketing Communication Manager in the Anglican Church Grammar School (Churchie)—Education House Finland, a Head of Academic Affairs of Alqalaa Primary School, a Principal of Um Ammar Kindergarten School, an Engineer from Mohammed Bin Rashid Space Centre, and a Technical Support Technician from Abu Dhabi Technical and Vocational Education and Training Centre (ACTVET). Both these means of collecting information would help in generating a more in depth understanding and explores more elaborate data for the study which is very important for distance learners.

The study would be focused on the descriptive and the evaluative approach where the variables of the study would be measured through the qualitative research. The data collected is non-numerical in nature which is suitable according to the nature of the study. The study design followed in the research is not experimental in nature suggesting that the variables in the study cannot be altered, instead they have to be observed more efficiently through the framework designed through the usual framework for the study.

The data observed in the study would be more descriptive in nature and would be collected in a systematic manner. This section of the report would be more focused on selection of the right research articles or the literature for the study and also to effectively generate the interviews for the study. Moreover, these sources selected for collecting the data would be secondary sources.

The research has focused on investigating the meaning, attributes and interactions of the right methods from different research phenomenon. The qualitative approach adopted has helped in collecting the evidence supporting whether the technology plays an important role in the process of distance learning, especially in facilitating the users under the present circumstances of COVID-19 (Al Khasawneh et al., 2021a, 2021b). The research will highlight how these strategies can be modified in order to sustain the provision of education for a longer period of time in relevance to the changing circumstances across the world.

Interviews were conducted online with the staff members. An interview was formulated with 20 questions, where the staff members were asked about the use of technology, how they consider that the distance learning can be an effective way of learning for the students. Also would be focused on how the organization can use its different strengths in overcoming the weaknesses of the organization or schools.

The interviewees were also asked to provide their recommendations regarding the distance learning and the risks associated. In addition, the interview questions were focusing on different potential risks involved associated with the usage of technology in implementing the distance learning.

4 Results

Based upon the literature generated from the literature review and the interview, the interviewees mentioned that the most significant risks associated with the distance learning or other online activities especially when the technology is involved included the Brute Force attack, MIMT attach, Cross site request, IP Spoofing, Masquerade, Session Hijacking, Session Prediction etc. the interviewees suggested that the most widely used protection measures included installing the firewalls, implementing the security management systems, improving the authenticity and accountability of the online sessions more effective and also ensuring the digital management and training professionals to be hired for the users to help them getting accustomed with the online sessions.

The interviews' outcomes show that it is extremely important to develop the understanding of the challenges faced especially in terms of unknown unknowns. With the global pandemic COVID-19, the education sector is facing number of issues where the technologies have made it even more important to prepare the schools and organizations against the unknown risks. Technology has significantly advanced over the period of time, although has now become a common practice to find technologically advanced solutions and to the pay high cost of the information and communication infrastructure.

The findings indicate that it is extremely important to understand the root cause of the issues arising. Although most of the issues might seem to be large and unsolvable, but it is important to investigate their root cause and to determine the productive results for these unknown unknowns to find the solutions. It is the responsibility of the management in every educational organization to communicate the risks involved, especially when the risk becomes a reality. It is important to provide all the stakeholders, especially the students with the suitable solutions to move forward and determine the right solutions. It is also important to keep a regular check on the communication plan and to keep an update on the progress.

The results can be more effectively presented in the diagram below. In the diagram the larger circles would represent the basic concepts and they are joined with each other in the form of the clusters and the dots. It has also been identified that the basic concept behind this learning process is that this helped the students in being able to learn more effectively and get acquainted according to the concept of online courses and the environment more effectively. This also helped the students and the educators to make use of the technology in the best possible way and help them with supporting and developing the access to the online learning mediums more effective.

The interviews have identified that it is extremely important to get a grasp of the unknown unknowns at the root and develop the technology skills and competencies to deal with the hitches involved. It is important for all the members involved to be effectively involved in the process of improving their skills and experience within upstream and downstream in being able to mitigate the possible risks.

While the interviewees also highlighted that there were number of risks involved which included protecting the online class sessions and also to ensure the confidentiality, integrity and the accessibility to be maintained for all the users involved through the effective means of communication.

5 Discussion of Results

Distance learning is the new trend that most of the educational organizations are aiming to adopt it. But some of the organizations are still working on that where the needed technologies are not in their scope. As the technology plays a big role in the success process of the distance learning.

The major patterns we have observed are that the risk management tool which needs to be raised as it fell short in the identification of risk on time. As discussed through the interviews, the organization's strategy considered highly during the current pandemic where it reflects on the way organization dealing with the pandemic. So, there are certain mechanisms that organizations must consider the unknown risks in their strategy such as planning for risk management, and analysis the quantitative and qualitative data. These mechanisms must provide alternative solutions, and adopt the latest tools/technologies as a lesson learnt from the COVID-19 pandemic (Hanaysha et al., 2021a, 2021b; Harahsheh et al., 2021). By mentioning the latest tools/technologies we must say that the availability of the online platforms and IT support as the main two factors of the distance learning; affects the overall process. One more point to discuss is the risks associated with distance learning are not only restricted to the risk associated with the technical system, but also cater to the teaching methodologies, examination, and evaluation of students' learning processes.

An important point to discuss is the significant risks that associated with the distance learning is the security and safety of the education system through the online platform which needs to be considered as threats to be solved to avoid all obstacles (Alnazer et al., 2017; Haitham Alzoubi and Ahmed, 2019). By search for an effective protection software to secure the safety of the education system or education online platform, and to improve the authenticity and the effectiveness and efficiency of the distance learning process.

We do agree on that the technology distance learning can mitigate unknown risk; considering COVID-19 as unknown risk must of the schools/universities shut down and closed. But, with the availability of great educational systems, online platforms, IT support and infrastructure of the internet. All together help the organizations in ensuring the continuity of the educational process which mitigated the negative expected occur outcomes in the education sector. Taking UAE as an example of supporting the above agreement; Prior to the announcement of the Ministry of Education (MoE) to starting the teaching and learning from a distance most of the educational organizations were able to start the day after due to the strong internet and communication infrastructure and the availability of the sources and technologies. On March 22, Sheikh Mohammed bin Rashid, the Vice President and Ruler of Dubai attended the first tele-school day in the state's schools (Alzoubi et al., 2020a, 2020b; Mehmood et al., 2019). His Highness wrote on Twitter that 1.200 Million students are enrolled in our government schools, private sector schools and out national universities" (The National, 2020). Moreover, HBMSU is a great example as well where being the first smart university in the UAE shows that there are already leading this initiative prior to others where it was inaugurated in 2009 (Alzoubi and Yanamandra, 2020; Alzoubi et al., 2020a, 2020b; Joghee et al., 2020). Briefly, we can say that there is a positive relationship between mitigation unknown risk and the availability of tools/technologies. Where adopting new process and technologies play a big role in dealing with unknown risks (Hamadneh et al., 2021; Hanaysha et al., 2021a, 2021b).

One of the common hypothesis that most of the interviewee referred to is the government and the educational organization's needs to study the adopting of the distance learning as an alternative way of teaching and learning process (Alnuaimi et al., 2021; Alshurideh et al., 2020a, 2020b, 2020c; Alzoubi et al., 2021). We also agree on that; where the new situation we are facing now is considering as the new norm, based on that our way of thinking, about building our future and innovative strategy should be redesigned based on planning for unknown risk.

Technology is the rescue; over the past few months we have seen how distance learning technology is helping the governments in the absence of the virus within the schools' facilities. According to Jiang and Ryan "Digitalized economies have lower epidemic risks and explain some channels through which risks can be mitigated". As follow (Fig. 1) the World Bank has shown that countries with wider internet access and safer internet servers tend to be more resilient to epidemics such as COVID-19 (Ali et al., 2021; Alzoubi and Aziz, 2021; Ghazal et al., 2021). Therefore, learning process through the online platforms connects millions of students to their teachers from home during the pandemic to contain the spread of the COVID-19 that's practically mitigated the current risk.

6 Conclusion

The world events are often an inflection point for rapid innovation—a clear example is the distance learning. The COVID-19 pandemic effected all organizations in varying proportions depending on organization's risk management strategy, on the adopted technologies and available tools, and on it are internal and external recourses. By now the successful organization must have prepared innovative plans that build to avoid or mitigate the unknown risks. In addition to that, the application of the distance learning is hypothetically limitless and could be the new norm of the education system (Al

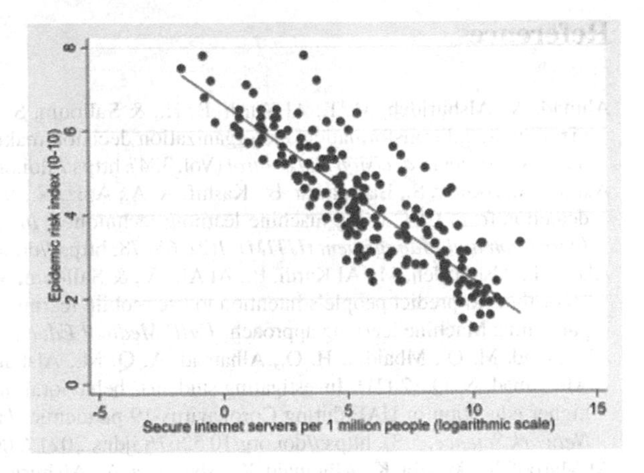

Fig. 1 World Bank Group's World Development Indicators

Kurdi et al., 2021; Alhamad et al., 2021). The technology nowadays exists to deliver an efficient and effective education system where it helps in overcoming barriers of distance, time, cost, and educational resources being continued refinement due to ensure continuity of education.

7 Recommendations

1. Government should consider keeping the distance learning and remote work as an alternative way of continuing business in the UAE.
2. Free online courses and e-library to fulfill the needs of all students.
3. COVID-19 pandemic has rearranged the education systems. The organizations must be reconsidered the infrastructure of the learning in the world, especially in the poor countries.
4. Implementation of an effective risk mitigation planning and enhanced risk analysis need to be followed to make early detection of the known and unknown risk factors.
5. Maintain the possible resources that the organization utilizes include critical pathway, workforce, memory and error budget.

References

Ahmad, A., Alshurideh, M. T., Al Kurdi, B. H., & Salloum, S. A. (2021). Factors impacts organization digital transformation and organization decision making during Covid19 pandemic. In *Studies in systems, decision and control* (Vol. 334). https://doi.org/10.1007/978-3-030-67151-8_6

Akhtar, A., Akhtar, S., Bakhtawar, B., Kashif, A. A., Aziz, N., & Javeid, M. S. (2021). COVID-19 detection from CBC using machine learning techniques. *International Journal of Technology, Innovation and Management (IJTIM), 1*(2), 65–78. https://doi.org/10.54489/ijtim.v1i2.22

Akour, I., Alshurideh, M., Al Kurdi, B., Al Ali, A., & Salloum, S. (2021). Using machine learning algorithms to predict people's intention to use mobile learning platforms during the COVID-19 pandemic: Machine learning approach. *JMIR Medical Education, 7*(1), 1–17.

Al-Hamad, M. Q., Mbaidin, H. O., Alhamad, A. Q. M., Alshurideh, M. T., Kurdi, B. H. A., & Al-Hamad, N. Q. (2021). Investigating students' behavioral intention to use mobile learning in higher education in UAE during Coronavirus-19 pandemic. *International Journal of Data and Network Science, 5*(3). https://doi.org/10.5267/j.ijdns.2021.6.001.

Al-Maroof, R., Ayoubi, K., Alhumaid, K., Aburayya, A., Alshurideh, M., Alfaisal, R., & Salloum, S. (2021a). The acceptance of social media video for knowledge acquisition, sharing and application: A comparative study among YouYube users and TikTok users' for medical purposes. *International Journal of Data and Network Science, 5*(3). https://doi.org/10.5267/j.ijdns.2021.6.013.

Al-Maroof, R. S., Alshurideh, M. T., Salloum, S. A., AlHamad, A. Q. M., & Gaber, T. (2021b). Acceptance of google meet during the spread of coronavirus by Arab university students. *Informatics, 8*(2). https://doi.org/10.3390/informatics8020024.

Al-Maroof, R. S., Alhumaid, K., Alhamad, A. Q., Aburayya, A., & Salloum, S. (2021c). User acceptance of smart watch for medical purposes: an empirical study. *Future Internet, 13*(5), 127.

Al-Mashaqbeh, I. F., & Al Hamad, A. (2010). Student's perception of an online exam within the decision support system course at Al al Bayt University. In *2010 Second International Conference on Computer Research and Development* (pp. 131–135).

Al Ali, A. (2021). The impact of information sharing and quality assurance on customer service at UAE banking sector. *International Journal of Technology, Innovation and Management (IJTIM), 1*(1), 01–17. https://doi.org/10.54489/ijtim.v1i1.10.

Al Khasawneh, M., Abuhashesh, M., Ahmad, A., Alshurideh, M. T., & Masa'deh, R. (2021a). Determinants of e-word of mouth on social media during COVID-19 outbreaks: An empirical study. In *Studies in systems, decision and control* (Vol. 334). https://doi.org/10.1007/978-3-030-67151-8_20.

Al Khasawneh, M., Abuhashesh, M., Ahmad, A., Masa'deh, R., & Alshurideh, M. T. (2021b). Customers online engagement with social media influencers' content related to COVID 19. In *Studies in systems, decision and control* (Vol. 334). https://doi.org/10.1007/978-3-030-67151-8_22.

Al Kurdi, B., Alshurideh, M., Nuseir, M., Aburayya, A., & Salloum, S. A. (2021). The effects of subjective norm on the intention to use social media networks: An exploratory study using PLS-SEM and machine learning approach. In *Advances in intelligent systems and computing* (Vol. 1339). https://doi.org/10.1007/978-3-030-69717-4_55.

Al Kurdi, B., Alshurideh, M., Salloum, S., Obeidat, Z., & Al-dweeri, R. (2020). *An empirical investigation into examination of factors influencing university students' behavior towards elearning acceptance using SEM approach.*

Alameeri, K. A., Alshurideh, M. T., & Al Kurdi, B. (2021). The effect of Covid-19 pandemic on business systems' innovation and entrepreneurship and how to cope with it: A theatrical view. In *Studies in systems, decision and control* (Vol. 334). https://doi.org/10.1007/978-3-030-67151-8_16.

AlHamad, A., Alshurideh, M., Alomari, K., Kurdi, B., Alzoubi, H., Hamouche, S., & Al-Hawary, S. (2022). The effect of electronic human resources management on organizational health of telecommuni-cations companies in Jordan. *International Journal of Data and Network Science, 6*(2), 429–438.

AlHamad, A. Q., & Al Qawasmi, K. I. (2014). Building an ethical framework for e-learning management system at a university level. *Journal of Engineering and Economic Development*, *1*(1), 11.

AlHamad, A. Q. M. (2020). Predicting the intention to use Mobile learning: A hybrid SEM-machine learning approach. *International Journal of Engineering Research & Technology*, *9*(3), 275–282.

Alhamad, A. Q. M., Akour, I., Alshurideh, M., Al-Hamad, A. Q., Kurdi, B. A., & Alzoubi, H. (2021). Predicting the intention to use google glass: A comparative approach using machine learning models and PLS-SEM. *International Journal of Data and Network Science*, *5*(3), 311–320. https://doi.org/10.5267/j.ijdns.2021.6.002

AlHamad, A. Q. M., & AlHammadi, R. A. (2018). Students' perception of E-library system at Fujairah University. In *International Conference on Remote Engineering and Virtual Instrumentation* (pp. 659–670).

Alhashmi, S. F. S., Alshurideh, M., Al Kurdi, B., & Salloum, S. A. (2020). A systematic review of the factors affecting the artificial intelligence implementation in the health care sector. In *Advances in intelligent systems and computing* (Vol. *1153*) AISC. https://doi.org/10.1007/978-3-030-44289-7_4.

Ali, N., Ahmed, A., Anum, L., Ghazal, T. M., Abbas, S., Khan, M. A., et al. (2021). Modelling supply chain information collaboration empowered with machine learning technique. *Intelligent Automation and Soft Computing*, *30*(1), 243–257. https://doi.org/10.32604/iasc.2021.018983.

Ali, N. M., Ghazal, T., Ahmed, A., Abbas, S., Khan, M. A., Alzoubi, H. et al. (2022). Fusion-based supply chain collaboration using machine learning techniques. *Intelligent Automation & Soft Computing*, *31*(3), 1671–1687. https://doi.org/10.32604/iasc.2022.019892.

Alnazer, N. N., Alnuaimi, M. A., & Alzoubi, H. M. (2017). Analysing the appropriate cognitive styles and its effect on strategic innovation in Jordanian universities. *International Journal of Business Excellence*, *13*(1), 127–140. https://doi.org/10.1504/IJBEX.2017.085799

Alnuaimi, M., Alzoubi, H. M., Ajelat, D., & Alzoubi, A. A. (2021). Towards intelligent organisations: An empirical investigation of learning orientation's role in technical innovation. *International Journal of Innovation and Learning*, *29*(2), 207–221. https://doi.org/10.1504/IJIL.2021.112996

Alsharari, N. (2021). Integrating blockchain technology with internet of things to efficiency. *International Journal of Technology, Innovation and Management (IJTIM)*, *1*(2), 1–13.

Alsharari, N. M., & Alshurideh, M. T. (2021). Student retention in higher education: the role of creativity, emotional intelligence and learner autonomy. *International Journal of Educational Management*, *35*(1). https://doi.org/10.1108/IJEM-12-2019-0421.

Alshurideh, M., et al. (2019a). Understanding the quality determinants that influence the intention to use the mobile learning platforms: A practical study. *International Journal of Interactive Mobile Technologies (IJIM)*, *13*(11), 157–183.

Alshurideh, M., Al Kurdi, B., & Salloum, S. A. (2020a). Examining the main mobile learning system drivers' effects: A mix empirical examination of both the expectation-confirmation model (ECM) and the technology acceptance model (TAM). In *Advances in intelligent systems and computing* (Vol. 1058). https://doi.org/10.1007/978-3-030-31129-2_37.

Alshurideh, M., Al Kurdi, B., Salloum, S. A., Arpaci, I., & Al-Emran, M. (2020b). Predicting the actual use of m-learning systems: A comparative approach using PLS-SEM and machine learning algorithms. *Interactive Learning Environments*. https://doi.org/10.1080/10494820.2020.1826982

Alshurideh, M., Gasaymeh, A., Ahmed, G., Alzoubi, H., & Kurd, B. A. (2020c). Loyalty program effectiveness: Theoretical reviews and practical proofs. *Uncertain Supply Chain Management*, *8*(3). https://doi.org/10.5267/j.uscm.2020.2.003.

Alshurideh, M. T., Al Kurdi, B., AlHamad, A. Q., Salloum, S. A., Alkurdi, S., Dehghan, A., et al. (2021a). Factors affecting the use of smart mobile examination platforms by universities' postgraduate students during the COVID-19 pandemic: An empirical study. *Informatics*, *8*(2). https://doi.org/10.3390/informatics8020032.

Alshurideh, M. T., Al Kurdi, B., Masa'deh, R., & Salloum, S. A. (2021b). The moderation effect of gender on accepting electronic payment technology: A study on United Arab Emirates consumers.

Review of International Business and Strategy, 31(3). https://doi.org/10.1108/RIBS-08-2020-0102.

Alshurideh, M., Salloum, S. A., Al Kurdi, B., & Al-Emran, M. (2019b). Factors affecting the social networks acceptance: an empirical study using PLS-SEM approach. In *Proceedings of the 2019b 8th International Conference on Software and Computer Applications* (pp. 414–418).

Alshurideh, M. T., Al Kurdi, B., Alzoubi, H. M., Ghazal, T. M., Said, R. A., AlHamad, A. Q., et al. (2022). Fuzzy assisted human resource management for supply chain management issues. *Annals of Operations Research*, 1–19.

Alzoubi, A. (2021a). The impact of process quality and quality control on organizational competitiveness at 5-star hotels in Dubai. *International Journal of Technology, Innovation and Management (IJTIM), 1*(1), 54–68. https://doi.org/10.54489/ijtim.v1i1.14.

Alzoubi, A. (2021b). Renewable Green hydrogen energy impact on sustainability performance. *International Journal of Computations, Information and Manufacturing (IJCIM), 1*(1), 94–110. https://doi.org/10.54489/ijcim.v1i1.46.

Alzoubi, H. M., Ahmed, G., Al-Gasaymeh, A., & Al Kurdi, B. (2020a). Empirical study on sustainable supply chain strategies and its impact on competitive priorities: The mediating role of supply chain collaboration. *Management Science Letters, 10*(3), 703–708. https://doi.org/10.5267/j.msl.2019.9.008.

Alzoubi, H. M., & Aziz, R. (2021). Does emotional intelligence contribute to quality of strategic decisions? The mediating role of open innovation. *Journal of Open Innovation: Technology, Market, and Complexity, 7*(2), 130. https://doi.org/10.3390/joitmc7020130

Alzoubi, H. M., Vij, M., Vij, A., & Hanaysha, J. R. (2021). What leads guests to satisfaction and loyalty in UAE five-star hotels? AHP analysis to service quality dimensions. *Enlightening Tourism, 11*(1), 102–135. https://doi.org/10.33776/et.v11i1.5056.

Alzoubi, H. M., & Yanamandra, R. (2020). Investigating the mediating role of information sharing strategy on agile supply chain. *Uncertain Supply Chain Management, 8*(2), 273–284. https://doi.org/10.5267/j.uscm.2019.12.004

Alzoubi, H., Alshurideh, M., Kurdi, B., Akour, I., & Aziz, R. (2022). Does BLE technology contribute towards improving marketing strategies, customers' satisfaction and loyalty? The role of open innovation. *International Journal of Data and Network Science, 6*(2), 449–460.

Alzoubi, H., & Ahmed, G. (2019). Do TQM practices improve organisational success? A case study of electronics industry in the UAE. *International Journal of Economics and Business Research, 17*(4), 459–472. https://doi.org/10.1504/IJEBR.2019.099975

Alzoubi, H., Alshurideh, M., Kurdi, B. A., & Inairat, M. (2020b). Do perceived service value, quality, price fairness and service recovery shape customer satisfaction and delight? A practical study in the service telecommunication context. *Uncertain Supply Chain Management, 8*(3), 579–588. https://doi.org/10.5267/j.uscm.2020.2.005

Amarneh, B. M., Alshurideh, M. T., Al Kurdi, B. H., & Obeidat, Z. (2021). The impact of COVID-19 on e-learning: Advantages and challenges. *The International Conference on Artificial Intelligence and Computer Vision*, 75–89.

Anon. (2020). *Types of risks: The known and the unknowns.*

Ashal, N., Alshurideh, M., Obeidat, B., & Masa'deh, R. (2021). The impact of strategic orientation on organizational performance: Examining the mediating role of learning culture in Jordanian telecommunication companies. *Academy of Strategic Management Journal, Special Is*(Special Issue 6), 1–29.

Aziz, N., & Aftab, S. (2021). Data mining framework for nutrition ranking: Methodology: SPSS modeller. *International Journal of Technology, Innovation and Management (IJTIM), 1*(1), 85–95.

Baber, H. (2020). Determinants of students' perceived learning outcome and satisfaction in online learning during the pandemic of COVID19. *Journal of Education and E-Learning Research, 7*(3), 285–292. https://doi.org/10.20448/JOURNAL.509.2020.73.285.292.

Bettayeb, H., Alshurideh, M. T., & Al Kurdi, B. (2020). The effectiveness of mobile learning in UAE universities: A systematic review of motivation, self-efficacy, usability and usefulness. *Int. J. Control Autom, 13*(2), 1558–1579.

Bijeesh, N. A. (2017). Advantages and disadvantages of distance learning. Retrieved from https://www.Indiaeducation.Net/Online-Education/Articles/Advantages-and-Disadvantages-Ofdistance-Learning.html (Дата Обращения: 10.06. 2020).

Bušelić, M. (2017). Distance Learning – concepts and contributions. *Oeconomica Jadertina, 2*(1), 23–34. https://doi.org/10.15291/OEC.209.

Cole, D. (2020). *70 cases of COVID-19 at French schools days after re-opening*. Associated Press.

Cruz, A. (2021). Convergence between blockchain and the internet of things. *International Journal of Technology, Innovation and Management (IJTIM), 1*(1), 35–56.

Dhawan, S. (2020). Online learning: A panacea in the time of COVID-19 crisis. *Journal of Educational Technology Systems, 29*(1), 5–22. https://doi.org/10.1177/0047239520934018

Eli, T. (2021). Students perspectives on the use of innovative and interactive teaching methods at the University of Nouakchott Al Aasriya, Mauritania: English department as a case study. *International Journal of Technology, Innovation and Management (IJTIM), 1*(2), 90–104.

Farouk, M. (2021). The universal artificial intelligence efforts to face coronavirus COVID-19. *International Journal of Computations, Information and Manufacturing (IJCIM), 1*(1), 77–93. https://doi.org/10.54489/ijcim.v1i1.47.

Folsom, J. (2020). *Distance learning during the COVID-19 pandemic.*

Ghazal, T. M., Hasan, M. K., Alshurideh, M. T., Alzoubi, H. M., Ahmad, M., Akbar, S. S., Al Kurdi, B., & Akour, I. A. (2021). IoT for smart cities: Machine learning approaches in smart healthcare—A review. *Future Internet, 13*(8), 218. https://doi.org/10.3390/fi13080218

Guergov, S., & Radwan, N. (2021). Blockchain convergence: Analysis of issues affecting IoT, AI and blockchain. *International Journal of Computations, Information and Manufacturing (IJCIM), 1*(1), 1–17. https://doi.org/10.54489/ijcim.v1i1.48.

Hamadneh, S., Pedersen, O., & Al Kurdi, B. (2021). An Investigation of the role of supply chain visibility into the scottish bood supply chain. *Journal of Legal, Ethical and Regulatory Issues, 24*(Special Issue 1), 1–12.

Hanaysha, J. R., Al-Shaikh, M. E., Joghee, S., & Alzoubi, H. (2021a). Impact of innovation capabilities on business sustainability in small and medium enterprises. *FIIB Business Review*, 1–12. https://doi.org/10.1177/23197145211042232.

Hanaysha, J. R., Al Shaikh, M. E., & Alzoubi, H. M. (2021b). Importance of marketing mix elements in determining consumer purchase decision in the retail market. *International Journal of Service Science, Management, Engineering, and Technology (IJSSMET), 12*(6), 56–72.

Harahsheh, A. A., Houssien, A. M. A., Alshurideh, M. T., & Mohammad, A. M. (2021). The effect of transformational leadership on achieving effective decisions in the presence of psychological capital as an intermediate variable in private Jordanian universities in light of the corona pandemic. In *Studies in Systems, Decision and Control* (Vol. 334). https://doi.org/10.1007/978-3-030-67151-8_13.

Jiang, N., & Ryan, J. (2020). *How does digital technology help in the fight against COVID-19?*

Joghee, S., Alzoubi, H. M., & Dubey, A. R. (2020). Decisions effectiveness of FDI investment biases at real estate industry: Empirical evidence from Dubai smart city projects. *International Journal of Scientific and Technology Research, 9*(3), 3499–3503.

Kashif, A. A., Bakhtawar, B., Akhtar, A., Akhtar, S., Aziz, N., & Javeid, M. S. (2021). Treatment response prediction in hepatitis C patients using machine learning techniques. *International Journal of Technology, Innovation and Management (IJTIM), 1*(2), 79–89. https://doi.org/10.54489/ijtim.v1i2.24

Kelly, M. (2020). *Lessons on COVID-19 risk management in higher education - Galvanize.*

Khan, M. A. (2021). Challenges facing the application of IoT in medicine and healthcare. *International Journal of Computations, Information and Manufacturing (IJCIM), 1*(1), 39–55. https://doi.org/10.54489/ijcim.v1i1.32.

Kim, S. D. (2017). Characterization of unknown unknowns using separation principles in case study on Deepwater Horizon oil spill. *Journal of Risk Research, 20*(1), 151–168. https://doi.org/10.1080/13669877.2014.983949

King, F. B. (The U. of C., Young, M. F., Drivere-Richmond, K., & Schrader, P. G.) (2001). Defining distance learning and distance education. *Educational Technology Review, 9*(1), 1–14.
Kurdi, B. A., Alshurideh, M., & Salloum, S. A. (2020). Investigating a theoretical framework for e-learning technology acceptance. *International Journal of Electrical and Computer Engineering, 10*(6). https://doi.org/10.11591/IJECE.V10I6.PP6484-6496.
Lake, D. (1999). Reducing isolation for distance students: An on-line initiative. *Open Learning, 14*(3), 14–23. https://doi.org/10.1080/0268051990140304
Lee, C., & Ahmed, G. (2021). Improving IoT privacy, data protection and security concerns. *International Journal of Technology, Innovation and Management (IJTIM), 1*(1), 18–33. https://doi.org/10.54489/ijtim.v1i1.12.
Lee, K., Azmi, N., Hanaysha, J., Alzoubi, H., & Alshurideh, M. (2022a). The effect of digital supply chain on organizational performance: An empirical study in Malaysia manufacturing industry. *Uncertain Supply Chain Management, 10*(2), 495–510.
Lee, K., Romzi, P., Hanaysha, J., Alzoubi, H., & Alshurideh, M. (2022b). Investigating the impact of benefits and challenges of IOT adoption on supply chain performance and organizational performance: An empirical study in Malaysia. *Uncertain Supply Chain Management, 10*(2), 537–550.
Leo, S., Alsharari, N. M., Abbas, J., & Alshurideh, M. T. (2021). From offline to online learning: A qualitative study of challenges and opportunities as a response to the COVID-19 pandemic in the UAE higher education context. *The Effect of Coronavirus Disease (COVID-19) on Business Intelligence, 334*, 203–217.
Liu, & (Lu), S. (2008). Student interaction experiences in distance learning courses: A phenomenological study. *Online Journal of Distance Learning Administration, 11*(1).
Masoud, N., & Bohra, O. P. (2020). Challenges and opportunities of distance learning during Covid-19 in UAE. *Academy of Accounting and Financial Studies Journal, 24*(1), 1–12.
Mehmood, T. (2021). Does information technology competencies and fleet management practices lead to effective service delivery? Empirical evidence from e-commerce industry. *International Journal of Technology, Innovation and Management (IJTIM), 1*(2), 14–41.
Mehmood, T., Alzoubi, H. M., Alshurideh, M., Al-Gasaymeh, & A., & Ahmed, G. (2019). Schumpeterian entrepreneurship theory: Evolution and relevance. *Academy of Entrepreneurship Journal, 25*(4), 1–10.
Miller, D. (2021). The best practice of teach computer science students to use paper prototyping. *International Journal of Technology, Innovation and Management (IJTIM), 1*(2), 42–63. https://doi.org/10.54489/ijtim.v1i2.17.
Mondol, E. P. (2021). The impact of block chain and smart inventory system on supply chain performance at retail industry. *International Journal of Computations, Information and Manufacturing (IJCIM), 1*(1), 56–76. https://doi.org/10.54489/ijcim.v1i1.30.
Mukhtar, K., Javed, K., Arooj, M., & Sethi, A. (2020). Advantages, limitations and recommendations for online learning during covid-19 pandemic era. *Pakistan Journal of Medical Sciences, 36*(COVID19-S4), S27–S31. https://doi.org/10.12669/pjms.36.COVID19-S4.2785.
Nazir, J., Rahaman, S., Chunawala, S., Ahmed, G., Alzoubi, H., Alshurideh, M., et al. (2022). Perceived factors affecting students academic performance. *Academy of Strategic Management Journal, 21*(Special Issue 4), 1–15.
Nuseir, M. T., Aljumah, A., & Alshurideh, M. T. (2021a). How the business intelligence in the new startup performance in UAE during COVID-19: The mediating role of innovativeness. In *Studies in systems, decision and control* (Vol. 334). https://doi.org/10.1007/978-3-030-67151-8_4.
Nuseir, M. T., Al Kurdi, B. H., Alshurideh, M. T., & Alzoubi, H. M. (2021b). Gender discrimination at workplace: Do Artificial Intelligence (AI) and Machine Learning (ML) have opinions about it. In *The International Conference on Artificial Intelligence and Computer Vision* (pp. 301–316).
Obaid, A. J. (2021). Assessment of smart home assistants as an IoT. *International Journal of Computations, Information and Manufacturing (IJCIM), 1*(1), 18–36. https://doi.org/10.54489/ijcim.v1i1.34.

Otair, M. A., & Hamad, A. Q. A. (2005). Expert personalized e-learning recommender system. *Retrieved June, 13*, 2010.

Qasim AlHamad, A., Salameh, M., & Al Makhareez, L. (2014). Identifying students' trends toward personalizing Learning Management System (LMS) at Zarqa University (Extended). *Journal of Information Technology & Economic Development*, *5*(1).

Radwan, N., & Farouk, M. (2021). The growth of Internet of Things (IoT) In the management of healthcare issues and healthcare policy development. *International Journal of Technology, Innovation and Management (IJTIM)*, *1*(1), 69–84. https://doi.org/10.54489/ijtim.v1i1.8.

Salloum, S. A., Alshurideh, M., Elnagar, A., & Shaalan, K. (2020). Machine learning and deep learning techniques for cybersecurity: A review. In *Advances in intelligent systems and computing* (Vol. 1153). AISC. https://doi.org/10.1007/978-3-030-44289-7_5.

Singh, V., & Thurman, A. (2019). How many ways can we define online learning? A systematic literature review of definitions of online learning (1988–2018). *American Journal of Distance Education, 33*(4), 289–306. https://doi.org/10.1080/08923647.2019.1663082

Sultan, R. A., Alqallaf, A. K., Alzarooni, S. A., Alrahma, N. H., AlAli, M. A., & Alshurideh, M. T. (2021). How students influence faculty satisfaction with online courses and do the age of faculty matter. In *The International Conference on Artificial Intelligence and Computer Vision* (pp. 823–837).

Taryam, M., Alawadhi, D., Aburayya, A., Albaqa'een, A., Alfarsi, A., Makki, I., et al. (2020). Effectiveness of not quarantining passengers after having a negative COVID-19 PCR test at arrival to dubai airports. *Systematic Reviews in Pharmacy, 11*(11), https://doi.org/10.31838/srp.2020.11.197.

The National. (2020). *The UAE has an opportunity to make distance learning a success.*

Totaro, M. W., Tanner, J. R., Noser, T., Fitzgerald, J. F., & Birch, R. (2005). Faculty perceptions of distance education courses: A survey. *Journal of College Teaching & Learning (TLC)*, *2*(7), 13–20. https://doi.org/10.19030/tlc.v2i7.1841.

Wetsman, N. (2020). *Here's who's most at risk from the novel coronavirus - The Verge.* Retrieved from https://www.theverge.com/2020/3/11/21173157/coronavirus-health-effects-age-covid-risk-diabetes-hypertension-disease-isolation.

Pharmacy Education and Conducting OSCE Exam During COVID-19: An Overview

Hamza Alhamad, Nazek Qasim Mohammad Al-hamad, Ahmad Qasim Mohammad AlHamad, and Muhammad Turki Alshurideh

Abstract Advances in pharmacy practice have transformed the role of a pharmacist from traditional dispensing to more patient-centered care practices. Apart from the traditional dispensing practices, pharmacists are currently assigned to provide pharmaceutical care services focusing on identifying medication-related problems, taking a medication history, patient interviewing, and designing an evidence-based care plan. The advancement of the pharmacist's role to be part of the health care team to maximise the health care provided to the patient should be met with restructuring the pharmacy education and assessment from being focused only on medicine compounding, selling, and dispensing to include pharmaceutical care provision. Undergraduate pharmacy education has changed tremendously over the years, evidenced by the shift from a customer-based approach to a patient-centered approach using different clinical learning and examinations models such as Objective Structured Clinical Examination (OSCE). OSCE exam is considered a powerful and valuable tool for evaluating pharmacy students' clinical performance. During the COVID-19 pandemic, pharmacy education and examination are globally affected. Many pharmacy schools are confronting surpassing challenges to sustain education

H. Alhamad (✉)
Department of Clinical Pharmacy, Faculty of Pharmacy, Zarqa University, Zarqa, Jordan
e-mail: halhamad@zu.edu.jo

N. Q. M. Al-hamad
Pharmacy College, Jadara University, Irbid, Jordan
e-mail: n.alhamad@jadara.edu.jo

A. Q. M. AlHamad
Department of Information Systems, College of Computing and Informatics, University of Sharjah, Sharjah, United Arab Emirates
e-mail: aalhamad@sharjah.ac.ae

M. T. Alshurideh
Department of Marketing, School of Business, The University of Jordan, Amman 11942, Jordan
e-mail: m.alshurideh@ju.edu.jo; malshurideh@sharjah.ac.ae

Department of Management, College of Business, University of Sharjah, 27272 Sharjah, United Arab Emirates

M. Alshurideh et al. (eds.), *The Effect of Information Technology on Business and Marketing Intelligence Systems*, Studies in Computational Intelligence 1056,
https://doi.org/10.1007/978-3-031-12382-5_30

during the COVID-19 pandemic. The global experiences of academics and healthcare instructors in delivering emergency remote teaching, ensuring purposeful experiential pharmacy student placements, communicating and supporting the displaced or isolated pharmacy students are considered an accelerating opportunity for new models of pharmacy education. Therefore, this chapter aims to have an overview of pharmacy education during the COVID-19 pandemic, the challenges to online pharmacy education, and the impact of the COVID-19 pandemic on pharmacy learning and education. Also, to have an overview about conducting OSCE clinical examination during COVID-19; the online OSCE exam preparation, set up, implementation, and assessment during COVID-19. Finally, to describe what can happen in the future to the online pharmacy student clinical education, OSCE assessment, and examination.

Keywords Pharmacy education · COVID-19 · OSCE exam

1 Overview of Pharmacy Education During COVID-19

The COVID-19 pandemic has impacted the educational sector globally (AlHamad et al., 2012; Alhamad et al., 2021; Lyons et al., 2020). As a result, many pharmacy schools were suspended their campus activities and quickly transitioned their teaching and assessment strategies to an emergency remote online courses delivery (i.e., no face-to-face interaction) (AlHamad et al., 2014; Ali et al., 2021; Hall et al., 2020; Kawaguchi-Suzuki et al., 2020; Lyons et al., 2020; Mohamed et al., 2020). In addition, the COVID-19 pandemic has driven pharmacy schools and higher institutions to adjust their teaching and learning strategies to comply with international and national safety measures, including reduced movements of staff and students due to lockdowns, movement control orders, and global border closures (Al Hamad, 2016; Hall et al., 2020; Mohamed et al., 2020; Tolsgaard et al., 2020). Hence, institutions employed online-based teaching and learning using various information technology platforms (Al Hamad, 2016; AlHamad et al., 2012, 2014).

Both academic staff and pharmacy students faced challenges to ensure smooth delivery of synchronous and asynchronous online teaching, including remote delivery, experiential placements, and academic integrity with assessments (Hall et al., 2020; Kawaguchi-Suzuki et al., 2020; Rajab et al., 2020; Tolsgaard et al., 2020). Also, academic staff should be agile and flexible while delivering online teaching considering the significance of showing empathy and patience to students and each other (AlHamad & Al Qawasmi, 2014; Brazeau & Romanelli, 2020; Kawaguchi-Suzuki et al., 2020; Mohamed et al., 2020; Tolsgaard et al., 2020). Fortunately, and in some aspects, many of these challenges can be faced. First, a curriculum transformation focusing on active learning and skill development should be achieved to ensure purposive experiential training and support remote students (Brazeau & Romanelli, 2020; Hall et al., 2020; Lyons et al., 2020). Second, the use of a mix of face-to-face, blended learning, and online based on course and program competencies (Fuller et al., 2020; Kawaguchi-Suzuki et al., 2020). Finally, there is a need for a comprehensive understanding of how the COVID-19 pandemic accelerates the need of having new teaching models for pharmacy students considering the need to conduct clinical examinations such as Objective Structured Clinical examinations (OSCE) (Hall et al., 2020; Lyons et al., 2020). This requires significant efforts from academics, healthcare instructors, and pharmacy students to become comfortable with various technological platforms and adjust to the educational challenges of the COVID-19 pandemic.

The OSCEs require a face-to-face assessment of clinical skills, which presents a significant challenge to schools and academic staff in the health profession, including pharmacy, especially during pandemic situations such as COVID-19 (Awaisu et al., 2007; Sepp & Volmer, 2021; Zayyan, 2011). As a result, medical and pharmacy schools switched to online-based clinical training and examination; for example, many schools used online-based OSCE (e.g., web-based, virtual, tele-OSCE). However, with a significant variation in the online platforms used, the patient's and academic staff skills, and ability to use the online technology, there is a need to evaluate these alternate methods of online OSCE in contrast with traditional face-to-face OSCE.

2 Challenges to Online Pharmacy Education

Many pharmacy schools and universities have transitioned from traditional face-to-face teaching methods to online or a combination of online and traditional education (i.e., blended learning) during the COVID-19 pandemic (Orleans, 2014; Rajab et al., 2020). Blended learning includes replacing part of the face-to-face interaction with online instruction (Edginton & Holbrook, 2010). This demonstrates how online education continues to grow in which face-to-face teaching is transformed into an online environment (Rajab et al., 2020). However, the subsequent enforcement of social distancing (i.e., increasing the physical space between people) due to the COVID-19 pandemic has forced pharmacy schools and universities to have an empty classroom as students kept away from their institutions (Rajab et al., 2020). Consequently, these institutions have switched automatically into distance learning using the most convenient conferencing platforms, email, and phone (Lederman, 2020).

The COVID-19 pandemic negatively impacted online learning in different ways. First, the transition from the traditional face-face into online education was challenging to many pharmacy schools and institutions as they were not ready (Esani, 2010; Rajab et al., 2020). Also, students become financially vulnerable because of the negative COVID-19 impact on the economy; therefore, they were worried about their ability to afford pharmacy school fees (Rajab et al., 2020). Finally, academic staff and students have other duties due to being at home, such as managing their children who are not in the school and other family members (Rajab et al., 2020).

Other challenges to online education were reported in the literature. For example, time management issues, the use of information technology tools, academic staff-student communication, and the lack of face-face (i.e., in-person) interaction and assessment tools (Esani, 2010; Rajab et al., 2020). The quality and access of online learning (i.e., poor or unavailability of the internet) is another issue (Rajab et al., 2020). Not all academic staff and students have access to laptops or high-speed internet at home (Rosen & Weil, 1995). Also, technophobia and the user age (academic staff and students) have negatively impacted the online learning process (Rosen & Weil, 1995). Other challenges such as not having a comfortable and quiet (i.e., noise-free) room and the additional tasks that come with online learning were reported by academic staff and students in the literature. Students reported that there is no campus atmosphere to create social interaction and may need to go back and listen to the recorded video (difficult and time-consuming for the students). Academic staff had reported difficulties following up and controlling students outside the country and were not satisfied with online exams and assessment as memory testing is not the best measure of learning in any environment (Ferri et al., 2020; Hall et al., 2020; Rajab et al., 2020). These challenges significantly affect the adoption of technology-enabled online learning (Chiasson et al., 2015).

3 The Impact of COVID-19 Pandemic on Pharmacy Learning and Education

The COVID-19 pandemic has created global challenges in pharmacy learning and education. Despite this, technology-enabled online learning and education adoption would be considered an advantage (Lederman, 2020). The COVID-19 situation facilitates the distribution and global acceptance of technology-enabled online learning (Hixon et al., 2012; Orleans, 2014). Also, the COVID-19 situation forced pharmacy schools and institutions to prepare and support their academic staff and students, which means implementing a new culture for online learning or blended learning (Rajab et al., 2020). Furthermore, online education was reported to be more convenient to students and academic staff. It requires not to travel to the pharmacy school campus, which saves them the hustle time with no or less travel and housing cost; students can work and still attend the lecture online (Ahmed et al., 2021; Almaghaslah et al., 2018).

Pandemics such as COVID-19 are a global health and economic crisis not just to learning and education (Alhamad et al., 2021). However, identifying learning and educational challenges would be primary in transforming them into innovative opportunities and brighter realities by preparing pharmacy students to impact their communities, country, and world (Mohamed et al., 2020). One of the significant opportunities is engaging pharmacy students and academic staff in transforming the current pandemic-imposed remote pharmacy learning into an evidence-based learning model (Al-Alami et al., 2021). Students and academic staff should learn from the COVID-19 pandemic and use their experiences and lessons to continue improving online pharmacy learning and education, advance the vital role of pharmacists in patient care and conduct the scholarly research needed for future advances in inpatient care (Strawbridge et al., 2021). The future of online learning education is uncertain. However, academic staff, students, pharmacy schools, and institutions should be ready and have an alternate plan in the future for such similar situations that may require switching partially (i.e., blended learning) or ultimately (i.e., online education) (Strawbridge et al., 2021).

Adapting to the pandemic situations such as COVID-19 should impose elaborative planning, creativity, flexibility, and collaboration by all key stakeholders (i.e., Countries and governments, pharmacy schools and institutions, academic staff, and students). Also, future research should be conducted to evaluate and provide an insight into how and to what extent pharmacy students' knowledge, skills, and attributes are affected as a result of the educational transformation during and post-pandemic situations such as COVID-19.

4 Conducting OSCE During COVID-19

The objective structured Clinical Examination (OSCE) can be defined as a clinical performance examination used in health education and medical disciplines such as physicians, pharmacy, nursing, and dentist (Harden et al., 1975, 2015; Khan et al., 2013). OSCE exam, developed in the mid-1970s, has become a standard gold examination to assess a wide range of students' clinical skills and competencies such as history taking, communicational and interpersonal skills, professionalism, critical thinking, clinical reasoning, and clinical examination (Blamoun et al., 2021; Harden et al., 1975; Majumder et al., 2019). Therefore, OSCE is a robust, extraordinary, functional, and significant tool of summative and formative clinical assessment of the competencies and highly effective psychometric assessment that require the learner not just to "know-how" but to "show-how" and demonstrate their competencies in practice (Alnasser, 2016; Blamoun et al., 2021; Miller, 1990). However, the non-verbal behaviour of the student, patient, examiner, and the environment in which the OSCE exam is implemented would affect the examiner's decision and the reliability of the OSCE exam (Alnasser, 2016; Blamoun et al., 2021). OSCE exam comprises a series of stations through which all candidates rotate on a timed basis (Harden & Gleeson, 1979; Newble, 2004). In each station, the candidate faces a simulated task or problem; these candidates are asked to perform a specific task or deal with the addressed problem (Almuqdadi et al., 2017; Majumder et al., 2019). As a performance-based assessment method, OSCE measures cognitive learning, essential practice skills, and the ability to communicate effectively using problem-solving skills (Blythe et al., 2021; Mohamed et al., 2020).

In the view of the COVID-19 pandemic, which led to the online transformation of teaching activities and assessment, academic staff (i.e., educators) and learners (students) of pharmacy schools and institutions globally faced several challenges in conducting realistic and face-face OSCE (Blythe et al., 2021; Craig et al., 2020). Online OSCE, sometimes referred to as remote, virtual, or tele-OSCE, is a new model of implementing OSCE (Blamoun et al., 2021; Blythe et al., 2021). Transforming into online OSCE necessitates changes to the standard face-face OSCE (Blamoun et al., 2021; Blythe et al., 2021). For example, educational objectives and assessment need to be reformed to be consistent with the online style (i.e., online properties need to be introduced) of conducting OSCE. Also, standardised training for student examiners and patients is required to translate in-person to an online OSCE environment (Blamoun et al., 2021; Blythe et al., 2021).

Online OSCE would be feasible and practical. No extra efforts, fewer costs, and logistics are required for exam preparation, setup, and implementation; for example, it is paperless and safer when it comes to pandemics (Shaban et al., 2021). However, online OSCE requires a higher number of examiners and patients, training and awareness of students, examiners, and patients, making online OSCE very impractical and time-consuming for many students (Blythe et al., 2021; Shaban et al., 2021). Also, online OSCE is affected by technology such as the need for Information Technology (IT) supporting staff presence, multiple computers, and other electronic devices, and requires generating multiple channels in the platform used and is highly affected by internet connection and device (e.g., microphone & camera) functionality (Blythe et al., 2021; Craig et al., 2020). Furthermore, in online OSCE, it is difficult to include a competency-based assessment of clinical examination (Blamoun et al., 2021; Shaban et al., 2021). Also, the lack of student in-person presence would affect the exam performance the ability of examiners to assess the student compared to the standardised face-face OSCE exam (Shaban et al., 2021).

The content of OSCE questions developed to assess the student clinical skills should be developed based on the nature of online OSCE compared to the standardised OSCE, which assesses a broader range of clinical examination skills; thus, a more concise OSCE blueprint is mandatory (Shaban et al., 2021).

To overcome the challenges of conducting online OSCE, some pharmacy schools and institutions used realistic simulation techniques in a three-dimensional (3D) virtual environment with an interactive simulator system of clinical cases or web-based immersive technology to assess clinical skills (Faria et al., 2021).

5 What Can Happen to Online Pharmacy Student Clinical Education, OSCE Assessment, and Examination in the Future?

The online assessment (paperless) may be considered an advantage over other traditional assessments where the student can have instant feedback on their education assessment from their academic instructors (Walsh, 2015). Also, the academic instructors can easily monitor pharmacy students learning progress and achievements (Craig et al., 2020). Despite this, the online OSCE assessment has costs of being dependent on technology, time-consuming and labour intensive, and the inability to assess Students' physical examinations and procedural skills would be a limitation (Walsh, 2015). Also, the logistics of the distribution and completion of examiner checklists would be another limitation (Craig et al., 2020).

The future of online clinical pharmacy education, assessment, and examination (e.g., OSCE) is vague and highly depends on the integrational success of virtual simulation-based technologies and implementing virtual clinical experience into the pharmacy training courses to achieve the clinical competencies and outcomes (Egarter et al., 2020; Kim, 2006; Walsh, 2015). Therefore, pharmacy schools and institutions should be prepared by having a proactive plan for a pandemic situation such as COVID-19 in the future. Currently, the standard face-to-face OSCE would still be the common practice, with online OSCE being an alternative in a pandemic or lockdown situation.

References

Ahmed, N. J., Alkhawaja, F. Z., Alrawili, A. S., & Alonazi, W. (2021). Pharmacy students' perceptions towards online learning. *Journal of Pharmaceutical Research International,* 88–93.

Al-Alami, Z. M., Adwan, S. W., & Alsous, M. (2021). Remote learning during Covid-19 lockdown: A study on anatomy and histology education for pharmacy students in Jordan. *Anatomical Science Education.*

Alhamad, H., Abu-Farha, R., Albahar, F., & Jaber, D. (2021). Public perceptions about pharmacists' role in prescribing, providing education and delivering medications during COVID-19 pandemic era. *International Journal of Clinical Practice, 75*(4), e13890.

AlHamad, A. Q., Yaacob, N., & Al-Omari, F. (2012). Applying JESS rules to personalize learning management system (LMS) using online quizzes. In *2012 15th International Conference on Interactive Collaborative Learning (ICL)* (pp. 1–4). IEEE.

AlHamad, A. Q., & Al Qawasmi, K. I. (2014). Building an ethical framework for e-learning management system at a university level. *Journal of English Economic Development, 1*(1), 11.

AlHamad, A. Q., Al Omari, F., & AlHamad, A. Q. (2014). Recommendation for managing patients' privacy in an integrated health information network. *Journal of Information Technology Economic Development, 5*(1).

Al Hamad, A. Q. (2016). Students' perception of implementing a smart learning system (SLS) based on Moodle at Fujairah college. In *2016 13th International Conference on Remote Engineering and Virtual Instrumentation (REV)* (pp. 315–318). IEEE.

Ali, M., Allihyani, M., Abdulaziz, A., Alansari, S., Faqeh, S., Kurdi, A., et al. (2021). What just happened? Impact of on-campus activities suspension on pharmacy education during COVID-19 lockdown–A students' perspective. *Saudi Pharmaceutical Journal, 29*(1), 59–66.

Almuqdadi, A., Yousef, A.-M., Majdalawi, K., Masoud, Z., Kalabani, R., & Al-Hadeed, H. (2017). Validation and evaluation of an OSCE in undergraduate doctor of pharmacy program. *Indian Journal of Pharmaceutical Education Research, 51*(3), 380–387.

Almaghaslah, D., Ghazwani, M., Alsayari, A., & Khaled, A. (2018). Pharmacy students' perceptions towards online learning in a Saudi pharmacy school. *Saudi Pharmaceutical Journal, 26*(5), 617–21.

Alnasser, S. S. A. (2016). *It does not need to be voiced to be counted. Non-verbal behaviour influences assessors' global marking when examining medical students using objective structured clinical examinations.* University of Leeds.

Awaisu, A., Mohamed, M. H. N., & Al-Efan, Q. A. M. (2007). Perception of pharmacy students in Malaysia on the use of objective structured clinical examinations to evaluate competence. *American Journal of Pharmaceutical Education, 71*(6).

Blythe, J., Patel, N. S. A., Spiring, W., Easton, G., Evans, D., Meskevicius-Sadler E, et al. (2021). Undertaking a high stakes virtual OSCE ("VOSCE") during Covid-19. *BMC Medical Education, 21*(1), 1–7.

Blamoun, J., Hakemi, A., & Armstead, T. (2021). A guide for medical students and residents preparing for formative, summative, and virtual objective structured clinical examination (OSCE): Twenty tips and pointers. *Advance Medical Education Practice, 12*, 973.

Brazeau, G., & Romanelli, F. (2020). Navigating the unchartered waters in the time of COVID-19. *American Journal of Pharmaceutical Education, 84*(3), 290–1.

Chiasson, K., Terras, K., & Smart, K. (2015). Faculty perceptions of moving a face-to-face course to online instruction. *Journal of College Teaching & Learning, 12*(3), 231–40.

Craig, C., Kasana, N., & Modi, A. (2020). *Virtual OSCE delivery-the way of the future?* Medical Education.

Egarter, S., Mutschler, A., Tekian, A., Norcini, J., & Brass, K. (2020). Medical assessment in the age of digitalisation. *BMC Medical Education, 20*(1), 1–8.

Edginton, A., & Holbrook, J. (2010). A blended learning approach to teaching basic pharmacokinetics and the significance of face-to-face interaction. *American Journal of Pharmaceutical Education, 74*(5).

Esani, M. (2010). Moving from face-to-face to online teaching. *American Social Clinical Laboratory Science, 23*(3), 187–90.

Faria, A. L., Perdigão, A. C. B., Marçal, E., Kubrusly, M., Peixoto, R. A. C., & Peixoto, A. A. (2021) OSCE 3D: a virtual clinical skills assessment tool for coronavirus pandemic times. *Revista Brasileira Education Medicine, 45*.

Ferri, F., Grifoni, P., & Guzzo, T. (2020). Online learning and emergency remote teaching: Opportunities and challenges in emergency situations. *Societies, 10*(4), 86.

Fuller, K. A., Heldenbrand, S. D., Smith, M. D., & Malcom, D. R. (2020). A paradigm shift in US experiential pharmacy education accelerated by the COVID-19 pandemic. *American Journal of Pharmaceutical Education, 84*(6).

Hall, A. K., Nousiainen, M. T., Campisi, P., Dagnone, J. D., Frank, J. R., Kroeker, K. I., et al. (2020). Training disrupted: Practical tips for supporting competency-based medical education during the COVID-19 pandemic. *Medical Teacher, 42*(7), 756–61.

Harden, R. M., & Gleeson, F. A. (1979). Assessment of clinical competence using an objective structured clinical examination (OSCE). *Medical Education, 13*(1), 39–54.

Harden, R. M., Lilley, P., & Patricio, M. (2015). *The definitive guide to the OSCE: The objective structured clinical examination as a performance assessment*. Elsevier Health Sciences.

Harden, R. M., Stevenson, M., Downie, W. W., & Wilson, G. M. (1975). Assessment of clinical competence using objective structured examination. *British Medical Journal, 1*(5955), 447–51.

Hixon, E., Buckenmeyer, J., Barczyk, C., Feldman, L., & Zamojski, H. (2012). Beyond the early adopters of online instruction: Motivating the reluctant majority. *Internet Higher Education, 15*(2), 102–7.

Kawaguchi-Suzuki, M., Nagai, N., Akonoghrere, R. O., & Desborough, J. A. (2020). COVID-19 pandemic challenges and lessons learned by pharmacy educators around the globe. *American Journal of Pharmaceutical Education, 84*(8).

Khan, K. Z., Gaunt, K., Ramachandran, S., & Piyush, P. (2013). The objective structured clinical examination (OSCE): AMEE Guide No. 81. Part II: An historical and theoretical perspective. *Medical Teacher, 35*, 1447–1463.

Kim, S. (2006). The future of e-learning in medical education: current trend and future opportunity. *Journal of Educational Evaluation for Health Professions, 3*.

Lederman, D. (2020). *Will shift to remote teaching be boon or bane for online learning.* Inside Higher Education.

Lyons, K. M., Christopoulos, A., & Brock, T. P. (2020). Sustainable pharmacy education in the time of COVID-19. *American Journal of Pharmaceutical Education, 84*(6).

Majumder, M. A. A., Kumar, A., Krishnamurthy, K., Ojeh, N., Adams, O. P., & Sa, B. (2019). An evaluative study of objective structured clinical examination (OSCE): Students and examiners perspectives. *Advance Medical Education Practice, 10*, 387.

Miller, G. E. (1990). The assessment of clinical skills/competence/performance. *Academic Medicine, 65*(9), S63-7.

Mohamed, M. H. N., Mak, V., Sumalatha, G., Nugroho, A. E., Hertiani, T., Zulkefeli, M., et al. (2020). Pharmacy education during and beyond COVID-19 in six Asia-Pacific countries: Changes, challenges, and experiences. *Pharmacy Education*, 183–195.

Newble, D. (2004). Techniques for measuring clinical competence: Objective structured clinical examinations. *Medical Education, 38*(2), 199–203.

Orleans, M. (2014). *Cases on critical and qualitative perspectives in online higher education*. IGI Global.

Rajab, M. H., Gazal, A. M., Alkawi, M., Kuhail, K., Jabri, F., & Alshehri, F. A. (2020). Eligibility of medical students to serve as principal investigator: an evidence-based approach. *Cureus, 12*(2).

Rajab, M. H., Gazal, A. M., & Alkattan, K. (2020). Challenges to online medical education during the COVID-19 pandemic. *Cureus, 12*(7).

Rosen, L. D., & Weil, M. M. (1995). Computer availability, computer experience and technophobia among public school teachers. *Computer Human Behaviour, 11*(1), 9–31.

Sepp, K., & Volmer, D. (2021). Use of face-to-face assessment methods in E-learning—An example of an objective structured clinical examination (OSCE) test. *Pharmacy., 9*(3), 144.

Shaban, S., Tariq, I., Elzubeir, M., Alsuwaidi, A. R., Basheer, A., & Magzoub, M. (2021). Conducting online OSCEs aided by a novel time management web-based system. *BMC Medical Education* [Internet], *21*(1), 508. https://doi.org/10.1186/s12909-021-02945-9.

Strawbridge, J., Hayden, J. C., Robson, T., Flood, M., Cullinan, S., Lynch M, et al. (2021). Educating pharmacy students through a pandemic: Reflecting on our COVID-19 experience. *Research Social Administrative Pharmacy*.

Tolsgaard, M. G., Cleland, J., Wilkinson, T., & Ellaway, R. H. (2020). How we make choices and sacrifices in medical education during the COVID-19 pandemic. *Medical Teacher, 42*(7), 741–3.

Walsh, K. (2015). Point of view: Online assessment in medical education–current trends and future directions. *Malawi Medical Journal, 27*(2), 71–2.

Zayyan, M. (2011). Objective structured clinical examination: the assessment of choice. *Oman Medical Journal, 26*(4), 219.

Business and Data Analytics

Using Logistic Regression Approach to Predicating Breast Cancer DATASET

Feras A. Haziemeh, Saddam Rateb Darawsheh, Muhammad Alshurideh, and Anwar Saud Al-Shaar

Abstract The aim of this study is to predicate breast cancer by using three approaches: Multilayer perceptron, multiple linear regression, and logistic regression, then make a comparison between the performance of these approaches to decide the best approach to analyses the Wisconsin breast cancer dataset. In this study, we used the Wisconsin breast cancer dataset (699 instances and 11 attributes). Three approaches (multilayer perceptron, multiple linear regression, and logistic regression) were applied to the dataset to predicate the Wisconsin breast cancer. Performance was calculated for each approach and a comparison was made between the three approaches to see which is better. From the performance comparison between multi-layer perceptron, multiple linear regression, and logistic regression, the Wisconsin breast cancer dataset is best to analyses using multi-layer perceptron, with 100% compared to logistic regression 97.6% and multiple linear regression 84.4%. The computerized and especially logistic regression model could be useful to the predicate for many fields of science. For this reason, it's used abundantly in the medical field specially to build a model to predicate breast cancer. The conclusion

F. A. Haziemeh
Department of Information Technology, Al-Balqa Applied University, Al-Salt, Jordan
e-mail: 1haziemh.feras@bau.edu.jo

Irbid - Hashemite Kingdom of Jordan, Irbid, Jordan

S. R. Darawsheh (✉)
Department of Administrative Sciences, The Applied College, Imam Abdulrahman Bin Faisal University, P.O. Box 1982, Dammam, Saudi Arabia
e-mail: srdarawsehe@iau.edu.sa

M. Alshurideh
Department of Management, College of Business Administration, University of Sharjah, Sharjah, United Arab Emirates
e-mail: malshurideh@sharjah.ac.ae; m.alshurideh@ju.edu.jo

Marketing Department, School of Business, The University of Jordan, Amman, Jordan

A. S. Al-Shaar
Deanship of Preparatory Year and Supporting Studies, Department of Self Development, Imam Abdulrahman Bin Faisal University, P. O. Box 1982, Dammam 43212, Saudi Arabia

M. Alshurideh et al. (eds.), *The Effect of Information Technology on Business and Marketing Intelligence Systems*, Studies in Computational Intelligence 1056,
https://doi.org/10.1007/978-3-031-12382-5_31

from this experiment showed that not all the independent variables have the coefficient effect with the dependent variable. However, the logistic regression approach is the best approach to analyses the Wisconsin breast cancer dataset rather than the multi-layer perceptron and multiple linear regression approach.

Keywords Logistic regression · Breast cancer · Multinomial logistic regression · Multi-layer perceptron · Multiple linear regression

1 Introduction

We address in this paper issues such as the global concept and interpretation of logistic regression and the process to apply those concepts is divided into a few sections with some illustrative, the first section containing the overview on the logistic regression which is used to mining the data in this study. The second section is the information about the datasets and methods used to analyze the data using logistic regression in SPSS tools. In the final section, the study explains the result and conclusion of the mining activity done to the datasets of Wisconsin breast cancer which been investigated. Logistic regression is part of a category of statistical models called generalized linear models. This broad class of models includes ordinary regression and ANOVA, as well as multivariate statistics such as ANCOVA and log-linear regression. An excellent treatment of generalized linear models is presented in Agresti (2018). Logistic regression allows one to predict a discrete outcome, such as group membership, from a set of variables that may be continuous, discrete, dichotomous, or a mix of any of these. Generally, the dependent or response variable is dichotomous, such as presence/absence or success/failure. A discriminate analysis is also used to predict group membership with only two groups. However, discriminate analysis can only be used with continuous independent variables. Thus, in instances where the independent variables are categorical, or a mix of continuous and categorical, logistic regression is preferred. The goal of logistic regression is to correctly predict the category of outcome for individual cases using the most parsimonious model. To accomplish this goal, a model is created that includes all predictor variables that are useful in predicting the response variable. Several different options are available during model creation. Variables can be entered into the model in the order specified by the researcher or logistic regression can test the fit of the model after each coefficient is added or deleted, called stepwise regression. There are two main uses of logistic regression. The first is the prediction of a group membership. Since logistic regression calculates the probability of success over the probability of failure, the results of the analysis are in the form of an odds ratio. For example, logistic regression is often used in epidemiological studies where the result of the analysis is the probability of developing cancer after controlling for other associated risks. Logistic regression also provides knowledge of the relationships and strengths among the variables (e.g., smoking 10 packs a day puts you at a higher risk for developing cancer than working in an asbestos mine).

2 Literature Review

The logistic regression approach is useful, researchers use it to predicate many diseases or to develop strategies for controlling these diseases, and it might be considered the cornerstone to improve these diseases. Several ancient studies have been used the logistic regression approach in many fields such as (Carpio et al., 2007; Choi et al., 2008; Giasson et al., 2006; Kemp & Alliss, 2007; Kyle et al., 2002; Larson, 2002; Olson & Olson, 1986; Rosa et al., 1981, 1992; Wang et al., 2004; Wolberg & Mangasarian, 1990; Zhang, 1992) studied the regression analysis and developed models to achieve predictive models. Their studies explained predictive models usage to predicate and improvement the systems output. In recent studies many researchers they studied breast cancer and they tried to build many of models to predicate and deal with the breast cancer. (Chaurasia & Pal, 2014) presented a diagnosis system for detecting breast cancer based on RepTree, RBF Network and Simple Logistic. To evaluate the proposed system performances, they applied tenfold cross validation method in test stage to the University Medical Centre, Institute of Oncology, Ljubljana, Yugoslavia database. The proposed system has a right classification rate of 74.5%. The study showed that Simple Logistics can be used to reduce the size of the feature space and the Rep Tree and RBF Network model proposed can be used to obtain fast automatic diagnostic systems for other diseases. According to the same researchers (Chaurasia & Pal, 2007). They said that Sequential Minimal Optimization (SMO) has higher prediction accuracy i.e. 96.2% than IBK and BF Tree methods.

The three methods were used for developing accurate prediction models for breast cancer by using data mining techniques. They used breast cancer data with a total 683 rows and 10 columns. Rajesh and Anand (2012) applied C4.5 classification algorithm to SEER breast cancer dataset to classify patients into three group (either "Carcinoma in situ" (beginning or pre-cancer stage) or "Malignant potential"). Rajesh and Anand (2012) Pre-processing techniques were applied to prepare the raw dataset and to identify relevant classification attributes. In order to get classification rules, random test samples were selected from the pre-processed data.

The rule set that was obtained was checked on the remaining data. In 2011, (Gupta et al., 2011) hey presented study paper summarized various review and technical articles on breast cancer diagnosis and prognosis. They presented an overview of the current research being carried out using the data mining techniques to enhance the breast cancer diagnosis and prognosis.

This study has the following objectives:

1. To study distribution of Wisconsin breast cancer data.
2. To examine the relationship independent variables (attributes) and the dependent variables (target or class).
3. To build logistic regression model for Wisconsin breast cancer data set.

3 Method

This section describes the used dataset in this paper and the acquisition of the dataset, which has been described in detail in the following subsection.

3.1 Data Description

The Wisconsin breast data has been acquired through UCI machine learning repositry. Several researchers have used the same data in their study. Based on the information provided at the above URL, there details description about Wisconsin breast cancer data (Table 1).

1. Number of Instances: 699.
2. Number of Attributes: 10 plus the class attribute.
3. Attribute Information: (class attribute has been moved to last column).

There are 26 missing values in Groups 1 to 6 that contain a single missing (i.e., unavailable) attribute value, now denoted by "?".

Class distribution of the Wisconsin breast cancer data is: Benign: 458 (65.5%). Malignant: 241 (34.5%). The distribution of data can also be represented in a bar char as shown in Fig. 1.

1. Logistic regression tools

The logistic regression model for Wisconsin breast cancer can be written as follows:

Table 1 Data description

#	Attribute	Code	Domain
1	Sample code number	CodeNum	id number
2	Clump thickness	CThick	1–10
3	Uniformity of cell size	CellSize	1–10
4	Uniformity of cell shape	CellShape	1–10
5	Marginal adhesion	MasAd	1–10
6	Single epithelial cell size	EpiCells	1–10
7	Bare nuclei	BareNuc	1–10
8	Bland chromatin	BlChr	1–10
9	Normal nucleoli	NormNuc	1–10
10	Mitoses	Mito	1–10
11	Class	Cl	2 for benign 4 for malignant

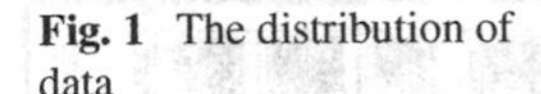

Fig. 1 The distribution of data

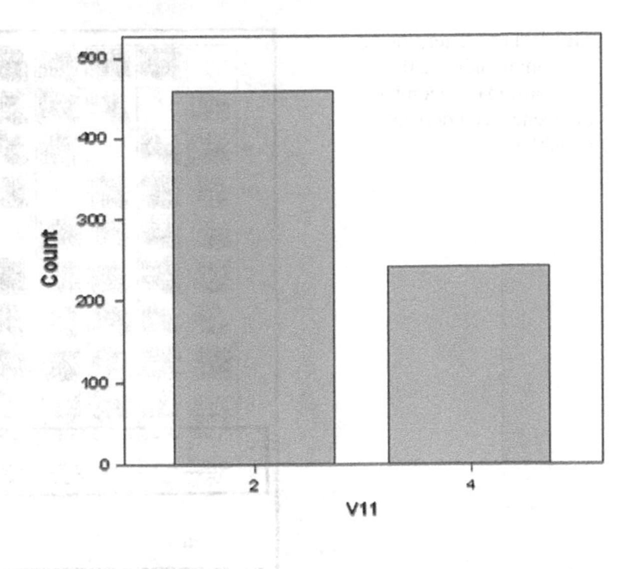

Cons. + B1 * CThick + B2 * CellSize + B3 * CellShape + B4 * MarAd + B5 * EpiCells + B6 * BareNuc + B7 * BlChr + B8 * NormNuc + B9 * Mito.

P = 1/1 + e.

Several steps should be performed to obtain the model:

1. Determine the relationship between the independent and dependent variables is liner. Draw a scatterplot diagram.
2. If the relationship between the independent and dependent variables is not liner, take a natural log of particular independent variables. In this study a Multinomial regression is used since the dependent variable has more than 2 classes of target (Wisconsin breast cancer).
3. Draw the scatterplot diagram again.
4. Get case processing summary and classification table and describe the distribution of data i.e. indicating how many of datasets represent each target class, and mention if there is any missing value.
5. Provide the model fitting information and form hypothesis for Chi-Square.
6. Get Wald-statistics table.
7. If the particular logit is not significant, get the correlation matrix.

First draw the scatterplot diagram to indicate the relationship between the independent and dependent variables, see Fig. 2.

The relationship between the dependent variable (V11) and the independent variables (V2 to V10) is not linear. Therefore, the attributes values should be transformed using logarithmic function (Naperian log or loge). After applying Naperian log to the dataset.

Once the logarithmic function has been applied all values (Except V1 and V11), the scatterplot should be obtained to determine the relationship of the variables for building the model. As shown in Fig. 3, the relationship between the dependent

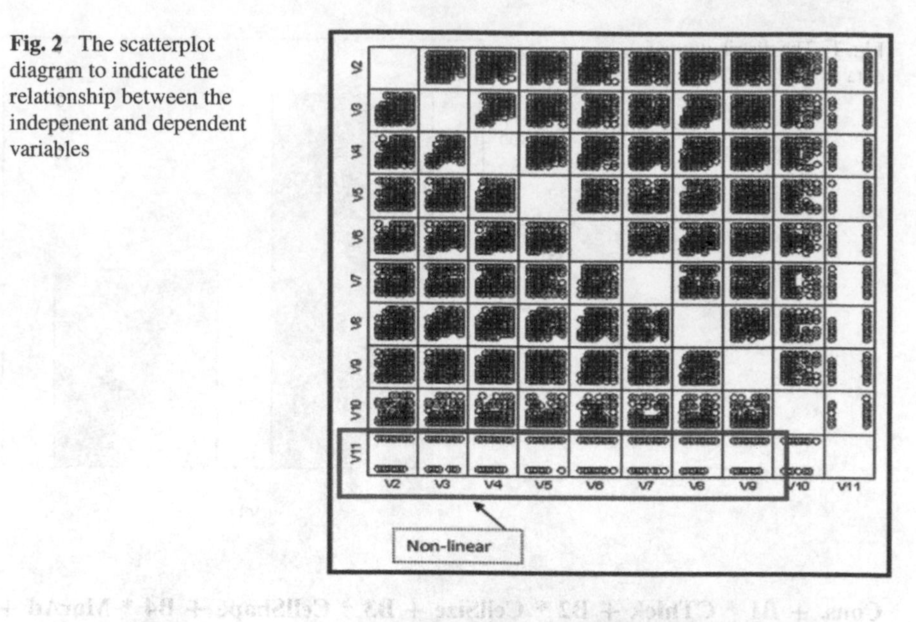

Fig. 2 The scatterplot diagram to indicate the relationship between the indepenent and dependent variables

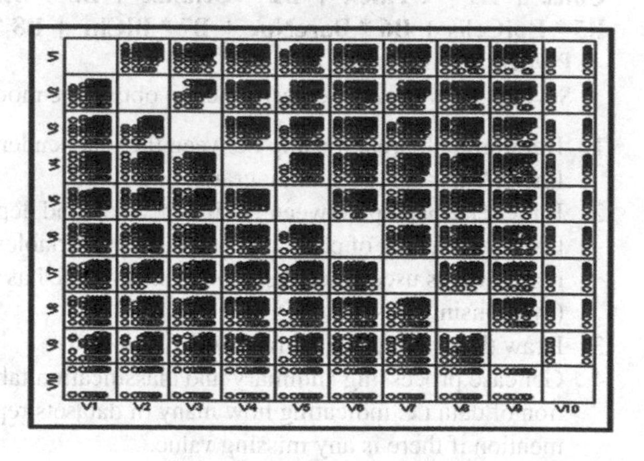

Fig. 3 Scatterplot for transformed dataset

and independent variables is not linear. Therefore, a logistic regression analysis is more suitable for this data. Multinomial Logistic Regression tool used is SPSS for processing the dataset.

4 Results

4.1 Distribution of Data

Table 2 show the case processing summary:

Table 2 Case processing summary

Unweighted cases[a]		N	Percent
Selected cases	Included in analysis	699	100.0
	Missing cases	0	0.0
	Total	699	100.0
Unselected cases		0	0.0
Total		699	100.0

[a]If weight is in effect, see classification table for the total number of cases

The minimum ratio of valid cases to independent variables for logistic regression is 10 to 1, with a preferred ratio of 20 to 1. In this analysis, there are 699 valid cases and 9 independent variables. The ratio of cases to independent variables is 77.6 to 1, which satisfies the minimum requirement. In addition, the ratio of 77.6 to 1 satisfies the preferred ratio of 20 to 1.

1. **Hypotheses**: there is no different between the model with only a constant and the model with independent variables.
2. **Significant value**: sine the significant value is < 0.05, therefore reject the hypothesis.
3. **Conclusion**: there is a different between the model with only a constant and the model with independent variables.

The presence of a relationship between the dependent variable and combination of independent variables is based on the statistical significance of the model chi-square at step 1 after the independent variables have been added to the analysis.

In this analysis, the probability of the model chi-square (783.516) was < 0.001, less than or equal to the level of significance of 0.05, see Table 3. The null hypothesis that there is no difference between the model with only a constant and the model with independent variables was rejected. The existence of a relationship between the independent variables and the dependent variable was supported.

As seen in Table 4, the independent variables could be characterized as useful predictors distinguishing survey respondents who have not seen an X-rated movie from survey respondents who have seen an X-rated movie if the classification accuracy rate was substantially higher than the accuracy attainable by chance alone. Operationally, the classification accuracy rate should be 25% or more high than the proportional by chance accuracy rate. The proportional by chance accuracy rate was computed by first calculating the proportion of cases for each group based on the

Table 3 Omnibus tests of model coefficients

Step 1	Chi-square	df	Sig
Step	783.516	9	0.000
Block	783.516	9	0.000
Model	783.516	9	0.000

Table 4 Classification

Observed		Predicted		
		V11		Percentage Correct
		2	4	
	V11			100.0
Step 0	2	458	0	
	4	241	0	0.0
Overall Percentage				65.5

[a]Constant is included in the model
[b]The cut value is 0.500

number of cases in each group in the classification table at Step 0. The proportion in the "2" group is 458/699 = 0.6552. The proportion in the "4" group is 241/699 = 0.3447. Then, we square and sum the proportion of cases in each group ($0.6552^2 + 0.3447^2 = 0.5481$). 0.5481 is the proportional by chance accuracy rate.

The accuracy rate computed by SPSS was 97.6% which was greater than or equal to the proportional by chance accuracy criteria of 68.5% (1.25 × 54.8% = 68.5%). The criteria for classification accuracy are satisfied.

4.2 *Pseudo R Square*

Among the three pseudo R-square Nagelkerke obtained the highest accuracy i.e. 93.1 as shown in Table 5.

1. While logistic regression does compute correlation measures to estimate the strength of the relationship (pseudo R square measures, such as Nagelkerke's R^2), these correlations measures do not really tell us much about the accuracy or errors associated with the model.
2. A more useful measure to assess the utility of a logistic regression model is classification accuracy, which compares predicted group membership based on the logistic model to the actual, known group membership, which is the value for the dependent variable.

Table 5 Model summary (Pseudo R–Square)

Step	−2 Log likelihood	Cox & Snell R Square	Nagelkerke R Square
1	117.012[a]	0.674	0.931

[a]Estimation terminated at iteration number 8 because parameter estimates changed by less than 0.001

Table 6 Parameter estimates

Step 1[a]	B	S.E	Wald	df	Sig	Exp(B)
CThick	0.531	0.132	16.237	1	0.000	1.701
CellSize	0.006	0.186	0.001	1	0.975	1.006
CellShape	0.333	0.208	2.567	1	0.109	1.395
MasAd	0.240	0.115	4.380	1	0.036	1.272
EpiCells	0.069	0.151	0.212	1	0.645	1.072
BareNuc	0.400	0.089	20.041	1	0.000	1.492
BlChr	0.411	0.156	6.918	1	0.009	1.508
NormNuc	0.145	0.102	2.003	1	0.157	1.156
Mito	0.551	0.303	3.311	1	0.069	1.734
Constant	−9.671	1.051	84.623	1	0.000	0.000

[a]Variable(s) entered on step 1: CThick, CellSize, CellShape, MasAd, EpiCells, BareNuc, BlChr, NormNuc, Mito

4.3 Wald Statistic

Variables in the equation as shown in Table 6.

Multicollinearity in the logistic regression solution is detected by examining the standard errors for the B coefficients. A standard error larger than 2.0 indicates numerical problems, such as multicollinearity among the independent variables, zero cells for a dummy-coded independent variable because all of the subjects have the same value for the variable, and 'complete separation' whereby the two groups in the dependent event variable can be perfectly separated by scores on one of the independent variables. Analyses that indicate numerical problems should not be interpreted. None of the independent variables in this analysis had a standard error larger than 2.0. (The check for standard errors larger than 2.0 does not include the standard error for the Constant).

1. **Hypothesis**: The coefficients for variables (nine variables) are zero.
2. **Significant value**: since the significant value for (9 value) are < 0.05. (e.g. p = 0.06, 0.000 and 0.0024)). Reject the hypothesis.
3. **Conclusion**: the coefficients for 9 values are not zero.

The probability of the Wald statistic for the variable CThick was 0.001, less than or equal to the level of significance of 0.05. The null hypothesis that the b coefficient for CThick was equal to zero was rejected. This supports the relationship that "survey respondents who were older were more likely to have not seen an X-rated movie."

The value of Exp(B) was 1.701 which implies that a one unit increase in CThick increased the odds that survey respondents have not seen an X-rated movie by 53.1%. This confirms the statement of the amount of change in the likelihood of belonging to the modeled group of the dependent variable associated with a one unit change in

Table 7 The performance of multi-layer perceptron, multiple linear regression and logistic regression

Method	Performance (%)
Multilayer perceptron	100.00
Multi Linear regression	84.4
Logistic regression	97.6

the independent variable, CThick. If variables are not significant, get the correlation table.

9.671-0.531CThick-0.006CellSize-0.333CellShape-0.240MarAd-0.060EpiCells-0.4BareNuc-0.411BlChr-0.145NormNuc-0.551Mito

5 Conclusion

The computerized and especially logistic regression model could be useful to predicate for many. fields of science. For this reason, it's used abundantly in medicine field especially to build a model to predicating breast cancer.

The conclusion from this experiment showed that not all the independent variables have the coefficient effect with dependent variable. However, from the performance comparison between: Multilayer perceptron, multiple linear regression and logistic regression, the Wisconsin breast cancer dataset is best to analyses using multi-layer perceptron, with 100% compared to logistic regression 97.6% and multi linear regression 84.4% as shown in Table 7.

References

Agresti, A. (2018). *An introduction to categorical data analysis*. John Wiley & Sons.

Carpio, C. E., Sydorovych, O., & Marra, M. C. (2007). Relative importance of environmental attributes using logistic regression.

Chaurasia, V., & Pal, S. (2014). Data mining techniques: To predict and resolve breast cancer survivability. *International Journal of Computer Science Mobile Computing IJCSMC, 3*(1), 10–22.

Choi, T., Schervish, M. J., Schmitt, K. A., & Small, M. J. (2008). A Bayesian approach to a logistic regression model with incomplete information. *Biometrics, 64*(2), 424–430.

Chaurasia, V., & Pal, S. (2017). A novel approach for breast cancer detection using data mining techniques. *International Journal of Innovative Research in Computer and Communication Engineering (An ISO 3297 2007 Certified Organisation), 2*(1), 2456–2465.

De la Rosa, D., Cardona, F., & Almorza, J. (1981). Crop yield predictions based on properties of soils in Sevilla, Spain. *Geoderma, 25*(3–4), 267–274.

De la Rosa, D., Moreno, J. A., García, L. V., & Almorza, J. (1992). MicroLEIS: A microcomputer-based Mediterranean land evaluation information system. *Soil Use and Management, 8*(2), 89–96.

Giasson, E., Clarke, R. T., Inda Junior, A. V., Merten, G. H., & Tornquist, C. G. (2006). Digital soil mapping using multiple logistic regression on terrain parameters in southern Brazil. *Scientific Agriculture, 63*(3), 262–268.

Gupta, S., Kumar, D., & Sharma, A. (2011). Data mining classification techniques applied for breast cancer diagnosis and prognosis. *Indian Journal of Computer Science Engineering, 2*(2), 188–195.

Kemp, E. M., & Alliss, R. J. (2007). Probabilistic cloud forecasting using logistic regression. *Northrop Grumman Information Technoogy A, 8*. Retrieved from https://ams.confex.com/ams/22WAF18NWP/techprogram/paper_122886.htm.

Kyle, R. A., et al. (2002). A long-term study of prognosis in monoclonal gammopathy of undetermined significance. *New England Journal of Medicine, 346*(8), 564–569.

Larson, R. R. (2002). A logistic regression approach to distributed IR. In *Proceedings of the 25th Annual International ACM SIGIR Conference on Research and Development in Information Retrieval* (pp. 399–400).

Olson, K. R., & Olson, G. W. (1986). Use of multiple regression analysis to estimate average corn yields using selected soils and climatic data. *Agricultural Systems, 20*(2), 105–120.

Rajesh, K., & Anand, S. (2012). Analysis of SEER dataset for breast cancer diagnosis using C4. 5 classification algorithm. *International Journal of Advanced Research Computer Communication Engineering, 1*(2), 1021–2278.

Wang, X., Summers, C. J., & Wang, Z. L. (2004). Large-scale hexagonal-patterned growth of aligned ZnO nanorods for nano-optoelectronics and nanosensor arrays. *Nano Letters, 4*(3), 423–426.

Wolberg, W. H., & Mangasarian, O. L. (1990). Multisurface method of pattern separation for medical diagnosis applied to breast cytology. *Proceedings of the National Academy of Sciences, 87*(23), 9193–9196.

Zhang, J. (1992). The mean field theory in EM procedures for Markov random fields. *IEEE Transactions on Signal Processing, 40*(10), 2570–2583.

The Usage of 3D Printing Technology by Small-Medium Sized Enterprise in Jordan

Ra'ed Masa'deh, Rand Al-Dmour, Raja Masadeh, Hani Al-Dmour, and Ahmed H. Al-Dmour

Abstract The purpose of the current research paper is to measure the effectiveness of the usage of 3DPT applications by small-medium sized enterprises (SME) in Jordan as a case from developing countries. It also aims to test whether the extent of the usage of 3DPT applications differs across different SME in terms of their type, size, and experiences. Primary data were collected from 250 respondents and 23 measures of 3D technology effectiveness were employed. A questionnaire was developed and extracted from previous studies. Six factors for measuring 3D technology effectiveness were extracted, namely: (1) time, (2) cost, (3) quality, (4) innovation, (5) competitiveness and (6) management. The findings indicate the extent of the 3DPT being used by small-medium sized enterprises (SME) in Jordan could be considered quite good at this stage. It also provides positive evidence that the 3DPT technology effectiveness indicators are significantly related to the level of 3DPT applications being used by SMEs, particularly, cost, time, quality and innovation. The findings revealed that the extent of using 3DPT technology applications were differ among

---

R. Masa'deh (✉) · R. Al-Dmour
Department of Management Information Systems, School of Business, The University of Jordan, Amman, Jordan
e-mail: r.masadeh@ju.edu.jo

R. Al-Dmour
e-mail: rand.aldmour@ju.edu.jo

R. Masadeh
Department of Computer Science, School of Information Technology, The World Islamic Sciences and Education University, Amman, Jordan
e-mail: raja.masadeh@wise.edu.jo

H. Al-Dmour
Department of Marketing, School of Business, The University of Jordan, Amman, Jordan
e-mail: dmourh@ju.edu.jo

A. H. Al-Dmour
Department of Accounting Information Systems, Faculty of Business, Al-Ahliyya Amman University, Amman, Jordan

M. Alshurideh et al. (eds.), *The Effect of Information Technology on Business and Marketing Intelligence Systems*, Studies in Computational Intelligence 1056,
https://doi.org/10.1007/978-3-031-12382-5_32

the target respondents based upon their size while there were not differ in terms of their types and business experiences.

Keywords 3DPT applications · Small-medium –sized enterprises · Effectiveness measures · Jordan

1 Introduction

Some common aspects of new industry are the high changing technologies, systems and processes that affect the effectiveness, efficiency, and cost of production in some or in another group (Al Kurdi et al., 2020; Alshurideh et al., 2019, 2021, 2022; Alzoubi et al., 2021). Additive manufacturing known as 3D printing, is a new industrial revolution (Altamony et al., 2012; Berman, 2012; Hunaiti et al., 2009; Qian, 2020; Weller et al., 2015). It is a disruptive technology because it can have a fundamental impact on production processes, supply chain, product life-cycle planning, and consumer behavior, among other things (Abu Zayyad et al., 2021; Abualoush et al., 2018; Masadeh et al., 2020; Obeidat et al., 2019; Schniederjans, 2017; Westerweel et al., 2019; Wu & Chen, 2018). According to several academic studies, AM technology has several advantages over traditional manufacturing such as cost, speed, quality and innovation/transformation, for example, (Attaran, 2017) indicated that the advantages of 3D-printing to manufacturing have become more prolific as the technology continues to advance inflexibility and capabilities. 3D-printing has seen an exponential growth in the number of materials a 3D-printer can use to build objects (Alzaqebah et al., 2019; Kietzmann et al., 2015; Saliba et al., 2020; Schniederjans, 2017) Other researchers claim that small company is going to adaption 3DPT much than the large company (Lopes da Silva, 2013; Masadeh et al., 2021; Mellor et al., 2014). So, this is the core of any business to grow and sustain, and because this technology increases the innovatively, so it creates and gains several opportunities and expandability to add extra market share (Weller et al., 2015). Furthermore, AM technology allows the flexible production of custom products without imposing fines on manufacturing costs. This is done by using direct digital manufacturing processes that convert 3D data directly into physical parts, without the need for tools or templates. In addition, the category manufacturing principle can also produce functionally integrated parts in a single production step, reducing the need for assembly activities. Thus, AM technology greatly affects flexibility, customization, capital costs and marginal production costs (Berman, 2012; Ford & Despeisse, 2016; Weller et al., 2015).

Understanding the factors affecting the adoption of 3D printing applications has received great attention from academic researchers and professionals around the world (e.g., (Guo & Leu, 2013; Martinsuo & Luomaranta, 2018). Other studies have emphasized several benefits of adopting AM technologies, including design freedom, efficiency and speed, customization of products, enabling of small batches, flexibility, adaptability, simplification of supply chains and reduction of waste (Weller et al.,

2015). Furthermore, literature shows that many studies concerning challenges to AM adoption have been carried out solely in large firms (Flores et al., 2016; Rylands et al., 2016).

However, studies that examine measures the effectiveness of the usage of 3DPT applications by small-medium sized enterprises (SME) in Jordan as a case from developing countries are rarely available. Digital manufacturing technologies are expected to have tremendous advantages for businesses and industries, but these benefits are rarely realized in practice. Therefore, this study has come to answer the following question: to which extent the use of 3D-printing technology applications enhancing the effectiveness of small-medium sized manufacturing enterprises operations in Jordan? Furthermore, this paper aims to elaborate a better understanding of the relationship between the extent of the implementation of 3DPT applications (i.e., H3DPT scope) and its effectiveness might be useful for both mangers and practitioners. The outline of this research has other six sections. Section 2 comprehensively presents the theoretical background and literature review. Section 3 establishes a research model and proposes the research hypotheses. Section 4 describes the research method and data collection. Section 5 discusses data analysis. Section 6 summarizes research conclusion and implications.

2 Theoretical Background and Literature Review

2.1 3DPT: Definition and Value

3D printing is a recognized broad term of Additive Manufacturing (AM), which known as, the processes so as to convert a digital file or information piece by piece, surface by surface and layer to layer into physical obsession (Thompson et al., 2016). Furthermore, Rapid Prototyping (RP) and Rapid Manufacturing (RM) are two extensively acknowledged synonyms which are incremental manufacturing. Compared to RP and RM, AM is a supplementary generic marquee that reveals the processing strategy of this superior industrial technology (Mognol et al., 2006; Obeidat et al., 2017; Schniederjans, 2017).

3DP technologies were patented and marketed in the late 1980s and early on 1990s by companies such as Stratasys, 3D Systems, and EOS. For several years these technologies have been referred to as rapid prototype implements. Stimulating technological maturity and termination of patents for progress and adoption with the rising certainty that 3DP technologies can be used for speedy manufacturing, not just prototypes (Cerdas et al., 2017; Ruffo & Hague, 2007). To date, on the other hand, the employ of 3DP for end-use products and components is mainly limited to dedicated industries. For example, while creating tools on the International Space Station is a logical application, the commercial viability of doing so on the ground is disputed (Khoo et al., 2015; Rayna & Striukova, 2016).

3DP operations vary from conventional manufacturing in several major ways. First of all, 3DP is an additive, that is, it generates solid objects by adding together materials in layers, somewhat than eliminating the materials to produce the required shape. This feature offers the ability to lessen waste of raw materials, while energy use may be superior than in conventional processes (Huang et al., 2013). Second, 3DP stirs directly from the digital model to a physical object, lacking the necessitate for tools. This facilitates on-demand production with extremely short lead times (Petrovic et al., 2011). It also assists distributed manufacturing, since resources could be shared throughout the digital transmission of designs for manufacture near their point of use. This offers apparent benefits when demand is low, discontinuous, or geographically disseminated, for example in parts supply chains (Khajavi et al., 2014). Third, 3DP manufacturing is not subject to economies of scale that sustain conventional manufacturing as there is no cost consequence coupled with low volume production (Mellor et al., 2014).

3D Printing provides several benefits ranging from customization to ease of designing, an elasticity of prototyping and production, shortening the lead time, reducing wastage, and enhancing efficiency in the supply chain. As compared to the conventional manufacturing, these major benefits considered as (1) it is ability to produce customized products in small batches (Ford & Despeisse, 2016), (2) as designs are in digital form, can be shared, and manufacture can be outsourced (Berman, 2012; Ford & Despeisse, 2016), (3) it offers speed, easiness, and flexibility in designing and adjusting products (Berman, 2012), (4) it assists in reusing material and dropping waste, therefore, leading to material savings (Berman, 2012; Weller et al., 2015), (5) it can lead to less reliance on high energy consuming manufacturing activities such as forging, casting, etc. (Khorram & Nonino, 2017), (6) it can lead to an enhanced and shorter value chain (Khorram & Nonino, 2017), and (7) 3D Printing has also engrossed many researches from sustainability community owing to the various environmentally friendly and societal paybacks it can present (Ford & Despeisse, 2016).

In spite of a number of benefits the 3D Printing be able to offer, there are several challenges which are acting as restraints or barriers for 3D Printing execution in diverse organizations across industries. Some of these challenges are: (1) The high charge is involved in 3D Printing functioning (Berman, 2012; Ford & Despeisse, 2016), (2) Speed of manufacture is not yet up to the mark (Berman, 2012), (3) Limited selection of materials to work with (Berman, 2012; Weller et al., 2015), (4) AM faces disputes regarding the quality of output which could be in the form of restricted strength, opposition to heat and moisture, and color stability (Berman, 2012; Thompson et al., 2016), (5) AM is appropriate for manufacturing products in smaller quantities but faces a vast confront with large volumes (Khorram & Nonino, 2017), (6) Difficulty in decisive the mindset and approach of the designers (Ford & Despeisse, 2016), (7) Lack of knowledge and awareness among the organizations (Martinsuo & Luomaranta, 2018), (8) Status quo and struggle to change in the organizations (Martinsuo & Luomaranta, 2018), (9) Lack of management and leadership maintain (Martinsuo & Luomaranta, 2018) and (10) Intellectual Property Rights issues (Ford & Despeisse, 2016; Martinsuo & Luomaranta, 2018).

2.2 Additive Manufacturing Technologies Applications

There are numerous systems and technologies for Additive Manufacturing, and over the years these technologies have developed and improved. Some of the most commonly discussed AM technologies are Selective Laser Sintering (SLS), Fused Deposition Modelling (FDM), Laser Melting (LM), Laminated Object Manufacturing (LOM), Stereo lithography (SL), Laser Metal Deposition (LMD), Electron Beam Melting (EBM), 3D Printing (3DP) among many others (Petrovic et al., 2011). According to Wong and Hernandez (2012), various technologies like Prometal, Polyjet, Laminated Engineered Net Shaping (LENS), and EBM were non-existent. The progress of new technologies or processes occurs like Stereolithography, Laser Beam Melting (LBM), MIT's 3D Printing process, laminated object manufacturing, Fused Disposition Modelling (FDM), and solid ground curing. In that period, not only there was a progress of technologies but also the thriving commercialization of these technologies, such as FDM, solid ground curing, Stereolithography, and laminated object manufacturing (Alananzeh et al., 2018; Thompson et al., 2016; Weller et al., 2015). New processes such as Electron Beam Melting (EBM) were improved and commercialized besides developing the existing technologies in 1990s and 2000s.

The enhancement in the technologies meant the processes were able to generate patterns, tooling and finishing parts. By the late 2000s, the 3D Printing industry became more effervescent as the patents pertaining to some of the 3D Printing technology processes perished which opened the doors for other players to contribute in the industry; the commoditization of AM processes which were commercialized before took place, together with growing AM hobby community, and invariable innovation (Chen et al., 2017). Also, the progression of AM can be characterized into three phases: Phase I- where AM processes mainly concerted on mock-ups of original designs, prototypes and models. Phase II- In this phase, AM processes enhanced and have had the ability to manufacture or generate finished products. 'Rapid Tooling' and 'Direct Digital Manufacturing' are other words that are occasionally used to refer to this phase. Phase III- At this phase, ultimate consumers will have the admission and be able to be the owner of the 3D Printers such as conventional Ink Jet and Laser Desktop Printers (Berman, 2012; Thompson et al., 2016).

From the applications view, 3D Printing applications used in various functions across diverse industries. Manufacturing, aerospace, transportation, medical, architecture, dental, hobbies, space exploration, education, military, art, and energy among others are extensively using the AM products and support services (Thompson et al., 2016; Weller et al., 2015; Wong & Hernandez, 2012). 3D Printing is predominantly used for prototyping and yet used for product development and innovation. Generally, it has been assured that 3D Printing is mainly used in applications concerning low production, small part sizes and having multifaceted printers (Berman, 2012).

2.3 Literature Review

The extant literature stresses on the importance of examining the value and the effectiveness of 3DP technology on manufacturing operations and functions. It has been recognized that one of the most significant challenges faced by managers and owners of small-medium sized enterprises today is assessing the effectiveness of using 3DPT applications in order to justify the value-added contribution of the 3DPT in achieving the organization's objectives (Thompson et al., 2016). Adopting and implementing of 3DPT applications may seem an important sub-system for many organizations, but unless it is an effective tool for manufacturing operations, its potential power would be significantly hindered. Recently, the relationship between using of digital technologies and its effectiveness has received great attention in the organizational and IT literature. Several researchers indicate that digital technologies, such as 3DP, can contribute to operation effectiveness in at least two relevant ways (e.g., Lu & Ramamurthy, 2011; Mishra et al., 2013). Firstly, digitally conduct business operation efficiency and assistance in overcoming uncertainty (Alshurideh et al., 2022; Alzoubi et al., 2021; Tariq et al., 2022). Secondly, IT helps to coordinate and synchronize business functions such as innovation, design product manufacturing and marketing (Bharadwaj et al., 2007; Candi & Beltagui, 2019; Kamaruddeen et al., 2022). Adopting digital tools in innovation can help enterprises achieve successful results from innovation projects (Marion et al., 2015). The relationship between flexible manufacturing systems, such as 3DP, and product innovation is a logical one, which was also empirical testing (Oke, 2013). As a result, the use of 3DP is expected to be associated with higher business performance. (Steenhuis & Pretorius, 2017) claims that the adoption of 3DP technology would affect business by improving their operational efficiencies objectives: quality, speed, flexibility and reliability cost.

According to Oettmeier and Hofmann (2017), operational efficiency is not considered as radical as strategic positioning effect of 3DP technology on business. The 3DP technology can have a big impact on poisoning (corporate image) on all players involved in supply chain i.e., suppliers, manufacturers and customers. The 3DP technology has the ability to make supply chains narrower as they provide an opportunity for integration and optimization product functions which to a certain extent exclude the need for sub-component suppliers. Thus, with the adoption of the company's 3DP technology, processes become more flexible due to the ability of technology to change product designs providing high quality services with decentralized production closer to consumer location. In addition, the 3DP technology not only affect the chain supply chain structure but also processes involved in the supply chain by grasping opportunities for product design through the new modified management and R & D processes. In this context, the adoption of 3DP technology can do changes in process planning and quality control processes.

According to Schniederjans (2017), the adoption of 3D printing technology allows fairly fast and fast prototypes enterprises, especially SMEs, to produce product models in a short time for quick testing and improving on-time parts that need to be improved. In this way, not only improves the production process but also enhances creativity in the enterprises in the development of their products. Moreover, with the adoption of 3D printing technology enterprises not only quickly prototype but also significantly reduced costs with traditional models are inevitable. Other potential benefits of the adoption of 3DPT include: reduction in time-to-market, material savings by reduction of subtractive manufacturing processes, reduced need for tools, molds or punches, enhanced density of final part and fabrication of free-form enclosed structures (Schniederjans, 2017). Besides, lead times can be with the three-dimensional printing system significantly lower while can produce different forms of products simultaneously without additional switching costs and time (Weller et al., 2015).

When we come to the cost in the 3DPT usually, the cost concept includes three dimensions: printer cost, materials cost, and operating cost. Some researchers also indicate that 3DPT helps to get an optimal design, which leads to a decreasing consumption in materials up to 40% (Achillas et al., 2015), and all of these variables can be control and changed based on the quality and precision needed of the printed products, also the type of 3DPT and the model of the printer effect the cost, in some cases the cost get affected by a level of finishing required such as sanding, coloring, or any post-process needed, In addition, 3DPT helps to reduce labor cost, because the 3D printing process is completely automated (Dedoussis & Giannatsis, 2009); that gives high flexibility in RP cycle for NPD and the cancelation of molds and patterns is a great benefit that saves a lot of money and also saves high lead times (Oettmeier & Hofmann, 2017; Thompson et al., 2016), also, this technology is lowering the cost needed to invest in inventory space and its management (Liu, 2014; Müller & Karevska, 2016), because the working method in 3DPT is near to Just-in-Time (JIT) method so no need for big factory space (Fawcett & Waller, 2014).

Furthermore, increasing adoption of additive manufacturing may lead to a reduction in raw material cost through economies of scale (Al-Dmour et al., 2020; Alshurideh & Alkurdi, 2012; Ford & Despeisse, 2016; Thomas, 2016). Other important advantage of adopting 3D printing technology can be gained from the supply chain perspective. Enterprises now can manage with the product warranties, repairs and upgrades by offering to the customers a downloadable 3D printing design of the parts needed for repair. Additionally, a recent search appears that there are already some logistics enterprises that already apply 3D and small technology part of the thinking is to start applying it in their business operations. The search shows that enterprises that are not considering applying the 3DP technology in their operations are mainly because of their lack of information about the potential advantages that this technology offers business (Müller & Karevska, 2016). This also means that production in the developing country is low states will no longer be necessary because enterprises can produce at the same low-cost point of consumption without bearing any transportation costs and logistics. What was said, besides these advantages, many new services are available to appear as in this case market barriers are relatively low (Müller & Karevska, 2016).

One of the most common things in small- medium sized enterprises is the desire and passion of creating really new creative, good, and well-developed ideas in order to convert it to be good products that will have a good opportunity of acceptance in the market, to gain a more loyal customer (Dimitrov et al., 2008). So many researchers said, that 3DPT is definitely improving product quality, on another side, it has many weak points such as poor surface quality, the accuracy of dimensions, a tolerance range of error, and some physical or mechanical issue (Khorram & Nonino, 2017). According to Oettmeier and Hofmann (2016), enterprises not only aim to achieve technological progress but also competitive advantages that distinguish them from competitors through highly innovative production customized products give value to the customer. Under this, the additives industry, also known as three-dimensional printing, was one of the latest industries Technological developments that Enterprises tend to adopt to improve them technological capabilities. The 3D printing was created with its rapid technological development these opportunities for Enterprises to reach and enhance the competitive advantage manufacturing capabilities (Rylands et al., 2016). It is clear that the 3DP technique is extensive adopted today as one of the subversive innovations that generate radical changes for one industry.

Additionally, many benefits from 3DPT drive to improve company's and products' competitiveness, but, developed countries face the problem of losing competitiveness in mass manufacturing while the 3DPT have it (Petrovic et al., 2011), so that what lead giant in mass production like China to adopt and invest great recourses in 3DPT (Bai et al., 2017), the results show growth in competitiveness of the early entrepreneurial startup's that adopt 3DPT in RM, especially with enterprises have vast experience in AM working for RM (Khorram & Nonino, 2017). 3DPT can also provide a competitive advantage, especially if the market is uncertain and demands a great variety of products and adaptability to varying customer needs (Weller et al., 2015), along with a shorter time to market (Weller et al., 2015). Petrovic et al. (2011) noted that there is a clear changing in management style and planning process for the SME company that adopting 3DPT, and that change happened in different areas such as needed infrastructure like electric power, material cost or wasting materials, space, inventory, labors or experts required, maintenance style, that facilitate logistics management and gives elasticity and a possible decrease in production costs. Also, AM affecting supply chain management by minimizing required stock holding (Fawcett & Waller, 2014), recently the concept of "Digital inventory" become popular in supply chain management, and the "Digital inventory" management model allows producing on demand for the users, like JIT model of providing for firms that can reduce storage expenses (Holmström et al., 2010), that increases inventory turnover (Tuck et al., 2006) and the ability to produce locally besides of that, AM requires fewer tools in the production process and employing fewer laborer's, also, developed operating capital management (Berman, 2012).

2.4 3DPT Status in Jordan

There is no doubt of 3DPT success, especially when we are talking about AM technology at the level of developed countries, there are very sophisticated uses, almost interference in all the simple and complex industries. From nuclear bombs (ExOne,

2014) and rockets used to travel through space to the uses of medical and therapeutic advances of heart surgery, arteries, brain and bones surgeries, also movie maker industry, fashion, and food. Moreover, its extensive uses of NPD that is technical or ordinary products such as kitchenware.

All of that because of the development has done in the entire ecosystem of entrepreneurship and innovation in developed countries. So, to look to the Jordanian ecosystem, the researchers follow the model adopted in the Brazilian entrepreneurial ecosystem of start-ups studies, which subtitled by Daniel Isenberg to determine domains of the entrepreneurship Ecosystem (Isenberg, 2011), and when he talks about the Entrepreneurship Ecosystem, he looks from a wide angle to see the broad picture and taking into his consideration all the major stakeholders in the ecosystem that affect adaptation of the new technologies and changes. So, that what we need in Jordan to guarantee a suitable environment that facilitates and incubate 3DPT in Jordan, to develop a comprehensive technological ecosystem that leverages the entrepreneurship ecosystem at this sector:

- **Policy and Laws Makers**: from governments and leaders and their role in enacting laws and legislation that facilitate and regulate the work of 3DPT sector.
- Financing System: from capital investments and lending policies for projects in 3DPT, and funding from international institutions to develop this sector.
- **Cultural System**: It means the way that society is seeing and focus on the success stories in this sector. Also, the social norms that have tolerance with mistakes, failures, and risk resulting from the core of RM and prototyping concepts that germane to the 3DPT and focus on the positive aspect of the development of creativity and innovation.
- **Support System**: about the infrastructure of energy, telecommunications, transportation, and Development Zones, in addition to non-governmental institutions in the private sector and its role in supporting entrepreneurship through conferences and courses and raising awareness about the nature of this type of ECO systems and professional support of experts in legal, accounting, banking issues, and Technical Advisors.
- **Human Capital**: means investing in human resources are both skilled or unskilled, to become dependable trained workers are familiar with 3DPT, and that transition should happen through educational institutions concerned of spreading awareness by education, training, human resources development, and leadership development, to increase the awareness of 3DPT capability and its usage and benefits.
- **Market System**: from local and international Enterprises that have production line intersected with 3DPT in order to make or develop a particular product and network that matter with all aspects of the market, also, customers who accept 3DPT by use product made from it, and adopt it in their works and Enterprises, and develop it by feedback from opinions, criticisms, and recommendations.

So, if the change going to happen in the ecosystem of Jordan to have an adoption of 3DPT, several steps need to activate in sequences and parallel sometimes, starting from creating an educational and cultural arm that comes with raising awareness of

3DPT, RM, and NPD in the parallel of that allocation of finance system to support and fund these initiatives and institutions. Then using these two steps to make pressure and motivate to the policymakers in order to lay down lists of laws, standards, limits, and regulations to organize this environment between all stakeholders in the ecosystem, so that could help the adoption of 3DPT in the society, which in turn will develop the technology and entrepreneurship ecosystem overall. After that, the coverage that comes from the laws with the opportunity that appears after developing the human resources and society to adopt this transform will help to create a culture based, and support system, that will courage the market mix from Enterprises and customers to deal and accept this adoption for 3DPT. Finally, after all of this effort, as a result of the development that occurs in the Jordanian ecosystem of entrepreneurship, this will impact the success or the entrepreneurial start-up success. The researcher reached to this conclusion after completing many brainstorming and interviews with several stakeholders in several institutions, initiatives, company, educational foundations, investment institutions, entrepreneurs, and some government organizations in Jordan.

3 The Study's Model and Research Hypotheses

Specific theories developed in the ground of Information Systems facilitate explicate how 3DP be able to contribute to business performance. Also, some researchers (e.g., Khorram & Nonino, 2017; Masa'deh et al., 2013; Masa'deh et al., 2014) argued that the notion of process virtualization assured that processes could be executed more efficiently as IT is performed to separate them from conventional physical or face to face methods. Also, based upon reviewing the related studies and literature and practical experience of the researchers (Khorram & Nonino, 2017; Martinsuo & Luomaranta, 2018; Masa'deh et al., 2018; Rylands et al., 2016; Schniederjans, 2017; Steenhuis & Pretorius, 2017), the study's model consists of two types of variables: independent variable (the extent of the adoption of 3DPT) and dependent variable (3DPT effectiveness indicators). Indeed, the research model for the current study was developed and proposed that the extent of 3DPT being used by entrepreneurial enterprises are positively likely to have a significant cause on their operational performance indicators in terms of time, cost, quality, competitiveness and management process.

In addition, three control variables were included to the research model: firm size (number of employees), firm experience (number of years from founding), and type of business. Firm size could be linked with a firm's ability to implement new technologies, such as 3DP. Also, firm size could be associated with the level of IT-manufacturing coordination required. In addition, in a very small firm, both functions possibly will exist in with the same people. Firm age could also be interrelated with the focal variables. Thus, in order to bring the distributions of these variables closer to being normally distributed, the logarithms of the numerical values were calculated.

The expected relationships between the independent and dependent variables were formulated in the following sub-hypotheses:

H1: The extent of the use of 3DPT applications directly and positively influences the perceived 3DPT effectiveness factors in SMEs' enterprises in Jordan.

H2: The extents of the use of 3DPT applications are differ amongst SME's enterprises due to their organizational demographic characteristics (type of business, size, and business experiences).

4 Research Methodology

4.1 Research Setting and Data Collection

The population of this study consists of small- medium sized industrial enterprises in Jordan, since the aim of the study is to examine the effect of the adoption of 3DPT on their operational effectiveness indicators: time, cost, quality, innovation, competitiveness, management process. Based on Ministry of Industry, the number of the actively registered small- medium sized industrial enterprises in Jordan is 4567 (Ministry of Industry Commerce, 2020). Because of the limited time and money, a connivance sample of 300 SME industrial enterprises located in Amman was selected for the purpose of this study. Taking this into account, a number of criteria were used to ensure that respondents were actually qualified to provide the information sought. The first criterion was that the enterprises they worked for should be engaged in developing products. The second criterion was that the enterprises should be actually used 3DPT. A total of 300 self-administered questionnaires were distributed to the respondents by e-mail and hand and the response rate was 72% (i.e., 216 respondents).

In this study, some variables are factual (for example Enterprises' profile such as type of SME enterprises, size, business experience, and type of the 3DPT applications), whereas others are perceptual (extent of the adoption of 3DPT application and type of operational effectiveness indicators: time, cost, quality, innovation, competitiveness and management). The extent of the adoption of 3DPT as well as the 3DPT effectiveness indicators were measured using Likert five-point scale, which ranged from 1 = strongly disagree, to five = strongly agree. The questionnaire contents were mainly selected from the previous studies (Khorram & Nonino, 2017; Martinsuo & Luomaranta, 2018; Rylands et al., 2016; Schniederjans, 2017; Steenhuis & Pretorius, 2017). For construct validation purpose, the questionnaire content was modified to the practice of Jordanian business context based on the results of a pilot study and feedback from five professional academic staff in this filed. However, the reliability of the study is tested by the Cronbach's Alpha coefficient. The value of Cronbach's alpha is found to be 0.87. It shows the stability and consistency of the scale are acceptable (Hair et al., 2010).

4.2 Respondents' Enterprises Profile

In this study, respondents' profile includes information about their characteristics such as: size, business experience, experience with the use of 3DP printing technology, the purpose of using 3DPT and the type of 3DPT applications. Table 1 summarizes the respondents' profile characteristics.

5 Data Analysis

The researchers tested the model adopting factor analysis and multiple regression analysis with the application of SPSS software version 25. The main rationale behind the use of factor analysis technique is to lessen the large number of variables that lie behind the perceived 3DPT effectiveness measures into orthogonal indices for additional analysis by the regression analysis.

Table 1 Respondents' enterprises profile

Respondents' characteristics	Frequency	Percent
Type of industry		
Automotive, Steel & Metal, Cement	46	0.21
Machinery, Power, Computer/Electronic	35	0.16
Automation, Plastics, Non-metallic Mineral	32	0.15
Food, Furniture, Wooden	47	0.22
Paper, Textile, Chemical	56	0.26
Business experience		
Less than 10 Years	86	0.40
15 Less than 20 years	78	0.36
More than 20 years	52	0.24
Number of employees (size)		
1–24 Employees	61	0.28
25–49 Employees	56	0.26
50–74 Employees	49	0.23
75–100 Employees	50	0.23
Experience with using 3DP applications		
1–5 years	121	0.56
5–10 years	68	0.31
More than 10 years	37	0.13

5.1 Validity and Reliability Test

By employing the principle component analysis techniques in order to explore the main patterns of factors that underlie these measures, the principle components analysis was used. It was considered an appropriate method to overcome the potential problems of multicollinearity among the variables that pertain to each construct. The results of the factor analysis were examined using multiple -criteria including, eigenvalues, interpretability, sampling adequacy (Kaiser–Meyer–Olkin measure > 0.5) and internal consistency, as recommended by Hair et al. (2010). Therefore, items with eigenvalues more than one and factor loadings less than (0.60) were determined. This means that the items had little or no relationship with each other; hence they were discarded (Hair et al., 2010). The results of the principle components analysis indicate that 6 factors can be extracted from 26 items. Cronbach's alpha reliabilities were also examined for each factor. Each coefficient greater than 0.40 for adapted and 0.70 for current scales was considered a reliable indicator of the constructs under study (Hair et al., 2010). Reliability analysis was carried out for each factor and the results score ranged from 0.85 to 0.91 as shown in Table 2. In summary, based on the preliminary analysis, the evaluation of the data by factor analysis and reliability estimates indicated that all scale items were appropriate and valid for further statistical analysis.

The results of the principal component analysis Table 4 indicate that six significant factors can be extracted from the 3DPT effectiveness measures. This construct composed of (26) items (variables) as presented in Table 2. The first factor, which accounts for (20.120%) of the variance with loadings ranging from 0.72 to 0.84, can be identified as a "Cost factor". The second factor, which explains 18.342% of variance with loadings range from 0.71 to 0.81, could be labelled as "Time" factor. The third factor which accounts for (15.121) can be identified as "Quality" factor and the fourth factor which account for (11.231) can be labelled as "Innovation" factor. The fifth factor can be named which account for 8.652 as "Competitiveness" factor and the sixth factor which account for 4.521can be called as "Management" factor. The combinations of these factors account for 78.987 of the total variances in the questionnaire data as can be shown in Table 3.

Table 2 A summary results of the principle component analysis

Factors	No. of items	Eigen-value	% of variance	Cumulative %
Cost	6	5.245	21.120	20.120
Time	4	5.177	18.342	39.462
Quality	5	4.793	15.121	54.583
Innovation	3	3.671	11.231	65.814
Competitiveness	3	2.345	8.652	74.466
Management	5	1.670	4.521	78.987

Table 3 Main factors underlying the perceived effectiveness of 3DPT

Code	Items (Variables)	Loadings	Communality
	Factor 1 cost		
C3	3D printing technology has minimized the product development cost	0.84	0.87
C4	3D printing technology has minimized the cost of maintenance	0.80	0.85
C2	3D printing technology has minimized the cost of tool for prototyping	0.77	0.82
C1	3D printing technology has minimized the cost of raw material	0.75	0.78
C6	3D printing technology has minimized the cost for transportation and delivery	0.73	0.76
C5	3D printing technology has reduced the cost of labor (workers)	0.72	0.74
	Factor 2 time		
T3	3D printing technology has minimized the time needed for prototyping cycle	0.81	0.85
T4	3D printing technology has minimized the time needed for production cycle	o.76	0.80
T1	3D printing technology has minimized the lead time	0.74	0.77
T2	3D printing technology has speeded time-to-market	0.71	0.76
	Factor 3 quality		
Q4	3D printing technology has increased the use of a new way to form materials for productions (e.g., Laser, UV, and Electric Beam)	0.78	0.81
Q5	3D printing technology has invented a new ability to get different properties of materials	0.75	0.77
Q3	3D printing technology has enhanced the optimization designs	0.72	0.78
Q2	3D printing technology has enhanced the precision and accuracy of the products output	0.69	0.74
Q1	3D printing technology has minimized the number of assembled components for products	0.67	0.76
	Factor 4 innovations		
N3	We played a leading role in market activities in our market	0.74	0.77
N2	We introduced more new products/services on the market than our competitors	0.67	0.69
N1	We adopted more technological innovations than our competitors	0.66	0.68
	Factor 5 competitiveness		

(continued)

Table 3 (continued)

Code	Items (Variables)	Loadings	Communality
O1	3D printing technology has improved our competitiveness capabilities	0.76	0.84
O3	3D printing technology has seriously improved our company's market positioning	0.71	0.75
O2	3D printing technology has speeded overall business operations more effectively better than our competitors	0.64	0.68
	Factor 6 management		
M4	3D printing technology has improved the management style and planning process of our company	0.75	0.79
M2	3D printing technology has decreased the risks in the decision-making process	0.68	0.76
M5	3D Printing technology has increased the privacy and security of new products in early stages	0/66	0.74
M3	3D printing technology has increased the health and safety standard in our company (smells, heats, noise)	0.64	0.68
M1	3D printing technology gives our employees greater control over their work	0.61	0.67

Table 4 The extent of the use of 3DP printing applications

No	Variables	Mean	%	Standard deviation	Sig. (2-tailed)
1	The company is fully integrated 3D printing technology in its business plan operations	3.781	0.7562	0.82783	0.000
2	The company has invested completely in IT infrastructure to adopt 3D printing technology	3.653	0.7306	0.8053	0.000
3	The company has 3DPT experts available to fully deploy 3D printing technology	3.681	0.7362	0.76469	0.000
4	The company top management is fully supported the decision to adopt 3D printing technology	3.541	0.7082	0.85610	0.000
5	The company has adopted 3D printing Technology in its manufacturing operations completely	3.432	0.6870	0.83426	0000
	Average practice	3.617	0.7235	0.75279	0.000

5.2 The Extent of the Use of 3DPT by SME in Jordan

The mean values, standard deviation and T-test are employed here to determine whether the 3DP printing applications are used by SMEs in Jordan. Finding shown in Table 4 indicate that the extent of 3DPT application being practiced by SME is conserved to be good (i.e., 72% or 3.61), since their mean are more than the mean of the scale, which is 3 (mean of the scale = Σ Degrees of the scale 5 = 1 + 2 + 3 + 4 + 5/5 = 3). This implies that there are some variations among SME in terms of their extent of use of 3DPT printing applications.

5.3 Hypotheses Testing Results

5.3.1 Testing the Relationship Between the Extent Use of 3DPT and the Perceived Effectiveness of 3DPT by SMEs in Jordan

The multiple regression analysis technique is used to examine the first hypothesis. Table 5 summarizes the results of multiple regression analysis, with the F-ratio test, for the study hypotheses (H1). The results indicate that there is a significant relationship between the extent of use of 3DPT applications and the perceived effectiveness of 3DPT factors at 0.000 level of significant. This result empirically proved that the extent of use of 3DPT has a positive and a significant impact on the perceived effectiveness of 3DPT. Accordingly, it may be concluded that the higher is extent use of 3DPT applications, the higher is the perceived effectiveness of 3DPT factors.

In order to determine the most important of perceived operational effectiveness of 3DPT factors which highly correlated with the use of 3DPT, the stepwise regression analysis was employed. The F value at (0.00) level of significance is used to determine the "goodness of fit" for the regression equation as well as the beta coefficient. The results of the beta coefficients in Table 6 indicate that the most important operational effectiveness factors indicators that highly associated with the use of 3DPT are four factors, namely and in order of the importance: cost, time, quality and innovations. It can be concluded that the higher the use of the 3DPT, the higher will be those operational effectiveness indicators, i.e., cost, time, quality and innovations indicators. Therefore, SMEs should be fully utilized and ingrate the 3DPT in their business process in order to enhance these operational effectiveness measures more effectively. However, the competitiveness and management effectiveness factors were found insignificantly related to the use of 3DPT. These results might be due to the

Table 5 Multiple regression for the (H1) hypothesis

Hypothesis	Multiple R	R. Square	Adjusted R Square	DF	F	Sign
Ho1	0.776	0.602	0.602	6	10.632	0.000

Table 6 Stepwise regression analysis results

Indictors	Step	R	R Square	Adjusted R Square	Beta	Sig
Cost	1	0.512	0.262	0.250	0.470	0.000
Time	2	0.556	0.314	0.308	0.366	0.000
Quality	3	0.590	0.348	0.331	0.238	0.000
Innovations	4	0.602	0.362	0.351	0.196	0.000

fact that management and competitiveness are long-term strategic effectiveness indicators, and as the majority of SMEs in Jordan have recently just adopted 3DPT applications on their business process, they may not yet grasp this other positive effectiveness of using 3DPT.

5.3.2 Testing the Variation on the Extent of Using 3DPT Application Based on the SMEs' Demographic Attributes

The ANOVA analysis technique is also used to examine the 2nd hypotheses (H2). To assess the differences among SMEs in terms of the extent of the use of 3DPT based on their organization's demographic characteristics such as size, type of business and business experience (age). As it is shown in Table 7, the results indicated that there was no variation among SMEs using 3DPT applications in terms of their types of industrial sector and business experience, while there was a significant difference amongst them in terms of their size of business. It may be concluded that type of SMEs and their business experience did not have any influence upon on their extent of using 3DPT applications while their business size has an important role in this respect. This result might be due to the fact that the larger size of company, the more ability to invest in the adoption of IT.

Table 7 The significance of the extent of using 3DPT among groups of SMEs base on their demographic attributes (type, experience and size)

Attributes		Sum of squares	df	Mean Square	F	Sig
Type	Between groups	0.170	4	0.085	0.124	0.884
	Within groups	167.660	212	0.687		
	Total	167.830	216			
Experience	Between groups	0.194	2	0.097	0.106	0.900
	Within groups	223.329	214	0.915		
	Total	223.522	216			
Size	Between groups	22.127	3	7.376	10.091	0.000
	Within groups	177.614	213	0.731		
	Total	199.741	216			

6 Discussions and Conclusion

The purpose of this study was to examine the effectiveness of using 3DPT applications by SMEs in Jordan and to determine whether using 3DPT among them differ in terms of their organizational demographic characteristics (type, experience and size). In order to achieve the study objectives, and to conduct the research using a systematic approach, hypotheses were developed and tested. The results of using factor analysis indicated that 26 items of effectiveness measures of 3DPT can be grouped into six orthogonal factors, namely; cost, time, quality, innovation, competitiveness and management. These patterns of factors are supported by previous studies (Martinsuo & Luomaranta, 2018; Schniederjans, 2017; Westerweel et al., 2019).

The findings also indicated that the extent of the 3DPT applications being used by SMEs was considered good (0.72). This result is reinforced by prior studies such as those by Martinsuo and Luomaranta (2018), Rylands et al. (2016), Khorram and Nonino (2017), Steenhuis and Pretorius (2017). The potential benefits of the adoption of 3DPT applications to SME include developing their competitiveness, in changes occurred at cost, time, and quality for their products, and management process. The analysis also provides empirical evidence regarding the effectiveness using 3DPT by SMEs as suggested in the first hypothesis (H1). The results have supported empirically the linkage between 3DPT utilization and the perceived operational effectiveness factors. The results also suggest that about 35% of the variation on the operational effectiveness factors s could be explained by the 3DPT being used by SMEs. This result is supported by Schniederjans (2017), Steenhuis and Pretorius (2017). Also, the stepwise regression analysis results indicated that the most important operational effectiveness factors that highly associated with using 3DPT were cost time, quality and innovation. Therefore, the managers of SMEs should be fully utilizing and ingrate the 3DPT in their business process in order to improve their operational effectiveness performance more effectively, particularly time, cost, quality and innovation measures.

Based on the above results, it is worthy to argue that digital technologies like 3DPT can be correlated to innovation performance in several ways. Conducting business processes digitally would advance efficiency and assist in coping with uncertainty (Lu & Ramamurthy, 2011; Mishra et al., 2013; Qian, 2020). Also, IT assists in coordinating and synchronizing business functions like innovation and product manufacturing (Bharadwaj et al., 2007). Innovation is an information prosperous activity, which benefits from the exploitation of IT to harmonize and deal with digital data, for instance offering access to information to all those concerned in a project. By assembly the innovation process more digital, 3DP be able to assist designers to distribute information at early stages of idea development, to rapidly examine working prototypes and allocate them with engineers in the development stage and to correspond to customers' requirements using physical models to extract customer inputs or during product commence (Jong & Bruijn, 2013).

Also, adoption of digital tools in innovation be able to assist firms attain thriving outcomes from innovation projects (Marion et al., 2015; Shannak et al., 2010; Tarhini et al., 2015). The association between elastic manufacturing systems, such as 3DP, and product innovation is a rational one, which has also been tested (Oke, 2013). Consequently, the commence of 3DP is likely to be related with superior innovation performance. A specific concern for innovation is the incorporation between innovation and manufacturing, as the ability to develop successful products entails both (Thompson et al., 2016). Innovation is gradually more reliant on the utilize of digital design tools, which in turn depend on the accessibility of appropriate IT infrastructures (Marion et al., 2015). This is reliable with the growing acknowledgment that IT should have a strategic rather than supporting part in business (Bharadwaj et al., 2007).

Furthermore, the results of the 2nd hypotheses testing revealed that the variation of using 3DPT applications among SMEs in Jordan could be mainly due to their size but not their type of business sector and experience This might be due the fact that more than half of target respondents (i.e., about 51% less than 5 years, Table (1) have just recently adopted the 3DPT and they were still in the early stage of the adoption, therefore, this short time of experience might not enough for the majority of them to take advantages of the potential benefits and opportunities using 3DPT application in their operational process.

6.1 Theoretical and Managerial Implications

The theoretical implications of this study are fourfold: first, the findings of this study contribute to literature on 3DPT utilizations which can be used to improve operational effectiveness of SMEs. Second, this study contributed to the 3DPT literature by identifying and examined six operational effectiveness factors measures of using 3DPT for the first time in Jordanian business context as a developing country. The present study has important implications for SMEs managers in the surveyed companies and in similar organizations. The authors believe that the decision-makers of these companies could benefit from this study's findings with a better understanding of potential benefits of using 3DPY on their business process.

Furthermore, the results of this study can aid managers in comprehending how 3DPT can improve operating process effectiveness and business performance. Managers or owners of SMEs could search for and adopt tools that can increase the use of 3DPT applications if they want to improve their operational effectiveness performances. Also, they should be fully aware of the importance of the effect of 3DPT applications on their operating process and business performances, so that they could make the appropriate change within their organizations. Second, the role of 3DPT applications to enhance company's operational effectiveness process should be considered in strategic choices for the future at every organization.

References

Abualoush, S., Obeidat, A., Masa'deh, R., & Tarhini, A. (2018). The role of employees' empowerment as an intermediary variable between knowledge management and information systems on employees' performance. *VINE Journal of Information and Knowledge Management Systems, 48*(2), 217–237.

Abu Zayyad, Z., Obeidat, Z., Alshurideh, M., Abuhashesh, M., Maqableh, M., & Masa'deh, R. (2021). Corporate social responsibility and patronage intentions: The mediating effect of brand credibility. *Journal of Marketing Communications, 27*(5), 510–533.

Achillas, C., Aidonis, D., Iakovou, E., Thymianidis, M., & Tzetzis, D. (2015). A methodological framework for the inclusion of modern additive manufacturing into the production portfolio of a focused factory. *Journal of Manufacturing Systems, 37*(1), 328–339.

Alananzeh, O. A., Jawabreh, O., Al Mahmoud, A., & Hamada, R. (2018). The impact of customer relationship management on tourist satisfaction: The case of radisson blue resort in Aqaba city. *Journal of Environmental Management and Tourism, 9*(2), 227–240. https://doi.org/10.14505/jemt.v9.2(26).02

Al-Dmour, R., Al Haj Dawood, E., Al-Dmour, H., & Masa'deh, R. (2020). The effect of customer lifestyle patterns on the use of mobile banking applications in Jordan. *International Journal of Electronic Marketing and Retailing, 11*(3), 239–258.

Al Kurdi, B., Alshurideh, M., & Salloum, S. A. (2020). Investigating a theoretical framework for e-learning technology acceptance. *International Journal of Electrical and Computer Engineering (IJECE), 10*(6), 6484–6496.

Alshurideh, M., & Alkurdi, B. (2012). The effect of customer satisfaction upon customer retention in the Jordanian mobile market: An empirical investigation. *European Journal of Economics, Finance and Administrative Sciences*, Issue 47, April, 69–78.

Alshurideh, M., Al Kurdi, B., & Salloum, S. A. (2019, October). Examining the main mobile learning system drivers' effects: A mix empirical examination of both the expectation-confirmation model (ECM) and the Technology Acceptance Model (TAM). In *International Conference on Advanced Intelligent Systems and Informatics* (pp. 406–417). Cham: Springer.

Alshurideh, M. T., Al Kurdi, B., & Salloum, S. A. (2021). The moderation effect of gender on accepting electronic payment technology: a study on United Arab Emirates consumers. *Review of International Business and Strategy, 31*(3), 375–396.

Alshurideh, M. T., Al Kurdi, B., Alzoubi, H. M., Ghazal, T. M., Said, R. A., AlHamad, A. Q., ... Al-kassem, A. H. (2022). Fuzzy assisted human resource management for supply chain management issues. *Annals of Operations Research*, 1–19.

Altamony, H., Masa'deh, R. M. T., Alshurideh, M., & Obeidat, B. Y. (2012). Information systems for competitive advantage: Implementation of an organisational strategic management process. *Paper presented at the Innovation and Sustainable Competitive Advantage: From Regional Development to World Economies—Proceedings of the 18th International Business Information Management Association Conference*, 1 (pp. 583–592)..

Alzaqebah, A., Al-Sayyed, R., & Masadeh, R. (2019, October). Task scheduling based on modified grey wolf optimizer in cloud computing environment. In *2019 2nd International Conference on new Trends in Computing Sciences (ICTCS)* (pp. 1–6). IEEE.

Alzoubi, H. M., Alshurideh, M., & Ghazal, T. M. (2021, June). Integrating BLE Beacon technology with intelligent information systems IIS for operations' performance: A managerial perspective. In *The International Conference on Artificial Intelligence and Computer Vision* (pp. 527–538). Cham: Springer.

Attaran, M. (2017). The rise of 3-D printing: The advantages of additive manufacturing over traditional manufacturing. *Business Horizons, 60*(5), 677–688.

Bai, X., Liu, Y., Wang, G., & Wen, C. (2017). The pattern of technological accumulation: The comparative advantage and relative impact of 3D printing technology. *Journal of Manufacturing Technology Management, 28*(1), 39–55.

Berman, B. (2012). 3-D printing: The new industrial revolution. *Business Horizons* (pp. 155–162).

Bharadwaj, S., Bharadwaj, A., & Bendoly, E. (2007). The performance effects of complementarities between information systems, marketing, manufacturing, and supply chain processes. *Information Systems Research, 18*(4), 437–453.

Candi, M., & Beltagui, A. (2019). Effective use of 3D printing in the innovation process. *Technovation, 80*, 63–73.

Cerdas, F., Juraschek, M., Thiede, S., & Herrmann, C. (2017). Life cycle assessment of 3D printed products in a distributed manufacturing system. *Journal of Industrial Ecology, 21*(1), 80–93.

Chen, Y. J., Lin, H., Zhang, X., Huang, W., Shi, L., & Wang, D. (2017). Application of 3D–printed and patient-specific cast for the treatment of distal radius fractures: initial experience. *3D Printing in Medicine, 3*(1), 11.

Dedoussis, V., & Giannatsis, J. (2009). Developing competitive products using stereolithography rapid prototyping tools. In *IEEE International Conference on Industrial Engineering and Engineering Management*, December, 154–158.

Dimitrov, D., Schreve, K., De Beer, N., & Christiane, P. (2008). Three dimensional printing in the South African industrial environment. *South African Journal of Industrial Engineering, 19*(1), 195–213.

ExOne. (2014). Case studies/US Navy. Retrieved from exone.com. https://www.exone.com/Portals/0/ResourceCenter/CaseStudies/X1_CaseStudies_All%202.pd.

Fawcett, S., & Waller, M. (2014). Supply chain game changers—mega, nano, and virtual trends—and forces that impede supply chain design (i.e., building a winning team). *Journal of Business Logistics, 35*(3), 157–164.

Flores, I., Khajavi, S., & Partanen, J. (2016). Challenges to implementing additive manufacturing in globalised production environments. *International Journal of Collaborative Enterprise, 5*(3/4), 232–247.

Ford, S., & Despeisse, M. (2016). Additive manufacturing and sustainability: An exploratory study of the advantages and challenges. *Journal of Cleaner Production, 137*, 1573–1587.

Guo, N., & Leu, M. C. (2013). Additive manufacturing: Technology, applications and research needs. *Frontiers of Mechanical Engineering, 8*(3), 215–243.

Hair, J., Black, W., Babin, B., & Anderson, R. (2010). *Multivariate data analysis* (7th ed.). Pearson.

Holmström, J., Partanen, J., Tuomi, J., & Walter, M. (2010). Rapid manufacturing in the spare parts supply chain: Alternative approaches to capacity deployment. *Journal of Manufacturing Technology*, 687–697.

Huang, S. H., Liu, P., Mokasdar, A., & Hou, L. (2013). Additive manufacturing and its societal impact: A literature review. *International Journal of Advanced Manufacturing Technology, 67*(5–8), 1191–1203.

Hunaiti, Z., Masa'deh, R.M., Mansour, M., & Al-Nawafleh, A. (2009). Electronic commerce adoption barriers in small and medium-sized enterprises (SMEs) in developing countries: The case of Libya. *Paper presented at the Innovation and Knowledge Management in Twin Track Economies Challenges and Solutions—Proceedings of the 11th International Business Information Management Association Conference, IBIMA 2009*, 1–3 (pp. 1375–1383).

Isenberg, D. (2011). Domains of the entrepreneurship ecosystem. Babson Park: MA: Babson College. Retrieved from http://entrepre-neurial-revolution.com/view-the-ecosystem-diagram/

Jong, D., & Bruijn, D. (2013). Innovation lessons from 3-D printing. *MIT Sloan Management Review, 54*(2), 42–53.

Kamaruddeen, A., Rui, L., Lee, S. K., Alzoubi, H., & Alshurideh, M. (2022). Determinants of emerging technology adoption for safety among construction businesses. *Academy of Strategic Management Journal, 21*(Special Issue 4), 1–20.

Kietzmann, J., Pitt, L., & Berthon, P. (2015). Disruptions, decisions, and destinations: Enter the age of 3-D printing and additive manufacturing. *Business Horizons, 58*, 209–215.

Khajavi, S., Holmström, J., & Partanen, J. (2014). Additive manufacturing in the spare parts supply chain. *Computers in Industry., 65*(1), 50–63.

Khoo, Z. X., Teoh, J. E. M., Liu, Y., Chua, C. K., Yang, S., An, J., ... Yeong, W. Y. (2015). 3D printing of smart materials: A review on recent progresses in 4D printing. *Virtual and Physical Prototyping, 10*(3), 103–122.

Khorram, N., & Nonino, F. (2017). Impact of additive manufacturing on business competitiveness: A multiple case study. *Journal of Manufacturing Technology Management, 28*(1), 56–74.

Liu, P. H. (2014). The impact of additive manufacturing in the aircraft spare parts supply chain: Supply chain operation reference model-based analysis. *Production Planning & Control, 25*(3–14), 1169–1181.

Lopes da Silva, J. (2013). 3D technologies and the new digital ecosystem: a Brazilian experience. *Proceedings of the Fifth International Conference on Management of Emergent Digital EcoSystems* (pp. 278–284.). Luxembourg: ACM.

Lu, Y., & Ramamurthy, K. (2011). Understanding the link between information technology capability and organizational agility: An empirical examination. *MIS Quarterly, 35*(4), 931–954.

Marion, J. T., Eddleston, K., Friar, J., & Deeds, D. (2015). The evolution of inter organizational relationships in emerging ventures: An ethnographic study within the new product development process. *Journal of Business Venturing, 30*(1), 167–184.

Martinsuo, M., & Luomaranta, T. (2018). Adopting additive manufacturing in SMEs: Exploring the challenges and solutions. *Journal of Manufacturing Technology Management, 29*(6), 937–957.

Masa'deh, R., Shannak, R., & Maqableh, M. (2013). A structural equation modeling approach for determining antecedents and outcomes of students' attitude toward mobile commerce adoption. *Life Science Journal, 10*(4), 2321–2333.

Masa'deh, R. M. T., Maqableh, M. M., & Karajeh, H. (2014). A theoretical perspective on the relationship between leadership development, knowledge management capability, and firm performance. *Asian Social Science, 10*(6), 128–137. https://doi.org/10.5539/ass.v10n6p128

Masa'deh, R., Al-Henzab, J., Tarhini, A., & Obeidat, B. Y. (2018). The associations among market orientation, technology orientation, entrepreneurial orientation and organizational performance. *Benchmarking, 25*(8), 3117–3142.

Masadeh, R., Sharieh, A., Jamal, S., Qasem, M. H., & Alsaaidah, B. (2020). Best path in mountain environment based on parallel hill climbing algorithm. *International Journal of Advanced Computer Science Application, 11*(9).

Masadeh, R., Alsharman, N., Sharieh, A., Mahafzah, B. A., & Abdulrahman, A. (2021). Task scheduling on cloud computing based on sea lion optimization algorithm. *International Journal of Web Information Systems*.

Mellor, S., Hao, L., & Zhang, D. (2014). Additive manufacturing: A framework for implementation. *International Journal of Production Economics, 149*, 194–201.

Ministry of Industry & Commerce. (2020). Retrieved from https://www.mit.gov.jo/Default.

Mishra, A. N., Devaraj, S., & Vaidyanathan, G. (2013). Capability hierarchy in electronic procurement and procurement process performance: An empirical analysis. *Journal of Operations Management, 31*(6), 376–390.

Mognol, P., Lepicart, D., & Perry, N. (2006). Rapid prototyping: Energy and environment in the spotlight. *Rapid Prototyping Journal, 12*(1), 26–34.

Müller, A., & Karevska, S. (2016). *How will 3D printing make your company the strongest link in the value chain: EY's global 3D printing report 2016*. Ernst & Young GmbH.

Obeidat, B., Hadidi, A., & Tarhini, A. (2017). Factors affecting strategy implementation: A case study of pharmaceutical companies in the middle east. *Review of International Business and Strategy, 27*(3), 386–408.

Obeidat, Z. M., Alshurideh, M. T., Al Dweeri, R., & Masa'deh, R. (2019). The influence of online revenge acts on consumers psychological and emotional states: Does revenge taste sweet? *Paper presented at the Proceedings of the 33rd International Business Information Management Association Conference, IBIMA 2019: Education Excellence and Innovation Management through Vision 2020* (pp. 4797–4815).

Oettmeier, K., & Hofmann, E. (2016). Impact of additive manufacturing technology adoption on supply chain management processes and components. *Journal of Manufacturing Technology Management, 27*(7), 944–968.

Oettmeier, K., & Hofmann, E. (2017). Additive manufacturing technology adoption: an empirical analysis of general and supply chain-related determinants. *Journal of Business Economics, 87*(9) 1, 97–124.

Oke, A. (2013). Linking manufacturing flexibility to innovation performance in manufacturing plants. *International Journal of Production Economics, 143*(2), 242–247.

Petrovic, V., Gonzalez, J., Ferrando, O., Gordillo, J., Puchades, J., & Grinan, L. (2011). Additive layered manufacturing: Sectors of industrial application shown through case studies. *International Journal of Production Research, 49*(4), 1061–1079.

Qian, M. (2020). Additive manufacturing. *The 2nd Asia–Pacific International Conference on Additive Manufacturing*. Royal Melbourne Institute of Technology, Melbourne, Australia. https://doi.org/10.1007/s11837-020-04034-6.

Rayna, T., & Striukova, L. (2016). From rapid prototyping to home fabrication: How 3D printing is changing business model innovation. *Technological Forecasting and Social Change, 102*, 214–224.

Ruffo, M., & Hague, R. (2007). Cost estimation for rapid manufacturing simultaneous production of mixed components using laser sintering. *Proceedings of the Institution of Mechanical Engineers, Part B. Journal of Engineering Manufacture, 221*(1), 1585–1591.

Rylands, B., Böhme, T., Gorkin, R., III., Fan, J., & Birtchnell, T. (2016). The adoption process and impact of additive manufacturing on manufacturing systems. *Journal of Manufacturing Technology Management, 27*(7), 969–989.

Saliba, S., Kirkman-Brown, J., & Thomas-Seale, L. (2020). Temporal design for additive manufacturing. *The International Journal of Advanced Manufacturing Technology, 106*, 3849–3857.

Schniederjans, D. G. (2017). Adoption of 3D-printing technologies in manufacturing: A survey analysis. *International Journal of Production Economics, 183*, 287–298.

Shannak, R. O., Masa'deh, R. M. T., Obeidat, B. Y., & Almajali, D. A. (2010). Information technology investments: A literature review. *Paper presented at the Business Transformation through Innovation and Knowledge Management: An Academic Perspective—Proceedings of the 14th International Business Information Management Association Conference, IBIMA 2010*, 2, 1356–1368.

Steenhuis, H. M., & Pretorius, L. (2017). The additive manufacturing innovation: A range of implications. *Journal of Manufacturing Technology Management, 28*(1), 122–143.

Tariq, E., Alshurideh, M., Akour, I., & Al-Hawary, S. (2022). The effect of digital marketing capabilities on organizational ambidexterity of the information technology sector. *International Journal of Data and Network Science, 6*(2), 401–408.

Thomas, D. (2016). Costs, benefits, and adoption of additive manufacturing: A supply chain perspective. *The International Journal of Advanced Manufacturing Technology, 85*(5–8), 1857–1876.

Thompson, M. K., Moroni, G., Vaneker, T., Fadel, G., Campbell, R. ... Martina, F. (2016). Design for additive manufacturing: trends, opportunities, considerations, and constraints. *CIRP Annals, 65*(2), 737–760.

Tuck, C., Hague, R., & Burns, N. (2006). Rapid manufacturing: Impact on supply chain methodologies and practice. *International Journal of Services and Operations Management, 3*(1), 1–22.

Tarhini, A., Mgbemena, C., Trab, M. S. A., & Masa'deh, R. (2015). User adoption of online banking in Nigeria: A qualitative study. *Journal of Internet Banking and Commerce, 20*(3). https://doi.org/10.4172/1204-5357.1000132.

Weller, C., Kleer, R., & Piller, F. (2015). Economic implications of 3D printing: Market structure models in light of additive manufacturing revisited. *International Journal of Production Economics, 164*, 43–56.

Westerweel, B., Basten, R., & Houtum, G. (2019). Preventive maintenance with a 3D printing option. Available at SSRN. https://ssrn.com/abstract=3355567. https://doi.org/10.2139/ssrn.3355567.

Wong, K. V., & Hernandez, A. (2012). A review of additive manufacturing. *Mechanical Engineering*, 1–10.

Wu, H. C., & Chen, T. C. T. (2018). Quality control issues in 3D-printing manufacturing: A review. *Rapid Prototyping Journal, 24*(3), 607–614.

Development of Market Analysis Study in Aqaba

Ali S. Hyasat and Ra'ed Masa'deh

Abstract This study aimed to understand the current conditions of the tourism product as well as tourist perceptions in Aqaba City. A quantitative approach methodology was developed to meet the research aims and objectives. The quantitative technique was used based upon the opinion of tourists who visited Aqaba to understand the current condition of tourism products and tourist perceptions. The result indicated that most of the tourists complained about the transportation and traffic. Therefore, there is a need for local institutions to overcome this problem by facilitating access to important tourist attractions areas. Moreover, these institutions should give more attention to the cleanliness issue in Aqaba. Tourists also indicated that Aqaba lacks good restaurants to use or café to enjoy in the city. This indicates that existing places are all popular initiating the need to construct new places with suitable entertainment for tourists. Aqaba Special Economic Zone (ASEZA) should encourage the private sector to establish more cafés, restaurants, and other facilities in Aqaba. Moreover, public institutions should control prices, as tourists complained about the manipulation of prices, especially with tourists using their ignorance of prices. Another weak element was raised by tourists is the poor activities included in the tourist programs in Aqaba. Tourists who visited Aqaba felt boring in particular during night times due to the poor program. The tourists expected to see the city with a wide range of tourist activities distributed in different places. Authorities and policymakers in Aqaba should create a suitable and enjoyable activates program for tourists. The study also recommended that tourism in Aqaba should be marketing along with Petra and Wadi Ram as standalone destinations. Furthermore, tourism in Aqaba needs a special

A. S. Hyasat (✉)
Tourism Management, Deanship of Scientific Research, Al-Balqa Applied University, Al-Salt, Jordan
e-mail: ali.hyasat@bau.edu.jo

R. Masa'deh
Department of Management Information Systems, School of Business, The University of Jordan, Amman, Jordan
e-mail: r.masadeh@ju.edu.jo

M. Alshurideh et al. (eds.), *The Effect of Information Technology on Business and Marketing Intelligence Systems*, Studies in Computational Intelligence 1056,
https://doi.org/10.1007/978-3-031-12382-5_33

Beach Management Board, which should be responsible for all tourism activities on the beach of Aqaba.

Keywords Market analysis · Tourism product · Tourist · Aqaba · Jordan

1 Introduction and Background

Tourism continues to grow fast, exceeding expectations despite the economic crises and geopolitical conflicts in many regions around the world. This growth has deepened the diversification of tourism offerings to become one of the fastest-growing economic sectors in the world (Alaali et al., 2021; ASEZA, 2015; EDP, 2011–2013). The UNWTO promotes tourism as a vehicle for economic development, social inclusion, and sustainable environmental development. The UNWTO recognized tourism for its ability to enhance peace between nations, create jobs, varied employment opportunities through youth and women inclusion in the host local communities, and contribute to poverty elevation, social development, cultural diversity, and a sustainable environment. Meanwhile, the UNWTO emphasizes the shared global responsibilities by tourism sector stakeholders in local communities to achieve sustainable tourism development in destinations (UNESCO, 2015; USAID, 2009).

In Jordan, tourism is a key driver of the national economy, it is the single largest employer, and the highest generator of foreign exchange. The potential for economic and social development stemming from tourism is high, particularly because Jordan has world-class historic and religious sites, fascinating cultural heritage, and stunning natural landscapes, including six UNESCO World Heritage sites. Tourism accounts for 13% of the Kingdom's income generation; it is a fundamental organ of employment operations and a vital contributor to the national economy. According to the Jordan National Tourism Strategy (JNTS), about 50,000 people were directly employed in the tourism sector. It is believed that employment generated by the indirect economic impact of tourism is several times greater than the direct employment generated. The Government of Jordan, through the execution of its Executive Development Program 2011–2013 (EDP), is trying to strengthen the competitiveness of the tourism industry by developing and diversifying tourism products, improving the quality of tourism services, exploring new markets, and increasing tourism-related investment. The Jordanian Ministry of Tourism and Antiquities (MoTA) is working on comprehensive tourism development through a public–private partnership to increase tourism revenues (Abu Zayyad et al., 2021; Alananzeh et al., 2018; Aljumah et al., 2021; Alshurideh et al., 2015; JDS, 2014; JNTS, 2010a, b; MOTA, 2011).

However, despite its advantages and its importance to the economy, Jordan's tourism sector was performing at a level far below its potential. Many factors contributed to lower visitor spending per arrival compared to the world average and some of Jordan's closest competitors (Egypt and Lebanon for example) (Deeb et al., 2020; European Union Development Programme (IS-ASEZA), 2005; Hamadneh et al., 2021), this disparity was largely due but not limited to three main reasons: The short length of stay, low visitor expenditure per day, and un-coordinated visitor servicing after arrival in Jordan.

According to the classification of the MoTA, tourist attraction resources in Jordan were divided into three main types in Jordan. These are historical and Cultural Heritage Sites., Natural Sites, and Medical Sites (Getz, 1997; Goldman et al., 1997; Jawabreh et al., 2020; Kamakura et al., 1991; Khwaldeh et al., 2020; Kim & Mauborgne, 1993; Kotler & Armstrong, 2007; Kotler et al., 2010; Lovelock et al., 2004; Luo, 2002). Under each type, there are several sites and tourist attractions including tourist services and facilities. As one of the major tourist sites in Jordan, Aqaba has the base ingredients for tourism success. In addition to its beautiful underwater environment and excellent climate, it has, unlike other Red Sea Resorts, its history, heritage, and character. Moreover, it benefits from proximity to a world-famous attraction—Petra—and to the desert splendor of Wadi Rum (Masa'deh et al., 2017, 2018, 2019; McDonald, 1999).

Aqaba, however, faces a difficult marketing environment. Worldwide economic and political problems have severely impacted the international tourism industry. Competition, particularly from other Red Sea resorts and also from Lebanon and Turkey, has intensified; earnings are under pressure. A further planned expansion of hotel rooms in Aqaba could damage occupancy and profitability unless strong and targeted marketing can deliver vigorous growth over the next five years. However, if Aqaba takes the necessary steps to improve its accessibility, enrich its product, strengthen its brand differentiation, and professionalize its marketing, then there is no reason why the spectacular success achieved in attracting resort investment cannot be matched in the future by the overall health of its incoming tourism industry (Ritchie & Crouch, 2003; Sweiss et al., 2021).

Aqaba has hitherto been held back by lack of accommodation capacity, but this phase is coming to a rapid end. Aqaba now urgently needs to improve its marketing performance (Tariq et al., 2022). According to the JNTS, Aqaba should improve and diversify its product offer, especially in the historic city center—visitors must be given good reasons to stay longer and spend more and to build demand. Both of these elements are vital to successful marketing, which needs to embrace the full cycle from research and product development through to promotion, with research then feeding back into product development and modification.

In conclusion, there was still some lack in analyzing and discussing all aspects of tourism in Aqaba to spot the light on the tourism market in Aqaba, therefore, this study was prepared to fill this gap (Hyasat et al., 2022). This study was conducted to understand the current condition of tourism products in Aqaba and tourist perceptions toward the City.

1.1 Materials and Methods

As previously mentioned, the main aims for this study were to understand the current conditions of the tourism product as well as tourist perceptions in Aqaba City through the following objectives:

- To identify the profile of tourists and their visits overview.
- To classify the type of activities and the amount of expenditure.
- To identify the perception of Aqaba (satisfaction and problems).

The study was based on applying quantitative research methodology. A quantitative method was used to meet the main aim and objectives for the survey based on developing a questionnaire that targeted tourists who visited Aqaba, to have a general understanding of their opinion about tourism in Aqaba and their state of participation. Questions were asked to tourists who visited Aqaba about their visits, activities, and amount of expenditure, whether their visit was arranged by a travel agency or individually carried out, and their perception of Aqaba, satisfaction, and problems. By using a random sample technique, the researchers distributed questionnaires to tourists, (105) out of those questionnaires have been filled up by tourists and collected by the researchers. However, the researchers found that there were only (80) questionnaires filled up properly in a reliable manner which enabled the researchers to conduct the research. According to researchers, the use of a questionnaire could contribute to mutual method prejudice; on the other hand, others stated that the reality of common method bias in self-reported surveys is unfounded (Podsakoff et al., 2003; Spector, 2006). To avoid this concern, the researchers ensured a rigorous methodology. A pilot study was used to address the concern. Moreover, the literature review, statistical data, and all documents about tourists and tourism in Aqaba have been collected by the researchers from the official institutions in Aqaba.

The list of samples was based on the random sampling technique. To ensure that the questionnaire reached most tourists from different nationally the questionnaire was distributed in most of the sites and tourists' attractions in Aqaba, which were expected to be seen by most of the tourists who visited Aqaba.

The researchers ensured the balanced rating scales to reduce the effect of bias. Descriptive analysis was then used to analyze the data. Data analysis included frequencies, percentages, means, and ranking of the results. The data was reflected on answering the fundamental questions of the research for the questionnaire, and it was analyzed to achieve the aims and objectives of this study. After analyzing the data collected for the survey, this study produced a framework of tourism-related business and tourist participation in Aqaba City. Some important factors such as a profile of the tourist, overview of the visit, types of activities, and any arrangements by travel agencies have been examined. Having presented the data, the researchers conducted a SWOT analysis to identify all strengths, weaknesses, opportunities, and threats points of the tourism market in Aqaba.

2 Data Presentation and Analysis

2.1 Sample Characteristics

The sample covered composed of a wide variety of nationalities. The highest number of tourists was from Russia with a percentage of 20.0%, followed by Italian tourists with a percentage of 15.0%. The medium percentage was recorded for German, French, American, and Sweden with percentages ranging from 6 to 8%. About 4% of the sample was recorded for Australian, Spanish, and Jordanian, while the lowest was recorded for Ukraine, Japan, Iraq, Israel, and UK (Table 1).

The majority of the sample was males with a percentage of 56.4%, while females composed 43.6%. The majority of sample ages ranged from 31–50 years with a percentage reaching 65.8% of the sample. The younger tourists formed about 21.1% of the sample, while the older tourists formed only 13.2% (Fig. 1).

Table 1 Sample structure

Nationality	Frequency	Percent
Russian	16	20.0
Italian	12	15.0
German	8	10.0
French	8	10.0
American	6	7.5
Sweden	6	7.5
Australian	4	5.0
Spanish	4	5.0
Jordanian	4	5.0
Hungarian	2	2.5
Okrine	2	2.5
Japan	2	2.5
UK	2	2.5
Iraq	2	2.5
Israel	2	2.5
Total	80	100.0

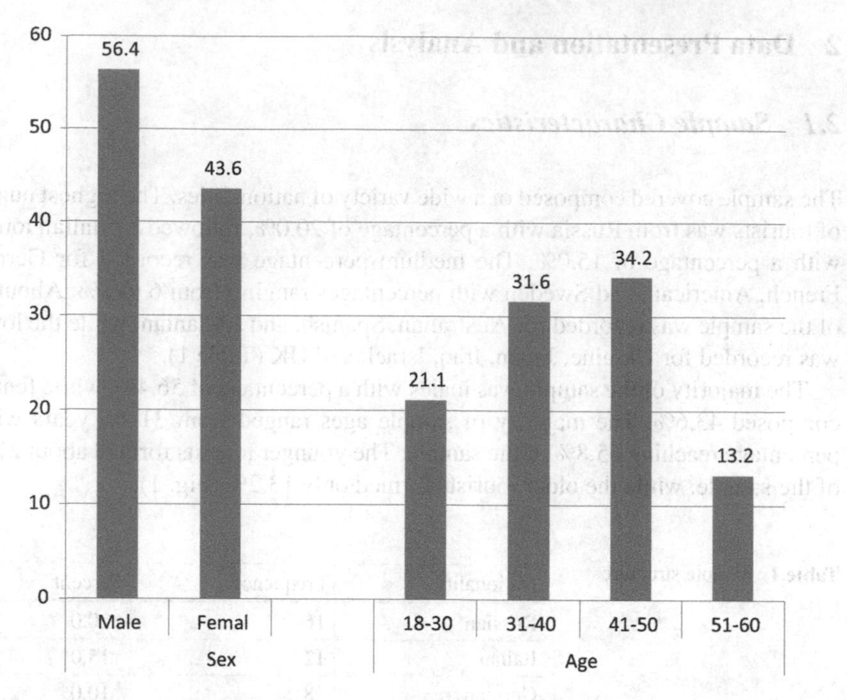

Fig. 1 Characteristics of sample tourists

2.2 *Visit Overview*

2.2.1 The Time of Stay in Aqaba

The majority of the sample spend two to three days in Aqaba with a percentage of 55.0%, while the number of tourists who spend more than three days was 42.5%, the lowest percentage of visitors stay for one day or less (Fig. 2). Most tourists accompanied husband/wife with a percentage of 42.5%, followed by a friend with a percentage of 25.0%. About 15.0% of tourists did not have any company in their visit to Aqaba. The least percentages had colleagues or family members as a company during their visit to Aqaba (Fig. 2).

2.2.2 Method of Transportation

Tour buses were the dominant used as a method of transportation (30.0%). The rented cars were the second method of transportation used to get to Aqaba followed by the usage of both private or rental cars and the same time using taxi and tour buses through their visit to Aqaba with percentages 22.5% and 20.0% respectively. The fourth tool used the use of a private car with a percentage of 12.5%. The same and

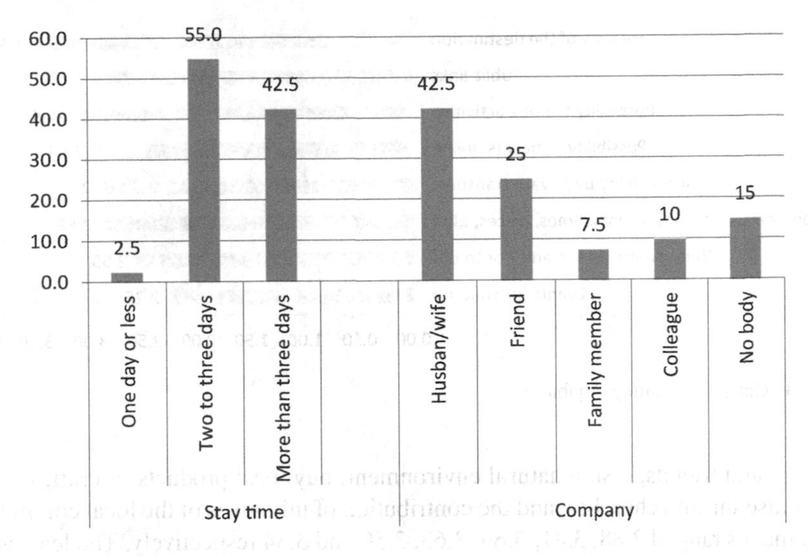

Fig. 2 Time of stay and tourist company

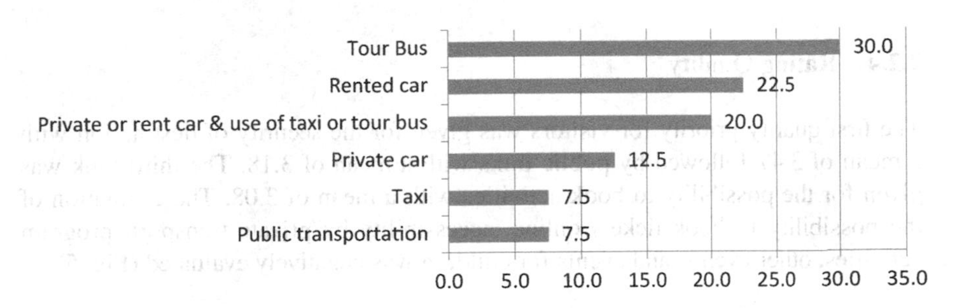

Fig. 3 Method of transportation

least percentage was recorded for Taxi and public transportation with a percentage of 7.5% (Fig. 3).

2.2.3 Causes of Choosing Aqaba

Means calculation was used to evaluate the rank of different causes of choosing Aqaba indicating that 5 is the highest rank and 1 is the lowest rank. The results in Fig. 4 show that break out of the routine and relaxation with a mean of 4.38 was the cause of selecting Aqaba. The second factor that affect the selection of Aqaba listed was the local product (Food and beverages) with a mean of 4.09. Meeting new people was ranked as a third cause of selecting Aqaba with a mean of 4.08 and curiosity about places/events unknown and unseen places was the fourth cause with a mean of 4.03. The causes living strong emotion and addictive, share the experience with

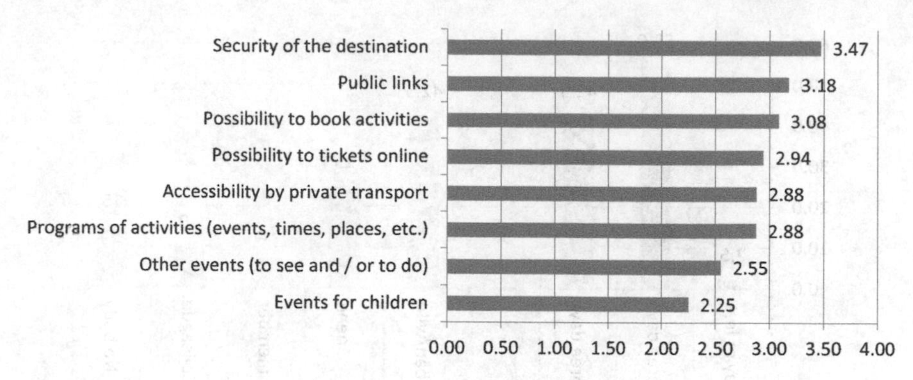

Fig. 4 Causes of visiting Aqaba

family and friends, visit a natural environment, buy local products or craft, visiting the museum, and churches, and the contribution of initiatives of the local community had means ranged 3.88, 3.81, 3.69, 3.63, 3.50 and 3.34 respectively. The least value was given for playing outdoor sports with a mean of 2.93 (Fig. 4).

2.2.4 Rating Quality

The first quality priority for visitors was given for the security of destination with a mean of 3.47 followed by public links with a mean of 3.18. The third rank was given for the possibility to book activities with a mean of 3.08. The evaluation of the possibility to book tickets online, accessibility by private transport, program activities, other events, and events for children was negatively evaluated (Fig. 5).

2.2.5 The Type of Accommodation Selected

Most of the sample indicated that they used hotels during their stay in Aqaba. The highest proportion used four starts and five starts hotels with a percentage 46.2% for each and only 7.7% used three start hotels. Only 2.5% of the sample stayed with their relatives or friend during their stay in Aqaba (Fig. 6).

2.2.6 Knowing About Aqaba

The majority of the sample (39.0%) knows about Aqaba through travel agents, followed by friends and family with a percentage of 26.3%. The guidebook was the third source of knowledge about Aqaba with a percentage of 13.6%. The rest of the methods of knowing about Aqaba were distributed according to their importance among websites (11.0%), visitors center in Amman (5.1%), brochures at hotels (3.4%), and other methods (1.7%) (Fig. 7).

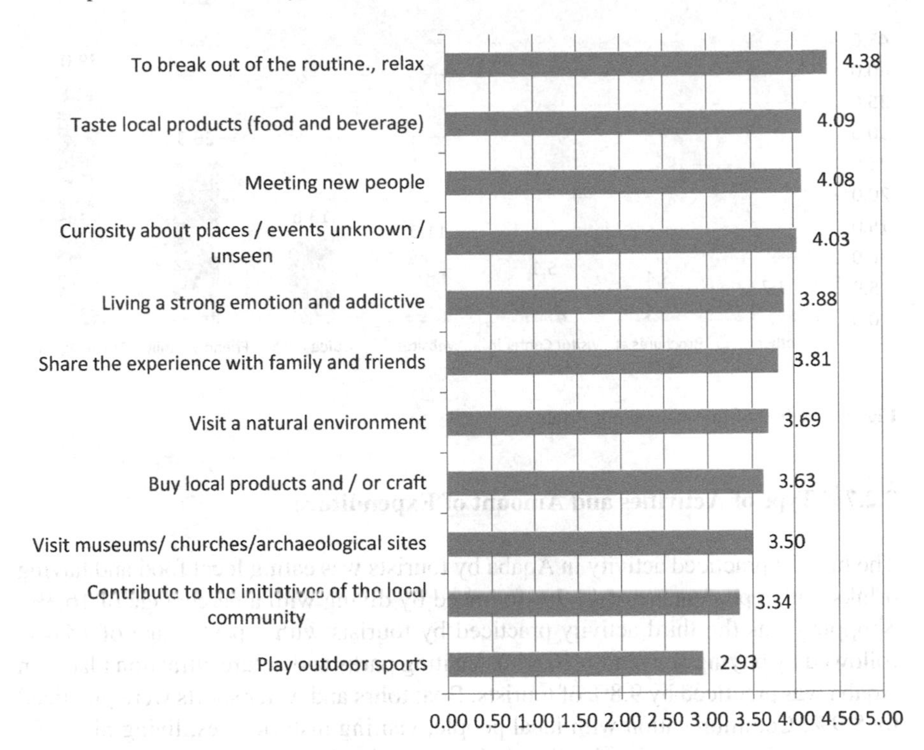

Fig. 5 Rating quality of different facilities

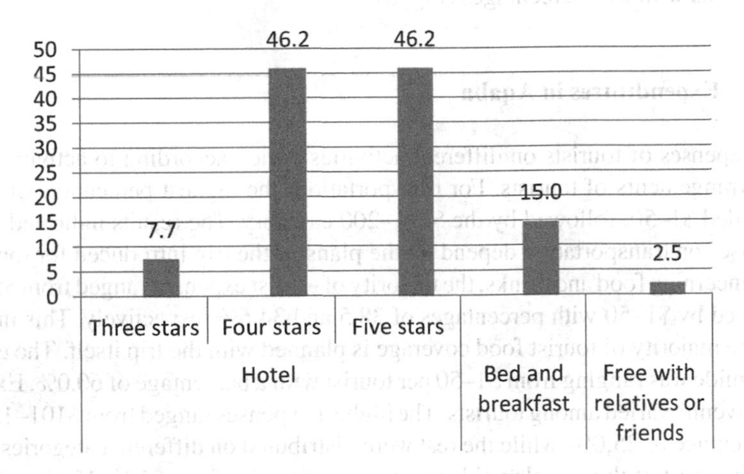

Fig. 6 Type of accommodation selected

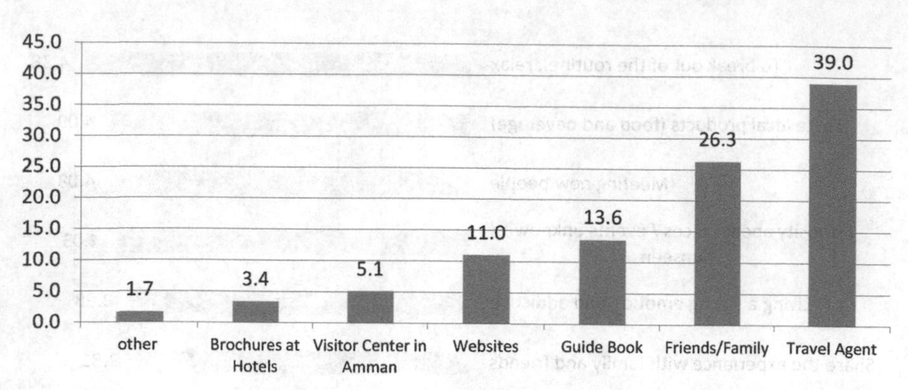

Fig. 7 Method of knowing about Aqaba

2.2.7 Type of Activities and Amount of Expenditure

The highest practiced activity in Aqaba by tourists was eating local food and having drinks with a percentage of 18.7% followed by diving with a percentage of 16.3%. Shopping was the third activity practiced by tourists with a percentage of 15.4%, followed by buying souvenirs (10.6%). Visiting parks and nature attraction places in Aqaba was practiced by 9.8% of tourists. Boat tours and water sports were practiced by 8.9%. Communication with local people, visiting historic sites, living nightlife, other activities, and visiting family relatives and friend house was activities practiced by tourists with low percentages (Fig. 8).

2.2.8 Expenditures in Aqaba

The expenses of tourists on different activities varied according to activity and the trip arrangements of tourists. For transportation, the highest percentage of tourists expended $1–50, followed by the $151–200 category. The results indicated that the expenses on transportation depend on the plans of the trip introduced for tourists.

Concerning food and drinks, the majority of tourist expenses ranged from $51–100 followed by $1–50 with percentages of 38.5 and 34.6% respectively. This indicates that the majority of tourist food coverage is planned with the trip itself. The expense for a guide was ranging from $1–50 per tourist with a percentage of 60.0%. Expenses on souvenirs varied among tourists. The highest expenses ranged from $101–150 with a percentage of 25.0%, while the rest were distributed on different categories almost equally. Most of the sample paid entrance fees ranging from $101–150 (see Figs. 9, 10, 11, 12, 13 and 14).

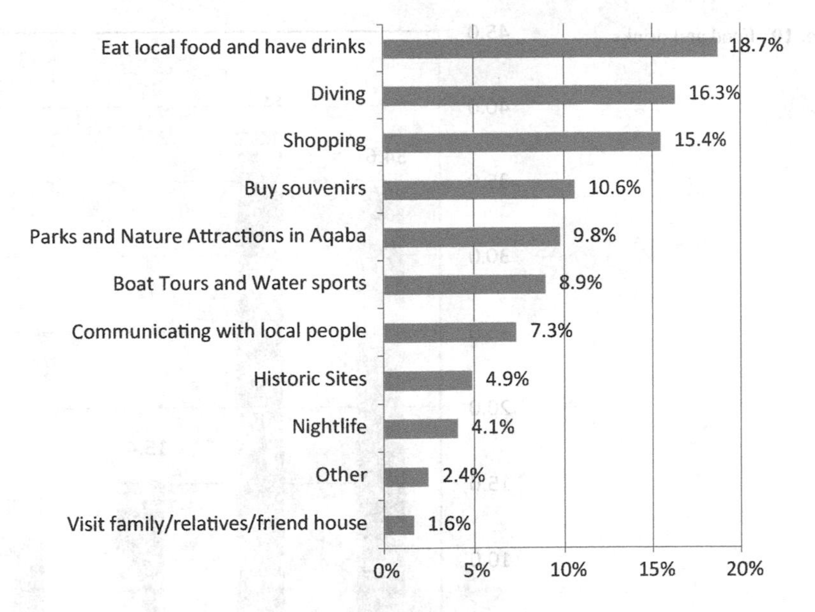

Fig. 8 Type of activities practiced in Aqaba by tourists

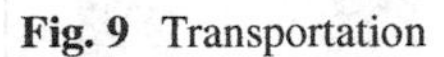

Fig. 9 Transportation

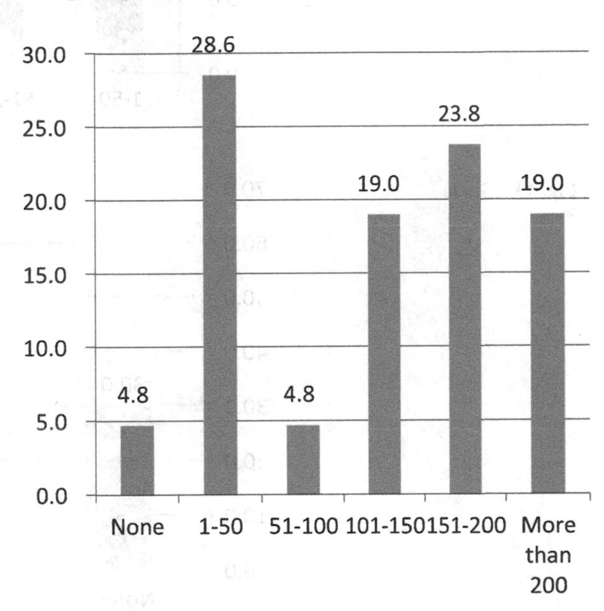

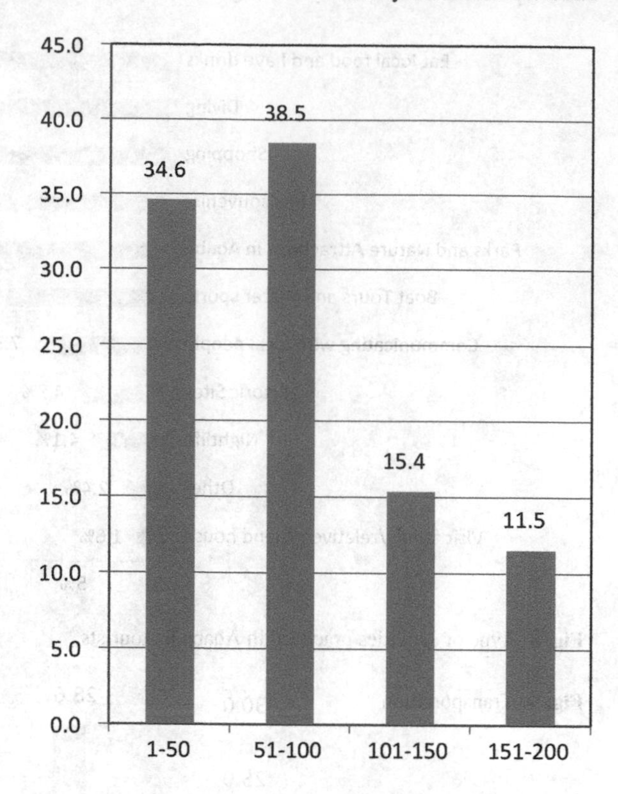

Fig. 10 Food and drinks

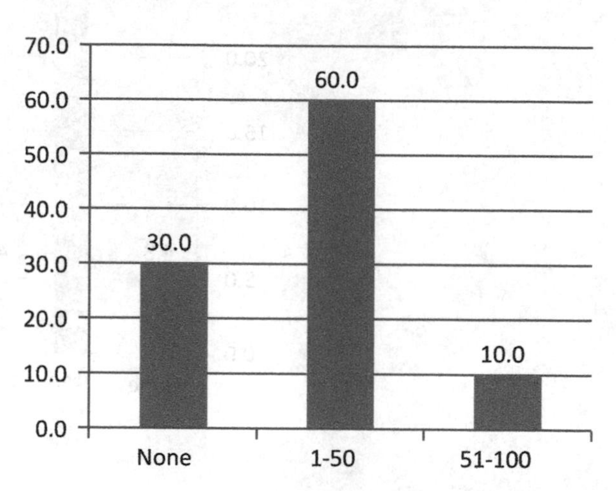

Fig. 11 Guide

2.2.9 Arrangement by a Travel Agency

The visit for the majority of tourists was arranged by a travel agent with a percentage of 74.4%. This indicates that the majority of tourists are committed to the travel

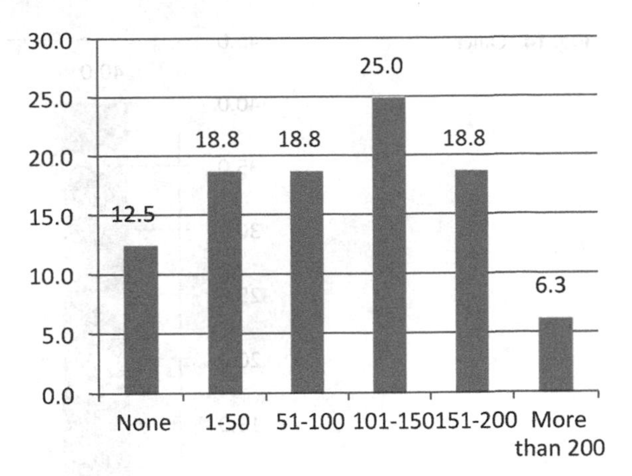

Fig. 12 Souvenir

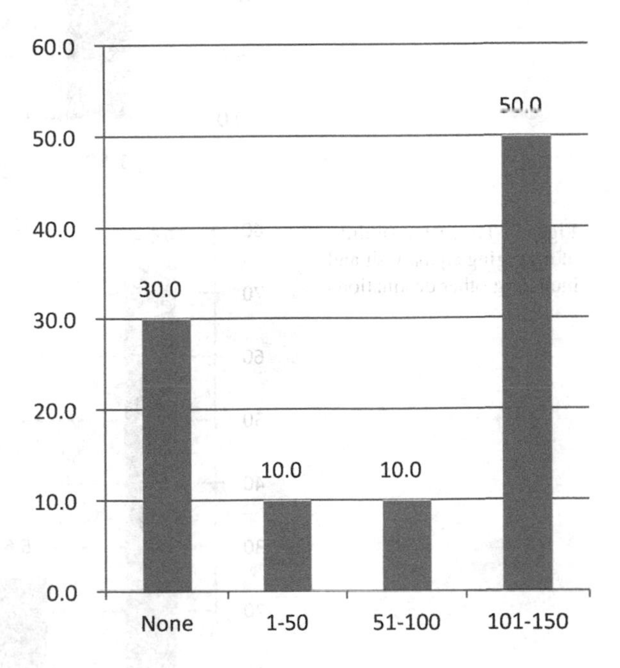

Fig. 13 Entrance fees

agent's plans to visit Aqaba. About 62.5% of the tourists indicated that other places are planned for a visit besides Aqaba (Fig. 15).

2.2.10 The Destinations Included in the Package

The first four places included in the package were Petra, the Dead Sea, Wadi Rum, and Mount Nebo with percentages of 23.0, 21.8, 19.5, and 12.6% respectively. The

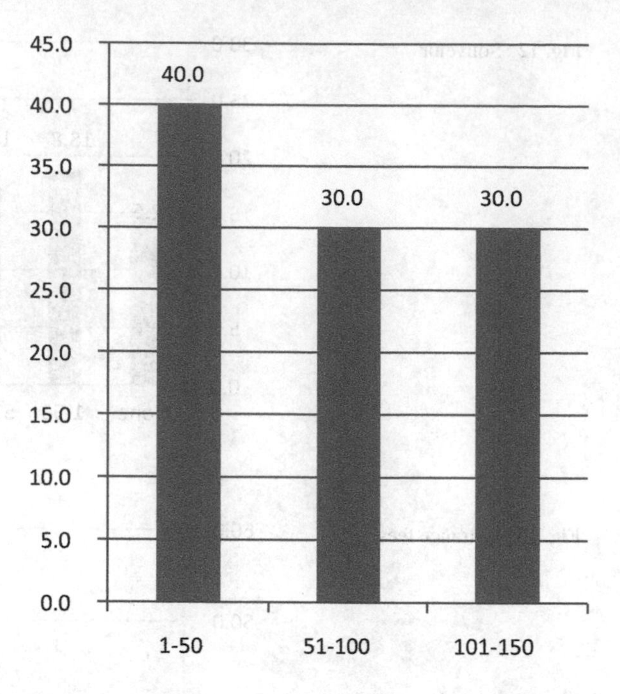

Fig. 14 Other

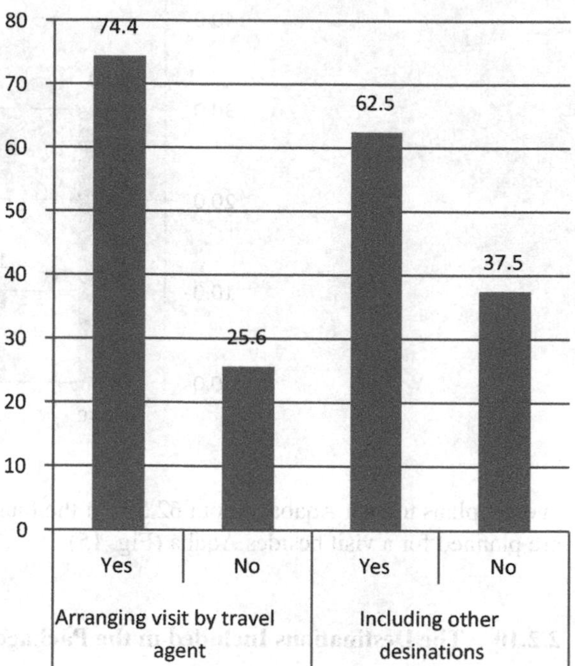

Fig. 15 The responsibility of arranging aqaba visit and including other destinations

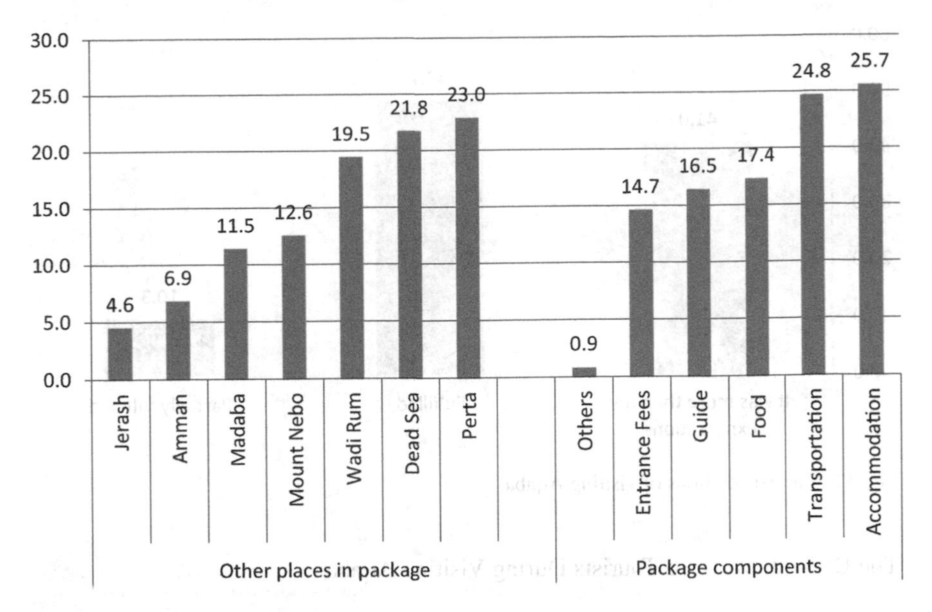

Fig. 16 Other places in package and its components

reset of places distributed on Madaba, Amman, and Jerash. The package included the accommodation with percentage 25.7%, transportation with percentage 24.8%, food (17.4%), guide (16.5%), and entrance fees (14.7%), (Fig. 16).

2.2.11 Average Expenses Paid

The average total expenses paid was $1675.3 with a standard deviation of 1188.97 indicating that there is a wide variation of the totals paid in visiting Aqaba. The high variation has resulted from the different sources of tourists included in the sample. The minimum expense was $204.55 while the highest recorded was $5000.

2.2.12 Perception of Aqaba (Satisfaction, Problems)

Expectations of Visiting Aqaba

The results showed positive attitudes of the sample for visiting Aqaba. The results showed that 41.0% of the sample was more than their expectation, while 48.7% were fulfilled in their visit. The results also showed that only 10.3% were partially fulfilled (Fig. 17).

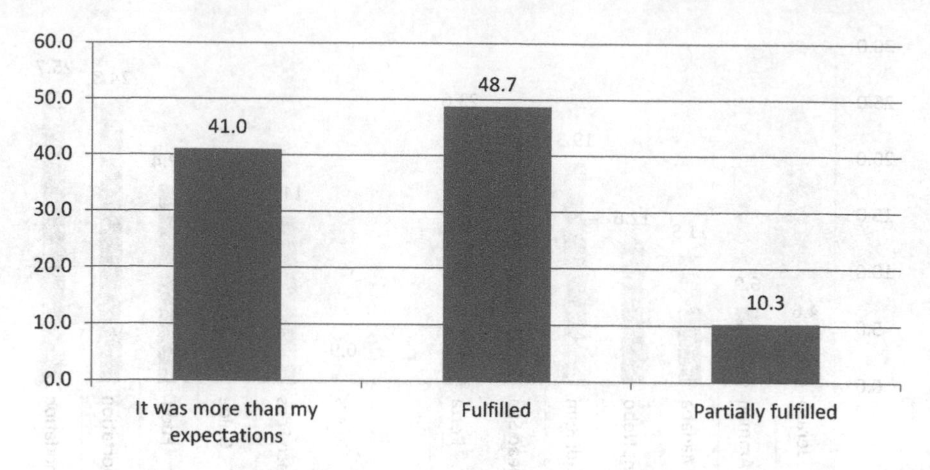

Fig. 17 The expectations of visiting Aqaba

The Difficulties Faced Tourists During Visiting Aqaba

The garbage and bad traffic were the first two difficulties faced tourists during their visit with a percentage of 24.5% for each. Poor and insufficient services were in the third rank with a percentage of 13.2%. Boring and no information for tourists were two difficulties faced by tourists with a percentage of 9.4% followed by no good café or restaurants (7.5%). The rest of the difficulties were minor in their classification (Fig. 18).

2.2.13 Recommending Others to Visit Aqaba

More than half of the sample indicated that they will recommend others to visit Aqaba, while 21.6% indicated that they perhaps recommend others to visit Aqaba. The rest of the sample were not sure, or do not recommend or never recommend visiting Aqaba (Fig. 19).

2.2.14 The Preferred Events in Aqaba According to Tourist's Perspectives

About 18.3% of tourists indicated that they prefer to have traditional food festivals, while a guided tour of walking trails on a specific story such as history, craftsman and sweats was the second activity preferred by tourists with a percentage of 17.3%. Traditional music and dance festival was the third preferred activity by tourists (Fig. 20).

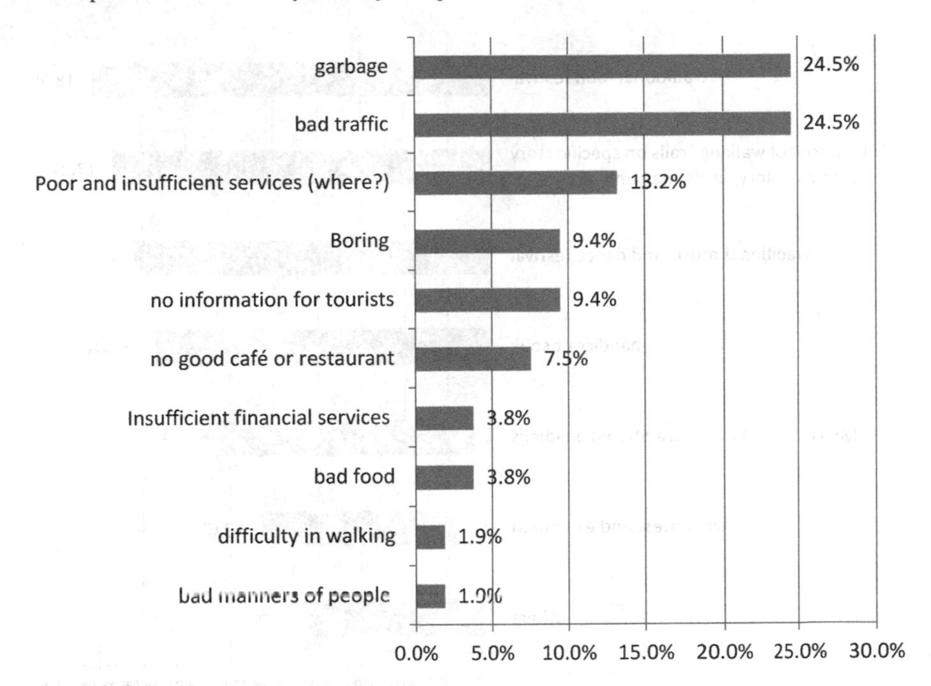

Fig. 18 The difficulties faced tourists during visiting Aqaba

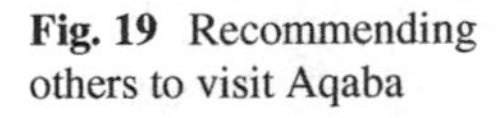

Fig. 19 Recommending others to visit Aqaba

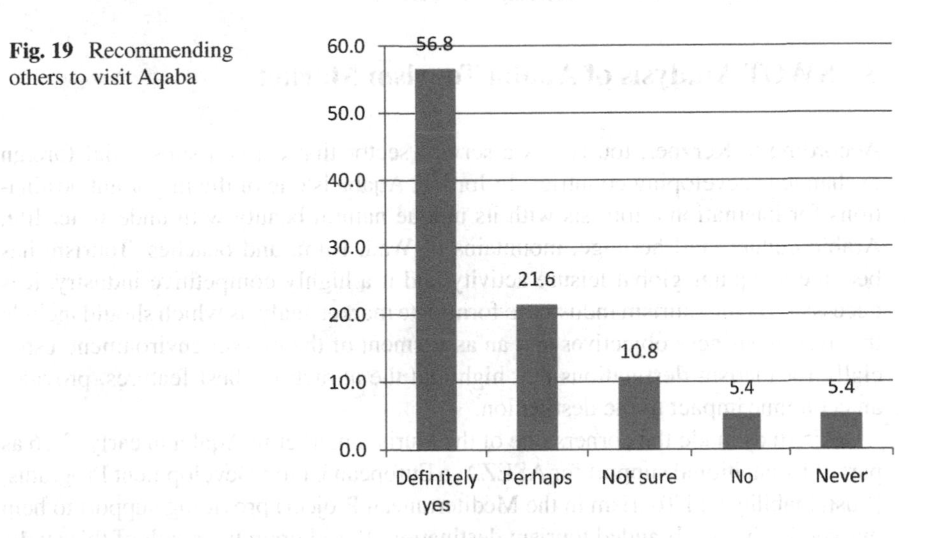

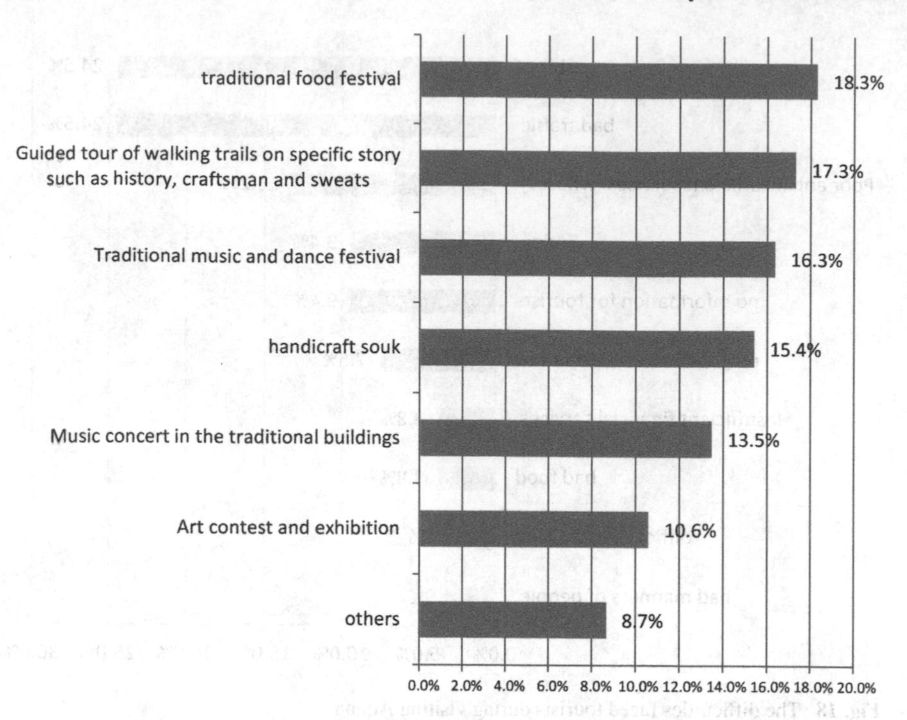

Fig. 20 The preferred activities according to tourists

3 SWOT Analysis of Aqaba Tourism Market

According to Kerzner, tourism is a service sector that earns a substantial foreign exchange to developing countries. In Jordan, Aqaba is one of the important destinations for international tourists with its unique natural beauty with underwater life, Arabic culture, and heritage, mountains of Wadi Rum, and beaches. Tourism has become a popular global leisure activity and it a highly competitive industry, it is necessary for the tourism industry to formulate market analysis which should include the overall business objectives and an assessment of the market environment, especially for tourism destinations that highlight the attractions best features, provides an economic impact to the destination.

This study made the cornerstone of the tourism market in Aqaba in early 2016 as part of its intuitional support for ASEZA, a European Union Development Programs, (Sustainability and Tourism in the Mediterranean Project) providing support to help market Aqaba as a branded tourism destination. Based upon the result of this study, tourists' expectations before visiting Aqaba depend on the information they receive from their agents, friends, or any other source of information they use. When the current situation does not meet the expectations, especially for foreign tourists, the long impact on promoting Aqaba as a tourist place will drop. The image of Aqaba as a tourist place will depend on different characteristics including transportation, what can be visited while getting there, and what is there. These three components were covered by the questionnaire, which will be used to analyze the market their using SWOT analysis.

Table 2 shows the SWOT analysis of tourism's current status in Aqaba. The matrix shows that there are many strong components of tourism in Aqaba. The most important of them those different attractive elements are there in Aqaba. The weather in winter times is considered one of the attractive elements for tourism, especially for local tourism. Another important element is the existing different historic places in Aqaba which represent different old ages as well as summarize part of Jordan's history. Moreover, the sea in Aqaba is a major tourism component that facilitates swimming activities as well as other related activities. Getting to Aqaba as the southern city of Jordan tourist needs to pass three governorates Karak, Tafeela, and Maan. The three governorates are considered rich in tourism places that are considered highly supportive in planning trips to Aqaba. The strength elements showed that the nature of Aqaba is considered encouraging and strong base for tourism in the city.

This result was online with the updating Tourism Market Strategy in (2013–2015). The strategy suggested that the traditional trend of international tourists coming to Aqaba for a relaxation-beach experience as part of a 'classic' Jordan tour is continuing. However, there is an increase in the number of visitors, particularly from Scandinavian countries, and to some extent, Eastern Europe, in search of good weather is using Aqaba as a base and extending the length of stay significantly for their segment.

Table 2 SWOT Analysis Matrix for Aqaba Market According to the Result of this Study

Strength	Weak
Wide variety of tourism components – Sea – Different weather – Historic places – Closeness to other tourist places – Wadi Rum, Petra – Summer sun (charter and domestic) – Winter sun (mainly charter and domestic) – Diving	– Weak transportation methods – Immature infrastructure (City – cleanness, traffic jam, local tourism promotion – Program activities – Children activities – Online ticketing – High Prices – Lack of well-arranged parks and attracted natural places
Opportunity	**Threats**
– Promoting historic places in Aqaba – Improving cultural activities including folklore, food, and other parties that reflect Aqaba – Improving nightlife activities – Improving infrastructure – Using empty places to construct new tourist activities to alleviate crowding – Promoting Aqaba as an integrating component with other tourism places in Jordan	– High prices – Excluding local area citizens from tourist activities – Lack of tourist entertainments places (such as cafés, restaurants, and night entertainment places) – Competitors from the near tourism destinations

However, overall Aqaba's length of stay is really low at 2.17 nights for three-four- and five-star hotels.

According to the Market Tourism Strategy (MTS), the main products being sold are as follows:

- Summer sun (charter and domestic)
- Winter sun (mainly charter and domestic)
- Diving
- Cruise tourism
- Heritage (base for Petra)
- Desert and activities (base for Wadi Rum)
- Other water sports and beaches (mainly domestic tourism).

On the other hand, the analysis of the situation in Aqaba in this study shows different weak elements in tourism infrastructure in the city which as pointed out by tourists. The major one was the lack of transportation management in the city, which makes it a weak point for tourists to rely on when touring in the city. The traffic jams in the city were indicated as other weak points. The crowding has resulted in the concentration of facilities in the limited area causing traffic jams and difficulties in pedestrian movement. The clean of the city is considered another weak point that should be handled. Even though, the efforts made to improve the cleanliness of the city, the culture of cleanness should be distributed among the tourists, particularly local ones to help improve the cleanliness of the city.

Another weak element was raised by tourists is the poor activities included in the tourist programs in Aqaba. These poor programs create boring for tourists in particular during night times. The tourists expected to see a city with a wide range of tourist activities distributed in different places. Moreover, the city lacks any activities for children's entertainment, which creates a shortage of services available for tourism in the city. Online ticketing was a difficulty faced by tourists when reserving in activities, indicating the presence physically to solve this issue. This creates an overburden for tourists. Some tourists complained of the high prices in Aqaba, which seems impediment for tourists to utilize a wide variety of activities. Some other tourists have shown that Aqaba lacks public parks and natural places that can be used for entertainment.

Nevertheless, some of the weaknesses points were obtained from this study were also online with the (MTS), these can be summarized as follow:

- Aqaba's hotels are not considered price-competitive. There is an understanding of the increase in the cost of utilities and overheads and that the benefits of Special Economic Zone status are not being passed on.
- There is a severe lack of hotels outside of the five-star category.
- Although the issue has been alleviated somewhat by the launch of the Berenice Beach Club which is considered a quality product, limited public access to the beach is a concern to operators.

- The availability of quality restaurants, other than hotel restaurants is severely limited and does not support extending the length of stay. In addition, there are no quality authentic restaurants or food experiences.
- The current range of activities other than water-based activities and excursions to Petra and Wadi Rum are extremely limited. There is an understanding that some desired experiences are being developed by the mega projects however, more daytime and nighttime activities are required in the city. The development of such activities is critical.
- There is a great lack of pedestrian-friendly areas and footpaths for tourists to explore. Requests were made for closing some streets to cars either for a 24 h period or at least from midafternoon until the evening.
- The increase in domestic tourism has increased the challenge that lack of parking presents. The result is considered to be having a negative effect that is both dangerous and time-consuming and greatly impacts the quality of the tourism experience.
- There is a great lack of pedestrian-friendly areas and footpaths for tourists to explore. Requests were made for closing some streets to cars either for a 24 h period or at least from midafternoon until the evening.

As a result of this study, there are a lot of opportunities in Aqaba that can be utilized to improve tourism and create a more attractive city. One major activity is promoting historic places in Aqaba. This activity to be included in tourists' programs will introduce added value for tourists. The visitor of Aqaba expected to see some folklore activities that reflect the culture there, but these activities have not existed anywhere in Aqaba city. The governorate of Aqaba should find special streets for talented people to practice these activities in. Moreover, the introduction of local food will be considered an encouraging tool for tourists to share and live while visiting Aqaba. Most tourists complained of boring because it has resulted in a lack of street activities that fill the tourist times.

Citizens in Aqaba should be trained to participate in tourism activities. Local vocational training programs should be including Aqaba citizens to practice activities that are considered a source of income and improve entertainment in the city. Folklore programs should be stated in continuous programs all over the year. On the other hand, tourists indicated that Aqaba lacks good restaurants to use or café to enjoy in the city. This indicates that existing places are all popular initiating the need to construct new places with suitable entertainment for tourists.

The core of the city suffers from crowding due to the concentration of most activities in this part of the city, while the other parts lack any attractive activities or trade movement that helps in attracting visitors out of crowding. Also, the traffic jams occurred in Aqaba as the concentration of trade movement is concentrated in a little number of streets which creates this crowding. The opportunity in this aspect is to expand activities to other parts of Aqaba which improve the entertainment of tourists in one aspect and solve the traffic jams in the middle of the city. The opportunity to succeed in arranging trips to Aqaba is to include it as part of the integrated program with other tourism places in Jordan such as Petra and Wide Ram.

According to the MTS many opportunities should the Jordanian tourism market in general and the tourism market in Aqaba in particular benefit from them, for example, Asia showed strong growth of 7% in outbound travel as incomes rise and consumers can travel more. China and Japan in particular performed strongly with double-digit growth rates in outbound travel. Prospects for Asian tourism in 2013 are even stronger.

The Market Tourism Strategy concluded that many opportunities are available for the tourism market in Aqaba; such opportunities can be as follows:

- Responsible Tourism: The trend in demand for responsible tourism is continuing to rise and greater account is being taken off the congestion tourism generates and of the negative effects on resources and host communities that can come with it. Although some developers (such as Saraya) are giving extensive emphasis to Corporate Social Responsibility (CSR), Aqaba has a long way to go in terms of responsibility. Green practices, such as reducing waste and emissions, using alternate energy sources, and producing natural products, have become something of a "me too "cause in recent years. Yet companies and destinations with a history in green innovation have reaped the most benefits and are making real cost savings, and will continue to do so through the economic recession and beyond. ASEZA has initiated a Green Key program for hotels, and this needs to be marketed.
- Community-Based Tourism: There is a growing demand for transparency in social and economic benefit to communities and tourism contribution to preservation and promotion of cultural heritage. There is growing support by the international tourism trade, including major international operators to recognize destinations and businesses who are actively addressing these sustainability themes. There is an increasing trend among tour operators in transparency in their choice of suppliers based on these themes. Based on its brand position and differentiation from other Red Sea resorts as a living city, the demand for community-based tourism is a good opportunity for Aqaba through the development of community-based tourism experiences and community engagement. To support the development of community-based tourism, the implementation of ongoing tourism awareness and training plan is required as a priority.
- Diving Tourism: Diving tourism is the main reason for travel and as an ancillary product continues to grow and divers are continuously looking for new and interesting destinations to dive. According to Professional Association of Diving Instructors (PADI) figures, in 2012, PADI cumulative diving certification figures alone reached more than 21 million worldwide – though some divers may hold more than one certificate. Two-thirds of PADI certified divers are men and though divers cover every age bracket, the median age of divers is 30. While this is an easy segment to target, consideration must be maintained in that the diving product worldwide needs particular environmental care, the highly informed diving community is extremely sensitive to coral deterioration and litter problems, and diving destinations that do not manage and regulate these issues face a backlash.

- Culinary Tourism: The tourism market internationally is increasingly being influenced and motivated by food and culinary experiences. Visitors are increasingly demanding a wide range of quality dining experiences and the availability of these experiences is influencing both destination choice and the perception of the quality of the overall holiday experience and satisfaction criteria. In addition, there is a growing demand for a variety of quality, authentic local food, and culinary specialties. Tourists want to try distinctive foods that reflect the tradition, heritage, and culture of a place, and which preserve traditional forms of agriculture and cultural heritage. As food and dining are a major component of Jordan's history and culture, from authentic dishes to seasonal dishes to celebratory dishes, Aqaba can benefit from this growing market if the range of experiences is increased and standards and quality are raised.
- Family Tourism: The family tourism market continues to grow internationally from both traditional tourism markets and new tourism markets. Aqaba has had growing success in attracting family business, particularly after the establishment of Tala Bay. The potential for family tourism from traditional markets such as Russia and new markets such as China and India are great and are increasing.
- Shopping: Shopping while on holiday has always been an income generator. While there is a significant increase in shopping destinations, shopping as an ancillary product remains one of the highest contributors for both traditional and developing tourism markets.
- Value for Money: While there will always be ultra-deluxe products catering for small but growing numbers of wealthy clients and special occasions, value-for-money has become critical in most current travel decisions, a growing trend as a result of the economic crisis, facilitated by numerous travel blogs, travel forums, and booking engines. Consumers now have much better cost-comparison tools. This is a key challenge for Aqaba; while some hotels are addressing this issue, many hotels need to change to room yield pricing rather than using fixed pricing.
- Safety and Security: Safety and security are important to tourism's continued growth. Tourism does however show increased resilience: travelers are better informed; they have acquired a more balanced perspective; they now include security concerns as just another consideration when selecting destinations. Aqaba and Jordan can benefit from this trend.

4 Conclusions

As mentioned earlier, this study was conducted to spot the light on the current conditions of the tourism market in Aqaba City. The survey was conducted to understand the current conditions of tourism products and tourist perceptions in Aqaba City by using a quantitative technique, mainly a questionnaire. The sample covered composed of a wide variety of nationalities. The average age for tourists was younger than expected which could indicate that Aqaba was getting more famous and recognized by young people in general.

It was concluded through the analysis that Aqaba is highly visited by both genders of various ages, indicating that tourists were diversified because they mainly felt comfortable being in Aqaba, and it was recommended by them to be visited by others. It could be also concluded that Aqaba was a family tourism destination as most tourists came to Aqaba along with their husband/wife, and the majority of them spend two to three days in Aqaba which indicated that tourists feel safe and comfortable coming to Aqaba.

Tour buses were the dominant used as a method of transportation, and the rented cars were the second method of transportation used to get to Aqaba followed by the usage of both private or rental cars and the same time using taxi and tour buses through their visit to Aqaba. However, it was noticed that the vast majority of tourists complained about the public transportation sector which did not meet their needs.

The results also showed that breaking out of the routine and relaxation was the cause of selecting Aqaba. The second factor that affect the selection of Aqaba listed was the local product (Food and beverages). Meeting new people was ranked as a third cause of selecting Aqaba and curiosity about places/events unknown and unseen places was the fourth cause.

The highest practiced activity in Aqaba by tourists was eating local food and having drinks with percentage followed by diving. Shopping was the third activity practiced by tourists, followed by buying souvenirs. Visiting parks and nature attraction places in Aqaba was also having the attention of tourists. Boat tours and water sports were practiced. The communication with local people, visiting historic sites, living nightlife, other activities, and visiting family relatives and friends' houses were also activated by tourists. The expenses of tourists on different activities varied according to activity and the trip arrangements of tourists. For transportation, the highest percentage of tourists expended $1–50, followed by the $151–200 category. The results indicated that the expenses on transportation depend on the plans of the trip introduced for tourists. Concerning food and drinks, the majority of tourist expenses ranged from $51–100 followed by $1–50 with percentages of 38.5 and 34.6% respectively. This indicates that the majority of tourist food coverage is planned with the trip itself. The expense for a guide was ranging from $1–50 per tourist with a percentage of 60.0%. Expenses on souvenirs varied among tourists. The highest expenses ranged from $101–150 with a percentage of 25.0%, while the rest were distributed on different categories almost equally. Most of the sample paid entrance fees ranging from $101–150.

The outcomes revealed that tourists were generally satisfied during their visit to Aqaba, and more than half of the sample indicated that they will recommend others to visit Aqaba. The visit for the majority of tourists was arranged by a travel agent. This indicated that the majority of tourists are committed to the travel agent's plans to visit Aqaba. The first four places included in the package were Petra, the Dead Sea, Wadi Rum, and Mount Nebo. The reset of places distributed on Madaba, Amman, and Jerash. The garbage and bad traffic were the first two difficulties faced tourists during their visit. Poor and insufficient services were in the third rank. Boring and no information for tourists were two difficulties faced by tourists followed by no good café or restaurants.

4.1 Practical Implications and Recommendations

Most of the tourists who visited Aqaba complained about the transportation and traffic. Therefore, there is a need for local institutions to overcome this problem by facilitating access to important tourist attractions areas. Moreover, these institutions should give more attention to the cleanliness issue in Aqaba. Tourists indicated that Aqaba lacks good restaurants to use or café to enjoy in the city. This indicates that existing places are all popular initiating the need to construct new places with suitable entertainment for tourists. ASEZA should encourage the private sector to establish more cafés, restaurants, and so on in Aqaba. The relative intuitions should control prices, as tourists complained about the manipulation of prices, especially with tourists using their ignorance of prices. Another weak element was raised by tourists is the poor activities included in the tourist programs in Aqaba. These poor programs create boring for tourists in particular during night times. The tourists expected to see the city with a wide range of tourist activities distributed in different places. Authorities and policymakers in Aqaba should create a suitable and enjoyable activates program for tourists. This program might include activities such as Beach and Marine Tourism—Red Sea Access, Cultural Tourism—Proximity to Petra, Desert and Tradition—Proximity to Wadi Rum, Special Interest Tourism—(cruising, diving, family, etc.), More day time and night time and Shopping and walking street.

As a result, the city lacks any activities for children's entertainment, which creates a shortage of services available for tourism in the city. Online ticketing was a difficulty faced by tourists when reserving in activities, indicating the presence physically to solve this issue. This creates an overburden for tourists. Some tourists complained of the high prices in Aqaba, which seems impediment for tourists to utilize a wide variety of activities. Some other tourists have shown that Aqaba lacks public parks and natural places that can be used for entertainment. One major activity that should have the attention of policymakers in Aqaba is promoting historic sites. This activity to be included in tourists' programs will introduce added value for tourists. The visitor of Aqaba expected to see some folklore activities that reflect the culture there, but these activities do not exist anywhere in Aqaba city. Most tourists complained of boring because it has resulted in a lack of street activities that fill the tourist times. Therefore, the governorate of Aqaba should find special streets for talented people to practice these activities in. The introduction of local food will be considered an encouraging tool for tourists to share and live while visiting Aqaba. Citizens in Aqaba should be trained to participate in tourism activities. Local vocational training programs should be including Aqaba citizens to practice activities that are considered a source of income and improve entertainment in the city. Folklore programs should be stated in continuous programs all over the year. The core of the city suffers from crowding due to the concentration of most activities in this part of the city, while the other parts lack any attractive activities or trade movement that helps in attracting visitors out of crowding. Also, the traffic jams occurred in Aqaba as the concentration of trade movement is concentrated in a little number of streets which creates this crowding.

The opportunity in this aspect is to expand activities to another part of Aqaba which improves the entertainment of tourists in one aspect and solve the traffic jams in the middle of the city. The opportunity to succeed in arranging trips to Aqaba is to include it as part of the integrated program with other tourism places in Jordan in the northern, middle, and southern parts. An updated marketing strategy is required to cope with the challenges of filling Aqaba's increasing number of rooms and the expansion of its holiday home portfolio, to increase current hotel occupancy, extend the length of stay, and support the increase of value and benefits of tourism to the wider community.

In Aqaba, while some achievements have been realized, there are significant weaknesses in human resources, visitor attractions and activities, beach quality, family activities, night entertainment, general service quality both indirect tourism services and supporting tourism services, marketing, and competitiveness, which need to be addressed. The launch of Turkish Airlines into Aqaba provides great opportunities to promote the destination to their wide network from Europe and beyond, however, a keen focus on increasing access to underpin the accommodation expansion now in hand is imperative. While several achievements have been made recently, one significant aspect of the 2002 strategy that has yet to be realized is the development of Aqaba as a meeting, incentives, conferences, and exhibitions (MICE) destination. This has been hampered to date by a limited product (the National Convention Centre was built near the Dead Sea), limited room supply. A great deal needs to be done to transform a destination with 5-star hotels into a fully functioning 5-star destination that attracts and meets all the requirements for this level of customer. Tourism in Aqaba should be marketing along with Petra and Wadi Ram as standalone destinations. Finally, tourism in Aqaba needs a special Beach Management Board, which should be responsible for all tourism activities on the beach of Aqaba.

References

Abu Zayyad, H. M., Obeidat, Z. M., Alshurideh, M. T., Abuhashesh, M., Maqableh, M., & Masa'deh, R. E. (2021). Corporate social responsibility and patronage intentions: the mediating effect of brand credibility. *Journal of Marketing Communications, 27*(5), 533–510.

Alaali, N., Al Marzouqi, A., Albaqaeen, A., Dahabreh, F., Alshurideh, M., Alrwashdh, S., … Aburayya, A. (2021) The impact of adopting corporate governance strategic performance in the tourism sector: A case study in the Kingdom of Bahrain. *Journal of Legal, Ethical and Regulatory, 24* (Special Issue 1), 1–18.

Alananzeh, O. A., Jawabreh, O., Al Mahmoud, A., & Hamada, R. (2018). The impact of customer relationship management on tourist satisfaction: The case of Radisson blue resort in Aqaba city. *Journal of Environmental Management and Tourism, 9*(2), 227–240. https://doi.org/10.14505/jemt.v9.2(26).02

Aljumah, A., Nuseir, M. T., & Alshurideh, M. T. (2021). The impact of social media marketing communications on consumer response during the COVID-19: Does the brand equity of a university matter. In *The Effect of Coronavirus Disease (COVID-19) on Business Intelligence* (pp. 367–384).

Alshurideh, M., Bataineh, A., Alkurdi, B., & Alasmr, N. (2015). Factors affect mobile phone brand choices–Studying the case of Jordan universities students. *International Business Research, 8*(3), 141–155.

ASEZA. (2015). Marketing and tourism directorate, *Tourism statistics and charter flights division aqaba tourism statistics.*

Deeb, A., Alananzeh, O. A., & Tarhini, A. (2020). Factors affecting job performance: The case of Jordanian hotels' kitchen staff. *International Journal of Public Sector Performance Management, 6*(3), 340–360. https://doi.org/10.1504/ijpspm.2020.107766

EDP, Executive Development Program. (2011–2013).

European Union Development Programme (IS-ASEZA). (2005). *Tourism marketing strategy for Aqaba 2005–2010.*

Getz, D. (1997). *Event management & event tourism.* Cognizant Communication Corporation.

Goldman, D., Leibowitz, A., Buchanan, J., & Keesey, J. (1997). Redistribution consequences of community rating. *Health Services Research, 32*(1), 71–86.

Hamadneh, S., Hassan, J., Alshurideh, M., Al Kurdi, B., & Aburayya, A. (2021). The effect of brand personality on consumer self-identity: The moderation effect of cultural orientations among British and Chinese consumers. *Journal of Legal, Ethical and Regulatory Issues, 24,* 1–14.

Hyasat, S., Al-Weshah, A., & Kakeesh, F. (2022). Training needs assessment for small businesses: The case of the hospitality industry in Jordan. *GeoJournal of Tourism and Geosites, 40*(1), 20–29.

Jawabreh, O., Mahmoud, R., & Hamasha, S. A. (2020). Factors influencing the employee's service performances in hospitality industry case study AQbA five stars hotel. *Geojournal of Tourism and Geosites, 29*(2), 649–661. https://doi.org/10.30892/gtg.29221-496

JDS. (2014). Jordan Department of Statistics, Annual Report.

JNTS. (2010a). Jordan National Tourism Strategy.

JNTS. (2010b). National Strategy for Tourism Handicraft Development in Jordan, Amman—Jordan.

Kamakura, W. A., Ramaswami, S., & Srivastava, R. K. (1991). Applying latent trait analysis in the evaluation of prospects for cross-selling of financial services. *International Journal of Research in Marketing, 8*(4), 329–349.

Khwaldeh, S., Alkhawaldeh, R. S., AlHadid, I., & Alrowwad, A. (2020). The impact of mobile hotel reservation system on continuous intention to use in Jordan. *Tourism and Hospitality Research, 20*(3), 358–371. https://doi.org/10.1177/1467358420907176

Kim, W. C., & Mauborgne, R. A. (1993). Effectively conceiving and executing multinationals' worldwide strategies. *Journal of International Business Studies, 24*(3), 419–448.

Kotler, P., & Armstrong, G. (2007). *Principles of marketing* (12th ed.). Prentice-Hall.

Kotler, P., Bowen, J., & Makens, J. (2010). *Marketing for hospitality and tourism.* Prentice-Hall.

Lovelock, C., Wirtz, J., & Chatterjee, J. (2004). *Services marketing people, technology strategy: A South Asian perspective.* Pearson Education Inc.

Luo, Y. (2002, January–February). Capability exploitation and building in a foreign market: Implications for multinational enterprises. *Organization Science, 13,* 48–63.

Masa'deh, R., Alananzeh, O., Algiatheen, N., Ryati, R., Albayyari, R., & Tarhini, A. (2017). The impact of employees' perception of implementing green supply chain management on hotel's economic and operational performance. *Journal of Hospitality and Tourism Technology, 8*(3), 395–416https://doi.org/10.1108/JHTT-02-2017-0011.

Masa'deh, R., Alananzeh, O., Tarhini, A., & Algudah, O. (2018). The effect of promotional mix on hotel performance during the political crisis in the Middle East. *Journal of Hospitality and Tourism Technology, 9*(1), 32-47https://doi.org/10.1108/JHTT-02-2017-0010

Masa'deh, R., Alananzeh, O., Jawabreh, O., Alhalabi, R., Syam, H., & Keswani, F. (2019). The association among employees' communication skills, image formation and tourist behavior: Perceptions of hospitality management students in Jordan. *International Journal of Culture, Tourism, and Hospitality Research, 13*(3), 257–272. https://doi.org/10.1108/IJCTHR-02-2018-0028

McDonald, M. (1999). *Strategic marketing planning: Theory and practice* (4th ed.). Butterworth Heinemann.

MOTA. (2011). Statistics. Ministry of Tourism and Antiquities. Retrieved February 15, 2016, from www. tourism.jo.

Podsakoff, M., MacKenzie, B., & Jeong, L. (2003). Common method biases in behavioral research: A critical review of the literature and recommended remedies. *Journal of Applied Psychology, 88*(5), 879–903.

Ritchie, J. R. B., & Crouch, G. I. (2003). *The competitive destination: A sustainable tourism perspective*. CABI.

Spector, P. (2006). Organizational research methods. *Method Variance in Organizational Research*: Truth or Urban Legend?, *6*(2), 221–232.

Sweiss, N., Obeidat, Z. M., Al-Dweeri, R. M., Mohammad Khalaf Ahmad, A., M. Obeidat, A., & Alshurideh, M. (2021). The moderating role of perceived company effort in mitigating customer misconduct within Online Brand Communities (OBC). *Journal of Marketing Communications*, 1–24.

Tariq, E., Alshurideh, M., Akour, E., Al-Hawaryd, S., & Al Kurdi, B. (2022). The role of digital marketing, CSR policy, and green marketing in brand development at the UK. *International Journal of Data and Network Science, 6*(3), 1–10.

UNESCO. (2015). World heritage tentative list, United Nations - UNESCO World Heritage Centre. Retrieved February 14, 2016, from http://whc.unesco.org/en/tentativelists/state=jo.

USAID. (2009). Filed report (Salt Rapid Appraisal) conducted through needs assessment methodology through participation; Assessment of tourism sector and the tourism economic potentials with local community in salt.

Classification Thyroid Disease Using Multinomial Logistic Regressions (LR)

Saddam Rateb Darawsheh, Anwar Saud Al-Shaar, Feras Ahmad Haziemeh, and Muhammad Turki Alshurideh

Abstract Classifying using Multinomial Logistic Regression is one of the techniques used in the statistical model. Binomial logistic regression is a form of regression that is used when the dependent is a dichotomy and the independents are continuous variables, categorical variables, or both. Multinomial Logistic Regression Model is used to handle the case of dependent variables with more classes and suitable for non-linear data by using Maximum Likelihood Estimation and Wald Statistics. This experiment aims to predict the Thyroid disease Dataset by using the Multinomial Logistic Regression model. In this experiment, SPSS software is used to run the Multinomial Logistic Regression by Thyroid disease data supplied by Randolf Werner, which was adopted from the UCI Repository of *Machine Learning Database*.

Keyword Multinomial logistic regression and application of logistic regression

S. R. Darawsheh (✉)
Department of Administrative Sciences, The Applied College, (Imam Abdulrahman Bin Faisal University), P.O. Box 1982, Dammam, Saudi Arabia
e-mail: srdarawsehe@iau.edu.sa

A. S. Al-Shaar
Department of Self Development, Deanship of Preparatory Year and Supporting Studies, Imam Abdulrahman Bin Faisal University, P.O. Box 1982, Dammam 43212, Saudi Arabia

F. A. Haziemeh
Department of Information Technology, Al-Balqa Applied University, Al-Salt, Jordan
e-mail: haziemh.feras@bau.edu.jo

M. T. Alshurideh
Department of Marketing, School of Business, The University of Jordan, Amman 11942, Jordan
e-mail: malshurideh@sharjah.ac.ae; m.alshurideh@ju.edu.jo

Department of Management, College of Business, University of Sharjah, 27272 Sharjah, United Arab Emirates

M. Alshurideh et al. (eds.), *The Effect of Information Technology on Business and Marketing Intelligence Systems*, Studies in Computational Intelligence 1056,
https://doi.org/10.1007/978-3-031-12382-5_34

1 Introduction

Binomial (or binary) logistic regression is a form of regression that is used when the dependent is a dichotomy and the dependents are continuous variables, categorical variables, or both. Multinomial Logistic Regression exists to handle the case of dependents with more classes. However, Logistic regression applies maximum likelihood estimation after transforming the dependent into a logit variable (the natural log of the odds of the dependent occurring or not). In this case, logistic regression estimates the probability of a certain event occurring. Moreover, logistic regression calculates changes in the log odds of the dependent, not changes in the dependent itself as OLS regression does. Logistic regression can be used to predict a dependent variable based on the independent variable. Logistic regression is also used to see the percentage of variance the independent variable is capable of explaining the dependent variable.

Moreover, Logistic regression does not make any assumptions of normality, linearity, and homogeneity of variance for the independent variables. Because it does not impose these requirements, it is preferred to discriminant analysis when the data does not satisfy these assumptions. Logistic regression is used to analyze relationships between a non-metric dependent variable and metric or dichotomous independent variables. SPSS now supports Multinomial Logistic Regression that can be used with more than two groups.

Logistic regression combines the independent variables to estimate the probability that a particular event will occur, i.e. a subject will be a member of one of the groups defined by the dichotomous dependent variable. In SPSS, the model is always constructed to predict the group with higher numeric code. If responses are coded 1 for Yes and 2 for No, SPSS will predict membership in the No category. If responses are coded 1 for No and 2 for Yes, SPSS will predict membership in the Yes category. We will refer to the predicted event for a particular analysis as the modelled event. Figure 1 below the general model of logistic regression. There are three components for logistic regression. An input layer, transfer function and output layer.

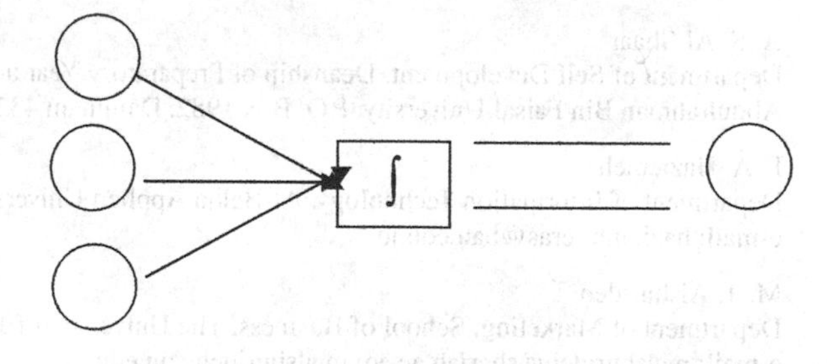

Fig. 1 The components for logistic regression

2 Literature Review

The review of literature established an ideal reference source of material and research writing concerning Logistic Regression (LR) it has been commonly used in various fields in the medical, Information Retrieval, Production and so on. This section showed how many ways Logistic Regression (LR) can give the best result in many fields?

2.1 Application of Logistic Regression in Medical

Logistic regression application has been commonly used to provide solutions to the medical field. According to Ng and Lai (2004), the Application of Anthropometric Indices in Childhood Obesity has been used the Logistic regression method to determine which anthropometric index can best predict childhood obesity and to investigate the interrelationship between these anthropometric indices and metabolic abnormalities. Data were obtained on the children's anthropometry, blood pressure, and serum lipid profiles. Body mass index (cutoff points by international age and sex-specific body mass index [BMI]) and weight-length index (WLI) were measured. The result of this study showed that there were 951 girls (47.43%) and 1,054 boys (52.57%).

The mean age was 7.3 _ 0.4 years. The prevalence of obesity was 4.21% in girls and 7.87% in boys using BMI definition, 12.83% in girls and 14.14% in boys using WLI definition. The prevalence of obesity revealed an increasing trend with age in both sexes, whether by BMI or WLI definition. Moreover, the prevalence of childhood obesity was evidently underestimated by using the BMI definition or by using the WLI definition. Childhood obesity exhibits a significant correlation with some metabolic abnormalities both indices, BMI and WLI, should be used together to define childhood obesity in clinical practice until a more appropriate and excellent index can be established. Another application, According to Steyerberg et al. (2002) state that the Application of Shrinkage Techniques in Logistic Regression Analysis has been used to develop a predictive model for a dichotomous medical outcome, such as short-term mortality. When the data set is small compared to the number of covariables studied, shrinkage techniques may improve predictions. This study has been compared the performance of three variants of shrinkage techniques a linear shrinkage factor, which shrinks all coefficients with the same factor; penalized maximum likelihood where a penalty factor is added to the likelihood function such that coefficients are shrunk individually according to the variance of each covariable and the Lasso, which shrinks some coefficients to zero by setting a constraint on the sum of the absolute values of the coefficients of standardized covariables.

Logistic regression models were constructed to predict 30-day mortality after acute myocardial infarction. Small data sets were created from a large randomized controlled trial, half of which provided independent validation data. The result of this study showed that all three shrinkage techniques improved the calibration of predictions compared to the standard maximum likelihood estimates. This study showed that shrinkage is a valuable tool to overcome some of the problems of overfitting in medical data.

2.2 Application of Logistic Regression Information Retrieval IR

Larson (2002) has applied the application of Logistic Regression to the problem of distributed Information Retrieval (IR). However, using a Logistic Regression algorithm for estimation of collection relevance. The algorithm is compared to other methods for distributed search using test collections developed for distributed search evaluation. As increasing numbers of sites around the world make their databases available through protocols such as OAI or Z39.50 the problem arises of determining, for any given query, which of these databases is likely to contain information of interest to a worldwide population of potential users. This is the central problem of distributed IR and there's much research have been active in this area such as (Becker et al., 2010; Callan, 2002) on GlOSS application of inference networks to distributed IR (CORI).

Along with several collaborators (Powell & French, 2003) enabled comparative evaluation of distributed IR by defining test collections derived from TREC data, where the TREC databases are divided into sub-collections representing virtual distributed collections. In addition, they defined some measures for evaluation of the performance of distributed IR used in this study to compare previously published results on collection selection with a probabilistic model based on logistic regression.

2.3 Application of Logistic Regression in E-rater

The linear logistic regression model is considered one of the most important and widely applicable techniques in analyzing. According to Qian (2003), logistic regression has been used widely currently in building e-rater models instead of linear regression. In this study, comparisons between logistic regressions with linear regression were made and tried to show the appropriateness of the use of logistic regression modelling in the analysis of *e-rater* data. There are several strategies in logistic regression modelling, which are:

(a) Forming suitable training data set and cross-validation data.
(b) Including high order interactions of features in model construction.
(c) Using three different types of estimators in prediction.
(d) Adjusting predicted scores.

As a result of this study, these factors played key roles in incorporating logistic regression models to *e-rater* modelling and help to outperform the linear regression. Using exact agreement as to the criterion of logistic regression in most situations, exceeded the regular linear regression model.

2.4 Objectives

Generally, the objective of this project is to classify thyroid data using the Neural Networks approach.

3 Methodology

3.1 Multinomial Logistic Regression Tool

The multinomial Logistic Regression tool used is SPSS Version 11.5 for processing the dataset. The dataset is taken from the raw data of the Thyroid Disease dataset in the UCI Repository of Machine Learning Databases to determine whether a patient referred to the clinic is hypothyroid. The data then was processed using the SPPS Version 11.5 according to the following process as shown in Fig. 2. Firstly, all the data are converted from Microsoft Excel format.

Next, the data are imported to the data source according to columns in the SPSS format by selecting Analyze menu, Regression and then Multinomial. All the variables are named according to its attribute in the variable view. A Multinomial Logistic Regression setup was used to configure the data according to the experiment required as in Fig. 3.

For the next step (refer to Fig. 4) insert the target in the dependent variable and insert the rest of the attributes in the covariates. To obtain the classification table as part of the regression analysis, select ***Statistics*** and check the Classification dialogue box as shown in Fig. 5.

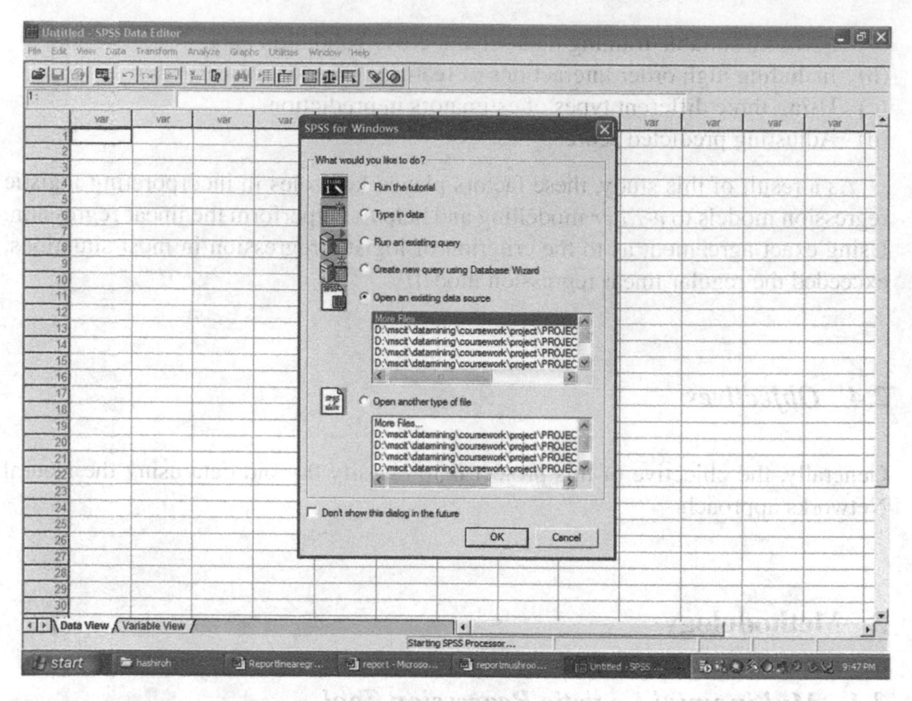

Fig. 2 Multinomial logistic regression process

4 Experimental Results

4.1 Multinomial Logistic Regressions

From Tables 1 and 2, both of the tables show the target has three classes to predict for the thyroid diseases dataset. For the first class, the percentage of correctly classified is 100%, for the second class the percentage correct is 100% and for the third class, the percentage correct is 100%. As a result, the total percentage of correct classification of the classes are 100%. There is no missing value from the dataset.

4.2 To Test the Significance of the Chi-Square

Model chi-square provides the usual significance test for a logistic model to see whether the hypothesis created is accepted or rejected. Model chi-square tests the null hypothesis that the independent variables are not linearly related to the log odds of the dependent variables. That is, model chi-square tests the null hypothesis that all population logistic regression coefficients except the constant are zero. It is thus an overall model test that does not assure that every independent is significant. From

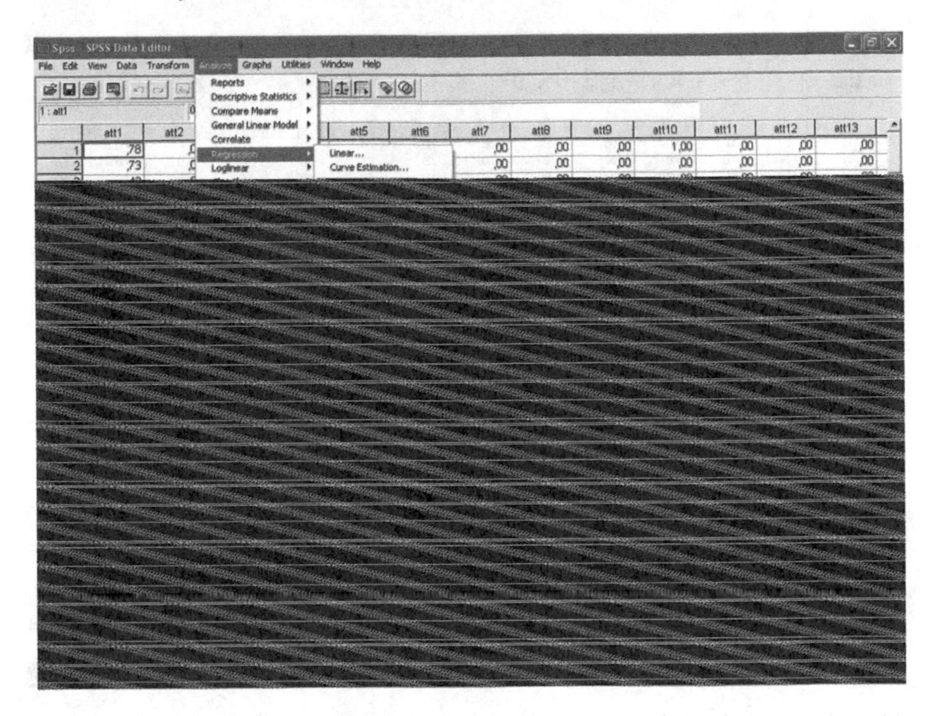

Fig. 3 Setting up multinomial logistic regression

Table 3, the probability of the model chi-square (662.322) is significant at 0.000 levels which is less than 0.05 (95%). Therefore, the hypothesis is rejected and the independent variables are linearly related to the log odds of the dependent variables (for each independent variable).

4.3 Pseudo-R-Square

Pseudo-R-Square is an Aldrich and Nelson's coefficient which serves as an analogue to the square contingency coefficient, with an interpretation like R square. Its maximum is less than 1. It may be used in either dichotomy or multinomial logistic regression. The Cox and Snell, Nagelkerke or McFadden in Table 4 attempt to provide a logistic analogy to R-square in the Multiple Linear Regression. The performance of this model using thyroid disease data set is 100%.

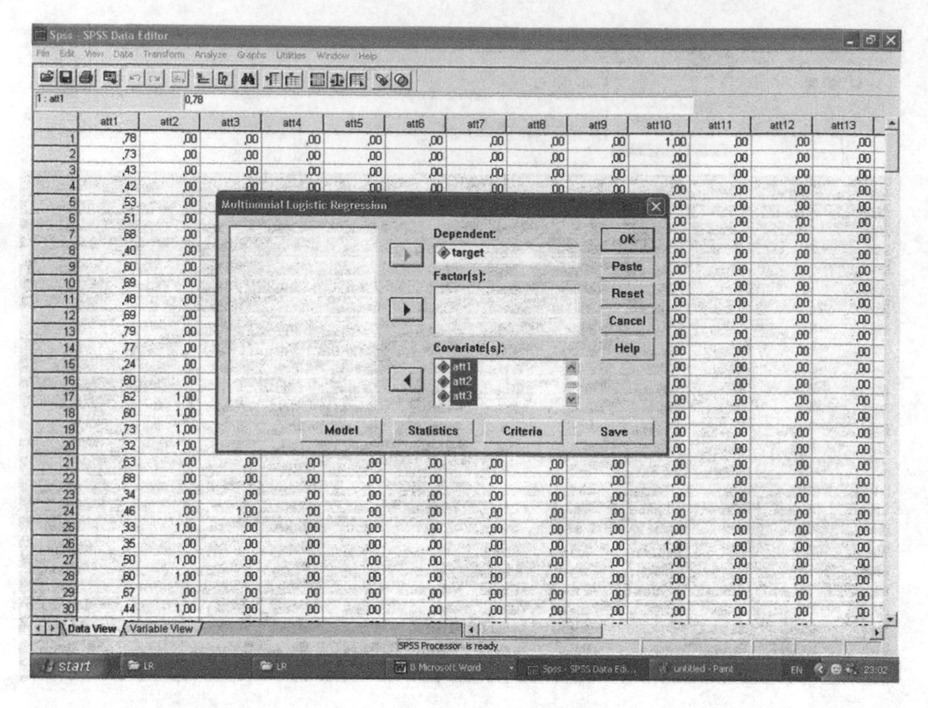

Fig. 4 Multinomial logistic regression process with dependent and independent variables

4.4 Likelihood Ratio Tests

The further analysis is by examining each of the independent variables using the –2 likelihood ratio tests or Likelihood Ratio Tests. The "–2 Log Likelihood" above is "model chi-square" which is used to test the significance of the logistic model the chi-square statistic is the difference in –2 log-likelihoods between the final model and a reduced model. Omitting and effect from the final model form the reduced model.

The reduced model is formed by omitting an effect from the final model. When probability (model chi-square) is less than or equal to 0.05, the null hypothesis that knowing independent in logistic regression is rejected. Thus, model chi-square needs to be significant at the 0.05 level or better. The null hypothesis is that all parameters of that effect are 0. Table 5 shows that the independent variable ATT16 is significant at below 0.05 levels. Therefore, the hypothesis is rejected. On the other hand, nineteen variables have a significance value of more than 0.05, which are ATT1, ATT2, ATT3, ATT4, ATT5, ATT6, ATT7, ATT8, ATT9A, TT10, ATT11, ATT12, ATT13, ATT14 ATT15, ATT17A, TT18, ATT19, ATT20 have the relationship between the dependent variable, TARGET. Therefore, further analysis needs to be considered.

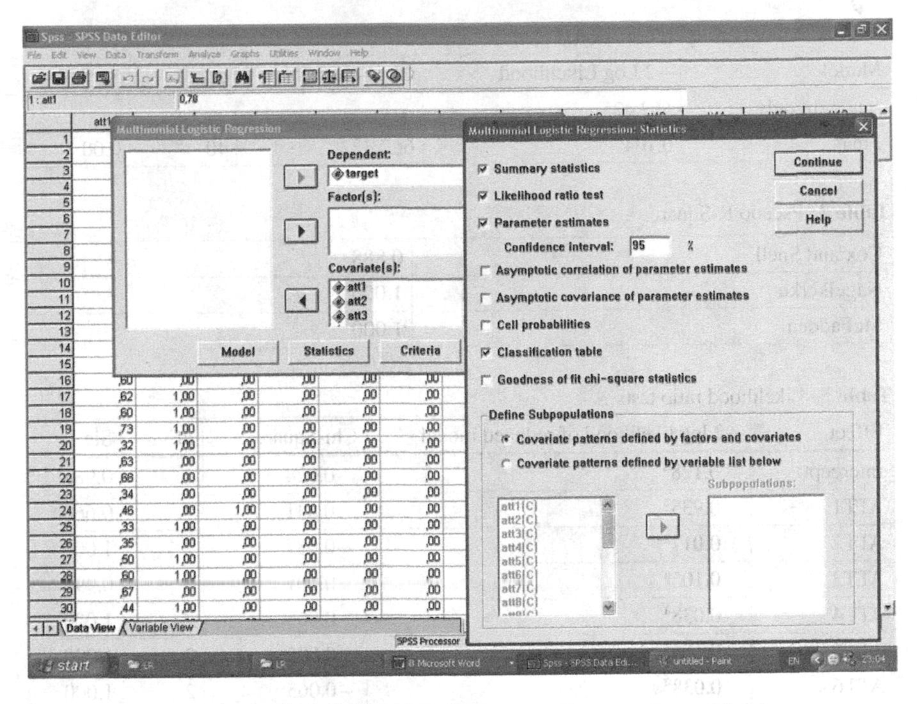

Fig. 5 Final process for producing the output view

Table 1 Case processing summary

		N
Target	1.00	92
	2.00	105
	3.00	105
Valid		302
Missing		0
Total		302

Table 2 Classification

Observed	Classification			
	1.00	2.00	3.00	Percent correct (%)
1.00	92	0	0	100
2.00	0	105	105	100
3.00	0	0	0	100
Overall percentage	30.5%	34.8%	34.8%	100

Table 3 Model fitting information

Model	−2 Log Likelihood	Chi-Square	df	Sig
Intercept only	662.425			
Final	0.104	662.322	40	0.00

Table 4 Pseudo R-Square

Cox and Snell	0.888
Nagelkerke	1.000
McFadden	1.000

Table 5 Likelihood ratio tests

Effect	−2 log likelihood of reduced model	Chi-Square	df	Sig
Intercept	0.128[a]	0.024	2	0.988
ATT1	0.935[a]	0.831	2	0.660
ATT2	0.017[a]	−0.087	2	1.000
ATT3	0.105[a]	0.001	2	0.999
ATT4	0.038[a]	−0.065	2	1.000
ATT5	0.104[a]	0.000	2	1.000
ATT6	0.038[a]	−0.065	2	1.000
ATT7	0.038[a]	−0.065	2	1.000
ATT8	0.039[a]	−0.665	2	1.000
ATT9	0.291[a]	0.187	2	0.911
ATT10	0.106[a]	0.002	2	0.999
ATT11	0.104[a]	0.000	2	1.000
ATT12	0.041[a]	−0.063	2	1.000
ATT13	0.038[a]	−0.066	2	1.000
ATT14	0.754[a]	0.650	2	0.722
ATT15	0.039[a]	−0.065	2	1.000
ATT16	228.462[a]	228.358	2	0.000
ATT17	2.085[a]	1.981	2	0.371
ATT18	0.102[a]	−0.002	2	1.000
ATT19	0.101[a]	−0.003	2	1.000
ATT20	0.015[a]	−0.089	2	1.000

The chi-square statistic is the difference in −2 log-likelihoods between the final model and a reduced model. The reduced model is formed by omitting an effect from the final model. The null hypothesis is that all parameters of that effect are 0

[a]Unexpected singularities in the Hessian matrix are encountered. There may be a quasi-complete separation in the data. Some parameter estimates will tend to infinity

4.5 Wald Statistic

The Wald statistic above and the corresponding significance level test the significance of the covariate and factors (dummy) independents in the model. The Wald statistic is commonly used to test the significance of individual logistic regression coefficients for each independent variable. It is the ratio of the unstandardized logit coefficient to its standard error. The null hypothesis is a particular effect coefficient is zero. From Table 6, the probability of the Wald statistic for all the independent variables is more than 0.05, therefore the variables are not significant which the null hypothesis cannot be rejected since the coefficient for all independent variables is zero. As result, all variables are not significant to predict Thyroid Disease Dataset.

5 Conclusion

The conclusion from this experiment showed that not all the independent variables have the coefficient effect with the dependent variable. However, from the performance between Multilayer Perceptron, Multiple Linear Regression and Logistic Regression, the Thyroid Disease Dataset is best to analyze using Logistic Regression, with 100% compared to Multi-Layer Perceptron 78.208% and Multi Linear Regression 65.5%. Table 7 concludes the result.

Table 6 Parameter estimates (Wald Statistics)

	Target	B	Std. Error	Wald	df	Sig	Exp(B)	95% confidence interval for Exp(B)	
								Lower bound	Upper bound
1.00	Intercept	111.101	2152.785	0.003	1	0.959			
	ATT1	73.048	209.239	0.122	1	0.727	5.3E + 31	4.168–147	6.7411 + 209
	ATT2	−6.508	68.344	0.009	1	0.924	1.491E−03	9.980E−62	2.2282E + 55
	ATT3	−24.762	304.057	0.007	1	0.935	1.761E−11	2.702–270	1.1480 + 248
	ATT4	30.202	32,679.454	0.000	1	0.999	1.3E + 13	0.000	a
	ATT5	69.855	681.283	0.011	1	0.918	2.2E + 30	0.000	a
	ATT6	397.605	3280.806	0.015	1	0.904	4.76 + 172	0.000	a
	ATT7	129.102	2149.969	0.004	1	0.952	1.17E + 56	0.000	a
	ATT8	5.442	0.000		1		230.907	230.907	230.907
	ATT9	−15.632	311.250	0.003	1	0.960	1.626E−07	1.883−272	1.4030 + 258
	ATT10	18.452	997.501	0.000	1	0.985	1.0E + 08	0.000	
	ATT11	−143.596	9149.724	0.000	1	0.987	4.336E−63	0.000	a
	ATT12	131.783	3515.224	0.001	1	0.970	1.71E + 57	0.000	a
	ATT13	9.353E−02	927.382	0.000	1	1.000	1.098	0.000	a
	ATT14	−107.399	1626.991	0.004	1	0.947	2.276E−47	0.000	a
	ATT15	27.356	585.486	0.002	1	0.963	7.6E + 11	0.000	
	ATT16	10,987.618	10,692.042	1.056	1	0.304		0.000	a
	ATT17	615.440	3179.081	0.037	1	0.846	1.91 + 267	0.000	a
	ATT18	−1283.398	39,510.774	0.001	1	0.974	0.000	0.000	a
	ATT19	496.005	24,682.969	0.000	1	0.984	2.58 + 215	0.000	

(continued)

Table 6 (continued)

	Target	B	Std. Error	Wald	df	Sig	Exp(B)	95% confidence interval for Exp(B)	
								Lower bound	Upper bound
	ATT20	−3068.229	34,542.601	0.008	1	929	0.000	0.000	a
2.00	Intercept	−108.300	523.991	0.043	1	0.836			
	ATT1	5.930	74.836	0.006	1	0.937	375.972	7.490E−62	1.8872E+66
	ATT2	−3.537	16.749	0.045	1	0.833	2.909E−02	1.610E−16	5.258E + 12
	ATT3	6.681	240.439	0.001	1	0.978	797.211	1.737−202	3.6587 + 207
	ATT4	18.814	32648.326	0.000	1	1.000	1.5E + 08	0.000	a
	ATT5	0.589	324.432	0.000	1	0.999	1.803	1.256−276	2.5872 + 276
	ATT6	22.557	3036.615	0.000	1	0.994	6.3E + 09	0.000	a
	ATT7	20.380	931.428	0.000	1	0.983	7.1E + 08	0.000	a
	ATT8	−8.143	749.070	0.000	1	0.991	2.908E−04	0.000	a
	ATT9	−13.090	309.786	0.002	1	0.966	2.066E−06	4.218−270	1.0123 + 253
	ATT10	7.671	974.787	0.000	1	0.994	2146.054	0.000	a
	ATT11	14.563	457.441	0.001	1	0.975	2111993	0.000	a
	ATT12	−29.420	3189.320	0.000	1	0.993	1.671 E−13	0.000	a
	ATT13	4.778	421.292	0.000	1	0.991	118.881	0.000	a
	ATT14	−71.627	2495.788	0.001	1	0.977	7.809E−32	0.000	a
	ATT15	16.880	220.326	0.006	1	0.939	2.1E + 07	6.160−181	7.4544 + 194
	ATT16	10854.727	1 0801.21 8	1.010	1	0.315	a	0.000	a
	ATT17	236.121	1874.495	0.016	1	0.900	3.51 + 102	0.000	a

(continued)

Table 6 (continued)

	Target	B	Std. Error	Wald	df	Sig	Exp(B)	95% confidence interval for Exp(B)	
								Lower bound	Upper bound
	ATT18	−401.588	5405.697	0.006	1	0.941	3.912−175	0.000	[a]
	ATT19	301.551	5062.226	0.004	1	0.952	9.16 + 130	0.000	[a]
	ATT20	526.839	5766.177	0.008	1	0.927	6.36 + 228	0.000	[a]

[a]Foating point occurred while computing this statistic, its value is therefore set to system missing

Table 7 Comparison between multilayer perceptron, multiple linear regression and logistic regression

Multi-layer perceptron (MLP)	Multiple linear regression (MLR)	Logistic regression
78.208%	65.5%	100%

References

Becker, H., Xiao, B., Naaman, M., & Gravano, L. (2010). Exploiting social links for event identification in social media.

Callan, J. (2002). Distributed information retrieval. In *Advances in information retrieval* (pp. 127–150). Springer.

Larson, R. R. (2002). A logistic regression approach to distributed IR. January 2002, p. 399. https://doi.org/10.1145/564376.564463.

Ng, K.-C., & Lai, S.-W. (2004). Application of anthropometric indices in childhood obesity. *Southern Medical Journal, 97*(6), 566–571.

Powell, A. L., & French, J. C. (2003). Comparing the performance of collection selection algorithms. *ACM Transaction on Information Systems, 21*(4), 412–456.

Qian, J. (2003). Application of logistic regression in analysis of e-rater data. *Educational Testing Service. ETS MS.*

Steyerberg, J. D. F., Eijkemans, E. W., & Habbema, M. J. C. (2002). Application of shrinkage techniques in logistic regression analysis: A case study. *Statistica Neerlandica, 55*(1), 76–88.

IT Governance and Control: Mitigation and Disaster Preparedness of Organizations in the UAE

Ismail Ali Al Blooshi, Abdulazez Salem Alamim, Raed A. Said, Nasser Taleb, Taher M. Ghazal, Munir Ahmad, Haitham M. Alzoubi, and Muhammad Alshurideh

Abstract The success of any business continuity today depends on its levels of preparedness to address any disasters that could emerge from the risks it faces in its operations. Most of the businesses in the UAE have adopted information technology and systems to aid them in their operations. Unfortunately, the adoption of such systems enhances the risks that businesses face, hence the need to ensure they adopt reliable disaster management strategies to address any potential challenges they could face. The aim of this research is to review and measure the levels of information system disaster preparedness in both the private and public sector organizations operating in the UAE. A study was conducted to measure the use of some of the popular disaster risk management models such as STRIDE, DREAD, and BCM. The study indicated that many of the firms in the UAE are not fully prepared to tackle disasters emanating from their information management systems but have

I. A. Al Blooshi · A. S. Alamim · R. A. Said · N. Taleb
Canadian University Dubai, Dubai, UAE

T. M. Ghazal
Center for Cyber Security, Faculty of Information Science and Technology, Universiti Kebansaan Malaysia (UKM), 43600 Bangi, Selangor, Malaysia
e-mail: taher.ghazal@skylineuniversity.ac.ae

School of Information Technology, Skyline University College, Sharjah, UAE

M. Ahmad
School of Computer Science, National College of Business Administration & Economics, Lahore, Pakistan

H. M. Alzoubi (✉)
School of Business, Skyline University College, Sharjah, UAE
e-mail: haitham.alzubi@skylineuniversity.ac.ae

M. Alshurideh
Department of Marketing, School of Business, University of Jordan, Amman, Jordan
e-mail: m.alshurideh@ju.edu.jo; malshurideh@sharjah.ac.ae

Department of Management, College of Business Administration, University of Sharjah, Sharjah, UAE

M. Alshurideh et al. (eds.), *The Effect of Information Technology on Business and Marketing Intelligence Systems*, Studies in Computational Intelligence 1056,
https://doi.org/10.1007/978-3-031-12382-5_35

taken some steps and have invested part of their resources in improving the state of their preparedness. The findings in this paper can benefit researchers in the field of IT, private and public organizations in the region, government entities, and other stakeholders investing in the gulf country.

Keywords Business continuity management · Disaster preparedness · IT governance · Risk management

1 Introduction

The United Arab Emirates (UAE) is undisputedly one of the most attractive investment destinations in the world. Studies on its economy indicate that the country has grown at an unprecedented scale and rate over the past few years pre Covid-19. The emergence of the UAE as an economic hub in the MENA (Middle East and North Africa) and GCC (Gulf Cooperative Council) regions and its profile as a leading oil and gas producer has attracted significant interest for investment in the country (Al-Gasaymeh et al., 2020; Alshubiri et al., 2019; Assad & Alshurideh, 2020a, b; Shah et al., 2020, 2021;). According to Mosteanu (2019), the country has combined strategic investment in business and the development of technology to position itself as an appealing destination for both local and international entrepreneurs and investors. The business sector of the UAE remains one of the most advanced and diversified, considering the large number of investors that have initiated operations in the country (De Jong et al., 2019). Technologies intended for improved decision-making strategies, enhancing operational efficiency, and guaranteeing the security of the resources, including data and information, have been among the most popular. Given such a trend, the levels of investment in technology in the businesses operating in the country have been reviewed in several studies (Alzoubi & Aziz, 2021). Today, the UAE enjoys foreign investment borne out of the significant investments both public and private sector stakeholders have made in the adoption and utilization of technology in their operations (Mahmah & Kandil, 2019). The high levels of reliance on technology for operations mainly in trade and industry in the UAE has led to concerns regarding the threat of IT-related disasters, which could destroy not just organizational-level activities and operations but also the delivery of services to the public (Haddad et al., 2020).

Unfortunately, businesses across the country have faced various disaster that has highlighted some of the weaknesses in the technologies in place. Ali et al. (2022) explains further that the concept of disaster recovery has not been fully mastered and incorporated with the technologies, specifically the information systems in place in most of the country's businesses.

The trend of disasters involving information systems across the world has been worsening each day (Geer et al., 2020). Businesses have recorded significant losses arising from the emergence of disasters affecting their operations directly across the country. Geer et al. (2020) indicated that the trend is getting even worse each considering the uncontrolled sprouting of technologies and their adoption even when the required approvals and certification is undertaken. The trend has affected the UAE directly considering that it is a leading hub for business and innovation, as shown in (Kashif et al., 2021). The country has invested heavily in encouraging innovators to come up with technologies that can help improve the trends of business performance in the country (Alzoubi & Ahmed, 2019). The rapid trend of technology development has made it impossible for UAE regulators to control the nature of technological ideas and innovations that end up being incorporated into the operations of various businesses (Al-Maroof et al., 2020; Alketbi et al., 2020; Alshurideh et al., 2021; Amarneh et al., 2021; Baadel et al., 2017; Leo et al., 2021; Turki et al., 2021). Unfortunately, unless the trend is controlled, the risk of disasters will continue to rise, thereby affecting the business continuity and ambitions of the UAE negatively.

The introduction of disaster and fault-preparedness systems in the field of business has emerged to be a positive aspect of global business operations. According to Li et al. (2019), recent innovations have enabled the creation of systems that can detect disasters before they happen, present warnings, recommend corrective measures to enable the prevention of major risks that could affect the business negatively. Li et al. (2019) further explain that in the event of a disaster, such systems can safeguard the resources of the business to enable the kickstarting of the recovery process once everything is under control from the effects of the disaster (Lee et al., 2022a, b). Disaster recovery solutions for information technology systems have not been developing at the same rate as the IT systems in place currently (Mendonca et al., 2019). The situation complicates the ability of business organizations to acquire information technology systems that fit the nature of their operations and to enable them to remain secure at all times considering that the environment of operation in the modern market is defined by unending uncertainty (Alzoubi et al., 2021). The findings of Mendonca et al. (2019) study make it necessary for scholars reviewing the UAE context of disaster recovery systems to consider reviewing the strategies in place in the businesses and the markets to enable an effective approach to guarantee the safety and resource security even after disasters (Alshurideh et al., 2022a, b; Radwan & Farouk, 2021).

The research undertaken herein focuses on the strategy on the development, adoption, and incorporation of disaster recovery systems in the UAE (Alzoubi et al., 2022). It has been indicated that the development of a disaster recovery system, their adoption and incorporation to business and organizational operations remains to be the most effective way to guarantee success globally (Al-Dmour et al., 2021; Alnuaimi et al., 2021; Li et al., 2019; Mashaqi et al., 2020). In the UAE, many companies still rely on information technology systems that are not effective in ensuring that they remain stable and helpful in the event of disasters and faults. Al Shamsi (2019) suggests that various business organizations in the UAE have not been able to update the information technology systems they implemented years ago to ensure that they conform to the needs of the modern age and market (Mehmood et al., 2019). The major challenges hinder the adoption of effective information technology systems capable of remaining stable, operations and efficient even in the face of exposure to risk is cost, lack of knowledge and challenges in linking them to the exact context of the operations of an organization (Gomez et al., 2019). Like most other corporations and business organizations across the world, those in the UAE are yet to come up with an effective approach to enable them to adopt effective disaster preparedness information systems in their operations.

The study herein reviews the trends and practices of adoption and implementation of a disaster preparedness information system in the UAE. The aim will be to focus mainly on the practices currently in place in most of the businesses across the country. Our research will also review some of the key performance indicators of disaster preparedness information systems i.e. models used for companies' operating within the UAE.

The rest of the paper is structured as follows: Sect. 2 provides the literature review on disaster preparedness and risk models. This is followed by the methodology in Sect. 3. In Sect. 4, the paper will analyze and discuss the results from the study. Finally, conclusion and future work is presented in Sect. 5.

2 Literature Review

The current review of literature considers the issues that scholars have focused on in their studies in their quest to understand IT-disasters, preparedness for adverse events and business continuity during and after IT emergencies in the UAE.

2.1 IT Disaster-Preparedness of Organizations in the UAE

The common feature of modernity all around the world is the reliance on technology systems to accomplish various day-to-day private and public activities (Al Barghuthi et al., 2019). The study indicated that popularity in the use of IT systems in the UAE has made the country vulnerable to attacks (Ali et al., 2022). While the study introduced the causes for concern that business firms should have as they operate in the UAE market, the authors did not address comprehensively the nature of attacks that firms are likely to face as they run their operations in the country. The study by Al Barghuthi et al. (2019) also failed to address the question of the nature of preparation that UAE firms should put in place to guarantee fault preparedness and resilience in the event of an attack (Miller, 2021).

Al Issa (2019) acknowledges that recent disasters have been a major cause for concern for the organizations that have invested heavily in the adoption and utilization of information technology systems in their operations. In his study, Al Issa (2019) indicates that recent disasters, which luckily have been on a limited scale, have demonstrated the nature of issues and complications that they could force firms to go through once they occur (Alshurideh et al., 2020). Unfortunately, Al Issa (2019) study is not clear of the specific steps that the UAE firms should take to ensure disaster preparedness in their IT systems to ensure business continuity during and after attacks (Aziz & Aftab, 2021). The question must be asked of the way forward including clear guidelines that must be created to guide IT disaster preparedness before an organization adopts specific IT systems in its UAE operations (Aasriya, 2021).

A recent report published by the EIU Digital Solutions (The Economist Intelligence Unit, 2018) rated the levels of preparedness to deal with cyberattacks among the countries that make up the GCC (Alhamad et al., 2021). In the report, it was indicated that Oman and Qatar are the only countries that have high preparedness. It was indicated that Oman and Qatar had a rating of 0, the UAE had a rating of one, Saudi Arabia and Bahrain had 2 and Kuwait had 3 using a scale of 0 to 4 with 0 being highest preparedness and 4 being lowest preparedness (Alzoubi, 2021a, b). At a score of 1, the UAE is somewhat highly prepared as per the findings of the report. However, recent trends have indicated that many businesses in the UAE, and indeed across the world, are rarely investing in the advancement of the disaster recovery systems in their operations (Alzoubi et al., 2020a, b). The question of the reluctance in increasing the budgetary allocations to IT disaster preparedness of the firms operating in the UAE market needs to be covered as extensively as possible, which is also the focus of the study being undertaken herein (AlHamad et al., 2022).

Sharma (2019) reported that most of the businesses have not been keen to increase the budgets for the security and disaster preparedness of their information technology systems. Such a trend indicates that many firms in the UAE are staring at a major risk, which will most likely influence negative outcomes in future (Alnazer et al., 2017). The study by Sharma (2019) leads to the question of whether the failure to increase IT disaster budgets increase the likelihood of emergencies with a firm's operations even as it pursues profitability through cost-saving (Joghee et al., 2020a). The research herein also focuses on the way firms in the UAE can be strategic in funding in disaster preparedness and preparedness of the systems that they use in their operations (Akhtar et al., 2021).

2.2 *Disaster Risk Management (DRM) Models*

DRM Model are developed to help with risk assessment when developing, deploying, and incorporating disaster preparedness and fault preparedness in information systems. Our study reviews and discusses four of such models (Alzoubi, 2021a, b).

Firstly, the STRIDE model (Spoofing, Tampering, Repudiation, Information disclosure, Denial of Service, Elevation of privilege) identifies the specific form of risk that an organization implementing an IT system is likely to face in its operations (Xin & Xiaofang, 2014). Poniatowski (2018) has criticized the STRIDE model arguing that experienced modelers will only rely on the STRIDE model when seeking to check and confirm if any potential attack is left unaddressed (Alzoubi & Yanamandra, 2020). However, he admits that the STRIDE model is still effective and can be applied to help an organization defend itself and its resources from attackers who could be keen on affecting its operations negatively (Ali et al., 2021). According to (Khan, 2017), the STRIDE model ought to be taken seriously in the UAE organizations seeking to identify the potential challenges their operations face following the various information technology systems they could be seeking to implement or could have adopted in their operations (Alshurideh, 2022; Alshurideh et al., 2022a, b; Alzoubi & Ahmed, 2019). The question of the applicability of the STRIDE model in UAE organizations and the effects of its application in the UAE is an issue that the current research aims at addressing through the data collected and analyzed (Alzoubi et al., 2020a, b).

Secondly, in IT disaster preparedness planning, it is important to consider a mitigation plan, which should be adopted to ensure that any adverse effects of an attack are addressed as comprehensively as required (Walker et al., 2010). The risk mitigation plan widely accepted in the planning for disaster preparedness involves four key types of actions; accepting the risk, avoiding it, transferring it, and reducing its effects (Herrera, 2013). Herrera (2013) explains that in most cases, IT disasters are often difficult to mitigate when they occur, which means that firms are accepting them by giving up whatever resources affected and creating ways to create new resources. According to Zahid et al. (2020), avoiding the risk involves taking measures to ensure the risk is evaded whenever possible. For instance, an organization facing the risk of its IT system being spoofed can take various measures such as enhancing the levels of security to ensure that attackers find it difficult to intrude and penetrate the layers of security that will be in place (Alzoubi et al., 2020a, b). According to Matthews (2020), the action of risk transfer involves an organization involving the third party to help it fight against the risk it could be facing. In the context of IT risks, an organization can choose to hand over its risks to an IT security firm to enable it to tackle the security and disaster preparedness challenges it could be facing (Mehmood et al., 2019).

Thirdly, the DREAD model has been presented to help firms plan for an effective response when responding to an attack (Suprihanto et al., 2018). The DREAD model has been under the application in various areas that have involved the deliberations of the disaster recovery framework to be adopted whenever an adverse challenge occurs (Czagan, 2014). Once an attack is executed, it is possible to rate various aspects of issues the organization experiences and then consider the strategy and levels of resource allocation to be delivered to help with the resolution of the challenges (Joghee et al., 2020b). The model is summarized the Table 1.

Lastly, the Business Continuity Management (BCM) model has been promoted as an effective approach to ensure an organization minimizes the levels of disruption in its operations following a disaster or a fault in its information technology systems (Torabi et al., 2016). The application of the BCM model allows an organization to address the challenges that emerge as it provides a framework, which involves all the various stakeholders in its operations. According to Vaidhyanathan (2010), the BCM model comprises the following steps:

Step 1—the solicitation of the support of the top leadership of the organization.

Step 2—communicating the business continuity plan to all the stakeholders.

Table 1 A summary of the DREAD model

Item	Description
Damage potential	How expansive is the damage relative to the entire security system of the organization?
Reproducibility	Can the attack be reproducible in the other systems of the organization not affected yet?
Exploitability	Who can exploit the attack? What skills do they have?
Affected users	Who among the stakeholders of the organization is affected by the attack?
Discoverability	How easily did the organization identify the attack? How much damage had been done before the attack was discovered?

Step 3—picking a sponsor to oversee the implementation of the business continuity plan in the operations of an organization.

Step 4—creating an incident management plan and an incident response structure, which will be effective in enabling the business recovery plan and a disaster recovery plan.

Step 5—developing an extensive check and review of the alternative strategy implemented in the previous stage to determine any limitations and gaps that are likely to hinder full operations in the organization.

Step 6—a plan-do-check-act cycle, which will focus on every single issue that could emerge, or be thought of, about the IT system that will be under implementation in the organization post the disaster or attack.

The organizations in the UAE, both public and private, which are at a significantly high risk of experiencing IT disasters and faults must be keen on implementing the BCM model as part of their recovery measures anytime they face such devastating issues in their operations (Ghazal et al., 2021).

3 Methodology

A questionnaire was the preferred tool used for the collection of the data that was analyzed in the current research. The research involved a sample of 62 respondents whose responses were the main basis of the analysis that was undertaken in the current study. The sample was drawn from employees working for both public and private firms in the UAE. The survey resulted in 51 complete responses with 9 incomplete surveys that were eventually discarded yielding an 82% response rate. Given the current state of business activity due to the Covid-19 pandemic, we deemed the response rate to be sufficient for our purpose of study.

The sample was selected using a purposeful sampling technique where we identified specific companies in the UAE to participate in the survey based on their business type and nature of the organization. Such techniques have been discussed in the literature and are being widely used for this type of study (Ghane & Esmaeili, 2020; Ghazal et al., 2020; Hodgetts et al., 2020).

Of the 62 respondents initially surveyed, 71% were drawn from the public sector organizations, while 29% were drawn from the private sector firms. The sample population indicate the reality in the UAE business industry where most of the organizations are public sector firms, which are the major contributors to economic growth (Farouk, 2021). The focus was mainly on employees who had information regarding the information systems that their firms were implementing (Lee & Ahmed, 2021).

The first part of the survey focused on the employee perception of whether their organization is prepared for disasters and have a plan in place. The second part of the survey had questions based on the STRIDE model framework and its adoption in the firm. The third part was designed to gauge the trends of application of the DREAD model in the organizations in which the respondents worked (Ali, 2021; Singh & Singh, 2021). It was the expectation of the research that the application of the DREAD model would be an indication of efforts to enhance the levels of disaster preparedness in the organizations in which the respondents worked (Hamadneh et al., 2021). The final part of the survey was designed to gauge the trends of application of the BCM model in the information systems of the organizations. The aim was to understand whether the BCM model, one of the core indicators of disaster preparedness levels of the information systems of an organization, was being employed in the firms that are currently operating in the UAE. Each of their scores were measured on a Likert scale of 1–5, where strongly agree = 5, agree = 4, neutral = 3, disagree = 2, and strongly disagree = 1.

The research was undertaken in strict conformity to the ethical guidelines and principles of scholarly investigation. All the respondents involved in the study were requested to volunteer to provide the required details with guaranteed anonymity. In the event they needed authorization from their firms to comment on the issues raised in the questionnaire, such permission was sought prior to their participation.

4 Results and Analysis

The following table provides a descriptive summary of what the respondents identified as their major observation on the need of having a strategy in place.

Two main observations are made from the summary statistics in Table 2 above. Firstly, all the indices have positive means with the average of 4.2. Secondly, the indices are negatively skewed with a Kurtosis average of 2.6 (less than the normal standard of 3) thus implying the indices behave linearly.

Figure 1 depicts their responses for each of the 5 indices measured.

Figure 1 indicate the factors driving the investments that organizations were putting in place in their disaster preparedness. Clearly, the cost of downtime and the objectives for organizational success, and the risk aversions of technology systems are major driving forces for investments of firms in UAE (Mondol, 2021).

Table 2 Summary statistics on the need for disaster preparedness strategy

	Mean	Std. Dev.	Skewness	Kurtosis
Cost of downtime	4.34	1.0059	−2.082	6.1333
Regulatory/Legal requirements	4.17	0.9399	−1.667	3.7892
Risks to the technology systems of the business	4.25	0.8960	−0.917	−0.1693
Priority of organization's leadership	4.15	1.0210	−1.206	0.9939
Objectives for organizational success	4.09	1.2305	−1.557	2.2645
Average	4.2	1.0187	−1.486	2.6023

Fig. 1 Mendonca et al., 2019

The study also investigated on the emphasis these firms have placed on the various stages of disaster awareness and mitigation. Majority of the respondents indicated they placed a lot of energy on prevention techniques and strategy as opposed to the recovery efforts. Figure 2 below illustrates this.

The study found that 88.00% of the respondents affirmed that they often consider the likelihood of an IT risk occurring and devastating the systems and operations of their organizations. Moreover, 92% of the respondents indicated that their organizations saw preparedness as an important strategy in their efforts to address the risks associated with IT disasters (Alnuaimi et al., 2021; Alzoubi et al., 2020a, b).

Additionally, the study analyzed some of the disaster risk management models utilized by the different organizations and measured some of the aspects of those models that were under-utilized or considered not effective by each firm. A summary of their responses is provided below.

With regards to the BCM model as highlighted in Fig. 3, many firms don't see the need of identifying a disaster recovery team or creating a new incident management plan that will oversee the implementation of the business continuity plan (Guergov & Radwan, 2021; Obaid, 2021). Instead, it seems that senior staff with their traditional

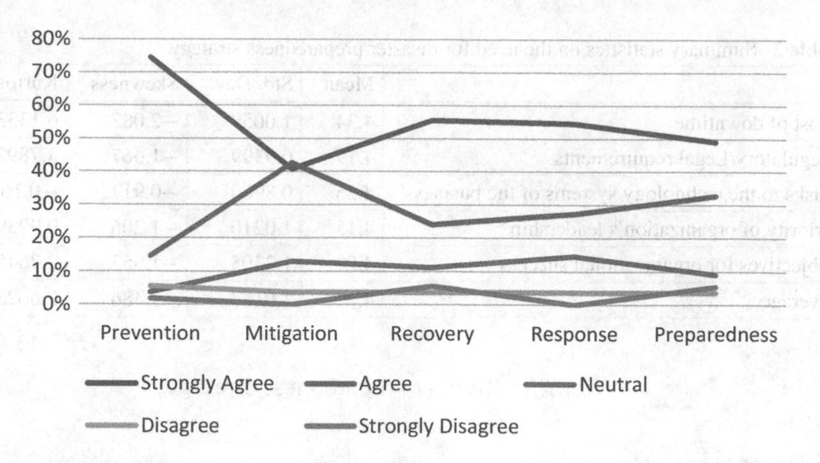

Fig. 2 Disaster awareness and strategy

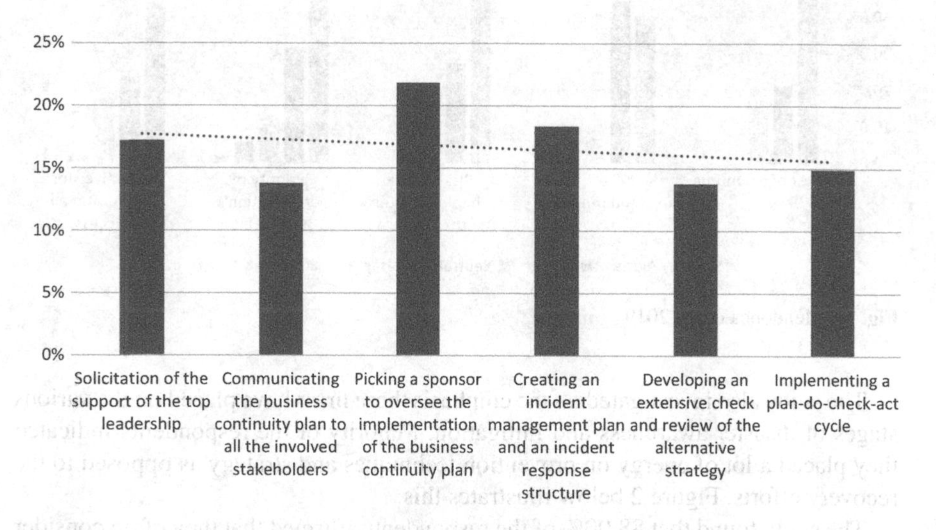

Fig. 3 Aspects of BCM model not followed in UAE firms

managerial roles also take charge of the recovery process. The study also indicated that 86% reported that their organizations were keen on integrating the BCM model to their recovery plan (Alzoubi & Aziz, 2021; Cruz, 2021).

Similarly, Figs. 4 and 5 look at the same issue with relation to DREAD and STRIDE models.

From the figures above, we note that most firms don't focus on the damage potential or exploitability of disasters, whereas information disclosure is highly ignored in the STRIDE model.

When it came to mitigation techniques, a lot of emphasis is put on either avoiding the risk or reducing the risk effect as illustrated in Fig. 6.

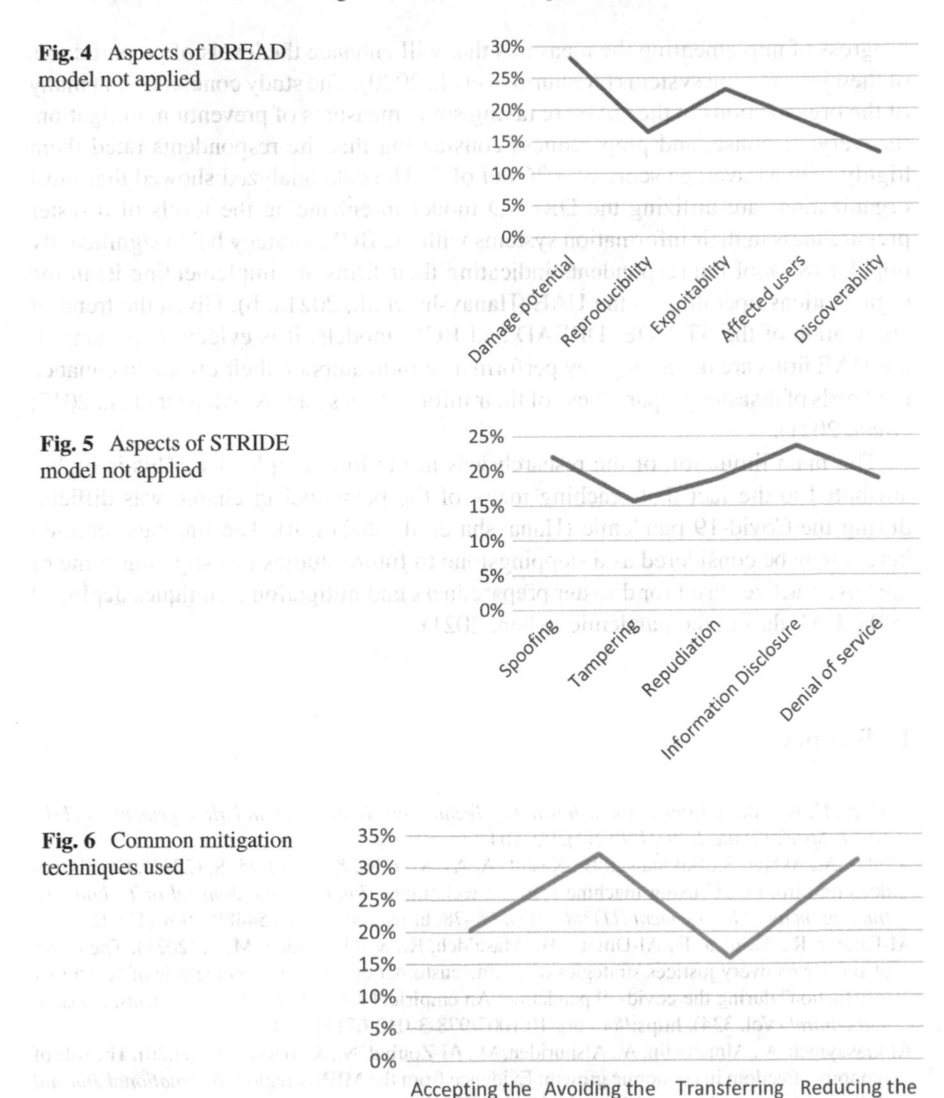

Fig. 4 Aspects of DREAD model not applied

Fig. 5 Aspects of STRIDE model not applied

Fig. 6 Common mitigation techniques used

5 Conclusion and Future Work

The objective of this study was to measure the levels of application of the various models of IT disaster preparedness and planning in the organizations operating in the UAE (Lee et al., 2022a, b). The levels of disaster preparedness in the information systems of an organization contribute substantively to the rest of the resources that the firm could be deployed for its business processes and operations (Mehmood, 2021). Some of the organizations have realized the need for such a step, and they are in the progress of implementing the measures that will enhance the levels of preparedness of their information systems (Alshurideh et al., 2020). The study concludes that many of the organizations in the UAE are taking some measures of prevention, mitigation, recovery, response, and preparedness considering that the respondents rated them highly with an average score of 4.26 out of 5. The data analyzed showed that most organizations are utilizing the DREAD model in enhancing the levels of disaster preparedness in their information systems with the BCM strategy being significantly popular (86% of the respondents indicating their firms are implementing it) in the organizations operating in the UAE (Hanaysha et al., 2021a, b). Given the trend of application of the STRIDE, DREAD and BCM models, it is evident that many of the UAE firms are deploying key performance indicators on their efforts to enhance the levels of disaster preparedness of their information systems (Alnazer et al., 2017; Khan, 2021).

The main limitation of the research was in the low sample size. This is highly attributed to the fact that reaching many of the personnel in charge was difficult during the Covid-19 pandemic (Hanaysha et al., 2021a, b). The findings reported herein may be considered as a steppingstone to future studies investigating some of the best practices used for disaster preparedness and mitigation techniques deployed in the UAE during the pandemic (Khan, 2021).

References

Aasriya, N. A. (2021). *International Journal of Technology, Innovation and Management (IJTIM), Vol. 1, Special Issue 1, 2021 90 1*(1), 90–104.

Akhtar, A., Akhtar, S., Bakhtawar, B., Kashif, A. A., Aziz, N., & Javeid, M. S. (2021). COVID-19 detection from CBC using machine learning techniques. *International Journal of Technology, Innovation and Management (IJTIM), 1*(2), 65–78. https://doi.org/10.54489/ijtim.v1i2.22.

Al-Dmour, R., AlShaar, F., Al-Dmour, H., Masa'deh, R., & Alshurideh, M. T. (2021). The effect of service recovery justices strategies on online customer engagement via the role of "customer satisfaction" during the covid-19 pandemic: An empirical study. In *Studies in systems, decision and control* (Vol. 334). https://doi.org/10.1007/978-3-030-67151-8_19.

Al-Gasaymeh, A., Almahadin, A., Alshurideh, M., Al-Zoubid, N., & Alzoubi, H. (2020). The role of economic freedom in economic growth: Evidence from the MENA region. *International Journal of Innovation Creativity and Change, 13*(10), 759–774.

Al-Maroof, R. S., Salloum, S. A., AlHamadand, A. Q., & Shaalan, K. (2020). Understanding an extension technology acceptance model of google translation: A multi-cultural study in United Arab Emirates. *International Journal of Interactive Mobile Technologies (IJIM), 14*(3), 157–178.

Al Ali, A. (2021). The impact of information sharing and quality assurance on customer service at UAE banking sector. *International Journal of Technology, Innovation and Management (IJTIM), 1*(1), 1–17. https://doi.org/10.54489/ijtim.v1i1.10.

Al Barghuthi, N. B., Ncube, C., & Said, H. (2019). State of art of the effectiveness in adopting blockchain technology-UAE survey study. In *2019 Sixth HCT Information Technology Trends (ITT)* (pp. 54–59).

Al Issa, A. (2019). Strategic data and cyber security management in the arab world: Running successful lives and businesses during the data Tsunami era. In *Strategic thinking, planning, and management practice in the Arab world* (pp. 182–196). IGI Global.

AlHamad, A., Alshurideh, M., Alomari, K., Kurdi, B. A., Alzoubi, H., Hamouche, S., & Al-Hawary, S. (2022). The effect of electronic human resources management on organizational health of telecommuni-cations companies in Jordan. *International Journal of Data and Network Science, 6*(2), 429–438. https://doi.org/10.5267/j.ijdns.2021.12.011

Alhamad, A. Q. M., Akour, I., Alshurideh, M., Al-Hamad, A. Q., Kurdi, B. A., & Alzoubi, H. (2021). Predicting the intention to use google glass: A comparative approach using machine learning models and PLS-SEM. *International Journal of Data and Network Science, 5*(3), 311–320. https://doi.org/10.5267/j.ijdns.2021.6.002

Ali, N., Ahmed, A., Anum, L., Ghazal, T. M., Abbas, S., Khan, M. A., ... Ahmad, M. (2021). Modelling supply chain information collaboration empowered with machine learning technique. *Intelligent Automation and Soft Computing, 30*(1), 243–257.

Ali, N., M. Ghazal, T., Ahmed, A., Abbas, S., A. Khan, M., Alzoubi, H., Farooq, U., Ahmad, M., & Adnan Khan, M. (2022). Fusion-based supply chain collaboration using machine learning techniques. *Intelligent Automation & Soft Computing, 31*(3), 1671–1687. https://doi.org/10.32604/iasc.2022.019892.

Alketbi, S., Alshurideh, M., & Al Kurdi, B. (2020). The influence of service quality on customers'retention and loyalty in the uae hotel sector with respect to the impact of customer'satisfaction, trust, and commitment: a qualitative study. *PalArch's Journal of Archaeology of Egypt/Egyptology, 17*(4), 541–561.

Alnazer, N. N., Alnuaimi, M. A., & Alzoubi, H. M. (2017). Analysing the appropriate cognitive styles and its effect on strategic innovation in Jordanian universities. *International Journal of Business Excellence, 13*(1), 127–140.

Alnuaimi, M., Alzoubi, H. M., Ajelat, D., & Alzoubi, A. A. (2021). Towards intelligent organisations: An empirical investigation of learning orientation's role in technical innovation. *International Journal of Innovation and Learning, 29*(2), 207–221.

Al Shamsi, A. A. (2019). Effectiveness of cyber security awareness program for young children: A case study in UAE. *International Journal of Information Technology and Language Studies, 3*(2).

Alshubiri, F., Jamil, S. A., & Elheddad, M. (2019). The impact of ICT on financial development: Empirical evidence from the Gulf Cooperation Council countries. *International Journal of Engineering Business Management, 11*, 1847979019870670.

Alshurideh, M. (2022). Does electronic customer relationship management (E-CRM) affect service quality at private hospitals in Jordan? *Uncertain Supply Chain Management, 10*(2), 1–8.

Alshurideh, M. T., Al Kurdi, B., Alzoubi, H. M., Sahawneh, N., & Al-kassem, A. H. (2022a). Fuzzy assisted human resource management for supply chain management issues. *Annals of Operations Research, 24*(1), 1–19.

Alshurideh, M, Al Kurdi, B., Alzoubi, H., Ghazal, T., Said, R., AlHamad, A., Hamadneh, S., Sahawneh, N., & Al-kassem, A. (2022b). Fuzzy assisted human resource management for supply chain management issues. *Annals of Operations Research*, 1–19.

Alshurideh, M., Gasaymeh, A., Ahmed, G., Alzoubi, H., & Kurd, B. A. (2020). Loyalty program effectiveness: Theoretical reviews and practical proofs. *Uncertain Supply Chain Management, 8*(3), 599–612. https://doi.org/10.5267/j.uscm.2020.2.003.

Alshurideh, M. T., Kurdi, B. Al, AlHamad, A. Q., Salloum, S. A., Alkurdi, S., Dehghan, A., Abuhashesh, M., & Masa'deh, R. (2021). Factors affecting the use of smart mobile examination platforms by universities' postgraduate students during the COVID 19 pandemic: An empirical study. *Informatics, 8*(2), 32.

Alzoubi, H., & Ahmed, G. (2019). Do total quality management (TQM) practices improve organisational success? A case study of electronics industry in the UAE. *International Journal of Economics and Business Research, 17*(4), 459–472. https://doi.org/10.1504/IJEBR.2019.099975

Alzoubi, A. (2021a). The impact of process quality and quality control on organizational competitiveness at 5-star hotels in Dubai. *International Journal of Technology, Innovation and Management (IJTIM), 1*(1), 54–68. https://doi.org/10.54489/ijtim.v1i1.14.

Alzoubi, A. (2021b). Renewable green hydrogen energy impact on sustainability performance. *International Journal of Computations, Information and Manufacturing (IJCIM), 1*(1), 94–110. https://doi.org/10.54489/ijcim.v1i1.46.

Alzoubi, H. M., Vij, M., Vij, A., & Hanaysha, J. R. (2021). What leads guests to satisfaction and loyalty in UAE five-star hotels? AHP analysis to service quality dimensions. *Enlightening Tourism. A Pathmaking Journal, 11*(1), 102–135.

Alzoubi, H., Alshurideh, M., Kurdi, B. A., Akour, I., & Azi, R. (2022). Does BLE technology contribute towards improving marketing strategies, customers' satisfaction and loyalty? The role of open innovation. *International Journal of Data and Network Science, 6*(2), 449–460. https://doi.org/10.5267/j.ijdns.2021.12.009

Alzoubi, H., Alshurideh, M., Kurdi, B. A., & Inairat, M. (2020a). Do perceived service value, quality, price fairness and service recovery shape customer satisfaction and delight? A practical study in the service telecommunication context. *Uncertain Supply Chain Management, 8*(3), 579–588. https://doi.org/10.5267/j.uscm.2020.2.005

Alzoubi, H. M., Ahmed, G., Al-Gasaymeh, A., & Al Kurdi, B. (2020b). Empirical study on sustainable supply chain strategies and its impact on competitive priorities: The mediating role of supply chain collaboration. *Management Science Letters, 10*(3), 703–708. https://doi.org/10.5267/j.msl.2019.9.008

Alzoubi, H. M., & Aziz, R. (2021). Does Emotional intelligence contribute to quality of strategic decisions? The mediating role of open innovation. *Journal of Open Innovation: Technology, Market, and Complexity, 7*(2), 130. https://doi.org/10.3390/joitmc7020130

Alzoubi, H. M., & Yanamandra, R. (2020). Investigating the mediating role of information sharing strategy on agile supply chain. *Uncertain Supply Chain Management, 8*(2), 273–284. https://doi.org/10.5267/j.uscm.2019.12.004

Amarneh, B. M., Alshurideh, M. T., Al Kurdi, B. H., & Obeidat, Z. (2021). The impact of COVID-19 on E-learning: Advantages and challenges. In *The International Conference on Artificial Intelligence and Computer Vision* (pp. 75–89).

Assad, N. F., & Alshurideh, M. T. (2020a). Financial reporting quality, audit quality, and investment efficiency: Evidence from GCC economies. *WAFFEN-UND Kostumkd. J, 11*(3), 194–208.

Assad, N. F., & Alshurideh, M. T. (2020b). Investment in context of financial reporting quality: A systematic review. *Waffen-Und Kostumkunde Journal, 11*(3), 255–286.

Aziz, N., & Aftab, S. (2021). Data mining framework for nutrition ranking Nauman Aziz. *International Journal of Technology, Innovation and Management, 1*(1), 90–100.

Baadel, S., Majeed, A., & Kabene, S. (2017). Technology adoption and diffusion in the Gulf: Some challenges. In *Proceedings of the 8th International Conference on E-Education, E-Business, E-Management and E-Learning*, (pp. 16–18).

Cruz, A. (2021). Convergence between blockchain and the internet of things. *International Journal of Technology, Innovation and Management (IJTIM), 1*(1), 35–56.

Czagan, D. (2014). *Qualitative risk analysis with the DREAD model*. InfoSecInstitute.Com.

De Jong, M., Hoppe, T., & Noori, N. (2019). City branding, sustainable urban development and the rentier state. How do Qatar, Abu Dhabi and Dubai present themselves in the age of post oil and global warming? *Energies, 12*(9), 1657.

Farouk, M. (2021). The universal artificial intelligence efforts to face coronavirus COVID-19. *International Journal of Computations, Information and Manufacturing (IJCIM), 1*(1), 77–93. https://doi.org/10.54489/ijcim.v1i1.47.

Geer, D., Jardine, E., & Leverett, E. (2020). On market concentration and cybersecurity risk. *Journal of Cyber Policy, 5*(1), 9–29.

Ghane, G., & Esmaeili, M. (2020). Nursing students' perception of patient-centred care: A qualitative study. *Nursing Open, 7*(1), 383–389.

Ghazal, T. M., Hasan, M. K., Alshurideh, M. T., Alzoubi, H. M., Ahmad, M., Akbar, S. S., Al Kurdi, B., & Akour, I. A. (2021). IoT for smart cities: Machine learning approaches in smart healthcare—A review. *Future Internet, 13*(8), 218. https://doi.org/10.3390/fi13080218

Ghazal, T. M., Hasan, M. K., Hassan, R., Islam, S., Abdullah, S., Afifi, M. A., & Kalra, D. (2020). Security vulnerabilities, attacks, threats and the proposed countermeasures for the Internet of Things applications. *Solid State Technology, 63*(1s), 2513–2521.

Gomez, S. R., Mancuso, V., & Staheli, D. (2019). Considerations for human-machine teaming in cybersecurity. In *International Conference on Human-Computer Interaction* (pp. 153–168).

Guergov, S., & Radwan, N. (2021). Blockchain convergence: Analysis of issues affecting IoT, AI and blockchain. *International Journal of Computations, Information and Manufacturing (IJCIM), 1*(1), 1–17. https://doi.org/10.54489/ijcim.v1i1.48.

Haddad, A., Ameen, A., Isaac, O., Alrajawy, I., Al-Shbami, A., & Chakkaravarthy, D. M. (2020). The impact of technology readiness on the big data adoption among UAE organisations. In *Data management, analytics and innovation* (pp. 249–264). Springer.

Hamadneh, S., Pedersen, O., Alshurideh, M., Kurdi, B. Al, & Alzoubi, H. (2021). An investigation of the role of supply chain visibility into The Scottish blood supply chain. *Journal of Legal, Ethical and Regulatory Issues, 24*(Special Issue 1), 1–12.

Hanaysha, J. R., Al Shaikh, M. E., & Alzoubi, H. M. (2021a). Importance of marketing mix elements in determining consumer purchase decision in the retail market. *Internation Journal of Service Science, 12*(6), 56–72.

Hanaysha, J. R., Al-Shaikh, M. E., Joghee, S., & Alzoubi, H. (2021b). Impact of innovation capabilities on business sustainability in small and medium enterprises. *FIIB Business Review*, 1–12.https://doi.org/10.1177/23197145211042232

Herrera, M. (2013). Four types of risk mitigation and BCM governance, risk and compliance. Retrieved from June 11, 2017.

Hodgetts, G., Brown, G., Batić-Mujanović, O., Gavran, L., Jatić, Z., Račić, M., Tešanović, G., Zahilić, A., Martin, M., & Birtwhistle, R. (2020). Twenty-five years on: Revisiting Bosnia and Herzegovina after implementation of a family medicine development program. *BMC Family Practice, 21*(1), 7.

Joghee, S., Alzoubi, H., & Dubey, A. (2020a). Decisions effectiveness of FDI investment biases at real estate industry: Empirical evidence from Dubai smart city projects. *International Journal of Scientific & Technology Research, 9*(3), 1245–1258.

Joghee, S., Alzoubi, H. M., & Dubey, A. R. (2020b). Decisions effectiveness of FDI investment biases at real estate industry: Empirical evidence from Dubai smart city projects. *International Journal of Scientific and Technology Research, 9*(3), 3499–3503.

Kashif, A. A., Bakhtawar, B., Akhtar, A., Akhtar, S., Aziz, N., & Javeid, M. S. (2021). Treatment response prediction in Hepatitis C patients using machine learning techniques. *International Journal of Technology, Innovation and Management (IJTIM), 1*(2), 79–89. https://doi.org/10.54489/ijtim.v1i2.24.

Khan, M. A. (2021). Challenges facing the application of IoT in medicine and healthcare. *International Journal of Computations, Information and Manufacturing (IJCIM), 1*(1), 39–55. https://doi.org/10.54489/ijcim.v1i1.32.

Khan, S. A. (2017). A STRIDE model based threat modelling using unified and-or fuzzy operator for computer network security. *International Journal of Computing and Network Technology, 5*(01), 13–20.

Lee, C., & Ahmed, G. (2021). Improving IoT privacy, data protection and security concerns. *International Journal of Technology, Innovation and Management (IJTIM), 1*(1), 18–33. https://doi.org/10.54489/ijtim.v1i1.12.

Lee, K. L., Azmi, N. A. N., Hanaysha, J. R., Alzoubi, H. M., & Alshurideh, M. T. (2022a). The effect of digital supply chain on organizational performance: An empirical study in Malaysia manufacturing industry. *Uncertain Supply Chain Management, 10*(2), 495–510. https://doi.org/10.5267/j.uscm.2021.12.002

Lee, K. L., Romzi, P. N., Hanaysha, J. R., Alzoubi, H. M., & Alshurideh, M. (2022b). Investigating the impact of benefits and challenges of IOT adoption on supply chain performance and organizational performance: An empirical study in Malaysia. *Uncertain Supply Chain Management, 10*(2), 537–550. https://doi.org/10.5267/j.uscm.2021.11.009

Leo, S., Alsharari, N. M., Abbas, J., & Alshurideh, M. T. (2021). From offline to online learning: A qualitative study of challenges and opportunities as a response to the COVID-19 pandemic in the UAE higher education context. *The Effect of Coronavirus Disease (COVID-19) on Business Intelligence, 334*, 203–217.

Li, B., Deng, C., Yang, J., Lilja, D., Yuan, B., & Du, D. (2019). Haml-ssd: A hardware accelerated hotness-aware machine learning based ssd management. In *2019 IEEE/ACM International Conference on Computer-Aided Design (ICCAD)* (pp. 1–8).

Mahmah, A. E. L., & Kandil, M. E. (2019). The balance between fiscal consolidation and non-oil growth: The case of the UAE. *Borsa Istanbul Review, 19*(1), 77–93.

Mashaqi, E., Al-Hajri, S., Alshurideh, M., & Al Kurdi, B. (2020). The impact of e-service quality, e-recovery services on e-loyalty in online shopping: Theoretical foundation and qualitative proof. *PalArch's Journal of Archaeology of Egypt/Egyptology, 17*(10), 2291–2316.

Matthews, G. (2020). *Managing third party risk*. EthicalBoardRoom.Com.

Mehmood, T. (2021). Does information technology competencies and fleet management. *International Journal of Technology, Innovation and Management, 1*(1), 14–41.

Mehmood, T., Alzoubi, H. M., Alshurideh, M., Al-Gasaymeh, A., & Ahmed, G. (2019). Schumpeterian entrepreneurship theory: Evolution and relevance. *Academy of Entrepreneurship Journal, 25*(4), 1–10.

Mendonca, J., Andrade, E., Endo, P. T., & Lima, R. (2019). Disaster recovery solutions for IT systems: A Systematic mapping study. *Journal of Systems and Software, 149*, 511–530.

Miller, D. (2021). The best practice of teach computer science students to use paper prototyping. *International Journal of Technology, Innovation and Management (IJTIM), 1*(2), 42–63. https://doi.org/10.54489/ijtim.v1i2.17.

Mondol, E. P. (2021). The impact of block chain and smart inventory system on supply chain performance at retail industry. *International Journal of Computations, Information and Manufacturing (IJCIM), 1*(1), 56–76. https://doi.org/10.54489/ijcim.v1i1.30.

Mosteanu, N. R. (2019). Intelligent foreign direct Investments to boost economic development–UAE case study. *The Business & Management Review, 10*(2), 1–9.

Obaid, A. J. (2021). Assessment of smart home assistants as an IoT. *International Journal of Computations, Information and Manufacturing (IJCIM), 1*(1), 18–36. https://doi.org/10.54489/ijcim.v1i1.34.

Poniatowski, K. (2018). *Is The STRIDE Approach Still Relevant for Threat Modeling?* Security Innovation Blog.

Radwan, N., & Farouk, M. (2021). The growth of internet of things (IoT) in the management of healthcare issues and healthcare policy development. *International Journal of Technology, Innovation and Management (IJTIM), 1*(1), 69–84. https://doi.org/10.54489/ijtim.v1i1.8.

Shah, S.F., Alshurideh, M. T., Al-Dmour, A., & Al-Dmour, R. (2021). Understanding the influences of cognitive biases on financial decision making during normal and COVID-19 pandemic situation in the United Arab Emirates. In *Studies in systems, decision and control* (Vol. 334). https://doi.org/10.1007/978-3-030-67151-8_15.

Shah, S. F., Alshurideh, M., Al Kurdi, B., & Salloum, S. A. (2020). The impact of the behavioral factors on investment decision-making: A systemic review on financial institutions. In *International Conference on Advanced Intelligent Systems and Informatics* (pp. 100–112).

Sharma, A. (2019). *Half of global businesses not ready to tackle cyber attacks, report says*. TheNational News.Com.

Singh, R., & Singh, P. K. (2021). Integrating blockchain technology with IoT. *CEUR Workshop Proceedings, 2786*(1), 81–82.

Suprihanto, D., Wardoyo, R., & Mustofa, K. (2018). Determination of weighting assessment on DREAD model using profile matching. *International Journal of Advanced Computer Science and Applications, 9*(10), 68–72.

The Economist Intelligence Unit. (2018). *Cyber attacks: is the GCC prepared?* Eiu.Com.

Torabi, S. A., Giahi, R., & Sahebjamnia, N. (2016). An enhanced risk assessment framework for business continuity management systems. *Safety Science, 89*, 201–218.

Turki, A. M., Barween, A. K., Ra'ed, M., & A., S. S. (2021). The moderation effect of gender on accepting electronic payment technology: a study on United Arab Emirates consumers. In *Review of International Business and Strategy: Vol. ahead-of-p* (Issue ahead-of-print). https://doi.org/10.1108/RIBS-08-2020-0102.

Vaidhyanathan, R. (2010). *Six business continuity management (BCM) lifecycle guidelines.* ComputerWeekly.Com.

Walker, J., Williams, B. J., & Skelton, G. W. (2010). Cyber security for emergency management. In *2010 IEEE International Conference on Technologies for Homeland Security (HST)* (pp. 476–480).

Xin, T., & Xiaofang, B. (2014). Online banking security analysis based on STRIDE threat model. *International Journal of Security and Its Applications, 8*(2), 271–282.

Zahid, M., Inayat, I., Daneva, M., & Mehmood, Z. (2020). A security risk mitigation framework for cyber physical systems. *Journal of Software: Evolution and Process, 32*(2), e2219.

Aircraft Turnaround Manager (ATM): A Solution to Airport Operations

Amber Aziz, M. Nawaz Brohi, Tariq Rahim Soomro, Taher M. Ghazal, Haitham M. Alzoubi, and Muhammad Alshurideh

Abstract The set of activities performed during the preparation of an inbound aircraft for the succeeding outbound flight, which are planned for the same aircraft is called turnaround operation. The turnaround operations consist of activities the exchange of passengers for both inbound and outbound, crew, baggage and cargo handling, catering services. The evaluation and control of activities at the present time, is being carried out manually, which lead to flight delays and unreliable data. The proposed solution is to show the aircraft turnaround operations in real time and is based on a rule based systems called Aircraft Turnaround Manager (ATM) that is used to manage and monitor the flight groundwork throughout the turnaround

A. Aziz
Department of Computer Science, SZABIST Dubai Campus, Dubai, UAE

M. N. Brohi
Bath Spa University RAK Campus, RAK, UAE
e-mail: mnbrohi@bathspa.ae

T. R. Soomro
Institute of Business Management, CCSIS, Karachi, Pakistan
e-mail: ariq.soomro@iobm.edu.pk

T. M. Ghazal
School of Information Technology, Skyline University College, Sharjah, UAE
e-mail: taher.ghazal@skylineuniversity.ac.ae

Center for Cyber Security, Faculty of Information Science and Technology, Universiti Kebangsaan Malaysia (UKM), Selangor, Malaysia

H. M. Alzoubi (✉)
School of Business, Skyline University College, Sharjah, UAE
e-mail: haitham.alzubi@skylineuniversity.ac.ae

M. Alshurideh
Department of Marketing, School of Business, University of Jordan, Amman, Jordan
e-mail: m.alshurideh@ju.edu.jo; malshurideh@sharjah.ac.ae

Department of Management, College of Business Administration, University of Sharjah, Sharjah, UAE

M. Alshurideh et al. (eds.), *The Effect of Information Technology on Business and Marketing Intelligence Systems*, Studies in Computational Intelligence 1056,
https://doi.org/10.1007/978-3-031-12382-5_36

process. The Critical Path Method (CPM) and Project Evaluation and Review Technique (PERT) are exhibited in the system. The study has shown, real time monitoring of turnaround operation can decrease delays occurring from operations belongs to airline and can optimize the aircraft ground time.

Keywords Airline operation · Aircraft turnaround process · Flight management · Delay propagation · Real-time monitoring

1 Introduction

Air transport is meant to connect people and cultures across the globe. The worldwide network of transportation that backs up tourism and supports international business is provided by the aviation industry. In addition to this, the world has been exponentially changed by the commercial aviation. It has not only been helpful in facilitating world trade but has also brought the people together. The World's gross domestic product was supported by the aviation industry, which is $2.7 trillion (3.5%). In 2017, airlines transported 4 billion passengers and 60 million metric tons of freight and the cities all over the world are connected with 36,000 routes at one point. The aviation industry has an essential role in boosting the economy as well as generating various financial and social advantages.

Airlines gain income only when the aircraft is flying. Thus, diminishing the non-profitable time spent on the ground is one of the greatest challenges, which aircraft confront nowadays particularly, when the market is oversaturated with business rivalry.

There are stages, where a traveler gone through one after another, and many of the aircraft's turnaround activities have to run in a row, whereas remaining may run in parallel. Theoretically, the aircraft is de-boarded, cleaned, boarded, refueled, supplied with drinking water and food and, if necessary, maintained before the boarding of upcoming passengers. Practically, the procedures are not just interrelated, they are likewise influenced by outside elements, for example, flight changes or climate conditions, which frequently brings about exorbitant changes to the first plan. The aviation business is not only a fast-growing industry, but also a highly competitive one. This is the very reason that it is much necessary for the airlines to use a proactive approach on the major factor of 'punctuality', which resultantly would lead to the satisfaction of passengers as well as specifically results in cost savings per minute.

This punctuality intends to enhance on-time performance. According to a few analysts; On-time performance is the sole largest performance metric in the airlines business. To deal with the turnaround activities, there are no comprehensive automated systems in use at airports at the moment. Generally, airlines have used paper-based procedures for the management of turnarounds. However, this caused inefficiencies. Complex and timely synchronization of a set of activities and services is required to deal with the turnaround of a wide-bodied passenger jet. The sending of key messages required dispatchers to leave the aircraft and run to the gate resulting in loss of valuable time. Likewise, the central dispatch group couldn't see a real-time status of their turnaround activities. Because of these issues, a framework would be made to adequately manage turnaround operations.

It additionally needed to guarantee that, with airport authorities hoping to better share data between airlines, air traffic controllers, and the airport management, it was in a situation to deliver real-time status info to these stakeholders. This study organizes in a way that first section is of introduction, Sect. 2 discussed the airport operations and airport background, along with current approaches for handling turnaround process. Materials gathered and used methods has been discussed along with the architecture of the system in Sect. 3. However, Sect. 4 presents the overall functionality of the system. The final section summarizes and presents the conclusions derived from this study and provides insight into the future implications.

2 Literature Review

A report released under the head of IATA in 2014 asserted that the number of passengers could touch the limits of 7.3 billion till 2034. This shows a 4.1% average growth rate per year. The increase in massive commute at airports and the surge to follow strict flight schedules will lead to airport congestion that will ultimately pose a challenge to optimum on-time performance (Akhtar et al., 2021; Al Ali, 2021; AlHamad et al., 2022; Alhamad et al., 2021; Ali et al., 2021). The issues will not only be of the congested airport lounges rather the broader airport process shall also bear the strain. Efficiency and effectiveness are the major challenges for the airlines on the day of operation. Efficiency is tested, when each day of the year is to be followed by the same routine, but of course, different schedules. The management has to make sure that the flight operations should run as smoothly as possible.

2.1 Airline Operational Procedures

There is a complicated network of various procedures, which are being performed whether in the air or on the ground. Most of the time, it becomes difficult to predict their effect on one another. As shown in Fig. 1, operational plans are to be developed in two different environments on the day of operations:

- The airline operations control center (OCC), also known as or the integrated operations control center (IOCC) or the systems operations control center (SOCC).
- The hub control center (HCC), also referred to as hub control (HC) or the airport control center (ACC).

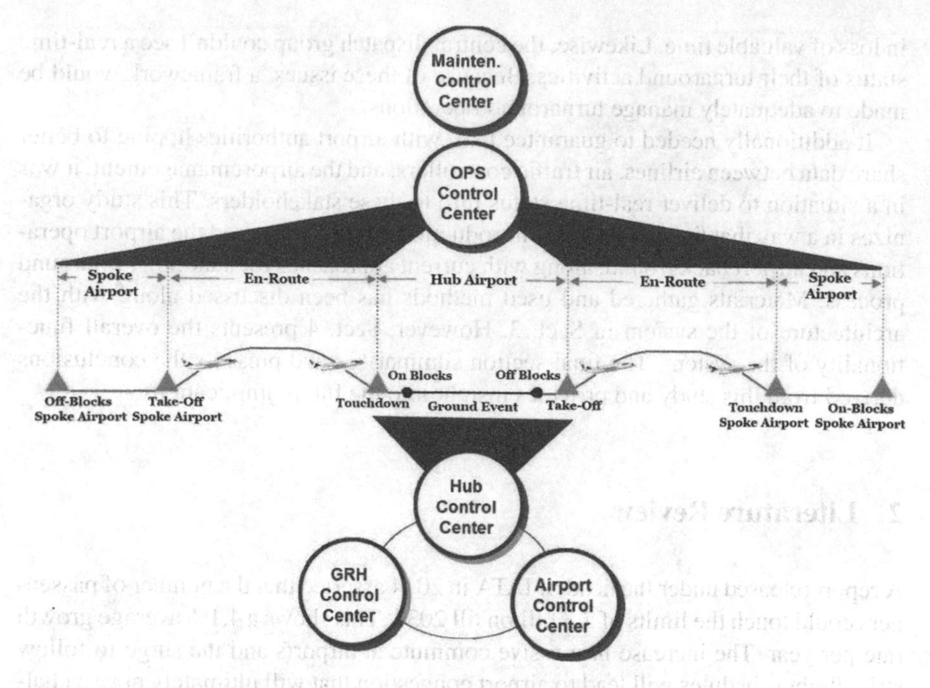

Fig. 1 The Hub management

The second major process deals with the control of on-ground turnaround processes (Ali et al., 2022; Alnazer et al., 2017). Aircraft turnaround details the procedure of servicing an aircraft at the time it is waiting between the two successive flights on the ground. The term of a turnaround states the quick sequence between an arrival and a departure (Alnuaimi et al., 2021; Alsharari, 2021; Alshurideh et al., 2022). However, specifically in long-haul flights, a long-time interval can be programmed between them for many air transport operations. This variation cab was seen in the following figure (ref Fig. 2).

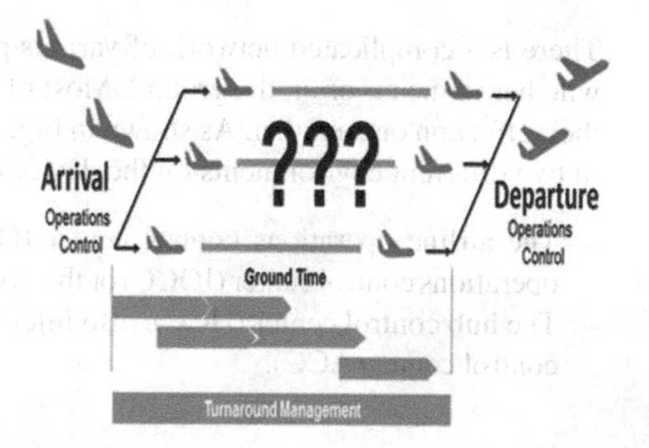

Fig. 2 Turnaround management

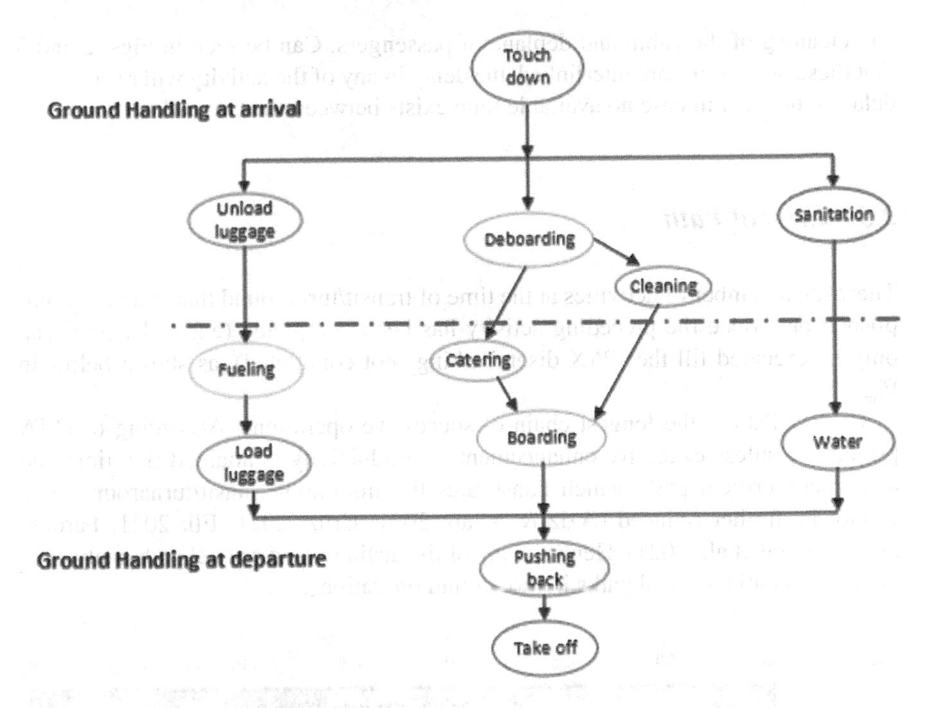

Fig. 3 Ground activities flow

Turnaround involves a set of a complicated process of general ground handling procedures, which an aircraft has to undergo. These activities include boarding/ landing, loading/unloading of luggage, catering, fueling, cleaning, sanitizing etc. The major ground handling activities and their precedence constraints around a grounded aircraft are described in Fig. 3. The ground handling tasks of arrival and departure are separately explained.

2.2 *Sequence of Ground Handling Activities*

While managing the sequence of aircraft ground handling procedures, some activities can be run on their own without considering other activities, but some are dependent on the other ones. The observance of interaction sequence of these ground handling activities is required for a smooth completion of the operations and on-time departure of the plane (Ali Alzoubi, 2021a, b; Asem Alzoubi, 2021a, b; Alzoubi et al., 2022; Alzoubi et al., 2020a, b). The independent ground handling activities may include the front and rear load off-loading with two lower deck loaders as these processes can be initiated and stopped without the involvement of any other activity (Alzoubi & Ahmed, 2019; Alzoubi & Aziz, 2021; Alzoubi & Yanamandra, 2020; Alzoubi et al., 2020a, b, 2021). Rest of the activities must take place in a predetermined set fashion e.g. cleaning of the cabin and deplane of passengers. Can be seen in Figs. 2 and 3 that these procedures are interlinked, the delay in any of the activity will result in the delay of the next in case no available time exists between the two tasks.

2.3 *Critical Path*

There are a number of activities at the time of transit/turnaround that can be accomplished only once the preceding activity has been completed (e.g. 'Cleaning' can only be executed till the 'PAX disembarking' not completed), as shown below in Fig. 4.

Critical Path is the longest chain of successive operations. According to IATA principles, unless excessive enhancement in productivity is attained (i.e. time and resources), critical path, which constitutes the minimum transit/turnaround time cannot be further reduced (Aziz & Aftab, 2021; Cruz, 2021; Eli, 2021; Farouk, 2021; Ghazal et al., 2021). Hence, many of the airlines used critical path analysis to identify potential critical paths in turnaround operations.

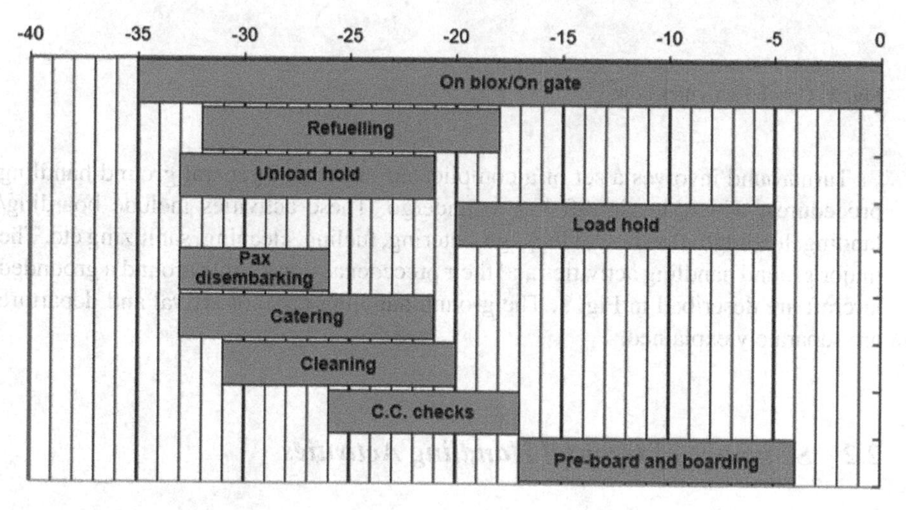

Fig. 4 Turnaround activities

2.4 *Current Approaches*

To deal with turnaround issues, different researchers' have presented different models. Few approaches have been used for the aircraft turnaround operation: Critical Path Method (CPM), Modeling technique and Monitoring System, Resource-Constrained Project Scheduling Problem (RCPSP). Table 1 depicts the benefits and shortcomings of these approaches.

3 Materials and Methods

This study derives as an applied research, based on an inductive approach. The review of previously published literature including journal and research papers has been performed to acquire the knowledge. This vast analysis aimed to detect maximum possible information about the Precision Time Schedule (PTS) at the current time, along with the determination to identify the sources of data and results that can help to support, enhance and validate the work being done in this study. As mentioned, the findings of this study can be used for the development of the tool, based on the proposed solution and design. ATM will be designed by taking the idea from the techniques of CPM and PERT model. CPM model is widely used to obtain the critical

Table 1 Comparison between different approaches (Author Created)

Approaches	Benefits	Shortcomings
Resource-Constrained Project Scheduling Problem (RCPSP)	– Focused on two operational targets, increasing passenger numbers and delivering punctual turnarounds	– For small aircrafts – Very limited activities – Dynamic handling of activities
CPM model	– Sub-operation activities handled simultaneously – Critical path analysis	– Do not capture the uncertainties of schedule punctuality – Operational uncertainties of aircraft turnaround operations
Markovian type model	– Can handle dynamic and stochastic transitions behavior – Can handle disruptions to aircraft turnarounds	– No buffer times – Very limited activities
Wu and caves analytical model	– Scheduling buffer times – Can handle operational uncertainties	– For small aircrafts – Very limited activities
PERT model	– Describe the interdependencies between some activities – Managing activities in chronological sequence	– Cannot handle dynamic and stochastic transitions behavior

path of the plan (Guergov & Radwan, 2021; Samer Hamadneh et al., 2021a, 2021b; Hanaysha et al., 2021b). It can be a project plan or a process, which has many tasks involved and the tasks are dependent on each other (Hanaysha et al., 2021a; Joghee et al., 2020; Kashif et al., 2021). The objective of using CPM model is to detect the critical path of the activities involved in turnaround process, the management of activities in a sequence has been inherited from the PERT model. Above, in the study, the activity sequence diagram, where a Path shows the arrangement of activities that must be executed in a sequence. A Use case methodology has been used for system analysis to be classifying and clarifying the complete system requirements.

The subsequent section will elaborate the structure and overall architecture of the system. The proposed architecture has been designed based on PERT and CPM model techniques. Further in the study includes more details on the model application using use cases.

3.1 ATM Structure and Flow Development

The control flow and structure has been elaborated in current section. Figure 5 demonstrates the proposed complete control flow and data flow in the different system functions.

The Aircraft Turnaround Monitor (ATM) is designed to gather functional data in real time through aircraft turnaround process. The system will also be used for real time monitoring by using the gathered information. The ATM framework is designed as open framework due to operational and different ground handling unit needs, which gives the capability to add on modules based on ADD¡ITIONAL requirement on the same structure and platform (Alshurideh et al., 2019; Khan, 2021; Lee & Ahmed, 2021; Lee et al., 2022a). According to most of aircraft PTS's, turnaround activities can be categorized in four process flows, specifically passenger, engineering, cargo and catering (Ghazal, 2021; Maasmi et al., 2021; Siddiqui et al., 2021). The framework includes the activities with respect to each process flow according to data collection needs and based on importance of specific activity involved in turnaround process. On the basis of this framework, each specific event handler will require to gather the timestamps of main activities during the process of turnaround that enables every specific process flow progress shared between control unit and the handler, e.g. catering center and catering loading staff. Meanwhile, the same information collected during turnaround operation can be shared amongst various handling units involved in turnaround process, the Airport Operation Control Center (AOCC) and the Network Operational Control Center (NOCC) at the airlines headquarters (Hamadneh et al., 2021a, 2021b; Lee et al., 2022b; Mehmood, 2021; Mehmood et al., 2019; Miller, 2021). The accessibility of real time operational data along with turnaround operations transparency significantly raises the situation awareness of all ground handling units and AOCC, NOCC involved in aircraft turnaround process. In the turnaround process some of the activities have operative sequence that needs to be followed, such as, passenger, cargo, and catering flows (Zitar, 2021; Zitar et al., 2021). The rest of

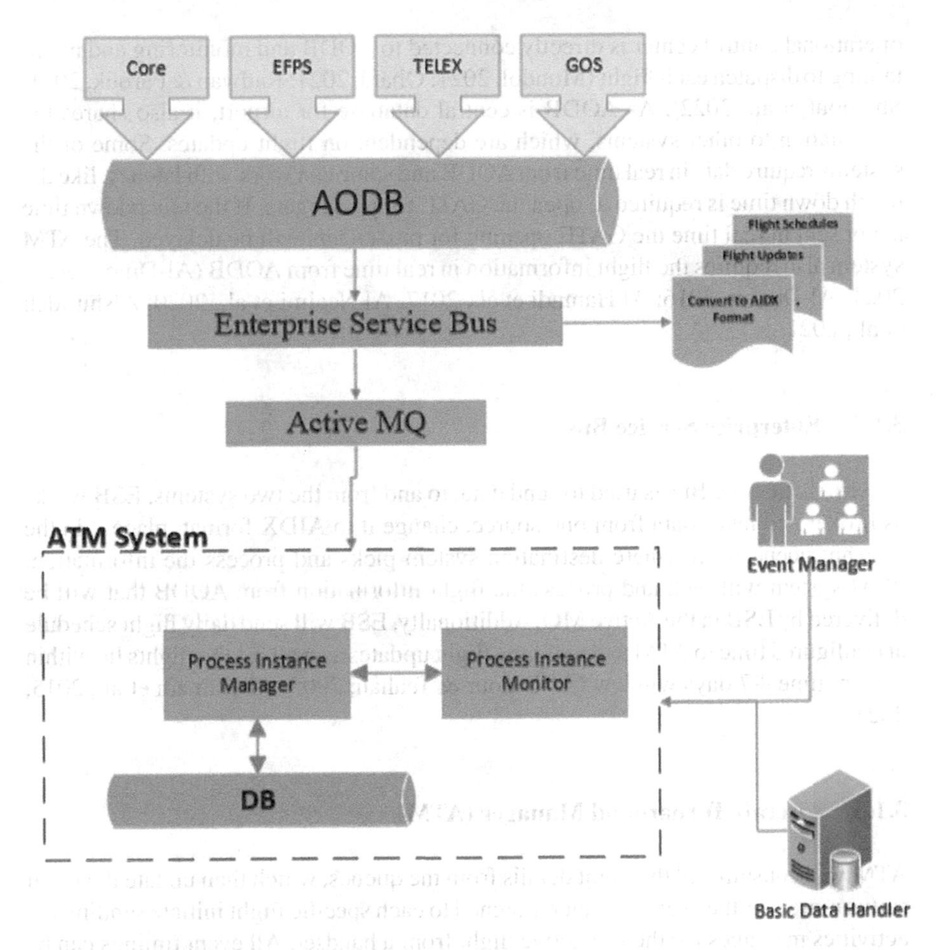

Fig. 5 Proposed ATM architecture

the activities, such as, engineering checks and refueling are executed autonomously from others.

External Systems: Core, EFPS, TELEX, GOS etc. these systems are the source of data, which updates information for each flight. Every system is responsible for specific type of data and information e.g. the Aircraft Actual Landing time is captured by EFPS system and it has been sent directly to AODB.

3.1.1 Airport Operational Database

Airport Operational Database (AODB) is called the heart of an airport. It stores all flights information and process each update received from different sources for specific flight. Airport has dozens of systems, which are connected to AODB. Airport operational control center is directly connected to AODB and monitoring and maintaining to dispatch each flight (Mondol, 2021; Obaid, 2021; Radwan & Farouk, 2021; Shamout et al., 2022). As AODB is central database for airport, it also shares the information to other systems, which are dependent on flight updates. Some of the systems require data in real time from AODB and some can work with latency, like the touch down time is required to open the GATE for passengers. If the touchdown time is not sent in real time the GATE opening for passengers will be delayed. The ATM system also requires the flight information in real time from AODB (Al-Dmour et al., 2021; Al-Dmour, 2016; Al Hamadi et al., 2017; Al Neaimi et al., 2020; Alshurideh et al., 2021).

3.1.2 Enterprise Service Bus

Enterprise Service Bus is used to send data, to and from the two systems. ESB works as broker and takes data from one source, change it to AIDX format, place it in the relevant queue from, where destination system picks and process the information. ATM system will pick and process the flight information from AODB that will be delivered by ESB in the Active MQ. Additionally, ESB will send daily flight schedule at configured time to ATM and real time flight updates as well for the flights lie within current time +7 days window (Al-Dmour & Teahan, 2005; Al-Hamadi et al., 2015, 2021).

3.1.3 Aircraft Turnaround Manager (ATM)

ATM will consume all the flight details from the queues, which then update the flight status moreover the event manager assigned to each specific flight initiate sending the activities messages for the respective flight from a handset. All event timings can be visible to all handling units and event managers, so that in case of any criticality they can take proactive measures (Al-Naymat et al., 2021; Ali & Dmour, 2021; Aslam et al., 2021; Bibi et al., 2021).

4 Results and Findings

After discussing the flow of the system, this section will emphasis on the functionality part of the system and will elaborate each function with the help of narrative use case.

Functionality Definition

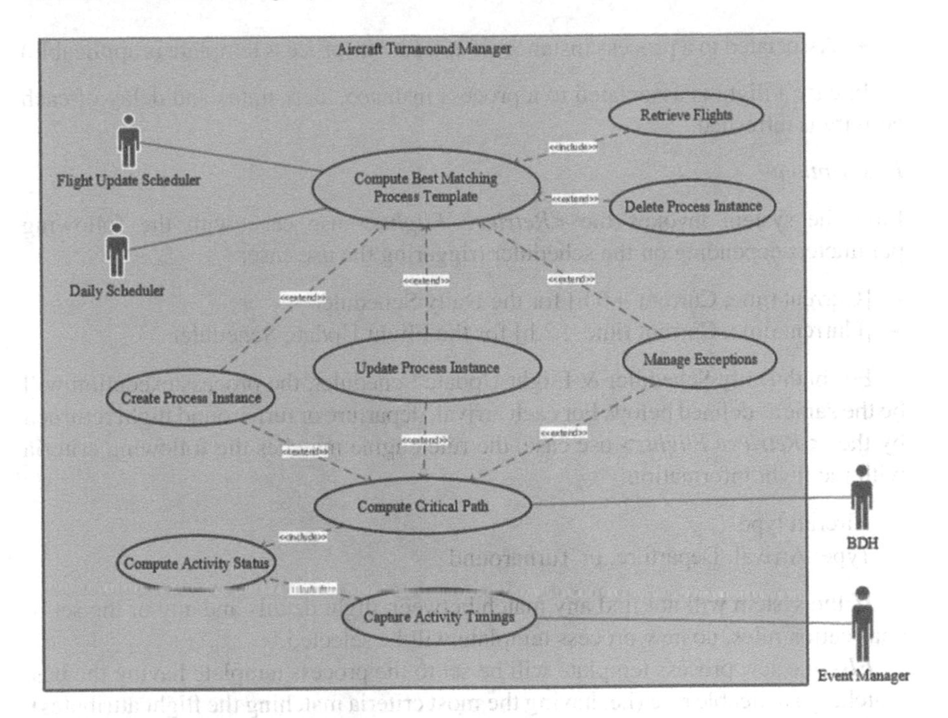

Fig. 6 Functionality diagram *(Author created)*

4.1 Manage Processes Instance—Use Case (High Level Use Case)

Figure 6 below shows proposed function diagram.

(a) Compute Best Matching Process Template—Use Case

Users

- Daily Scheduler
- Flight Update Scheduler

Pre-Conditions

- The configured Daily Scheduler threshold is reached: {Daily Instantiation Time} {Daily Instantiation Time Zone}
- The Flight Update Scheduler threshold is reached (every {Flight Update Frequency})

Post-Conditions

- Flights are either:
  - Not associated to a process instance (if no process template is applicable)

- Associated to a process instance (if at least one process template is applicable)

In case a flight is associated to a process instance, alert status and delay of each activity is refreshed.

Description

First the system invokes the <*Retrieve Flights*> use case with the following parameters depending on the scheduler triggering the use case:

- [Current time, Current +6 h] for the Daily Scheduler
- [Current time, Current time +2 h] for the Flight Update Scheduler

For both Daily Scheduler & Flight Update Scheduler, the process execution will be the same as defined below. For each arrival, departure or turnaround flight returned by the < *Retrieve Flights*> use case, the rule engine matches the following criteria with the flight information:

- Aircraft type
- Type: Arrival, Departure, or Turnaround

If the system will not find any match between flight details and any of the set of application rules, no new process template will be selected.

Else the new process template will be set to the process template having the best matching applicable rule (i.e. having the most criteria matching the flight attributes).

If no new process template is selected and a current process instance is set on the flight, the < *Delete Process Instance*> use case will be invoked.

Else if the system detects one of an exception based on the flight data, the < *Manage Exception* > and < *Compute Critical Path* > use cases will be invoked.

Else if a new process template is selected and the flight has no associated process instance, the < *Create Process Instance* > and < *Compute Critical Path* > use cases will be invoked.

Else If a new process template is selected and the current process instance set on the flight derives from a different process template the < *Update Process Instance* > and < *Compute Critical Path* > use cases will be invoked.

Since the computation of the critical path entails the update of activities earliest start/end time (against which delay value is computed) and latest start/end time (against which the severity of the delay is evaluated), the system then invoke < *Compute Activities Delay* > to refresh delay and associated severity for each activities.

(b) Retrieve Flights—Use Case

Users: N/A

Pre-Conditions: N/A

Post-Conditions: A set of selected flights containing arrival, departure and turnaround flight will be returned.

Description

The system selects arrival and departure flights based on:

- Best arrival timing (Arrival Flights)
- Best departure timing (Departure Flights)
- The time window as provided in input of the use case

For each selected flight the following processing will be performed:

The aircraft split time is defined as the sum of the Minimum Turnaround Time (MTTT), as defined in Basic Data Handler (BDH) and {Split Time Buffer}, as configured in PTS per aircraft type.

If a flight linkage exists, and departure flight's best departure timing minus arrival flight's best arrival timing is less than the aircraft split time, the system adds the *turnaround* in the set of selected flights.

Else if a flight linkage exists, and departure flight's best departure timing minus arrival flight's best arrival timing is greater than or equal to the aircraft split time, the system respectively adds the *arrival* and *departure* flight to the set of selected flights as separate entities.

Else If no flight linkage exists, the system adds the *arrival* or *departure* flight to the set of selected flights as separate entities.

The system then excludes the flights for which no further processing is needed (cancelled, off-block flight) as follow:

- Departure flights that are either cancelled or off-block
- Arrival flight that are either cancelled
- Turnaround flight that are fully cancelled (i.e. both arrival and departure in cancelled status) or off-block
- Turnaround flight for which either its arrival or its departure is cancelled

In the last case, the non-cancelled part of the turnaround flight will be added to the list of selected flight as a separate entity (*arrival or departure).*

Ground flight movements (tows) are not considered as flights or part of a turnaround.

(c) Create Process Instance—Use Case

Users: N/A

Pre-Conditions: A new process template has been selected for the flight

Post-Conditions: The flight is associated with a process instance

Description: The system instantiates the process template for the flight. Each activity and their characteristics are associated to the flight.

(d) Update Process Instance—Use Case

Users: N/A

Pre-Conditions

- A new process template has been selected for the flight.
- An existing process instance exists for the flight.

Post-Conditions

- The flight is associated with a process instance.
- Actual start time and actual end time of common activities between current and new process instance will be preserved.

Description

The system looks up for any existing activity present both on process instance associated the flight and on the process template to be applied. Selected activities actual timings will be preserved. Other existing activities and relationship are deleted.

Note that this copy of an activity actual timing(s) also applies in the case a turnaround is split into an arrival flight and a departure flight, and vice versa:

- In case an activity is present in both the turnaround existing process instance and in the new process instance of the arrival and/or the departure, actuals will be copied for this activity.
- In case an activity is present in both the process instance of arrival and/or the departure flight and in the turnaround new process instance, actuals will be copied for this activity.

The associated actual timing information are thus preserved for common activities between the existing and new process template.

(e) Delete Process Instance—Use Case

Users: N/A

Pre-Conditions

- No new process template has been selected for the flight.
- An existing process instance exists for the flight.

Post-Conditions: The flight is not associated to any process instance.

Description

The system deletes any process instance associated to the flight.

(f) Manage Exceptions—Use Case

Users: N/A

Pre-Conditions

- A new process template has been selected for the flight
- An exception has been detected for the flight (Aircraft Change, Bay Change or Gate Change)

Post-Conditions

- The flight will be associated with a process instance.

- Actual start time and actual end time each activity will be updated as per system configuration
- Former activities in reset status will be attached to the process instance

Description

In case an exception is detected, the system first instantiates the new process template for the flight. Each activity and their characteristics are associated to the flight.

In a second step, each activity actual will be revisited according to the system configuration (exception triggered, activity action). For each exception, the system allows the user to specify one of the following behaviors per exception:

- Reset
  - If the activity was also present in the former process instance, and if any actual time was present (activity started/finished), the former activity will be attached to the new process instance, without link to any of the associated activities of new process instance. Its status will be flagged as 'STOPPED'.
  - The activity present in the new process instance is not updated with any actual (status remains 'planned')
- Fetch from flight:
  - If the activity was also present in the former process instance, and if any actual time was present, the actual times are transferred to the new activity. Previous activity is discarded.
- Fetch from aircraft:
  - If the activity was also present in the former process instance, and if any actual time was present (activity started/finished), the activity is stored against the former registration for further re-use in a pool.
  - The system looks for any activity stored in the pool or, on a flight associated to the same registration which is still on ground. This only applies if the activities actuals are fully included in the exception window (between current time minus {Exception Window Start} and current time plus { Exception Window End})

The last step of exception management consists in applying exceptional activities according to the system configuration (exception triggered, exception activity, exception condition). These are templated activities that apply on specific exceptions and additional conditions. These additions to the process template include the new activities needed to carry out flight processes as per exception procedure.

- E.g. 'Passenger disembark, 'Passenger transfer from previous gate'
- E.g. 'Transport GPU from previous bay', Etc.

(g) Compute Critical Path—Use Case

Users: NA.

Pre-Conditions

- A process instance is applied to the flight.
- A best departure time is set for the turnaround or departure flight
- A best arrival time is set for the turnaround or arrival flight

Post-Conditions

- A earliest start and end time will be set on each activity
- A latest start and end time will be set on each activity

Description

Computing the critical path consists in identifying, for each activity:

- The earliest start and end time achievable depending on the flight arrival time based on all predecessor activity duration
- The latest start and end time of the activity has to meet in order to enable the flight to depart on time.
- In order to achieve this computation, the system sets the reference end time of the process as follow:
- Reference end time is set to best departure time for turnaround and departure flights.
- Reference start time is set to the best arrival time for turnaround and arrival flights

The system then computes the *earliest* start and end times of activities in sequence from the *start* to the *end* of the process for all the activities linked directly or indirectly to the start of the process (flight arrival) as follow:

- Earliest start time of an activity is the earliest achievable time for this activity to start considering all predecessor activities earliest end times and associated link lag.
- Earliest end time is set to the earliest start time plus activity duration.

The system also computes the *latest* start and end times of activities in sequence from the *end* to the *start* of the process for all the activities linked directly or indirectly to the end of the process (flight departure) as follow:

- Latest end time of an activity is the latest possible time for this activity to end not endangering any successor activities latest start times and associated link lag.
- Latest start time is set to the earliest start time plus activity duration.

For the activities that are not connected to the start of the process, the above mentioned calculation does not provide any earliest time. The systems by default set earliest start/end to latest start/end time. This case typically happens for departure flights, where no reference start time is set by the system but might also happen for turnaround activities not related to the start of the process. For the activities that are not connected to the end of the process, the above mentioned calculation does not provide any latest time. The systems by default set latest start/end to earliest start/end time. This case typically happens for arrival flights, where no reference end time is set by the system but might also happen for turnaround activities not related to the end of the process.

The system does not support activities that are not connected neither to the start nor the end of the process.

(h) Capture Activity Timings—Use Case

Users

- Event Manager

Pre-Conditions

- An event is detected by the Event Manager.
- The concerned flight exists and the corresponding activity is associated to the flight.

Post-Conditions

- The current activity is set with an actual start and/or end time.
- Successor activities are set with estimated earliest start and end time.

Description

The Event Manager parses incoming external interfaces messages and identifies:

- The concerned flight
- The concerned activity
- The actual start or end time of the activity (if a condition start or condition end is triggered for the activity
- If a reset condition is triggered for the activity

In case the concerned flight is not found, or if no process instance exists for this flight, or if the activity is not part of the process instance of the flight, the message is discarded.

If a reset condition is detected, actual timings pertaining to the activity (start, end or both as identified in the message) are removed from ATM system.

Else if a start condition is detected, the actual start time is added to the activity and system then re-evaluates the current activity estimated end time satisfying its duration.

Else an end condition is detected; the actual end time is added to the activity.

In any of the above mentioned case, the system then re-evaluates all successor activities with an estimated start and estimated end, satisfying their successive relationship offset and duration.

The system then invokes the < *Compute Activity Status* > use case for all updated activities.

(i) *Compute Activity Status*—Use Case

Users: N/A

Pre-Conditions

- Earliest Start and End Time are set on the activity
- Latest Start and End Time are set on the activity

Post-Conditions

- Delay value will be set on the activity
- Alert status will be set on the activity

Description

The system sets the alerts status and associated delay for each activity depending on the timings associated to the activity. As a general rule, the delay value of an activity is always computed versus the earliest possible times. The alert severity depends on the latest times (critical if latest time is breached, non-critical otherwise).

The following section describes the applicable scenarios for each alert status:

- Not Set (PTS in the future)
- Started (PTS started and no issue)
- Finished (PTS finished)
- Non-critical alert (PTS is delayed but no impact to an on-time departure)
- Critical alert (PTS is delayed and impacts an on-time departure)

The alert status is *'Not Set'* (PTS in the future) in the following scenarios, as shown in Table 2:

The alert status is *'Started–No Alert'* (PTS started and no issue) in the following scenario, as shown in Table 3.

The alert status is *'Finished–No Alert'* (PTS finished) in the following scenario, as shown in Table 4.

The alert status is *'Non–Critical Alert'* (PTS is delayed but no impact to an on-time departure) in the following scenarios, as shown in Table 5.

Table 2 Alert Not Set *(Authors created)*

Scenario	Conditions	*Delay*
PTS instance is present on the flight but not started (no actual timings received for any activity)	– No actual start time is present on the activity – No actual end time is present on the activity – No estimated start time is present on the activity – No estimated end time is present on the activity	0
PTS instance started (at least one actual timing is received for one activity), and the current activity is estimated as not delayed	– Estimated end time is set on the activity – Estimated end time is before earliest Start time	0

Table 3 Started–No Alert *(Authors created)*

Scenario	Conditions	Delay
Activity is started on time	– Actual start time is set on the activity – Actual end time is not set on the activity – Actual start time is before earliest start time	0

Table 4 Finished–No Alert *(Authors created)*

Scenario	Conditions	Delay
Activity is completed on time	– Actual end time is set on the activity – Actual end time is before earliest end time	0

The alert status is '*Critical Alert*' (PTS is delayed and impacts an on-time departure) in similar scenarios, except that flight departure time is impacted, as shown in Table 6.

5 Discussion and Future Work

This research unveils that main objective of an Operation Control Center of any airline to uphold the declared flight timetable, so that the flight departs at its scheduled timing diminishing delays and obstacles if occurred during the flight operations. Therefore, transparency of all activity status during flight turnaround operations is to give the complete picture to Operation Control Center. It empowers the OCC to efficiently take swift and suitable decisions about to face any complication to the flight schedule, detect responsibilities and asses' performance regarding all involved activities. In the past days the assessment and control of these activities has been performed by telephone communication and manual procedures, due to human interference it made the data doubtful and lead to delays and unreliable information. Hence, to achieve high OTP the real time turnaround activity monitoring is necessary. Airlines do their best to achieve good OTP, a flight depart on the announced time means a good OTP, which has a major impact on service quality to passengers, airline reputation and economy of the airline. The solution offered in this study has presented a system for aircraft turnaround process management and monitoring. CPM and PERT models were used to manage turnaround activities by obtaining critical path and critical activities including passenger handling, cargo, engineering checks and catering services.

For the later stage, the proposed system can accept more functionality and extend to numerous directions. Currently the alerts are generated and are shown on the monitoring screen; this can be extended to send SMS messages to the concerned flight coordinator or handler. With current solution every handler has to update the start and finish time of activity from handheld device, this can be extended to use barcode reader to update the timings, which will switch to full automation of activity

Table 5 Non−Critical Alert *(Authors created)*

Scenario	Conditions	Delay
PTS instance started (at least one actual timing is received on any activity), and current activity is not started but estimated as delayed. However, it does not impact the departure time	– Actual start time is not set on the activity – Actual end time is not set on the activity – Estimated end time is after earliest end time – Estimated end time is before latest end time or activity is not on the critical path	Estimated end time minus earliest end time
Activity is not started and the current time is after estimated start time. However, it does not impact the departure time	– Actual start time is not set on the activity – Actual end time is not set on the activity – Current time is after the earliest start time – Current time is before the latest start time or activity is not on the critical path	Previous value
Activity is started and completion is estimated as delayed. However, it does not impact the departure time	– Actual start time is set on the activity – Actual end time is not set on the activity – Actual start time is after earliest start time – Actual start time is before latest Start time	Estimated end time minus earliest end time
Activity is started, not completed and the current time is after estimated end time. However, it does not impact the departure time	– Actual start time is set on the activity – Actual end time is not set on the activity – Current time is after the earliest end time – Current time is before the latest end time or activity is not on the critical path	Previous value
Activity is completed and delayed. However, it does not impact the departure time	– Actual end time is set on the activity – Actual end time is after earliest start time – Actual end time is before latest Start time	Actual end time minus earliest end time

times. Even though the mobile devices can work to get the operational data during turnaround process still it needs human interaction for transmission of data, which add additional burden to agents which are handling the ground events. To minimize this manual intervention automated procedure can be added using RFID tags.

Table 6 Critical Alert *(Authors created)*

Scenario	Conditions	Delay
PTS instance started (at least one actual timing is received for any activity), and current activity is not started but estimated as delayed. It does impact the departure time	– Actual start time is not set on the activity –Actual end time is not set on the activity – Estimated end time is after latest end time	Estimated end time minus earliest end time
Activity is not started and the current time is after estimated start time. It does impact the departure time	– Actual start time is not set on the activity – Actual end time is not set on the activity – Current time is after the latest start time	Previous value
Activity is started and completion is estimated as delayed. It does impact the departure time	– Actual start time is set on the activity – Actual end time is not set on the activity – Actual start time is after earliest start time – Actual start time is before latest Start time	Estimated end time minus earliest end time
Activity is started, not completed and the current time is after estimated end time. It does impact the departure time	– Actual start time is set on the activity – Actual end time is not set on the activity – Current time is after the latest end time	Previous value
Activity is completed and delayed. It does impact the departure time	– Actual end time is set on the activity – Actual end time is after latest end time	Actual end time minus earliest end Time

References

Akhtar, A., Akhtar, S., Bakhtawar, B., Kashif, A. A., Aziz, N., & Javeid, M. S. (2021). COVID-19 detection from CBC using machine learning techniques. *The International Journal of Technology, Innovation and Management, 1*, 65–78.

Al-Dmour, A., Al-Dmour, H., Al-Barghuthi, R., Al-Dmour, R., & Alshurideh, M. T. (2021). Factors influencing the adoption of E-payment during pandemic outbreak (COVID-19): Empirical evidence. *Studies in Systems, Decision and Control.*

Al-Dmour, N. (2016). Using unstructured search algorithms for data collection in IoT-based WSN. *International Journal of Engineering Research & Technology*. ISSN 974-3154.

Al-Dmour, N. A., & Teahan, W. J. (2005). Peer-to-peer protocols for resource discovery in the grid. In *Parallel and Distributed Computing and Networks* (pp. 319–324). Citeseer.

Al-Hamadi, H., Gawanmeh, A., & Al-Qutayri, M. (2015). An automatic ECG generator for testing and evaluating ECG sensor algorithms. In *2015 10th International Design & Test Symposium (IDT)* (pp. 78–83). IEEE.

Al-Hamadi, H., Nasir, N., Yeun, C. Y., & Damiani, E. (2021). A verified protocol for secure autonomous and cooperative public transportation in smart cities, In *2021 IEEE International Conference on Communications Workshops (ICC Workshops)* (pp. 1–6). IEEE.

Al-Naymat, G., Hussain, H., Al-Kasassbeh, M., & Al-Dmour, N. (2021). Accurate detection of network anomalies within SNMP-MIB data set using deep learning. *International Journal of Computer Applications in Technology, 66*, 74–85.

Al Ali, A. (2021). The impact of information sharing and quality assurance on customer service at UAE banking sector. *The International Journal of Technology, Innovation and Management, 1*, 01–17.

Al Hamadi, H., Gawanmeh, A., & Al-Qutayri, M. (2017). Guided test case generation for enhanced ECG bio-sensors functional verification. *International Journal of E-Health and Medical Communications, 8*, 1–20.

Al Neaimi, M., Al Hamadi, H., Yeun, C. Y., & Zemerly, M. J. (2020). Digital forensic analysis of files using deep learning. In *2020 3rd International Conference on Signal Processing and Information Security (ICSPIS)* (pp. 1–4). IEEE.

AlHamad, A., Alshurideh, M., Alomari, K., Kurdi, B., Alzoubi, H., Hamouche, S., & Al-Hawary, S. (2022). The effect of electronic human resources management on organizational health of telecommuni-cations companies in Jordan. *International Journal of Data and Network Science, 6*, 429–438.

Alhamad, A. Q. M., Akour, I., Alshurideh, M., Al-Hamad, A. Q., Kurdi, B. A., & Alzoubi, H. (2021). Predicting the intention to use google glass: A comparative approach using machine learning models and PLS-SEM. *International Journal of Data and Network Science, 5*.

Ali, L., & Dmour, N. (2021). The shift to online assessment due to COVID-19: An empirical study of university students, behaviour and performance, in the region of UAE. *International Journal of Information and Education Technology, 11*, 220–228.

Ali, N., Ahmed, A., Anum, L., Ghazal, T. M., Abbas, S., Khan, M. A., Alzoubi, H. M., & Ahmad, M. (2021). Modelling supply chain information collaboration empowered with machine learning technique. *Intelligent Automation and Soft Computing, 30*, 243–257.

Ali, N. M., Ghazal, T., Ahmed, A., Abbas, S., Khan, M. A., Alzoubi, H., Farooq, U., Ahmad, M., & Adnan Khan, M. (2022). Fusion-based supply chain collaboration using machine learning techniques. *Intelligent Automation and Soft Computing ,31*, 1671–1687.

Alnazer, N. N., Alnuaimi, M. A., & Alzoubi, H. M. (2017). Analysing the appropriate cognitive styles and its effect on strategic innovation in Jordanian universities. *International Journal of Business Excellence, 13*, 127–140.

Alnuaimi, M., Alzoubi, H. M., Ajelat, D., & Alzoubi, A. A. (2021). Towards intelligent organisations: An empirical investigation of learning orientation's role in technical innovation. *International Journal of Innovation and Learning, 29*, 207–221.

Alsharari, N. (2021). Integrating blockchain technology with internet of things to efficiency. *International Journal of Technology, Innovation and Management., 1*, 1–13.

Alshurideh, M., Alsharari, N., & Al Kurdi, B. (2019). Supply chain integration and customer relationship management in the airline logistics. *Theoretical Economics Letters, 9*, 392–414.

Alshurideh, M., Gasaymeh, A., Ahmed, G., Alzoubi, H., & Kurd, B. A. (2020). Loyalty program effectiveness: Theoretical reviews and practical proofs. *Uncertain Supply Chain Management, 8*.

Alshurideh, M. T., Al Kurdi, B., Alzoubi, H. M., Ghazal, T. M., Said, R. A., AlHamad, A. Q., Hamadneh, S., Sahawneh, N., & Al-kassem, A. H. (2022). Fuzzy assisted human resource management for supply chain management issues. *Annals of Operations Research*, 1–19.

Alshurideh, M. T., Al Kurdi, B., Masa'deh, R., Salloum, & S. A. (2021). The moderation effect of gender on accepting electronic payment technology: A study on united arab emirates consumers. *Review of International Business and Strategy. 31*.

Alzoubi, A. (2021a). The impact of process quality and quality control on organizational competitiveness at 5-star hotels in dubai. *International Journal of Technology, Innovation and Management, 1*, 54–68.

Alzoubi, A. (2021b). Renewable green hydrogen energy impact on sustainability performance. *International Journal of Computations, Information and Manufacturing, 1*, 94–110.

Alzoubi, H., & Ahmed, G. (2019). Do TQM practices improve organisational success? a case study of electronics industry in the UAE. *International Journal of Business and Economics Research, 17*, 459–472.

Alzoubi, H., Ahmed, G., Al-Gasaymeh, A., & Kurdi, B. (2020a). Empirical study on sustainable supply chain strategies and its impact on competitive priorities: The mediating role of supply chain collaboration. *Management Science Letters, 10*, 703–708.

Alzoubi, H., Alshurideh, M., Kurdi, B. A., & Inairat, M. (2020b). Do perceived service value, quality, price fairness and service recovery shape customer satisfaction and delight? a practical study in the service telecommunication context. *Uncertain Supply Chain Management, 8*, 579–588.

Alzoubi, H., Alshurideh, M., Kurdi, B., Akour, I., & Aziz, R. (2022). Does BLE technology contribute towards improving marketing strategies, customers' satisfaction and loyalty? The role of open innovation. *International Journal of Data and Network Science, 6*, 449–460.

Alzoubi, H. M., & Aziz, R., (2021). Does emotional intelligence contribute to quality of strategic decisions? the mediating role of open innovation.

Alzoubi, H. M., Vij, M., Vij, A., & Hanaysha, J. R. (2021). What leads guests to satisfaction and loyalty in UAE five-star hotels? AHP analysis to service quality dimensions. *Enlightening Tourism, 11*, 102–135.

Alzoubi, H. M., & Yanamandra, R. (2020). Investigating the mediating role of information sharing strategy on agile supply chain. *Uncertain Supply Chain Management, 8*, 273–284.

Aslam, M. S., Ghazal, T. M., Fatima, A., Said, R. A., Abbas, S., Khan, M. A., Siddiqui, S. Y., & Ahmad, M. (2021). Energy-efficiency model for residential buildings using supervised machine learning algorithm.

Aziz, N., & Aftab, S. (2021). Data mining framework for nutrition ranking: Methodology: SPSS modeller. *International Journal of Technology, Innovation and Management, 1*, 85–95.

Bibi, R., Saeed, Y., Zeb, A., Ghazal, T. M., Rahman, T., Said, R. A., Abbas, S., Ahmad, M., & Khan, M. A. (2021). Edge AI-based automated detection and classification of road anomalies in VANET using deep learning. *Computational Intelligence and Neuroscience, 2021*.

Cruz, A. (2021). Convergence between blockchain and the internet of things. *International Journal of Technology, Innovation and Management., 1*, 35–56.

Eli, T. (2021). Students perspectives on the use of innovative and interactive teaching methods at the university of nouakchott Al Aasriya, Mauritania: English department as a case study. *International Journal of Technology, Innovation and Management, 1*, 90–104.

Farouk, M. (2021). The universal artificial intelligence efforts to face coronavirus COVID-19. *International Journal of Computations, Information and Manufacturing, 1*, 77–93.

Ghazal, T. M. (2021). Internet of things with artificial intelligence for health care security. *Arabian Journal for Science and Engineering*, 1–12.

Ghazal, T. M., Hasan, M. K., Alshurideh, M. T., Alzoubi, H. M., Ahmad, M., Akbar, S. S., Al Kurdi, B., & Akour, I. A. (2021). IoT for smart cities: Machine learning approaches in smart healthcare—a review. *Future Internet, 13*, 218.

Guergov, S., & Radwan, N. (2021). Blockchain convergence: Analysis of issues affecting IoT, AI and blockchain. *International Journal of Technology, Innovation and Management, 1*, 1–17.

Hamadneh, S., Pedersen, O., & Al Kurdi, B. (2021a). An investigation of the role of supply chain visibility into the scottish bood supply chain. *Journal of Legal, Ethical and Regulatory Issues, 24*, 1–12.

Hamadneh, S., Pedersen, O., Alshurideh, M., Kurdi, B. A., & Alzoubi, H. M. (2021b). An investigation of the role Of supply chain visibility into the scottish blood supply chain. *Journal of Legal, Ethical and Regulatory Issues, 24*.

Hanaysha, J. R., Al-Shaikh, M. E., Joghee, S., & Alzoubi, H. (2021a). Impact of innovation capabilities on business sustainability in small and medium enterprises. *FIIB Business Review*, 1–12.

Hanaysha, J. R., Al Shaikh, M. E., & Alzoubi, H. M. (2021b). Importance of marketing mix elements in determining consumer purchase decision in the retail market. *International Journal of Service Science, Management, Engineering, and Technology, 12*, 56–72.

Joghee, S., Alzoubi, H. M., & Dubey, A. R. (2020). Decisions effectiveness of FDI investment biases at real estate industry: Empirical evidence from Dubai smart city projects. *International Journal of Scientific & Technology Research, 9*, 3499–3503.

Kashif, A. A., Bakhtawar, B., Akhtar, A., Akhtar, S., Aziz, N., & Javeid, M. S. (2021). Treatment response prediction in hepatitis C patients using machine learning techniques. *International Journal of Technology, Innovation and Management, 1*, 79–89.

Khan, M. A. (2021). Challenges facing the application of IoT in medicine and healthcare. *International Journal of Computations, Information and Manufacturing, 1*, 39–55.

Lee, C., & Ahmed, G. (2021). Improving IoT privacy, data protection and security concerns. *International Journal of Technology, Innovation and Management, 1*, 18–33.

Lee, K., Azmi, N., Hanaysha, J., Alzoubi, H., & Alshurideh, M. (2022a). The effect of digital supply chain on organizational performance: An empirical study in malaysia manufacturing industry. *Uncertain Supply Chain Management., 10*, 495–510.

Lee, K., Romzi, P., Hanaysha, J., Alzoubi, H., & Alshurideh, M. (2022b). Investigating the impact of benefits and challenges of IOT adoption on supply chain performance and organizational performance: An empirical study in Malaysia. *Uncertain Supply Chain Management., 10*, 537–550.

Maasmi, F., Morcos, M., Al Hamadi, H., & Damiani, E. (2021). Identifying applications' state via system calls activity: A pipeline approach. In *2021 28th IEEE International Conference on Electronics, Circuits, and Systems (ICECS)* (pp. 1–6). IEEE.

Mehmood, T. (2021). Does information technology competencies and fleet management practices lead to effective service delivery? empirical evidence from E-commerce industry. *International Journal of Technology, Innovation and Management, 1*, 14–41.

Mehmood, T., Alzoubi, H. M., & Ahmed, G. (2019). Schumpeterian entrepreneurship theory: Evolution and relevance. *Academy of Entrepreneurship Journal, 25*.

Miller, D. (2021). The best practice of teach computer science students to use paper prototyping. *International Journal of Technology, Innovation and Management, 1*, 42–63.

Mondol, E. P. (2021). The impact of block chain and smart inventory system on supply chain performance at retail industry. *International Journal of Computations, Information and Manufacturing, 1*, 56–76.

Obaid, A. J. (2021). Assessment of smart home assistants as an IoT. *International Journal of Computations, Information and Manufacturing, 1*, 18–36.

Radwan, N., & Farouk, M. (2021). The growth of internet of things (IoT) in the management of healthcare issues and healthcare policy development. *International Journal of Technology, Innovation and Management, 1*, 69–84.

Shamout, M., Ben-Abdallah, B., Alshurideh, M., Alzoubi, H., Al Kurdi, B., & Hamadneh, S. (2022). A conceptual model for the adoption of autonomous robots in supply chain and logistics industry. *Uncertain Supply Chain Management, 10*, 1–16.

Siddiqui, S. Y., Haider, A., Ghazal, T. M., Khan, M. A., Naseer, I., Abbas, S., Rahman, M., Khan, J. A., Ahmad, M., & Hasan, M. K. (2021). IoMT cloud-based intelligent prediction of breast cancer stages empowered with deep learning. *IEEE Access, 9*, 146478–146491.

Zitar, R. A. (2021). A review for the genetic algorithm and the red deer algorithm applications. In *2021 14th International Congress on Image and Signal Processing, BioMedical Engineering and Informatics (CISP-BMEI)* (pp. 1–6). IEEE.

Zitar, R. A., Abualigah, L., & Al-Dmour, N. A. (2021). Review and analysis for the Red Deer Algorithm. *Journal of Ambient Intelligence and Humanized Computing*, 1–11.

Information Systems Solutions for the Database Problems

Nidal A. Al-Dmour, Liaqat Ali, Mohammed Salahat, Haitham M. Alzoubi, Muhammad Alshurideh, and Zakariya Chabani

Abstract This paper has attempted to build on an explanation of solutions for database problems in information systems. The main focal area for this paper is majorly on enterprise resource planning systems and their effects and characteristics that make them a primary software tool which is used by businesses worldwide whether they be at large scales or work at smaller scales compared to the big-name enterprises out there. The paper has given a basic introduction into what an enterprise resource planning system is and defined its key characteristics that make it exactly an ERP. It also points out how businesses and enterprises make use of such systems and the methodologies that they use to implement ERP systems into their business practices. This paper also attempts to characterize ERP in relation to the solutions that they provide for the specific industry that they are used in. This is

N. A. Al-Dmour
Department of Computer Engineering, College of Engineering, Mutah University, Mutah, Jordan

L. Ali · M. Salahat
College of Engineering and Information Technology, University of Science and Technology of Fujairah, Fujairah, UAE
e-mail: l.ali@ustf.ac.ae

M. Salahat
e-mail: m.salahat@ustf.ac.ae

H. M. Alzoubi (✉)
School of Business, Skyline University College, Sharjah, UAE
e-mail: haitham.alzubi@skylineuniversity.ac.ae

M. Alshurideh
Department of Marketing, School of Business, University of Jordan, Amman, Jordan
e-mail: m.alshurideh@ju.edu.jo; malshurideh@sharjah.ac.ae

Department of Management, College of Business Administration, University of Sharjah, Sharjah, UAE

Z. Chabani
Faculty of Management, Canadian University Dubai, Dubai, UAE
e-mail: zakariya.chabani@cud.ac.ae

M. Alshurideh et al. (eds.), *The Effect of Information Technology on Business and Marketing Intelligence Systems*, Studies in Computational Intelligence 1056,
https://doi.org/10.1007/978-3-031-12382-5_37

mostly done through categorizing the ERP through characteristics such as according to size, technology, risk factor, cost effectiveness and more.

Keywords Information systems solutions for the database problems · Enterprise resource planning systems

1 Introduction

Enterprise resource planning (ERP) is a core enterprise software and solution stack comprised of a central database surrounded by multiple plug-in services and software packages, each of which are designed and made to solve specific enterprise problems and address specific challenges. There are a multitude of various processes that are essential to the running of an enterprise business. These processes include skills and techniques such as inventory management, customer relationship management, human resources, accounting and more (Akhtar et al., 2021; Al Ali, 2021; AlHamad et al., 2022; Alhamad et al., 2021; Alshurideh, 2022). To put it simply, ERP management software integrates these various functions into one complete software package system. This is done in order to streamline processes and information across the entire business and enterprise organization (Al Kurdi et al., 2020a, b, c; Hamadneh et al., 2021a, b; Tariq et al., 2022). For and ERP system to function to its maximum and full capability, the ERP system needs to involve a shared database that supports multiple functions used by different divisions (Ali et al., 2021, 2022; Alnazer et al., 2017; Alnuaimi et al., 2021). An example of this would be the human resources department only being reliant on the information that is specific needs.

ERP is valuable as ERP systems are able to tie together a multitude of different business processes and enable the flow of all the data that is required for the business between them. In this day and age ERP systems are critical for medium to large size businesses (Alsharari, 2021; Alshurideh et al., 2020a, b, c, 2022; Alzoubi, 2021a) and enterprises if they wish to manage the thousands of data driven systems that they require to function properly and efficiently. To the businesses and enterprises that make use of ERP systems, it is an indispensable tool for them. Being able to align separate departments and improve day to day workflow is almost viewed as an essential practice for the enterprise market (Alzoubi & Yanamandra, 2020; Alzoubi et al., 2020a, 2022; Alzoubi, 2021b). This is due to the fact that ERP is a critical enterprise software that is able to collect information from different departments in a common database, making it easy to monitor and understand. ERP systems are able to generate major financial saving and as well as time savings as they are able to provide the organization insight into the processes and reveal opportunities for growth. There are many ways for ERP systems to be implemented into business practices (Alzoubi et al., 2021; Alzoubi & Aziz, 2021; Alzoubi et al., 2020b; Alzoubi & Ahmed, 2019). These include cloud based systems, onsite premises systems and it is also possible to include a hybrid of both the cloud an onsite system in one. In recent years cloud based ERP system services has seen the larger and more substantial rise in practice and popularity (Aziz & Aftab, 2021; Cruz, 2021; Eli, 2021; Farouk, 2021), however which approach an enterprise or business adopt into their working environment depends on the certain and specific needs of the company and their work culture. Enterprises should take importance to understand the integration needs and requirements, the methods of implementation, the different capabilities and benefits of each and the total cost of each different system before choosing to which one they wish to take into effect (Ghazal et al., 2021; Guergov & Radwan, 2021; Hamadneh et al., 2021a, b; Hanaysha et al., 2021a, b).

Companies across every industry, with diverse business models, have come to term with the benefits that come with utilizing ERP systems. ERP has been seen to be used in a multitude of industries and businesses such as educations, health and beauty, manufacturing, IT services, retail, food and beverage and many more. There are a few fundamental features that make an ERP system distinguishable from other types of software (Maasmi et al., 2021; Siddiqui et al., 2021; Zitar et al., 2021). These traits, in their most basic form, are having a common database, a consistent UI, business process integration, automation and data analysis. As an all-inclusive source of data, ERP systems also provide a number of reports and analytics that can be the clear difference makers in the working of a business. Turning the influx of data into comprehensible models makes ERP capabilities invaluable.

2 Characterization of ERP Solutions

Numbers of ways are available for characterization for ERP solutions-by the supplier's stake in the market, by the solution's size, and by the specific industry, the system is made for the sustenance of, or according to its working technological stage.

According to Size: Oracle and SAP are two other leading examples of ERP systems based on size-they're responsible for more than half of the total investments on ERP, expenditure, globally. Trailed by Consona, CDC, Microsoft (Dynamics AX and Dynamics GP), Epicor, QAD, Infor (VISUAL, SyteLine, and many other solutions) among others (Hanaysha et al., 2021a, b; Joghee et al., 2020; Kashif et al., 2021; Khan, 2021).

According to Technology: Positioned on operational setting (database/ hardware/ operating system), previously there were considerable distinctions in ERP solutions they worked on. Due to the global application of Windows and web-based architecture, there has been a notable loss of division (Alshurideh et al., 2019, 2020a, b, c; Al Kurdi et al., 2020a, b, c). Nevertheless, database and server problems can be classified using ERP solutions, although many solutions can be implemented in different environments. Mainframe systems, intel/window platforms, UNIX are worked on by versions of products of SAP, for example.

3 Application Analysis

Certain factors must be kept in mind when choosing a proper ERP system specifically:
Benefit Factor: These are those features of the ERP system which widely manage the system's ability to understand and align various departments' operations across various locations. Since the UN has an office in each country that is a member, it's particularly related in this case. Comprehending these departments' operations and synchronizing them can have an effect on its successful action that is a negative (Aslam et al., 2021; Bibi et al., 2021; Ghazal, 2021).

Satisfaction of Executive: It discusses the fulfillment of expectations relating to ERP system's expectations as described by the top-level managers or the executives of an organization.
Satisfaction of Employee: This requirement also discusses the ERP system's ability to fulfill the needs of the employees, like the previous point, given works relating to the chosen systems of ERP. It's essential to remember that this factor could either make or destroy the ERP system's success rates. This is inclusive of user-friendliness and the easy operation of the system.

Risk Factor of Business: Recognizing different factors connected with the application of the project of ERP (enterprise resource planning). A few of the challenges are of the business procedures regarding restructuring to the procedures followed by the ERP software, investing in recruiting and employing professionals of IT, the hurdle of seeking external individuals, and conforming their technical proficiency and knowledge with the fundamental teams and the possibility of technological bottlenecks (Lee & Ahmed, 2021; Lee et al., 2022a, b; Mehmood, 2021).
Cost-benefit and cost-effectiveness: If ERP can be indicated by business risk factors and the benefit factor which is essential to highlight, the different prices laid out for testing, implementation, customization, and purchase of an ERP system, with least investments monetarily is known as the cost-effectiveness of ERP (Zitar, 2021).

User friendliness: The next thing observed was that the user-friendliness of the ERP system can be indicated by the satisfaction of executives and employees. The user-friendliness of an ERP system is studying the ease in its operations or interaction in its solutions. When the answers are easy to configure, intuitive, integrated, graphical, and visual, it is always more beneficial. Hence the picture below gives us an idea of the present scenario of ERP (Fig. 1).

ERP Solution Satisfaction/
Benefits Realization

Tier I and Tier II ERP Solutions

	SAP	Oracle	Microsoft	Tier II	Industry Average
Benefits Factor	72.2%	58.0%	68.0%	68.6%	65.3%
Executive Satisfaction	76.4%	75.9%	65.4%	67.7%	70.7%
Employee Satisfaction	73.6%	60.3%	76.9%	76.5%	67.4%
Business Risk Factor	50.0%	56.9%	57.7%	61.8%	54.0%

Source: Panorama's 2010 ERP Report
Three-year study of over 1,600 ERP implementations across the globe
Full report available at http://www.panorama-consulting.com/2008ERPReport.html

PANORAMA CONSULTING GROUP

Fig. 1 Choosing the right ERP. *Source* Panorama's 2010 ERP Report. Accessed on Dec-2021. Available at https://www.panorama-consulting.com/wpcontent/uploads/2016/07/2010-ERP-Vendor-Analysis-Report.pdf

3.1 SAP ERP

One of the main organizations to develop the possibility of individual projects was SAP while getting information from normal databases that can be procured, introduced, and implemented autonomously. With an extended past of more than 4 decades and around 50,000 establishments in roughly 50% of 1000 organizations of the world's Fortune, SAP is a solid supplier of administrations that has advanced and adjusted to alterations (Mehmood et al., 2019; Miller, 2021; Mondol, 2021).

Following are the key focal points related to SAP programming:

- SAP can undoubtedly tweak their product to fit the requirements of any industry because of its broad involvement with the field. For example, it tends to be implemented to assembling, acquirement, administrations, accounts, and even intergovernmental associations.
- The significant advantage of SAP ERP is its exceptional coordination with different business modules that aren't there for contender programming projects like BAAN and Oracle.

1. It empowers less complex worldwide combination conquering issues like language, cash conversion scale, and culture
2. It gives continuous data
3. It diminishes the likelihood of having excess data

3.2 Oracle ERP

If Consultancy offerings, industry verticals, functional and technical are combined by the Oracle ERP systems by ERP solutions, into one understanding organization aka Oracle E-business Suite as shown in Fig. 2. It merges and assembles offshore, onsite and offsite activities of a firm to shorten the ease of the burden of operations that are integrated. Various functions like business intelligence analysis, aids in the management of CRM, and HR and financial calculations are performed by it. Technical functions are carried out and format for forms and reports are provided that help the organization's workflow systems and QA Frameworks (Alhashmi et al., 2020; Almaazmi et al., 2020; Alshurideh et al., 2020a, b, c; Obaid, 2021; Radwan & Farouk, 2021; Shamout et al., 2022). It aids in the backup of media functions, retail, insurance, and manufacturing and maintenance.

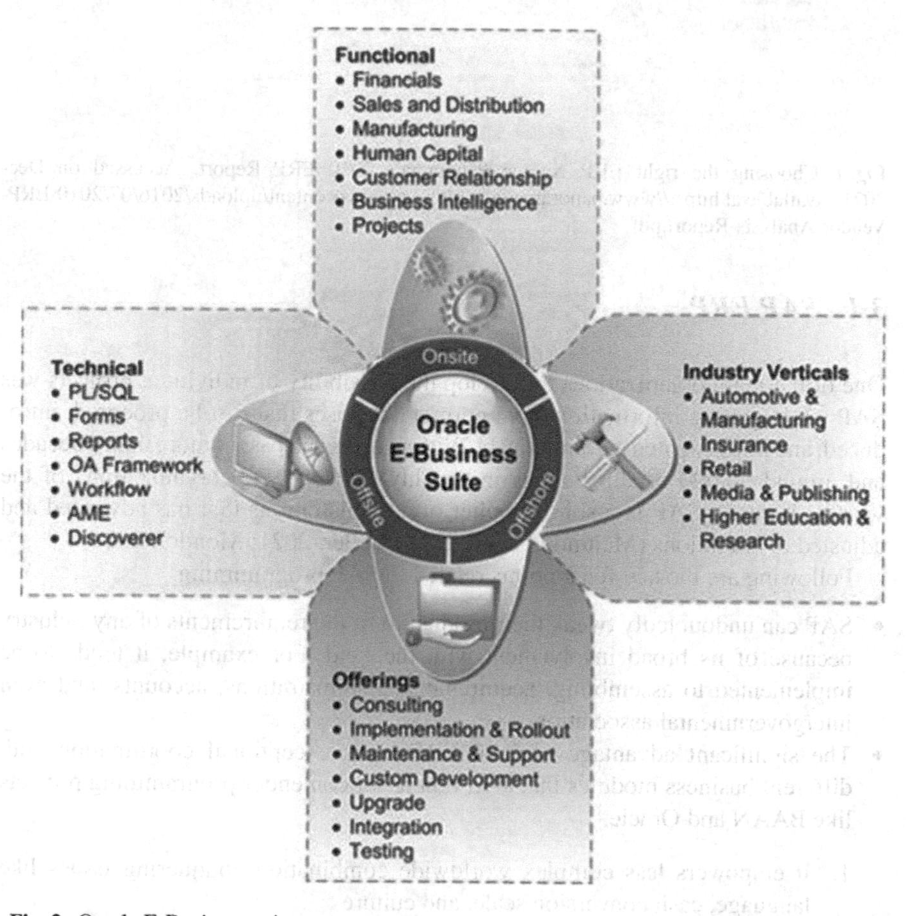

Fig. 2 Oracle E-Business suite

3.3 *Microsoft ERP System*

A wide variety of advantageous features for integrations provided by the Microsoft AX ERP system can be seen from the Fig. 3 shared below. The functions of all the departments in the organization are joined and the personalized solutions are provided with interfaces that are user friendly and very advantageous to organizations that have employee amounts ranging from 1000 to 10,000 (Al-Dmour, 2016; Al-Dmour & Teahan, 2005; Al Hamadi et al., 2017; Al Neaimi et al., 2020).

1. Reduced risk of business
2. All ERP problems were joined with a single platform solution
3. Cost-effective
4. Effective decreases the demands of manpower.
5. Cloud backup and memory storage
6. Omitted maintenance
7. Higher production.

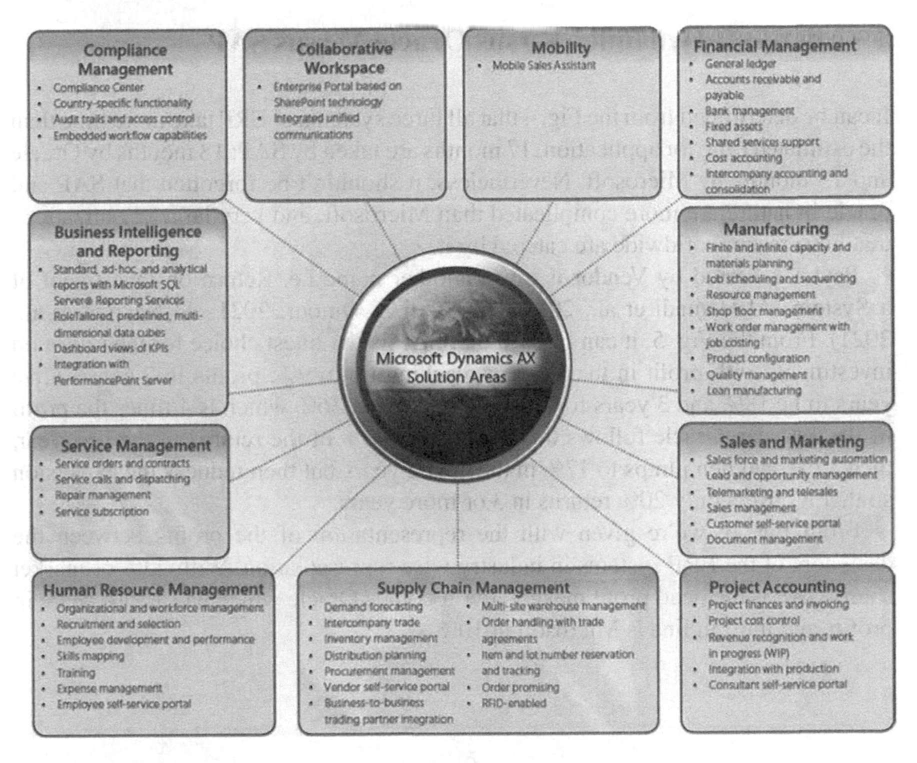

Fig. 3 Integration features of Microsoft AX ERP

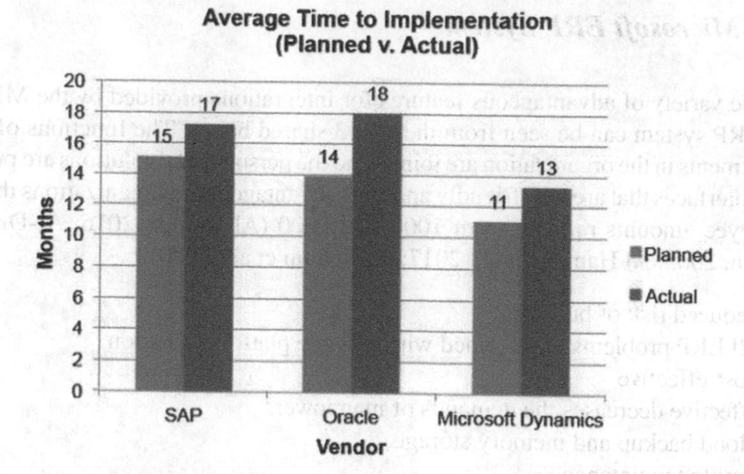

Fig. 4 Average implementation time for SAP, Oracle, and Microsoft Dynamics

4 Microsoft Dynamics Versus Oracle Versus SAP

It can be determined from the Fig. 4 that all three systems of ERP take more time than the estimated time for application. 17 months are taken by SAP; 18 months by Oracle and 13 months by Microsoft. Nevertheless, it shouldn't be forgotten that SAP and oracle in nature are more complicated than Microsoft, and very large organizations that have reach worldwide are catered by it.

Payback Period by Vendor is given another name i.e. Return on Investment of a System (Al-Hamadi et al., 2015, 2021; Ali & Dmour, 2021; Al-Naymat et al., 2021). From the Fig. 5, it can be seen that SAP is the finest choice for the return on investments with profit in the 1st year reaching up to 9%, profits in 2 consecutive years to be 18% and 3 years for the profits to reach 36% which is 4 times the profit of the 1st year. Oracle follows this which gives 5% of the returns in the first year, then 12%, and hen jumps to 17% in the next 2 years but then reduces the expansion so that it covers only 20% returns in 3 or more years.

From Fig. 6, we're given with the representation of the profits between the designers of the ERP systems in industry wise representation. With 22% of market share, SAP is the head profit maker, after it comes Oracle which has a 15% share in profits and third in line is Microsoft at 10%.

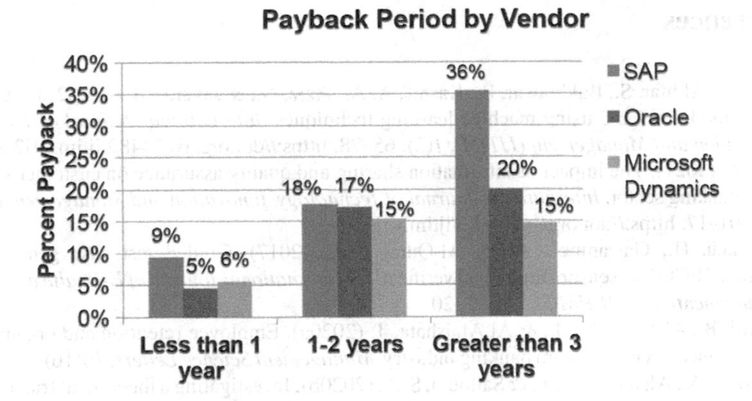

Fig. 5 Payback period by vendor

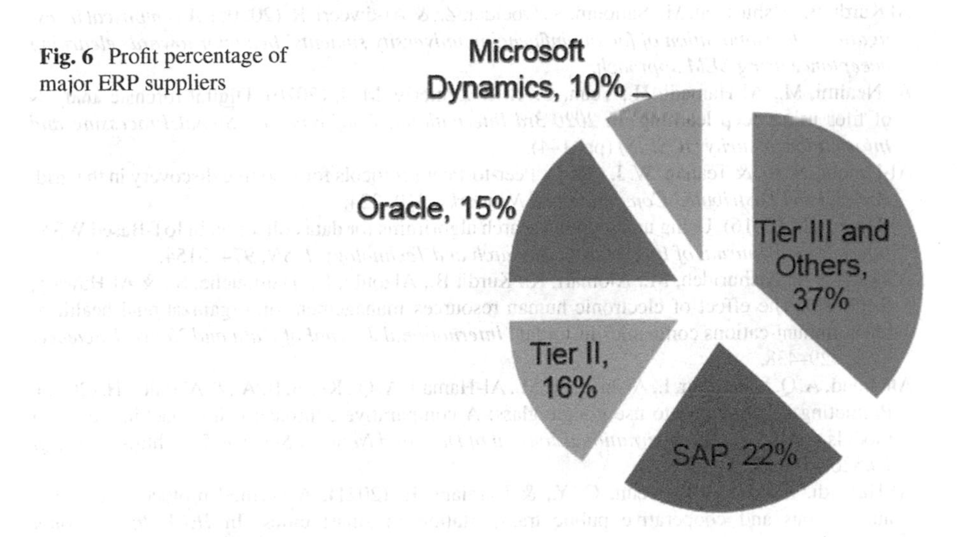

Fig. 6 Profit percentage of major ERP suppliers

5 Conclusions

This paper presented a basic introduction into what an enterprise resource planning system is and defined its key characteristics that make it exactly an ERP. This paper compared Average Implementation Time for SAP, Oracle, and Microsoft Dynamics. It also presented Payback Period by Vendor and Profit Percentage of Major ERP Suppliers.

References

Akhtar, A., Akhtar, S., Bakhtawar, B., Kashif, A. A., Aziz, N., & Javeid, M. S. (2021). COVID-19 Detection from CBC using machine learning techniques. *International Journal of Technology, Innovation and Management (IJTIM), 1*(2), 65–78. https://doi.org/10.54489/ijtim.v1i2.22

Al Ali, A. (2021). The impact of information sharing and quality assurance on customer service at UAE banking sector. *International Journal of Technology, Innovation and Management (IJTIM), 1*(1), 01–17. https://doi.org/10.54489/ijtim.v1i1.10

Al Hamadi, H., Gawanmeh, A., & Al-Qutayri, M. (2017). Guided test case generation for enhanced ECG bio-sensors functional verification. *International Journal of E-Health and Medical Communications (IJEHMC), 8*(4), 1–20.

Al Kurdi, B., Alshurideh, M., & Al Afaishata, T. (2020a). Employee retention and organizational performance: Evidence from banking industry. *Management Science Letters, 10*(16), 3981–3990.

Al Kurdi, B.A., Alshurideh, M., & Salloum, S. A. (2020b). Investigating a theoretical framework for e-learning technology acceptance. *International Journal of Electrical and Computer Engineering, 10*(6). https://doi.org/10.11591/IJECE.V10I6.PP6484-6496

Al Kurdi, B., Alshurideh, M., Salloum, S., Obeidat, Z., & Al-dweeri, R. (2020c). *An empirical investigation into examination of factors influencing university students' behavior towards elearning acceptance using SEM approach.*

Al Neaimi, M., Al Hamadi, H., Yeun, C. Y., & Zemerly, M. J. (2020). Digital forensic analysis of files using deep learning. In *2020 3rd International Conference on Signal Processing and Information Security (ICSPIS)* (pp. 1–4).

Al-Dmour, N. A., & Teahan, W. J. (2005). Peer-to-Peer protocols for resource discovery in the grid. *Parallel and Distributed Computing and Networks*, 319–324.

Al-Dmour, N. (2016). Using unstructured search algorithms for data collection in IoT-Based WSN. *International Journal of Engineering Research and Technology. ISSN*, 974–3154.

AlHamad, A., Alshurideh, M., Alomari, K., Kurdi, B., Alzoubi, H., Hamouche, S., & Al-Hawary, S. (2022). The effect of electronic human resources management on organizational health of telecommuni-cations companies in Jordan. *International Journal of Data and Network Science, 6*(2), 429–438.

Alhamad, A. Q. M., Akour, I., Alshurideh, M., Al-Hamad, A. Q., Kurdi, B. A., & Alzoubi, H. (2021). Predicting the intention to use google glass: A comparative approach using machine learning models and PLS-SEM. *International Journal of Data and Network Science*, *5*(3). https://doi.org/10.5267/j.ijdns.2021.6.002

Al-Hamadi, H., Nasir, N., Yeun, C. Y., & Damiani, E. (2021). A verified protocol for secure autonomous and cooperative public transportation in smart cities. In *IEEE International Conference on Communications Workshops (ICC Workshops),* (Vol. 2021) (pp. 1–6).

Al-Hamadi, H., Gawanmeh, A., & Al-Qutayri, M. (2015). An automatic ECG generator for testing and evaluating ECG sensor algorithms. In *2015 10th International Design & Test Symposium (IDT)* (pp. 78–83).

Alhashmi, S. F. S., Alshurideh, M., Al Kurdi, B., & Salloum, S. A. (2020). A systematic review of the factors affecting the artificial intelligence implementation in the health care sector. In *Advances in Intelligent Systems and Computing: Vol. 1153 AISC*. https://doi.org/10.1007/978-3-030-44289-7_4

Ali, L., & Dmour, N. (2021). The shift to online assessment due to COVID-19: An empirical study of university students, behaviour and performance, in the region of UAE. *International Journal of Information and Education Technology, 11*(5), 220–228.

Ali, N., Ahmed, A., Anum, L., Ghazal, T. M., Abbas, S., Khan, M. A., Alzoubi, H. M., & Ahmad, M. (2021). Modelling supply chain information collaboration empowered with machine learning technique. *Intelligent Automation and Soft Computing, 30*(1), 243–257. https://doi.org/10.32604/iasc.2021.018983

Ali, N., M. Ghazal, T., Ahmed, A., Abbas, S., A. Khan, M., Alzoubi, H., Farooq, U., Ahmad, M., & Adnan Khan, M. (2022). Fusion-based supply chain collaboration using machine learning techniques. *Intelligent Automation & Soft Computing, 31*(3), 1671–1687. https://doi.org/10.32604/iasc.2022.019892

Almaazmi, J., Alshurideh, M., Al Kurdi, B., & Salloum, S. A. (2020). The effect of digital transformation on product innovation: A critical review. In *International Conference on Advanced Intelligent Systems and Informatics* (pp. 731–741).

Al-Naymat, G., Hussain, H., Al-Kasassbeh, M., & Al-Dmour, N. (2021). Accurate detection of network anomalies within SNMP-MIB data set using deep learning. *International Journal of Computer Applications in Technology, 66*(1), 74–85.

Alnazer, N. N., Alnuaimi, M. A., & Alzoubi, H. M. (2017). Analysing the appropriate cognitive styles and its effect on strategic innovation in Jordanian universities. *International Journal of Business Excellence, 13*(1), 127–140. https://doi.org/10.1504/IJBEX.2017.085799

Alnuaimi, M., Alzoubi, H. M., Ajelat, D., & Alzoubi, A. A. (2021). Towards intelligent organisations: An empirical investigation of learning orientation's role in technical innovation. *International Journal of Innovation and Learning, 29*(2), 207–221. https://doi.org/10.1504/IJIL.2021.112996

Alsharari, N. (2021). Integrating blockchain technology with internet of things to efficiency. *International Journal of Technology, Innovation and Management (IJTIM), 1*(2), 1–13.

Alshurideh, M. (2022). Does electronic customer relationship management (E-CRM) affect service quality at private hospitals in Jordan? *Uncertain Supply Chain Management, 10*(2), 1–8.

Alshurideh, M., et al. (2019). Understanding the quality determinants that influence the intention to use the mobile learning platforms: A practical study. *International Journal of Interactive Mobile Technologies (IJIM), 13*(11), 157–183.

Alshurideh, M., Al Kurdi, B., & Salloum, S. A. (2020a). Examining the main mobile learning system srivers' effects: A mix empirical examination of both the Expectation-Confirmation Model (ECM) and the Technology Acceptance Model (TAM). In *Advances in Intelligent Systems and Computing* (Vol. 1058). https://doi.org/10.1007/978-3-030-31129-2_37

Alshurideh, M., Al Kurdi, B., Salloum, S. A., Arpaci, I., & Al-Emran, M. (2020b). Predicting the actual use of m-learning systems: A comparative approach using PLS-SEM and machine learning algorithms. *Interactive Learning Environments*. https://doi.org/10.1080/10494820.2020.1826982

Alshurideh, M., Gasaymeh, A., Ahmed, G., Alzoubi, H., & Kurd, B. A. (2020c). Loyalty program effectiveness: Theoretical reviews and practical proofs. *Uncertain Supply Chain Management*, 8(3). https://doi.org/10.5267/j.uscm.2020c.2.003

Alshurideh, M. T., Al Kurdi, B., Alzoubi, H. M., Ghazal, T. M., Said, R. A., AlHamad, A. Q., Hamadneh, S., Sahawneh, N., & Al-kassem, A. H. (2022). Fuzzy assisted human resource management for supply chain management issues. *Annals of Operations Research*, 1–19.

Alzoubi, A. (2021a). The impact of process quality and quality control on organizational competitiveness at 5-star hotels in Dubai. *International Journal of Technology, Innovation and Management (IJTIM), 1*(1), 54–68. https://doi.org/10.54489/ijtim.v1i1.14

Alzoubi, A. (2021b). Renewable Green hydrogen energy impact on sustainability performance. *International Journal of Computations, Information and Manufacturing (IJCIM), 1*(1), 94–110. https://doi.org/10.54489/ijcim.v1i1.46

Alzoubi, H., & Ahmed, G. (2019). Do TQM practices improve organisational success? A case study of electronics industry in the UAE. *International Journal of Economics and Business Research, 17*(4), 459–472. https://doi.org/10.1504/IJEBR.2019.099975

Alzoubi, H. M., & Yanamandra, R. (2020). Investigating the mediating role of information sharing strategy on agile supply chain. *Uncertain Supply Chain Management, 8*(2), 273–284. https://doi.org/10.5267/j.uscm.2019.12.004

Alzoubi, H. M., & Aziz, R. (2021). Does emotional intelligence contribute to quality of strategic decisions? The mediating role of open innovation. *Journal of Open Innovation: Technology, Market, and Complexity, 7*(2), 130. https://doi.org/10.3390/joitmc7020130

Alzoubi, H., Ahmed, G., Al-Gasaymeh, A., & Kurdi, B. (2020a). Empirical study on sustainable supply chain strategies and its impact on competitive priorities: The mediating role of supply chain collaboration. *Management Science Letters, 10*(3), 703–708.

Alzoubi, H., Alshurideh, M., Kurdi, B. A., & Inairat, M. (2020b). Do perceived service value, quality, price fairness and service recovery shape customer satisfaction and delight? A practical study in the service telecommunication context. *Uncertain Supply Chain Management, 8*(3), 579–588. https://doi.org/10.5267/j.uscm.2020.2.005

Alzoubi, H. M., Vij, M., Vij, A., & Hanaysha, J. R. (2021). What leads guests to satisfaction and loyalty in UAE five-star hotels? AHP analysis to service quality dimensions. *Enlightening Tourism, 11*(1), 102–135. https://doi.org/10.33776/et.v11i1.5056

Alzoubi, H., Alshurideh, M., Kurdi, B., Akour, I., & Aziz, R. (2022). Does BLE technology contribute towards improving marketing strategies, customers' satisfaction and loyalty? The role of open innovation. *International Journal of Data and Network Science, 6*(2), 449–460.

Aslam, M. S., Ghazal, T. M., Fatima, A., Said, R. A., Abbas, S., Khan, M. A., Siddiqui, S. Y., & Ahmad, M. (2021). *Energy-efficiency model for residential buildings using supervised machine learning algorithm.*

Aziz, N., & Aftab, S. (2021). Data mining framework for nutrition ranking methodology: SPSS modeller. *International Journal of Technology, Innovation and Management (IJTIM), 1*(1), 85–95.

Bibi, R., Saeed, Y., Zeb, A., Ghazal, T. M., Rahman, T., Said, R. A., Abbas, S., Ahmad, M., & Khan, M. A. (2021). Edge AI-based automated detection and classification of road anomalies in VANET using deep learning. *Computational Intelligence and Neuroscience, 2021.*

Cruz, A. (2021). Convergence between blockchain and the internet of things. *International Journal of Technology, Innovation and Management (IJTIM), 1*(1), 35–56.

Eli, T. (2021). StudentsPerspectives on the use of innovative and interactive teaching methods at the University of Nouakchott Al Aasriya, Mauritania: English Department as a Case Study. *International Journal of Technology, Innovation and Management (IJTIM), 1*(2), 90–104.

Farouk, M. (2021). The universal artificial intelligence efforts to face coronavirus COVID-19. *International Journal of Computations, Information and Manufacturing (IJCIM), 1*(1), 77–93. https://doi.org/10.54489/ijcim.v1i1.47

Ghazal, T. M., Hasan, M. K., Alshurideh, M. T., Alzoubi, H. M., Ahmad, M., Akbar, S. S., Al Kurdi, B., & Akour, I. A. (2021). IoT for smart cities: Machine learning approaches in smart healthcare—a review. *Future Internet, 13*(8), 218. https://doi.org/10.3390/fi13080218

Ghazal, T. M. (2021). Internet of things with artificial intelligence for health care security. *Arabian Journal for Science and Engineering*, 1–12.

Guergov, S., & Radwan, N. (2021). Blockchain convergence: Analysis of issues affecting IoT, AI and blockchain. *International Journal of Computations, Information and Manufacturing (IJCIM), 1*(1), 1–17. https://doi.org/10.54489/ijcim.v1i1.48

Hamadneh, S., Keskin, E., Alshurideh, M., Al-Masri, Y., & Al Kurdi, B. (2021a). The benefits and challenges of RFID technology implementation in supply chain: A case study from the Turkish construction sector. *Uncertain Supply Chain Management, 9*(4), 1071–1080.

Hamadneh, S., Pedersen, O., & Al Kurdi, B. (2021b). An Investigation of the role of supply chain visibility into the scottish bood supply chain. *Journal of Legal, Ethical and Regulatory Issues, 24*(Special Issue 1), 1–12.

Hanaysha, J. R., Al-Shaikh, M. E., Joghee, S., & Alzoubi, H. (2021a). Impact of innovation capabilities on business sustainability in small and medium enterprises. *FIIB Business Review*, 1–12. https://doi.org/10.1177/23197145211042232

Hanaysha, J. R., Al Shaikh, M. E., & Alzoubi, H. M. (2021b). Importance of marketing mix elements in determining consumer purchase decision in the retail market. *International Journal of Service Science, Management, Engineering, and Technology (IJSSMET), 12*(6), 56–72.

Joghee, S., Alzoubi, H. M., & Dubey, A. R. (2020). Decisions effectiveness of FDI investment biases at real estate industry: Empirical evidence from Dubai smart city projects. *International Journal of Scientific and Technology Research, 9*(3), 3499–3503.

Kashif, A. A., Bakhtawar, B., Akhtar, A., Akhtar, S., Aziz, N., & Javeid, M. S. (2021). Treatment response prediction in hepatitis C patients using machine learning techniques. *International Journal of Technology, Innovation and Management (IJTIM), 1*(2), 79–89. https://doi.org/10.54489/ijtim.v1i2.24

Khan, M. A. (2021). Challenges facing the application of IoT in medicine and healthcare. *International Journal of Computations, Information and Manufacturing (IJCIM), 1*(1), 39–55. https://doi.org/10.54489/ijcim.v1i1.32

Lee, K., Azmi, N., Hanaysha, J., Alzoubi, H., & Alshurideh, M. (2022a). The effect of digital supply chain on organizational performance: An empirical study in Malaysia manufacturing industry. *Uncertain Supply Chain Management, 10*(2), 495–510.

Lee, K., Romzi, P., Hanaysha, J., Alzoubi, H., & Alshurideh, M. (2022b). Investigating the impact of benefits and challenges of IOT adoption on supply chain performance and organizational performance: An empirical study in Malaysia. *Uncertain Supply Chain Management, 10*(2), 537–550.

Lee, C., & Ahmed, G. (2021). Improving IoT privacy, data protection and security concerns. *International Journal of Technology, Innovation and Management (IJTIM), 1*(1), 18–33. https://doi.org/10.54489/ijtim.v1i1.12

Maasmi, F., Morcos, M., Al Hamadi, H., & Damiani, E. (2021). Identifying Applications' state via system calls activity: A pipeline approach. In *2021 28th IEEE International Conference on Electronics, Circuits, and Systems (ICECS)* (pp. 1–6).

Mehmood, T., Alzoubi, H. M., & Ahmed, G. (2019). Schumpeterian entrepreneurship theory: evolution and relevance. *Academy of Entrepreneurship Journal, 25*(4).

Mehmood, T. (2021). Does information technology competencies and fleet management practices lead to effective service delivery? empirical evidence from e-commerce industry. *International Journal of Technology, Innovation and Management (IJTIM), 1*(2), 14–41.

Miller, D. (2021). The best practice of teach computer science students to use paper prototyping. *International Journal of Technology, Innovation and Management (IJTIM), 1*(2), 42–63. https://doi.org/10.54489/ijtim.v1i2.17

Mondol, E. P. (2021). The impact of block chain and smart inventory system on supply chain performance at retail industry. *International Journal of Computations, Information and Manufacturing (IJCIM), 1*(1), 56–76. https://doi.org/10.54489/ijcim.v1i1.30

Obaid, A. J. (2021). Assessment of smart home assistants as an IoT. *International Journal of Computations, Information and Manufacturing (IJCIM), 1*(1), 18–36. https://doi.org/10.54489/ijcim.v1i1.34

Radwan, N., & Farouk, M. (2021). The growth of internet of things (IoT) in the management of healthcare issues and healthcare policy development. *International Journal of Technology, Innovation and Management (IJTIM), 1*(1), 69–84. https://doi.org/10.54489/ijtim.v1i1.8

Shamout, M., Ben-Abdallah, B., Alshurideh, M., Alzoubi, H., Al Kurdi, B., & Hamadneh, S. (2022). A conceptual model for the adoption of autonomous robots in supply chain and logistics industry. *Uncertain Supply Chain Management, 10*, 1–16.

Siddiqui, S. Y., Haider, A., Ghazal, T. M., Khan, M. A., Naseer, I., Abbas, S., Rahman, M., Khan, J. A., Ahmad, M., & Hasan, M. K. (2021). IoMT Cloud-based intelligent prediction of breast cancer stages empowered with deep learning. *IEEE Access, 9*, 146478–146491.

Tariq, E., Alshurideh, M., Akour, I., & Al-Hawary, S. (2022). The effect of digital marketing capabilities on organizational ambidexterity of the information technology sector. *International Journal of Data and Network Science, 6*(2), 401–408.

Zitar, R. A. (2021). A review for the Genetic Algorithm and the Red Deer Algorithm Applications. In *2021 14th International Congress on Image and Signal Processing, BioMedical Engineering and Informatics (CISP-BMEI)* (pp. 1–6).

Zitar, R. A., Abualigah, L., & Al-Dmour, N. A. (2021). Review and analysis for the Red Deer Algorithm. *Journal of Ambient Intelligence and Humanized Computing*, 1–11.

How Drones Can Mitigate Unknown-Unknown Risks Case of Covid-19

Mounir El khatib, Alaa Al-Shalabi, Ali Alamim, Hanadi Alblooshi, Shahla Alhosani, Elham Al-Kaabi, Haitham M. Alzoubi, and Muhammad Alshurideh

Abstract This paper shows the importance of technology in mitigating the spread of Covid-19 which has significantly changed the lives of people, affecting health, livelihood, and the global economy due to the closure of international borders. It has promoted social distancing and increased the need for measures that can eliminate the impact of COVID-19 related issues and outcomes. The research investigates the ways technology can benefit to support teams in preventing Covid-19. It also aims to examine the main advantages of drones during Covid-19. In addition to suggesting the use of drones to fight the Covid-19 pandemic. The inability of some countries to cope with the virus and the effective use of drones to deal with it highlights its importance. However, there is a need to adjust some cultures to the use of drones during the coronavirus crisis. Based on the review of literature, it indicated that people couldn't eliminate the virus impact yet, but they could alleviate the damage they caused. The current research has used interviews to collect data from five organizations involved in the mitigation of unknown-unknown risks, including the Ministry of Health & Community Protection, Dubai Executive Council, DEWA, Melaaha Drones, and ArabDrones. Moreover, a face-to-face interview was conducted by phones and e-mail conversations. As well as a literature review has helped in exploring the unknown-unknown risk method along with the use of Artificial Intelligence such as drones to assist in combating the Covid-19 pandemic. The descriptive qualitative research design has helped to test the hypothesis regarding its ability of drones to mitigate the unknown-unknown risk of Covid-19. The research results part indicated that drones

M. El khatib · A. Al-Shalabi · A. Alamim · H. Alblooshi · S. Alhosani · E. Al-Kaabi
Hamdan Bin Mohamad Smart University, Dubai, UAE

H. M. Alzoubi (✉)
School of Business, Skyline University College, Sharjah, UAE
e-mail: haitham.alzubi@skylineuniversity.ac.ae

M. Alshurideh
Department of Marketing, School of Business, University of Jordan, Amman, Jordan
e-mail: m.alshurideh@ju.edu.jo; malshurideh@sharjah.ac.ae

Department of Management, College of Business Administration, University of Sharjah, Sharjah, UAE

M. Alshurideh et al. (eds.), *The Effect of Information Technology on Business and Marketing Intelligence Systems*, Studies in Computational Intelligence 1056,
https://doi.org/10.1007/978-3-031-12382-5_38

are vital constituents of contingency planning. They have proven their value by going for the extra mile in saving costs and lives by incorporating drones technology in minimizing the effects of the crisis which was illustrated with some real examples. Also, the paper has proven the hypothesis by showing another real example on a drone company like ArbaDrone and how they assess, plan and mitigate unknown-unknown risks by following such a specialized program called RIO Project Program to measure and minimize the crisis impacts in the running company. Thus, drones have significantly helped countries such as the UAE to deal with the risks against Covid-19 through the elimination of infection risks with about 96% accuracy via computerized tomography scans. Finally, since the drones have contributed to international success in mitigating the Covid-19 pandemic and having an essential role in fighting the spread of the virus, so, it is very necessary to re-enforce the drone's experience to enhance the fight against coronavirus attracting other tools. Furthermore, extra studies still needed to give more bigger pictures about the impact of this coronavirus statistically to have more information about its influences on the global market.

Keywords Drones · Unknown risks · Risk mitigation · Project risk management · Covid-19

1 Introduction

With technological development around the world, the drones have taken an especial place regarding aerial vehicles technology during the Covid-19. It has gained more recognition. As far as the Covid-19 situation is concerned, it is more about social distancing and avoiding direct contact with the people and effected things around in order to avoid the spread of viral disease (Akour et al., 2021; Alshurideh et al., 2021a, b; Taryam et al., 2020). During the crisis, the drones were an effective medium to avoid all these emergencies where these were used to deliver medical supplies and to monitor the social distancing among the people at public places (Ali et al., 2022; Alshurideh et al., 2022; Hanaysha et al. 2021a, b; Khan, 2021). Along with these drones were helpful in keeping and monitoring the infected places besides, all these drones have a special place in the disinfection programs for sprays.

As UAE has taken keen measures to control COVID-19 spread and this plan is facilitated with the help of drone technology where during the lockdown hours' drones were used to sanitize the city in the sterilization and announcements were made to be delivered around the huge buildings through drones (Spires, 2020). It was easy to track areas and people with the COVID-19 symptoms where drones to manage the procedures, which enhance the speed and efficiency of the tracking areas and managing sterilization operations.

With the help of drones, it was easy to detect the people and separate healthy from sock as these are helpful in measuring the temperature, heartbeat, and breathing rate with their vital intelligence (Farouk, 2021; Mondol, 2021; Obaid, 2021). Socially these were used for the provision of medical supplies and food during lockdown to follow social distancing rules as well.

During the pandemic situation, the businesses have faced greater losses where drones were helpful in providing them a safe way to approach their customers (Asem Alzoubi, 2021b; Guergov & Radwan, 2021). Technology being the greatest allies in order to fight the pandemic has provided greater ease to continue fighting the battle by keeping the supplies possible to public drones have shown success with versatility (Gascueña, 2020).

2 Literature Review

2.1 Drones

As Earls (2019) mention that drone is unmanned aircraft technology which is formally known as unmanned aerial vehicles technology UAVs or unmanned aircraft systems (UASes). Also, Earls (2019) added that a drone is a flying robot that is remotely controlled. Furthermore, Spires (2020) emphasis that drone's technology proving it's extremely beneficial to reach places that humans cannot reach. This is the reason why it is adopted worldwide particularly, by four sectors: commercial, personal, military and future technology, which leads to the minimum amount of effort and time and energy (AlHamad et al., 2022; Alnuaimi et al., 2021; Alzoubi et al., 2022; Joghee et al., 2020). The use of drones has rapidly changed during the past years, Earls (2019) asserts that drone was used first in the military plus Its contain a technological element and they are: electronic speed controllers, flight controller, GPS module, antenna, cameras, ultrasonic sensors and collision avoidance sensors. In contrast, nowadays drone technology has rapidly developed so the Individuals and governments realized that drones have various uses.

2.2 Unknown–Unknown Risks

All businesses facing risks and everything we do have an attached with the risk. We cannot eliminate risks, but we can mitigate the damage that may cause (Ahmad et al., 2021; Odeh et al., 2021; ONeal, 2018). Besides, As Kim (2012) mention that the classification of the risks is based on the level of knowledge about the impact plus there are four categories and they are: Known–knowns, Unknown–knowns (impact is unknown), Known–unknowns (risks), Unknown–unknowns (risk and impact uncertainty). In addition, BBSCommunications (2019) mention that the business leaders have to make choices that are impacted by factors known and unknown, also the most prevalent risks are unknown-unknown category. Besides, BBSCommunications

(2019) asserts that the challenge is for looking to minimize risks is to better understand risk. Consequently, BBSCommunications (2019) emphasis to be comfortable with all types of risk it's important to prioritize developing the skills and competencies to be adaptable and responsive to the face of unknown unknowns. On the other hand, ONeal (2018) mention that Unknown unknowns are the most dangerous types of risks plus The danger here is that since the organization is unaware of the risk presence so it cannot manage the risk which results in disaster, to mitigate those risks the organization must look for a solution for every risk and control it.

2.3 How the Drone Technology Can Mitigate Unknown–Unknown Risk of Covid19

Undoubtedly, technology plays a critical role which can mitigate unexpected risks like Covid-19. The question is how the drone technology can mitigate unknown–unknown risk of covid19? It is surprised during the COVID-19 case with the unknown–unknown risk, but building scenarios depend on visibility into the future to mitigate the unknown–unknown risk (Wunker, 2020). In the United Arab Emirates, there were serious action is being taken to keep people safe. According to ONeal (2018), UAE used drones to send across the most public message which is used in the current situation (stay home). On the other hand, the drone is an easy solution to minimize the number of Covid-19 cases. Moreover, drone technology is a part of fighting the spread of Covid-19, according to ONeal (2018) during the lockdown hours, drones were used to sanitize the city in the sterilization program and its start from 8:00 pm tell 6:00 am every day. On the other hand, Staff Report (2020) mention that using drones to manage the procedures which enhance the speed and efficiency of the sterilization operations.

2.4 Use Cases of Drone Technology in Mitigating Unknown Risks

Several attempts have been reported on the use of drones in response to the current global COVID-19 pandemic (Xiao & Torok, 2020). However, drone technology was only occasionally utilized in different situations and not on a scale (Akhtar et al., 2021; Ali Alzoubi, 2021). Based on the analysis and review of research papers and publicly available information on the use of drones in mitigating unknown risks, here are some drone use cases (Martins et al., 2021). According to recent research reports, there are three main use cases of drone technology in mitigating unknown risk, taking the COVID-19 pandemic as a case study. These are (UNICEF, 2020):

1. Transportation: Drone technology has been adopted for transporting (pick-up and/or delivery) medical supplies and lab samples. This does not only minimize

the exposure to infection and unknown risks but also reduces the transportation times.

2. Aerial Spraying: Drone technology has been adopted for aerial spraying of public areas to disinfect potentially contaminated areas.
3. Close Monitoring: Drone technology has been adopted to monitor public areas and ensure guidance during quarantine and lockdown.

2.5 *Transportation*

According to a recent report, about 18 nations have adopted drone technology for transportation purposes during the COVID-19 pandemic—this includes pick-up and delivery (Al Kurdi et al., 2021; Alhamad et al., 2021). Though some of these countries are still in their testing and experimentation phase, others like Rwanda, Ghana, and Malawi have fully adopted regular drone transportation operations in response to the COVID-19 pandemic, using drones to deliver medical samples, regular medical commodities, and COVID-19 supplies since the outbreak of the pandemic. This has helped the three countries to meet up with the increased demand for medical commodities and COVID-19 supplies (Greenwood, 2021).

While the use of drone technology for transporting samples and supplies during the pandemic period was notable (Sedov et al., 2020), the cost-effectiveness of using drone technology for transportation operations (pick-up and delivery) operations compared to using other established means of transport was not well documented, and has to be considered (Lee et al., 2022a, b). However, the notable and well-documented benefits of drone technology for transportation during the COVID-19 pandemic include reduced risk of infection transmission, limited physical contact, extended transportation network (reaching a wider area), increased speed of delivery of supplies and samples (Kentenyingi, 2021).

2.6 *Aerial Spraying*

There have been several reports on the use of drone technology for aerial spraying of disinfectants in outdoor spaces and public areas to control and prevent the spread of the Coronavirus. According to reports, using drones for aerial spraying was attempted in China, South Korea, Spain, UAE, and other countries (Ali et al., 2021; Ghazal et al., 2021). In some use cases, drones were able to cover up to about 3km^2 of an area with spraying. However, there hasn't been substantial scientific evidence suggesting the efficiency and effectiveness of this application.

2.7 Close Monitoring

Several countries across the world including the UK, France, the US, Italy, China, India, Rwanda, Sierra Leone, Spain, and others have deployed drone technology to track non-compliant citizens and monitor public spaces to gain better situational awareness (Kentenyingi, 2021), and if necessary, enforce lockdown by broadcasting messages over a loudspeaker attached to the drone (Alzoubi et al., 2021; Alzoubi & Aziz, 2021). The surveillance, as well as the message broadcasting using drone technology, is expected to reduce the increase in the number of offenders and the possibility of responders having direct contact with potentially infected people (Alsharari, 2021; Eli, 2021; Hamadneh et al., 2021). More so, some academics have recently begun experimenting with drone technology to carry out symptom tracing that is enabled by AI and thermal imagery. According to reports, the use of drone technology for close monitoring is the most widespread use case in response to the COVID-19 pandemic (Khanam et al., 2019). However, some human rights activists have criticized this move and termed it as a potential abuse of civil rights. This led to the suspension of this drone program in some areas (Al Ali, 2021; Hanaysha et al., 2021a, b; Mehmood, 2021).

3 Research Method

3.1 Data Collection Methodology

In order to meet the aims of this project, exploratory and empirical evidence were determined based on **interview questioners** on the data and experience with five particular organizations in mitigating unknown-unknown risks (Al Kurdi, 2016; Alnuaimi et al., 2021). The unknown-unknown risk approach implementation was provided through a contribution by the following organizations:

1. The Ministry of Health & Community Protection
2. DEWA
3. Dubai Executive Council
4. AeabDrones
5. Melaaha Drones.

A face-to-face interview has been conducted for that, along with emails and phone conversations. In support, various papers in journal articles were also explored on unknown-unknown risk and in using Artificial Intelligence like Drones to help combat Covid-19.

The research deployed was a descriptive qualitative design that was adopted to provide answers to the questions associated with a particular research problem in order to obtain information concerning the hypothesis such as *"How the drones can mitigate the unknown-unknown risk of Covid-19"*.

3.2 Data Analysis

This Coronavirus is considered as one of the greatest business management and most challenging we have ever faced, even more bigger than the financial crisis occurred in 2008. For that reason, by evaluating ways how these modern organizations prepared themselves from any un-identified risk (Covid-19 crisis) and its unknown impacts, you can see that their main strategic goal is to understand the risk itself and the probability tails that comes along with it (Alshurideh et al., 2020; Alshurideh et al., 2021; Harahsheh et al., 2021).

But, since the start of the coronavirus outbreak, drones have acquired a new role where it proved it's essential in response to the pandemic, especially in UAE. According to Dr. Ahmed Al Mansoori the Director of International Health Relations Department in the Ministry of Health & Community Protection, the DJI's partnerships with Dubai Municipality, Dubai Police and Sharjah Police have seen the companies of drones were deployed in many applications to mitigate the crisis risk and for battling the spread of the virus by Table 1.

Since the need to reduce the human contact as a health precaution in the time of Covid-19 has provided a boost to the use of drones, According to Mr. Abdel Rahman the founder and the CEO of ArabDrones, many drone companies had got their share from economic repercussions at the beginning, especially the ones used for sports, and shooting commercials & movies, however, the success and failure of a business

Table 1 Types of applications that used drones (UNICEF, 2020)

Medical drones deliveries	
Situational awareness lockdown/curfew enforcement	
Broadcasting useful information	
Spraying streets with water and disinfectant	
Body temperature scanning	

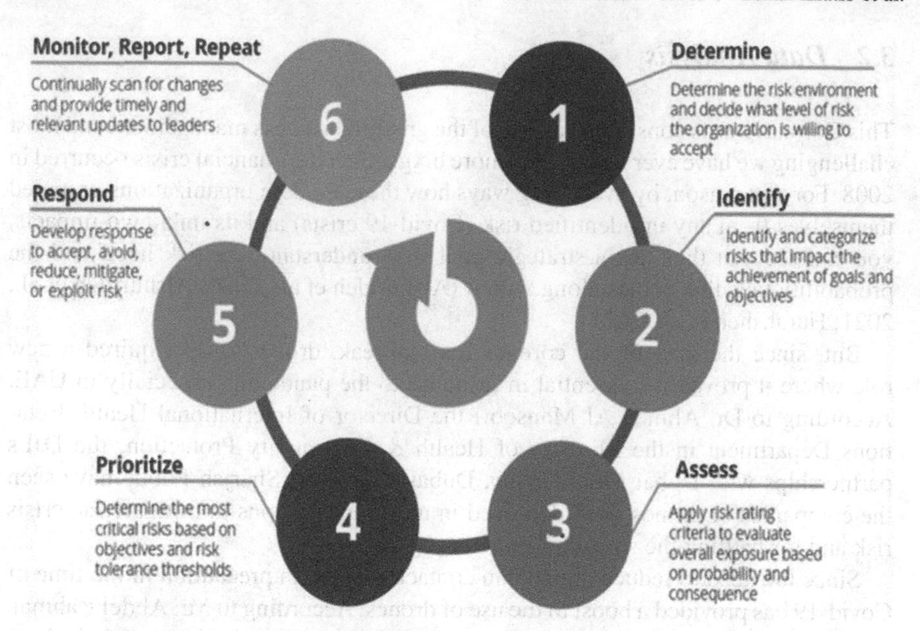
Monitor, Report, Repeat
Continually scan for changes and provide timely and relevant updates to leaders
6
1
Determine
Determine the risk environment and decide what level of risk the organization is willing to accept
Respond
Develop response to accept, avoid, reduce, mitigate, or exploit risk
Identify
Identify and categorize risks that impact the achievement of goals and objectives
5
2
Prioritize
Determine the most critical risks based on objectives and risk tolerance thresholds
4
3
Assess
Apply risk rating criteria to evaluate overall exposure based on probability and consequence

Fig. 1 Risk management process (ONeal, 2018)

depends on risk management to identify, assess and control the threats to an organization's capitals and earnings (Alshurideh, 2022; Alzoubi et al., 2020a, b; Svoboda et al., 2021). For instance, ArabDrones in mitigating their own unknown-unknown risk, they are following the RIO management program guidance for proper decision making (Kashif et al., 2021). The following figure show the process steps which are followed in ArabDrones Company for mitigating risk management approach (Fig. 1).

3.3 SWOT Analysis

SWOT analysis is one of the strategic plans which are used by organizations to exploit their core competencies along with the background and the environmental influences in order to initiate competitive advantage among the business world (Alzoubi & Yanamandra, 2020).

As stated by our interview questioner, the main factors for helping a drone business owner to identify risks based on SWOT related to the business competition are summarized in the Table 2.

Table 2 SWOT analysis

Strength and Opportunities (Positive risks)	Weeknesses and Threats (Negative risks)
– New businesses are being expected due to Covid-19 – Regulations are coming and implemented faster due to Covid-19 – Businesses are leaning into other means of delivery due to Covid-19 – The resources and the networks are supported by the government of UAE – Having many strategic alliances with very important suppliers and clients – Having clear country regulations that can be followed easily – Great low enforcement that can increase drones' company's chances of reducing any legal risk	– Country rules and regulations implications – Lockdown with other businesses due to Covid-19 – Less products supply and resources – Lack of UAE nationals in the industry – Technology is not penetrating all sectors yet – Competitors offering similar products quicker – Competitors with more flexibility in pricing – Lack of commercial knowledge among. Consumers about drone's architectures – The changes in rules and regulations to cut drones operations in UAE skies to be opened sometimes for military usage only

4 Results

Autonomous drones are essential part of the contingency planning for the support of business continuity. The multi mission and on-site autonomous drones have been proven during Covid-19 crisis, the increase efficiency in saving cost and lives by;

1. Covering large areas quicker than the foot patrols.
2. Offering a cheaper alternative to helicopter call-outs.
3. Providing crucial and urgent real time situational awareness aided by thermal and zoom cameras.

Consequently, Dr. Ahmed Al Mansoori noted that drones have been participating with big steps in combating the risk against coronavirus by eliminating the risk of infection with 96% accuracy via computerized tomography scans. In additional to that, according to a recent survey of financial lenders, 42% of respondents in the UAE said that their companies could go back to 'Business as usual' within 3–6 months, thanks to the help from drones and other AI robots over doing what they can do in mitigating the crisis (Alhashmi et al., 2020; AlShamsi et al., 2021; Euronews, 2020; Yousuf et al., 2021). Through Drones IT Technology, a disinfectant study provided essential information to protect those responding to the COVID-19 pandemic, including first responders and health care providers by 96.4% of the cases, as well as provided best practices for individuals to highly reduce the potentials for the contamination on the most effective streets and materials to clean their surfaces to rid them of the virus (Euronews, 2020).

5 Discussion of Results

Key findings

- Results indicate that in addition to others parts which helps the contingency planning which keep continuity of the business as AI, Robots, etc. drones are highly considered as essential and effective method to keep our business and life going on during any crises will happen to our globe (Mehmood et al., 2019). Drones have a speed, short ways, cheap cost transportation, and others benefits generally, which make us depends on it in a large area of usage, which already proved its ability of handling issues regarding to Dr. Ahmed Al Mansoori say. That makes us think more and more to develop extra methods which allow increasing useful capacity of drones.
- The study shows by SWOT analysis the positive and negative risks. Whereas positive risks get the advantage of negative risks, as we can see in the Table 2, almost negative risks are about governmental management issues, which can solve easier than move drones away from the scene (Alzoubi et al., 2020a, b). Others risks are about training and commercial issues, only one risk must be taken in the consideration, which makes the skies open, we can solve it by making the drones traceable with some AI control which declares intension and permissions control (Cruz, 2021; Lee & Ahmed, 2021).
- The Analysis confirm that for the time being, we are in most ever crises we've ever faced, so all organizations must take a steps to improve its ability for dealing with coronavirus, with a new rules and policies in order to mitigate the impact of the crises, as UAE, we have already applied a drones in many fields, as shown in Table 1, but we think it's not enough (Aziz & Aftab, 2021; Miller, 2021; Radwan & Farouk, 2021).
- The data shows that we have to take in the consideration the risk management to identify, and we should mitigate unknown-unknown risk.

6 Conclusion

We've found that UAE is already uses drones in many sides, Medical, Movement control, broadcasting alerts and information, sterilization of public facilities and streets and Body temperature scanning etc. So, the issue we have shown here is related to existing culture, but we need to improve that culture to be useful in coronavirus crises (Abuhashesh et al., 2021; Alzoubi & Ahmed, 2019; Hamadneh & Al Kurdi, 2021). Generally, the basic culture of drones is already existing, we won't need any more training or awareness campaigns.

7 Implications

- Societies will increase using drones in new sectors in future we can see that drones' services are ascending and will get more influence in the markets, that because drones will have the ability to enter areas where human being cant, to mitigate unknown risks such as firefighting, or risks with tangible aspects. Using the drowns will allow authorities to provide the right mitigation of the risks. It might also be a crucial tool for military industry as well.
- Regarding To Dr. Ahmed Al Mansoori declaration which says that drones clear the risk of infection around 96%, which mean 4% out of the total persons who provides the services to infected persons or unknown infected or not persons, the rest of 96% of services providers persons will be decreased out breaking coronavirus, it's good indicator which lead to benefits of drones fighting against the crises.
- But still yet some challenges must be defeated as shown in the Table 2. How much we decrease the challenges how much drones will useful.

8 Limitation

We are still yet not able generalization using drones in a vast side, we need some extra studies with bigger sample size, especially we don't have the time especially what we are in coronavirus crises, any new practical experience will be useful, and we will find many positive outcomes, but still in short term, not for long term.

9 Recommendation

In our opinion, the golden era of the drones has just begun and we think that we are about to see many other companies are going to adopt more effective and efficient ways to use this technology to minimize, monitor and control the impacted risks related to coronavirus. In our recommendations, Drone utilization should be applied to more opportunities across all country sectors to return this huge investment with humanity benefits, as well as technology benefits (Alnazer et al., 2017). We reflect that the need to emphasize drones' capabilities is mandatory to play a stronger role during the pandemic period such as Covid-19 Crisis. Besides that, more studies need to be implemented to introduce detailed tangible results to reflect the impact of drones in mitigating Covid-19 in statistical perspectives. High technology drones' leaders should balance between the short-term preparation while developing new capabilities and new regulations for the long-term. However, customers need to be more educated about this technology and entrepreneurs need to work harder in penetrating drones through social media, Internet information, and advertising into the local market. Moreover, they need to build a strong connection to the community via creative clubs or earned media generated which will increase awareness from the number of potentially loyal customers. This can help pursuing an outstanding opportunity for drone companies to target other markets such as agriculture, mining, wildlife monitoring, sports coverage, real estate, science, and education.

References

Abuhashesh, M. Y., Alshurideh, M. T., & Sumadi, M. (2021). The effect of culture on customers' attitudes toward Facebook advertising: The moderating role of gender. *Review of International Business and Strategy.*

Ahmad, A., Alshurideh, M., Al Kurdi, B., Aburayya, A., & Hamadneh, S. (2021). Digital transformation metrics: A conceptual view. *Journal of Management Information and Decision Sciences, 24*(7), 1–18.

Akhtar, A., Akhtar, S., Bakhtawar, B., Kashif, A. A., Aziz, N., & Javeid, M. S. (2021). COVID-19 Detection from CBC using machine learning techniques. *International Journal of Technology, Innovation and Management (IJTIM), 1*(2), 65–78. https://doi.org/10.54489/ijtim.v1i2.22

Akour, I., Alshurideh, M., Al Kurdi, B., Al Ali, A., & Salloum, S. (2021). Using machine learning algorithms to predict people's intention to use mobile learning platforms during the Covid-19 pandemic: Machine learning approach. *JMIR Medical Education, 7*(1), 1–17. https://doi.org/10.2196/24032

Al Ali, A. (2021). The impact of information sharing and quality assurance on customer service at UAE banking sector. *International Journal of Technology, Innovation and Management (IJTIM), 1*(1), 01–17. https://doi.org/10.54489/ijtim.v1i1.10

Al Kurdi, B. (2016). *Healthy-food choice and purchasing behaviour analysis: an exploratory study of families in the UK*. Durham University.

Al Kurdi, B., Elrehail, H., Alzoubi, H., Alshurideh, M., & Al-adaileh, R. (2021). The interplay among HRM practices, job satisfaction and intention to leave: An empirical investigation. *Journal of Legal, Ethical and Regulatory, 24*(1), 1–14.

AlHamad, A., Alshurideh, M., Alomari, K., Kurdi, B., Alzoubi, H., Hamouche, S., & Al-Hawary, S. (2022). The effect of electronic human resources management on organizational health of telecommuni-cations companies in Jordan. *International Journal of Data and Network Science, 6*(2), 429–438.

Alhamad, A. Q. M., Akour, I., Alshurideh, M., Al-Hamad, A. Q., Kurdi, B. A., & Alzoubi, H. (2021). Predicting the intention to use google glass: A comparative approach using machine learning models and PLS-SEM. *International Journal of Data and Network Science, 5*(3), 311–320. https://doi.org/10.5267/j.ijdns.2021.6.002

Alhashmi, S. F. S., Alshurideh, M., Al Kurdi, B., & Salloum, S. A. (2020). A systematic review of the factors affecting the artificial intelligence implementation in the health care sector. In *Advances in Intelligent Systems and Computing: Vol. 1153 AISC*. https://doi.org/10.1007/978-3-030-44289-7_4

Ali, N., Ahmed, A., Anum, L., Ghazal, T. M., Abbas, S., Khan, M. A., Alzoubi, H. M., & Ahmad, M. (2021). Modelling supply chain information collaboration empowered with machine learning technique. *Intelligent Automation and Soft Computing, 30*(1), 243–257. https://doi.org/10.32604/iasc.2021.018983

Ali, N., M. Ghazal, T., Ahmed, A., Abbas, S., A. Khan, M., Alzoubi, H., Farooq, U., Ahmad, M., & Adnan Khan, M. (2022). Fusion-based supply chain collaboration using machine learning techniques. *Intelligent Automation & Soft Computing, 31*(3), 1671–1687. https://doi.org/10.32604/iasc.2022.019892

Alnazer, N. N., Alnuaimi, M. A., & Alzoubi, H. M. (2017). Analysing the appropriate cognitive styles and its effect on strategic innovation in Jordanian universities. *International Journal of Business Excellence, 13*(1), 127–140. https://doi.org/10.1504/IJBEX.2017.085799

Alnuaimi, M., Alzoubi, H. M., Ajelat, D., & Alzoubi, A. A. (2021). Towards intelligent organisations: An empirical investigation of learning orientation's role in technical innovation. *International Journal of Innovation and Learning, 29*(2), 207–221. https://doi.org/10.1504/IJIL.2021.112996

Alshamsi, M., Salloum, S. A., Alshurideh, M., & Abdallah, S. (2021). Artificial intelligence and blockchain for transparency in governance. In *Artificial Intelligence for Sustainable Development: Theory, Practice and Future Applications* (pp. 219–230). Springer.

Alsharari, N. (2021). Integrating blockchain technology with internet of things to efficiency. *International Journal of Technology, Innovation and Management (IJTIM), 1*(2), 1–13.

Alshurideh, M. (2022). Does electronic customer relationship management (E-CRM) affect service quality at private hospitals in Jordan? *Uncertain Supply Chain Management, 10*(2), 1–8.

Alshurideh, M., Gasaymeh, A., Ahmed, G., Alzoubi, H., & Kurd, B. A. (2020). Loyalty program effectiveness: Theoretical reviews and practical proofs. *Uncertain Supply Chain Management, 8*(3). https://doi.org/10.5267/j.uscm.2020.2.003

Alshurideh, M.T., Al Kurdi, B., AlHamad, A. Q., Salloum, S. A., Alkurdi, S., Dehghan, A., Abuhashesh, M., & Masa'deh, R. (2021a). Factors affecting the use of smart mobile examination platforms by universities' postgraduate students during the COVID-19 pandemic: An empirical study. *Informatics, 8*(2). https://doi.org/10.3390/informatics8020032

Alshurideh, M. T., Hassanien, A. E., & Masa'deh, R. (2021b). *The effect of coronavirus disease (Covid-19) on business intelligence*. Springer.

Alshurideh, M. T., Al Kurdi, B., Alzoubi, H. M., Ghazal, T. M., Said, R. A., AlHamad, A. Q., Hamadneh, S., Sahawneh, N., & Al-kassem, A. H. (2022). Fuzzy assisted human resource management for supply chain management issues. *Annals of Operations Research*, 1–19.

Alzoubi, A. (2021a). The impact of process quality and quality control on organizational competitiveness at 5-star hotels in Dubai. *International Journal of Technology, Innovation and Management (IJTIM), 1*(1), 54–68. https://doi.org/10.54489/ijtim.v1i1.14

Alzoubi, A. (2021b). Renewable green hydrogen energy impact on sustainability performance. *International Journal of Computations, Information and Manufacturing (IJCIM), 1*(1), 94–110. https://doi.org/10.54489/ijcim.v1i1.46

Alzoubi, H., & Ahmed, G. (2019). Do TQM practices improve organisational success? A case study of electronics industry in the UAE. *International Journal of Economics and Business Research, 17*(4), 459–472. https://doi.org/10.1504/IJEBR.2019.099975

Alzoubi, H. M., & Yanamandra, R. (2020). Investigating the mediating role of information sharing strategy on agile supply chain. *Uncertain Supply Chain Management, 8*(2), 273–284. https://doi.org/10.5267/j.uscm.2019.12.004

Alzoubi, H. M., & Aziz, R. (2021). Does emotional intelligence contribute to quality of strategic decisions? The mediating role of open innovation. *Journal of Open Innovation: Technology, Market, and Complexity, 7*(2), 130. https://doi.org/10.3390/joitmc7020130

Alzoubi, H. M., Ahmed, G., Al-Gasaymeh, A., & Al Kurdi, B. (2020a). Empirical study on sustainable supply chain strategies and its impact on competitive priorities: The mediating role of supply chain collaboration. *Management Science Letters, 10*(3), 703–708. https://doi.org/10.5267/j.msl.2019.9.008

Alzoubi, H., Alshurideh, M., Kurdi, B. A., & Inairat, M. (2020b). Do perceived service value, quality, price fairness and service recovery shape customer satisfaction and delight? A practical study in the service telecommunication context. *Uncertain Supply Chain Management, 8*(3), 579–588. https://doi.org/10.5267/j.uscm.2020.2.005

Alzoubi, H. M., Vij, M., Vij, A., & Hanaysha, J. R. (2021). What leads guests to satisfaction and loyalty in UAE five-star hotels? AHP analysis to service quality dimensions. *Enlightening Tourism, 11*(1), 102–135. https://doi.org/10.33776/et.v11i1.5056

Alzoubi, H., Alshurideh, M., Kurdi, B., Akour, I., & Aziz, R. (2022). Does BLE technology contribute towards improving marketing strategies, customers' satisfaction and loyalty? The role of open innovation. *International Journal of Data and Network Science, 6*(2), 449–460.

Aziz, N., & Aftab, S. (2021). Data mining framework for nutrition ranking methodology: SPSS modeller. *International Journal of Technology, Innovation and Management (IJTIM), 1*(1), 85–95.

BBSCommunications. (2019). *Identifying risk–Known knowns, known unknowns and unknown unknowns. BBS Communications Group.*

Cruz, A. (2021). Convergence between blockchain and the internet of things. *International Journal of Technology, Innovation and Management (IJTIM), 1*(1), 35–56.

Earls, A. (2019). *What is a Drone? Definition from WhatIs.com.*

Eli, T. (2021). Studentsperspectives on the use of innovative and interactive teaching methods at the University of Nouakchott Al Aasriya, Mauritania: English department as a case study. *International Journal of Technology, Innovation and Management (IJTIM), 1*(2), 90–104.

Euronews. (2020). *Coronavirus: UAE moves to mitigate economic fallout|Euronews.*

Farouk, M. (2021). The universal artificial intelligence efforts to face coronavirus Covid-19. *International Journal of Computations, Information and Manufacturing (IJCIM), 1*(1), 77–93. https://doi.org/10.54489/ijcim.v1i1.47

Gascueña, D. (2020). *Drones to stop the COVID-19 epidemic | BBVA.*

Ghazal, T. M., Hasan, M. K., Alshurideh, M. T., Alzoubi, H. M., Ahmad, M., Akbar, S. S., Al Kurdi, B., & Akour, I. A. (2021). IoT for smart cities: Machine learning approaches in smart healthcare—a review. *Future Internet, 13*(8), 218. https://doi.org/10.3390/fi13080218

Greenwood, F. (2021). *Assessing the impact of drones in the global Covid response.*

Guergov, S., & Radwan, N. (2021). Blockchain convergence: Analysis of issues affecting IoT, AI and Blockchain. *International Journal of Computations, Information and Manufacturing (IJCIM), 1*(1), 1–17. https://doi.org/10.54489/ijcim.v1i1.48

Hamadneh, S., & Al Kurdi, B. (2021). The effect of brand personality on consumer self-identity: The moderation effect of cultural orientations among British and Chinese consumers. *Journal of Legal, Ethical and Regulatory Issues, 24*(Special Issue 1), 1–14.

Hamadneh, S., Pedersen, O., & Al Kurdi, B. (2021). An investigation of the role of supply chain visibility into the scottish bood supply chain. *Journal of Legal, Ethical and Regulatory Issues, 24*(Special Issue 1), 1–12.

Hanaysha, J. R., Al-Shaikh, M. E., Joghee, S., & Alzoubi, H. (2021a). Impact of innovation capabilities on business sustainability in small and medium enterprises. *FIIB Business Review*, 1–12,. https://doi.org/10.1177/23197145211042232

Hanaysha, J. R., Al Shaikh, M. E., & Alzoubi, H. M. (2021b). Importance of marketing mix elements in determining consumer purchase decision in the retail market. *International Journal of Service Science, Management, Engineering, and Technology (IJSSMET), 12*(6), 56–72.

Harahsheh, A. A., Houssien, A. M. A., & Alshurideh, M. T. (2021). The effect of transformational leadership on achieving effective decisions in the presence of psychological capital as an intermediate variable in private Jordanian. In *The Effect of Coronavirus Disease (COVID-19) on Business Intelligence* (pp. 243–221). Springer Nature.

Joghee, S., Alzoubi, H. M., & Dubey, A. R. (2020). Decisions effectiveness of FDI investment biases at real estate industry: Empirical evidence from Dubai smart city projects. *International Journal of Scientific and Technology Research, 9*(3), 3499–3503.

Kashif, A. A., Bakhtawar, B., Akhtar, A., Akhtar, S., Aziz, N., & Javeid, M. S. (2021). Treatment response prediction in hepatitis c patients using machine learning techniques. *International Journal of Technology, Innovation and Management (IJTIM), 1*(2), 79–89. https://doi.org/10.54489/ijtim.v1i2.24

Kentenyingi, R. (2021). *Leveraging drone technology in the Covid-19 rapid response in rural areas of Uganda.* UN Capital Development Fund (UNCDF).

Khan, M. A. (2021). Challenges facing the application of iot in medicine and healthcare. *International Journal of Computations, Information and Manufacturing (IJCIM), 1*(1), 39–55. https://doi.org/10.54489/ijcim.v1i1.32

Khanam, F. T. Z., Al-Naji, A., & Chahl, J. (2019). Remote monitoring of vital signs in diverse non-clinical and clinical scenarios using computer vision systems: A review. *Applied Sciences (Switzerland), 9*(20). https://doi.org/10.3390/APP9204474

Kim, S. D. (2012). *Characterizing unknown unknowns.*

Lee, C., & Ahmed, G. (2021). Improving IoT privacy, data protection and security concerns. *International Journal of Technology, Innovation and Management (IJTIM), 1*(1), 18–33. https://doi.org/10.54489/ijtim.v1i1.12

Lee, K., Azmi, N., Hanaysha, J., Alzoubi, H., & Alshurideh, M. (2022a). The effect of digital supply chain on organizational performance: An empirical study in Malaysia manufacturing industry. *Uncertain Supply Chain Management, 10*(2), 495–510.

Lee, K., Romzi, P., Hanaysha, J., Alzoubi, H., & Alshurideh, M. (2022b). Investigating the impact of benefits and challenges of IOT adoption on supply chain performance and organizational performance: An empirical study in Malaysia. *Uncertain Supply Chain Management, 10*(2), 537–550.

Martins, B. O., Lavallée, C., & Silkoset, A. (2021). Drone use for COVID-19 related problems: Techno-solutionism and its societal implications. *Global Policy.* https://doi.org/10.1111/1758-5899.13007

Mehmood, T. (2021). Does information technology competencies and fleet management practices lead to effective service delivery? empirical evidence from e-commerce industry. *International Journal of Technology, Innovation and Management (IJTIM), 1*(2), 14–41.

Mehmood, T., Alzoubi, H. M., Alshurideh, M., Al-Gasaymeh, A., & Ahmed, G. (2019). Schumpeterian entrepreneurship theory: Evolution and relevance. *Academy of Entrepreneurship Journal, 25*(4), 1–10.

Miller, D. (2021). The best practice of teach computer science students to use paper prototyping. *International Journal of Technology, Innovation and Management (IJTIM), 1*(2), 42–63. https://doi.org/10.54489/ijtim.v1i2.17

Mondol, E. P. (2021). The impact of block chain and smart inventory system on supply chain performance at retail industry. *International Journal of Computations, Information and Manufacturing (IJCIM), 1*(1), 56–76. https://doi.org/10.54489/ijcim.v1i1.30

Obaid, A. J. (2021). Assessment of smart home assistants as an IoT. *International Journal of Computations, Information and Manufacturing (IJCIM), 1*(1), 18–36. https://doi.org/10.54489/ijcim.v1i1.34

Odeh, R., Obeidat, B. Y., Jaradat, M. O., & Masa'deh, R., & Alshurideh, M. T. (2021). The transformational leadership role in achieving organizational resilience through adaptive cultures: The case of Dubai service sector. *International Journal of Productivity and Performance Management.* https://doi.org/10.1108/IJPPM-02-2021-0093

ONeal, B. (2018). *Risk management for different types of risk.* 360 Factors.

Radwan, N., & Farouk, M. (2021). The growth of internet of things (IoT) in the management of healthcare issues and healthcare policy development. *International Journal of Technology, Innovation and Management (IJTIM), 1*(1), 69–84. https://doi.org/10.54489/ijtim.v1i1.8

Sedov, L., Krasnochub, A., & Polishchuk, V. (2020). Modeling quarantine during epidemics and mass-testing using drones. *PLoS ONE, 15*(6 June). https://doi.org/10.1371/JOURNAL.PONE.0235307

Spires, J. (2020). *Drones capture Dubai's message honoring COVID-19 frontline heroes. DroneDJ.*

Staff Report. (2020). *Video: Drones disinfect Dubai in UAE Covid-19 sterilisation program-News. Khaleej Times.*

Svoboda, P., Ghazal, T. M., Afifi, M. A. M., Kalra, D., Alshurideh, M. T., & Alzoubi, H. M. (2021). Information systems integration to enhance operational customer relationship management in the pharmaceutical industry. In *The International Conference on Artificial Intelligence and Computer Vision* (pp. 553–572).

Taryam, M., Alawadhi, D., Aburayya, A., Albaqa'een, A., Alfarsi, A., Makki, I., Rahmani, N., Alshurideh, M., & Salloum, S. A. (2020). Effectiveness of not quarantining passengers after having a negative COVID-19 PCR test at arrival to dubai airports. *Systematic Reviews in Pharmacy, 11*(11). https://doi.org/10.31838/srp.2020.11.197

UNICEF, S. D. (2020). *How drones can be used to combat covid-19* (pp 1–4).

Wunker, S. (2020). *Business strategy through four phases of the coronavirus crisis*. Forbes.

Xiao, Y., & Torok, M. E. (2020). Taking the right measures to control COVID-19. *The Lancet Infectious Diseases, 20*(5), 523–524. https://doi.org/10.1016/S1473-3099(20)30152-3

Yousuf, H., Zainal, A. Y., Alshurideh, M., & Salloum, S. A. (2021). Artificial intelligence models in power system analysis. In *Studies in Computational Intelligence* (Vol. 912). https://doi.org/10.1007/978-3-030-51920-9_12

The Role of Remote Work in Mitigating Unknown-Unknown Risk During Covid-19

Mounir El khatib, Hamda Al Rais, Hessa Al Rais, Hessa Al Rais, Khawla AlShamsi, Latifa AlKetbi, Reem AlBanna, Haitham M. Alzoubi, and Muhammad Alshurideh

1 Introduction

From management perspective, understanding the strength of leadership, process, system, operation, and risks is essential to lead the organization to success. Focusing and studying the risks factors impact is important to take the organization to a new level of business excellency and give the management the chance to enhance and improve the performance and efficiency levels of the organization and give the management to understand the position of the organization and what it needs to have a better place among the rivals considering all possible challenges that the organization may have in the future. Which also give the management the opportunity to have other options and plans and be ready to deal with the risks and unknown challenges that any organization may face. The aim of this paper is to know how the risks of the COVID-19 crisis, as an unknown unknown risk, affects the framework of Dubai Ports World, Dubai Economy Department and RTA and how the management at these companies deal with such unknown risk and discuss the outcomes.

M. El khatib · H. Al Rais · H. Al Rais · H. Al Rais · K. AlShamsi · L. AlKetbi · R. AlBanna
Hamdan Bin Mohamad Smart University, Dubai, UAE

H. M. Alzoubi (✉)
School of Business, Skyline University College, Sharjah, UAE
e-mail: haitham.alzubi@skylineuniversity.ac.ae

M. Alshurideh
Department of Marketing, School of Business, University of Jordan, Amman, Jordan
e-mail: m.alshurideh@ju.edu.jo; malshurideh@sharjah.ac.ae

Department of Management, College of Business Administration, University of Sharjah, Sharjah, UAE

M. Alshurideh et al. (eds.), *The Effect of Information Technology on Business and Marketing Intelligence Systems*, Studies in Computational Intelligence 1056,
https://doi.org/10.1007/978-3-031-12382-5_39

2 Literature Review

Today's business world has changed recently due to COVID-19 and associated risks (Ahmad et al., 2021a, b; Alshurideh et al., 2021a,b; Leo et al., 2021). The pandemic was an unknown uknown risk for companies back in 2019, which made many businesses change their processes fully (Alshurideh et al., 2021a,b; Nuseir et al., 2021). For instance, the pandemic made employers around the world switch to remote working and protect their employees from the virus (Alshurideh et al., 2019a,b,c; Alshurideh et al., 2020a,b; Sultan et al., 2021). Remote working is one of the types of social isolation measures, which has proven effective to reduce the disease spread as employees do not gather in groups and, thus, do not infect one another. At the same time, remote working works differently in each industry (Alshurideh et al., 2019a,b,c; Sweiss et al., 2021). The example of Dubai Ports World company, Dubai Economy and RTA experience sheds light on how effective remote working is in mitigating risk during COVID-19 can be organized (Al Shebli et al., 2021; Alshurideh, 2022; Taryam et al., 2020).

Risks can be categories into three categories: known risks, unknown risks and unknown-unknown risks (Alnazer et al., 2017; Alzoubi & Ahmed, 2019; Mehmood et al., 2019). Unknown risks are defined as "dangers which can be calculated in terms of their probability and impact based on past records" (Kinnvall et al., 2020). Unknown-unknown risks should be differentiated from known risks and mitigated remembering that their consequences cannot be fully foreseen (AlSuwaidi et al., 2021; Alzoubi et al., 2020a, 2020b; Joghee et al., 2020). Inability to deal with unknown unknown risks leads to insecurity. This research will focus on the remote working tool in mitigating the unknown unknown risk at different companies in the UAE.

Risk management is the crucial part of the company's strategy allowing it to remain stable, profitable and efficient (Alshurideh et al., 2020; Alzoubi et al., 2020a, b). Based on the interview results with Eng. Sulaiman AlBloushi Software Department Manager at DP world, Mrs Suhaila Haji General Budget Manager at RTA and Laila AlRamsi, Revenue Section Manager at Dubai Economy, unknown risk mitigation is organized here via the so-called Enterprise Risk Management Framework, implementation of which is carefully monitored by the Audit Committee. Overall, these companies mitigate their unknown risks through continuous market analysis and monitoring of the macroeconomic indicators of the markets of influence, total quality management, big data forecasting analytics.

According to Tysiac (2012), the volume and complexity of risks has increased considerably during the last years and continues growing. Therefore, risk mitigation remains crucial for projects and operational activity (Alzoubi et al., 2021; Hanaysha et al., 2021a, b; Odeh et al., 2021). Based on the literature review in the field, uknown risk mitigation involves proper classification of the risks, crisis management, "reverse stress testing," uncertainty modelling, focusing on both pros and cons of the risk, enabling company-wide responses to emerging threats, and integrating risk management and strategic management as well as remembering about unpredictable outcomes (Firoozye & Ariff, 2016; Kim, 2012; Segal, 2011; Tysiac, 2012). It is important to continuously reevaluate and back-track perceived loss vs. actual loss from each identified risk because actual loss may be different (Firoozye & Ariff, 2016).

Overall, remote working played a significant role in sustaining work productivity through COVID-19 ensuring that all employees are working safely from home mitigating all potential risks that could possibly occur (Alameeri et al., 2021; Ali et al., 2021; Ghazal et al., 2021). In order to improve it and decrease the chances of unknown uknown risks in the future while working remotely, new strategies need to be applied to avoid low band-width communication, loss job motivation and productivity as well as loss of team collaboration, increased employees' conflicts and inability to resolve some of them remotely (Al Kurdi et al., 2020; Guergov & Radwan, 2021; Khan, 2021; Mondol, 2021). Therefore, companies should implement strategies to avoid possible risks that might occur because of these challenges and problems of remote working through hiring more technological leaders and project managers who can identify how to tackle the most common risks associated to remote teamwork (Al Kurdi et al., 2020; Alzoubi, 2021a; Farouk, 2021; Obaid, 2021; Tariq et al., 2022).

3 Research Methodology and Limitations

The secondary data method is used in this paper. Data that is related to risk management at Dubai Ports World, Dubai Economy and RTA was searched in order to find the main methods that these insitutions use in order to avoid any unexpected losses or risks that threatens the company's framework.

4 Findings

The key findings of Dubai Ports World, Dubai Economy and RTA remote working had significant relationship between the positive and negative outcomes, there was an noticeable increase in the productivity level of employees and had the highest rates of engagement due to the flexibility and since there was less distractions and the majority found that it's easier to concentrate at home, another positive impact was reducing employee turnover, since it was indicated that offering remote working is a valuable retention tool for all organizations (Alhamad et al., 2021; Alnuaimi et al., 2021; Hanaysha et al., 2021a, b). On the other hand, the negative outcomes had unknown unknown risks, since some companies have a good enough cyber-security system in place, while some are not aware of the risks involved in connecting remotely (Al Ali, 2021; Alzoubi, 2021b; Salloum et al., 2020). Thus, Home Wi-Fi Security as opposed to the office's internet connection, where IT support can control the security of all the Wi-Fi networks system, employee's home networks may probably have weaker protocols than office's internet connection which allows hackers to easily access the network's system (Akhtar et al., 2021; Al Kurdi et al. 2021; Eli, 2021; Kashif et al., 2021). The current COVID-19 pandemic led to a transformation in remote work which inevitably brings a new set of risks and challenges (Rubinstein, 2020). Since working from home became mandatory in the previous period, organizations must rapidly ensure the security of every laptop, it was found that the best practice to protect departments from unknown uknown risks mitigation is carrying out virtual training for employees to educate them on how to protect their IT system from hackers and the potential risks they should know about (Alsharari, 2021; Alshraideh et al., 2017; Mehmood, 2021; Miller, 2021).

Regarding DP world, RTA and Dubai Economy cases, it seems that they maintain an effective system of internal control which is embedded in the main key operations that is designed to provide reasonable assurance of the organization's strategic business objectives and to mitigate unknown unknown risks (Ahmad et al., 2021a, b ; Al-Dmour et al., 2021; Aziz & Aftab, 2021; Cruz, 2021; Radwan & Farouk, 2021; Ben-Abdallah et al., 2022). In addition, the key components of the internal controls are listed below:

1. Risk Assessment: identification and the analysis of risks that prevent from achieving their objectives, by identifying the risks in a proper manner the management will be able to easily determine how to mitigate and manage the risks effectively. Risk factors may consist of internal/external factors, which shall be evaluate risk on a regular basis, like changes in the company from new policies, staffing and software application that can all impact on the risk assessment of the company (Hamadneh et al., 2021; Lee & Ahmed, 2021).
2. Control activities: consisting of the procedures and guide of policies that help ensure that directives are implemented. One of the main points of control activities is segregation of duties, by having different individuals responsible for authorizing the organization transactions, performing reconciliations and having custody of assets (Alshurideh et al., 2022; Alzoubi & Aziz, 2021).
3. Information and communication: relating identification of pertinent information in a timely manner that permits personnel to perform responsibilities (Ali et al., 2022; Alzoubi et al., 2022). Like for example, having the finances report in a timely manner allow management to identify the differences in its operations in terms of high reserves and drops in margin.
4. Monitoring: the key element to internal control when it to comes to the management, since it's their responsibility to monitor all the controls and determine if they're operating as they were intended. If the operation of the controls is not effective, it's then the management responsibility to amend the internal control (AlHamad et al., 2022; Lee et al., 2022a, b).

It was found that if all these components of internal control are implemented and are operating effectively, they can ensure that the company is mitigating unknown uknown risks while achieving their organizational objectives, please see Table 1.

Table 1 SWOT analysis

Strengths:	**Weaknesses**:
– Solid Ground for the system – Strong Management Policies & Procedures – Widespread Awareness Program	– Strong Risk Management System which obstructs the internal work and progress – High level of security affecting performance
Opportunities: – World leading in trade innovation, global presence and entering new markets	**Threats**: – Out of control with continuance changing in laws and regulation

5 Recommendations and Conclusion

After understanding and analyzing the Dubai Ports World, Dubai Economy and RTA' management perspectives, the reserachers noticed some areas that might need few adjustment or enhancement that would give the company a push to a better position at the market in difficult times, such as the recent pandemic, and how to deal with the challenges and consequences of the crisis that they might suffer from in the future taking remote work as an option or solution as the following:

- Remote working is a new thing for many employees as theys haven't seen such situation before in which they have to work from their homes, they have a habit of going to offices but due to the current cercumstanses it is mandatory to do the usual tasks online. It semms that employers face difficulties in this, so the company should give their employees a complete online training in which they will be guided and explained how actually remote working requirements are and how it should be currectly done in order to make it more effective rather than wasting time doing tasks that add no value to the required objectives, and how to overcome the possible technical issues that they may face while using such technologies. The management also should run a technical tests on the systems and platforms the employees are using as well as examining the employees after the training is done to check whether they have learned about their remote working and if they have any confusion related to remote working it can be cleared out with them.
- Communication gap is a key of many problems, if there is no proper communication between the employees and company then the company will face huge loss. It is necessary for the management to develop such a system in which the communication gap can be reduced to minimum, employees should be in daily contact with the system so the management is aware of what is happening and can predict the risk factors and make suitable arrangements to solve them otherwise simple communication would not be effective. The company should conduct live sessions for their employees. It is suggested to conduct effective Microsoft teams training sessions through a webinar on MS teams; to extend the knowledge and demonstrate the key advanced features, which will help in improving the overall communication and collaboration between the employees.

- It is essential for the management to monitor their system at regular intervals especially in such situation where they are not able to deal with employees face to face. If a management keeps a check on the system after several months and witness any issue, then it will be too late to solve it. Reports should be generated and compared with previous ones to see the improvements in the company. If the previous reports are better than the new one then the management should immediately call a meeting of all the employees on MS teams and find where the actual problem is. So, through this daily communication and check the chances of loss for the company through remote working will be minimized.
- Hacking is the main issue of remote working; hackers can steal information and miss use them which attackes the company's confedintial database and create many serious problems the company. The management team should tackle this issue by installing security system in the devices of their employees on which they work. The system indicates any threat in the work and the employ can immediately tackle with it to avoid any data breaching. The companies should also hire an IT experts who should keep checking on the overall system so in case if the data is hacked then the expert can immediately recognize the problem and prevent the system from the data breaching and inform the management so they will be aware of such threats and work to find a solution to avoid it.

Due to the situation of Covid-19, remote working is basic need instead of shutting down the production line at such situations. It is an ulternative solution to keep the business running. It has its own pros and cons; therefore, the management team should find and create methods to make it as much as effective and effeciant as possible.

To conclude, the main issue regarding remote working is the increase of unknown uknown risks. The research was conducted in this regard with reference to different companies in the UAE and it was found that they have implemented many strategies regarding this situation which are very effective that helped to grow but these are only effective when the management keeps proper check on all implementations rather than making the policies and then not taking any action on them. Measures such as a complete training of employees, providing a complete security system to them so the data could not leak and ensuring that security is provided to all employees, along with daily communication and monitoring of all the activities can improve the operations to a greater limit.

Appendix

Questionnaire for interview

1. Organization name, interviewee name, date of interview
2. Brief about the organization
3. Brief about risk in the organization
4. Any experiences in unknown risks? Risks which taught you lessons

5. Do you have a risk strategy in your organization?
6. Do you have a risk methodology in your organization?
7. How do deal with unknown risks?
8. Do you have a strategy and/or methodology for unknown risks identification?
9. Can you tell us about this strategy and/or methodology?
10. Do you use any of these technology tools?

No	Tools	
1	Artificial intelligence	
2	Blockchain	
3	Telehealth technologies	
4	Three-dimensional printing	
5	Nanotechnology	
6	Drones	
7	Robots	
8	Remote Work	
9	Distance Learning	
10	Supply Chain 4.0	

11. Are you using any other tools?
12. If (Yes) please specify

No	Tool	Planning Risk management	Risk Identification	Qualitative analysis	Quantitative analysis	Planning risk responses	Implement Risk responses	Monitor risks
1	Artificial intelligence							
2	Blockchain							
3	Telehealth technologies							
4	Three-dimensional printing							
5	Nanotechnology							
6	Drones							
7	Robots							
8	Remote Work							
9	Distance Learning							
10	Supply Chain 4.0							

13. If (yes) Give us a brief about the tool
14. In which risk process do you use these technology tools
15. Based on Q14, In which risk process do these technologies mitigate unknown risk (Tick ✓ all applied)

No	Tool	Planning Risk management	Risk Identification	Qualitative analysis	Quantitative analysis	Planning risk responses	Implement Risk responses	Monitor risks
1	Artificial intelligence							
2	Blockchain							
3	Telehealth technologies							
4	Three-dimensional printing							
5	Nanotechnology							
6	Drones							
7	Robots							
8	Remote Work							
9	Distance Learning							
10	Supply Chain 4.0							

16. Based on Q15, Identify the tools and/or processes that mitigate unknown risks. What, how, what results, his personal judgement, how to improve it.
17. Any other tools or processes that lead to unknown risk mitigation?
18. What are the strengths and weaknesses in the organization based on the interviewee opinion?
19. What are the opportunities and threats in the organization based on the interviewee opinion?
20. Final recommendations or any additions the interviewee likes to add.

Instructor Feedback for the research components:

	Excellent	V. Good (h)	V. Good (L)	Remarks
1. Abstract			✓	
2. Introduction			✓	
3. Literature review			✓	
4. Hypothesis			✓	
5. Research method			✓	
6. Data gathering			✓	
7. questionnaire			✓	
8. Analysis			✓	
9. Limitations			✓	
10. Recommendations			✓	
11. Conclusion			✓	
12. Depth			✓	
13. Logic and flow			✓	
14. Others			✓	
15. Overall work			B	

More Details: Work is done to a Good level to meet the minimum requirements. Work can be improved in almost all aspects especially literature review and data gathering to analysis and conclusions.

Work structure and the flow and logic is good.

References

Ahmad, A., Alshurideh, M. T., Al Kurdi, B. H., & Salloum, S. A. (2021a). Factors impacts organization digital transformation and organization decision making during Covid19 pandemic. In *Studies in Systems, Decision and Control* (Vol. 334). https://doi.org/10.1007/978-3-030-67151-8_6

Ahmad, A., Alshurideh, M. T., Al Kurdi, B. H., & Alzoubi, H. M. (2021b). Digital strategies: A systematic literature review. In *The International Conference on Artificial Intelligence and Computer Vision* (pp. 807–822).

Akhtar, A., Akhtar, S., Bakhtawar, B., Kashif, A. A., Aziz, N., & Javeid, M. S. (2021). COVID-19 Detection from CBC using Machine Learning Techniques. *International Journal of Technology, Innovation and Management (IJTIM), 1*(2), 65–78. https://doi.org/10.54489/ijtim.v1i2.22

Al-Dmour, R., AlShaar, F., Al-Dmour, H., Masa'deh, R., & Alshurideh, M. T. (2021). The effect of service recovery justices strategies on online customer engagement via the role of "customer satisfaction" during the covid-19 pandemic: An empirical study. In *Studies in Systems, Decision and Control* (Vol. 334). https://doi.org/10.1007/978-3-030-67151-8_19

Al Ali, A. (2021). The impact of information sharing and quality assurance on customer service at UAE banking sector. *International Journal of Technology, Innovation and Management (IJTIM), 1*(1), 01–17. https://doi.org/10.54489/ijtim.v1i1.10

Al Kurdi, B., Alshurideh, M., & Al Afaishata, T. (2020). Employee retention and organizational performance: Evidence from banking industry. *Management Science Letters, 10*(16), 3981–3990.

Al Kurdi, B., Alshurideh, M., Nuseir, M., Aburayya, A., & Salloum, S. A. (2021). The effects of subjective norm on the intention to use social media networks: An exploratory study using PLS-SEM and machine learning approach. In *Advances in Intelligent Systems and Computing* (Vol. 1339). https://doi.org/10.1007/978-3-030-69717-4_55

Al Shebli, K., Said, R. A., Taleb, N., Ghazal, T. M., Alshurideh, M. T., & Alzoubi, H. M. (2021). RTA's employees' perceptions toward the efficiency of artificial intelligence and big data utilization in providing smart services to the residents of Dubai. In: *The International Conference on Artificial Intelligence and Computer Vision* (pp. 573–585).

Alameeri, K. A., Alshurideh, M. T., & Al Kurdi, B. (2021). The effect of covid-19 pandemic on business systems' innovation and entrepreneurship and how to cope with it: a theatrical view. In *Studies in Systems, Decision and Control* (Vol. 334). https://doi.org/10.1007/978-3-030-67151-8_16

AlHamad, A., Alshurideh, M., Alomari, K., Kurdi, B., Alzoubi, H., Hamouche, S., & Al-Hawary, S. (2022). The effect of electronic human resources management on organizational health of telecommuni-cations companies in Jordan. *International Journal of Data and Network Science, 6*(2), 429–438.

Alhamad, A. Q. M., Akour, I., Alshurideh, M., Al-Hamad, A. Q., Kurdi, B. A., & Alzoubi, H. (2021). Predicting the intention to use google glass: A comparative approach using machine learning models and PLS-SEM. *International Journal of Data and Network Science, 5*(3). https://doi.org/10.5267/j.ijdns.2021.6.002

Ali, N., Ahmed, A., Anum, L., Ghazal, T. M., Abbas, S., Khan, M. A., Alzoubi, H. M., & Ahmad, M. (2021). Modelling supply chain information collaboration empowered with machine learning technique. *Intelligent Automation and Soft Computing, 30*(1), 243–257. https://doi.org/10.32604/iasc.2021.018983

Ali, N., M. Ghazal, T., Ahmed, A., Abbas, S., A. Khan, M., Alzoubi, H., Farooq, U., Ahmad, M., & Adnan Khan, M. (2022). Fusion-Based Supply Chain Collaboration Using Machine Learning Techniques. *Intelligent Automation & Soft Computing, 31*(3), 1671–1687. https://doi.org/10.32604/iasc.2022.019892

Alnazer, N. N., Alnuaimi, M. A., & Alzoubi, H. M. (2017). Analysing the appropriate cognitive styles and its effect on strategic innovation in Jordanian universities. *International Journal of Business Excellence, 13*(1), 127–140. https://doi.org/10.1504/IJBEX.2017.085799

Alnuaimi, M., Alzoubi, H. M., Ajelat, D., & Alzoubi, A. A. (2021). Towards intelligent organisations: An empirical investigation of learning orientation's role in technical innovation. *International Journal of Innovation and Learning, 29*(2), 207–221. https://doi.org/10.1504/IJIL.2021.112996

Alsharari, N. (2021). Integrating Blockchain Technology with Internet of things to Efficiency. *International Journal of Technology, Innovation and Management (IJTIM), 1*(2), 1–13.

Alshraideh, A. T. R., Al-Lozi, M., & Alshurideh, M. T. (2017). The impact of training strategy on organizational loyalty via the mediating variables of organizational satisfaction and organizational performance: An empirical study on Jordanian agricultural credit corporation staff. *Journal of Social Sciences (COES&RJ-JSS), 6*(2), 383–394.

Alshurideh, M., Al Kurdi, B., & Salloum, S. (2019a). Examining the Main Mobile Learning System Drivers' Effects: A Mix Empirical Examination of Both the Expectation-Confirmation Model (ECM) and the Technology Acceptance Model (TAM). In *International Conference on Advanced Intelligent Systems and Informatics* (pp. 406–417).

Alshurideh, M., Salloum, S. A., Al Kurdi, B., Monem, A. A., & Shaalan, K. (2019b). Understanding the quality determinants that influence the intention to use the mobile learning platforms: A practical study. *International Journal of Interactive Mobile Technologies, 13*(11). https://doi.org/10.3991/ijim.v13i11.10300

Alshurideh, M., Salloum, S. A., Al Kurdi, B., & Al-Emran, M. (2019c). Factors affecting the social networks acceptance: an empirical study using PLS-SEM approach. In *Proceedings of the 2019c 8th International Conference on Software and Computer Applications* (pp. 414–418).

Alshurideh, M., Al Kurdi, B., Salloum, S. A., Arpaci, I., & Al-Emran, M. (2020a). Predicting the actual use of m-learning systems: A comparative approach using PLS-SEM and machine learning algorithms. *Interactive Learning Environments.* https://doi.org/10.1080/10494820.2020.1826982

Alshurideh, M., Gasaymeh, A., Ahmed, G., Alzoubi, H., & Kurd, B. A. (2020b). Loyalty program effectiveness: Theoretical reviews and practical proofs. *Uncertain Supply Chain Management, 8*(3). https://doi.org/10.5267/j.uscm.2020.2.003

Alshurideh, M.T., Al Kurdi, B., AlHamad, A. Q., Salloum, S. A., Alkurdi, S., Dehghan, A., Abuhashesh, M., & Masa'deh, R. (2021a). Factors affecting the use of smart mobile examination platforms by universities' postgraduate students during the COVID-19 pandemic: An empirical study. *Informatics, 8*(2). https://doi.org/10.3390/informatics8020032

Alshurideh, M. T, Hassanien, A. E., & Masa'deh, R. (2021b). *The Effect of Coronavirus Disease (COVID-19) on Business Intelligence.* Springer.

Alshurideh, M. (2022a). Does electronic customer relationship management (E-CRM) affect service quality at private hospitals in Jordan? *Uncertain Supply Chain Management, 10*(2), 1–8.

Alshurideh, M. T, Al Kurdi, B., Alzoubi, H. M., Ghazal, T. M., Said, R. A., AlHamad, A. Q., Hamadneh, S., Sahawneh, N., & Al-kassem, A. H. (2022b). Fuzzy assisted human resource management for supply chain management issues. *Annals of Operations Research,* 1–19.

Alsuwaidi, S. R., Alshurideh, M., Al Kurdi, B., & Aburayya, A. (2021). The main catalysts for collaborative R&D projects in Dubai industrial sector. In *The International Conference on Artificial Intelligence and Computer Vision* (pp. 795–806).

Alzoubi, A. (2021a). The impact of process quality and quality control on organizational competitiveness at 5-star hotels in Dubai. *International Journal of Technology, Innovation and Management (IJTIM), 1*(1), 54–68. https://doi.org/10.54489/ijtim.v1i1.14

Alzoubi, A. (2021b). Renewable Green hydrogen energy impact on sustainability performance. *International Journal of Computations, Information and Manufacturing (IJCIM), 1*(1), 94–110. https://doi.org/10.54489/ijcim.v1i1.46

Alzoubi, H. M., & Aziz, R. (2021). Does emotional intelligence contribute to quality of strategic decisions? The mediating role of open innovation. *Journal of Open Innovation: Technology, Market, and Complexity, 7*(2), 130. https://doi.org/10.3390/joitmc7020130

Alzoubi, H. M., Vij, M., Vij, A., & Hanaysha, J. R. (2021). What leads guests to satisfaction and loyalty in UAE five-star hotels? AHP analysis to service quality dimensions. *Enlightening Tourism, 11*(1), 102–135. https://doi.org/10.33776/et.v11i1.5056

Alzoubi, H. M., & Yanamandra, R. (2020). Investigating the mediating role of information sharing strategy on agile supply chain. *Uncertain Supply Chain Management, 8*(2), 273–284. https://doi.org/10.5267/j.uscm.2019.12.004

Alzoubi, H., Ahmed, G., Al-Gasaymeh, A., & Kurdi, B. (2020a). Empirical study on sustainable supply chain strategies and its impact on competitive priorities: The mediating role of supply chain collaboration. *Management Science Letters, 10*(3), 703–708.

Alzoubi, H., Alshurideh, M., Kurdi, B., Akour, I., & Aziz, R. (2022). Does BLE technology contribute towards improving marketing strategies, customers' satisfaction and loyalty? The role of open innovation. *International Journal of Data and Network Science, 6*(2), 449–460.

Alzoubi, H., & Ahmed, G. (2019). Do TQM practices improve organisational success? A case study of electronics industry in the UAE. *International Journal of Economics and Business Research, 17*(4), 459–472. https://doi.org/10.1504/IJEBR.2019.099975

Alzoubi, H., Alshurideh, M., Kurdi, B. A., & Inairat, M. (2020b). Do perceived service value, quality, price fairness and service recovery shape customer satisfaction and delight? A practical study in the service telecommunication context. *Uncertain Supply Chain Management, 8*(3), 579–588. https://doi.org/10.5267/j.uscm.2020.2.005

Aziz, N., & Aftab, S. (2021). Data mining framework for nutrition ranking methodology: SPSS Modeller. *International Journal of Technology, Innovation and Management (IJTIM), 1*(1), 85–95.

Ben-Abdallah, R., Shamout, M., & Alshurideh, M. (2022). Business development strategy model using EFE, IFE and IE analysis in a high-tech company: An empirical study. *Academy of Strategic Management Journal, 21*(Special Issue 2), 1–9.

Cruz, A. (2021). Convergence between blockchain and the internet of things. *International Journal of Technology, Innovation and Management (IJTIM), 1*(1), 35–56.

Eli, T. (2021). studentsperspectives on the use of innovative and interactive teaching methods at the University of Nouakchott Al Aasriya, Mauritania: English department as a case study. *International Journal of Technology, Innovation and Management (IJTIM), 1*(2), 90–104.

Farouk, M. (2021). The universal artificial intelligence efforts to face coronavirus Covid-19. *International Journal of Computations, Information and Manufacturing (IJCIM), 1*(1), 77–93. https://doi.org/10.54489/ijcim.v1i1.47

Firoozye, N., & Ariff, F. (2016). *Managing uncertainty, mitigating risk: Tackling the unknown in financial risk assessment and decision making*. Springer.

Ghazal, T. M., Hasan, M. K., Alshurideh, M. T., Alzoubi, H. M., Ahmad, M., Akbar, S. S., Al, B., & Akour, I. A. (2021). IoT for smart cities: Machine learning approaches in smart healthcare—a review. *Future Internet, 13*(8), 218. https://doi.org/10.3390/fi13080218

Guergov, S., & Radwan, N. (2021). Blockchain convergence: Analysis of issues affecting IoT, AI and Blockchain. *International Journal of Computations, Information and Manufacturing (IJCIM), 1*(1), 1–17. https://doi.org/10.54489/ijcim.v1i1.48

Hamadneh, S., Pedersen, O., & Al Kurdi, B. (2021). An investigation of the role of supply chain visibility into the Scottish bood supply Chain. *Journal of Legal, Ethical and Regulatory Issues, 24*(Special Issue 1), 1–12.

Hanaysha, J. R., Al-Shaikh, M. E., Joghee, S., & Alzoubi, H. (2021a). Impact of innovation capabilities on business sustainability in small and medium enterprises. *FIIB Business Review*, 1–12,. https://doi.org/10.1177/23197145211042232

Hanaysha, J. R., Al Shaikh, M. E., & Alzoubi, H. M. (2021b). Importance of marketing mix elements in determining consumer purchase decision in the retail market. *International Journal of Service Science, Management, Engineering, and Technology (IJSSMET), 12*(6), 56–72.

Joghee, S., Alzoubi, H. M., & Dubey, A. R. (2020). Decisions effectiveness of FDI investment biases at real estate industry: Empirical evidence from Dubai smart city projects. *International Journal of Scientific and Technology Research, 9*(3), 3499–3503.

Kashif, A. A., Bakhtawar, B., Akhtar, A., Akhtar, S., Aziz, N., & Javeid, M. S. (2021). Treatment response prediction in Hepatitis C patients using machine learning techniques. *International Journal of Technology, Innovation and Management (IJTIM), 1*(2), 79–89. https://doi.org/10.54489/ijtim.v1i2.24

Khan, M. A. (2021). Challenges facing the application of IoT in medicine and healthcare. *International Journal of Computations, Information and Manufacturing (IJCIM), 1*(1), 39–55. https://doi.org/10.54489/ijcim.v1i1.32

Kim, S. D. (2012). *Characterizing unknown unknowns*.

Kinnvall, C., Manners, I., & Mitzen, J. (2020). *Ontological insecurity in the European Union*. Routledge.

Lee, C., & Ahmed, G. (2021). Improving IoT privacy, data protection and security concerns. *International Journal of Technology, Innovation and Management (IJTIM), 1*(1), 18–33. https://doi.org/10.54489/ijtim.v1i1.12

Lee, K., Azmi, N., Hanaysha, J., Alzoubi, H., & Alshurideh, M. (2022a). The effect of digital supply chain on organizational performance: An empirical study in Malaysia manufacturing industry. *Uncertain Supply Chain Management, 10*(2), 495–510.

Lee, K., Romzi, P., Hanaysha, J., Alzoubi, H., & Alshurideh, M. (2022b). Investigating the impact of benefits and challenges of IOT adoption on supply chain performance and organizational performance: An empirical study in Malaysia. *Uncertain Supply Chain Management, 10*(2), 537–550.

Leo, S., Alsharari, N. M., Abbas, J., & Alshurideh, M. T. (2021). From offline to online learning: A qualitative study of challenges and opportunities as a response to the Covid-19 pandemic in the UAE higher education context. In *Studies in Systems, Decision and Control* (Vol. 334). https://doi.org/10.1007/978-3-030-67151-8_12

Mehmood, T. (2021). Does information technology competencies and fleet management practices lead to effective service delivery? Empirical evidence from e-commerce industry. *International Journal of Technology, Innovation and Management (IJTIM), 1*(2), 14–41.

Mehmood, T., Alzoubi, H. M., & Ahmed, G. (2019). Schumpeterian entrepreneurship theory: Evolution and relevance. *Academy of Entrepreneurship Journal, 25*(4).

Miller, D. (2021). The best practice of teach computer science students to use paper prototyping. *International Journal of Technology, Innovation and Management (IJTIM), 1*(2), 42–63. https://doi.org/10.54489/ijtim.v1i2.17

Mondol, E. P. (2021). The impact of block chain and smart inventory system on supply chain performance at retail industry. *International Journal of Computations, Information and Manufacturing (IJCIM), 1*(1), 56–76. https://doi.org/10.54489/ijcim.v1i1.30

Nuseir, M. T., Aljumah, A., & Alshurideh, M. T. (2021). How the business intelligence in the new startup performance in UAE during COVID-19: The mediating role of innovativeness. In *Studies in Systems, Decision and Control* (Vol. 334). https://doi.org/10.1007/978-3-030-67151-8_4

Obaid, A. J. (2021). Assessment of smart home assistants as an IoT. *International Journal of Computations, Information and Manufacturing (IJCIM), 1*(1), 18–36. https://doi.org/10.54489/ijcim.v1i1.34

Odeh, R., Obeidat, B. Y., Jaradat, M. O., Masa'deh, R., & Alshurideh, M. T. (2021). The transformational leadership role in achieving organizational resilience through adaptive cultures: The case of Dubai service sector. *International Journal of Productivity and Performance Management.* https://doi.org/10.1108/IJPPM-02-2021-0093

Radwan, N., & Farouk, M. (2021). The growth of internet of things (IoT) in the management of healthcare issues and healthcare policy development. *International Journal of Technology, Innovation and Management (IJTIM), 1*(1), 69–84. https://doi.org/10.54489/ijtim.v1i1.8

Rubinstein, C. (2020). *Beware: Remote work involves these 3 cyber security risks.*

Salloum, S. A., Alshurideh, M., Elnagar, A., & Shaalan, K. (2020). Machine learning and deep learning techniques for cybersecurity: A review. In *Advances in Intelligent Systems and Computing* (Vol. 1153) AISC. https://doi.org/10.1007/978-3-030-44289-7_5

Segal, S. (2011). *Corporate value of Enterprise risk management: The next step in business management*. John Wiley & Sons.

Sultan, R. A., Alqallaf, A. K., Alzarooni, S. A., Alrahma, N. H., AlAli, M. A., & Alshurideh, M. T. (2021). How students influence faculty satisfaction with online courses and do the age of faculty matter. In *The International Conference on Artificial Intelligence and Computer Vision* (pp. 823–837).

Sweiss, N., Obeidat, Z. M., Al-Dweeri, R. M., Ahmad, M. K., & A., M. Obeidat, A., & Alshurideh, M. (2021). The moderating role of perceived company effort in mitigating customer misconduct within Online Brand Communities (OBC). *Journal of Marketing Communications*, 1–24,. https://doi.org/10.1080/13527266.2021.1931942

Tariq, E., Alshurideh, M., Akour, I., & Al-Hawary, S. (2022). The effect of digital marketing capabilities on organizational ambidexterity of the information technology sector. *International Journal of Data and Network Science, 6*(2), 401–408.

Taryam, M., Alawadhi, D., Aburayya, A., Albaqa'een, A., Alfarsi, A., Makki, I., Rahmani, N., Alshurideh, M., & Salloum, S. A. (2020). Effectiveness of not quarantining passengers after having a negative COVID-19 PCR test at arrival to dubai airports. *Systematic Reviews in Pharmacy*, *11*(11). https://doi.org/10.31838/srp.2020.11.197

Tysiac, K. (2012). *Five emerging strategies for coping with unknown risks.*

Impact of Remote Work on Project Risks Management: Focus on Unknown Risks

Mounir El khatib, Ahmed Al-Nakeeb, Abdulla Alketbi, Ayesha Al Hashemi, Fatma Mustafawi, Roudha Almansoori, Shamma Alteneiji, Haitham M. Alzoubi, and Muhammad Alshurideh

Abstract Unidentified risks, also called unknown unknowns, have been outside the scope of organization risk management for a long time. Many leaders believe uncertainties are difficult to find or imagine in advance. For instance, when COVID-19 spread quickly to different parts of the world in early 2020, the sudden shift of tens of millions of workers from on-site to remote work locations was experienced across the globe. Many businesses sent their staff to work at home, and most enterprises adopted technology to maintain their operations. Remote work has been aided by new technological platforms, including virtual private networks (VPNs), voice over Internet protocols (VoIPs), virtual meetings, and other collaboration tools. During this COVID-19 pandemic, these technologies have reduced the spread of the virus, enabled employers to decrease overhead expenses, helped workers to gain flexibility, and abridged overall organization risks. This study gathers critical information about remote work amidst unknown risks, such as COVID-19 pandemic, by relying on qualitative and quantitative data. The study develops a model to characterize unidentified risks by gathering employees' perspectives on remote working. Data collected will be compared in terms of the benefits and challenges of remote work. The findings and discussion will be used to suggest and recommend ways organizations can rely on remote work to improve performance and mitigate sudden, unknown risks.

M. El khatib · A. Al-Nakeeb · A. Alketbi · A. Al Hashemi · F. Mustafawi · R. Almansoori · S. Alteneiji
Hamdan Bin Mohamad Smart University, Dubai, UAE

H. M. Alzoubi (✉)
School of Business, Skyline University College, Sharjah, UAE
e-mail: haitham.alzubi@skylineuniversity.ac.ae

M. Alshurideh
Department of Marketing, School of Business, University of Jordan, Amman, Jordan
e-mail: m.alshurideh@ju.edu.jo; malshurideh@sharjah.ac.ae

Department of Management, College of Business Administration, University of Sharjah, Sharjah, UAE

M. Alshurideh et al. (eds.), *The Effect of Information Technology on Business and Marketing Intelligence Systems*, Studies in Computational Intelligence 1056,
https://doi.org/10.1007/978-3-031-12382-5_40

Keywords Remote work · Unknown risks · Risk mitigation · Covid-19 · Project risk management

1 Introduction

With many risks yet unknown, organizations' ability to conduct business as usual will constantly be tested. Finding many unimaginable risks have been beyond the scope of project risk management in various institutions. The outbreak of coronavirus disease in 2019 revealed that employees can significantly experience disruptions in their work environments. As soon as the COVID-19 pandemic was announced by the World Health Organization and many countries outlined containment measures, organizations turned to technologies to help meet employees' needs and manage workloads remotely. Many people turned to remote work environments to a scale never experienced before. For example, as of mid-February, nearly 4.7 million employees in the United States worked remotely (PWC, 2021). When we mention remote work, various software and solution are included "such as video conference tools "Zoom, Cisco Webex, and Google Meet", enterprise collaboration tools such as Microsoft Teams, Slack, and Workplace from Facebook, and ephemeral messaging applications such as Snapchat, WhatsApp, Telegram, and Signal". Containment measures, such as social distancing, increased these numbers from 33% in February to 61% in March (PWC, 2021). The change of work environment also demanded a transformation of skillsets and attitudes for both organizations and workers. Remote working through technologies, such as VPNs, VoIPs, or cloud technology, requires certain skills to adopt (Xio & Fan, 2020). With unknown unknowns, organizations struggle to balance between security and flexibility for remote work employees. As a result, the role of technologies in mitigating unidentified risks has emerged as one of the critical concerns for organizational leaders. Many employees have embraced remote work because it provides flexibility and eliminates commute time. However, there have also been negative impacts of such arrangements on productivity, creativity, information security and privacy, loneliness, and lack of work-life balance (Xio & Fan, 2020). To learn comprehensively about the impact of technology-aided remote work on the productivity of employees during unidentified uncertainties, studies have turned to remote work benefits, challenges, and best practices (Ali et al., 2021). Therefore, remote work, with the help of technologies, can mitigate unknown risks if they can help organizations to visualize the real-time state of the workforce to make informed decisions, assess and bridge security gaps in a remote environment, and secure distributed work models for the long-term.

2 Literature Review

Unknown unknown risks are considered to be very rare in the IT industry but once they have been found either intentionally or non-intentionally, they could be mitigated because they might harm the system of the organization (Hamadneh et al., 2021). From the perspective of risk management, the unknown risks are not acceptable. The individual has the advantage that from the known risks they could easily make multiple solutions.

Considering the example of known risks, the current situation of the world, millions of workers are staying home due to COVID-19 (Wressell et al., 2018; Aljumah et al., 2021a, b; Alshurideh et al., 2019a, b, 2021; Harahsheh et al., 2021). Somehow, some of the industries allow their workers to work from home due to the fact that those organizations want to continue their operations. Therefore, it could be easily said that COVID-19 has built the remote working infrastructure for multiple businesses (Ahmad et al., 2021; Alameeri et al. 2021a, b; Leo et al., 2021; Nuseir et al., 2021a, b). The remote working is an attractive benefit for the employees as it provides ease for the workers to work remotely (Al Kurdi et al., 2020; Alameeri et al., 2021a, b; Kurdi et al., 2020). But at the same time, it presents some of the security challenges which are hard to break (AlShehhi et al., 2020; Alsuwaidi et al., 2021; Hanaysha et al., 2021a, b). It also facilitates a decentralized team in which the security will be top of the mind which is concerned with the IT department which does not increase with the VPN capacities, beefing up the safety protocols which ensure remote teams follow practices.

2.1 Strategic Plan to Mitigate Risks Associated with Risks

As per the study conducted by Rapp et al. (2020), there is a strategic plan in order to mitigate the unknown risks from working remotely that organizations should follow to increase their productivity insecure environment (Alnuaimi et al., 2021). The strategic plan for mitigating the risk is to provide access for employees to access various software applications that would manage and facilitate them to work remotely (Alshraideh et al., 2017; Alshurideh, 2022; Ben-Abdallah et al., 2022). It includes accommodation to employees to enhance the remote VPN workers. From the study conducted by Wressell et al. (2018), it has been witnessed that globally, the use of corporate VPNs is increasing. OpenVPN licenses could be purchased at the price of each connected device (Alhamad et al., 2021). However, as Ben Popper of Stack Overflow pointed out, for many organizations, their VPN infrastructure is not built to manage all internal organizations, so the need to speed up may be a challenge. It requires infrastructure and at the same time, SC Magazine recommends working hours to distribute with existing VPN functions.

In this same consideration, the study by Rapp et al. (2020) also recommends the organizations that they should ensure to keep the devices patched and up to date. Considering this strategic method, a study by Yeboah-Ofori and Opoku-Akyea (2019), addresses that the National Cyber Security Center's government guidelines on remote services urge IT, workers, to ensure that employees understand the importance of keeping software (and their equipment) up-to-date and should know how to perform the best from this equipment which ultimately enhance the level of productivity (Joghee et al., 2020). Besides, the other strategic option is to boost security awareness with mandatory training. According to the study conducted by Wressell et al. (2018), 45% of employers have not received safety training from their employers. Consumer safety training can help them understand how to prevent phishing attacks and other frauds. The same study found when organizations invest in cybersecurity training; they can reduce security threats by 70%. A short training will encourage employees to be vigilant and avoid risky behaviors, such as clicking unfamiliar links and accidentally deleting infected documents. The other way to mitigate the risks is the encouragement of the digital basic hygiene (Hanaysha et al., 2021a, b). This simply means that the individual is using a fast-paced security system to reduce online threats to the individual and its business.

The other strategic method to reduce the remote risks is to become pro-active in risk management. In this manner, the study conducted by Yeboah-Ofori and Opoku-Akyea (2019), states that it is important for the individual to follow and apply Murphy's Law and think that anything that might go wrong will be damaged. This may not be the time to invest or implement a comprehensive risk management system, but it is still important to formulate policies and strategies to minimize risks (Alnazer et al., 2017). According to Rapp et al. (2020), although each organization has its strategy, an important part of this process is clear and consistent in this activity so that it can cover all employees, outsiders, or others. In addition to this, the other strategic option to mitigate the security risks is the usage of the cloud software solutions which is concerned with the file management (Ghazal et al., 2021).

2.2 *Methods to Mitigating Risks*

2.2.1 Cloud-Based Environment

The unknown risk in this consideration is also the cloud-based environment risk which highlights that the remote workers often need technology solutions that rely heavily on the cloud. The use of the cloud can save the cost of implementing customized projects for key users; however, cloud computing carries its own set of network security threats (Ali et al., 2022). In such a way, it has third party control issue which is initiated due to the usage of multiple applications via adding a vector for the cyber security risk practices which sometimes become out of control. This means that the third party will initiate the intentionally hardware issue and breached it malware which could be potentially vulnerable to control (Yeboah-Ofori & Opoku-Akyea, 2019). In cloud-based environment the users also face other issues such as increased the potential for the risk breaches, this occurs when the information is presented by the third party (Alshurideh et al., 2020). In addition to this, the third issue faced by the remote workers is the account hijacking and insider threats which brought multiple hindrances in the work which is related to the cyber security.

2.2.2 Cloud Access Security Brokers (CASBs) and Protection Tools

CASBs such as MVISION Cloud, Bitglass and Microsoft Cloud App Security are tools or services that work as part of the gateway infrastructure of third-party cloud providers, thereby achieving security and early management. The study conducted by Firoozye and Ariff (2016), addresses that the individuals should use a third-party cloud. All CASBs provide network and firewall, test equipment, and data loss prevention functions to prevent the transmission of sensitive data outside authorized channels (Mehmood et al., 2019). The Cloud Access Security Foundation (CASB) is the state or entry point of a cloud security strategy that is set between the cloud service provider and the cloud provider to integrate the two (Alshurideh et al., 2022). Enter the security policy strategy as security. Get resources from the cloud. CASB brings together various security policy frameworks Examples of additional security include authentication, group identification, authorization, credentials, device profiles, encryption, tokenization, registration, input, and detection/prevention (Radwan & Farouk, 2021). CASB ensures that the network traffic between the local device and the cloud provider complies with the organization's security policy. According to Zhang et al. (2018), the user's advantage in protecting cloud access comes from its ability to provide information about the use of cloud applications and unauthorized use on the cloud platform. This is very important in the management industry.

As per Firoozye and Ariff (2016), CASB performs automatic detection to detect cloud applications, and then uses them to identify high-risk, high-risk users and other high-risk sources. Cloud recipients can apply administrative access rights to various securities, including encryption and authentication (Aziz & Aftab, 2021). When they are not logged in, they can also provide other services, such as certificate mapping. In addition to knowledge-based security, there must be a safe security tool system to prevent data leakage caused by human errors: firewalls, VPNs, security software, and all anti-malware software that plays a role in protecting security.

3 Research Methods

To understand more on how we can create a resilient society after the effects of the unknown risk, we decide to conduct a research on one of the current technological trends which is remote working and how did it mitigate and manage risks aspects. As this tool became a norm in today's world, a world that has been hugely altered by the pandemic. A pandemic that created an atmosphere of fear and unpredictable because the world has not expectation or knowledge of what's to coming. Using COVID-19, as an example it has been ranked for both known and unknown- unknown risk, known risk because organizations are aware of it and how dangerous it can be on a person's health since it spreads easily and there is no known cure for it yet (Lee et al., 2022a, b). It is an unknown risk because organizations do not know the extent of it and are unaware of the size and effects that this risk will have on their organization's performance and future.

The aim of our research was to get a better understanding of what remote working has done to organizations, has it helped the organizations to grow in facing risks or helped them to stabilize or did it led to their failure? Is remote working the way to go post-COVID-19? To gain this understanding, we looked at both qualitative and quantitative data. The benefits of qualitative research are that it provides an insight of the social reality that the subject is living in at a certain point of time. As Denzin and Lincoln (1994) have stated that qualitative research is a multi-method in focus, attempting to make sense of, or interpret, phenomena in terms of the meanings people bring to them. The benefit of incorporating quantitative research is that it provides statistical data which can be compared and analyzed, thus presenting straightforward findings (Cruz, 2021). Combing both qualitative and quantitative research helps improving the reliability of the data because both of these data balance out any data limitations. In our research for the qualitative research we have conduct interviews, one interview per member group, so 5 interviews in total to these specific companies:

National Search and Rescue Center: organization cooperate with federal authorities in all aid and emergency processes.

Emirates Central Cooling Systems Corporation (EMPOWER): largest district cooling Service Provider in the Middle East.

Emirates Airlines: one of the biggest most efficient Airlines organizations in the world.

Zayed University: Educational channel.

Emergency Crisis and disaster Operational Center: organization mitigate unknown risk as one of their daily core tasks.

Interviews are a great method to use to understand the situation better and explore the interviewee's, the research subject, opinion, experience and behavior around the research topic which remote working as for the quantitative research we have looked at secondary data because it is time effective since we has limited time for the research (Lee et al., 2022a, b). Another benefit of using secondary data is that has allowed us to evaluate and analyze a high volume of data in a short amount of time, to create a more rounded understanding on how remote working has led to the mitigation of unknown risk in our pandemic reality.

The interview questions that we used were the same questions that the professor has shared. It had a total of 20 questions, divided in to 3 sections which are general and background information about the employee and organization they work in, the second section is the main body questions that revolve around risk produced and risk methodology of the organization in questions. The third section is all about the strengths and weakness of the organization (AlHamad et al., 2022). The third section is different than the other sessions, in that it has 3 open-ended questions, this type of question allows the interviewee's to include more information and express their opinions and feelings more openly (Lee & Ahmed, 2021). The secondary data we used were surveys that were conducted both locally and internationally has a response and evolution on how remote working has helped in reducing the unknown risk. The analysis method that we used to analyze the interviews was looking at the language that was used by the interview (Shamout et al., 2022). Also look for patterns and themes within the 5 interviews. The analysis of the secondary data was based on comparing the tables and chart with each other and interprets them into words to link them to our primary data.

4 Analysis of Results

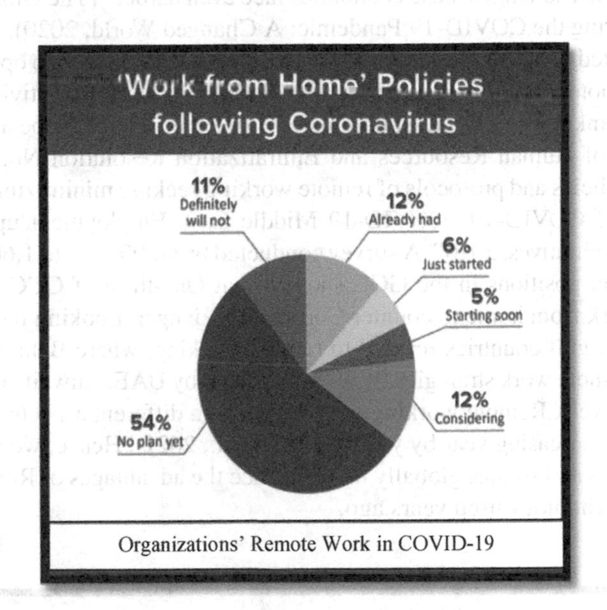

Organizations' Remote Work in COVID-19

4.1 Quantitative Method Analysis

Several critical questions were asked such as *does Remote Work mitigate any kind of risk including knowns and unknowns and is that measurable? Does it really mitigate unknown risk?* As we noticed, related to the last period, organizations dramatically expand adapting remote work in their strategy. However, the statistic showed how business in the GCC and internationally moved to use remote working in various risk cases from the secondary resources we collected, where these resources gathered statistic evidence how companies shift their system to remote work directly and succeed mitigate the pandemic risk (Al Ali, 2021). The most known case that organization measures their unknown risk mitigation system nowadays is "COVID-19" because it has been the current case situation labeled as unknown risk; accordingly, the world is still suffering from, Thus after COVID-19 announcement in the world a study was done to indicate remote work implementation in organizations. As the above chart clarify that 35% of the organizations are raising remote work in their business plan process (Alzoubi et al., 2022). Therefore, these GCC business organizations are processing the accommodation of remote working, where a big percentage didn't plan since the beginning to enhance this method, which made them weakened their business cycle process and face several issues, which is one of the reason that affected economic growth nowadays "Many emerging and developing economies were already experiencing weaker growth before this crisis; the shock of COVID-19 now makes the challenges these economies face even harder" (The Global Economic Outlook During the COVID-19 Pandemic: A Changed World, 2020). Many organizations adapted and bought systems to build new IT infrastructure to operate a valued usage of remote work for the sake of controlling their business activities under the unexpected unknown risks (Kashif et al., 2021). Especially after the announcement by ministry of Human Resources and Emiratization Resolution No. 281 of 2020 regarding policies and protocols of remote working seeking minimizing and limiting the spread of COVID-19 "COVID-19 Middle East: Employment update on new government initiatives, 2020". A survey conducted by GulfTalent to 1,600 employees from different positions in the GCC showed that One-third of GCC organizations arranged work from home to counter Coronavirus danger. Looking more closely on how different gulf countries reacted to remote working, where Bahrain formed the top rate of remote work strategies by 38%, followed by UAE, Kuwait, and Qatar with 37% respectively. Remote working got high rates in different aspects; numbers and growth were increasing year by year (Akhtar et al., 2021). Hence, we will represent some numbers and studies globally that evidence the advantages of Remote working as modern technology used years ago.

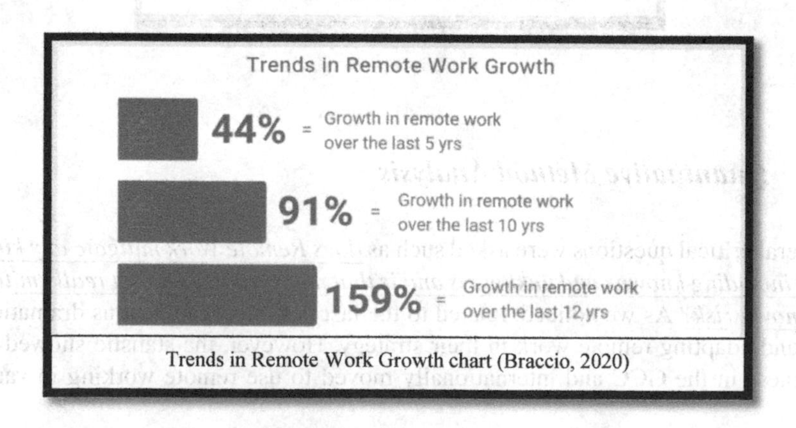

Trends in Remote Work Growth chart (Braccio, 2020)

The chart clarify how the trend of remote work is expanding year by year as people got more knowledge wise of its significance benefits and gains that is originating to both party organizations and employees (Alzoubi & Aziz, 2021). Depending on the result obtained in this chart, a huge increased rate of remote work growth has been noted; the growth rate over the last 12 years has increased dramatically to reach a 159% which means that people are looking forward and having a great business continuity plan included in their entities strategy. Moreover *Is it measurable,* productivity is a good measurement to judge on specific technology and here comes "FlexJobs' annual survey which found that 65% of respondents are more productive in their home office than at a traditional workplace" (Braccio, 2020), which is a big rate that can assure employee productivity off-site, which lead us to determine that employees have the ability, equipment, knowledge and skills to perform outside the office under different risks and circumstances.

4.2 Qualitative Method Analysis

The interviews we conducted has clarified that most organizations in the UAE tend to use remote work today to mitigate the pandemic risk. Few of them stopped in the UAE after the announcement of allowing employee perusing work in their office, were others added remote work in their new strategies to react for any unknown and known risk that can occur suddenly in the future. Organizations' Business Continuous Plans updated the remote work as a must method to mitigate coming risks (Alzoubi et al., 2021). As Ayesha Al-Hashemi mentioned in her interview Empower as UAE organization, had the plan to prepare new physical place in case any thing happened to the main work office. Where what happened actually was moving in the path of adapting remote work suddenly because it is more efficient and effective in facing the unknown risk that organization faced. Mr. Murugesan Vasanthan the director in Empower mentioned that business plan should be updated yearly and restructures more details aspects, which seem to be a strength that affects business process on adapting efficient initiatives such as remote work (Eli, 2021). That statement support experts' recommendations mentioned in the literature review in updating strategies methods and devices to enhance better risk plan mitigation. Furthermore, Dr. Amina Ali Abdulla the Risk Management and Business Continuity Project Manager from MOH recommended in her interview to adapt the implementation of remote work forever because of its usefulness (Alsharari, 2021). As she stated in her interview that today we can realize that remote work technology would proceed to what's behind COVID-19, where the GulfTalent, 2020 supported that by stating that "75% of current teleworkers say they plan to work remotely for the rest of their career" "GulfTalent, 2020". Besides, Maryam AlMarar from Zayed University stated that in this period everyone recognized how remote work had the value and the power to stand against unexpected threat arising anytime (Mehmood, 2021). So far we could value remote work and empower it to mitigate risks; yet, its importance was not stressed clear enough before facing previous unknown risks (Al Alzoubi et al., 2020a, b; Awadhi et al., 2021; Al.Khayyal et al., 2021). Emirates Airline merged the use of remote work and artificial intelligence in their process as they supported that by stating "we ensure to use latest IT infrastructure and secure technology to exchange the viable information securely without any breach or delay" (Miller, 2021). One of the most disappointing things we knew was that the National Search and Rescue Center; one of the most important organization are not using remote work benefits and advantages. Especially that they deal with unknown risk in a daily base.

Remote work seems to be a must today, not a method that may be implemented one day, because obviously that day came and everyone had to expose their planning effort and preparedness (Al Kurdi et al., 2021; Alshurideh et al. 2019a, b; Khan, 2021). Today, it isn't only about flexibility and employee satisfaction only by making them work from home which remote work can offer; it is more of keeping the business cycle going by applying one of the greatest effective methods that mitigates unknown risk (Agha et al., 2021; Al Batayneh et al., 2021; Al Shebli et al., 2021; AlMehrzi et al., 2020; Alyammahi et al., 2020; Nuseir et al., 2021a, b).

4.3 Challenges

Challenges that occurred to new remote work implemented strategy could be avoided in case of trainings presence to employees occurred. Maryam AlMarar mentioned the difficulties rose in adapting new systems, where they were not familiar with, which could be avoided if proper training courses were given to employees (Mondol, 2021). It is surprising how big developed entities were eliminating remote work from their plan; especially that it is not a new aspect that was introduced few years ago. E-government services were developed in our current region to support the idea of performing and producing through using technologies and internet. Wressell et al. (2018) stated that employees lack of safety protocols and cybersecurity knowledge increase threats of frauds and scams (Alzoubi & Ahmed, 2019). Challenges and difficulties rise with any new initiative; remote work has its own as well. There is no doubt of major problems that this tool can derive with, but in the same time organizations can't avoid enhancing advantages and values this method that can arise because of their fair of future coming problems. Somehow, remote work can be one of the best solutions to the coming future unknown risks.

A study was done in Doist, an organization that works fully remotely where it creates Applications. This study was indicted to clarify challenges alongside remote work. Communication, unplugging and loneliness were the major issues that face employees in their production by 58%, which is big number (Guergov & Radwan, 2021). Amir Salihefendic, the organization CEO mentioned that these are issues to be considers where it effects on employee performance and motivation toward producing, where people do not take into consideration the less seen side that result in really complex issue "State of Remote Work 2019, 2020".

Literature review as well, mentioned challenges such as cyber and security matters that can affect organizations and their privacy (Farouk, 2021). The loss in communication is a serious issue that can raise other misunderstanding in sharing information, delivering news and knowledge limitations (Obaid, 2021).

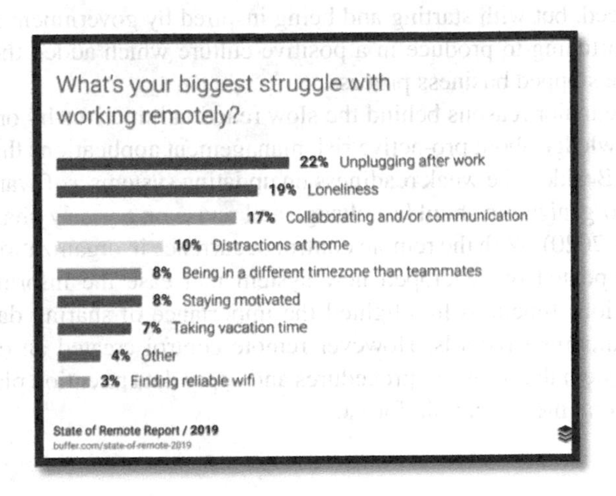

5 Discussion

Experts expected that remote work will be taken more seriously even after COVID-19 crises. As business owners, CEOs, managers and employees went through it, got the needed training and adapted new working lifestyle that can be used anytime and every time (Alzoubi et al., 2020a, b). Thus, the importance and advantages this technology derives tend to become a major way of mitigating unknown risk. Introducing a change and training employee to accept it may be difficult and takes time, while the case was different this time, employees and organizations had to collaborate in making remote working a success tool to overcome this unknown risk.

Organizations that adapted Remote Work way age, had the ability to develop, fix and prepare their technology for any coming risks. However, others who just listed Remote Work in the "In Process to Implement One day" faced the bigger issue. I believed that we were step behind mitigating unknown risk when we had the Remote Work strategy to be considered and most of the organizations did not take it seriously. Studies were done years ago to highlight importance of this strategy in proceeding work and how it derives many benefits any organization can achieve. However, the fear of failure is the most common issue that builds organization limitations in adapting a new technology, where it is not that new actually going back to Remote Work establishment. As we go over the literature review, we can view ways, solutions and issues that this tool generates. As employees who worked remotely just with the

COVID-19 presence, were experienced a modern working system and succeed in. Being able to reach work from home increase our ability to perform and flexibility to exchange information. Pros and Cons of remote work vary in different aspects and as per experience technologies has both good and bad sides. The resistance of adapting new way showed in the beginning merging the fair of the unknown risk that we experienced, but with starting and being inspired by government statements we had the opportuning to produce in a positive culture which added the productivity instead of the stopped business process.

One of the major reasons behind the slow reaction initiatives by organizations is the poor knowledge about pro-active risk management applications that this method can provide. Besides, the weak readiness on updating systems, software and training courses that organization should of change and revise on a yearly phase (Alzoubi & Yanamandra, 2020). With the remote control occurrence in organization, some structure where updated or developed new system that ease the information sharing system. Previous functions highlighted the importance of sharing data or keeping updates of important records. However remote control created an easier more e-functional system that simplify procedures and apply data retention plans that report faster ephemeral messaging platforms.

6 Matrix SWOT Analysis

	Strengths Support from the management Regular updates and check ins between the manager and employees Improve flexibility of employees work schedule Higher autonomy	**Weakness** Capacity Employees have low technological skills Over time the power of the employees will decrease with the increase stress of IT issues Lack of social contact Lack of research and preparation on remote working
Opportunities Improve risk management strategy New research and training course Improve work-life balance	**Strengths-Opportunities** Introduction of a new business model strategy that combines both remote working and onsite working (S1, S4, O1) New and increase method of communication to maintain a health work-life balance at home (S2, O3)	**Weakness-Opportunities** Invest more in training that focus on improve employee's IT skills and how to trouble shoot (W2, W3, O2) Regular check-ins with employee to check on their health, workload and get feedback. (W3) New researches and preparation of implementing remote work in the organization (W5, O2)

(continued)

(continued)

Threats	Strengths-Threats	Weakness-threats
Shortage of IT specialist Lack of shared introduction throughout the organization Unsecure network, thus increase phishing attacks	Hire more IT specialist that have experience with Risk management (S1, T1) Improve systems for the future (T3)	Low IT skills that might result in attacking organization's systems (W1, W2, T3)

7 Recommendations

Based on the literature review analysis and interviews conducted previously, we would recommend the governmental and private organizations the followings to improve role of remote work in risk mitigation.

7.1 Implement Remote Working

The organization should have a clear strategic plan in remote work and implement it precisely based on their criteria. Organizations can set specific procedures and techniques for implementing Remote Working to ensure the efficiency and productivity. There should be a clear strategy that adapts the Remote Working requirements to suit the nature of work and the job categories with emphasizing on the importance of committing to their duties and responsibilities. In addition, the organization should ensure that most of its services are available to its employees and customers through online communications, smart applications, etc. Moreover, learning from other entities or organization around the world about their experience with remote work in such circumstances will enhance the organization's dynamic work as well.

7.2 Remote Work and Business Continuity Planning

The pandemic of the COVID-19 was an imbalance of opportunities. Some organizations across the world had shut; other had failed in production whereas the rest did not even feel the change and the productivity were still at its same level. COVID-19 which is considered as unknown risk was the reason the organizations had to notify the importance of updating their business continuity plan. Essentially, the most effective elements of BCPs are the teleworking or working from home that the organizations were lacking them in their BCP's and strategic plans. Few organizations had the BCP's but were lacking their annual updating which resulted in not being able to enhance the productivity during the remote working (Alzoubi, 2021a).

Therefore, we would recommend all the organizations to follow the strategy that is followed by the Emergency Crisis and Disaster Operational Center (MOH) in which they are used to revise their BCP's annually even if nothing happened and update them when necessary with adding the new tools added around the world such as artificial intelligence, robots and other tools. By applying this strategy, the organizations will ensure the lower risks and harms possible in case of crises or risks appearance and will be able to maintain the organization's functions or resume them quickly. Whether the COVID-19 pandemic lasts for weeks, months or years, the organizations method of communicating with their capitals will be critically important for their continues success. In fact, the organizations that are more flexible and responsive to the modern BCP's will fare better in our current environment.

7.3 *Strategic Initiatives*

Based on the literature Review, other recommendations are preferred and prioritized to be implemented in every organization after the appearance of COVID-19 pandemic for their preparedness of any future unknown risks. The Strategic plan to mitigate risks associated with risks is to implement plans that would allow the employees to access the required software applications that would manage working remotely. This strategy includes the adaptation of employees to enhance the remote VPN workers. Another strategic plan is to become pro-active in risk management. Within this strategy it is important for the organization to follow and implement Murphy's Law to think in advance in case something might go wrong. Although each organization has its strategy, an important part of this process is clear and consistent in this activity so that it can cover all employees, outsiders, or others.

7.4 *Cloud Programs*

According on the literature Review mentioned above, other recommendations such as Cloud-Based Environment and Cloud Access Security Brokers (CASBs) and protection tools are chosen to implement. The Cloud-Based Environment saves the cost of implementing customized projects for the employees although we know that cloud computing carries its own set of network security threats such as having third party control issue which starts as a result of using multiple applications by adding cybersecurity risk practices which is sometimes out of control. Whilst the Cloud Access Security Brokers (CASBs) and protection tools such as Microsoft cloud are tools and services that work as part of the infrastructure of third-party cloud providers, they achieve security and early management to the organizations in order not to face any hackings (Alzoubi, 2021b). CASB brings together several security policy frameworks, such as authentication, group identification and authorization.

NSRC consider that one whole research should be done to the National Search and Rescue Center with strong recommendations and suggestion to implement remote work method.

8 Conclusion

To conclude with, individuals can work remotely in many ways. That is the magnificence of inaccessible work; individuals can select to work in a way that creates the foremost sense for their lives. Remote work is solving nowadays a big work issue due to the pandemic caused by covid-19 virus. In the future, we believe that most of the entities, government and companies around the world will have a separate remote work strategy to ensure the business continuity plans of any organization. This paper discussed remote work that mitigate unknown risk, provided strong literature review, implemented qualitative and quantitative research method, analyzed results of both methods and discussed them with providing recommendations and suggestions.

References

Agha, K., Alzoubi, H. M., & Alshurideh, M. T. (2021). Measuring reliability and validity instruments of technologically driven cognitive intrusion towards work-life balance. In *The International Conference on Artificial Intelligence and Computer Vision* (pp. 601–614).

Ahmad, A., Alshurideh, M. T., Al Kurdi, B. H., & Salloum, S. A. (2021). Factors impacts organization digital transformation and organization decision making during Covid19 pandemic. In *Studies in Systems, Decision and Control* (Vol. 334). https://doi.org/10.1007/978-3-030-67151-8_6

Akhtar, A., Akhtar, S., Bakhtawar, B., Kashif, A. A., Aziz, N., & Javeid, M. S. (2021). COVID-19 detection from CBC using machine learning techniques. *International Journal of Technology, Innovation and Management (IJTIM), 1*(2), 65–78. https://doi.org/10.54489/ijtim.v1i2.22

Al Ali, A. (2021). The impact of information sharing and quality assurance on customer service at UAE banking sector. *International Journal of Technology, Innovation and Management (IJTIM), 1*(1), 01–17. https://doi.org/10.54489/ijtim.v1i1.10

Al Batayneh, R. M., Taleb, N., Said, R. A., Alshurideh, M. T., Ghazal, T. M., & Alzoubi, H. M. (2021). IT governance framework and smart services integration for future development of dubai infrastructure utilizing AI and Big Data, its reflection on the citizens standard of living. In *The International Conference on Artificial Intelligence and Computer Vision* (pp. 235–247).

Al Khayyal, A. O., Alshurideh, M., Al Kurdi, B., & Salloum, S. A. (2021). Women empowerment in UAE: A systematic review. In *Advances in Intelligent Systems and Computing: Vol. 1261 AISC*. https://doi.org/10.1007/978-3-030-58669-0_66

Al Kurdi, B., Alshurideh, M., Nuseir, M., Aburayya, A., & Salloum, S. A. (2021). The effects of subjective norm on the intention to use social media networks: an exploratory study using PLS-SEM and machine learning approach. In *Advances in Intelligent Systems and Computing* (Vol. 1339). https://doi.org/10.1007/978-3-030-69717-4_55

Al Kurdi, B., Alshurideh, M., & Al Afaishata, T. (2020). Employee retention and organizational performance: Evidence from banking industry. *Management Science Letters, 10*(16), 3981–3990.

Al Shebli, K., Said, R. A., Taleb, N., Ghazal, T. M., Alshurideh, M. T., & Alzoubi, H. M. (2021). RTA's employees' perceptions toward the efficiency of artificial intelligence and big data utilization in providing smart services to the residents of Dubai. In *The International Conference on Artificial Intelligence and Computer Vision* (pp. 573–585).

Alameeri, K. A., Alshurideh, M. T., & Al Kurdi, B. (2021a). The Effect of Covid-19 pandemic on business systems' innovation and entrepreneurship and how to cope with it: A theatrical view. In *Studies in Systems, Decision and Control* (Vol. 334). https://doi.org/10.1007/978-3-030-67151-8_16

Alameeri, K., Alshurideh, M., Al Kurdi, B., & Salloum, S. A. (2021b). The Effect of Work Environment Happiness on Employee Leadership. In *Advances in Intelligent Systems and Computing: Vol. 1261 AISC*. https://doi.org/10.1007/978-3-030-58669-0_60

AlHamad, A., Alshurideh, M., Alomari, K., Kurdi, B. A., Alzoubi, H., Hamouche, S., & Al-Hawary, S. (2022). The effect of electronic human resources management on organizational health of telecommuni-cations companies in Jordan. *International Journal of Data and Network Science, 6*(2), 429–438. https://doi.org/10.5267/j.ijdns.2021.12.011

Alhamad, A. Q. M., Akour, I., Alshurideh, M., Al-Hamad, A. Q., Kurdi, B. A., & Alzoubi, H. (2021). Predicting the intention to use google glass: A comparative approach using machine learning models and PLS-SEM. *International Journal of Data and Network Science, 5*(3), 311–320. https://doi.org/10.5267/j.ijdns.2021.6.002

Ali, N., Ahmed, A., Anum, L., Ghazal, T. M., Abbas, S., Khan, M. A., Alzoubi, H. M., & Ahmad, M. (2021). Modelling supply chain information collaboration empowered with machine learning technique. *Intelligent Automation and Soft Computing, 30*(1), 243–257. https://doi.org/10.32604/iasc.2021.018983

Ali, N., M. Ghazal, T., Ahmed, A., Abbas, S., A. Khan, M., Alzoubi, H., Farooq, U., Ahmad, M., & Adnan Khan, M. (2022). Fusion-based supply chain collaboration using machine learning techniques. *Intelligent Automation & Soft Computing, 31*(3), 1671–1687. https://doi.org/10.32604/iasc.2022.019892

Aljumah, A., Nuseir, M. T., & Alshurideh, M. T. (2021). The impact of social media marketing communications on consumer response during the Covid-19: Does the brand equity of a university matter? in *Studies in Systems, Decision and Control* (Vol. 334). https://doi.org/10.1007/978-3-030-67151-8_21

AlMehrzi, A., Alshurideh, M., & Al Kurdi, B. (2020). Investigation of the key internal factors influencing knowledge management, employment, and organisational performance: A qualitative study of the UAE hospitality sector. *International Journal Innovation Creative Change, 14*(1), 1369–1394.

Alnazer, N. N., Alnuaimi, M. A., & Alzoubi, H. M. (2017). Analysing the appropriate cognitive styles and its effect on strategic innovation in Jordanian universities. *International Journal of Business Excellence, 13*(1), 127–140. https://doi.org/10.1504/IJBEX.2017.085799

Alnuaimi, M., Alzoubi, H. M., Ajelat, D., & Alzoubi, A. A. (2021). Towards intelligent organisations: An empirical investigation of learning orientation's role in technical innovation. *International Journal of Innovation and Learning, 29*(2), 207–221. https://doi.org/10.1504/IJIL.2021.112996

Alsharari, N. (2021). Integrating Blockchain Technology with Internet of things to Efficiency. *International Journal of Technology, Innovation and Management (IJTIM), 1*(2), 1–13.

AlShehhi, H., Alshurideh, M., Al Kurdi, B., & Salloum, S. A. (2020). The impact of ethical leadership on employees performance: A systematic review. *International Conference on Advanced Intelligent Systems and Informatics*, 417–426.

Alshraideh, A. T. R., Al-Lozi, M., & Alshurideh, M. T. (2017). The impact of training strategy on organizational loyalty via the mediating variables of organizational satisfaction and organizational performance: An empirical study on Jordanian agricultural credit corporation staff. *Journal of Social Sciences (COES&RJ-JSS), 6*(2), 383–394.

Alshurideh, M. (2022). Does electronic customer relationship management (E-CRM) affect service quality at private hospitals in Jordan? *Uncertain Supply Chain Management, 10*(2), 1–8.

Alshurideh, M., Salloum, S. A., Al Kurdi, B., Monem, A. A., & Shaalan, K. (2019a). Understanding the quality determinants that influence the intention to use the mobile learning platforms: A practical study. *International Journal of Interactive Mobile Technologies, 13*(11). https://doi.org/10.3991/ijim.v13i11.10300

Alshurideh, M., Salloum, S. A., Al Kurdi, B., & Al-Emran, M. (2019b). Factors affecting the social networks acceptance: an empirical study using PLS-SEM approach. In *Proceedings of the 2019b 8th International Conference on Software and Computer Applications* (pp. 414–418).

Alshurideh, M., Gasaymeh, A., Ahmed, G., Alzoubi, H., & Kurd, B. A. (2020). Loyalty program effectiveness: Theoretical reviews and practical proofs. *Uncertain Supply Chain Management, 8*(3). https://doi.org/10.5267/j.uscm.2020.2.003

Alshurideh, M.T., Al Kurdi, B., AlHamad, A. Q., Salloum, S. A., Alkurdi, S., Dehghan, A., Abuhashesh, M., & Masa'deh, R. (2021). Factors affecting the use of smart mobile examination platforms by universities' postgraduate students during the COVID-19 pandemic: An empirical study. *Informatics, 8*(2). https://doi.org/10.3390/informatics8020032

Alshurideh, M. T., Al Kurdi, B., Alzoubi, H. M., Ghazal, T. M., Said, R. A., AlHamad, A. Q., Hamadneh, S., Sahawneh, N., & Al-kassem, A. H. (2022). Fuzzy assisted human resource management for supply chain management issues. *Annals of Operations Research*, 1–19.

Alsuwaidi, M., Alshurideh, M., Al Kurdi, B., & Salloum, S. A. (2021). Performance appraisal on employees' motivation: A comprehensive analysis. In *Advances in Intelligent Systems and Computing: Vol. 1261 AISC*. https://doi.org/10.1007/978-3-030-58669-0_61

Alyammahi, A., Alshurideh, M., Kurdi, B. Al, & Salloum, S. A. (2020). The impacts of communication ethics on workplace decision making and productivity. In *International Conference on Advanced Intelligent Systems and Informatics* (pp. 488–500).

Alzoubi, A. (2021a). The impact of process quality and quality control on organizational competitiveness at 5-star hotels in Dubai. *International Journal of Technology, Innovation and Management (IJTIM), 1*(1), 54–68. https://doi.org/10.54489/ijtim.v1i1.14

Alzoubi, A. (2021b). Renewable green hydrogen energy impact on sustainability performance. *International Journal of Computations, Information and Manufacturing (IJCIM), 1*(1), 94–110. https://doi.org/10.54489/ijcim.v1i1.46

Alzoubi, H., & Ahmed, G. (2019). Do TQM practices improve organisational success? A case study of electronics industry in the UAE. *International Journal of Economics and Business Research, 17*(4), 459–472. https://doi.org/10.1504/IJEBR.2019.099975

Alzoubi, H., Alshurideh, M., Kurdi, B. A., Akour, I., & Azi, R. (2022). Does BLE technology contribute towards improving marketing strategies, customers' satisfaction and loyalty? The role of open innovation. *International Journal of Data and Network Science, 6*(2), 449–460. https://doi.org/10.5267/j.ijdns.2021.12.009

Alzoubi, H., Alshurideh, M., Kurdi, B. A., & Inairat, M. (2020a). Do perceived service value, quality, price fairness and service recovery shape customer satisfaction and delight? A practical study in the service telecommunication context. *Uncertain Supply Chain Management, 8*(3), 579–588. https://doi.org/10.5267/j.uscm.2020.2.005

Alzoubi, H. M., Ahmed, G., Al-Gasaymeh, A., & Al Kurdi, B. (2020b). Empirical study on sustainable supply chain strategies and its impact on competitive priorities: The mediating role of supply chain collaboration. *Management Science Letters, 10*(3), 703–708. https://doi.org/10.5267/j.msl.2019.9.008

Alzoubi, H. M., & Aziz, R. (2021). Does emotional intelligence contribute to quality of strategic decisions? *The Mediating Role of Open Innovation*. https://doi.org/10.3390/joitmc7020130

Alzoubi, H. M., Vij, M., Vij, A., & Hanaysha, J. R. (2021). What leads guests to satisfaction and loyalty in UAE five-star hotels? AHP analysis to service quality dimensions. *Enlightening Tourism, 11*(1), 102–135. https://doi.org/10.33776/et.v11i1.5056

Alzoubi, H. M., & Yanamandra, R. (2020). Investigating the mediating role of information sharing strategy on agile supply chain. *Uncertain Supply Chain Management, 8*(2), 273–284. https://doi.org/10.5267/j.uscm.2019.12.004

Awadhi, J., Obeidat, B., & Alshurideh, M. (2021). The impact of customer service digitalization on customer satisfaction: Evidence from telecommunication industry. *International Journal of Data and Network Science, 5*(4), 815–830.

Aziz, N., & Aftab, S. (2021). Data mining framework for nutrition ranking: methodology: SPSS modeller. *International Journal of Technology, Innovation and Management (IJTIM), 1*(1), 85–95.

Ben-Abdallah, R., Shamout, M., & Alshurideh, M. (2022). Business development strategy model using EFE, IFE and IE analysis in a high-tech company: An empirical study. *Academy of Strategic Management Journal, 21*(Special Issue 2), 1–9.

Cruz, A. (2021). Convergence between blockchain and the internet of things. *International Journal of Technology, Innovation and Management (IJTIM), 1*(1), 34–53.

Denzin, N, & Lincoln, A. (1994). *Lincoln.* Sage Publications Inc.

Eli, T. (2021). Students perspectives on the use of innovative and interactive teaching methods at the University of Nouakchott Al Aasriya, Mauritania: English department as a case study. *International Journal of Technology, Innovation and Management (IJTIM), 1*(2), 90–104.

Farouk, M. (2021). The Universal artificial intelligence efforts to face coronavirus Covid-19. *International Journal of Computations, Information and Manufacturing (IJCIM), 1*(1), 77–93. https://doi.org/10.54489/ijcim.v1i1.47

Firoozye, N., & Ariff, F. (2016). *Managing Uncertainty, Mitigating Risk: Tackling the Unknown in Financial Risk Assessment and Decision Making.* Springer.

Ghazal, T. M., Hasan, M. K., Alshurideh, M. T., Alzoubi, H. M., Ahmad, M., Akbar, S. S., Al Kurdi, B., & Akour, I. A. (2021). IoT for smart cities: Machine learning approaches in smart healthcare—a review. *Future Internet, 13*(8), 218. https://doi.org/10.3390/fi13080218

Guergov, S., & Radwan, N. (2021). Blockchain convergence: Analysis of issues affecting IoT, AI and blockchain. *International Journal of Computations, Information and Manufacturing (IJCIM), 1*(1), 1–17. https://doi.org/10.54489/ijcim.v1i1.48

Hamadneh, S., Pedersen, O., Alshurideh, M., Kurdi, B. Al, & Alzoubi, H. (2021). An investigation of the role of supply chain visibility into the Scottish blood supply chain. *Journal of Legal, Ethical and Regulatory Issues, 24*(Special Issue 1), 1–12.

Hanaysha, J. R., Al-Shaikh, M. E., Joghee, S., & Alzoubi, H. (2021a). Impact of innovation capabilities on business sustainability in small and medium enterprises. *FIIB Business Review*, 1–12,. https://doi.org/10.1177/23197145211042232

Hanaysha, J. R., Al Shaikh, M. E., & Alzoubi, H. M. (2021b). Importance of marketing mix elements in determining consumer purchase decision in the retail market. *International Journal of Service Science, Management, Engineering, and Technology (IJSSMET), 12*(6), 56–72.

Harahsheh, A. A., Houssien, A. M. A., & Alshurideh, M. T. (2021). The effect of transformational leadership on achieving effective decisions in the presence of psychological capital as an intermediate variable in private Jordanian. In *The Effect of Coronavirus Disease (COVID-19) on Business Intelligence* (pp. 243–221). Springer Nature.

Joghee, S., Alzoubi, H. M., & Dubey, A. R. (2020). Decisions effectiveness of FDI investment biases at real estate industry: Empirical evidence from Dubai smart city projects. *International Journal of Scientific and Technology Research, 9*(3), 3499–3503.

Kashif, A. A., Bakhtawar, B., Akhtar, A., Akhtar, S., Aziz, N., & Javeid, M. S. (2021). Treatment response prediction in hepatitis C patients using machine learning techniques. *International Journal of Technology, Innovation and Management (IJTIM), 1*(2), 79–89. https://doi.org/10.54489/ijtim.v1i2.24

Khan, M. A. (2021). Challenges facing the application of IoT in medicine and healthcare. *International Journal of Computations, Information and Manufacturing (IJCIM), 1*(1), 39–55. https://doi.org/10.54489/ijcim.v1i1.32

Kurdi, B., Alshurideh, M., & Alnaser, A. (2020). The impact of employee satisfaction on customer satisfaction: Theoretical and empirical underpinning. *Management Science Letters, 10*(15). https://doi.org/10.5267/j.msl.2020.6.038

Lee, C., & Ahmed, G. (2021). Improving IoT privacy, data protection and security concerns. *International Journal of Technology, Innovation and Management (IJTIM), 1*(1), 18–33.

Lee, K. L., Azmi, N. A. N., Hanaysha, J. R., Alzoubi, H. M., & Alshurideh, M. T. (2022a). The effect of digital supply chain on organizational performance: An empirical study in Malaysia manufacturing industry. *Uncertain Supply Chain Management, 10*(2), 495–510. https://doi.org/10.5267/j.uscm.2021.12.002

Lee, K. L., Romzi, P. N., Hanaysha, J. R., Alzoubi, H. M., & Alshurideh, M. (2022b). Investigating the impact of benefits and challenges of IOT adoption on supply chain performance and organizational performance: An empirical study in Malaysia. *Uncertain Supply Chain Management, 10*(2), 537–550. https://doi.org/10.5267/j.uscm.2021.11.009

Leo, S., Alsharari, N. M., Abbas, J., & Alshurideh, M. T. (2021). From Offline to online learning: A qualitative study of challenges and opportunities as a response to the Covid-19 pandemic in the UAE higher education context. In *Studies in Systems, Decision and Control* (Vol. 334). https://doi.org/10.1007/978-3-030-67151-8_12

Mehmood, T. (2021). Does information technology competencies and fleet management practices lead to effective service delivery? Empirical evidence from e-commerce industry. *International Journal of Technology, Innovation and Management (IJTIM), 1*(2), 14–41.

Mehmood, T., Alzoubi, H. M., Alshurideh, M., Al-Gasaymeh, A., & Ahmed, G. (2019). Schumpeterian entrepreneurship theory: Evolution and relevance. *Academy of Entrepreneurship Journal, 25*(4), 1–10.

Miller, D. (2021). The best practice of teach computer science students to use paper prototyping. *International Journal of Technology, Innovation and Management (IJTIM), 1*(2), 42–63. https://doi.org/10.54489/ijtim.v1i2.17

Mondol, E. P. (2021). The impact of block chain and smart inventory system on supply chain performance at retail industry. *International Journal of Computations, Information and Manufacturing (IJCIM), 1*(1), 56–76. https://doi.org/10.54489/ijcim.v1i1.30

Nuseir, M.T., Aljumah, A., & Alshurideh, M. T. (2021a). How the business intelligence in the new startup performance in UAE during Covid-19: The mediating role of innovativeness. In *Studies in Systems, Decision and Control* (Vol. 334). https://doi.org/10.1007/978-3-030-67151-8_4

Nuseir, Mohammed T, Al Kurdi, B. H., Alshurideh, M. T., & Alzoubi, H. M. (2021b). Gender Discrimination at Workplace: Do Artificial Intelligence (AI) and Machine Learning (ML) Have Opinions About It. In *The International Conference on Artificial Intelligence and Computer Vision* (pp. 301–316).

Obaid, A. J. (2021). Assessment of smart home assistants as an IoT. *International Journal of Computations, Information and Manufacturing (IJCIM), 1*(1), 18–36. https://doi.org/10.54489/ijcim.v1i1.34

PWC. (2021). *COVID-19: Making remote work productive and secure.*

Radwan, N., & Farouk, M. (2021). The growth of internet of things (IoT) in the management of healthcare issues and healthcare policy development. *International Journal of Technology, Innovation and Management (IJTIM), 1*(1), 69–84. https://doi.org/10.54489/ijtim.v1i1.8

Rapp, J., Niehaus, E., Ribó, A., Mej\'\ia, R., Quinteros, E., & Fath, A. (2020). Using Space Technology to Mitigate the Risk Caused by Chronic Kidney Disease of Unknown Etiology (CKDu). In *In Space Fostering Latin American Societies, Cham* (pp. 55–71).

Shamout, M., Ben-Abdallah, R., Alshurideh, M., Alzoubi, H., Kurdi, B. A., & Hamadneh, S. (2022). A conceptual model for the adoption of autonomous robots in supply chain and logistics industry. *Uncertain Supply Chain Management, 10*(2), 577–592. https://doi.org/10.5267/j.uscm.2021.11.006

Wressell, J. A., Rasmussen, B., & Driscoll, A. (2018). Exploring the workplace violence risk profile for remote area nurses and the impact of organisational culture and risk management strategy. *Collegian, 25*(6), 601–606.

Xio, Y., & Fan, Z. (2020). *10 technology trends to watch in the COVID-19 pandemic*. World Economic Forum.

Yeboah-Ofori, A., & Opoku-Akyea, D. (2019). May. Mitigating Cyber Supply Chain Risks in Cyber Physical Systems Organizational Landscape. In *2019 International Conference on Cyber Security and Internet of Things (ICSIoT)*, (pp. 74–81).

Zhang, N., Mi, X., Feng, X., Wang, X., Tian, Y., & Qian, F. (2018). *Understanding and mitigating the security risks of voice-controlled third-party skills on amazon alexa and google home.*

The Impediments of the Application E-Business to Classified the Restaurants in the Aqaba Special Economic Zone

Omar Jawabreh, Ra'ed Masa'deh, Tamara Yassen, and Muhammad Alshurideh

Abstract The study aims research paper seeks to study the status of tourism promotion in Jordan, in particular via the Internet, with a future plan to develop this type of promotion based on the needs of the country of this research paper is to draw conclusions that help to know and understand this type of tourism promotion, And to know how to develop it in Aqaba Special Economic Zone Disseminatethe maximum information about the restaurant in the Aqaba, this research paper is designed to determine the conditions of websites used in tourism promotion of the restaurants to find out the obstacles that face this type of promotion via websites, and the factors that affect its the promotion is to be a primary tool for economic development in tourism destinations, The objectives are to increase the contribution of the tourism sector to the obstacles of e-business in the classification of restaurants and to benefit economically from the tourism industry through employment opportunities and investment, product Development and tourism experience, promotion of tourism in restaurants, between the public and private sector in Aqaba.

Keywords Financial problems · Technical problems · Security problems · Management problems · Legal problems

O. Jawabreh (✉)
Department of Hotel Management, Faculty of Tourism and Hospitality, The University of Jordan, Amman, Jordan
e-mail: o.jawabreh@ju.edu.jo

R. Masa'deh
Department of Management Information Systems, School of Business, The University of Jordan, Amman, Jordan

T. Yassen
Department of Hotel Management, Faculty of Tourism and Hospitality, The University of Jordan, Amman, Jordan

M. Alshurideh
Department of Marketing, School of Business, The University of Jordan, Amman, Jordan
e-mail: m.alshurideh@ju.edu.jo

University of Sharjah, Sharjah, UAE

M. Alshurideh et al. (eds.), *The Effect of Information Technology on Business and Marketing Intelligence Systems*, Studies in Computational Intelligence 1056,
https://doi.org/10.1007/978-3-031-12382-5_41

1 Introduction

With the growth of tourism, the industry has become the world's biggest (Goeldner et al., 2000; Masa'deh et al., 2018; Yannopoulos & Rotenberg, 1999). A Social, Cultural, and Economic Phenomenon: Tourism is described by the World Tourism Organization as an activity involving individuals going to nations or locations outside of their normal surroundings for personal or commercial purposes. Consequently, tourism focuses on visitor behavior, which necessitates a deep understanding of how technology and consumer behavior impact travel-related data dissemination and accessibility. Understanding the content of information that may be available to internet visitors is a key step in developing successful tourist marketing campaigns, better information systems (Al Khasawneh et al., 2021; Fesenmaier et al., 2006; Jawabreh, 2017; Jawabreh & Al Sarayreh, 2017; Masadeh et al., 2019; Xiang et al., 2008); and acquiring superior innovations (Alrowwad et al., 2020; Mahmoud et al., 2021; Obeidat et al., 2019; Qandah et al., 2020). Though it seems that social media is on the rise in the online travel information search industry, there is no scientific data to support or explain this trend.

Websites that utilize a search engine to find appropriate travel information may expose tourists to social media content to a greater extent. When each seller focuses on nurturing and increasing their various channels of information and persuasion, this is referred to as promotion, the marketing of a place is intertwined with cultural aspects. (Herbig, 1998) says customers' needs vary between cultures. One conclusion can be drawn: all destination images and marketing materials produced by the local tourism industry should include a level of authentic destination identity (Go et al., 2004; Noordman, 2004; Onians, 1998; Rekom & Go, 2003; Riel, 1996). This tourism development strategy consists of two stages: creating a tourism "product," and commercializing the offering using that identity and the authenticity of the place (whether staged or real). Then, a projected tourism destination image is created through planned marketing and communication, or through vicarious experiences (Cohen-Hattab & Kerber, 2004). Destination image formation agents (Gartner, 1993), and Internet interactivity may allow these destination photos to be presented in novel ways. If a destination's brand identity and tourist product presentation don't match, there may be a tourism development strategy mismatch. Jordan's tourism promotion, namely the Internet, was studied by Alananzeh et al. (2019); Jahmani et al., 2020; Jawabreh et al., 2020), with the hope of extending this approach in the future to suit the needs of the tourism sector.

This study is intended to uncover facts that will help us to better understand this type of tourism marketing and to assist the Hashemite Kingdom of Jordan in their tourism promotion (Jawabreh, 2021). This coastal city has an excellent temperature all year. During the summer, it is hot and invigorating, while in the winter it is warm and comfortable. The city is designed to meet the requirements and wants of the whole population, including both summer and winter. There are many things to choose from while in Aqaba, including watersports, sailing, diving, and a variety of other exciting desert-based activities, including camelback riding and exploring Petra. Aqaba is a desert oasis where these activities occur.

Because of the location's unique aquatic environment, which includes 140 different kinds of coral reefs and colorful and unusual fish species found only in the seas of Aqaba, the ocean scenery in Aqaba is a breathtaking spectacle. Aqaba has archaeological sites that date back at least 5,500 years, which provide evidence of the early human presence in the area. By placing Aqaba in this context, it highlights the importance of Aqaba's role in the military. Land and sea connections link Asia, Africa, and Europe through this city. This newly discovered structure, which is widely believed to be the world's oldest church, comes from the Mamluk period, and it is the grandfather of His Majesty King Abdullah II. Aqaba is a unique place where the red of the sky combines with the blue of the sea to create a vivid, underwater scene. It is also a wonderful destination for conferences, holidays, and business retreats, especially Aqaba. If you are here for pleasure or business, you'll find enough to do, see, and enjoy, as the absence of monotony and boredom are also guaranteed. When in Aqaba, it leaves you with many wonderful memories and urges you to return. This place is a veritable Mecca for people from all over the world. It is a city in the desert. Think about how much you'll enjoy spending time in this beautiful oasis, and plan a vacation to Aqaba where you can relax in the sun (ASEZA, Aqaba Special Economic Zone Authority).

2 Literature Review

Managers and business owners in the hotel and tourism sectors spend the majority of their time concentrating on changing each of these components to provide a competitive offer to customers (Kotler, 2003). The researchers in this study examined the challenges faced by the online commerce sector while using the idea of the internet. Pillars of Content Performance, developed by the Arab country (and available on N.D.), laid forth a framework for identifying factors that affect website success. The main actors in website success are quality of information and service, system use, interactivity, and design. To achieve success, an e-commerce website should contain these criteria: a user-friendly online interface, superior customer and supplier assistance, a robust search engine to fuel the website, ensuring that the site's customers are satisfied, and providing up-to-date information. The articles in this review demonstrate that technology is utilized in the design, planning, production, and development of manufacturing. Also referred to as electronic commerce and the economic and financial consequences of such activity, this phrase relates to the term "electronic commerce" (Hoffman & Novak, 1996).

The goal of this study is to investigate Aqaba's unique economic zone, the progress made in online commerce, and the difficulties restaurants have with websites. Specifically, 36 restaurants were surveyed using questionnaires. These surveys discovered that the majority of these restaurants' websites were not used for electronic commerce. Internet business is difficult in Aqaba because of security, trust, and a lack of understanding about technical problems. This study looked at the speed of information and communication development, which allowed for the wide-ranging transmission of information. The current report said that it saw new discoveries in this area as a key component of economic development and online commerce. It included topics such as e-tourism and mobile tourism, as well as the issues that arise when tourists claim that there is competition among them. as well as how electronic services such as e-books, photos, videos, e-booking, and e-cards can serve tourists in Arab countries. Despite the efforts of the tourism industry, nothing has been written on the monetary worth of shopping as a tourist attraction (Heung & Qu, 1998; Law & Norman, 2000). Because tourism retailing research is just starting, it seems that we have a clear need to find out what tourists buy when they visit tourist attractions. The rise in domestic tourism observed in recent years may be attributed to Jordan.

Due to the increasing wealth of the Jordanian people and the growing interest of tourist organizations in promoting domestic tourism to assist in reducing the Kingdom's seasonal demand for tourism, this was a primary factor behind the rise in domestic tourism (Magablih, 2005). According to this theory, in order to increase the amount of tourist activity, "push" and "pull" motivations must be used (Josiam et al., 1999; Kim & Lee, 2000). Push motivations are the external societal and psychological forces that push a person to travel. The web also provides greater possibilities for regional tourist promotion than traditional marketing and advertising channels, and it is very cheap (Standing & Vasudavan, 2000).

Another significant aspect is the information included on tourist websites, and one study shows that this is what travelers are searching for (Ho & Liu, 2005). The chance of return visits is increased (Rosen & Purinton, 2004). For these reasons, when tourism businesses have shifted from simply giving information to interactive designs, they may uncover consumers' interests and help encourage their return. Personalizing service offerings empowers these businesses to better understand customer needs, which allows them to engage with customers on a personal level and to provide tailored services (Doolin et al., 2002). As a consequence, the information found on government tourism websites affects the destination's gastronomic image. It also offers a realistic tasting experience for food-oriented visitors, and the design of a website influences how people react (Rosen & Purinton, 2004).

The reorganization of municipal administrations in response to the change brought on by globalization and economic restructuring may be linked to the concept of place promotion. The future of post-industrial towns depends on attracting inward investment and migration, and in order to do so, places have been conceptualized in new ways. The achievement gap exists even among minority children who attend majority-minority schools. While according to Ashworth and Voogd (1994), the promotion of a particular bundle of facilities or the selling of a place as a whole may be referred to as marketing, promotion in the context of this research refers to targeting key employees to advance the sales team's objectives. Placement of an advertisement is a facet of location marketing, which is concerned with linking "local activities" to the wants of targeted. A wide range of techniques, not just advertising campaigns, are used to promote places, depending on the product, its surroundings, customers, and manufacturers. The purpose of place promotion is to create an appealing identity which will entice target groups such as investors and tourists (Palmer & McCole, 2000; Young & Lever, 1997). Increasingly, the Internet (with the most current web technology) offers a wide range of information in an incredibly rapid manner, necessitating it as a tourism-sector marketing and promotion tool (particularly for hotels and travel companies, e.g., DMOs and national tourist organization NTOs) (Abuhashesh et al., 2019; Brey et al., 2007; Choi et al., 2007; Kim et al., 2007).

The web also provides greater possibilities for regional tourist promotion than traditional marketing and advertising channels, and it is very cheap (Standing & Vasudavan, 2000). Because your website is accessible 24 h a day, 7 days a week, from anywhere on the globe, it may reach a global audience. Website material is essential, and it must be kept up to date (Lin & Huang, 2006). In order to find out why tourists frequent shopping malls, the purpose of this study was to investigate why tourists buy. To what extent are Jordanians motivated to purchase property in Aqaba? It may be concluded, based on the quantity of shopping malls in the location, that what attracts visitors is having the freedom to select among various shopping malls.

The rise in domestic tourism observed in recent years may be attributed to Jordan. Due to the increasing wealth of the Jordanian people and the growing interest of tourist organizations in promoting domestic tourism to assist in reducing the Kingdom's seasonal demand for tourism, this was a primary factor behind the rise in domestic tourism (Magablih, 2005). The goal of this research is to offer advice to business managers who wish to do a better job of marketing to meet the needs of their consumers.

3 Research Methodology

This section provides the methodology applied in the current study. It consists of the research operational definitions of the study's variables, besides data collection tool and research population and sample.

3.1 Research Operational Definitions

The current research considers five variables regarding (Jawabreh et al., 2010). These includes Financial problems, Technical problems, Security problems, Management problems, and Legal problems; which were measured in the research questionnaire through seven, seven, seven, six, and seven items respectively.

3.2 *Population and Sampling*

The targeted population of this study consisted of all staff that is responsible of disability services in Aqaba city hotels located in Jordan, specifically hotels 5, 4 and 3 stars. Thus, a judgment sampling technique was conducted. A total of 142 questionnaires were returned and applicable for statistical analysis. Indeed, the primary data was collected through a drop-and-collect survey technique. The surveys were distributed to the targeted staff working in the hotels that agreed to participate in the study.

The questionnaire consisted of two sections; the first section in questionnaire presents general personal information about a respondent, the gender, age, educational level, career level and ranking of the hotel. The second section includes questions to measure the ENAT standards based on their operational definitions.

4 Data Analysis and Results

In order to explore the degree to which Aqaba city hotels apply ENAT standards towards disabled tourists, in which the items for these variables have been measured using 5-points Liker scale that varies between strongly disagree = 1; 2 = disagree; 3 = neutral; 4 = agree and strongly agree = 5; reliability and validity analyses were conducted, descriptive analysis was used to describe the characteristic of sample and the respondents to the questionnaire's items.

4.1 *Validity and Reliability*

Validity and reliability are two important measures to determine the quality and usefulness of the primary data. Validity is about accuracy and whether the instrument measures what it is intended to measure while reliability is about precision; it is used to check the consistency and stability of the questionnaire. Indeed, the researchers depended on scales and items that were previously developed and used by Jawabreh et al. (2010).

Also, a draft of the questionnaire was formulated, and then it was reviewed by three academic lecturers–who have a sufficient knowledge and experience in this scope-to ensure that each item is measuring what is intended to be measured, and to avoid the ambiguity and complexity in the phrasing of questions. The reliability of the instrument was measured by the Cronbach's alpha coefficient. Further, some scholars (e.g., (Bagozzi & Yi, 1988; Creswell, 2009)) suggested that the values of all indicators or dimensional scales should be above the recommended value of 0.60. Table 1 represents the results of Cranach's alpha for the research variables. Cronbach's alpha

Table 1 The Cronbach's alpha coefficients of study variables

Variables	Number of items	Cronbach alpha
Financial problems	7	0.715
Technical problems	7	0.722
Security problems	7	0.730
Management problems	6	0.786
Legal problems	7	0.883

coefficients of all the tested variables are above 0.60 which suggesting the composite measure is reliable.

4.2 Respondents Demographic Profile

As indicated in Table 2, the demographic profile of the respondents for this study showed that they are typically females work in private sector, most of them between 20 and 40 years old, the majorities hold bachelor degrees; and most of them got a monthly income 350JD and more.

4.3 Descriptive Analysis

In order to describe the responses and thus the attitude of the respondents toward each question they were asked in the survey, the mean and the standard deviation were estimated. While the mean shows the central tendency of the data, the standard deviation measures the dispersion which offers an index of the spread or variability in the data (Pallant, 2005; Sekaran & Bougie, 2013). In other words, a small standard deviation for a set of values reveals that these values are clustered closely about the mean or located close to it; a large standard deviation indicates the opposite. The level of each item was determined by the following formula: (highest point in Likert scale - lowest point in Liker scale) / the number of the levels used = (3–1) / 5 = 0.66, where 1–1.66 reflected by "low", 1.67–2.33 reflected by "moderate", and 2.34–3 reflected by "high". Then the items were being ordered based on their means. Tables 3 and 4 show the results.

As presented in Table 3, data analysis results have shown that all problems in Aqaba hotels are applied to a high level in which the range of the mean score is 2.34–2.45. Table 4 demonstrates the mean, standard deviation, level, and order scores for items for each variable/standard.

Table 2 Description of the respondents' demographic profiles

Category	Category	Frequency	Percentage (%)
Gender	Males	38	38.4
	Females	61	61.6
	Total	99	100
Age	Less than 20	20	20.2
	20 years—less than 30	41	41.4
	30 years—less than 40	22	22.2
	40 years—less than 50	8	8.1
	50 years—less than 60	3	3.0
	More than 60 years old	5	5.1
	Total	99	100
Monthly income	Less than 250JD	15	15.2
	250—less than 350 JD	25	25.3
	350—less than 450 JD	31	31.3
	More than 450 JD	28	28.3
	Total	99	100
Occupation	Private sector	53	53.5
	Public sector	46	46.5
	Total	99	100
Educational level	High school	24	24.2
	Bachelor	54	54.5
	Masters	13	13.1
	PhD	8	8.1
	Total	99	100

Table 3 Overall mean and standard deviation of the study's variables

Type of variable	Mean	Std. Dev	Level	Order
Financial problems	2.4026	0.64419	High	3
Technical problems	2.4556	0.69395	High	1
Security problems	2.3997	0.75039	High	4
Management problems	2.4394	0.85165	High	2
Legal problems	2.3478	0.67073	High	5

5 Discussion and Conclusions

When confronted with the findings of the research, hypotheses and questions are answered. Thirty-six restaurants that provide websites and utilize them for Internet trading performed the research and found that of those, less than one-third have not

Table 4 Mean and standard deviation of the study's variables

Financial problems	**Mean**	**SD**	**Level**	**Order**
High prices of computer equipment and networking tools and software	2.04	1.339	Moderate	6
High wages charged by information professionals	2.34	1.451	High	5
The high financial cost of establishing and managing an electronic trading site	2.35	1.487	High	4
Banks are not cooperating in online transactions	2.54	1.431	High	1
There is no financial intermediary to guarantee payment in electronic payments	2.53	1.417	High	2
Our customers do not use electronic payment cards in their dealings	2.49	1.501	High	3
The transition to electronic commerce will not entail additional financial income	2.53	1.561	High	2
Technical problems	**Mean**	**SD**	**Level**	**Order**
Our organization lacks the equipment to link and share information (networks, servers)	2.38	1.426	High	6
Our database system is not configured to connect with the Internet	2.43	1.458	High	3
We do not have people who are specialized in computer programming	2.72	1.519	High	1
The difficulty of following up and maintaining the website of electronic commerce	2.42	1.512	High	4
Lack of specialized companies interested in the establishment and management of sites for electronic rent	2.41	1.471	High	5
Lack of proliferation of computers connected to the Internet when our clients	2.29	1.443	Moderate	7
Rapid and ongoing developments in hardware and software	2.53	1.548	High	2
Security problems	**Mean**	**SD**	**Level**	**Order**
Dealing with electronic payment system is not safe and offered to steal stocks	2.29	1.394	Moderate	7
Linking enterprise systems with the Internet presents them for information piracy	2.38	1.448	High	3
Electronic transactions in the Internet are not free of fraud and forgery	2.46	1.560	High	2
There are many cases of fraud and fraud in electronic transactions	2.63	1.588	High	1
The difficulty of protecting the data transmitted on the Internet from manipulation and switching	2.37	1.454	High	4
Connecting to the Internet increases the likelihood of a computer system being compromised	2.35	1.380	High	5

(continued)

Table 4 (continued)

The fear of the control of service providers and the telecommunications sector and its reflection on the work	2.30	1.432	Moderate	6
Management problems	**Mean**	**SD**	**Level**	**Order**
We do not have knowledge of how to launch online trading services	2.48	1.466	High	3
We have no plans to change the course of our business at this time	2.52	1.548	High	2
Our organization does not need to increase the number of its customers at the moment	2.41	1.471	High	4
Difficulty dealing with the status of cancellation of the service request after payment for electronic transactions	2.27	1.470	Moderate	6
There are no successful e-selection experiments at similar institutions	2.62	1.595	High	1
There is no government encouragement for enterprises to move towards e-commerce	2.33	1.392	Moderate	5
Legal problems	**Mean**	**SD**	**Level**	**Order**
The request for the service requires the physical presence of the customer first	2.86	1.597	High	1
Requesting the service electronically by different types is not binding on us	2.38	1.489	High	2
The verification document must be verified for the customer who benefits from the service	2.23	1.427	Moderate	6
Lack of information available about the legal aspects of electronic commerce	2.35	1.487	High	4
Electronic contracts pose a challenge in their legal catharsis	2.27	1.470	Moderate	5
Weak legislative tax policy in the trading environment	2.39	1.406	High	3
Lack of legal legislation regulating electronic commerce	1.94	1.284	Moderate	7

been used for online trading, but problems arose in implementing the Internet trade securely and confidently in Aqaba. This mostly resulted from the increase in the net worth of Jordanians and an increased focus on domestic tourism as a means of reducing seasonality in the Kingdom (Magablih, 2005).

Tourists have different requirements, desires, and preferences. Therefore, in order to succeed, marketing strategists must understand and fulfill their needs and preferences while competing (Magablih, 2005). The idea of tourist motivation is outlined in a well-documented theoretical framework, which is known as the "increase motivation through inspiring via drawing inspiration" (Josiam et al., 1999; Kim & Lee, 2000). When considering travel motives, you must also look at the socio-psychological requirements that drive a person to go on a trip. This opportunity to help rural tourism thrive will be unique, and much less costly compared to other marketing and advertising methods (Standing & Vasudavan, 2000). Global audiences may be reached by using an excellent website: it is available around the clock from anywhere on the globe. Website content is critical, and must be updated frequently in the tourist marketing industry.

The primary subject of this study was the Aqaba special economic zone and how e-commerce impacts the economy in the area. A smart question about electronic commerce and restaurants in the Aqaba special economic zone: What effect has electronic commerce had, and what complications have businesses in the Aqaba special economic zone faced when utilizing the internet to market their products? E-commerce websites, on the whole, are not typically utilized. With online commerce, the Aqaba has the most trust, security, and understanding of technical problems as obstacles. In this study, the rise in information access was shown to be mostly due to the massive expansion in the quantity of information in the information and communication industry.

The research also addressed the reality of e-tourism in Arab nations and what the e-services may offer visitors, such as e-books, pictures, video, e-booking and e-cards. It is now very impossible to get much information about shopping as a tourist attraction (Alshurideh et al., 2021; Heung & Qu, 1998; Law & Norman, 2000). Research on tourism retailing is a new area of study, and there seems to be a need for researching tourists' buying habits while visiting tourist destinations. The number of domestic tourists in Jordan has grown in the past few years. When it comes to meeting the varied wants, interests, and tastes of tourists, marketing experts must be very knowledgeable about the industry and able to offer goods and services in a cost-effective manner (Jawabreh, 2017).

Push and pull provide a well-established theoretical framework for assessing tourist motivation excitement (Josiam et al., 1999; Kim & Lee, 2000). Indeed, (Josiam et al., 1999; Kim & Lee, 2000) argued that for the most part, travel motives are motivated by social and psychological motivations. Additionally, it also serves as a useful tool for marketing the area's tourism. Coupled with more modern means of advertising, including the internet, it is less expensive than conventional media like television and radio (Standing & Vasudavan, 2000). To be successful, a website must target the whole world's population, and it may be done from any place at any time of day. Website content is essential, and it should be updated regularly in the industry (Lin & Huang, 2006).

Of the many resources on the internet, one of the best sources of travel information is websites like this (Ho & Liu, 2005). Once again, invite individuals back to the website and over (Rosen & Purinton, 2004). To enhance tourist returns, there are two possible routes. One way to provide information is through interactive design, while the other is to use interactive design to offer information. Additionally, these businesses are in a better position to understand their visitors' preferences, which allows them to provide personalized services to individuals rather than generic packages (Doolin et al., 2002). Gastronomic image and virtual culinary experience are both greatly affected by the content of government tourist websites. People's interests and preferences change (Rosen & Purinton, 2004).

Promotion is said to be connected to local government shifting to meet the changing nature of the economy because of globalization. New place identities that have been created to promote investment and in-migration are important tools in attracting inward investment and migration to local post-industrial areas (Law & Norman, 2000). In regards to location promotion, the idea of promoting a certain bundle of amenities was explored by Ashworth and Voogd (1994), and their research indicates that location promotion is the process of marketing a location as a whole by using pictures containing numerous different characteristics linked to it. Placement of promotional forms is inadequate when used alone. The requirements of targeted consumers should be taken into consideration while considering local activities. A question by Ashworth and Voogd (1994) is left open: Advertising is not the only strategy that places and goods use to get attention. Some of them include the context of the product, the intended audience, and the creator. Creating an appealing identity that draws a particular set of investors and tourists is key in a location marketing strategy (Young & Lever, 1997). This is typically done by deconstructing pictures of the location and then rebuilding them, as described by Jawabreh (2020). Electronic commerce has revolutionized the tourism sector by providing a case study (Palmer & McCole, 2000). New online technology makes the Internet a potent marketing and promotion tool for tourism.

References

Abuhashesh, M., Al-Khasawneh, M., & Al-Dmour, R. (2019). The impact of Facebook on Jordanian consumers' decision process in the hotel selection. *IBIMA Business Review, 2019*. https://doi.org/10.5171/2019.928418

Al Khasawneh, M., Abuhashesh, M., Ahmad, A., & Alshurideh, M. T. (2021). Customers online engagement with social media influencers' content related to Covid 19. https://doi.org/10.1007/978-3-030-67151-8_22

Alananzeh, O., Al-Badarneh, M., Al-Mkhadmeh, A., & Jawabreh, O. (2019). Factors influencing MICE tourism stakeholders' decision making: The case of Aqaba in Jordan. *Journal of Convention & Event Tourism, 20*(1), 24–43. https://doi.org/10.1080/15470148.2018.1526152

Alrowwad, A., Abualoush, S. H., & Masa'deh, R. (2020). Innovation and intellectual capital as intermediary variables among transformational leadership, transactional leadership, and organizational performance. *Journal of Management Development, 39*(2), 196–222. https://doi.org/10.1108/JMD-02-2019-0062

Alshurideh, M. T., Al Kurdi, B., AlHamad, A. Q., Salloum, S. A., Alkurdi, S., & Dehghan, A. (2021). Factors affecting the use of smart mobile examination platforms by universities' postgraduate students during the COVID-19 pandemic: An empirical study. *Informatics, 8*(2). https://doi.org/10.3390/informatics8020032

Ashworth, G., & Voogd, H. (1994). Marketing and place promotion. In J. R. Gold, & S.V. Ward (Eds.), *Place promotion: The use of publicity and marketing to sell towns and regions* (pp. 39–52). Wiley.

Bagozzi, R., & Yi, Y. (1988). On the evaluation of structural evaluation models. *Journal of the Academy of Marketing Science, 16*(1), 74–94.

Brey, E. T., So, S.-I., Kim, D.-Y., & Morrison, A. M. (2007). Web-based permission marketing segmentation for the lodging industry. *Tourism Management, 28*(6), 1408–1416.

Choi, S., Lehto, X. Y., & Morrison, A. M. (2007). Destination image representation on the web: Content analysis of Macau travel-related websites. *Tourism Management, 28*(1), 118–129.

Cohen-Hattab, K., & Kerber, J. (2004). Literature, cultural identity and the limits of authenticity: A composite approach. *International Journal of Tourism Research, 6*(2), 57–73.

Creswell, J. (2009). *Research design: Qualitative, quantitative, and mixed methods approaches* (3rd ed.). Sage Publications.
Doolin, B., Burgess, L., & Cooper, J. (2002). Evaluating the use of the web for tourism marketing: A case study from New Zealand. *Tourism Management, 23*(5), 557–561.
Fesenmaier, D. R., Wober, K., & Werthner, H. (2006). Introduction recommendation systems in tourism. In D. R. Fesenmaier, K. Wo¨ber, & H. Werthner (Eds.), *Destination recommendation systems: Behavioral foundations and applications*. CAB.
Gartner, W. C. (1993). Image formation process. *Journal of Travel and Tourism Marketing, 2*(2/3), 191–215.
Go, F. M., Lee, R. M., & Russ, A. P. (2004). E-Heritage in the globalizing society: Enabling cross-cultural engagement through ICT. *Informat Technology & Tourism, 6*(1), 55–68.
Goeldner, C. R., Ritchie, J. R. B., & MacIntosh, R. W. (2000). *Tourism: Principles, practices, philosophies*, (8th edn, pp. 615–665). John Wiley & Sons, Inc.
Herbig, P. A. (1998). *Handbook of cross-cultural marketing*. The Haworth Press.
Heung, V., & Qu, H. (1998). Tourism shopping and its contributions to Hong Kong. *Tourism Management, 19*(4), 383–386.
Ho, C.-I., & Liu, Y.-P. (2005). An exploratory investigation of web-based tourist information search behavior. *Asia Pacific Journal of Tourism Research, 10*(4), 351–360.
Hoffman, D. L., & Novak, T. P. (1996). Marketing in hypermedia computer mediated environments: Conceptual foundations. *Journal of Marketing, 60*(3), 50–68.
Jahmani, A., Bourini, I., & Jawabreh, O. A. (2020). The relationship between service quality, client satisfaction, perceived value and client loyalty: A case study of Fly Emirates. *Cuadernos De Turismo, 45*, 219–238.
Jawabreh, O., & Al-Sarayreh, M. (2017). Analysis of job satisfaction in the hotel industry: A study of hotels five- Stars in Aqaba special economic zone authority (AZEZA). *International Journal of Applied Business and Economic Research, 15*(26), 389–407.
Jawabreh, O., Tallal, H., & Harazneh, A. (2010). Challenges the application of e-business to the travel and tourism companies in the Aqaba Special Economic Zone. *Journal of Association of Arab Universities for Tourism and Hospitality, 5*(1), 223–231.
Jawabreh, O. A. (2017). An exploratory study of the motives of Jordanian out bound tourism and its impact on the development of tourism in Jordan. *International Journal of Applied Business and Economic Research, 15*(19 Part-II), 443–467.
Jawabreh, O. A. (2020). Innovation management in hotels industry in aqaba special economic zone authority; hotel classification and administration as a moderator. *GeoJournal of Tourism and Geosites, 32*(4), 1362–1369. https://doi.org/10.30892/gtg.32425-581.
Jawabreh, O. A. (2021). Tourists and local community of the case study aqaba special economic zone authority (ASEZA). *GeoJournal of Tourism and Geosites, 35*(2), 490–498. https://doi.org/10.30892/gtg.35229-676
Jawabreh, O., Masa'deh, R., Mahmoud, R., & Hamasha, S.A. (2020). Factors influencing the employees service performances in hospitality industry case study aqba five stars hotel. *GeoJournal of Tourism and Geosites, 29*(2), 649–661. https://doi.org/10.30892/gtg.29221-496
Josiam, B., Smeaton, G., & Clements, C. (1999). Involvement: Travel motivation and destination selection. *Journal of Vacation Marketing, 5*(2), 161–175.
Kim, E., & Lee, D. (2000). Japanese tourists' experience of the natural environments in North OLD Region-Great barrier reef experience. *Journal of Travel and Tourism Marketing, 9*(1/2), 93–113.
Kim, D. J., Kim, W. G., & Han, J. S. (2007). A perceptual mapping of online travelagencies and preference attributes. *Tourism Management, 28*(2), 591–603.
Kotler, P. (2003). *Marketing management* (11th ed.). Prentice Hall.
Law, R., & Norman, A. (2000). Relationship modeling in tourism shopping: A decision rules induction approach. *Tourism Management, 21*(3), 241–249.
Lin, Y.-S., & Huang, J.-Y. (2006). Internet blogs as a tourism marketing medium: A case study. *Journal of Business Research, 59*(10–11), 1201–1205.

Magablih, K. M. (2005). Evaluation of the nature and dimensions of domestic tourism in the golden tourism triangle: A survey of a selected sample of Jordanian tourists. *Mu'tah, 20*(1), 75–95.

Mahmoud, R., Al-Mkhadmeh, A., & Alananzeh, O. (2021). Exploring the relationship between human resources management practices in the hospitality sector and service innovation in Jordan: The mediating role of human capital. *Geojournal of Tourism and Geosites, 35*(2), 507–514. https://doi.org/10.30892/gtg.35231-678

Masa'deh, R., Alananzeh, O., Tarhini, A., & Algudah, O. (2018). The effect of promotional mix on hotel performance during the political crisis in the Middle East. *Journal of Hospitality and Tourism Technology, 9*(1), 32–47. https://doi.org/10.1108/JHTT-02-2017-0010

Masadeh, R., Alananzeh, O., Jawabreh, O., Alhalabi, R., Syam, H., & Keswani, F. (2019). The association among employee's communication skills, image formation and tourist behaviour: Perceptions of hospitality management students in Jordan. *International Journal of Culture, Tourism and Hospitality Research, 13*(3), 257–272.

Noordman, T. B. J. (2004). *Culture in de city marketing*. Elsevier.

Obeidat, Z. M., Alshurideh, M. T., & Al Dweeri, R. (2019). The influence of online revenge acts on consumers psychological and emotional states: Does revenge taste sweet? Paper presented at the *Proceedings of the 33rd International Business Information Management Association Conference, IBIMA 2019: Education Excellence and Innovation Management through Vision 2020* (pp. 4797–4815).

Onians, D. (1998). The real England. *RSA Journal (royal Society for the Encouragement of Arts, Manufactures & Commerce), 145*(5484), 34–40.

Pallant, J. (2005). *SPSS survival manual: A step guide to data analysis using SPSS for windows version 12*. Open University Press.

Palmer, A., & McCole, P. (2000). The role of electronic commerce in creating virtual tourism destination marketing organizations. *International Journal of Contemporary Hospitality Management, 12*(3), 198–204.

Qandah, R., Suifan, T. S., & Obeidat, B. Y. (2020). The impact of knowledge management capabilities on innovation in entrepreneurial companies in Jordan. *International Journal of Organizational Analysis*. https://doi.org/10.1108/IJOA-06-2020-2246

Van Rekom, J., & Go, F.M. (2003). Cultural Identities in a Globalizing World: Conditions for Sustainability of Intercultural Tourism. In P. Burns (Ed.), *Conference Proceedings: Global Frameworks and Local Realities: Social and Cultural Identities in Making and Consuming Tourism*. University of Brighton.

Van Riel, C. B. M. (1996). *Identiteiten Imago: Grondslagen Van corporate communication (Identity and image: Foundations of corporate communication* (2nd ed). Academic Service.

Rosen, D. E., & Purinton, E. (2004). Website design: Viewing the web as a cognitive landscape. *Journal of Business Research, 57*(7), 787–794.

Sekaran, U., & Bougie, R. (2013). *Research methods for business: A skill-building approach* (6th ed.). Wiley.

Standing, C., & Vasudavan, T. (2000). The marketing of regional tourism via the Internet: Lessons from Australian and South. *Marketing Intelligence & Planning, 18*(1), 45–48.

Xiang, Z., Wöber, K., & Fesenmaier, D. R. (2008). Representation of the online tourism domain in search engines. *Journal of Travel Research, 47*(2), 137–105.

Yannopoulos, P., & Rotenberg, R. (1999). Benefit segmentation of the near-home tourism market: The case of the Upper New York State. *Journal of Travel and Tourism Marketing, 8*(2), 41–55.

Young, C., & Lever, J. (1997). Place promotion, economic location and the consumption of city image. *Economist En Social Geography*. https://doi.org/10.1111/j.1467-9663.1997.tb01628.x

Adoption Factors of Digitize Facilities: A Management Review

Alaa Ahmad and Muhammad Turkih Alshurideh

Abstract This research study specified the Building Information Model has structures such as formation and sharing of the data with other associates of the project. Building Information model was used as a tool for the improvement of sustainability of constructions by consuming the energy effectively and upholding a waste organisation process in the post constructed level in the business highlighting the facility management. The objective of this systematic review is to provide the digitalization technology to have a better understanding which accomplishes the development of electrical devices, approaches and organisations that make the statement more precise and flexible by focusing on previous studies. Surveyed the strategies of showing the systematic review for which the investigators have focused on two main records which are google scholar and emerald insights from which the case studies were selected by clarifying the scholars based on year (2018–2020) and concentrating on eligibility standards. The three studies that were certain of which 2 were from google scholar and one was from emerald insights. These were choose based on sustaining the aims of the present study. The outcomes of the current study revolve the significance of combination between technological digitalization, capabilities, business, and organisation of digital Facility Management. To conclude, the digital practices of the sustainable facility management are experiencing appraisal and transformation.

A. Ahmad (✉)
DBA Student at University of Sharjah, 27272 Sharjah, United Arab Emirates
e-mail: U19106242@sharjah.ac.ae

M. T. Alshurideh
Department of Management, College of Business, University of Sharjah, 27272 Sharjah, United Arab Emirates
e-mail: malshurideh@sharjah.ac.ae

M. Alshurideh et al. (eds.), *The Effect of Information Technology on Business and Marketing Intelligence Systems*, Studies in Computational Intelligence 1056,
https://doi.org/10.1007/978-3-031-12382-5_42

1 Introduction

With the increasing digitalization of operations, facilities digital transformation creates challenges, the most crucial challenge present organizations adaptation (Tilson et al., 2010; Yoo et al., 2012). As a result, organizations must constantly monitor and assess the capabilities and risk posed by evolving digital technology (Bharadwaj et al., 2013). Scholars and practitioners alike have acknowledged the value of digital technology in organizations. This comprehension of how new digital technology practices affect businesses in remarkable ways has an influence on internal operations and how they interact with their external environment. Scholars and practitioners alike have acknowledged the value of digital technology in organizations. This comprehension of how new digital technology practices affect businesses in remarkable ways has an influence on internal operations and how they interact with their external environment (Carbonari et al., 2015). There is an added incentive for customers and facility managers since they stand to gain the most financially from the digitization of facility management services and use (IoT) (Eadie et al., 2013). More precisely, the research seeks to provide the groundwork that will lead to understand factors that influence facilities managers acceptance of digitalization facility management operations and utilize (IoT) projects.

2 Digitization of Facilities Management

The evolution of building technology has altered the way FM functions. Building technologies are rapidly being used in buildings (Wong et al., 2005). Many analogue, electrical, and mechanical equipment are gradually being converted to digital technologies, which has had a significant influence on industry (Alaloul et al., 2020). To comprehend digitization significance, In our study, we must look at the term digitization, Chapco-Wade (2018) defines *digitization* as "the conversion of analogue to digital" and *digitization* as "the use of digital technology" and digitized data to affect how work gets done, revolutionize how organisations connect and interact, and generate new digital income streams. The shift will be unlike anything civilization has ever seen in terms of its scale, range, and complexity (Fomunyam, 2019). This is causing significant disruption in the facilities management (FM) sector where "technology is an enabler of the workplace" and "smarter buildings are just around the corner" (Atta & Talamo, 2020; Cudmore, 2017; Kazado et al., 2019). Building technologies are rapidly being used in buildings, the rise of intelligent users' where knowledge is gathered in integrating building technology with building operations to provide an optimal building environment. As a result, there is a need to modify the way buildings are managed (Lee et al., 2001; Wong et al., 2018). Many researchers and facility managers have recommended leveraging data from building technologies to help with building and operations administration (Kensek, 2015).

3 Methodology

The study focused on providing all existing evidence digitalization on FM adoption and acceptability from the perspectives of facilities managers via a systematic review from the perspectives of various stakeholder. A literature review is defined as a "complete examination and interpretation of literature on a certain topic" (Grant & Booth, 2009). Given the study's objectives and questions in the first phase, a systematic review was conducted: first, to present robust evidence of the status of digitalization in FM service adoption and acceptance with specific and focused questions such as (Population, Intervention (or focus of Interest), Context, and Outcome); second, to identify the knowledge gap in the literature; and finally, to inform the study's subsequent phases. We intend to Using an explicit method, such as a systematic review, has proven to bring some significant benefits (Gopalakrishnan & Ganeshkumar, 2013), such as: reducing bias, producing accurate and reliable conclusions, making information easier to deliver to healthcare providers, researchers, and policymakers, developing new hypotheses about subgroups of study populations, and, most importantly, increasing the performance and quality of the results (Greenhalgh & Donald, 2000). Google Scholar, ScienceDirect, and Scopus were the three electronic databases included. These sites were chosen because of their reputation for covering a wide range of digitization and technology articles from scientific and academic journals. The search was limited to research done in English because this has been demonstrated to be the predominant language for digital facilities management literature. The following search terms were applied: ["Digitize Facilities management" OR "Digitize FM"] AND ["Facilities manager" OR "Facilities professionals"] AND [adoption OR acceptance]. The PRISMA checklist for reporting scope search results was used, as illustrated in (Fig. 1). Furthermore, an alert was set up in all databases to notify us of any freshly published articles that met the search parameters.

4 Data Extraction Table

This data extraction table is shown by aiming on the study style, whether it contains state, members, objective, limitations, results, along with the methodology of each study (see Outcomes section) (Table 1).

5 Discussion

The existing section will deliver the literature knowledge gap by aiming on studies that will support and have a contradictor effect on the studies that have been involved. This will help the researchers in focusing on other facility Management which is a productive device to find the positive and negative ascribes of the home-grown help,

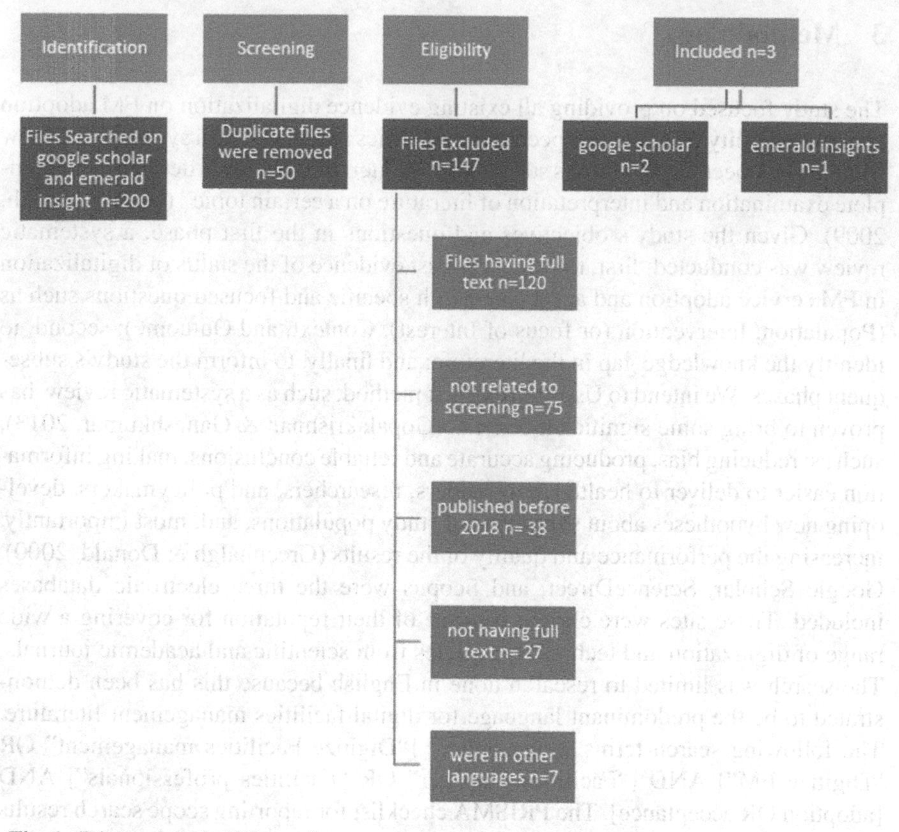

Fig. 1 Prisma chart (self-formed)

assist the important navigation, function as a source of perspective highlight balance past execution levels with the current ones and recognise mistakes and backing with reasonable solutions to be attempted. The literature review established that main execution indicators Key Performance Indicators are, for the most part, utilised in Facility Management for appraising the performance of the facilities as one of the important estimation devices. It is also maintained that the contribution of the facility management managers is significant in the study of value management, specifically with respect to answering the questions presented below:

6 Customer Satisfaction

The facility management managers should coordinate with the customers to recognise the need of the users and management businesses. The connection between the facility management and the society influences the sharing of skills appropriate to several aspects of the facility management with both the customers and the inventors. Earlier

Table 1 Data extraction table (Self-formed)

Reference	Country of Publishing	Participant Involvement	Aim	Methods	Outcomes	Limitations
Charef et al. (2018)	Europe	Studies contain stakeholders, businessman	The aim of the research is to proof the adoption in facility management and adequacy from the outlook of the management facility managers through the overview of various shareholder's outlook	The studies from databases cconcentrated on 28 European states as clear in the EU Commission website (2018) containing quantitative study	The data analysis raises the lack of consensus, and it is also highlighted on the digitization of the facility management	Limitations of having the great chance that the nature of operation where the information collection method is weak
Lu et al. (2018)	UK	Studies contain Businessman	The objective of the research is to determine the facility management manager during the building process can ensure sustainable policies which are not affected by following the facilities of delivery that plans and strategies which is kept up to Date	systematic literature searches of articles business literature from the user and organizational perspectives data and hand-searched the reference lists of retrieved articles containing qualitative	The results of this article give rise to advanced new claims and facilities	Limitation in having problems with models and selection Insufficient sample size for information measurements

(continued)

Table 1 (continued)

Reference	Country of Publishing	Participant Involvement	Aim	Methods	Outcomes	Limitations
Koch et al. (2019)	UK	Studies contain directors and patients	The aim of the research is to focus and highlighted the the significance of combination between the mechanical digitalization, abilities, association, and organisation of digital Facility Management	A community-based descriptive study	The outcome of the studies external advisors is involved, posing a problem of the in-house outsourced HR in the upcoming digital facility Management operation	The limitations of the study two case articles provide understandings, but those studies have the limited access on generalizability

studies have primarily shown the notion of significant creator interest in structuring all the activities at the occupancy stage. In many other studies, Wang et al. (2020) found that facility management helps in the training of the plan requirement. All the data are removed from the previous plans and buildings, which must be evaluated to grow design concerns for new services. The post-occupancy evaluations accordingly with the design of the facility which includes three main strategies (the high quality effect of organisational structures on the preferred results, the contribution of the organisation designs to cut off the non-relevant prices and increase the profits, and in last the effects of the organisational structure to improve the growth of human resources. Facility managers use the post-occupancy evaluation to improve building activities, and according to Stojanovic et al. (2018), the customer needs are frequently stated through the service level contract. The facility management services control the decision making of the facility management, which contain the growth of the administration, expenses, and human resources.

7 Sustainability

It has been decisively shown that numerous roles are required to involve and integrate sustainability into the value management study. It is also stated by Proskurin and Vorobeva (2019) that facility management manager must give the response of growth, including the sustainable issues. Though focusing too heavily on sustainable development, it might reduce problems at the macro-scale stage during the operational level, then the designs should have the capacity to lessen maintenance prices and ensure security. Pishdad-Bozorgi et al. (2018) reported that the active contribution of the facility management manager during the building process could also ensure sustainable policies which are not affected by following the facilities of delivery that plans and strategies which is kept up to Date. The usage of new materials and technological advancement could assist the control of energy consumption in the setting of adding up all the value of the current structures (Koch et al., 2019). The primary expert of the facility management is undoubtedly the facility director, away from the authorities stated above. According to Moretti et al. (2018) a facility director reports straight to the manager or top management of the organisation on facility matters and the performance. It is also received the official training in various skills like "human resource management skills", economic, organisational skills, and practical skills to achieve the firm for the purpose of increasing business goals. It is more likely to guide and accomplish a facility organisation team, which helps the daily operations process in the office while focusing on the strategic office development and organising matters (Lu et al., 2018). Furthermore, a facility director is also called the business director, who is accountable for endorsing the administration's efficiency and being protective of the workers' excellence of life. Output is approved by the facility director by guaranteeing that each facility that is significant in manufacturing functions correctly and successfully, and as detailed above, each space in the office gives in output. A facility director defends the excellence of life on workers by making comfortable and secure services and offices that assist workers doing the task efficiently. A facility director guarantees that no skill can generate harm, risk or threat to workers through preservation and regular examination of services. The business director, by accomplishing duties not only at the functioning level but also the corporate level, that has also performed jobs purposefully to make development and give the company a reasonable advantage. At the corporate level, a facility director needs to guarantee that facility policy plans match business planned strategies, capital costs are calculated and measured, worker productivity is exploited, and prices are minimalised and projected.

8 Digitalization and Performance Measurement in Facility Management

Immense information, internet effects and block cable are measured as digital technical tools in any business. Facility Management has been emerging through the close link with high technical developments in instant data and flexible statements. When the Digitalization Technology method started in the building, manufacturing, and the construction business, it was clear that workplace design was consecutively changed. Nowadays, the skill-driven variations in organisational work are moderately minor, though the full effects of current data and Public Services Skills on lessening the office service continue to be seen. An information management organisation for advantage care that can generate information and share data which is provided by the plan of the team associates who have a significant result on the activities of buildings. In the earlier units, the contributions of organization claims in Facility Management operations are argued with the previous case studies (Levo, 2021). The main aim of facility management is actually to support an organisation in increasing office management. The office, on the other hand, includes the individuals, the physical and fiscal resources, and the procedures and processes of a commercial; thus, the facility management activities should have an optimistic influence on these assets. Moreover, the planned aim of facility management is the actual organisation of facility assets and services in providing the primary support to all the operational processes of the organisations (Charef et al., 2018).

9 Building Information Model

Building Information Model is the procedure of data design and organisation for buildings through the design growth and process. The idea of the Building Information Model times back the previous year the 1970 and far along in the year 2000; it was effectively used mainly by the industry. Building Information Management was created from computer-assisted projects using Information and Communication

Technology and progressively established to generate BIM representations using the software equipment. The Building Information Model the idea was clear and studied in various research objects with different outlooks. This kind of research study specified the Building Information Model has features like the formation and sharing includes the data with other associates of the project. BI model was used as a tool for the improvement of sustainability of constructions by consuming the energy effectively and upholding a waste organisation procedure in the post created level in the organization. Building Information Model can be used for operational stage purposes with the great possibility, separately, for procedures and care management, the energy organisation, emergency organisation and quality organisation, along with the space organisation. Furthermore, BI Model is rarely applied in the operational phase, such as renovations and alterations. According to Kazado et al. (2019), the essential purpose of applying the Building Information Model is to make a facility notion that is nearly applicable to the complete development of a plan. BI Model used in the whole lifecycle of a project is applied Skill of producing, upholding, and using structure data for dealing with the operations and the strength preservation in their complete growth operations. The literature evaluation of study articles linked to Digitalization Technology of Facility Management specified that Facility Management has been developing with different certainty capture knowledge, such as involuntary ID, laser scanning, point influence and more, which can assist managers in gathering data about the operations of a created facility. Also, this expertise has enormous contributions to Facility Management as it allows facility executives and engineers to directly capture end users' requirements and preferences. Hosseini et al. (2018) claimed that knowledge of certainty charging deliver support by making digital models observing performance and certification. Evaluating Facility Management execution provides office managers with a clear summary of the strategy and plan of Facility Management. Decision taken by office managers for the policies generally vast effect on the effectiveness of facilities. Moreover, PM in Facility Management empowers resource managers to recognise the over a wide span of time operational productivity of the facilities and settle on compelling decisions for key preparation.

10 Project Stakeholder Management

Project "Stakeholder" Management includes the ways needed to identify the overall population, meetings, or connections that might disturb or be affected by the given task, to examine partner needs and their outcome on the responsibility, and to generate proper management techniques for effectively appealing partners in task selections and implementation. Partner management likewise focuses on constant communication with the partners to understand their needs and desires, inclining to matters as it happens, controlling clashing interests, and inspiring proper partner meetings in task picks and training. This contentment must be supervised as the main project aim. It is additionally characterised in the article that Preventive Management could be a proportion of how sufficient Facility Management is at recognising the necessities of the clients and its capacity to set up appropriate arrangements to fulfil such requirements and upgrade their worth. On the other hand, Ermolli (2019), represented PM as the most common way of evaluating progress towards acquiring pre-decided objectives, which remembers data for the proficiency as a way to such an extent that assets can be changed into labour and products, the nature of these results and yields and the adequacy of hierarchical destinations. Evaluating Facility Management execution provides office managers with a clear summary of the strategy and plan of Facility Management. Decision taken by office managers for the policies generally vast effect on the effectiveness of facilities. Moreover, PM in Facility Management empowers resource managers to recognise the over a wide span of time operational productivity of the facilities and settle on compelling decisions for key preparation. Emich (2020) opposed that PM refers to the ways of developing the sustainability of facilities by contributing the basic approaches to getting client fundamentals with least expenses and helps the course of asset shares and relocations. In the short-term, Dixit et al. (2019), placed that PM in The Key Performance Indicators can provide office managers with a precise examination to evaluate the performances of an office by utilising fitting boundaries produced from reviews and field studies. In their examination work, Che-Ani and Ali (2019) utilised a device of Key Performance Indicators coordinated from maintenance of board model to measure support of public consideration.

11 Conclusions

The study conducted mainly focused on the digitalization technology proof for its adoption in facility management and adequacy from the outlook of the management facility managers through the overview of various shareholder's outlook. In the first place the digitalized technology of the facility management has been discussed and the knowledge gap along with the service played an important role in this whole literature review. To conclude, the digital practices of the sustainable facility management are experiencing appraisal and transformation.

References

Alaloul, W. S., Liew, M. S., Zawawi, N. A. W. A., & Kennedy, I. B. (2020). Industrial Revolution 4.0 in the construction industry: Challenges and opportunities for stakeholders. *Ain Shams Engineering Journal*, 11(1), 225–230.

Atta, N., & Talamo, C. (2020). Digital transformation in facility management (FM). IoT and big data for service innovation. In *Digital Transformation of the Design, Construction and Management Processes of the Built Environment* (pp. 267–278). Springer, Cham.

Bharadwaj, A., El Sawy, O. A., Pavlou, P. A., & Venkatraman, N. V. (2013). Digital business strategy: toward a next generation of insights. *MIS Quarterly*, 471–482.

Carbonari, G., Ashworth, S., & Stravoravdis, S. (2015). How Facility Management can use Building Information Modelling (BIM) to improve the decision making process. *Journal of Facility Management*, 10(2015).

Chapco-Wade, C. (2018). Digitization, digitalization, and digital transformation: What's the difference. https://colleenchapcowadesafina.medium.com/digitization-digitalization-and-digital-transformation-whats-the-difference-eff1d002fbdf (Дата Звернення: 01.02.2020).

Charef, R., Alaka, H., & Emmitt, S. (2018). Beyond the third dimension of BIM: A systematic review of literature and assessment of professional views. *Journal of Building Engineering, 19*, 242–257.

Che-Ani, A. I., & Ali, R. (2019). Facility management demand theory: Impact of proactive maintenance on corrective maintenance. *Journal of Facilities Management*.

Cudmore, T. (2017). Wwhole life performance plus defining the relationship between indoor environmental quality and workplace productivity.

Dixit, M. K., Venkatraj, V., Ostadalimakhmalbaf, M., Pariafsai, F., & Lavy, S. (2019). Integration of facility management and building information modeling (BIM): A review of key issues and challenges. *Facilities*.

Eadie, R., Browne, M., Odeyinka, H., McKeown, C., & McNiff, S. (2013). BIM implementation throughout the UK construction project lifecycle: An analysis. *Automation in Construction, 36*, 145–151.

Emich, T. (2020). KIT-fachbereiche-facility management-research-digital transformation.

Ermolli, S. R. (2019). Digital flows of information for the operational phase: The Facility Management of Apple Developer Academy. *TECHNE-Journal of Technology for Architecture and Environment*, 235–245.

Fomunyam, K. G. (2019). Education and the fourth industrial revolution: Challenges and possibilities for engineering education. *International Journal of Mechanical Engineering and Technology (IJMET), 10*, 23–25.

Gopalakrishnan, S., & Ganeshkumar, P. (2013). Systematic reviews and meta-analysis: Understanding the best evidence in primary healthcare. *Journal of Family Medicine and Primary Care, 2*(1), 9.

Grant, M. J., & Booth, A. (2009). A typology of reviews: An analysis of 14 review types and associated methodologies. *Health Information & Libraries Journal, 26*(2), 91–108.

Greenhalgh, T., & Donald, A. (2000). Papers that summarize other papers (systematic reviews and meta-analysis). In U. K. London (Ed.), *Evidence based health care workbook* (pp. 111–129). BMJ Publishing Group.

Hosseini, M. R., Roelvink, R., Papadonikolaki, E., Edwards, D. J., & Pärn, E. (2018). Integrating BIM into facility management: Typology matrix of information handover requirements. *International Journal of Building Pathology and Adaptation*.

Kazado, D., Kavgic, M., & Eskicioglu, R. (2019). Integrating building information modeling (BIM) and sensor technology for facility management. *Journal of Information Technology in Construction (ITcon), 24*(23), 440–458.

Kensek, K. (2015). BIM guidelines inform facilities management databases: A case study over time. *Buildings, 5*(3), 899–916.

Koch, C., Hansen, G. K., & Jacobsen, K. (2019). Missed opportunities: Two case studies of digitalization of FM in hospitals. *Facilities*.

Lee, M. P. L., Kua, J. S. W., & Chiu, W. K. Y. (2001). The use of remifentanil to facilitate the insertion of the laryngeal mask airway. *Anesthesia & Analgesia, 93*(2), 359–362.

Levo, H. (2021). Real estate investors guide for sourcing property management function of facility management service in Finland.

Lu, Y., Papagiannidis, S., & Alamanos, E. (2018). Internet of Things: A systematic review of the business literature from the user and organisational perspectives. *Technological Forecasting and Social Change, 136*, 285–297.

Moretti, N., Dejaco, M. C., Maltese, S., & Re Cecconi, F. (2018). An information management framework for optimised urban facility management.

Pishdad-Bozorgi, P., Gao, X., Eastman, C., & Self, A. P. (2018). Planning and developing facility management-enabled building information model (FM-enabled BIM). *Automation in Construction, 87*, 22–38.

Proskurin, D., & Vorobeva, Y. (2019). Digitalization of energy facility management processes in the Voronezh region. In *E3S Web of Conferences* (Vol. 110, p. 2123). EDP Sciences.

Stojanovic, V., Trapp, M., Richter, R., Hagedorn, B., & Döllner, J. (2018). Towards the generation of digital twins for facility management based on 3D point clouds. *Management, 270*, 279.

Tilson, D., Lyytinen, K., & Sørensen, C. (2010). Research commentary—Digital infrastructures: The missing IS research agenda. *Information Systems Research, 21*(4), 748–759.

Wang, D., Khoo, T. J., & Kan, Z. (2020). Exploring the application of digital data management approach for facility management in Shanghai's high-rise buildings. *Progress in Energy and Environment, 13*, 1–15.

Wong, J. K. W., Li, H., & Wang, S. W. (2005). Intelligent building research: A review. *Automation in Construction, 14*(1), 143–159.

Wong, Johnny Kwok Wai, Ge, J., & He, S. X. (2018). Digitisation in facilities management: A literature review and future research directions. *Automation in Construction*, 92, 312–326.

Yoo, Y., Boland, R. J., Jr., Lyytinen, K., & Majchrzak, A. (2012). Organizing for innovation in the digitized world. *Organization Science, 23*(5), 1398–1408.

A Review of Civilian Drones Systems, Applications, Benefits, Safety, and Security Challenges

Khalifa Al-Dosari, Ziad Hunaiti, and Wamadeva Balachandran

Abstract Among the plethora of technologies revolutionising modern life, Unmanned Aerial Systems, also known as self-piloting drones, offer scope for profound changes in a variety of applications across many sectors. Commercially available drones of various shapes and sizes serve numerable practical user needs. Drones are expected to achieve great economies and improve sustainability, but significant legislative and technical challenges face their adoption and deployment in many counties. This chapter reviews civilian drone systems, and their applications, benefits, safety, and security challenges.

Keywords Drones · Civilian drones systems · Applications · Benefits · Safety · Security challenges

1 Introduction

A drone is a vehicle capable of flying or propelling itself into the air without the need for an on-board human pilot. Airborne drones can be controlled from the ground using flight controllers, remote control, or a ground control centre (GCC). A flight controller or remote control can be used by civilians to control their drone, often (for visual applications) with a small mounted camera so that the operator can see what the drone sees (i.e., the surrounding environment). A GCC is usually used in large-scale projects, such as military research, to access a variety of information (e.g., for aerial monitoring or surveillance).

K. Al-Dosari (✉) · Z. Hunaiti · W. Balachandran
Department of Electronic and Electrical Engineering, College of Engineering, Design and Physical Sciences, Brunel University London, Uxbridge, UK
e-mail: Khalifa.Al-Dosari@brunel.ac.uk

Z. Hunaiti
e-mail: ziad.hunaiti@gmail.com

W. Balachandran
e-mail: Wamadeva.Balachandran@brunel.ac.uk

M. Alshurideh et al. (eds.), *The Effect of Information Technology on Business and Marketing Intelligence Systems*, Studies in Computational Intelligence 1056,
https://doi.org/10.1007/978-3-031-12382-5_43

This chapter focuses on civilian drones, which come in various shapes and sizes, of which the four main commercially available civilian types are multi-rotor, fixed-wing, single-rotor helicopter, and fixed-wing hybrid VTOL drones (Smith, 2017). They all share essentially similar functional designs in terms of their flight properties, such as a fast-spinning propeller to generate enough thrust energy to lift the drone off the ground and stay afloat in the airspace. The selection of particular properties and specifications depends on the type or size of the drone and the intended application. For example, a relatively larger propeller with increased mass on a smaller drone would increase structural weight, which could actually inhibit drone flight. Therefore, a smaller propeller or the use of lightweight materials is more appropriate for smaller drone.

The most popular image of the application of drones is for localised aerial photography and recreational use, and potential delivery to consumers. This is the culmination of a long period of mainly military-led research and development, during which unmanned aerial vehicles and drones have evolved significantly. Drone research was particularly pioneered by the US Navy, which incorporated Sperry's three-axis gyroscopic flight control system (Altawy & Youssef, 2016) in a plane to test if it was able to fly on its own (unmanned). The first drone or unmanned plane was able to fly a distance of 1,000 yards. This time period was dominated by the prospect of world wars, and drone research thus had little interest in potential civilian applications. The prospect of the integration of drones into civilian life emerged during recent decades, commensurate with research into making drones safe for civilians, ensuring safety and security. The US Federal Aviation Administration (FAA) and the UK Fleet Air Arm are instrumental in the formulation of legislation and regulation governing drone production and use, for the safety of human operators, drones themselves and surrounding infrastructure and environments, and the general public.

While most commercial applications of drones nowadays relate to recreational use and aerial photography, there are many serious purposes and functions of drones, such as engineering inspections (e.g., to view structural integrity in accessible structures, such as offshore rigs or high-rise buildings, where direct human access would be expensive and difficult), monitoring sea level changes, and preparing evacuation plans for high-intensity rainfall events. Drones can also be used to help perform large-scale tasks productively, limiting the risk of human error. Aerial drones have applications that can benefit society by enabling the development of mapping technology, aerial photography, animal tracking, and delivery. These benefits can increase labour efficiency, produce aesthetic photographs or videos that add social value to product, and which can also provide political and economic value, by developing forecasting and mitigation methods for hazardous natural events to protect humans.

However, drones also raise challenges and concerns, which can be split into four categories: data and technical issues, privacy and ethics (mainly relating to personal data and privacy), security, and safety issues. The most direct potential threats of drones are unwanted surveillance and potential collisions with humans or infrastructure, causing injuries or damages. Cyber threats such as malware infecting the drone and allowing for unauthorised personnel to take control of the drone are issues that human operators need to be cognisant of, along with physical threats to drones themselves, such as malicious vandalism or theft. Due to such issues, security, safety, and privacy measures are being enforced to keep the surrounding areas, communities, and drones safe. There are different mitigation strategies pertaining to the application of security and safety measures that can prevent or reduce risks.

Sense-and-avoid mechanisms are excellent for the prevention of collision with obstacles such as buildings or tress, especially if the human operator does not react fast, or visualisation and communication with the drone is cut or disrupted. In such contexts, drones can have artificial intelligence (AI) capabilities to make appropriate autonomous decisions in the field. Such strategies are useful and beneficial in maintaining the security and safety of the drone as well as its surroundings. However, there are also disadvantages associated with potential drone add-ons, most notably the financial, time, and training costs associated with the purchase, installation, and use of advanced programmes or mechanisms. Furthermore, drones with such defences still face costly mitigation strategies, which could deter human operators from buying them, and instead choose to purchase a cheaper drone that may not have effective mitigation strategies; put simply, drone purchasing decisions involve cost-safety trade-offs relative to particular applications.

2 History of Drones

2.1 Military Research and Development

In modern society, drones embody the peak of technological advancements in many fields, located at the convergence of aeronautical engineering, smart technologies, AI, digital photography, and renewable energy and sustainable solutions. However, the history of drones is long and extensive, and is firmly rooted in military applications, as mentioned previously. The first recognised drone was documented by the US Navy (Parsch, 2005), resulting from a research project led by Elmer Sperry, the founder of the flight navigation control firm Sperry Corporation (Altawy & Youssef, 2016). The drone ("three-axis gyroscopic flight control system") successfully flew down the River Seine in France a month before WWI began (Flight Journal, 2018). Based on this success, Peter Cooper Hewitt began the Automatic Airplane or Aerial Torpedo partnership with Sperry in late 1916, which was fitted with the Sperry system and was able to fly a distance of 1,000 yards autonomously, following pre-set mechanical commands. This event, which occurred on March 6, 1918, is considered the first substantive unmanned aircraft vehicle (UAV) flight.

The next invention was the Sperry-Curtiss "Pilotless Flying Bomb" (Fig. 1). This attempt at a cruise missile had minor success in 1918, combining cutting-edge aeroplane and munitions technology, both of which were being fast-tracked by WWI. It was never produced or used again, and by WWII jet engines and rocket technology became the main concerns of military research for the areas tentatively addressed by

Fig. 1 "Pilotless flying bomb" (Parsch, 2005)

"drones" during WWI. There was a tentative resurgence of drones during the 1970s during the Vietnam War, for surveillance of Vietcong fighters and US prisoners of war, but more substantive use was evident in the widespread use of weaponised drones (in "drone strikes") by the US following the 2001 invasion of Afghanistan (Flight Journal, 2018). However, it should be noted that this was presaged by the use of drone reconnaissance during the US invasions of Panama (1989) and the Gulf War (1991) (O'Hanlon, 2011).

The main rationale for drone use was to minimise risks and casualties among US combat personnel (Satia, 2014). This lack of consequence for aggressors in drone strikes, and their technical imprecision, means that civilian massacres by drones have been one of the most egregious cases of US defence policy in Afghanistan, Iraq, and elsewhere, including the extra-judicial assassination of US citizens abroad, and legal provision for military drone strikes against US citizens within the US itself (posing legal and constitutional issues beyond the scope of this study) (Thompson, 2013). However, from the defence industry, drone development has been accompanied by conventional platitudes about the surgical nature of drone strikes and minimising civilian casualties, including the direct potential of drone observation of areas targeted for airstrikes to minimise civilian casualties (US Department of Defence, 2011).

2.2 *Civilian Issues*

Naturally, in the context of the carnage of the Western Front, these early military applications had no consideration of civilian safety, and they were in fact so limited in application that they posed no substantive threats. More recent military research and drone tests took place in US military testing grounds, with minimal human populations to get in the way (Altawy & Youssef, 2016). However, in the modern age, with the rapid deployment of commercial drones, civilian safety issues are paramount. The FAA and UK Fleet Air Arm are driving forces in reshaping the laws surrounding drones, with mandatory registrations of purchase and use of drones, to help identify any criminal uses (violating national airspace or conventional criminal laws). Drones are becoming a normalised commodity, accessible and serviceable to the public. Drones could be used for many purposes (mainly for aerial photography, mapping, and video recording), and may be deployed in commercial uses (such as delivery and transportation) and recreational applications (e.g., drone flying per se, racing with friends, entertainment, leisure, and photography) (Jones, 2017).

3 Different Types of Drones

There are four main types of drones commonly used by the general public, as mentioned previously: multi-rotor, fixed-wing, single-rotor helicopter, and fixed-wing hybrid VTOL drones (Smith, 2017).

Multi-rotor drones are widely used by creative industries and the academic sector, most commonly for aerial photography. These can be used in science research when observing inaccessible areas, such as densely forested green spaces, and to produce aerial digital content of urban landscapes. This type of drone is easier and cheaper to manufacture and use in comparison to other market-leading products. Multi-rotor drones are available in a number of different shapes and sizes, including a tricopter (with three rotors). However, these models have restricted flying time, rendering them unsuitable for long-term applications like mapping extensive areas and surveillance (Smith, 2017).

Fixed-wing drones have more particular design parameters than multi-rotor models, resembling the structure of a plane, with wing-like structures adorning the drone's main body to aid flight. They do not utilise energy to stay afloat, instead needing a remote unit for synchronous operation or implanted asynchronous guided control to remain airborne. This type of drone is more proficient for long-distance uses, and can be operated for long time periods, unlike the multi-rotor drone. These drones are more suitable for projects that need a huge distance or long hours, thus they are commonly used for aerial mapping or surveillance. However, they cannot be used for precise photography, due to their velocity of motion and inability to hover, although they are suitable for video-recording. This drone is relatively costly and requires expertise to operate (Smith, 2017).

Single-rotor drones resemble helicopter morphology. Whereas multi-rotor drones have multiple rotors, single-rotor drones have one eponymous rotor (accompanied by a smaller ancillary one on the tail, to control the heading). Much like the fixed-wing drones, this drone has long flying time, and it is more stable than the multi-rotor drone. However, it is complex and difficult to understand and operate. In addition, fatal injuries have been recorded with the use of this drone, and it could be very hazardous for inexperienced users (Smith, 2017).

Finally, hybrid VTOL is a drone model specialising in manual gliding and automation. Manual control or remote control or programmed flight course in the drone are some ways to control this drone. This drone has been used for commercial uses such as delivery services (Smith, 2017).

These four types are the most commonly used drones among the general public, and each has its particular advantages and disadvantages. They can all be used effectively for specific purposes, making them attractive to a customer in need of aerial support to execute various tasks. However, drone services come at the cost of dangers, for operators, infrastructure, and the general public. Drones are "airborne machines" that depend greatly on "cyber capabilities," which may pose many risks (Altawy & Youssef, 2016). The main risk of drones is collision, with people, property, or livestock etc. Drones could collide with airplanes when used (knowingly or unwittingly) in restricted airspace. Some units weigh over 10 kg (22 lbs), and any collision (particularly at high speed) can be catastrophic (Cerbair, 2020). While malfunctions are relatively rare in drones, bad weather or loss of battery are fundamental risks that can reduce operational efficiency and safety.

4 UAV/Drone System Architecture

Civilian drone technology utilises software to aid flight, ensuring its safety and security. Drones are available for purchase in a number of different sizes and shapes, depending on operator needs. A child could want a drone for entertainment purposes, and therefore purchase a small model, while a researcher who needs drones to map out the elevation of hills or mountains can purchase larger, high-specification models, equipped with different cameras and sensors to collect and generate large data sets. Figure 2 shows the design of a UAV in detail, illustrating the remote control and computer-based GCC methods of control. It shows the different vehicle properties and data links, and the individual components of a UAV (consisting of a propeller, compass, motor, and landing gear).

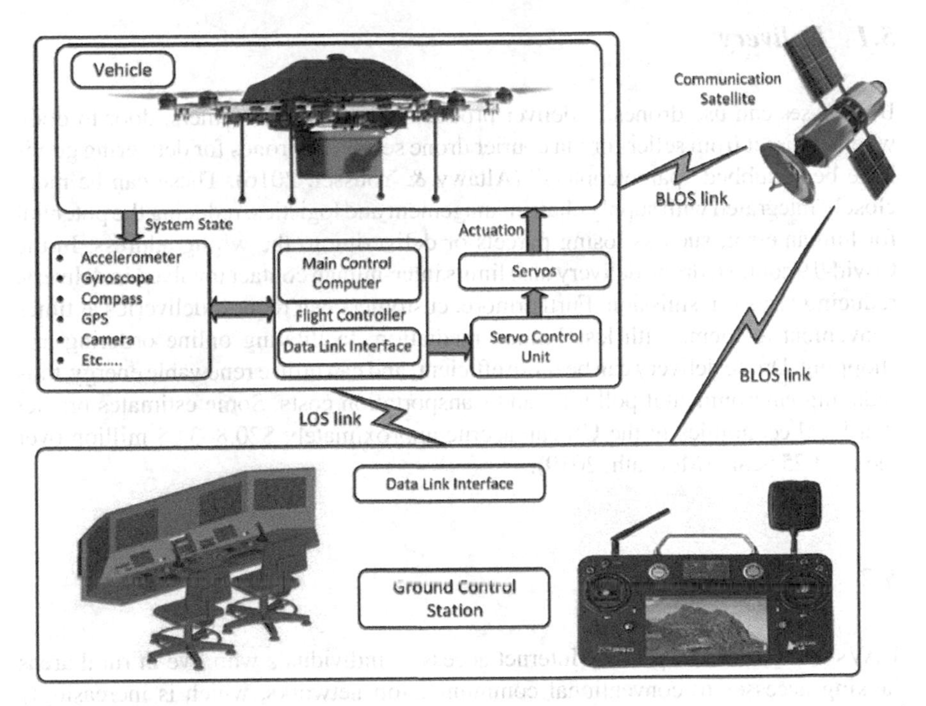

Fig. 2 Detailed overview of drone communication with human operator (Altawy & Youssef, 2016)

5 Civilian Drone Applications and Benefits

Technology advances relative to time and society moves forward with it. With drones evolving from one purpose to another, inevitably public access demand will increase. Previously, drones were very large and needed a GCC, but modern commercial and civilian drones are small, and only require a flight or remote controller. They have many applications that are beneficial to owners. The functions and purposes of drones are also increasing, which makes them more attractive to civilians who can afford to purchase them. These functions can vary from commercial to non-commercial sectors. The FAA suggests that the sales of civilian or commercial drones could approximately reach $2.7 million by 2020, which has important economic implications (Chang et al., 2017). Drone purchase and usage have the potential to improve and transform business, creating new industries and businesses, and cutting financial and environmental costs inherent in the current commercial delivery industry (Association for Unmanned Vehicle Systems International, 2013). The following subsections explore some of the applications of drones in more detail.

5.1 Delivery

Businesses can use drones to deliver products directly to customers, door-to-door, whether direct from sellers or via courier drone services. Drones for delivering goods have been dubbed "parcelcopters" (Altawy & Youssef, 2016). These can be more closely integrated with supply chain management and logistics, reducing the potential for human error, such as losing parcels or delivering to the wrong address. In the Covid-19 context, drone delivery also limits inter-human contact involved in delivery, reducing virus transmission. Furthermore, customers can request deliveries at times convenient to them, with less human mediation, facilitating online ordering and shopping. Drone delivery can be more efficient, and can utilise renewable energy, thus reducing environmental pollution and transportation costs. Some estimates predict that local economies in the US can accrue approximately $20.8–34.8 million over the next 25 years (McIlrath, 2019).

5.2 Internet Access

UAVs can be used to provide Internet access to individuals who live in rural areas lacking accesses to conventional communication networks, which is increasingly essential in the era of smart technologies and global interconnectivity.

5.3 Cinematography and Aerial Photography

Filmmakers can use aerial drones to film areas that are hard to reach or inaccessible to humans. For example, drones can be used to video tape an area from the sky, should the scene need the tops of buildings. These drones create countless possibilities for creative scenes and give the opportunity for the filmmaker or photographers new platforms to expand their artworks (traditionally, helicopters were used for aerial shots used in films, which is extremely expensive). Drones can also be used to document intangible cultural heritage for 3D reconstruction and preservation. They are also used by individual photographers to capture pictures with a bird's eye view (Rango et al., 2006). Aerial photography is popular for both recreational and business purposes. Indeed, one of the main advantages of drones for many commercial users is their high-quality imaging and live streaming. Drones have cameras that can take high-quality photos and videos, and a wide range of high-resolution data, such as the exact coordinates of the location. This helps if a person wants to go back to the area. Drones can capture video footage in high resolution, and they are commonly used for live steaming events, important occasion, or memories. Drones can be used of live streaming for political events, such as presidential election debates, or for entertainment events such as music concerts.

5.4 Mapping

Drones can collect a variety of high-resolution data that can be used to create accurate 3D maps and models, which has numerous applications. For instance, this can discover erosion or conservation needs, and map disaster areas to streamline relief, ensuring that rescue teams understand the terrain and are equipped with pertinent knowledge before entering hazardous zones.

5.5 Agriculture and Crop Management

Drones can be used to perform specific tasks after system customisation, such as to fertilise, irrigate, or monitor crops, warning operators of particular water or nutritional deficiencies (Altawy & Youssef, 2016).

5.6 Construction

Construction activities or tasks can be aided by the use of civilian drones to accelerate data collection, improve data quality, and reduce the costs of data collection. Drones can help with different aspects of construction projects, such as in geological survey (McIlrath, 2019). This is beneficial as drones' multipurpose functions can assist a wide range of project stakeholders. The usage of drones for this purpose can also help with local economies, with potential benefits of $690.1–1,116.7 million over the next 25 years (McIlrath, 2019).

5.7 Ambulance and Medical Services

Drones can be used to deliver emergency response equipment or medication to different patients, which is especially important when human paramedics or doctors are unable to reach patients in a timely fashion. Traditional delivery methods may be hindered by a variety of reasons, particularly during emergencies such as natural disasters, so this drone function is highly useful. In terms of everyday healthcare services, "medical parcelcopter" drones are already used to deliver medication to patients. Ambulance drones can provide emergency medication when external roads are blocked by debris. They can also be equipped with monitoring sensors to assess the state of the survivors and report back to human paramedics or medical services the injuries or trauma of survivors, as well as being used to assess the general damage done by disasters.

5.8 *Ensuring a Safe Environment Exists*

Drones can be used to promote safe environments, enabling local authorities to monitor areas clearly and in live moments, instead of relying on CCTV that could be old, broken, or missing required coverage and footage. Drones can monitor situations and potentially be able to report potential threats to local authorities and warn people of potential dangers and unsafe conditions (Altawy & Youssef, 2016). For instance, drones equipped with cameras that can evaluate traffic patterns. This is beneficial in helping drivers avoid high risk areas and can reduce the risk of traffic collisions or accidents, and consequently casualties and deaths. Adding to the military uses discussed previously, drones also help to scope target sites and check areas prior to the deployment of human troops.

6 Civilian Drone Safety Challenges

Civilian drone challenges include those to application or programme used for the safety of the drone itself, or those to surrounding people and areas.

6.1 *Weather*

Different weather conditions, particularly storms, can hinder drone manoeuvrability and navigation systems. Different systems such as a sense-and-avoid or situational awareness must be implemented to ensure drone safety.

6.2 *Friendly Drone Collision*

It is not uncommon for multiple drones to be in use simultaneously, raising the potential for collision if operators are unable to react fast enough and steer their drones to safety. Sense-and-avoid systems or AI can be beneficial to avoid this issue.

6.3 *Theft and Vandalism*

Drones can be stolen or vandalised by humans if in close range. AI or a facial recognition systems can be used to avoid this issue.

6.4 Protective Programmes

Numerous protective programmes may be installed within the UAV to protect the surrounding environment, including people, property, and other aircraft from unexpected consequences of dribe operation, as described below.

6.4.1 Sense-and-Avoid

Sense-and-avoid systems allow aerial drones to autonomously detect potential collisions with other aircraft or aerial objects and to take evasive actions, dodging aerial hazards like a human pilot would, ensuring drone safety. This reduces the risk of the drone being damaged, and therefore minimises any repair costs. However, sense-and-avoid does not give a blanket guarantee of collision avoidance; the FAA regulation states that drones must deploy an automated sense and-avoid intelligent system that provides safety levels equal to or even exceeding that of manned aircraft (Altawy & Youssef, 2016). One example of how a drone senses and avoids a collision is a safety parachute system being activated when it senses that the drone is out of batter or terminated. This aids in minimising the damage when a collision is unavoidable.

6.4.2 Situational Awareness

Drones must possess a minimum level of situational awareness to ensure they are able to deal with any situation that may occur in their location. This system can predict real-time information and data about the location or situation to the human operator concerning weather and human population size (according to need). For example, weather data is useful to determine if it is an important factor in erosional landforms looking the way they do (when using drones for geological survey). The system could also have a thermal camera (Altawy & Youssef, 2016). Benefits of this awareness system are that it mitigates any risk of collision. It is able to sense and observe any collision obstacles and set a trajectory course to avoid them safely.

6.4.3 Artificial Intelligence

AI drones operate autonomously, without direct supervision by human operators. Algorithms allow drones to fly and react to their situation in a quick and responsive manner. AI mathematical features and control theory-based algorithms allow quick autonomous decisions and reactions. Civilian drones need to have some sort of AI capable of sensing danger or scanning for threats in the vicinity. After they sense this, they will then need the AI to out a reaction plan or different ways to avoid the threat. This type of drone could have tactical sensors to observe the area, or any changes that could potentially harm the drone, and facial recognition software to allow only authorised personnel to access the system.

A localisation algorithm and technology could also be installed in AI drones to ensure they know where they are located, and where to go, should a human operator install co-ordinates for the drone to land at a specific time. This technology is beneficial as it can tell the drone where buildings are located, or which properties forbid the possibility of landing (e.g., should the batteries run out). Another advantage of this technology is that it is also able to tell the drone which airspace is illegal to enter, thereby reducing the risk of human owners or operators facing fines or criminal charges for trespass.

General benefits of AI algorithms installed in a drone mainly relate to improved capability to fly safely due to mathematical algorithms for collision avoidance and decision-making to avoid or respond to collisions. It can recognise operators or different locations, using facial recognition. AI capabilities are also essential in tasks in agriculture and military deployment due to faster reaction times than humans.

General disadvantages of this system include its expensive cost price, and objections to the use of facial recognition, and related privacy and security issues. Experienced programmers are also required to safely and successful implant AI programmes in drones, involving complex programming and long installation times. Additionally, the very efficiencies of AI drone systems can reduce employment in related fields rendered redundant by advanced technologies.

7 Cyber-Physical Security Challenges

7.1 General System Challenges

Cyber-physical security is created and then implemented in drones to protect different technologies and processes. The main components of a drone are a propulsion, autopilot, and recovery systems, aside from the physical structural airframe. Drones can be vulnerable to cyber attacks because of their dispersed physical systems. This can cause defective operation of the control loop, which is why cyber-physical security is necessary to ensure the security and effectiveness of the drone. Unauthorised personnel or attackers can manipulate drone processes and functions and misuse, such as denial of service attacks, whereby human operators lose all control of the drone. They can also sabotage operations, deny process or system functions, and corrupt data (Siddappaji et al., 2020).

However, drone cyber-physical security systems are quite efficient, and are able to deal with various real-world processes, adapting themselves to deal with changes in the surrounding area. They can perform calculations quickly, without wasting time, ensuring the safety and effectiveness of the drone, while avoiding obstacles or crashes (Siddappaji et al., 2020).

The main disadvantage of such systems is their cost. For an operator to ensure its drone is safe, they would need to have money on hand to install different security programmes to tackle different areas. They also only target certain areas, so again the operator would need to buy and install an ancillary system to deal with other processes that the cyber-physical security system does not address. These systems are relatively new to civilian drones, and therefore little research has studied their impacts, or potential installation in smaller drones (Altawy & Youssef, 2016). Also, the security community has started to identify categories and spectrums of emerging threats a civilian could encounter, and a lot of areas have been barely looked at (Altawy & Youssef, 2016).

7.2 Civilian Drone Security Challenges

7.2.1 Authentication

Aerial drones must have the means to ensure that only the authorised personnel can access them. They must be protected from hackers or cyber threats. One threat to a UAV is the possibility of it being hacked by unauthorised personnel who can then take control of the UAV, or inject falsified flight controller sensor data. A core security requirement of every drone is the guarantee that only authorised personnel can access the drone and its resources. Authentication mechanisms can implement distance protocols for authentication, determining the physical proximity of communicators within a system. This extra measure can mitigate spoofing attacks, as it is difficult for attackers to get close to the their target (Altawy & Youssef, 2016). This mechanism must be implemented in aerial drones to reduce chances of unauthorised personnel from either accessing the aerial drone or the GCC. It mitigates the risk of hacking or unauthorised personnel stealing the drone, or committing criminal acts with it.

This mechanism is beneficial as it allows the drone to always be safe from external threats. It reduces the risk of the authorised personnel accessing or stealing drones. This reduces the chance that the owner of the drone will need to pay for a new drone, or worry about the data such as photos taken on the drone being lost. It could also reduce the risk of insurance being paid for ether the stolen item or damages that the unauthorised personnel is creating with the drone. Other examples of mitigation techniques include cross-verification with readings from alternative sensors or firewalls. Cross-verification with readings from alternative sensors allows authorised personnel to ensure that the drone's main sensor has not been hijacked or is not displaying falsified data. Firewalls are beneficial and effective if they are implanted in drones, filtering threating from authorised communications. They are capable of excluding unauthorised communications, but they could be ineffective against malware, hence additional protection for the drone is required. This includes additional costs, and it can take time to install the new anti-malware or protection systems and programmes.

7.2.2 System Integrity

Common cyber threats of concern include unauthorised disclosure of the GCC, telemetric signals from the remote controller, and data stored within the drone. Aerial drones should be able to guarantee the authenticity of their software and hardware components, which consumer trust about their data being stored in hardware. Some trusted techniques that can help with this include memory curtaining, sealed storage, and remote attestation (Altawy & Youssef, 2016). Other ways to mitigate risks include the encryption of telemetric channels and data links, which guarantees the authenticity of system firmware and sensitive data. For extra protection, especially against viruses or malware, anti-virus software and firewalls can be used (Altawy & Youssef, 2016). These mitigation strategies are beneficial in reducing the risk of data falling into the wrong hands, and the costs of covering data breaches. If a business uses a drone for research or purposes, data breaches can cost an average of approximately $3.86 million (Ponemon Institute, 2018).

7.2.3 Accountability of Actions

The aerial drone should have multiple mechanisms to ensure operators of the drone are accountable and responsible for their actions. To do this an aerial drone can have a digital signature algorithm to authenticate operators and then to bind them to an issued action (Altawy & Youssef, 2016). In addition to this, logging procedures should also be implemented to track the sequence of actions in a chronological manner, tracking who executed what actions, and when. However, the possibility of holding the owner accountable for their actions using this mechanism is relatively low, as the drone is not able to physically stop the owner, who may have authorised remote accesses. On the other hand, drones could send alerts or warning messages to nearby police stations if owner actions are detected as potentially criminal.

Furthermore, laws and registration schemes are becoming more prevalent. In the UK for example, drone users must take an online test and pay a £9 fee licence per year if they wish to fly a drone weighing between 250 g and 20 kg. Other requirements are not exceeding an altitude of 400 ft above the ground or surface level (Cunliffe et al., 2017). If a user is caught flying a drone without a three-year flyer's ID or one-year operator ID they could face a fine of £1,000. This registration scheme has become mandatory due to civilian fears about the possibility of drone misuse, such as stalking and unauthorised voyeur photography. New laws and requirements brought by the UK's Civil Aviation Authority are a safety measure to identify the owners of a drone if unacceptable actions occur, making it easier to hold owners accountable. However, there is no way of knowing if everyone registers, particularly as drones become more widely available in supermarkets etc. and online, and therefore many illegal actions could go unnoticed. However, *effective* registration schemes can be advantageous in guaranteeing accountability in drone use.

7.2.4 Information Confidentiality

The confidentiality of telemetric information and GCC signals is a safety requirement that an aerial drone should protect, without which data can be disclosed to unauthorised personnel. One mitigation technique is the authentication of communication links or the encryption of data links. Aerial drones should be able to control the information being placed in them and keep it confidential, with systems and mechanisms to reduce the risk of unauthorised disclosure of telemetric and control information. Different encryption standards such as AES can be used for data link encryption, to enhance data security and strengthen protection and firewalls used to keep drone data confidential. Auto encryption of data is generally used for additional protection against unauthorised access, which avoids the costs of manual encryption and potential encryption errors.

Disadvantages of these methods include the minimal protection of data in transit, such as during transfer to a computer, and they slow down device speed to view data (e.g., possibly demanding an authentication key every time a user tries to access data). These factors also complicate data recovery.

7.2.5 Spoofing Attacks

Spoofing is a method by which unauthorised personnel or attackers disguise their manipulation or hacking into a system as something else, exploiting fake data or normal system aviation readings. This ensures that the owner or operator is not aware of the breach, and rather just thinks it is a normal system. Spoofing attacks render a variety of sensors redundant, further allowing the drone to be breached, whereby attackers can gain access to the server or networks, and to data or information stored within a drone. This threatens the integrity of telemetric information, control signals being sent form controller or control centre to the drone, and potentially to GPS locations. Some mitigation strategies against this include the installation of anti-malware, firewall, intrusion detection system, and defences against sensor readings being fabricated.

8 Key Challenges for Civilian Drone Safety and Security Applications

Civilian drones face challenges that can hinder deployment and cause serious harm to those around them, or to the drone itself. Drones are becoming increasingly popular, due to natural market development and improved commercial viability of drone purchasing, and the uptake of alternative hobbies following the Covid-19 pandemic and lockdown experiences. Business and education have started to move online or find new ways to adapt to modern technologies, and drone delivery to online customers is a trailblazer in this process. The increasing proliferation of drone usage creates many challenges, which fall under four categories: design and technical, privacy and ethics, drone security, and drone safety challenges.

8.1 Design and Technical Challenges

Delayed responses to instructions can cause technical issues. The control of civilian drones is dependent on control signals from either the GCC or the flight controller/remote controller. Sometimes delays can occur; if the drone is far away from the main control signal, then information being sent form that point could delay reaching the drone. Consequently, the operator might instruct a drone to swerve to avoid an obstacle, but the lag time in communication can result in collisions and damage.

Injecting falsified sensor data is a challenge that can create technical issues and destabilisation for drones, by comprising some drone sensors by inserting false and fabricated readings in the sensors and flight controllers. This undermines the authorised personnel's control of the drone and causes the drone to be easily accessible by unauthorised personnel. The design of the civilian drone can make it easier for unauthorised personnel to manipulate the drone, such as through external influence on sensors, such as radar.

Malevolent and malicious hardware and software can create a lot of issues in the drone. The civilian flight controller can be vulnerable to software or hardware Trojans, which are viruses deigned to disrupt service, steal data, and otherwise inflict damaging actions on data or networks. In 2011 a Trojan was detected at a ground control unit in Creech (UK), which infected Royal Air Force systems with viruses and malware (Hartmann and Steup, 2018). One of the soldiers, a key-logger, found traces of malware in the operators of drones that flew over Iraq and Afghanistan. There was no consequence of this incident, but was security increased due to the worry that sensitive data was copied. This could happen to civilian drones if they are not properly protected. One example of a malware or virus that can impact negatively or infect a drone is a software known as Maldrone (Altawy & Youssef, 2016), which allows unauthorised personnel to control the drone by opening a back-door connection whereby it acts as a proxy for flight control, sensors, and all communications. The malware opens a back-doored connection with its botmaster to receive commands, enabling unwanted access to communication and data. Overall, this malware is able to grant the unauthorised personnel the opportunity to control and land the drone as they wish (Altawy & Youssef, 2016).

A more basic design challenge is the size of drones; relatively large drones can be intrusive and objectionable for the public. The visual perception of drones can lead to people feeling threatened, and the reality of the drone can create new notion and perceptions relating to security and privacy. Small drones generally do not scare people, and can make them feel at ease, particularly when there is no evidence of the presence of cameras. However, some people may also react in a negative manner or feel suspicious to smaller drones and potential stealthy surveillance activities.

People could worry that small drones could sneak into air vents or other small places and restricted locations within and outside buildings. Areas that are not accessible to humans can be accessed by drones undetected; a study on understanding human perception on drones found that people thought that drones smaller than the Parrot AR were preferable (Chang et al., 2017).

8.1.1 Privacy Ethics and Cultural Challenges (Personal Data)

The general reflexive public opinion of drones is that they do not feel safe with the idea of having civilian drones flying in the sky (Chang et al., 2017). There is a high possibility of such drones taking photos or videos of them, and the difficulty of knowing if drones are present (e.g., drones with long-range cameras) can make people scared and paranoid. People are worried about their privacy being violated by drones capable of surveillance, photography, and video-recording, which frightens people. The challenge therefore is to educate the public in understanding or accepting drones. They need to understand the ethical and legal frameworks prepared for responsible drone use. Drones have multiple functions that can assist human life, and understanding the different purposes could put people at ease. Legislation and regulation of drone ownership and use can also comfort the public, particularly with regard to data protection. Images or videos obtained from drones should not be used without the consent of subjects, and conventional data protection legislation should apply to drone footage and other data.

Ultimately, responsible drone use depends on the responsible actions of owners and operators, and the basic presumption of innocence and trust. However, this is not a realistic proposition given the easy potential for misuse and increasing drone usage makes it harder to detect and monitor drone activity. Unregistered users unknown to the government or other authorities can be harder to identity. Ineffective laws and regulation make it easier for civilian drones to be used in harmful and illegal ways that genuinely threaten civilians. Common ways drones can be misused for illegal purposes are unauthorised tracking and surveying people and property without consent, but criminal uses of drones can be as severe as terrorist attacks or targeted assassination (Altawy & Youssef, 2016).

8.1.2 Security of Drone Itself

The main challenges to drone security come from cyber threats. One way to endanger the security of a drone is through a cyberattack on the wireless link. Drones can be operated by computers or a remote control, which can take over on-board multiple networks of sensors and actuators, which normally feedback or communicate to the operator on the ground via a wireless link (Altawy & Youssef, 2016). This communication channel is vulnerable to cyber-attacks. An example of this occurring was when military operators lost control of a drone due to being taken over by unauthorised personnel employing a mix of cyber-attacks on unmanned aerial vehicles.

All communications from the GCC to the drone were shut off by jamming both the ground control signals and the satellite. A GPS spoofing attach was employed by the unauthorised personnel into the feed of the unmanned aerial vehicle with a modified GPS data, which forced the drone to land in a new destination, while making the drone think it was landing back to its original destination that the military coded into it. This attack was carried out on a military drone, but the technology in this drone can be similar to civilian ones. The approach employed by the unauthorised personnel can be used to hijack a civilian drone.

A physical or cyber attack on the flight control or GCC causes the drone to be unmanned. This can create malfunctions within the drone, causing it to crash and potentially cause harm, due to the drone's dependence on information being fed to it from the flight controller or GCC. This is achieved via a data link and a sensor feeding back information about the surrounding environment. If unauthorised personnel are able to manipulate the internal communication system or data link they can falsify data, making the operator think they are in control, or making the drone think it has not been hijacked, therefore safety measures are not activated to counter the attack.

8.1.3 Safety of Drone Itself

Physical damage to the drone interacting with its surroundings is a safety issue. Collison is the main risk threatening drones and objects with which they collide. Inexperienced or amateur operators can easily accidently collide with other drones or with objects, and the drone may lose connectivity with the flight controller in the hands of the operator. Collision with people or animals can cause injuries or even death, and collision with infrastructure (e.g., buildings) can completely destroy the drone, and potentially cause criminal damage (Chang et al., 2017). For a successful and safe drone flight, the operator needs to see what the drone sees, and thus when to dodge or turn left or right if an obstacle is in the flightpath. Video footage captured by drones mounted with cameras is beneficial for collision avoidance and navigation, keeping the drone safe and reducing the risk of damage. An operator requests live video footage from the kernel of the operating system via a computer by issuing a system call. Unauthorised personnel or attackers who have knowledge of the system and its parameters are also able to gain access to the flight controller via this route. They can intercept the system call that the operator issues to the kernel, and then fabricate the footage that the operator sees. With a GPS spoofing attack, the attacker has full control of the drone.

This can consequently lead to the operator thinking they have control of the drone, when in reality the attacker has the control. The attacker can then land the drone with no consequence from the operator. Having the drone stolen or vandalised is an issue that operators could face. Flying low to the ground or at a distance that is easily seen by people makes drones attractive targets to thieves, and targets of vandals. There are multiple ways in which a drone can be brought down, such as dart guns and rifles, or physically grabbing it (e.g., with a net) if the drone is low enough. Police are issued with specialist anti-drone rifles, to shoot down drones in violation of laws (e.g., unlawful surveillance of civilians). These rifles are designed and created to disable drones immediately, without damaging them. They use radio pulses to disrupt the data link communication within a drone, forcing the drone to activate failsafe protocols, whereby it hovers closer to the ground in preparation to land safely. As mentioned above in relation to security challenges, the use of malware and software can be used to bring a drone down. The intentional infection of a software virus, such as Maldrone, can bring a drone down, whereby attackers can retrieved the drone from the landing site.

9 Summary and Conclusion

This chapter explains the history of UAVs, the architecture of drones, different applications and mechanisms within drones, and associated benefits and challenges. We have also classified the different types of cyber and physical attacks that can affect private and public sector drones, as well as the different techniques used for this to occur. Some case studies and examples have been explored to evidence common attack methods. Mitigation strategies have been provided to inform operators of the different safety techniques available. Finally, we have also discussed the consequences of drone use on civilian privacy. In conclusion, drones have great potential to modernise and digitise many industries, but there are number of challenges that need to be seriously addressed in order to facilitate the appropriate deployment of drones applications, in order to ensure the safety and security of drones and people.

References

Altawy, R., & Youssef, A. M. (2016). Security, privacy, and safety aspects of civilian drones: A survey. *ACM Transactions on Cyber-Physical Systems, 1*(2), 1–25.

Association for Unmanned Vehicle Systems International. (2013). The economic impact of unmanned aircraft systems integration in the United States. *Association for Unmanned Vehicle Systems International.* Retrieved October 8, 2021, from https://higherlogicdownload.s3.amazonaws.com/AUVSI/958c920a-7f9b-4ad2-9807-f9a4e95d1ef1/UploadedImages/New_Economic%20Report%202013%20Full.pdf.

Cerbair. (2020). Drone risks. *CERBAIR.* Retrieved October 8, 2021, from https://www.cerbair.com/drone-risks/.

Chang, V., Chundury, P., & Chetty. M. (2017). Spiders in the sky: User perceptions of drones, privacy, and security. In *Proc. 2017 CHI Conference on Human Factors in Computing Systems* (pp. 6765–6776).

Cunliffe, A. M., Anderson, K., DeBell, L., & Duffy, J. P. (2017). A UK Civil Aviation Authority (CAA)-approved operations manual for safe deployment of lightweight drones in research. *International Journal of Remote Sensing, 38*(8–10), 2737–2744.

Flight Journal. (2018). Drones: Technology changes the face of combat. *Flight Journal.* Retrieved October 8, 2021, from https://www.flightjournal.com/drones-technology-changes-face-combat/#visitor_pref_pop

Hartmann and Steup. (2018). Drone Pro Skills, Drone calibration: The complete guide. *Drone Pro Skills*. Retrieved October 8, 2021, from https://droneproskills.com/drone-calibration-guide-dji-mini-drone-calibration-trimming/.

Hydro Systems. (2018). What you should know about… landing-gear, *Hydro Systems*. Retrieved October 8, 2021, from https://www.hydro.aero/en/newsletter-details/what-you-should-know-aboutlanding-gear.html.

Jones, T. (2017). *International Commercial Drone Regulation and Drone Delivery Services*. Rand Corporation.

McIlrath, L. (2019). Drones benefit study: High level findings. *Me.consulting*. Retrieved October 8, 2021, from https://www.transport.govt.nz/assets/Uploads/Report/04062019-Drone-Benefit-Study.pdf.

O'Hanlon, M. E. (2011). *Technological Change and The Future of Warfare*. Brookings Institution Press.

Parsch, A. (2005). Curtiss/Sperry 'Flying Bomb'. *Designation Systems*. Retrieved October 8, 2021, from http://www.designation-systems.net/dusrm/app4/sperry-fb.html.

Perez, D., Maza, I., Caballero, F., Scarlatti, D., Casado, E., & Ollero, A. (2013). A ground control station for a multi-UAV surveillance system. *Journal of Intelligent & Robotic Systems, 69*(1–4), 119–130.

Ponemon Institute. (2018). Cost of a data breach study: Global overview. *IBM*. Retrieved October 8, 2021, from https://www.intlxsolutions.com/hubfs/2018_Global_Cost_of_a_Data_Breach_Report.pdf.

Rango, A., Laliberte, A., Steele, C., Herrick, J. E., Bestelmeyer, B., Schmugge, T., & Jenkins, V. (2006). Using unmanned aerial vehicles for rangelands: Current applications and future potentials. *Environmental Practice, 8*(3), 159–168.

Satia, P. (2014). Drones: A history from the British Middle East. *Humanity: An International Journal of Human Rights. Humanitarianism, and Development, 5*(1), 1–31.

Siddappaji, B., Akhilesh, K. B. (2020). Role of cyber security in drone technology. In *Smart Technologies*, K. Akhilesh and D. Möller, Eds. Singapore: Springer, pp. 169–178. https://doi.org/10.1007/978-981-13-7139-413.

Smith, E. (2017). Types of drones: Explore the different types of UAV's. *Circuits Today*. Retrieved October 8, 2021, from https://www.circuitstoday.com/types-of-drones.

Thompson, M. (2013). Legality of armed drone strikes against US Citizens within the United States. *BYU Law Review*, 153–182.

US Department of Defence. (2011). Rise of the drones: UAVS after 9/11. *Defence Talk*. Retrieved October 8, 2021, from https://www.defencetalk.com/rise-of-the-drones-uavs-after-911-37432/.

How to Build a Risk Management Culture that Supports Diffusion of Innovation? A Systematic Review

Mohammad N. Y. Hirzallah and Muhammad Turki Alshurideh

Abstract This research paper aimed to produce a systematic review of two of the critical terms in any organisation: risk management (RM) and diffusion of innovation (DOINV). The research sought to examine the relationship between RM and DOINV. The systematic review comprised 27 articles published from 2007 to 2020 sourced from different databases, such as; ProQuest, Emerald, ScienceDirect, IEEE, Scopus, and Google Scholar. A set of inclusion and exclusion criteria were used to filter the results, which were then assessed using a quality assessment process. Factors that appeared more frequently were highlighted as the main findings for future study. A proposed framework for future study was also defined at that point. Next, a comprehensive analysis was conducted on the 27 selected articles to answer the five research questions. This researches' main finding of this research was that a risk management culture positively supported the diffusion of innovation in firms. Several factors affected this relationship, such as; organisation learning, intellectual capital, human resources management, information technology, soft total quality management, knowledge management practice, strategy and structure.

Keywords Systematic review · Risk management · Firm performance · Diffusion of innovation

M. N. Y. Hirzallah (✉)
College of Business Administration, University of Sharjah, 27272 Sharjah, United Arab Emirates
e-mail: mhirzallah@sharjah.ac.ae

M. T. Alshurideh
Department of Management, College of Business, University of Sharjah, 27272 Sharjah, United Arab Emirates
e-mail: malshurideh@sharjah.ac.ae

M. Alshurideh et al. (eds.), *The Effect of Information Technology on Business and Marketing Intelligence Systems*, Studies in Computational Intelligence 1056,
https://doi.org/10.1007/978-3-031-12382-5_44

1 Introduction

Establishing something new is the essence of product innovation. Since this process necessarily involves early risk identification and management are required in innovative firms. Therefore, this paper aims to explore methods for managing risk in the diffusion of innovation. At the same time, the proposed method for managing risk in a prespecified innovation will be further explained. Risk is inherent to the activities of innovation companies; managing such risks represents an opportunity for companies to improve their capability to achieve their goals. The market of innovation development has been increasing, year by year, encompassing a greater and greater share of the economy. It is different from other industries because it consists of many small companies (Lowman et al., 2012). Together with other countries with more developed economies, the UAE has conducted a great deal of research related to this topic. In contrast, the number of such studies in developing countries has been limited (Nagano et al., 2014).

1.1 Research Questions

This systematic review has focused on the relationship between risk management culture and diffusion of innovation and has studied the factors and variables that affect this relationship. More precisely, this review performed its analysis by answering the research questions below:

RQ1: What are the primary research purposes of the studies selected?
RQ2: What are the main research methods of the studies selected?
RQ3: What are the participating countries in the context of the studies selected?
RQ4: What is the context of the studies selected?
RQ5: What is the study distribution by publication year regarding RM and DOINV?

1.2 Research Objectives

This systematic review has focused on developing many skills for new DBA students, such as:

1. Using different databases to develop research skills and techniques.
2. Learning the systematic review process.
3. Expanding knowledge concerning the search topic.
4. Discovering the factors affecting the relationship between RM and DOINV.
5. Creating ideas about future studies concerning the same topic.

1.3 Research Importance

Many firms have used RM as a competitive advantage. If risks are managed correctly between all parties in an organisation, everyone will act positively to achieve the organisation's goals and objectives. This study will help researchers see what other researchers have already achieved in this field, build from where others have stopped, and understand the various issues related to this topic: RM and DOINV—in addition to exploring the factors that affect the relationship between RM and DOINV. Finally, learning how to conduct a systematic review.

2 Literature Review

2.1 Risk Management

For companies to launch new products speedily and successfully, taking risks is essential. The ability to identify and manage risk is vitally important in risky innovation.

Definition

There has not been a single, universally accepted definition of the word risk (Green and Serbein, 1983). Its definition has changed as it has become interwoven with innovation and the rapidly globalising world. Companies that wish to survive need to innovate at a previously unparalleled rates and within a framework of greater uncertainty. This situation means their risks are deepening (Taplin, 2005). In more technical and specialised literature, as Kalvet and Lember (2010) pointed out, the word risk implies the measurement of the chance of an outcome, the size of the outcome, or a combination of both. According to the standard definition of risk, it is "the combination of the frequency or probability of occurrence and the consequence of a specified hazardous event"(Edwards et al., 2005). Some former writers in this field have distinguished between uncertainty and risk.

A risk situation is defined as a probability distribution for consequences is made on a meaningful basis, agreed upon by a set of relevant experts, and therefore it is 'known'. Uncertain situations arise when an agreement among experts cannot be gained, so there will be an undefined probability distribution on the set of outcomes (Vargas-Hernández, 2011).

Sources of Risk

Any factor affecting project performance can cause risk, and when this effect is both significant and uncertain on project performance, risk arises (Chapman & Ward, 1997). Fort-Rioche and Ackermann (2013) argued that categorising risk in a simple manner could be very unhelpful since such categories could be viewed independently. Thus, when considering a wider range of risk categories, it is important to consider more than just the risks themselves and their impact on one another. Risk should be considered systemic to represent the different aspects of risk accurately. According to their research, categorisation comprised; political, partner and supplier, customer, people, reputation, financial and market.

Following Vargas-Hernández (2011), an enterprise's risk aspects may be placed under the following major categories: property and personnel, marketing, finance, personnel and production and environment. Therefore, by considering the different sources of risk and the purpose of this paper, the best risk categorisation for this research work was as follows:

- Environment (exchange rates, availability of skilled labour, government policy, culture, weather).
- Technical (materials, new methods, technologies).
- Resources (materials, staff, finance).
- Integration (old and new systems, software modules).
- Management (experience, multiple parties, use of project management techniques, product transition management, set tight goals, HRM, organisation behaviour, organisation structure).
- Marketing (competitors, customers).
- Strategy.

Risk Management System

Risk management is acknowledged as 'the process of understanding the nature of uncertain future events and making positive plans to mitigate them where they present a threat or take advantage of them where they present opportunities' (Taplin, 2005). Considering that one of the main features of innovation is 'risk', risk management should facilitate innovation rather than stifle it (Taplin, 2005). Systematic risk management enhances an organisation's ability to manage risks. Risk management aims to improve project performance using; systematic identification, appraisal, and management of project-related risk (Chapman & Ward, 1997).Systematic risk management encourages decision-making inside an organisation that is more controlled, consistent, and flexible (Edwards et al., 2005). Following Edwards et al. (2005), it can be said that good risk management for a project should encompass these processes:

- Establishing appropriate context(s).
- Recognising project risk that faced by a stakeholder organisation.
- Analysing identified risks.
- Developing risk responses.
- Supervising and monitoring risks throughout a project.
- Permitting post-project collection of risk knowledge.

2.2 *Innovation Diffusion*

According to Mokyr (2002), innovation is the main source of economic growth, a key source of new employment opportunities, and has the potential to realise environmental benefits (Foxon et al., 2005). It is widely argued that in the current global economy, economic actions can be more cheaply conducted in low-wage economies, such as China. Therefore the main way in which other economies can compete and survive is for them to discover new and improved processes and products, in other words, they must innovate (Salaman & Storey, 2005).

Definition

The Oxford Dictionary of Economics states that innovation refers to the economic application of a new idea. Product innovation involves a modified or new product; process innovation involves a modified or new process of making a product (Black et al., 2006). Afuah (2003) stated that innovation involves the employment of new knowledge to provide a new product or service that the customers want; in other words, it is invention plus commercialisation. Van de Ven (1986) designated innovation as a new idea, which may be; a recombination of old ideas, a plan that challenges the present order, a formula, or an exclusive method that is perceived as new by individuals involved.

Different Innovation Types

The existing literature provides different innovation categories classified by; competence, degree, type, ownership and impact (Narvekar & Jain, 2006). Innovation can be considered in service and manufacturing sectors of different sizes (large, medium, and small). Although these two sectors differ, the general definition and innovation process is the same. The services sector has different characteristics from manufacturing. For instance, services are; perishable, intangible and heterogeneous (Salaman & Storey, 2005; Alshurideh et al., 2012; Alshurideh et al., 2021; Alshurideh, 2022).

Tidd and Hull (2006) stated that innovation was not just about opening new markets; it can also present new ways of serving older and established ones. He classified innovation into four groups (Product, Process, Position, and Paradigm), where each can happen along an axis, running from incremental through radical change. Incremental product innovation entails introducing an improved product, which, compared with its predecessor, has at least one additional desirable characteristic or is more efficient with the same characteristics. In contrast, radical or fundamental product innovation occurs when a new market has opened up, and the innovator begins to satisfy a hidden demand (Arundel et al., 2019).

3 Methods

This assessment used a systematic review of risk management and innovation. The research went through many stages, including; defining exclusion and inclusion criteria, selecting data sources and the search strategy, a quality assessment process, and data coding and analysis. These stages are detailed further in the subsequent subsections.

3.1 *Exclusion/Inclusion Criteria*

The exclusion and inclusion criteria were set up to support conducting this research. Firstly, the searched articles needed to involve risk management and innovation for the inclusion criteria. Secondly, the reviewed articles were required to show the relationship between RM and innovation. Thirdly the included articles needed to have been published between 2007 and 2020. Fourthly the articles needed to be written in the English language. Finally, the included articles should not include sector indicators.

Regarding the exclusion criteria, firstly, articles with technical backgrounds were excluded. Secondly, older articles published before 2007 concerning risk management and innovation were excluded. Finally, articles not written in English were excluded. Table 1 presents the exclusion and inclusion criteria used by this research work.

3.2 *Research Strategies and Data Sources*

Several databases and research engines were used to conduct this systematic literature review, such as; ProQuest, Emerald, Google Scholar, IEEE, Scopus, and ScienceDirect. The Search terms used were "risk management and innovation. In the identification stage, the results presented 1361 articles from the mentioned search engines and databases. In the screening stage, the results were filtered, and any duplicated articles were removed; most of the duplicates came from Google Scholar. The results were

Table 1 Exclusion and inclusion criteria

Inclusion criteria	Exclusion criteria
• Should involve risk management and innovation • Should show the relationship between innovation and RM • No sector indicators • Published between 2007 and 2020 • Must use the English language	• Articles with technical backgrounds • Non-English language • Articles published before 2007

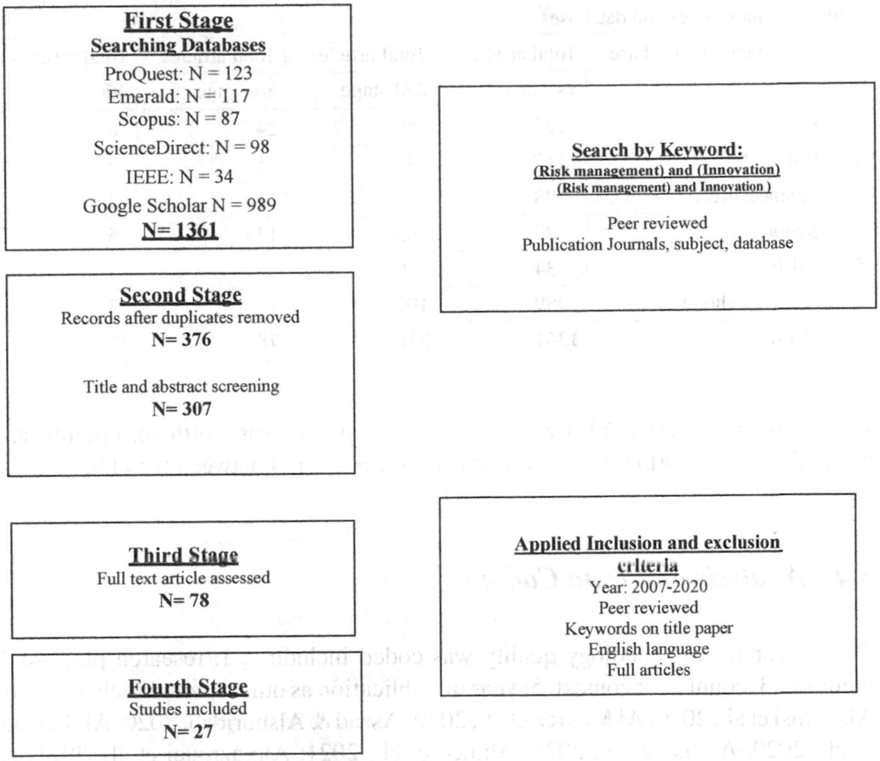

First Stage
Searching Databases
ProQuest: N = 123
Emerald: N = 117
Scopus: N = 87
ScienceDirect: N = 98
IEEE: N = 34
Google Scholar N = 989
N= 1361

Search by Keyword:
(Risk management) and (Innovation)
(Risk management) and Innovation)
Peer reviewed
Publication Journals, subject, database

Second Stage
Records after duplicates removed
N= 376
Title and abstract screening
N= 307

Third Stage
Full text article assessed
N= 78

Applied Inclusion and exclusion criteria
Year: 2007-2020
Peer reviewed
Keywords on title paper
English language
Full articles

Fourth Stage
Studies included
N= 27

Fig. 1 Flowchart for the selected studies

limited to only show peer-reviewed articles published from 2007 to 2020; the result comprised 376 articles. Following that, the titles and abstracts of all articles were checked, reducing to 307 the number of shown articles. Then the full articles and keywords included in the title were requested, which reduced the resulting articles to 78 in the eligibility stage. The exclusion and inclusion criteria were next applied for all articles, which reduced the shown articles to 27 in the final stage. The flowchart of the selected study is illustrated in Fig. 1 (Tables 2, 3, 4, 5, 6, 7, and 8).

3.3 *Quality Assessment*

To check the quality of the research studies in this research, "Quality assessment." was used similarly to how Da Silva Etges and Cortimiglia (2019) used it in his paper "A systematic review of Risk Management in Innovation-oriented Firms" as it's easy to use and gave an excellent way to evaluate the articles. The quality assessment contained nine questions to investigate the quality of the selected articles. There

Table 2 Data sources and databases

#	Search engine/database	Total articles 1st stage	Total articles 2nd stage	Total articles 3rd stage	Total articles 4th stage
1	ProQuest	123	89	24	6
2	Emerald	117	42	5	3
3	ScienceDirect	98	71	11	2
4	Scopus	87	63	14	5
5	IEEE	34	9	5	4
6	Google Scholar	989	102	19	7
	Total	**1361**	**376**	**78**	**27**

was a three-point scale. YES was worth one point, NO was worth zero points, and particularly was worth 0.5 points. Each study was scored between 0 and 9.

3.4 Analysis and Data Coding

This research's methodology quality was coded including; 1: research purpose 2: methods. 3: country. 4: context. 5: year of publication as other studied such as (Aisha Alshamsi et al., 2020; Al Mehrez et al., 2020; Assad & Alshurideh, 2020; Al Suwaidi et al., 2020; Ahmad et al., 2021; Ahmed et al., 2021; Almazrouei et al., 2021; Al Khayyal et al., 2021; Alshamsi et al., 2021; AlShehhi et al., 2021) did. The answers from this group of questions provided a precise representation of the paper's quality. A researcher can, at this point, include many articles by answering the five following questions:

RQ1: What are the primary research purposes of the studies selected?
RQ2: What are the main research methods of the studies selected?
RQ3: What are the participating countries in the context of the studies selected?
RQ4: What is the context of the studies selected?
RQ5: In terms of RM and innovation, what is the study distribution by year of publication?

Based on the selected 27 research articles concerning risk management and diffusion of innovation from 2007 to 2020. The systematic review results were analysed related to the answers to the above five questions.

RQ1: Research purpose distribution

The country of origin was developed as a crucial indicator to analyse the articles in this systematic review. This procedure helped to analyse the articles according to their purpose.

Table 3 Analysis of included research articles

No.	Authors	Year	Place	Context	Data collection methods	Sample size
S1	Da Silva Etges and Cortimiglia (2019)	2019	Brazil	Innovation oriented firms	Systematic review and empirical study	115 works
S2	Jin and Navare (2010)	2010	UK	Liverpool Victoria	Case study approach	One organisation
S3	Kwak et al. (2018)	2018	South Korea	Manufacturers and logistics intermediaries	Quantitative approach	
S4	Borgelt and Falk (2007)	2007	Australia	Three large organisations	Qualitative and case study methodology	60 leaders and managers
S5	Andersen (2009)	2009	Denmark	Different manufacturing industries	Quantitative approach	896 companies
S6	Chiambaretto et al. (2020)	2020	Sweden	Swedish firms	Quantitative approach	786 innovative Swedish firms with 50 to 1000 employees
S7	Wehn and Evers (2015)	2015	Netherlands	Citizen observations	Case study approach	2 Case Studies
S8	Migliori et al. (2020)	2020	Italy	Small and medium-sized family firms	Quantitative approach	1093 Italian small and medium-sized family firms
S9	Koru et al. (2018)	2018	United States	Home health agencies (HHAs)	Qualitative analysis	347 reports entered by HHA clinicians
S10	Games and Rendi (2019)	2019	Indonesia	Small and medium enterprises (SMEs)	Quantitative approach	165 small business owners in creative industries
S11	Ab Rahman et al. (2015)	2015	Malaysia	AIM participants	Quantitative approach	10 participant companies

(continued)

Table 3 (continued)

No.	Authors	Year	Place	Context	Data collection methods	Sample size
S12	Izumi et al. (2019)	2019	Japan	Academia, government, NGOs, and the private sector	Quantitative approach	A total of 228 answers were received from; universities and research institutes (145), government (30), NGOs (24), international and regional organisations (16), the private sector (6)
S13	Da Silva Etges et al. (2018)	2018	USA & Brazil	Hospitals in the United States and Brazil	Qualitative analysis	This study incorporated interviews with 15 chief risk officers (8 from the United States and 7 from Brazil)
S14	Wu and Wu (2014)	2014	China	Chinese firms	Quantitative approach	Panel data from a survey of 1178 Chinese firms
S15	Puente et al. (2019)	2019	Spain	Large pharmaceutical laboratories	Systematic analysis	31 large pharmaceutical laboratories
S16	Röth and Spieth (2019)	2019	Germany	Innovation projects	Semantic priming	455 participants
S17	Barczak et al. (2019)	2019	Poland	Micro, small, medium, and large enterprises	Quantitative approach	360 full interviews
S18	Ali et al. (2017)	2017	USA	Academic papers	Qualitative literature review and systematic review	31 academic papers
S19	Hock-Doepgen et al. (2021)	2021	Germany	Small and medium-sized enterprises (SMEs)	Qualitative comparative analysis	197 small and medium-sized enterprises (SMEs)
S20	Etges et al. (2017)	2017	Brazil	Innovative enterprises	Quantitative analysis	13 innovative companies

(continued)

Table 3 (continued)

No.	Authors	Year	Place	Context	Data collection methods	Sample size
S21	Bowers and Khorakian (2014)	2014	Iran and the UK	Large companies	Qualitative analysis	40 interviews
S22	Gurd and Helliar (2017)	2017	Australia	Zeta and Eta companies	Case study approach	Two case studies
S23	Vargas-Hernandez and Garcia-Santillan, (2011)	2011	Mexico	Innovation projects in lightweight medical	Case study and qualitative approach	One case
S24	Hölttä-Otto et al. (2015)	2015	United States	Kickstarter and Indiegogo customers	Quantitative Approach	127 consumers
S25	Berglund (2007)	2007	Sweden	Large corporations	Case study approach	Two cases
S26	O'Connor et al. (2008)	2008	USA	Large firms	Literature sharing and meetings with the IRI membership. Two longitudinal cross-case qualitative studies	85 high-level individuals
S27	Kalvet and Lember (2010)	2010	Estonia	Public procurement	Case study approach	Five case studies

Table 4 Articles included in this study with purpose, findings, and limitations

No.	Third term	Purpose	Finding	Limitations
S1	Ffinancial impact	The proposed framework was the first necessary state for future development. this study identified that innovation-oriented firms would provide a first empirical test of the interpretative framework of risk events, particularly regarding the comprehensive completeness of the proposed framework as perceived by practitioners	The main contribution of this paper was the proposal of a synthesising interpretative framework of risk events that could support the future development of payment ERM models for innovation-oriented firms. Additionally,	The empirical study signalled which risk events were more relevant in the Brazilian context, limited to discussing specific aspects of operational implementation
S2	Global sustainability (GS)	Determining the relationship between risk and innovative adoptive behaviour was examined to guide managers on managing risk and uncertainty under the different circumstances of their innovative practices	The study developed a conceptual framework, which consisted of; risk behaviour, environmental conditions, and innovation. Two key elements determined the performance of such behaviour	The research was only exploratory and used a single case study to illustrate the applicability of the REAI model The boundary and conditions of such applications were not specified. Therefore, there remains a need to test the REAI model using a large representative sample. Given that it takes a relatively long period to observe environmental turbulence and organisational behaviour changes, it is also essential to carry out such a study from a longitudinal perspective
S3	Supply chain	This paper proposed and validated a theoretical model to investigate whether supply chain (SC) innovation positively affected risk management capabilities, such as robustness and resilience in global SC operations. It also examined how these capabilities may improve competitive advantage	It was found that innovative SCs had a discernible positive influence on all dimensions of risk management capability, which in turn had a significant impact on enhancing competitive advantage Therefore, this research provided evidence on SC innovation and risk management capabilities in supporting competitive advantage	This research did not use objective data to measure the firm's competitive advantage but employed the respondents' perceptions of their competitive advantage. The impact of risk management capability on competitive advantage might take a long time. Data were collected from a single country, South Korea

(continued)

Table 4 (continued)

No.	Third term	Purpose	Finding	Limitations
S4	Leadership	This paper presented a data-driven discussion concerning whether effective leadership and innovation were being stifled in contemporary organisational environments of continuous change	The key finding was that the accurate application of interaction between leadership and management released social capital and related identity and knowledge resources, helping address tension between risk management and innovation	This paper only reported on one component of the original study. The findings and implications reported here are regarded as tentative
S5	Capital structure	This paper argued that strategic responsiveness was of paramount importance for effective risk management outcomes and demonstrated this by introducing an empirical study	The study revealed that risk management effectiveness combined the ability to exploit opportunities and avoid adverse economic impacts and had a significant positive relationship with performance. This effect was moderated favourably by investment in innovation and lower financial leverage	The analysis was based on a sample of large firms, which may have affected the generalisability of the results. Nonetheless, the study showed that effective RM capabilities differentiated firms and determined success and failure. It further underscored the importance of combined innovation policy and capital structure decisions as firms deal effectively with risk and uncertainty
S6	Small and large firms	This research investigated the extent to which small, and large firms differed when assessing the benefits and risks provided by competitors as partners in innovation	This paper's results confirmed that small and large firms valued the benefits and risks associated with competitors differently. It showed that small firms were less reluctant to compete than large firms, especially if the competition allowed them to reduce their costs and learn from their competitor. In contrast, it showed that large firms' agreed to cooperate with their competitors to enable them to reduce their time-to-market	The method used did not allow the researchers to investigate other attributes that might be equally or even more important. Second, the participants were forced to select a competitor for innovation based on only a few limited attributes, which were limited by the number of respondents
S7	Flood risk management	This paper analysed the social innovation potential of ICT-enabled citizen observations to increase e-participation in local flood risk management	This study analysed the social innovation potential of ICT-enabled citizen observations to increase e-participation in local flood risk management	The limitation of this research was its focus on only two cases. Further investigation is required to confirm the findings based on more cases

(continued)

Table 4 (continued)

No.	Third term	Purpose	Finding	Limitations
S8	Family management	This paper investigated the relationship between family management and innovation investment propensity in family firms by analysing the effect of two innovation impulses: demand-pull and technology-push. Extending the technology-push/demand-pull framework in the context of family firms, and adopting a direct measure of firms' innovation investment propensity	The findings suggested to practitioners and policymakers that family firm innovation impulses were important contingencies that needed to be considered when making innovation investment decisions. Other findings suggested that the effect of FTMT on INN_PRO was statistically significant. Additionally, the findings suggested that the TP impulse was less decisive than the DP impulse	First, the investigation was focused on Italian family firms. Second, the study adopted a cross-sectional analysis given that the dataset covered the period 2007–2009. Third, the study measured the DP and TP variables following some prior studies. Fourth, the sample and the family business definition was adopted. Finally, the focus was on family SMEs rather than medium or large firms
S9	Technology	The study performed a detailed examination of fall and near-fall incidents in an HHA to identify potential IT-based innovations that could be used in process improvement to achieve better fall risk management during home care	The results provided data about fall risk management during home care. The results suggested methods that could positively respond to the challenges and opportunities identified in the current study to impact fall risk management during home care positively	The perspectives of patients and caregivers related to fall incidents were not captured
S10	Financial performance	This study examined the effects of knowledge management and risk-taking on financial performance where negative innovation outcomes were chosen as a mediator variable	The study results indicated the importance of knowledge management and risk-taking to reduce negative innovation outcomes. Further, it was found that negative innovation outcomes did not mediate the links between knowledge management, risk-taking, and SMEs' financial performance	The study offered some insights from Indonesia as an emerging market economy, which may have had different characteristics from other contexts. Secondly, most respondents were newly in business and perhaps saw business innovation imperative in creative industries. Thirdly, it may also not be sufficient to see the result of the implementation of risk management. Fourthly, this research did not include positive innovation outcomes as a mediator variable. This study saw negative innovation outcomes as sufficient to see the impact of knowledge management and risk-taking

(continued)

Table 4 (continued)

No.	Third term	Purpose	Finding	Limitations
S11	Sustainability of microfinance	This study clarified issues around whether measures had a positive effect on different aspects of innovation and best practices	The study findings showed that performance measurement positively affected innovation and best practice issues. The measurements for several aspects of innovation and best practices had good potential in the microfinance industry	The study's main limitation descended from its exploratory nature because the objective was to gain theoretical clarification regarding the product innovation process and best practices. The findings were generalised, which limited the study
S12	Disaster management	This paper identified three key issues that could help overcome the following barriers: networking, eco-production of knowledge, and a stronger role played by academia	This paper clearly showed that innovations were not required to be completely new or high-tech products, as approaches and frameworks could lead to changes and influence people's thinking and behaviour	It was necessary to consider both innovative products and approaches in DRR strategies when working to develop them further and make them more effective and useful. CBDRR was still a relatively new approach, even among governments
S13	Health care	This study proposed a conceptual ERM framework specifically designed for health care organisations. The study explored how hospitals in the United States and Brazil structured and implemented ERM processes within their management structures	The study confirmed that adopting ERM for health care organisations had gained momentum and become a priority. The demand for risk economic assessment orientation was common among health care risk managers	Given the complexity of the variables affecting health care organisations in other countries: political, financial, and environmental, the results of this study may not be generalisable
S14	Board of directors	This study explored the role of the board of directors in the relationship between integrated risk management and product innovation. The study focused on the board's direct involvement in risk oversight and its use of external audits in risk oversight. It also examined their moderating effects on the relationship between integrated risk management and product innovation	The finding demonstrated that a board's risk involvement and its use of external audits in risk oversight negatively moderated that relationship. A board's direct involvement in risk oversight negatively moderated the positive relationship between integrated risk management and product innovation success. The use of external audits in risk oversight similarly weakened the relationship	First, this study examined only the relationship between IRM and product innovation. It would perhaps be fruitful to extend the theoretical framework to other innovative activities, such as process innovation or organisational innovation. Furthermore, it would be interesting to break down product innovation into radical and incremental innovations and examine any differences in the relationship with IRM and board involvement. Since the entire sample was from China, future research needs to validate the applicability of these findings elsewhere. Finally, this analysis used time-lagged variables to address this issue

(continued)

Table 4 (continued)

No.	Third term	Purpose	Finding	Limitations
S15	Strategic choices	This study developed a fuzzy evaluation model that provided managers at different responsibility levels in pharmaceutical laboratories with a rich picture of their innovation risk and that of competitors. This situation helped them make better decisions about managing their present and future portfolio of clinical trials in an uncertain environment	The main finding of this work was the development of an innovative fuzzy evaluation model that was useful for analysing the innovation risk characteristics of large pharmaceutical laboratories given their strategic choices	Small pharmaceutical laboratories are a relevant part of the pharmaceutical industry but were not covered in this paper. Other restrictions imposed in this work could be relaxed, allowing each expert to evaluate the model's variables by proposing their independently of the rest of the evaluators
S16	Resistance to change	This paper investigated why different individuals struggled to achieve a shared understanding of innovation projects. The primary focus was on the individuals' functional backgrounds as the cause of systematic differences in assessing innovativeness, neglecting dispositional traits. So the paper investigated how a decision's context and an individual's resistance to change influenced the assessment of an innovation project's risk and innovativeness	The paper showed that an individual's resistance to change moderated the positive relationship between an innovation project's innovativeness and its perceived risk. It also indicated that; department orientation, mental models, or thought worlds caused systematic differences when individuals evaluated innovation projects	First, the experimental design sought to clarify the specific relationship between the constructs, limiting it in terms of complexity. Second, it reduced complexity and neglected to include recent research avenues during innovation portfolio management. Third, recent research has not been able to clarify the influence of innovation project portfolio's innovativeness on agility. Finally, testing the generalisability of the findings with an investigation of practitioner behaviours could offer further novel insights
S17	Logistics management	This study determined the impact of the risk of implementing digital technologies for logistics management	The literature review proved that this issue still had a research gap in management theory, including in logistics. Analysis of the research results showed that the implementation and use of digital technologies implied changes in logistics management. However, this impact was diversified, depending on the digital technology type	The limitations of this study were that the implementation and use of digital technologies implied changes in logistics management. However, this impact was diversified, depending on the digital technology type. Therefore, it could be assumed that not all digital technologies arouse equal interest among logistics specialists, bringing with them the yet unknown effects of the risk of implementing digital technologies

(continued)

Table 4 (continued)

No.	Third term	Purpose	Finding	Limitations
S18	Cloud-based business	This study uncovered the myriad challenges managers confront as they seek to leverage cloud technology in the ongoing transformation of their organisation's service offerings. Combining this systematic literature analysis with relevant theory also helps managers identify their organisation's general risk profile and links that profile to a specific configuration of resolution	This study synthesised an integrated model for managing risk during the innovation of cloud-based business services. The model identified three types of risks (services, technology, and process risks) and four types of resolutions (stakeholder engagement, technology development, innovation planning, and innovation control)	One area for future research is to explore whether the risk-strategy model works differently for public, private, and hybrid cloud environments. The research did not differentiate between these categories; hence, further insights may be gained from investigating the differences and commonalities across these contexts. is how cloud-based business services providers might deal with the risks of transitioning all or some of their services to the cloud
S19	Knowledge management	This study examined the impact of internal and external KM capabilities on BMI and how risk-taking tolerance moderated these effects	The results from this study indicated that external KM capabilities stimulated BMI. This relationship was strengthened for firms with a high risk-taking tolerance. Internal knowledge was only effective for firms with a low risk-taking tolerance	It was important to consider the nature of the data basis for empirical analysis when interpreting the data. The study relied on key informants in each organisation. Multiple informants in each organisation might be preferable to capture different facets of BMI. Although this issue was lower for SMEs, future studies are encouraged to collect data with respondents at different levels in a firm. The study only focused on external and internal KM capabilities from a theoretical perspective
S20	Enterprise management	This paper showed that it was possible to measure the risks that an innovation company faces and that such risks can be managed with the support of a company strategy. An economic analysis was applied based on MCS and an indicator of CFaR to measure innovation risks. A strategic performance model for innovation companies was proposed, and the benefit of implementing risk management practices in innovation organisations was validated	The main findings indicated the link between analysing the risks to which a company was exposed, its origin, and their relationship to decision making to manage the company strategically. In brief, the probabilistic analysis of the risks indicated the variables that caused the greatest financial impact	The region limited this research. The case study was from Brazil and cannot be assumed as a global conclusion. Similarly, by only examining one case, the findings were specific to one company and its business sector

(continued)

Table 4 (continued)

No.	Third term	Purpose	Finding	Limitations
S21	Project management	This paper helped fill the knowledge gap based on a practical basis for designing the most appropriate form of risk management in different innovation-based industries. It also highlighted the possibilities for better integration of the theories of innovation and project risk management	The main finding of this paper was to provide a framework that diverse companies could appreciate. The framework offers a basis for discussing the most appropriate form of risk management in different innovation-based industries	It would be useful to extend this research by examining more case studies from other countries and industries
S22	Management control systems	This study used the concept of institutional leadership to explore how leaders balanced creativity and product innovation against administrative arrangements, such as risk management and management control	Using two case companies, distinct contrasts to this balance were produced. Neither showed the ambidexterity needed to 'secure the value of innovation and exploration and developed risk management and control systems for exploitative activities	A limitation of this study was that both enterprises were highly focused on engineering and dominated by engineers
S23	Project management	This research identified the sources of risk in innovation projects and determined whether they could be managed better	This research attempted to provide a system for managing the risk in innovation projects. It also created a method for prioritising different risks factors and managing the most important ones in the second stage of this risk management system for innovation	The proposed method for analysing the risk in risk the management system will be applied for one case
S24	Crowdfunding	This paper filled the research gap of understanding the innovation and risk balance strategy for new product development (NPD) at small companies. It also explored innovation versus risk for small companies using crowdfunding products as a proxy for analysis	The main result suggested a preliminary innovation and risk balance framework for crowdfunding NPD success. A statistical model was developed to correlate the amount of crowdfunding raised with 64% predictability. These results may contribute to better understanding and balancing risk and innovation in crowdfunding and small company contexts	The limitations are worth mentioning. The adapted RWW metric might require further refinement before being applied to a wider selection of products and crowdfunding platforms. Also, the current rating process was relatively slow to generate a large dataset for analysis. The results from this paper could be used to generate experience to design better machine learning tools for a faster and more rigorous data collection and analysis. Finally, the implications from this paper were constrained to the context of crowdfunding. Caution shall be taken before interpreting the results for other areas

(continued)

Table 4 (continued)

No.	Third term	Purpose	Finding	Limitations
S25	Risk conception	This paper explored the risk conceptions of innovators in two large corporations and identified three themes that illuminated the relationship between risk and innovation in the corporate setting	There were three main findings The first related risk to the issues of boundaries and control over parts of the innovation process. The second showed how risk was primarily related to innovation as a process and not an output. The third showed how a flexible view of business models could manage risk in corporate innovation	The two units came from different industries and differed in corporate history and performed work. Second, there was a tendency to discuss both innovation risks and the concept of quality about processes instead of outputs. Finally, given the small sample size, the results were not generalisable in a statistical sense
S26	Learning	The study investigated a subset of management practices that might have contributed to the success of radical innovation efforts by large established firms. The study focused on that subset of practices concerned with learning-oriented approaches to risk management	The study found that the development of dynamic capabilities for managing highly uncertain phenomena, such as radical innovation, included a risk management-by learning capability. Further research is necessary to replicate the findings and extend them to other high uncertainty venues	The limitations were that the interaction between industry clock speed and harvesting strategy negatively affected activity and output, suggesting that a strategy of harvesting early and often was more valuable for firms operating in less dynamic environments. Thus, the interaction between the use of experimental processes and industry clock speed did not significantly correlate with any of the RI success measures
S27	Public procurement	This study provided the public sector with what was necessary to deal with different kinds of process risks. The study took an exploratory approach in determining the state of practice regarding risk management in public procurement for innovation at the local level	The study found that cities were, for the most part, actively engaged in risk identification; the risks were being met mainly with mixed solutions in contracting strategies rather than comprehensive risk-management tools	There were no signs of addressing these risks via comprehensive risk management strategies aimed at reducing providers' technology or innovation-related risks. Further studies are needed to document how the public sector meets the risks when using public procurement to promote innovation. Also, further research is needed for building coherent theoretical models addressing this issue

Table 5 Independent and dependents factors

No.	Independent factors	Dependent factors
1	• Innovation output • Innovation input • Source of innovation • Innovation strategy • Technology • External orientation	• Innovation • Risk management
2	• Environmental turbulence • Adoptive innovator behaviour • Appetite for risk-taking • Innovation performance	• Innovation • Risk management
3	• Supply chain innovation (SCI) • Robustness capability (RB) • Resilience capability (RS) • Competitive advantage (CA)	• Innovation • Risk management
4	• Leadership in a Leadership Configuration (LLC) • Management in a Leadership Configuration (MLC) • Leading in a Management Configuration (LMC) • Managing in a Management Configuration (MMC)	• Innovation • Risk Management
5	• Risk Management Effectiveness (RME) • Investment in Innovation (II) • Financial Leverage (FL) • Performance (PER)	• Innovation • Risk management
6	• Partner's intensity of competition • Partner's time-to-market • Innovation degree • Partner's assumed level of risk • Partner's allowed cost reduction • Partner' delivered learning opportunities • Strategic importance of resources accessed through the partner	• Innovation • Risk management
7	• Individual education • Communicative influence • Advise/Consult • Co-govern • Direct authority • Technical expertise • Deliberate and negotiate • Aggregate and bargain • Develop performance • Express performance • Explicate data collection • Listen as a spectator • Implicate data collection	• Innovation • Risk management

(continued)

Table 5 (continued)

No.	Independent factors	Dependent factors
8	• INN PRO • AGE • PUB • Fam CEO • CEO AGE • EMPL • FTMT • DEM HETER • HIGH QUALITY	• Innovation • Risk management
9	• Patient engagement • Caregiver involvement • Expectation management • Care coordination • Sociocultural awareness	• Innovation • Risk management
10	• Risk-taking	• Innovation • Risk management
11	• Suitability • Best practices • Business performance • Organisational change	• Innovation • Risk management
12	• Networking • Coproduction of knowledge • Academia	• Innovation • Risk management
13	• Process and Strategy • Risk culture and education • Risk Identification • The Action plan and KPI's	• Innovation • Risk management
14	• Constant • R&D expenditure • Foreign • Domestic • Firm size • Sales • Industry dummy	• Innovation • Risk management
15	• Avg sales • Avg dept • Constant • R-squared • No observation	• Innovation • Risk management
16	• Resistance to change • Perceived Innovation • Market-related facets • Technology related facets • Perceived risk	• Innovation • Risk management

(continued)

Table 5 (continued)

No.	Independent factors	Dependent factors
17	• Three-dimensional printing • Artificial intelligence • Blockchain • Drones • Augmented reality • Self-propelled vehicles	• Innovation • Risk management
18	• Stakeholder engagement • Technology development • Innovation planning • Innovation control	• Innovation • Risk management
19	• Internal KM capabilities • External knowledge management capabilities • Organisational risk-taking tolerance	• Innovation • Risk management
20	• Physical change in space • Delivery date products and equipment • Important delay • Dibbing • Quality of workmanship • Technology changes • Use • Process management	• Innovation • Risk management
21	• Customer feedback • Employees' opinion • Competitors • Market • Profit • Internal control	• Innovation • Risk management
22	• Organisational • Products • Strategic	• Innovation • Risk management
23	• Learning • Creativity • Selection • Incubation • Implementation	• Innovation • Risk management
24	• Voice of customer • Functional feasibility • Tangible and intangible benefits • Vulnerability risk evaluation and growth strategy	• Innovation • Risk management

(continued)

Table 5 (continued)

No.	Independent factors	Dependent factors
25	• Quality assurances • Customer relations management • Relations to partners • Dated support systems • Organisational efficiency • Customer contracts • Human capital risk • Business culture • Development	• Innovation • Risk management
26	• Options thinking • Experimental processes • Harvest strategy	• Innovation • Risk management
27	• Nature of innovation • Procurer's level of risk • Risks identified by cities • Risk management tools applied by cities • Deviation from regular tendering procedures • Technology standards • Performance agreements • In-depth market study	• Innovation • Risk management

Articles from China:

There were five articles from China. The first article highlighted the research's purpose to mediate the role of decision quality in the relationship between risk management and firm performance (Tidd & Hull, 2006) The second article examined the impact of risk management strategies on high technology firms in China (Yang, 2008). While Zheng et al. (2010) conducted his study by examining the mediation role on risk management and organization Innovation culture, structure, strategy, and effectiveness. The fourth article asserted the relationship between risk management orientation, organisation innovation, and organisational performance (Darroch, 2005). The last article from China focused on the fit between risk management strategies and intellectual capital and their impacts on firm performance (Hsu & Sabherwal, 2012) .

Articles from Spain:

In this study, three articles from Spain were covered. The first article showed a model of strategic human resource practices as a catalytic mechanism influencing the effectiveness of risk management (Inkinen et al., 2015). The second paper aimed to evaluate structural and relational capital's roles as the conclusions of codification and personalisation of risk management strategies in renewal capital and innovation in low- and high-tech companies (Buenechea-Elberdin et al., 2018) . The third article showed the relationship between SMEs' R&D internationalisation and innovation outcomes (Ren et al., 2015). Edwards et al. (2005) defined his research paper

Table 6 Frequency variables according to the databases

No.	Databases	Culture	Structure	Strategy	Technology	Intellectual capital	Leadership
1	ProQuest	7	6	4	5	2	2
2	Emerald	2		2	6	2	2
3	Science Direct	1	1	2	1	1	1
4	Tylor & Francis						
5	Google Scholar	1		1		1	
Total		**11**	**7**	**9**	**12**	**6**	**5**
1	2015	X			X		X
2	2017					X	
3	2018		X	X			
4	2017	X		X	X	X	
5	2011			X			
6	2010				X		
7	2013	X	X				X
8	2016				X	X	
9	2014	X		X			
10	2019			X			
11	2014	X	X	X			
12	2015				X		
13	2019	X					
14	2017		X				
15	2017	X		X	X		
16	2013						X
17	2018				X	X	
18	2019	X	X		X		X
19	2015			X			
20	2019						
21	2019				X	X	
22	2015	X	X				
23	2016				X		
24	2019				X	X	
25	2015	X	X				X
26	2010			X	X		
27	**2013**						
Total	**11**	**7**	**9**	**12**	**6**	**5**	

Table 7 Checklist for quality assessment

#	Question
1	Were the research aims specified clearly?
2	Did the study design achieve these aims?
3	Were the variables considered in the study specified clearly?
4	Was the study context/discipline specified clearly?
5	Were the data collection methods detailed adequately?
6	Did the study explain the validity/reliability of the measures?
7	Were the statistical techniques used to analyse the data described adequately?
8	Did the results add to the literature?
9	Did the study add to your understanding or knowledge?

by assessing the impact of risk management RM strategies on a firm's corporate performance and innovation.

Articles from India:

Three articles from India were covered. The first article evaluated organisational learning, firm performance, and risk management (Fort-Rioche & Ackermann, 2013) The second article proposed a conceptual model comprising six knowledge management and two performance constructs (Joshi & Chawla, 2019). Narvekar and Jain (2006) studied the influence of efficient risk management practices on firm performance.

Articles from Iran:

There were four articles from Iran. The first one evaluated risk management value and organisational innovation (Salaman & Storey, 2005). Wehn and Evers (2015) defined his research to determine relationships between; organisational learning, transformational leadership, organisational innovation, knowledge management, and organisational performance. The third article studied the influence of organisational innovation and risk management practices in SMEs using structural equation modelling SEM (Calvo-Mora et al., 2016). The last study presented a universal approach to evaluating RM practices on FP (Douthwaite & Ward, 2005).

Articles from Taiwan and Japan:

Gurd and Helliar (2017) examined the relationship between RM strategy and FP performance. While Tseng and Lee (2014) showed how enterprises could effectively apply their risk management capabilities and develop uniquely dynamic capabilities. The article from Japan assessed explicit and implicit risk's role in translating management innovation into firm performance in Japanese companies (Van de Ven, 1986).

Table 8 Quality assessment results

Study	Q1	Q2	Q3	Q4	Q5	Q6	Q7	Q8	Q9	Total	Percentage (%)
S1	1	1	1	1	1	0.5	1	1	1	8.50	94
S2	1	1	1	1	1	0.5	1	0	1	7.50	83
S3	1	0.5	1	1	0.5	0.5	0	0	1	5.50	61
S4	1	1	1	1	1	0.5	1	0	0.5	7.00	78
S5	1	1	1	1	1	1	1	0	1	8.00	89
S6	1	1	1	1	0	0.5	0	0	0.5	5.00	56
S7	1	0.5	1	0.5	0.5	0.5	0	1	0.5	5.50	61
S8	1	1	1	1	1	1	1	0.5	1	8.50	94
S9	1	1	1	1	1	1	1	0.5	0.5	8.00	89
S10	1	1	1	1	1	1	1	0	0.5	7.50	83
S11	1	1	1	1	1	1	0.5	1	0.5	8.00	89
S12	1	1	1	1	1	1	1	0.5	0.5	8.00	89
S13	1	0.5	1	1	0.5	1	1	0.5	0.5	7.00	78
S14	1	0.5	1	0.5	1	1	1	0.5	1	7.50	83
S15	1	1	1	1	1	1	1	0.5	0.5	8.00	89
S16	1	0.5	1	1	1	1	1	0	0	6.50	72
S17	1	0.5	1	1	1	1	1	0.5	0.5	7.50	83
S18	1	1	1	1	1	1	1	1	1	9.00	100
S19	1	1	1	1	1	1	1	0.5	0.5	8.00	89
S20	1	1	0.5	0.5	0.5	1	1	1	1	7.50	83
S21	1	1	1	1	0.5	1	1	0.5	0.5	7.50	83
S22	1	0.5	1	1	1	1	1	0.5	0.5	7.50	83
S23	1	1	0.5	1	1	1	1	0	0.5	7.00	78
S24	1	1	1	1	1	0	1	0.5	0.5	7.00	78
S25	1	1	1	1	1	1	1	0.5	0.5	8.00	89
S26	1	1	1	1	1	1	1	1	1	9.00	100
S27	1	1	1	1	1	1	1	1	1	9.00	100

Articles from the US and Brazil:

Bowers and Khorakian (2014) in his research tested the relationship between strategic intent, including; the goals, vision and mission of intellectual capital (i.e., structural capital, relational capital, human capital), risk management, and. While Vargas-Hernandez and Garcia-Santillan (2011) reviewed a conceptual view of the relationship between; business innovation and risk management, supply chain technology investments, and overall firm performance. The author from Brazil classified the type and intensity of relationships between; organisational performance, innovation performance, and strategic knowledge management (SKM) practices (Afuah, 2003).

Articles from Bahrain and Egypt:

The Bahraini article evaluated a comprehensive model comprising various relationships between; risk management (RM) processes, transformational and transactional leadership, and organisational performance (Birasnav, 2014). While Jin and Navare (2010) from Egypt, in his study, highlighted the role of risk management to enhance organisational performance and, consequently, developed an integrated risk management capabilities framework for assessing organisational innovation.

Articles from Malaysia and Indonesia:

Three articles were from Malaysia, where the first author examined the indirect relationships between Soft TQM factors and risk management with firm performance (Etges et al., 2017). The second author evaluated the relationship between organisation performance and risk management capabilities (Berglund, 2007). The last author highlighted the critical role of the risk creation process in demonstrating the risk management enabler's T-shaped skill to create creativity in the organisation, which wand, thus, firm performance (Azmi et al., 2018). There were two articles from Indonesia. The first investigated the environmental hostility contingencies on the relationship between firm innovation and risk management strategy (Ambad & Wahab, 2013). The second article, Hock-Doepgen et al. (2021), studied the application of risk management and intellectual capital as variables and tested their effect on firm performance.

Articles from Finland, the UK, Italy, Slovenia, and Serbia:

The author from Finland covered the relationship between risk management and intellectual capital and firm performance (Hock-Doepgen et al., 2021). While Hölttä-Otto et al. (2015) researched to show the effect of the company performance on risk management with information technology support. The author from Italy assessed the link between; problem-solving processes, knowledge management practices, and organisational innovation (Giampaoli et al., 2017). The Slovenian article showed that through; creating, collecting, organising, and utilising knowledge, that organisations could enhance organisational performance (Rasula et al., 2012). The article from Serbia focused on the effect of risk management on organisational performance and innovation in Serbia's economy (Dwyer et al., 2016).

RQ2: Distribution of research methods

Regarding the study distribution by research method used, the chart below shows that numerous research papers depended on the case study approach in their studies as it was the most prevalent method for collecting data. Several other studies used interview methods. The framework, quantitative and qualitative, literature review, mail survey, and conceptual study methods were also used in the examined studies (Fig. 2).

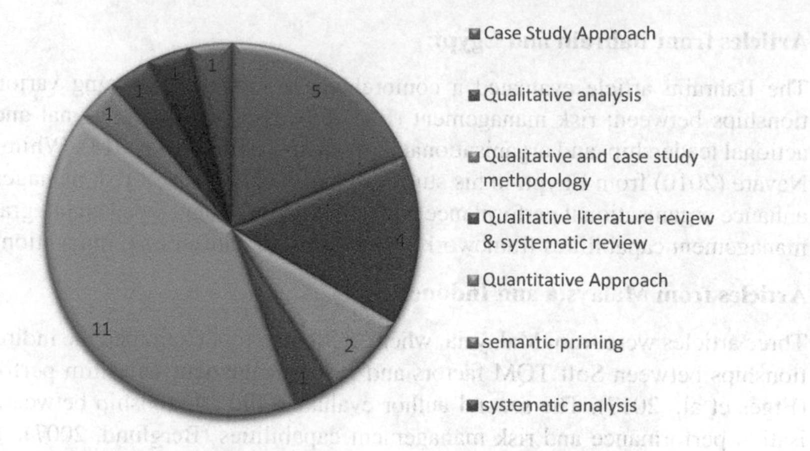

Fig. 2 Study distribution by research method

RQ3: Country distribution of research studies

The chart below shows that the USA produced the most research articles concerning risk management and innovation in the examined studies. This situation may be because the USA is a leading industrial country globally. Brazil ranked in second place with three articles. The UK, Germany, Sweden, and Australia were ranked third with two articles. The rest of the countries each had one article, as shown in Fig. 3.

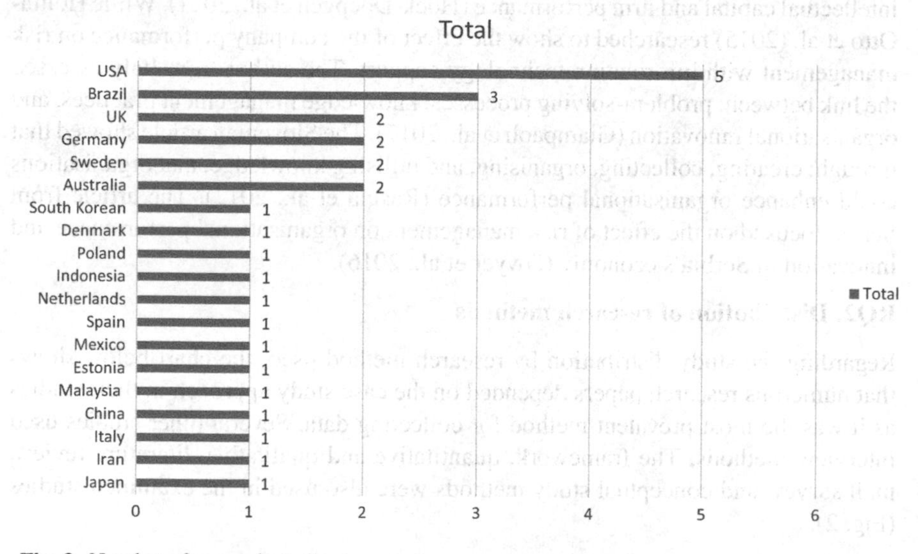

Fig. 3 Number of research studies by country

RQ4: Distribution of research context

This systematic review covered many contexts: investment, large firms, academia, government, NGOs, the private sector, different manufacturing industries, hospitals, innovation projects, high technology firms, electronic and electric firms, hospitality services, domestic companies, and manufacturing organisations. Firms, SMEs, technology and IT cross-sectional industries, textile industries, family businesses, PLU-SMEs, engineering firms, and firms adopted the risk management system. The chart below illustrates that large firms were ranked highest with eight articles, small-medium enterprises had three articles, and manufacturing firms came next in the context of this research. Technology and IT firms came after innovation projects (Fig. 4).

RQ5: Distribution of studies by year of publication

Listed by the year of publication, Fig. 5 presents the distribution of the examined studies investigating risk management and innovation from 2007 to 2020. The number of research papers covered in this research totalled 27. It can be seen that there was one article in 2008, 2009, 2011, 2013, 2014, and 2016, respectively. There were two articles in 2007, 2010, and 2018. There were three articles in 2015 and four articles in 2017. In 2019 the number of articles totalled eight.

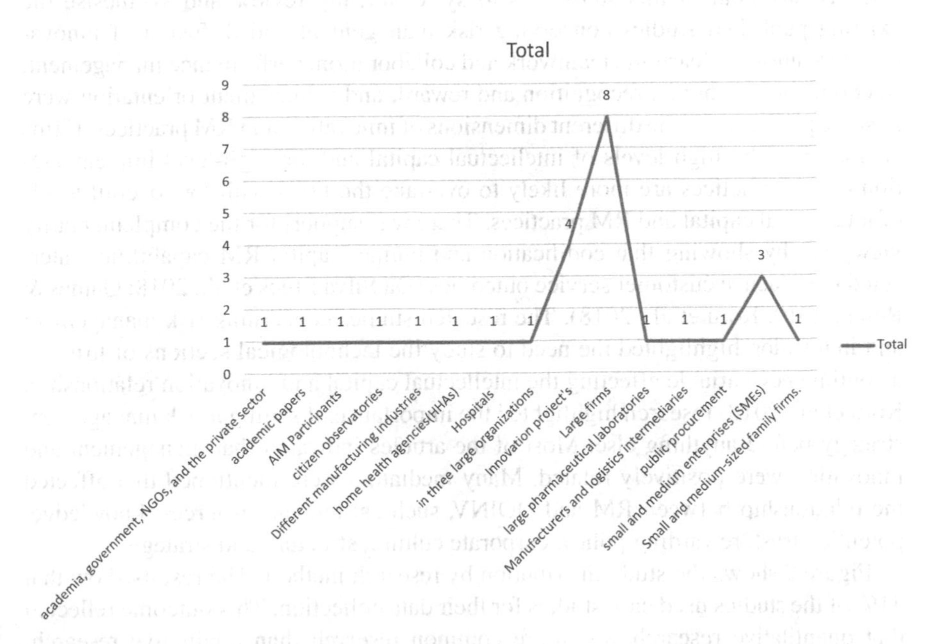

Fig. 4 Distribution of studies by context/discipline

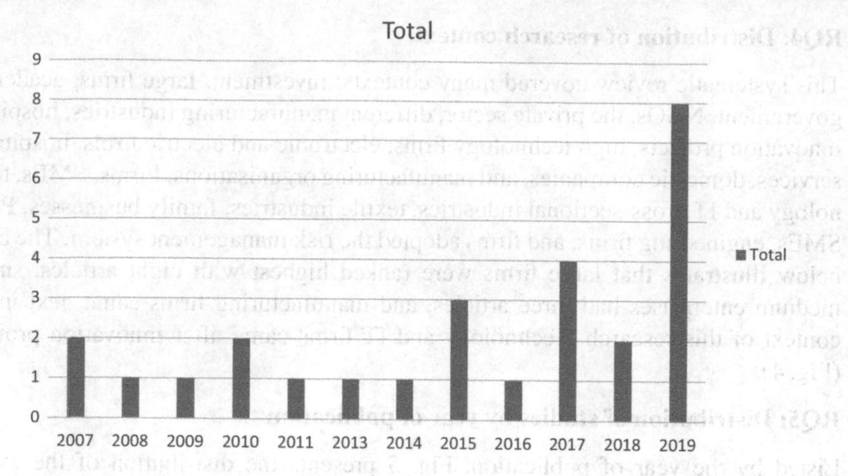

Fig. 5 Number of articles by year

4 Discussion

The primary goal of this study was to systematically review and synthesise the existing published studies concerning risk management and diffusion of innovation. Organisation learning, teamwork and collaboration, performance management, freedom and autonomy, recognition and reward, and achievement orientation were positive predictors of the different dimensions of innovation and RM practices. Firms characterised by high levels of intellectual capital and the high-level implementation of RM practices are more likely to overtake the firms with low overall levels of intellectual capital and RM practices. There was support for the complementarity viewpoint by showing that codification and human capital RM capabilities interacted to influence customer service outcomes (Da Silva Etges et al., 2018; Games & Rendi, 2019; Koru et al., 2018). The research studies concerning risk management and innovation highlighted the need to study the technological sections of firms as a contingency variable affecting the intellectual capital and innovation relationship. Koru et al. (2018) research highlighted the importance of setting a risk management strategy before anything else. Most of the articles indicated that management and innovation were positively related. Many mediators were mentioned that affected the relationship between RM and DOINV, such as; human resources, knowledge, potential, trust, rewarding policy, corporate culture, structure and strategy.

Figure 2 shows the study distribution by research method. The results show that 41% of the studies used case studies for their data collection. This outcome reflected that quantitative research was more common research than qualitative research. According to Figure 3, most of the examined studies came from the USA, comprising 12% of the research articles, as the USA is one of the leading manufacturing countries. Brazil came second with 9.7% of the research articles. Figure 5 shows the research studies in terms of the context. 14 % of the research articles were from the context of large firms, 9.7% from manufacturing firms, 7.2% from SMEs, and 2.4% from IT and technology firms. Figure 5 illustrates the year of publication.

After checking the dependent and independent factors, it could be seen that the most frequent factors were (culture, structure, technology, intellectual capital, and leadership). This comprehensive research highlights paths for future studies concerning the internal drivers that affect risk management and diffusion of innovation. The research model is shown below: culture, structure, strategy, knowledge sharing, and technology were used as internal factors that affect risk management and diffusion of innovation. The model shown below was drawn based on the hypotheses of the paper (Fig. 6).

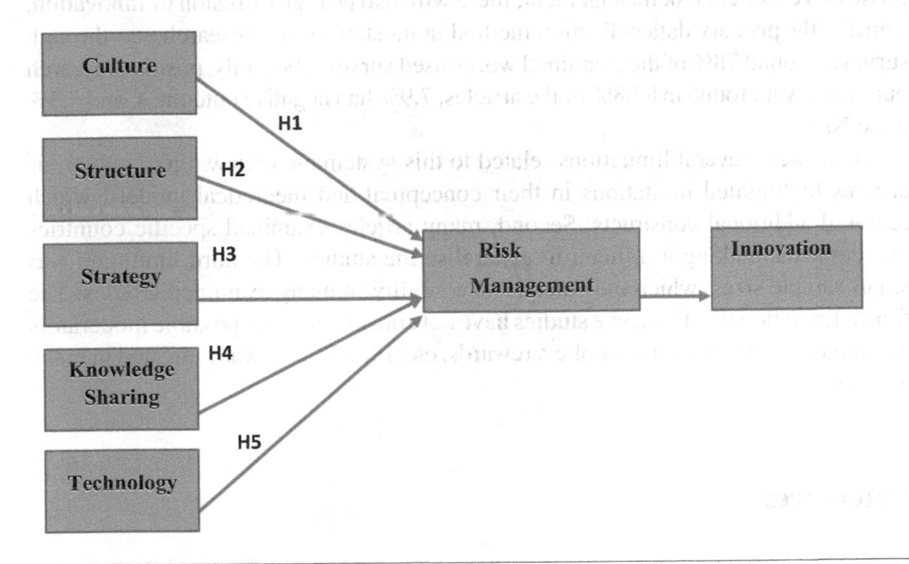

H1: Culture positively impacts risk management

H2: Organisation structure negatively impacts risk management

H3: Great organisation strategy positively impacts risk management

H4: Knowledge sharing positivel impacts risk management

H5:A higher degree of technology has a positive impact on risk management.

H6: Risk management has a positive impact on firm performance

H7: Risk management mediates the relationship between culture and organisation performance

Fig. 6 Research model

5 Conclusion

Many findings can be drawn from this systematic review concerning risk management and diffusion of innovation. In this research, several factors reoccurred, such as; culture, structure, strategy, technology, intellectual capital, leadership, process, and rewards. The majority of the repeated elements were located within firms and organisations. This outcome will lead any future studies regarding the internal drivers affecting RM and DOINV. Secondly, the majority of the articles indicated a positive relationship between risk management and diffusion of innovation. An increasing relationship between RM and DOINV was discovered. Thus, if a firm possesses excellent risk management, there will also be high diffusion of innovation. Thirdly, the primary data collection method in most examined research was through surveys; around 78% of the examined works used surveys. Fourthly, positive research outcomes were found in 87.8% of the articles, 7.9% had negative outcomes, and 4.3% were N/A.

There were several limitations related to this systematic review. First, numerous articles highlighted limitations in their conceptual and theoretical models, which required additional constructs. Second, many articles examined specific countries and contexts, making it difficult to generalise the studies. The third limitation was small sample sizes, which indicated low reliability in many examined articles. The fourth limitation was that some studies have not considered other possible moderators and mediators, such as; trust, policy, rewards, etc. Finally, bias was indicated in many articles.

References

Ab Rahman, N. A., Hassan, S., & Said, J. (2015). Promoting sustainability of microfinance via innovation risks, best practices and management accounting practices. *Procedia Economics and Finance, 31*, 470–484.

Afuah, A. (2003). Redefining firm boundaries in the face of the internet: Are firms really shrinking? *Academy of Management Review, 28*(1), 34–53.

Ahmad, A., Alshurideh, M. T., Al Kurdi, B. H., & Alzoubi, H. M. (2021). Digital Strategies: A Systematic Literature Review. *The International Conference on Artificial Intelligence and Computer Vision*, 807–822.

Ahmed, A., Alshurideh, M., Al Kurdi, B., & Salloum, S. A. (2021). Digital Transformation and Organizational Operational Decision Making: A Systematic Review. In *Advances in Intelligent Systems and Computing: Vol. 1261 AISC*. https://doi.org/10.1007/978-3-030-58669-0_63

Al Khayyal, A. O., Alshurideh, M., Al Kurdi, B., & Salloum, S. A. (2021). Women Empowerment in UAE: A Systematic Review. In *Advances in Intelligent Systems and Computing: Vol. 1261 AISC*. https://doi.org/10.1007/978-3-030-58669-0_66

Al Mehrez, A. A., Alshurideh, M., Al Kurdi, B., & Salloum, S. A. (2020). Internal Factors Affect Knowledge Management and Firm Performance: A Systematic Review. *International Conference on Advanced Intelligent Systems and Informatics*, 632–643.

Al Suwaidi, F., Alshurideh, M., Al Kurdi, B., & Salloum, S. A. (2020). The impact of innovation management in SMEs performance: A systematic review. *International Conference on Advanced Intelligent Systems and Informatics*, 720–730.

Ali, A., Warren, D., & Mathiassen, L. (2017). Cloud-based business services innovation: A risk management model. *International Journal of Information Management, 37*(6), 639–649.

Almazrouei, F. A., Alshurideh, M., Al Kurdi, B., & Salloum, S. A. (2021). Social Media Impact on Business: A Systematic Review. In *Advances in Intelligent Systems and Computing: Vol. 1261 AISC*. https://doi.org/10.1007/978-3-030-58669-0_62

Alshamsi, A., Alshurideh, M., Kurdi, B. A., & Salloum, S. A. (2021). The Influence of Service Quality on Customer Retention: A Systematic Review in the Higher Education. In *Advances in Intelligent Systems and Computing: Vol. 1261 AISC*. https://doi.org/10.1007/978-3-030-58669-0_37

Alshamsi, Aisha, Alshurideh, M., Al Kurdi, B., & Salloum, S. A. (2020). The Influence of Service Quality on Customer Retention: A Systematic Review in the Higher Education. *International Conference on Advanced Intelligent Systems and Informatics*, 404–416.

AlShehhi, H., Alshurideh, M., Kurdi, B. A., & Salloum, S. A. (2021). The Impact of Ethical Leadership on Employees Performance: A Systematic Review. In *Advances in Intelligent Systems and Computing: Vol. 1261 AISC*. https://doi.org/10.1007/978-3-030-58669-0_38

Alshurideh, M. (2022). Does electronic customer relationship management (E-CRM) affect service quality at private hospitals in Jordan? *Uncertain Supply Chain Management, 10*(2), 1–8.

Alshurideh, M. T., Hassanien, A. E., & Masa'deh, R. (2021). *The Effect of coronavirus disease (COVID-19) on business intelligence*. Springer.

Alshurideh, Masa'deh, R., & Al Kurdi, B. (2012). The effect of customer satisfaction upon customer retention in the Jordanian mobile market: An empirical investigation. *European Journal of Economics, Finance and Administrative Sciences, 47*(12), 69–78.

Ambad, S. N. A., & Wahab, K. A. (2013). Entrepreneurial orientation among large firms in Malaysia: Contingent effects of hostile environments. *International Journal of Business and Social Science, 4*(16).

Andersen, T. J. (2009). Effective risk management outcomes: exploring effects of innovation and capital structure. *Journal of Strategy and Management*.

Arundel, A., Bloch, C., & Ferguson, B. (2019). Advancing innovation in the public sector: Aligning innovation measurement with policy goals. *Research Policy, 48*(3), 789–798.

Assad, N. F., & Alshurideh, M. T. (2020). Investment in context of Financial Reporting Quality: A Systematic Review. *Waffen-Und Kostumkunde Journal, 11*(3), 255–286.

Azmi, Z., Misral, M., & Maksum, A. (2018). Knowledge Management, the Role of Strategic Partners, Good Corporate Governance and Their Impact on Organizational Performance. *Prosiding CELSciTech, 3*, 20–26.

Barczak, A., Dembińska, I., & Marzantowicz, Ł. (2019). Analysis of the risk impact of implementing digital innovations for logistics management. *Processes, 7*(11), 815.

Berglund, H. (2007). Risk conception and risk management in corporate innovation: Lessons from two Swedish cases. *International Journal of Innovation Management, 11*(04), 497–513.

Birasnav, M. (2014). Knowledge management and organizational performance in the service industry: The role of transformational leadership beyond the effects of transactional leadership. *Journal of Business Research, 67*(8), 1622–1629.

Black, J., Lodge, M., & Thatcher, M. (2006). *Regulatory innovation: A comparative analysis*. Edward Elgar Publishing.

Borgelt, K., & Falk, I. (2007). The leadership/management conundrum: innovation or risk management? *Leadership & Organization Development Journal*.

Bowers, J., & Khorakian, A. (2014). Integrating risk management in the innovation project. *European Journal of Innovation Management*.

Buenechea-Elberdin, M., Sáenz, J., & Kianto, A. (2018). Knowledge management strategies, intellectual capital, and innovation performance: a comparison between high-and low-tech firms. *Journal of Knowledge Management*.

Calvo-Mora, A., Navarro-García, A., Rey-Moreno, M., & Periañez-Cristobal, R. (2016). Excellence management practices, knowledge management and key business results in large organisations and SMEs: A multi-group analysis. *European Management Journal, 34*(6), 661–673.

Chapman, C. B., & Ward, S. C. (1997). Managing risk management processes: navigating a multidimensional space. *Managing Risks in Projects*, 109–118.
Chiambaretto, P., Bengtsson, M., Fernandez, A.-S., & Näsholm, M. H. (2020). Small and large firms' trade-off between benefits and risks when choosing a coopetitor for innovation. *Long Range Planning, 53*(1), 101876.
da Silva Etges, A. P. B., & Cortimiglia, M. N. (2019). A systematic review of risk management in innovation-oriented firms. *Journal of Risk Research, 22*(3), 364–381.
da Silva Etges, A. P. B., Grenon, V., de Souza, J. S., Neto, F. J. K., & Felix, E. A. (2018). ERM for Health Care Organizations: An Economic Enterprise Risk Management Innovation Program (E2RMhealth care). *Value in Health Regional Issues, 17*, 102–108.
Darroch, J. (2005). Knowledge management, innovation and firm performance. *Journal of Knowledge Management*.
Douthwaite, M., & Ward, P. (2005). Increasing contraceptive use in rural Pakistan: An evaluation of the Lady Health Worker Programme. *Health Policy and Planning, 20*(2), 117–123.
Dwyer, L., Dragićević, V., Armenski, T., Mihalič, T., & Knežević Cvelbar, L. (2016). Achieving destination competitiveness: An importance-performance analysis of Serbia. *Current Issues in Tourism, 19*(13), 1309–1336.
Edwards, T., Delbridge, R., & Munday, M. (2005). Understanding innovation in small and medium-sized enterprises: A process manifest. *Technovation, 25*(10), 1119–1127.
Etges, A. P. B. da S., Souza, J. S. de, & Kliemann, F. J. (2017). Risk management for companies focused on innovation processes. *Production, 27*.
Fort-Rioche, L., & Ackermann, C.-L. (2013). Consumer innovativeness, perceived innovation and attitude towards "neo-retro"-product design. *European Journal of Innovation Management*.
Foxon, T. J., Gross, R., Chase, A., Howes, J., Arnall, A., & Anderson, D. (2005). UK innovation systems for new and renewable energy technologies: Drivers, barriers and systems failures. *Energy Policy, 33*(16), 2123–2137.
Games, D., & Rendi, R. P. (2019). The effects of knowledge management and risk taking on SME financial performance in creative industries in an emerging market: The mediating effect of innovation outcomes. *Journal of Global Entrepreneurship Research, 9*(1), 1–14.
Giampaoli, D., Ciambotti, M., & Bontis, N. (2017). Knowledge management, problem solving and performance in top Italian firms. *Journal of Knowledge Management*.
Gurd, B., & Helliar, C. (2017). Looking for leaders:'Balancing'innovation, risk and management control systems. *The British Accounting Review, 49*(1), 91–102.
Hock-Doepgen, M., Clauss, T., Kraus, S., & Cheng, C.-F. (2021). Knowledge management capabilities and organizational risk-taking for business model innovation in SMEs. *Journal of Business Research, 130*, 683–697.
Hölttä-Otto, K., Otto, K., Song, C. Y., Luo, J. X., & Seering, W. (2015). Risk and innovation balance in crowdfunding new products. In *International Conference on Engineering Design*.
Hsu, I., & Sabherwal, R. (2012). Relationship between intellectual capital and knowledge management: An empirical investigation. *Decision Sciences, 43*(3), 489–524.
Inkinen, H. T., Kianto, A., & Vanhala, M. (2015). Knowledge management practices and innovation performance in Finland. *Baltic Journal of Management*.
Izumi, T., Shaw, R., Djalante, R., Ishiwatari, M., & Komino, T. (2019). Disaster risk reduction and innovations. *Progress in Disaster Science, 2*, 100033.
Jin, Z., & Navare, J. (2010). Exploring the relationship between risk management and adoptive innovation: A case study approach. *World Journal of Entrepreneurship, Management and Sustainable Development*.

Joshi, H., & Chawla, D. (2019). How knowledge management influences performance?: Evidences from Indian manufacturing and services firms. *International Journal of Knowledge Management (IJKM), 15*(4), 56–77.

Kalvet, T., & Lember, V. (2010). Risk management in public procurement for innovation: the case of Nordic–Baltic Sea cities. *Innovation–the European Journal of Social Science Research, 23*(3), 241–262.

Koru, G., Alhuwail, D., Jademi, O., Uchidiuno, U., & Rosati, R. J. (2018). Technology innovations for better fall risk management in home care. *Journal of Gerontological Nursing, 44*(7), 15–20.

Kwak, D.-W., Seo, Y.-J., & Mason, R. (2018). Investigating the relationship between supply chain innovation, risk management capabilities and competitive advantage in global supply chains. *International Journal of Operations & Production Management.*

Lowman, M., Trott, P., Hoecht, A., & Sellam, Z. (2012). Innovation risks of outsourcing in pharmaceutical new product development. *Technovation, 32*(2), 99–109.

Migliori, S., De Massis, A., Maturo, F., & Paolone, F. (2020). How does family management affect innovation investment propensity? The key role of innovation impulses. *Journal of Business Research, 113*, 243–256.

Mokyr, J. (2002). Innovation in an historical perspective: tales of technology and evolution. In *Technological innovation and economic performance* (pp. 23–46). Princeton University Press.

Nagano, M. S., Stefanovitz, J. P., & Vick, T. E. (2014). Innovation management processes, their internal organizational elements and contextual factors: An investigation in Brazil. *Journal of Engineering and Technology Management, 33*, 63–92.

Narvekar, R. S., & Jain, K. (2006). A new framework to understand the technological innovation process. *Journal of Intellectual Capital.*

O'Connor, G. C., Ravichandran, T., & Robeson, D. (2008). Risk management through learning: Management practices for radical innovation success. *The Journal of High Technology Management Research, 19*(1), 70–82.

Puente, J., Gascon, F., Ponte, B., & de la Fuente, D. (2019). On strategic choices faced by large pharmaceutical laboratories and their effect on innovation risk under fuzzy conditions. *Artificial Intelligence in Medicine, 100*, 101703.

Rasula, J., Vuksic, V. B., & Stemberger, M. I. (2012). The impact of knowledge management on organisational performance. *Economic and Business Review for Central and South-Eastern Europe, 14*(2), 147.

Ren, S., Eisingerich, A. B., & Tsai, H.-T. (2015). How do marketing, research and development capabilities, and degree of internationalization synergistically affect the innovation performance of small and medium-sized enterprises (SMEs)? A panel data study of Chinese SMEs. *International Business Review, 24*(4), 642–651.

Röth, T., & Spieth, P. (2019). The influence of resistance to change on evaluating an innovation project's innovativeness and risk: A sensemaking perspective. *Journal of Business Research, 101*, 83–92.

Salaman, G., & Storey, J. (2005). Achieving "fit": Managers' theories of how to manage innovation. *Strategic human resource management: Theory and practice. A reader*, 91–115.

Taplin, R. (2005). *Risk management and innovation in japan, britain, and the united states.* Routledge New York.

Tidd, J., & Hull, F. M. (2006). Managing service innovation: The need for selectivity rather than 'best practice.' *New Technology, Work and Employment, 21*(2), 139–161.

Tseng, S.-M., & Lee, P.-S. (2014). The effect of knowledge management capability and dynamic capability on organizational performance. *Journal of Enterprise Information Management.*

Van de Ven, A. H. (1986). Central problems in the management of innovation. *Management Science, 32*(5), 590–607.

Vargas-Hernández, J. G. (2011). Modeling risk and innovation management. *Journal of Competitiveness Studies, 19*(3/4), 45.

Vargas-Hernandez, J. G., & Garcia-Santillan, A. (2011). Management in the innovation project. *Journal of Knowledge Management, Economics and Information Technology, 7*(1), 1–24.

Wehn, U., & Evers, J. (2015). The social innovation potential of ICT-enabled citizen observatories to increase eParticipation in local flood risk management. *Technology in Society, 42*, 187–198.
Wu, J., & Wu, Z. (2014). Integrated risk management and product innovation in China: The moderating role of board of directors. *Technovation, 34*(8), 466–476.
Yang, J. (2008). Antecedents and consequences of knowledge management strategy: The case of Chinese high technology firms. *Production Planning and Control, 19*(1), 67–77.
Zheng, W., Yang, B., & McLean, G. N. (2010). Linking organizational culture, structure, strategy, and organizational effectiveness: Mediating role of knowledge management. *Journal of Business Research, 63*(7), 763–771.

9783031123818VOL01